SUBTERRANEAN HYDROLOGY

by G. Kovács
and Associates

WATER RESOURCES PUBLICATIONS

For information and correspondence:

WATER RESOURCES PUBLICATIONS
P.O. Box 2841
Littleton, Colorado 80161, USA

SUBTERRANEAN HYDROLOGY

by G. Kovács and Associates

This book is a product of lectures at the International Post-Graduate Course in Hydrology, within the Research Centre for Water Resources Development (VITUKI) Budapest, Hungary. Dr. G. Kovacs and his associates are the lecturers at the International Post-Graduate Course in Hydrology and the staff members of the Research Centre.

ISBN-0-918334-35-7
U.S. Library of Congress Catalog Card Number 80-54120

This publication is printed and bound by BookCrafters, Inc., Chelsea, Michigan, USA

PREFACE

The growing problems related to volume and quality of surface water supplies have increasingly directed attention towards the exploration of subterranean water resources. In response to such demands, many scientists, engineers and professionals in related fields have devoted intensive efforts to investigations of hydrological processes which take place below the terrain, in the vicinity of the surface, as well as in deeper layers, and which affect the groundwater supplies at attainable depths, or the subterranean part of the hydrologic cycle.

The objective of this book is to meet the urgent demand of needed investigations, by presenting to the interested reader the state-of-the-art of subterranean hydrology. This includes the most recent advances attained in this field by incorporating the results achieved in related disciplines. In this way it is endeavored to provide methodological guidance to professionals engaged in practical work and to recently graduated hydrologists alike in broadening their knowledge and in solving their routine problems, as well as in the introduction of new methods.

The task of hydrology is to explore the interrelations between the various forms of water on the continental surfaces and to analyze its continuous movement along the hydrological cycle. When investigating the subsurface water from a hydrological point of view, its connection with atmospheric and surface waters as well as the flow of water along the subterranean branch of the cycle must be studied. Considering this definition, it is quite evident that the topic of subterranean hydrology covers the bordering fields of earth sciences, physics and engineering sciences. The role of geology is to determine the structure composed of various layers where the investigated processes develop. The principles of physics are applied to describe the transport and storage processes through the diverse system of water-conveying formations. The solution of practical problems is the task belonging to the field of interest of engineering sciences. Considering the multidisciplinary character of the topic, it was intended, when writing this book, to give a more detailed explanation in connection with geological, hydrodynamical and geophysical problems related to subterranean hydrology, than that usually available in other manuals. The reader interested in specific details can also obtain welcome information from the bibliography which was compiled specifically for this purpose and attached to each part of this book.

This book, entitled "Subterranean Hydrology," has been developed from the lecture notes compiled and first published in 1966 for the International Post-Graduate Course on Hydrology, Budapest, Hungary, sponsored by UNESCO, subsequently sponsored also by WMO, under the training program adopted for the International Hydrological Decade and later for the International Hydrological Program.

The revised and enlarged material of lecture notes published in book form is expected to be of interest not only to participants of the Budapest Post-Graduate Course, but also to other hydrologists striving to improve their knowledge of the subject.

The management of the Budapest International Post-Graduate Course on Hydrology is to be credited with having made possible the publication of this work in its present form. Thanks are due also to the co-authors, further to the scientific staff members of the Research Center for Water Resources Development (VITUKI), participating in the editorial work for their devoted efforts, suggestions and recommendations. The work of all others involved in the final preparation of the manuscript is also appreciated.

Dr. G. Kovács
Corresponding member of the
Hungarian Academy of Sciences

TABLE OF CONTENTS

Part | Page

Part | Page

Part Page

Part		Page

PART I

HYDROLOGICAL INVESTIGATION OF SUBSURFACE WATERS

by

Dr. J. Gálfy and Dr. G. Kovács

SECTION 1

GENERAL INTRODUCTION

1-1 Objectives and Methods of Groundwater Hydrology

A considerable amount - about 20 percent - of the freshwater resources on the globe is stored below the surface. In most cases this water is protected from the pollutants produced by human activities on the land surface. The utilization of groundwater is, therefore, an important part of water resources development. Studies which aim to insure both the quantitative control of this source and its protection against pollution have high economic importance. There are, however, many scientific aspects suitable to analyze the versatile character of the water stored below the surface and the reservoirs containing this water. In this case, the seepage hydraulics investigates the flow through porous media; geology describes the position as well as the texture and structure of water-bearing layers; geochemistry deals with the chemical composition of subsurface waters and its change as a result of chemical interaction between the solid matrix and the water; the field of interest of soil science is limited to the upper shallow zone of the crust, investigating the combined system of soil and water, including the influence of both the chemical components and the plants.

The purpose of the present work is to summarize the objectives of hydrological investigations of subsurface waters and to survey the methods generally applied in these studies. This analysis requires the accurate definition of hydrology to clarify the aspects to be considered in the investigation of the various types of water stored below the surface. According to the most commonly accepted definition:

> Hydrology is not an independent science, but an interdisciplinary synthesis of the bordering fields of the other water-oriented sciences, which includes and harmonizes those investigations carried out on the basis of the different principles of the related sciences but aiming at the characterization of the complete hydrological cycle (including its stretches in the atmosphere, on and below the surface), and thus the common feature of hydrological studies is that they deal with the water that is a part of the continuous movement along the cycle.

According to this definition, every investigation independent of the scientific aspects applied in research are regarded as hydrological studies if their purpose includes such details as:

- the description of the mechanism maintaining the movement along the cycle;

- the survey of interactions between the water propagating along the cycle and its environment;

- the analysis of the interrelation between the different branches of the cycle;

- the determination of the parameters characterizing the water as a medium, and its movement as a process at various stretches of the hydrological cycle, including the investigation of the relationships between these parameters.

Although this list is not complete, it does provide a guideline to understand and outline the range of the hydrological investigations in general and to determine the objectives of groundwater hydrology in our special case. The research may be considered hydrological if the conditions summarized in the definition above are satisfied, i.e., if the water is studied as a material moving along the cycle, if processes are investigated and the exploration of interactions existing between the water and its environment and those creating contacts between the investigated process and the hydrological cycle as a whole are studied.

Considering the explanations given in the previous paragraphs, the investigation of subsurface stretches of the hydrological cycle may be considered the most general objective of groundwater hydrology. The hydrological character is independent of the scientific aspects applied in the research, since the research may be based on physical (dynamic), geological, biological or chemical (physicochemical, colloidal) principles. In addition, the analysis may use the methodology of any other related science, sometimes using a combination of various methods. The hydrological investigations may extend to the entire system of the interconnected reservoirs storing water below the surface, or they may be concentrated only to one well-defined part of this system. The basic precondition for the use of the adjective hydrological is that the continuity of the cycle should be considered in any case.

Within the broad framework outlined by the general scope of groundwater hydrology, there remains a large variety of subjects, from which the direct purpose of the investigation may be selected. It follows from consideration of the dynamic aspects of the system that processes should always be studied, although the investigations aimed at the characterization of a single process may be regarded as a part of hydrology, similar to those dealing with the interconnected series of various processes. The interdisciplinary analyses applying different aspects of the basic sciences listed previously assist greatly in the understanding of the versatile processes, but the studies using the principles of a sole branch of science may also produce hydrological results. The only criterion to be satisfied is that the process investigated should influence the movement of water along the hydrological cycle and the space where the process takes place, and that the process should not be analyzed as a closed system separate from its natural environment.

There is no clear distinction between the hydrological investigations and other groundwater studies, and some cases may even be regarded as a transition form between the two groups. The methods and basic principles applied for the research may be common and only the scope of hydrological studies is larger, requiring that the role of the investigated process in its interrelation with the hydrological cycle be analyzed. The correctness of this explanation may be demonstrated with an example of groundwater exploitation. Since the water demand can be small related to the available resource, knowledge of the geological structure and application of a relatively simple model providing the yield of a well is sufficient to solve the hydrogeological problem, since the extraction of the required amount of water in this case does not significantly alter the water regime of the aquifer.

The exploitation of higher yield creates large drawdown, causing a change of flow conditions in the system. The result of this process is the modification of water exchange between the drained aquifer and the neighboring layers. The interactions between surface waters and groundwaters may be altered as well, and in some cases even the natural recharge and discharge developing through the land surface may be changed. It is necessary, therefore, that the contact with the hydrological cycle be analyzed and thus the research has hydrological characteristics. The water demand increases rapidly and therefore, the aquifers are drained generally at several points and thus the interaction of wells must be considered. In general, the objective of modern investigations is the optimum utilization of a groundwater system and not the design of a single well. In many cases the conjunctive use of surface water and groundwater insures the most economical solution, which further emphasizes the importance of the hydrological analysis of a large system. Thus, consideration of the role of the hydrological cycle in groundwater studies is unavoidable, and the hydrological investigation of subsurface water may be regarded as the needed way of research.

One aspect used to classify the processes investigated in groundwater hydrology is the distinction made between natural processes and those created and maintained by human activities. Although the differences between these two groups may appear fundamental, in most cases the natural and man-influenced processes must be investigated in a combined form and therefore the distinction between them serves no real practical purpose. Scientific activities should serve human actions, and should result in practical products. The natural regime of subsurface waters is never analyzed for itself only, but the determination of the parameters describing the behavior of the system is aimed at making possible the prediction of the interactions between hydrological processes and some planned activities in this way. There are only a few cases when groundwater data are needed for planning the utilization of surface-water resources (determination of the base flow of rivers). In general, the practical purpose of a hydrological investigation of subsurface reservoirs is the prediction of man-induced changes in the regime of groundwater and soil moisture. It is advisable, therefore, to group the problems investigated in applied hydrology according to the character of the processes influenced by various human activities, including water utilization, change of regime caused by hydraulic structures, and activities influencing soil moisture conditions. Since the purpose of the first part of this book is to explain the fundamentals of groundwater hydrology, further explanation of this grouping will be given in the second part of this text.

As it was stressed in the definition, the movement of water along the subsurface branch of the hydrological cycle should be investigated in groundwater hydrology. Geology describes the medium (its material and geometry) where the movement takes place, and the basic subject of hydrodynamics is the determination of flow characteristics. For this reason, the principles of these two sciences play a dominant role in the hydrological investigation of subsurface waters, and are dealt with separately, summarizing only the aspects important in hydrological analysis. The principles of other related sciences, i.e., geophysics, geochemistry, soil science, will be discussed in the course of their application.

Although the methods applicable to the hydrological analysis of subsurface waters differ considerably depending on the type of water and its reservoir (structure and position of the layer; pressure condition, temperature and chemical composition of the water; contacts prevailing between the system and its surroundings), some fundamental rules to be followed in all cases can be determined using the definition and objectives of groundwater hydrology. Let us recall the main tasks of the investigations and in this way determine the common principles on the basis of which general methodological aspects can be summarized. The requirements arising in connection with any kind of research are as follows:

- to investigate the subsurface stretches of the hydrological cycle and to describe the phenomena governing the regime of the groundwater and soil moisture;

- to analyze the dynamic and time-dependent processes influencing the movement and the change in amount of water stored below the surface by considering the special conditions caused by the fact that the investigated space is partly occupied by solid particles and only the remaining part is free for the flow and storage of water;

- to explore the interrelations between the subsurface water and the phenomena, including both hydrological and meteorological events, occurring on and above the land surface;

- to investigate the contact and water exchange between the various water-bearing horizons;

- to determine numerical parameters characterizing the flow, storage and interactions mentioned in the previous points, giving the complete quantitative description of the regime of subsurface waters;

- to predict the changes expected as a result of the modified natural conditions or those of planned artificial effects (human activities).

The system of subsurface waters is one of the natural systems - and perhaps the most important one - where the amount of available water resources is fundamentally influenced by the process of storage. Similar to other storage systems, the calculation of the amount available from groundwater depends on the operation of the reservoirs. In this case, however, the operation is not controlled by man, or at least is only influenced by man's activities to a very limited extent, but the process of storage depends to a large extent on uncontrolled natural phenomena. Thus, groundwater hydrology can be regarded as a special branch of systems hydrology which offers an excellent field for the hydrological application of systems analysis. The basic principles of systems analysis should therefore be utilized as a guideline, and the methods developed for this concept can be used to derive and solve mathematical models by application of analytic, numerical or analog techniques. The application of models requires the proper consideration of boundary conditions, which are governed by the continuity between the system being investigated and its surroundings. This technique involves, therefore, the analysis of the contacts and interrelations within the hydrological cycle and insures the hydrological character of the investigations.

Dynamically balanced conditions should always be investigated within the storage systems. All groundwater catchments or parts of the interconnected water horizons can be regarded as independent systems, and the equation of continuity can be applied to describe the balance of the investigated quantities, e.g., heat, chemical substances and water, if the processes acting along the borders of the separated space are duly considered. It can be stated, therefore, that the difference between input and output summarized for an investigated period should be equal to the change in the amount of the quantity in question stored in the system. It is evident that the water balance is investigated in most cases in hydrological studies. It is necessary to emphasize, however, that the determination of the balance of other substances such as dissolved chemicals, isotopes, and heat may provide us with important supplementary information on the hydrological regime of subsurface waters.

The fluctuation of the water table (commonly observed under natural conditions), the piezometric level, and to a smaller extent the change of moisture content, indicates the internal operation of the system (the storage in wet seasons or periods and the drainage during dry periods). It can also be stated, however, that the wet and dry periods counterbalance each other under natural conditions, if the interval being investigated is long enough. The system is, therefore, in a state of equilibrium; the average of the natural change in the amount of stored water tends toward zero as the length of the investigated period increases. The validity of this statement is evident, because the natural effects influencing the balance of subsurface waters have acted since geological ages without causing considerable modification of the stored amount, the latter realizing only periodic fluctuation. The average position of the water table developing under natural conditions indicates, therefore, the level where the recharge and drainage are in equilibrium.

It was also mentioned that the balanced condition should be regarded dynamically. The adjective "dynamic" means the consideration of two facts; namely

- the investigation of the forces acting is a basic requirement in hydrological studies;

- the balance of the energy available in the system must be analyzed.

A considerable part of the subsurface water is in continuous movement. The expression of the equilibrium in the form of water amount (recharge, drainage, and change in storage) does not provide us, therefore, with sufficient information. The balanced condition of the forces acting should be investigated by comparing the accelerating and retarding forces, or by considering that the difference in the energy contents of the recharged and drained masses must be equal to the sum of two quantities, i.e., the change in the energy content of the stored water and the energy consumed by the resistances acting against the flow of water.

The second important dynamic aspect originates from the first one. The realization of the dynamic concept requires the investigation of the energy content of water, apart from that of the changes in the stored amount of water. In systems where the flow is hindered by impervious layers, the movement does not start until the energy content has achieved the upper limit of the latent energy; that is, the energy gradient is smaller than the threshold gradient of the system. The limit value is determined by the resistance of the system belonging to the static condition. The accumulation of energy is a continuous process in most cases. It is only a question of time until the stored energy surpasses the limit and flow will develop in the system. Thus, the imperviousness of a layer is not an absolute term, but is related to the instantaneous energy content of the entire system.

When man's activities begin to influence the groundwater, the water table or piezometric level (in general the energy content) will change continuously until the development of a new equilibrium of the natural and artificial phenomena. Naturally, this modified balance should be interpreted in a dynamic sense as well since its parameters also depend on the available energy content increased by man's activities that are in question. Thus, investigating the expected new conditions created by man, the change in time of both the moving and the stored amount of water as well as that in energy content should be predicted to determine whether a new water regime exists. This is generally described by a new position of either the water table or the piezometric level, which satisfies the required conditions of equilibrium, or indicates that a continuous change of the situation should be foreseen for a long period.

The study of natural and artificial balances requires knowledge of the various processes recharging and draining the groundwater space and the soil-moisture zone, such as infiltration, evaporation and evapotranspiration, percolation to surface waters and recharge of groundwater from this direction, as well as the water exchange between water-bearing layers.

For this purpose, the recharging and draining areas of the groundwater should be distinguished and the stretches through which the movements create contacts between the two types of areas should be determined, insuring the overall mass balance within the entire basin.

The aspects described here again emphasize the importance of the movement, the characterization of which requires investigation of the entire structure of the system and consideration of hydrodynamic principles. The basic data needed for such investigations can be summarized, and the list can be used as a general guideline in any hydrological study dealing with subsurface waters:

- the geometry of the water transporting layers;

- the physical parameters of the system composed of different layers, the parameters being sufficient to characterize the resistance against the movement of water either under static conditions or in the presence of flow with given velocity already developed in the systems;

- the dynamic conditions of the system, i.e., the forces and pressure values prevailing at the various parts of the system or along its different stretches, considering the accumulation of the energy content which results from the development of the forces and pressure;

- the numerical values describing the flow through the system, i.e., velocity, flow rate, change in energy content, calculated from the physical and dynamic parameters.

The application of the general principles and the determination of the characteristic data listed as fundamental information will be discussed in detail in connection with various types of subsurface waters.

SECTION 2

DEVELOPMENT AND GEOLOGICAL FEATURES OF WATER-BEARING SYSTEMS

The subsurface branch of the hydrological cycle takes part in the outer rocky shell of the earth called the lithosphere. The description of the subsurface features, underground structures and rock types, i.e., the determination of the lithological units of the lithosphere is the task of the geologist. If particular attention is directed towards the hydrology or water-bearing properties of rocks, then hydrogeology is utilized.

The rock as a lithological unit may be considered a system consisting of a solid, rigid framework and of water or other liquid or gaseous material filling out the interstices of the framework. The characteristic quantities used for the description of a rock type should be related to the framework, i.e., to the reservoir, as well as to the water contained in it.

The duty of the hydrogeologist is the study and determination of geometry, including size, shape, and spatial extension, and the petrographical description of hydrologically important rocks in the sense explained above, using parameters suitable for hydrological work such as porosity and permeability.

If it is useful or necessary for the explanation of the present hydrogeological situation, then the development of events and processes of the geological history of rocks should be studied.

At this point, a few words should be said about the definition of hydrogeology. In conformity with the common usage of the term, hydrogeology can be defined as the study of groundwater with particular emphasis given to its chemistry, mode of migration, and relation to the geological environment. Practically, a wide variety of activities are undertaken under the name hydrogeology, ranging from water prospecting to water-well perforation. This variation can be eliminated by proper distinction between the science itself and the application of it to different problems.

The term hydrogeology as defined above will be used in the following discussion in a somewhat restricted sense: it constitutes an applied branch of the geology proper, applied not to the study of groundwater, but to the study of rocks in their geologic environment. This definition is in agreement essentially with the term first used by Lamarck and Powell (1885), according to whom "hydric geology" is defined as a study of the "phenomena of degradation (erosion) and deposition by aqueous agents." Hydrogeology, as applied geology, in this sense includes the bordering fields of other sciences, such as geomorphology, geophysics, and geochemistry as is the case with geology proper in water-oriented forms of application. It utilizes petrography, stratigraphy, structural geology, and geomorphology and other geologic specializations to a lesser extent.

The difference between conventional geologic work and a hydrogeological study is well illustrated by a pair of maps in which the geology and the hydrogeology of the same hypothetical area are shown. On the geological map (Fig. 1-1) the age or geologic time of the rocks is given in detail. The periods and the subdivisions (e.g., Amran formation) are marked on the map. The petrophysical description of the formations is roughly outlined. It should be noted that the Quaternary, the youngest sediment, is represented as the uniform medium.

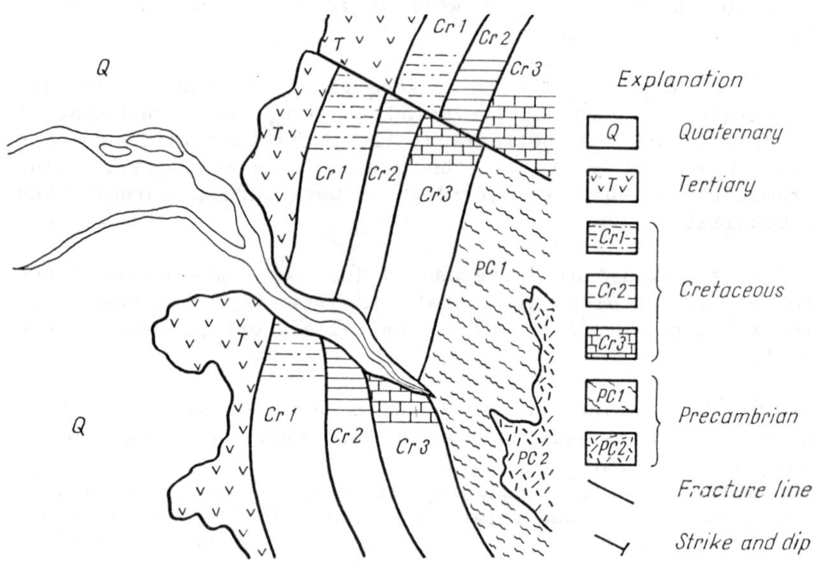

Fig. 1-1. Geological map of the hypothetical area representing the gorge of a wadi and its surrounding. Q: alluvial sand and silt with some gravel. T: trap, dark gray basalt. Cr1: Kohlan group, siltstone with limestone beds. Cr2: Amran group, massive dark limestone. Pc1: gneiss, quartzite and feldspar. Pc2: Granite, abundant pink orthoclase.

On the basic hydrogeological map (Fig. 1-2) units such as granite, gneiss and quartz are divided into groups according to their water-bearing properties, and shales and siltstones as aquicludes are grouped together. On the other hand, many deposits of clay, sand, and gravel which would be mapped geologically as a single unit, i.e., the Quaternary, are differentiated. The geological time scale is missing, pointing to the fact that the lithological or stratigraphical description of rocks is more important than the geological age.

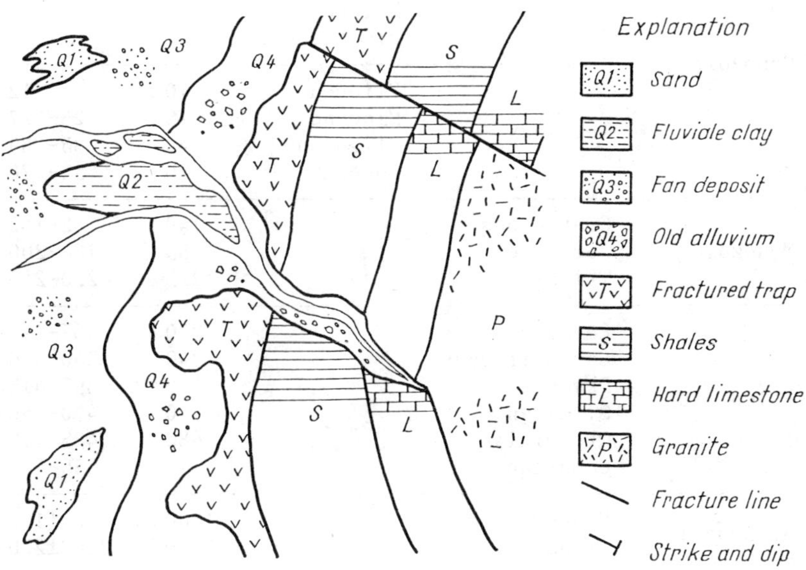

Fig. 1-2. Hydrogeological map of the area shown in Fig. 1-1. Note the contrast in detail that reflects the difference in purpose between the two maps (after Davis, 1966). Q1: sand, 1-10 m thick. Q2: fluvial clay, 1-5 m thick. Q3: alluvial fan deposit, 1-100 m thick. Q4: older alluvial and fan deposit, an aquifer. T: trap basalt, fractured, may be permeable. S: shale and siltstone, eroded, probably not an aquifer. P: unweathered granite and gneiss, impermeable.

When the geological age is given sometimes in lithological description, it is used as short expression of a group of petrographical parameters, e.g., "Devonian limestone" means chalky limestone of 29.5 percent porosity and 37.8 mdarcy permeability or 37.8 x 10^{-7} m/sec hydraulic conductivity. Moreover, taking into account the use of different names in different countries for the subdivisions in the geological time scale, the lithological description of rocks is preferred over the geological description. The geological time scale is shown in Table 1-1 for orientation.

Table 1-1. Geologic time scale.

Era (sequence)	Period (system)	Epoch (series)	Begin.	Limits
Cenozoic	Quaternary	Holocene		
		Pleistocene	1	1- 2
	Tertiary	Pliocene	10	6- 12
		Miocene	25	23- 27
		Oligocene	40	35- 39
		Eocene	70	54- 70
Mesozoic	Cretaceous		140	132-142
	Jurassic		185	180-200
	Triassic		225	220-240
Paleozoic	Permian		270	265-295
	Carboniferous		340	340-360
	Devonian		400	395-415
	Silurian		440	430-450
	Ordovician		480	485-515
	Cambrian		570	555-585
Archeozoic	Proterozoic		1800	1900±100
	Archaic		3400	3200±200

All the time data, given im million years, relate to the beginning of each geological time interval. The most probable values are listed in column Begin., the maximal and minimal ones determined by different dating methods may be found in column Limits.

The petrographical (lithological) description of different types of rocks, as well as the determination of petrographical parameters such as porosity, permeability (or hydraulic conductivity), and grain size distribution, will be given in Section 2.

Discussing the development and geological structure of water-bearing systems, the objective of this section, let us return to the definition of hydrogeology given by Lamarck and Powell and consider the lithosphere and the erosional processes taking part in it.

On the surface of the earth, various agents are engaged in rock destruction: rivers, waves, wind, etc. These are all processes in which the agent is moving while in operation. There are other kinds of destruction in which the agent is essentially motionless: alternate heating and cooling, expansion of rock moisture in freezing, etc. Whether accomplished by moving or motionless agents, the rock destruction is called erosion. Mechanical wear by water or wind is termed corrasion, that by chemical agents is corrosion and, when the agent is motionless, the process is called weathering.

The fragments and debris formed by erosion ordinarily are borne away from their source, partially in solution and partially transported mechanically and eventually accumulate as sedimentary material. Deposits of this kind are transported sediments, and those products of erosion which remain in situ are called residual deposits. Transported deposits are chemical if they are precipitated from solutions, and they are clastic (mechanical, fragmental) if they consist of particles and fragments of the parent rock transported by mechanical means. All of these deposits, organic or inorganic, including clay, silt or mud, sand, gravel, soil and the like are known collectively as mantle rock. Everywhere below the mantle rock a solid rock, bedrock or substratum exists. When the bedrock projects through the overlying mantle rock, the protruding portions are called outcrops or exposures. The solid-rock basis may consist equally of the three main rock types:

- sedimentary;

- igneous, which are consolidated from igneous conditions;

- metamorphic, i.e., rocks altered due to the changes of conditions under which they were formed.

Description and characterization of sedimentary rocks in their original forms, called internal structure, will be given in Subsection 2-1.

From the time of their origin, rocks are more or less disturbed by forces acting within the lithosphere. Forces outside of the lithosphere such as the erosional forces were discussed above. Evidence of the internal forces are furnished by joints, cleavage, folds, and faults which are often seen in consolidated rocks. The deformation represented by these or other allied structures may be brought about through flowage or fracture. Flowage means a gradual change in the form and internal structure of a rock mass by microscopic fractures, while the rock remains essentially rigid. Under very great pressure and temperature, such as exists in the deeper part of the lithosphere, all rocks are deformed by flowage. Nearer to the surface, where rocks are more apt to break, rocks deform by fracture. Thus, the main structural forms due to tectonic movements are foled forms, e.g., anticline and faulted forms, e.g., overthrust. It should be noted, however, that fracture elements may be originated by other causes, such as the removal of overburden by erosion. This process is called exfoliation.

The word structure is used in the geologic literature in two senses: the internal structure of a rock, as well as the forms caused by tectonic forces. In the following discussion the latter meaning will be used unless the misunderstanding is excluded.

The study of the structural forms of sedimentary strata and the description of the tectonic processes in the lithosphere will be discussed in Subsection 2-2.

The hydrogeological properties of igneous and metamorphic rocks show considerable similarity to that of indurated sediments and in lesser extent to non-indurated ones, e.g., volcanic ash. Having discussed the properties of sedimentary rocks, the igneous and metamorphic rocks will be characterized hydrogeologically in Subsectin 2-3.

With full knowledge of stratigraphy and structural geology of the principal water-bearing formations, aquifers and aquicludes of different origin can be divided into well-defined groups, according to their setting and geometric form. These types may be considered as the building blocks of more complicated hydrogeological systems. There are many gradational types of rocks and there are different aspects of classification such as genetic, lithologic, geographic, etc. The geological classification combined with lithological aspects is a relatively simple grouping which provide us with sufficient information for hydrogeological purposes. The basic hydrogeological forms classified according to this system are quoted and illustrated in Subsection 2-4.

The hydrogeological studies are practically performed by using the different methods of field geology. The objective of this exploration includes the determination of areas where water-bearing formations are likely to occur and the location of potential aquifer systems, in short, the delineation of the basic hydrogeological forms as explained above.

The fundamental principle of the hydrogeological exploration is: hydrogeological units can be detected by hydrogeological exploration, where marked difference exists in parameters between the unit to be studied and its surrounding. The parameters may have stratigraphic character, e.g., unconformity, morphological character (dislocation on the surface), or physical character, e.g., differences in electric resistivity, etc.

The combinations of the basic hydrogeological forms are realized in nature in unlimited number of different geological formations. This is the reason why exploration problems in hydrogeology are unique with flexible solution techniques and why the applied methods in the hydrogeological exploration are very versatile and manifold.

The hydrogeological exploration includes primarily field methods of geology proper such as petrography, stratigraphy, and structural geology and secondarily forms of bordering sciences such as geomorphology, geochemistry, geophysics, and recently in increasing degree remote sensing (photogeology) which itself is a mixture of applied sciences. The application of the practical methods in solution of exploration problems is very diverse, according to the problem to be solved. On areas where the water-bearing strata are rather uniform (depth and thickness) over many hundreds of square kilometers the advantage of the hydrogeological exploration may be very slight. In mountainous or hilly lands, the hydrogeological situation may be determined generally by means of geomorphological (air photography) studies without the extensive use of other methods. In the plains or any case at great depths, the hydrogeologist cannot give answers to questions as where to drill and to what depth without the use of surface geophysics.

Exploration methods used in hydrogeology and the preparation of hydrogeological maps will be discussed in Subsection 2-5. The reconnaissance or regional types of exploration will be explained thoroughly, the methods applied during detailed survey will be dealt with in the course of their main application in the following parts and their sections.

Knowing the geometry of the water-bearing formations and that of water transporting structures, the main path of the subsurface branch of the hydrological cycle can be determined. Although the definition of hydrogeology is not completely unanimous, as it was quoted earlier, it is generally agreed that it should include not only the reservoirs storing the water but that of the fluids stored in the interstices (pores, fissures or fractures) of the formations.

It is necessary, therefore, that the characterization of subsurface waters should also be discussed apart from the petrographical, stratigraphical, and structural analysis of geological formations. Since there are several aspects according to which the various types of water stored and transported below the surface can be grouped, the method of classification should be suited to the character of the hydrological investigation. This topic will be dealt in Subsection 2-6.

2-1 Development and Lithological Description of Sedimentary Rocks

The most favorable formations for water accumulation are predominantly found in sedimentary rocks. About 80 percent of known subsurface water supply comes from non-indurated sediments. The hydrogeological terminology suitable for the characterization of sedimentary rocks can be simply obtained considering the development of sediments.

During the process of sedimentation the agent carrying the rock particles may have the power of separating the lighter from heavier particles. The difference in weight is caused mostly by the size of the particles because they hardly differ in specific weight, although the differences of the latter may influence the particle transport as well. This process is called sorting and it is characteristic of the transport by wind or running water. Where the transporting current is weak, light particles are moved and deposited, where it becomes stronger, heavier particles are transported and distributed as a coarse layer over the earlier, finer material. The deposit may evolve to have a layered structure called bedding, or stratification. Beds (layer, strata) are sedimentary units of stratified rocks of uniform or varied composition and thickness. The bedding plane separates two strata of differing character. The position of the bedding plane is determined by strike and dip angles (Fig. 1-3). The absence of bedding might be expected in deposits accumulated in very uniform conditions.

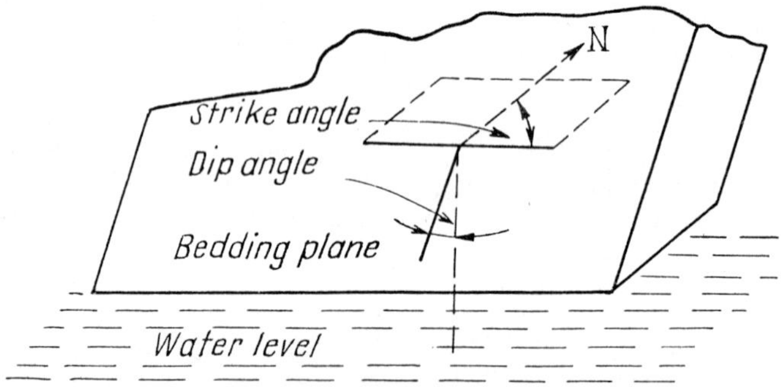

Fig. 1-3. Definition of dip and strike.

A bed or an assemblage of beds with well-marked upper and lower boundaries constitute the geological formation. When sediments are being laid down in uninterrupted sequence, the beds are usually parallel though the strata may differ in texture and composition. This relation between beds is called conformity. Unconformity is a buried erosion surface which separates two strata in a well-defined manner. Mutual relations of beds or bedding types and forms of unconformity are illustrated in Fig. 1-4 and Fig. 1-5, respectively.

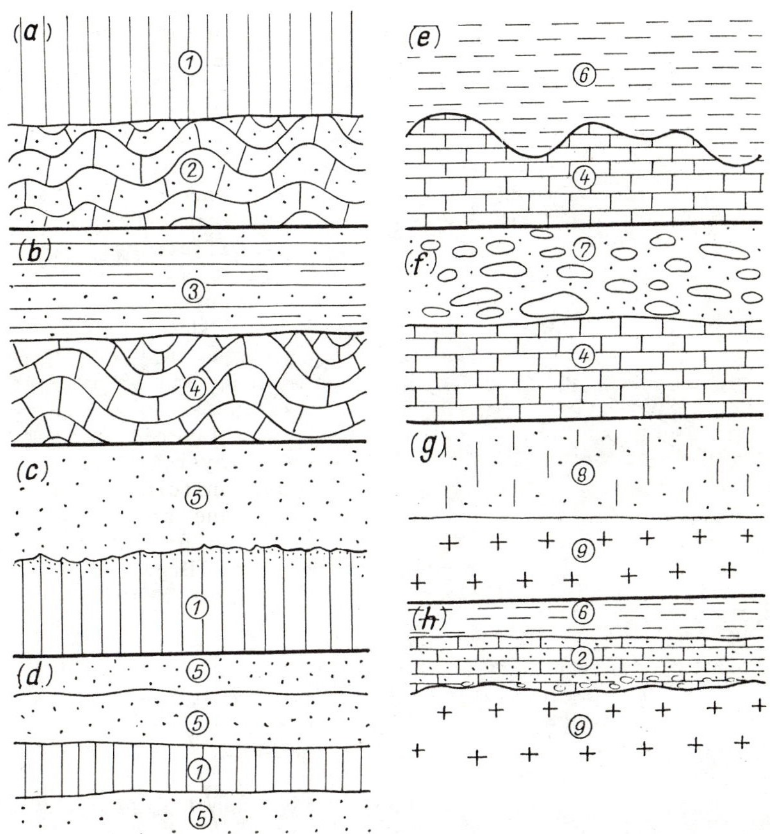

Fig. 1-4. Various types of unconformities. a: lava flow (1) resting on folded sandstone (2). b: loose unconsolidated layer (3) resting on carbonate rocks (4). c: ash deposit (5) overlying a lava flow (1) with erosional surface of its upper vesicular part. d: lava flow (1) with unconformable ash bed (5). e: eroded limestone (4) surface overlain by shale (6). f: bouldery till (7) resting on a limestone surface (4). g: loess (8) overlying a granite substratum (9). h: clastic elements (6) and (2) lying upon igneous rock (9).

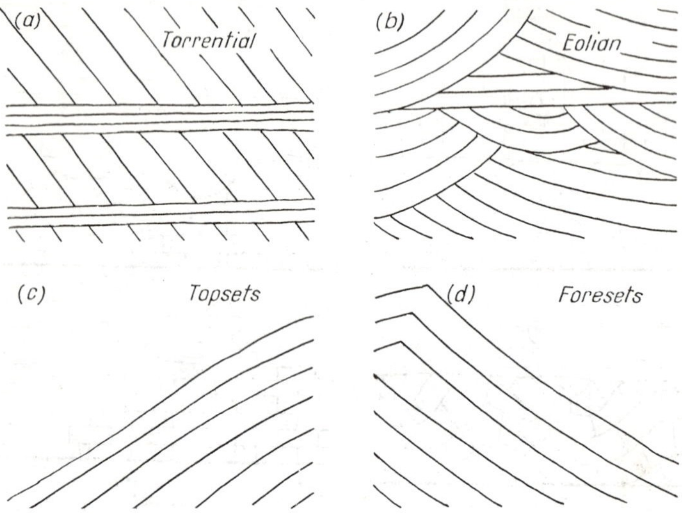

Fig. 1-5. Various forms of bedding. a: torrential cross-bedding. b: eolian cross-bedding. c: topset bedding. d: foreset bedding of a dune.

A very important type of unconformity is produced during the course of sedimentation when older sediments are overlapped by a younger one. Two kinds of overlapping are recognized: transgressive and regressive. Transgressive overlapping is illustrated by the case in which the sea is advancing or transgressing upon a low shelving land. A vertical section through this formation displays finer clastics over coarser ones. Contrary to this, marine regressive overlapping is produced where the sea recedes or regresses from the land, thus lying coarser clastics over the finer sediments. Both cases are illustrated in Fig. 1-6.

These are the most important phenomena controlling the development of sedimentary formations during the process of depositing transported materials. The evolution of the sedimentary rocks has not yet ceased, post-depositional geologic processes are in many cases of greater significance to aquifers (from the hydrogeological aspect) than the original rock features. The effect of tectonic forces will be considered in Subsection 2-2. Here the diagenesis, i.e., the process leading to lithification of rocks, will be discussed.

Fig. 1-6. Forms of regressive (1) and transgressive (2) overlaps with ideal sections to illustrate the sequence of deposit.

Silt or mud, sand, and gravel are unconsolidated mechanical sediments. Under certain conditions these may become mudstone, sandstone, and conglomerate, respectively. In like manner, all clastics have their consolidated equivalent.

At greater depth, the consolidation by cementation is a very common process. The cementing material may have been transported in or more often has been derived within the deposit. The most important difference between the consolidated and unconsolidated clastic sediment lies in the presence of cement in the former and its absence in the latter.

During long continued sedimentation the older strata are covered and compressed by increasingly thick and heavy overburden. The compression, although produced by gravitational force acting downward, is transmitted in all directions such that the sedimentary material becomes more compact both across as well as along the bedding. The reduction in volume, the compaction, may vary considerably in sedimentary materials. Sandstones and limestones are practically incompressible. Most of the compaction occurring in sandstones and limestones takes place before cementation begins. A mud, clay or shale may show considerable compaction, which may be over 40 percent of the original volume. The volume of the average clay decreases to about 13 percent when indurated into shale.

It was stated earlier that petrography is the first and most important consideration in the study of hydrogeological formations. The main petrographical parameters used in hydrogeological description of rocks are porosity and permeability. The porosity determines the amount of water which can be held in storage, and the permeability determines the ease of withdrawing the water for use. In general, a distinctive porosity and permeability may be attached to a given type of rock. Although the lithological description of rocks and not a petrographical one is given earlier (this will be discussed later), mentioning data of porosity and permeability cannot be avoided. The exact definition of porosity and permeability as well as the detailed description of rocks using petrographical terms will be found partly in Subsection 2-2 and partially in Parts IV and V. The term lithological facies is frequently used in geological descriptions. It is defined as the lithological character of a stratum or formation.

The search for groundwater commonly starts with an investigation of non-indurated sediments. Among the many reasons for this preference, let us mention the following:

- these deposits are easy to drill or dig;
- they are most likely found in valleys, where the groundwater level is close to the surface;
- the specific yield is generally higher than in other materials.

Non-indurated sediments can be subdivided, on the basis of origin, into a large number of categories such as alluvium, till, loess, etc., which are quoted in Subsection 2-4. Residual sediments, although not transported, have many hydrogeological characteristics in common with alluvium.

The non-indurated sediments consist of clastic rocks transported by erosion. Small changes in the origin of transported material and in the transporting process itself cause changes in the deposit, thus creating the typical flat-lying bedding. The thickness of strata depends on the rate of supply of material and the duration of the uniform condition of sedimentation. Beds may vary from very thin to several tens of meters in thickness, though beds of a few meters are most common. The total thickness of the whole sedimentary formation may vary from a few meters to thousands of meters and the lateral extent may range a few to hundreds of kilometers.

Regarding the petrographical parameters, porosities range from a minimum of 20 percent to about 90 percent, the former in coarse alluvium, the latter in soft mud. Permeabilities vary in an extremely wide range of nine orders of magnitude.

The material transported by water is sorted according to the weight and size of particles: the coarse, more permeable clastic material generally occurs near the base of a continuous formation and closer to the source area. Sediments deposited in the same time period may be permeable nearer the source and continuously less permeable the farther away from it.

The matrix, i.e., the solid framework, of a clastic sediment consists of particles of different size. The small particles fill partially or entirely the voids between the larger grains. Almost all sediments have a small amount of cementing material yet they may give the appearance of being completely non-indurated. Clay and colloidal materials, which are present in all but the purest sediments, tend to form coatings on larger particles and open networks next to smaller particles. Inasmuch as clay will expand with hydration, the permeability of clay-cemented material will be strongly influenced by the moisture present. The effect of clay coating in geophysical exploration will be discussed in Part IV.

Among the forms of non-indurated sediments to be discussed in detail are: the river valleys, sedimentary basins, coastal plains, regions of eolian deposits, and glaciated terrains.

River valleys are topographic depressions that are open, although in some cases their floor may be diversified by small basins. The distribution of clay, silt, sand, and gravel within river valleys are exceedingly complicated. General patterns or forms are nevertheless quite regular and predictable. In wide valleys where the river channel occupies a restricted part of the valley, the deposit can be classified on the basis of its topographic form and position in relation to the channel (see Subsection 2-4). Coarsest deposits providing the best aquifers are sands and gravels, and these are deposited directly within the channel. Point bar deposits are special channel deposits formed during high water at river bends. The migration of the river, together with continuous flood deposition, will build up a thicker layer of fine sediments that grades from coarse silt at the natural levee into fine silt and clay within the backswamp deposit. The fine-grained flood deposits reach their maximal thickness where they have filled former cutoff meanders of the river channel. Not all river valleys display the entire sequence of floodplain deposits. Wadis carrying large bed loads of coarse sediment during flood period develop broad floodplains of sand and gravel with numerous interconnected channels (braided streams).

Despite the great lateral variability of river valley sediments, most of the valley deposits have a simple vertical succession from coarse sands or gravels near the bottom of the riverbed to silts and clays at the top. The relative thickness of the coarse and fine units depend on the type of sediments carried by the river and on its geologic history.

Coarse glacial outwash filling a mountain valley and deposits of a wadi near to the origin have only the slightest amount of fine-grained material at the top. On the other hand, rivers suddenly blocked near to their mouths may fill their valleys with silt and clay and will have only slight amounts of coarse material at the bottom of the sedimentary fills.

Large valleys generally are originated by tectonic movements rather than by erosion. Some of these valleys are bounded by long fault system, e.g., the Jordan River, others constitute broad alluvial plains such as those of the Po River and Ganges River. Large valleys are characteristically filled by rivers depositing vast amounts of fragments eroded from surrounding mountains. Other types of deposits may, however, be locally important. The total thickness of such deposits commonly exceeds 500 meters and may in some cases be more than 1500 meters thick.

The effect of compaction of sedimentary fill in great river valleys and in sedimentary lowlands is of importance, although it is often overlooked by hydrogeologists. According to laboratory tests, the compaction of sands and gravels is relatively minor and that most of the volume change is related to compaction of fine-grained material. Subsidence due to the compaction of aquifers and associated sediments and produced not by tectonic movements was demonstrated first in the Santa Clara Valley, California. The regional study of the Great Hungarian Plain suggests that flexures observed in the soft to semi-consolidated thick basin fill are the results of a dual process, i.e., compaction and faulting in the hard-rock basement complex.

Regarding the problems of hydrogeological exploration, the conditions are very similar whether the valleys are either of tectonic origin or of an erosional one. Few geologic clues exist on the surface as to the position of the buried inner valleys. Test drilling and geophysical measurements can be used to obtain reliable hydrogeological information relating to the thickness and types of alluvial deposits in river valleys.

Sedimentary basins. The term basin is used here for topographic depressions which are rimmed around all sides. They may be deep or shallow, large or small and they may or may not contain water.

Two great units of sedimentary basins may be distinguished: structural and depositional. Where the two forms are merged, they are called composite basins. The basin found to result from the direct effect of faulting or of tilting of the underlying rocks is a structural (tectonic, diastrophic) basin. When deposition by wind, water, ice, or vulcanism is irregular, or where natural barriers are thrown up across valleys, depositional basins may result.

Regional sedimentary basins are found in bowls formed by continental bucklers. Famous aquifers occurring in this type of setting are the continental sandstones of North Africa (Desert d'Ouest), the Dakota Sandstones of the United States and the aquifers of the Great Australian Artesian Basin (Fig. 1-7).

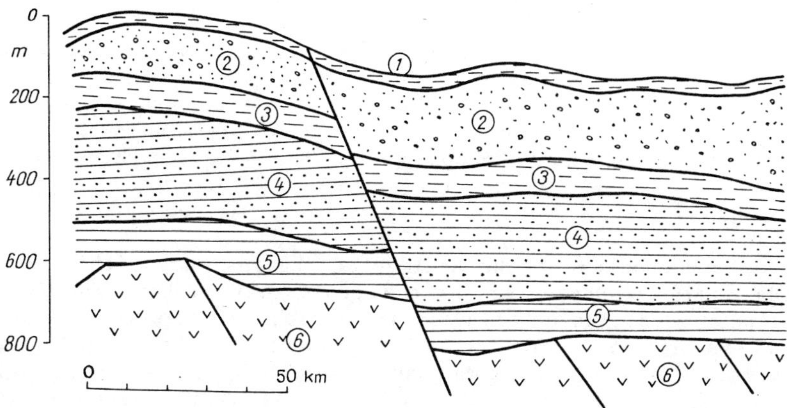

Fig. 1-7. Continental basin with folded sandstone aquifer. (1) sandy-silty cover, (2) sand and gravel, (3) impermeable clay layer, (4) water-bearing sandstone, (5) impermeable shale, (6) crystalline bedrock.

Barrier basins are produced by natural damming. Landslide debris or mudflow debris may choke up a narrow valley and a lake or swamp may be formed. Another type of barrier basin may be made by a sheet of lava which blocks the valley of a river. Barrier basins include lagoons, inland seas and bays which have been closed in by bay-mouth bars. Along a coast, where the mainland is very low, the waves break some little distance offshore, thus churning up the bottom sands and depositing them seawards as an offshore bar (Fig. 1-8). Between the bar and the mainland, a lagoon is formed containing clear saltwater. Streams carry mud and sand into the lagoon from the landward side and wind blows the dry sands of the bar into it, filling up the lagoon gradually. If there is an abundant supply of sand on the bar, and if winds blowing from definite directions are prevailing strong, dunes (Fig. 1-9) may be constructed due to the erosional effect of wind.

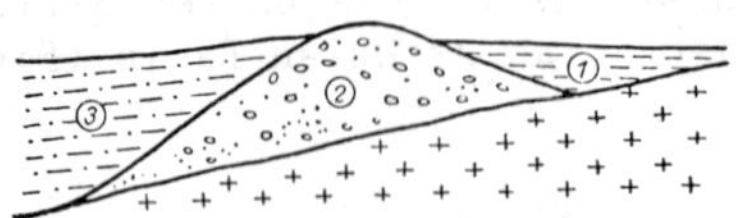

Fig. 1-8. Cross section of an offshore bar with fluvial deposit (1) at the lagoon side of the bar (2) opposite to the sea (3).

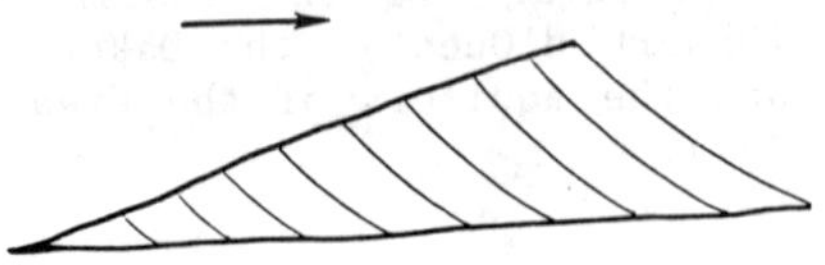

Fig. 1-9. Cross section of a dune migrating by the blowing sand from windward side to lee side, in direction of the arrow.

The deposits of the sedimentary basins vary within very wide limits. Depending on the geological history, beds of marine, lacustrine and fluviatile origin, pyroclastic, eolian, and even glacial materials are present in the sedimentary column. The basin fill may contain glacial deposits in cool zones (Fig. 1-10), clastics and evaporites of marine origin (Fig. 1-11), debris eroded from the surrounding rocks (Fig. 1-12) in temperate zones, and weathering products in tropics (Fig. 1-13). The total thickness of the sedimentary formation will range from about 500 meters to about 2000 meters.

The lithological characteristics and the process of compaction of deposits of sedimentary basins are quite similar to those of large river valleys. This problem was discussed above.

Geological and geophysical logs together with surface studies may be used to depict a useful picture of subsurface geological conditions. Although individual aquifers are hard to trace, reliable stratigraphic information, relating to water-bearing formations, can be obtained.

Coastal plains. Coastal plains vary in size from small isolated valley deposits to vast almost featureless plains of thick sedimentation. The sediments generally represent both alluvial as well as marine environments. In some areas alluvial and deltaic types (Fig. 1-14) are common. Many stratigraphic units along coastal plains grade oceanwards from partly alluvial deposits into entirely marine units, accompanied by the tendency to become progressively fine-grained. The depositional zone nearest to the shore is characterized by wide variations in the type of deposits such as sands, silts or mud, clays, shales, and organic remains (Fig. 1-15). In

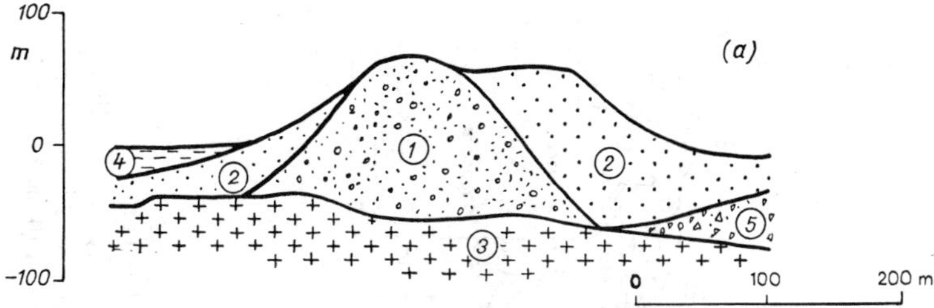

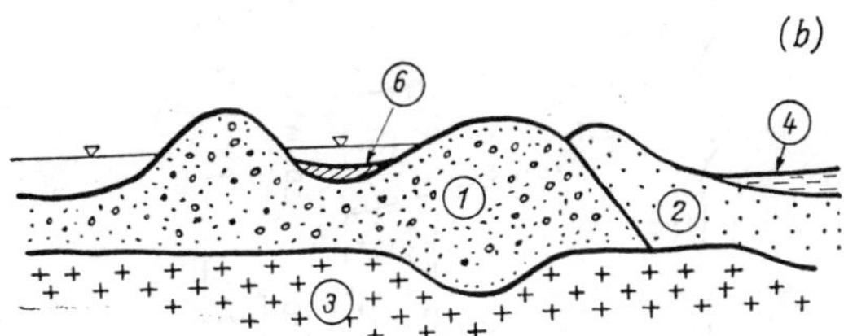

Fig. 1-10. An anticline-type esker on flat terrain and a kame field. The water table in the kame field (b) may be on a rather high level owing to the fine-grained material (6) accumulated on the bottom of the kettles (after Lahermo, 1973). (1) coarse glacial deposit, (2) fine-grained sediment, (3) granite bedrock, (4) peat, (5) till.

the hydrogeological exploration, test drilling and geophysical studies are the suitable methods.

Regions of eolian deposits. Deposits transported by wind are less abundant than either fluvial or glacial deposits. These deposits occur throughout many parts of the world such as Nebraska, Middle China, the littoral zone of the Red Sea, etc. This kind of sediment can be divided into two types: dune sand and loess.

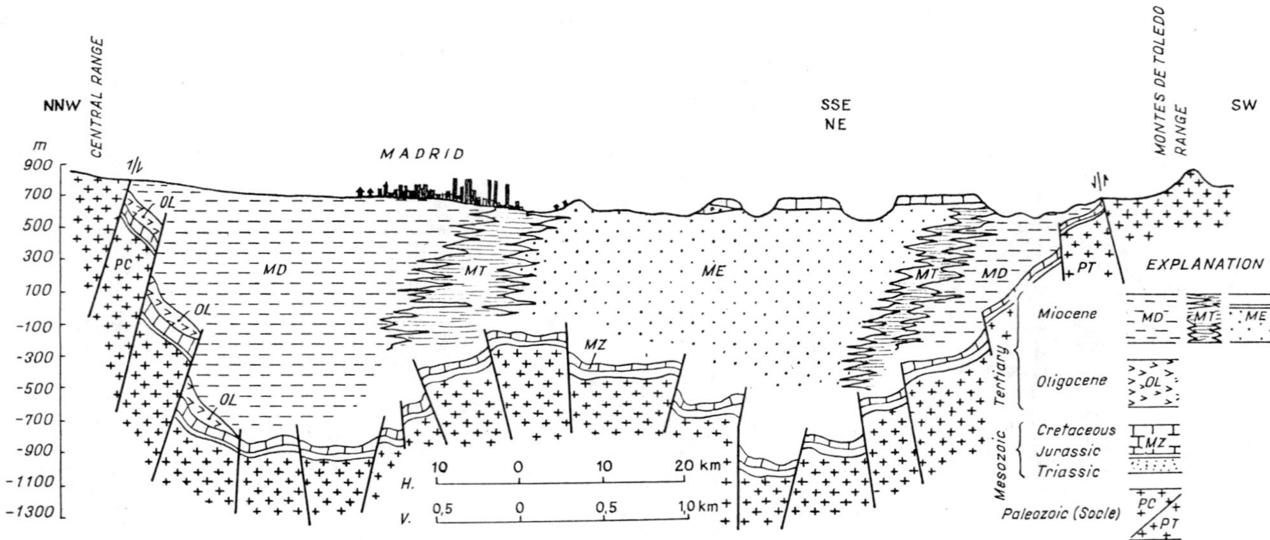

Fig. 1-11. Graben-like granitic structure filled with sediments of varying facies. MD: detrital facies, clay, and sand. MT: transitional facies, clay, marl, and gypsum. ME: evaporitic facies, gypsum, marl, and limestone. OL: marl, gypsum, and conglomerate. MZ: limestone, marlous limestone, marl, and sand. PC/PT: igneous and metamorphic basement (after Llamas, 1976).

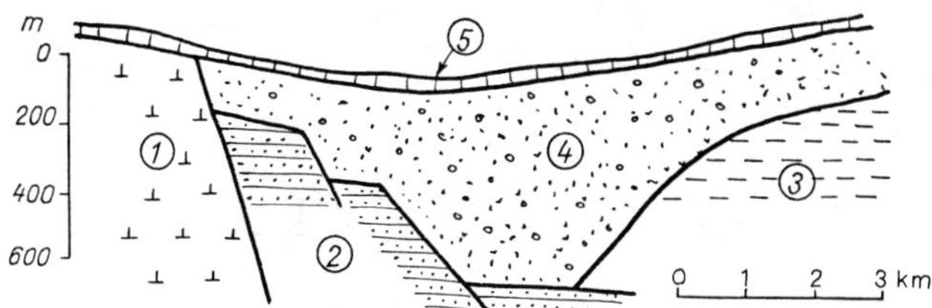

Fig. 1-12. Cross section of a structural basin in South Mongolia. (1) impermeable crystalline bedrock, (2) water-bearing sandstone, (3) impermeable clay, (4) clayey sand with fragments of sandstones and boulders, (5) sand overburden with silt spots.

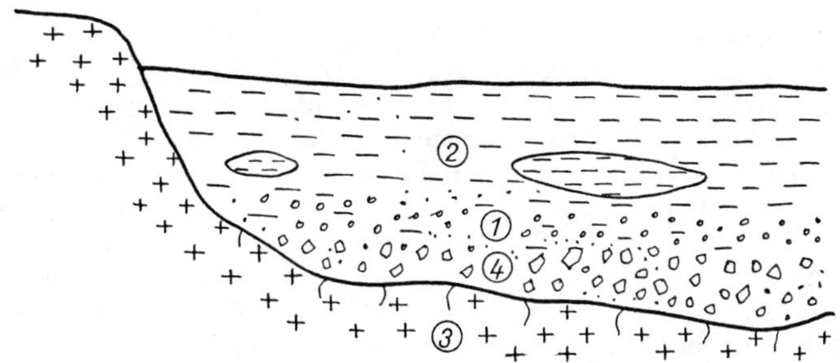

Fig. 1-13. Continental shield area. Cross section of a lateritic basin. The only aquifer of the area, the so-called flowing formation, consisting of clay and fine fragments of mica (1) was developed in the lateritic basin fill (2) of the bedrock (3) depression at the contact of the laterite with the fragmentary part (4) of the bedrock (after Barneman, 1973).

In arid regions and along the shore of seas and lakes (somewhere along or near to the courses of streams), dunes may be piled up if there is an abundant supply of sand and if wind is strong enough, blowing from certain definite directions, e.g., onshore in case of seas and lakes (Fig. 1-9). Regarding the grain size, the dune sand belongs to sands of medium to very fine grades. It is well-sorted having porosity between 35 and 40 percent, and permeability varying from 5 darcys to

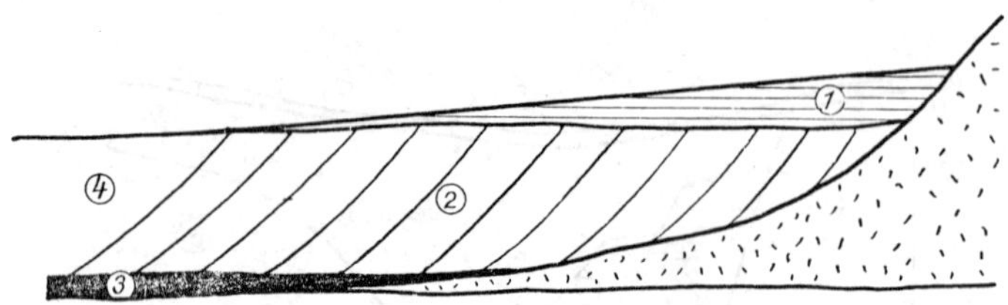

Fig. 1-14. Delta formation showing topset (1) foreset (2) prodelta clay (3) at the seashore (4).

50 darcys, or, in hydraulic conductivity from 5 x 10^{-5} to 5 x 10^{-4} m/sec.

The particles of loess belong to the coarser range of silts. The particles are transported by wind but the mode of deposition influences decisively the packing of particles.

The coarse loess is deposited directly on arid areas forming strata of considerable thickness, e.g., the loess-walls in China. It has porosities of 40 to 55 percent. These high porosities are possible because individual silt grains tend to be held in rather open networks by the cementing action of small amounts of clay always present in the loess.

Contrary to this, where the particles of loess are deposited in water, e.g., in lakes, the clay will expand by hydration and both porosity and permeability of loess strata decreases. Values of porosity of less than 15 percent should be characteristic for fine-grained or lacustrine-type loess. The permeability or hydraulic conductivity of this type of loess displays 10^{-2} darcy or 10^{-7} m/sec on the average. Loess is not regarded commonly as an aquifer because of its low permeability, though the porosity of coarse loess is comparatively high. Where coarse loess overlies impermeable stratum, or consolidated rock, water can be accumulated only within the lower part of the loess formation owing to the great and rapid capillary rise characteristic for loess strata.

Regarding the problems of exploration, the eolian cover makes the application of hydrogeological surveys difficult, especially the use of geophysical measurements.

Glaciated terrain. The lithological character of glacial deposits will be better understood if the climatic conditions of the Quaternary period are considered. The climatic changes in the Quaternary had been of global consequences. The accumulation of ice during the glaciations in the higher latitudes

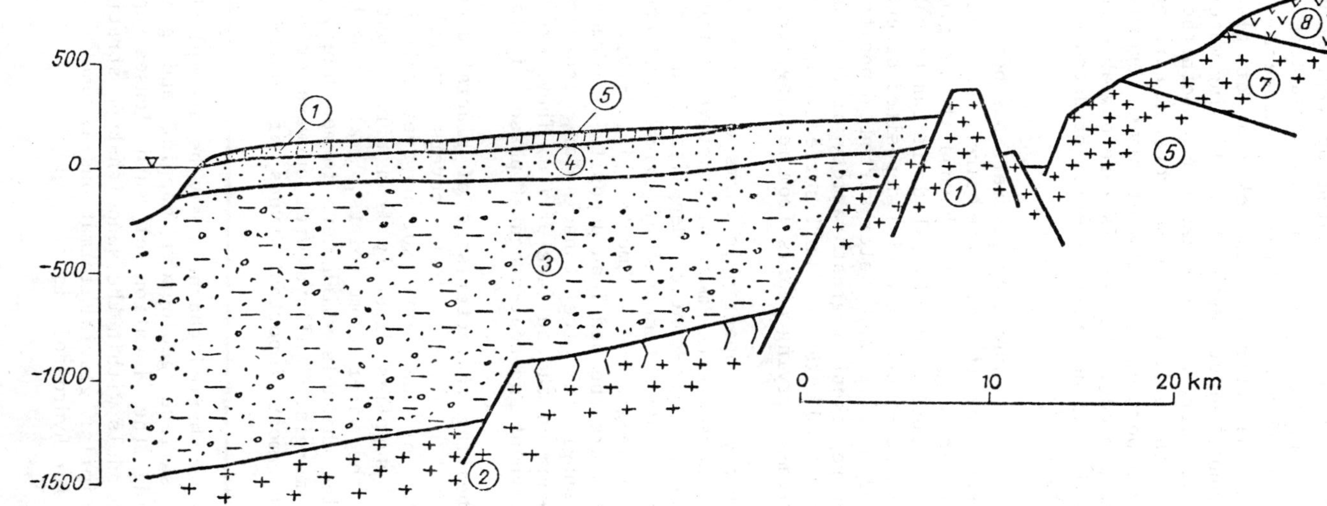

Fig. 1-15. Cross section of the southeastern coastal plain of the Red Sea. The graben structure built up of volcanic rocks (2) is overlain by a thick layer of impervious clay (3) covered by a fluvial deposit consisting of sand and gravel (4). The surface is blanketed by flowing sand (1) at the seashore and by silt (5) near to the mountain.

of the continents extended the ice cover to about 30 percent of the land surface of the earth. Within this period, large amounts of coarse detritus, originated by weathering, were made ready for transportation not only on the surface, covered by ice sheets, but also on the broad belt bordering the ice caps. The arid character of the climate of glaciations decreased the amount and strength of the flowing waters which were thus unable to carry away the detritus. Both the melting of ice as well as the uplift of areas covered earlier by ice increased the transporting capacity of rivers enormously in the non-glaciated periods. The rivers were now able to transport and re-deposit the detrital material produced during the preceding glaciation.

Much of the debris removed by glaciers has been deposited near to the margins of the glaciated areas or laid out as outwash along the drainage channels beyond the margin of the glacier. Glacial deposits or glacial drift can be subdivided into poorly sorted non-bedded material deposited directly from glacial ice called till and water-sorted deposits. Ice-contact deposits represent a great variety of textural types ranging from silt and clay to well-sorted coarse boulders deposited by streams carrying debris from melting ice.

A large variety of topographic forms were also produced among which the eskers are the most distinctive of the various landforms composed if ice-contact deposits. They are typically long sinuous ridges composed of coarse sand and gravel interbedded with fine-grained sediments (Fig. 1-10). Regarding the origin, eskers are the bed load deposits of former streams that occupied subglacial ice tunnels or, less commonly, deposits of streams on the surface of ice. They were probably formed during stagnant and near-stagnant phases of glaciation.

The water-bearing characteristic of glacial deposits is rather complicated. Glacial drift covers several million square kilometers of the earth surface yet relatively few wells draw their supplies directly from till. Most of them probably obtain the water from joints, small sand lenses or from gravel-filled channels which were formed during retreats of the glacial margin. Valleys with outwash or preglacial alluvium are the most important types of aquifers within glaciated regions.

Exploration of buried channels may be accomplished by a combination of surface geological mapping and geophysical methods. Finding the location of sand lenses and small gravel-filled channels within the sandwich-type stratification of eskers and similar glacial formations is one of the most difficult tasks for hydrogeologists.

The indurated types of the sedimentary rocks are of no less importance in water supply of large territories, for example the Nubian Sandstone Reservoir or the karst in Italy. Shale, claystone, siltstone, and other fine-grained detrital rocks account for roughly 50 percent of all sedimentary rocks. Next in abundance are sandstones, then the carbonate rocks.

The lithological character of sandstones is determined mainly by the processes affecting them during their post-depositional history. Among these effects the cementation is the most important. Common cementing materials are clay minerals, calcite, dolomite, and quartz. Clay minerals may be present as original constituents or as a product of diagenesis. Rocks cemented with clay are usually less firm than other types. The porosity of clay-cemented sandstones tends to be quite high, for the clay itself has considerable porosity. On the other hand, silica-cementation can produce orthoquartzites of very low porosity and of extreme hardness. The porosity of sandstones ranges from less than 5 percent to a maximum of about 30 percent. Permeabilities of sandstones are one to three orders of magnitude lower than permeabilities of corresponding non-indurated sediment, e.g., sand.

Considering porosities and permeabilities of extended sandstone formations, firmly cemented sandstones with low porosities and permeabilities will yield water along fractures, which are in general the most favorable areas for development of groundwater. In the hydrogeological exploration of sandstones special attention should be payed to this secondary porosity.

Among the carbonate rocks, limestone and dolomite are the two common types. Clastic limestones and dolomites originate from a large number of different sedimentary deposits such as shells, shell fragments, talus deposit, calcite sand, etc. The chemical precipitates consist of mineral matters precipitated at the place where the rock was formed and not transported as in the case of clastic carbonates. Chemically precipitated carbonate aquifers are the crystalline limestones and dolomites with more or less siliceous material. Clastic carbonate rocks are always more or less cemented with re-crystallized calcite. When the cementation has gone far, it is hard to tell what proportion of a carbonate rock is clastic and what is chemical precipitate.

The original porosity and permeability of many of these sediments are modified rapidly after burial. The more important changes are caused by compaction, solution of aragonite or calcite, re-precipitation of calcite cement, and formation of mineral dolomite. As a result, the original porosity is reduced in all except some young limestones. The permeability decreases due to these effects. The original pore space sometimes retained in old limestones is of less importance from the standpoint of water exploitation, but other forms of porosity are very considerable ones. These are fractures and secondary solution openings, though the post-depositinal change from calcite to dolomite also creates considerable pore space (without increasing the permeability).

Erosional and depositional processes are slow and continuously changing with time and space. As a result, alternating beds of shale, limestone, and sandstone are characteristic of most indurated sedimentary sequences. The individual beds may be thin but sometimes they are of considerable thickness. With the aid of hydrogeological study of the depositional processes, i.e., their history, it is possible to predict their water-bearing possibilities. In further studies, the location of fractured zones and solution openings (cavities) are the most important tasks. The horizontal openings, which tend to remain open are better developed near faults, so that traces of vertical faults at the surface can also be used to locate the openings. The location of cavities or caverns in limestone terrain is rather difficult either by geological or geophysical methods.

A special case of (young) limestones is found in atolls and in some seashores where the seawater moves in and out of the sediments through solution openings and channels in response to tidal fluctuations. The position of the channels, the seawater intrusion through them, and the geometry of the freshwater lenses can be traced by complex hydrogeological prospecting.

Most of the fine-grained detrital rocks have relatively high porosity but very low permeability; they serve therefore as reservoirs for artesian systems. Most commonly, the fine-grained rocks constitute barriers to the movement of water. This is the hydrogeological significance of the fine-grained sediments.

2-2 Movement of the Crust Resulting in Special Structures

Very conspicuous manifestations of the forces acting in the lithosphere are the living faults, e.g., the San Andreas fault in California. Within the lithosphere, enormous numbers of similar forms can be found which are dead at present but at one time, as results of tectonic forces, formed the structure of the strata.

In geometrical description of structural forms, the same terminology can be used as was for characterization of bedding planes (Fig. 1-3); this is because parts of structural forms can be approximated and described using planes and straight lines, e.g., fault plane, axis of anticline, etc.

Importance of tectonic processes must be emphasized because they may modify the hydrogeological conditions in hard rocks, e.g., producing secondary porosity, and also in the soft and semi-consolidated deposits. Faults or folds originating in the basement or in rigid substratum of a thick, loose sedimentary sequence may create permeable zones within the covering strata thus allowing the localized upward flow of deep groundwater. Movement along a fault may bring an aquifer against an aquiclude, forming in this way an impermeable barrier in a permeable formation. Secondary porosity due to tectonic forces and processes also represent a hydrogeological factor, though of minor importance, in non-indurated sediments.

Tilting and folding of strata can be induced by processes such as: settling due to weighting, differential compaction, tangential compression, igneous intrusion and intrusion of rock salt, secondary changes including chemical alteration and unequal weathering, and faulting.

The bending of strata due to unequal compaction is now discussed in detail. Tectonic movement of long duration and affecting large areas are continuously changing the position of parts of continents, thus causing processes of transgression and regression of the sea (quoted earlier in Subsection 2-1 and illustrated in Fig. 1-6). There are many large continental slopes and basins open to the sea where regression is an uninterrupted one-way process of very long duration. The consequences of this long lasting regression are manifested in a thick sequence of loose clastic material which is deposited in fluviatile and deltaic environment covering more and more areas following the retreating sea.

Considering now the average distribution of sediments along the coast, where sands predominate near-shore and mud or silt farther-out, in the process of transgression of the sea the mud phase will migrate over the basal sand layer. Between the principal sand body and mud body there will be a transitional zone of considerable thickness in which the mud phase is increasing upward and the sand phase downward in the series. At any given time during the transgression, the proportion of mud to sand, measured vertically from top to bottom in the growing sedimentary formation, will be greater seaward. The same effect may be observed in compaction of the whole sedimentary column which is different seaward, for the mud will suffer about 15 times higher vertical shrinkage than the sand. Considering the proportion of mud to sand, the result is that the vertical shrinkage of the total formation is increasing seaward from 4 percent to 12 percent, which means the bending of the sedimentary deposit.

A fault is defined as a fracture along which there has been slipping of the contiguous masses against each other. There are situations where the difference between faulting and folding is hardly discernible. The amount of displacement along a fault, measured between points formerly in contact at successively higher or lower levels in the disrupted mass, displays considerable variation, for the broken strata or other rock materials may vary in rigidity and elasticity. In the levels where the displacement is indiscernible, the fault dies out and passes into a flexure. Two terms important in regional structural hydrogeology should be mentioned here. An upthrown block between two downthrown blocks forms a horst and a downthrown block between two upthrown blocks constitutes a graben. In both structures, the movement of blocks is relative, e.g., in a graben the middle block may have gone-down, or the walls on both sides may have gone-up. The principal kinds of structural forms are enumerated and classified in Subsection 2-4.

Many of the hydrogeologically important forms owe their origin to tectonic movements as was quoted earlier in connection with sedimentary formations. The large valleys are generally faulted forms. The narrow types are bounded by a long fault system and the broad ones are characterized by downwarping of the earth's crust. Some of the so-called valleys are divided by ridges and grabens. The structural basin is formed by a block lowered on all sides by faulting, or by downwarping, or by both processes. The deposition of materials in coastal plains is generally controlled by faulting on areas affected by tectonic forces. It is characteristically represented by the Red Sea Shore and by the Gulf of Aden in the Arabian Peninsula where a graben and horst structure is filled by sediments in great thickness (Fig. 1-15).

It was mentioned in the introduction of this section that faulting and folding belong to the post-depositional history of sedimentary formations. Many structural features of large sedimentary basins, however, cannot be explained by post-depositional processes, consequently contemporaneous geological effects must be considered in some cases.

The geological structures are the most suitable objectives of the hydrogeological exploration. The reasons for this are as follows:

- geological clues exist generally on the surface as to the position of the structures which may be located by morphological studies, e.g., air photography;

- there are marked differences in physical parameters within the structure suitable for geophysical exploration;

- the structures are characterized through altered petrophysical parameters of the affected rock, e.g., broken material along the fault plane, or by-products, e.g., breccia, differing in parameters from the surrounding rock.

2-3 Development and Lithological Description of Igneous and Metamorphic Rocks

Igneous and metamorphic rocks are at or near to the earth's surface in more than 20 percent of the land surface of the world. Where it is covered by coarse sediments, as it is the situation in North Europe and in North America, this mantle rock, consisting of glacial deposits of variable thickness, yields most of the groundwater used in these regions. Great areas, however, must rely more heavily on small supplies of water exploited directly from igneous and metamorphic rocks.

Common metamorphic rocks, such as phyllite, slate, and schist, as well as common plutonic rocks, such as granite, diorite, and gabbro, are characterized by very small porosities and permeabilities of small scale. This means that solid pieces of these rocks can be considered practically of zero permeability. On large scale, however, great masses of these rocks show other petrographical parameters, i.e., sometimes considerable porosity and permeability. This secondary petrophysical characteristics is determined by joints, faults, fissures, and solution cavities produced in the rock body by tectonic movement, weathering, and erosion. From a hydrogeological point of view, the most interesting parts are the well-developed fault zones. In general, the most favorable water-bearing zones are in marble and dolomite which have been fractured and partly removed by solution.

In some tropical areas, in regions of intense weathering in the geological past, the weathered surface of granite basement and the overlain product of decomposition form the best aquifer under impermeable clastic cover (Fig. 1-13). Fractures that are not associated with pronounced faults, cavities due to the solution of carbonate minerals contained in many metamorphic rocks and in a small number of igneous rocks, and sheets caused by erosional unloading do not alter significantly the petrographical characteristics of the unfractured fresh rocks.

Volcanic rocks include materials having a wide range of lithographic and petrographic properties. Though the porosity and permeability of the unfractured effusive volcanic rocks (basalt and andesite) varies in larger range than that of the plutonic rocks, the petrographical parameters are mainly functions of secondary structures within the rock. Joints due to cooling, lava tubes, intersecting vesicles, fractures produced by buckling of partly congealed lava, and voids left between successive flows are some of the internal structures that give recent basalt and andesite its high permeability. Owing to this fact, groundwater may flow freely in fractured basalt and andesite.

In the exploration for groundwater, the zones of interest are those which will impede the loss of water and cause the water table to rise to the surface.

Effusive basalt can flow very easily from the place of eruption, covering great areas in comparatively uniform thickness. Where this basalt-sheet occurs, the trap is partly dislocated, owing to tectonic forces, and the characteristic staircase-structure, well known in South Arabia, will develop. In the valleys lying near to the volcanic eruption, the lava flow will bury any alluvium which may be present. Where the valleys contain streams from extensive drainage systems, large thicknesses of gravel and sand may be present, which on burial forms by chance important aquifers. Rivers blocked by lava will form lakes which, if filled by silt, clay, or volcanic ash, represent confining beds for the underlying stream gravel. In areas of extensive volcanism, basins may contain complicated sequences of alluvial, volcanic, and lacustrine material. In addition to this, there are silica-rich lavas, such as rhyolite, of higher viscosity which will be erupted in thick and dense flow or commonly as pyroclastics (fragments).

Unaltered pyroclastics show similar properties as alluvium characterized by poor sorting and abundance of fine material. They consist of volcanic dust, ash, bombs, and broken pieces of older rocks that were shattered by the force of eruption. Consolidated masses of volcanic dust, ash, and lapilli are known as tuff. The agglomerates are consolidated coarser pyroclastic debris with tufaceous matrix. Volcanic ash may decompose to clay, known as bentonite, which represents aquicludes in volcanic formations.

In hydrogeological exploration, the first step will be the reconstruction of the geomorphological history of the area using aerial photographs together with a study of outcrops as an aid in the determination of the position of ancient valleys, buried alluvium, and other features which may influence the flow of groundwater. Dikes are the major vertical barriers in young volcanic rocks representing, in general, impermeable beds offset by faults against permeable layers. The thorough study of outcrops is very important. The distribution of the fissures and openings determined on outcrops will serve as a basis for the calculation of average permeabilities.

2-4 Hydrogeological Classification of Lithological Forms with Special Reference to Water-Bearing Formations

The terms used here for the characterization of particles of loose sedimentary rocks, such as gravel, sand, rock flour or mo, silt or mud, clay, etc., will be explained in Section 3. The major water-bearing formations can be classified for the:

Aquifers Consisting of Sedimentary Rocks:

Non-indurated sediments. Based mainly on the origin of sedimentary rocks, the following typical formations can be distinguished:

- sand formations of coastal plain, sand bar, offshore bar, barrier islands along seashores (Figs. 1-16 and 1-8);

- stream alluvium, alluvial terraces and channel fill, occurring mainly in deltal distributary and in other stream channels composed of gravel and sand (Figs. 1-17 and 1-18);

- dunes and other eolian deposits (Fig. 1-9);

- alluvial deltas and alluvial fans consisting of a mass of ill-sorted alluvial deposits ramified in numerous stringers of well-sorted gravels (Figs. 1-19 and 1-14);

- talus slope and debris (Fig. 1-20);

- glacial drift: till plains and moraines consisting of till deposited directly from glacial ice (Figs. 1-21 and 1-22);

- glacial drift: eskers, kame terraces and outwash plains consisting of ice contact and glacial outwash, respectively, both composing materials are well-sorted, alluvium like sediments deposited in close proximity to glacial ice (ice contact) or somewhat shifted by meltwater streams (glacial outwash) (Fig. 1-10);

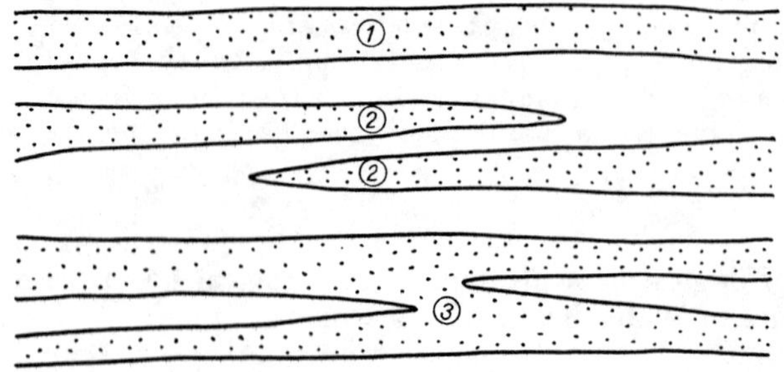

Fig. 1-16. Different forms of sand formations, in cross section, representing continuous sand beds (1), wedge-out (2), and coalescing sand beds (3).

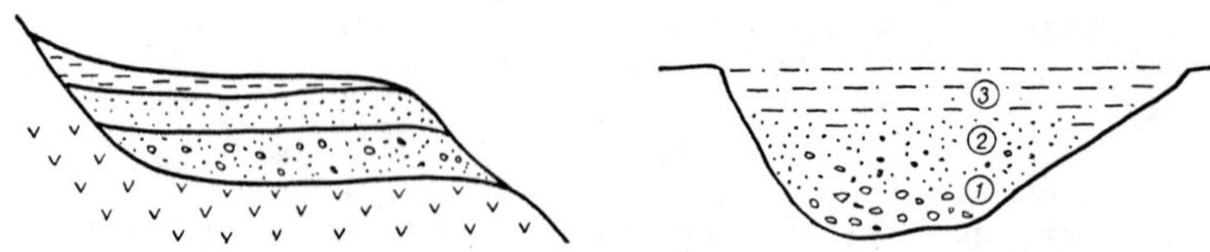

Fig. 1-17. Cross section of an alluvial terrace.

Fig. 1-18. Cross section of a channel-fill containing coarse sand and gravel (1), sand (2), and silt (mud) with fine sand (3).

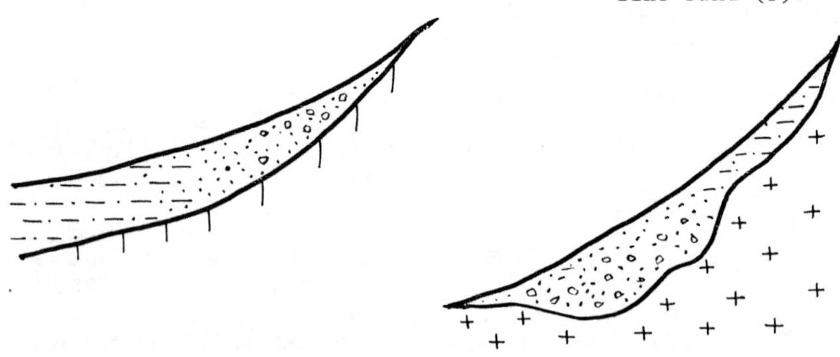

Fig. 1-19. Alluvial fan.

Fig. 1-20. Talus slope.

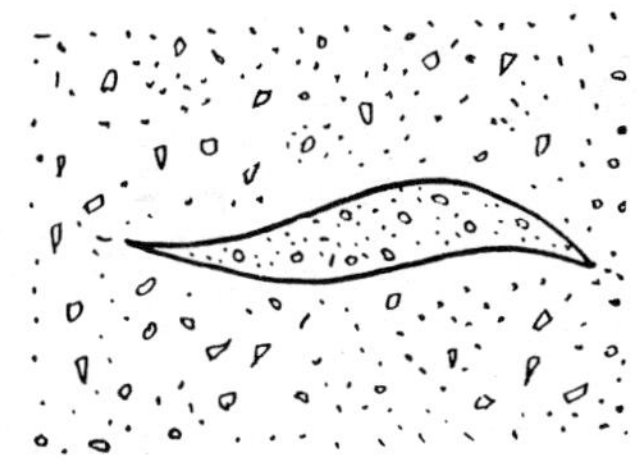

Fig. 1-21. Till containing a lens of gravel.

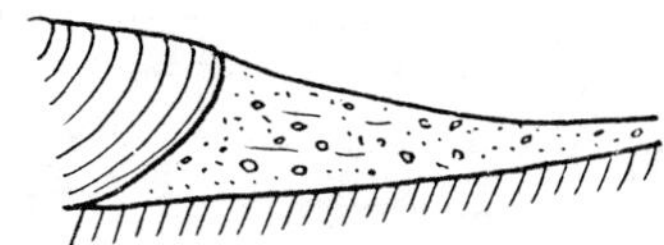

Fig. 1-22. Terminal moraine.

- reefs or organic reefs (bioherms), embedded organic debris in marine sediments having different sizes.

Indurated sediments can be divided into two main groups as follows:

- clastic indurated sediments, such as sandstones, siltstones, shales, argillites, and conglomerates (or intraformational breccias), the latter are sediments containing fragments and particles loosened from an underlying bed;

- chemical indurated sediments, such as limestones, dolomites, chalks, marls, gypsum, anhydrites, and rock-salt.

For the Aquifers Consisting of Igneous and Metamorphic Rocks:

- intrusive igneous rocks, such as granites, gabbros, diorites, etc.;

- effusive volcanic rocks, such as basalts, andesites, rhyolites, etc. (Fig. 1-23);

- pyroclastic volcanic rocks, such as volcanic dust, ash, agglomerates, tuff, breccia, etc. (Fig. 1-24);

- metamorphic rocks, such as quartzites, slates, schists, gneisses, marbles, phyllites, etc. (Fig. 1-25).

A classification according to structural forms and structures of rocks can also be determined.

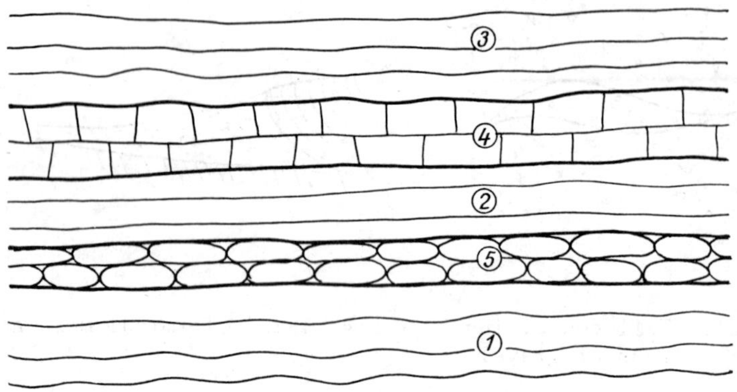

Fig. 1-23. Different flow units of basaltic flow (1), (2), and (3) with various kind of contacts: over jointed crust of flow (4) and below pillow lavas (5).

Fig. 1-24. Breccia structure.

For the Principal Folded Forms:

- synclines represent a downfold opening upwards (Fig. 1-26);

- anticlines represent an upfold opening downwards (Fig. 1-26);

- a monocline is a steplike bend in otherwise horizontal or gently dipping beds (Fig. 1-27);

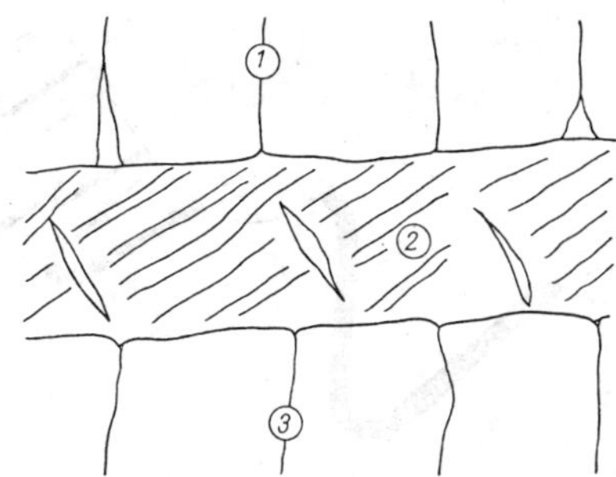

Fig. 1-25. Jointed massive quartzite (1) with fracture-cleavage partings (2) surrounded by a slate stratum with tension joints.

- structural terrace displays a steplike or shelflike flattening of the dip in more steeply inclined strata (Fig. 1-28);

- salt-rock structure (Fig. 1-29).

For the Principal Faulted Structures:

- faults (normal, reverse, dying types), horst, and graben (Fig. 1-30);

- joints (Fig. 1-25);

- thrusts, overthrust and underthrust are faults due to compression, the overlying block is termed a nappe in case of overthrust (Fig. 1-26);

- breccias may be classed among structures as indications of tectonic forces, these are unsorted rocks consisting of fragments contained in a matrix of the same or different material (Fig. 1-24);

- fracture cleavages may be regarded as miniature jointings (Fig. 1-25).

According to the geological setting, various physiographic units may be distinguished. There are single, multi-units, and composite aquifers or an aquifer system composed of the "elements" described above. The major physiographic units of an aquifer system in the lithosphere are as follows: continental shields and platforms, orogenic or mountain-folded areas, great sedimentary basins, intermontane areas, karstic areas, glaciated areas, deserts, lowlands, river valleys, seashores, and piedmont areas.

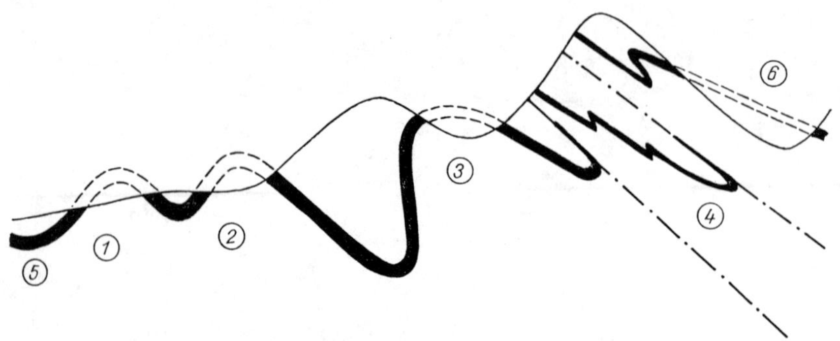

Fig. 1-26. Various types of folding and thrusting, showing symmetrical anticline (1), inclined anticline (2), recumbent fold (3), overthrust fold (4), syncline (5), and nappe (6).

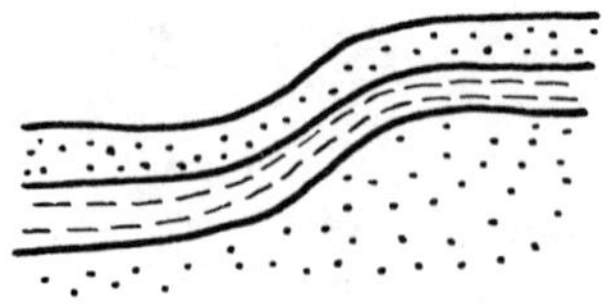

Fig. 1-27. Monocline.

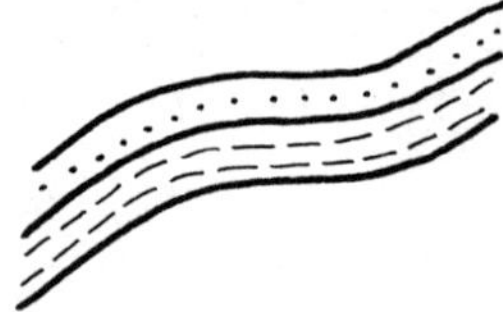

Fig. 1-28. Structural terrace.

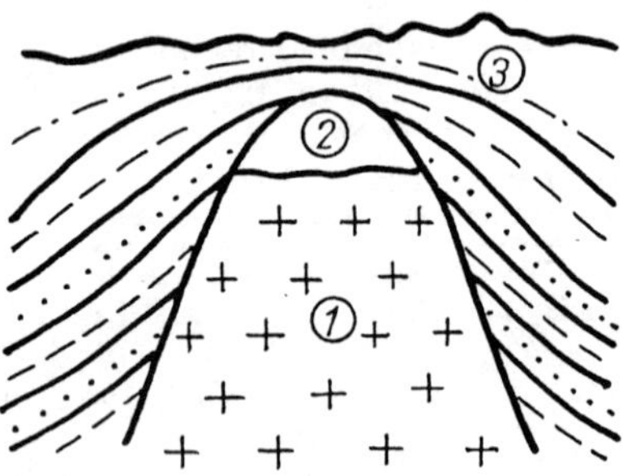

Fig. 1-29. Cross section of a salt dome structure showing rock-salt (1), cap rock (2), and folded strata (3).

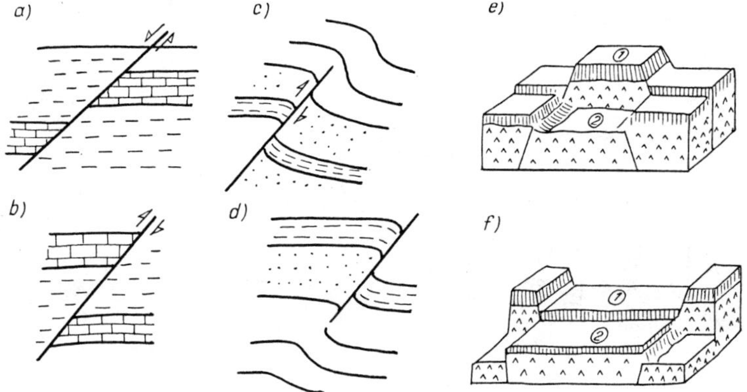

Fig. 1-30. Various faulted forms. a: normal fault. b: reverse fault. c: fault dying out upwards. d: that of downwards. e: horst. f: graben both after faulting (1) and after erosion (2).

These physiographic units, representing different situations from a hydrogeological point of view, can be grouped into a number of categories when taking into consideration other principles of classification. The similarity among different categories, e.g., pyroclastic sediments are similar to glacial drift and this to coarse alluvium, suggests a classification according to the exploration problems of hydrogeology.

Although there is an enormous diversity in the exploration problems of hydrogeology, a few arrangements of basic geological formations, common throughout the world, can be selected as types to study by means of a well-defined system of survey methods. From the point of view of exploration the most frequently encountered, typical regional groundwater schemes can be listed as follows:

- regional sedimentary basins (Fig. 1-7);
- continental shield regions (Figs. 1-10, 1-11, 1-12, 1-13);
- coastal plains, deltaic areas, large alluvial plains of river systems (Fig. 1-15);

- karst, limestone and dolomite terranes (Fig. 1-31);
- volcanic plateaus (Fig. 1-32).

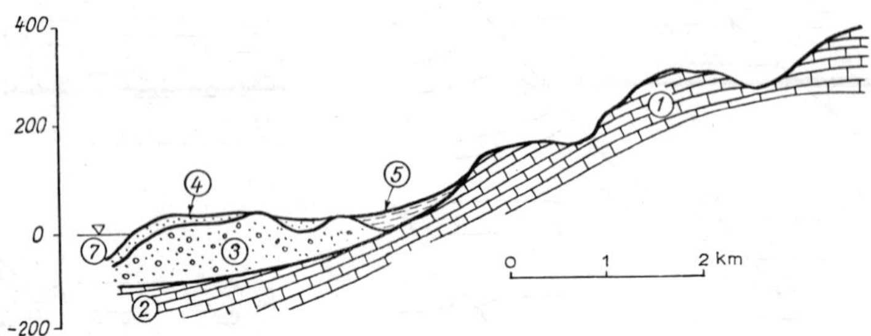

Fig. 1-31. Karstic area at the mediterranean seashore near to Beyrouth. (1) fissured dolomitic limestone, (2) Cretaceous limestone, (3) littoral sandstone called Ramlah, (4) sandy-gravelly layer produced by decementation process of Ramlah, (5) silty-sandy soil (after Mijatovic, 1975).

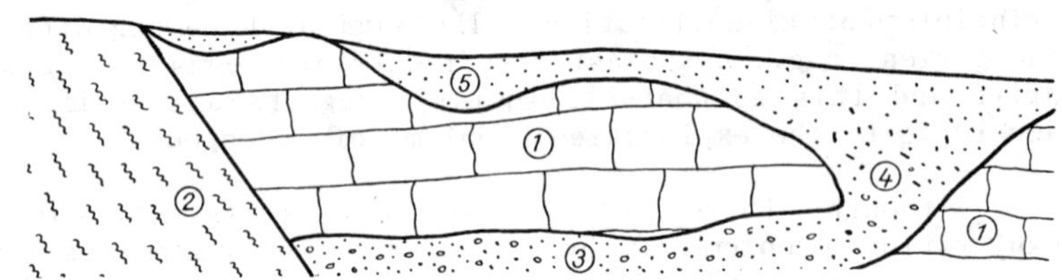

Fig. 1-32. Volcanic area. The basalt flow spreading out (1) from a quartzite block (2) covers the older alluvium (3) forming a window (4) and a buried valley (5) which are the suitable areas for water wells.

When the hydrogeological formations or system of formations should be classified according to the main types of subsurface waters, as it will be discussed in Subsection 2-6, the above mentioned five types will be grouped practically into two categories: loose clastic sedimentary layers and fractured and fissured solid rocks.

2-5 Hydrogeological Exploration and Mapping

The objectives of the hydrogeological exploration, i.e., the various hydrogeological formations (Subsection 2-4), are generally hidden in the lithosphere but with manifestations on the earth's surface. Surface hydrogeological surveying is intended to study these manifestations, to describe the forms observed and the effects measured on the surface, with explanations relating to the composition shape and attitude and possibly to the development of the formations to be explored. The surface hydrogeological surveying is of reconnaissance type, when the regional hydrogeological circumstances should be clarified. In recent hydrogeological explorations, airborne methods are used increasingly which, in the proper sense of the definition, belong to the reconnaissance type of surveying. In the course of a typical hydrogeological project, the reconnaissance type investigation will be followed by detailed surface surveying and this, as the last step in the field work, will be continued by subsurface geological exploration. During the subsurface exploration, data are generally obtained from drilling holes with the intentions: to determine the stratigraphic sequence of rocks penetrated by the hole, to give petrographical evaluation of rocks from a hydrogeological point of view, and to provide data for correlation from hole to hole within the lithological sequence penetrated. Among the types of hydrogeological explorations, the detailed investigations and the subsurface methods will be discussed in Part IV.

The Principles of Reconnaissance Hydrogeological Survey are:

Photomaps. It should be axiomatic that adequate topographic contour maps, which serve as base maps, be available in advance of all hydrogeological (or geological) exploration and mapping. If adequate maps do not exist for an area, photomaps, which are simply controlled air-photo mosaics, can be used. Hydrogeologists as well as those concerned with other mineral resources prefer photomaps to the more conventional line maps because the former assist them considerably with transferring photo-interpreted data to the map.

Geomorphology. Every hydrogeological exploration program begins with the study of topographic forms, i.e., with the application of geomorphology. Geomorphology is the science which treats the surface features of the earth, their form, nature, origin, development, and interrelationships. The pure description and interpretation of landforms as mere isolated phenomena are generally not sufficient. The various relief features of a region as a whole, the general appearance of a land area called topographic expression, is the product of some common cause. The topographic expression is determined by four important factors: rock materials, geological structure, erosion process, and stage of topographic development. By this way, geomorphology can yield lithological, stratigraphical, and structural information of interest.

The presence of permeable glacial sediments such as kames, eskers, and outwash sediments (Fig. 1-10), which are well-defined lithological formations, can commonly be mapped through the study of surface forms in relation to the environment. The topographical expression of a horst (Fig. 1-30) or graben displays one of the most conspicuous phenomena.

Air reconnaissance. The advantage of the use of air photographs in geomorphology is obvious when only considering that large geological units can be distinguished on air photographs which, when inspected in the field, disintegrate into an irregular mosaic. Air reconnaissance is defined generally as the detection of variations of the earth's surface or its properties from a distance. Here, in a restricted sense, air reconnaissance is limited to the extension of the electromagnetic spectrum from the visible region, and the ultraviolet region containing wavelengths shorter than the human eye can utilize. Sectors of longer wavelength, the infrared and microwave regions, are found to be increasingly useful in providing geological data. Air photography includes the use of black and white, color, solar infrared (0.7 - 1.1 μ), thermal infrared (3.5 - 5.5 μ, and 8 - 14 μ) emulsions. Waves used in microwave imagery are longer than 50 μ.

Air photography may help the geomorphological fieldwork in many aspects. The integrating character of the black and white photos was mentioned above. Using stereoscopic examination on paired overlapping air photos, the topographical features, especially dips, can be easily evaluated. This is because the dips observed under stereoscope are enhanced, they appear steeper than their actual inclination. Anticlinal and synclinal folds are clearly shown on air photos and faults as well as joints, which perhaps may be inconspicuous on the ground, may show up very strikingly.

Exploration in areas of consistently poor illumination due to meteorological conditions (clouds, fog) can be accomplished with microwave imagery, e.g., by side-looking radar equipment. This produces a photo-like image which could be described as an instant relief map. It is almost equivalent to an air photograph taken on a clear day when the sun is at a very low angle. The slightest morphological features are accentuated by the shadow they cast, thus facilitating the interpretation of faults and folds. Frequently the surface expression of these features are so subtle that they would not be recognizable on stereo pairs of conventional black and white or color air photos.

The black and white air photos, the color air photos, and the radar imagery furnish information about the structural geomorphology, but the information about the quality of the strata is given only scarcely. Depending on hydrogeological conditions, there are areas where surficial deposits may be depicted when considering different shades on photographs.

Additional data for the structural geomorphology and data for the stratigraphical one can be gained by infrared imagery. Infrared imagery indicates the heat distribution on the surface of the earth. This geothermal map is controlled in one part by the transmission of heat from depth, in the other part by the hydrogeological properties of rocks, the most important of which is the water content of the surficial layers. Infrared imagery has been successfully applied to various hydrogeological problems such as determination of rainfall distribution, localization of zones of infiltration and groundwater discharge, measuring the temperature of water surfaces, etc. (in Part V).

During recent years, evidence has accumulated which indicates that air photos made with passive radiometers, in the 21 cm band of the microwave range (characteristic of the radiation from interstellar hydrogen), show very good correlation with the moisture content of the upper 2.5 cm layer of the soil. The determination of the moisture content may be accomplished by aircraft as well as from satellites (EREP program).

Appraisal of infrared and microwave imagery with regard to investigation of the soil moisture zone will be given in Part II.

Airborne geophysics. There are geophysical methods of the reconnaissance type which may serve as aids in geomorphological studies. The result of these geophysical explorations are represented on maps which, after the necessary corrections, may be interpreted in terms of geomorphology.

The regional gravity map, or more exactly the regionl Bouguer-anomaly map, yields information about the natural variations of the force of gravity which is controlled primarily by the varying density of rocks as well as by their positions in the lithosphere.

The regional magnetic map shows the spatial variations of the magnetic force or sometimes the variation of its vertical or horizontal components. The magnetic field of the earth is modified by rocks of different magnetic properties.

Other airborne electromagnetic maps reflect, in rather complicated manner, the electric conductivities of rocks lying near to the earth's surface (see Part IV).

The effects, measured by airborne geophysical methods, originate within the lithosphere, at various depths. In other words, these mapping methods penetrate the earth's surficial cover thus affording a more continuous picture than that achieved by surficial examination of geomorphological forms and outcrops. The depth penetration of air photography is very small or even zero, for the waves used are reflected from the surface of the geomorphological elements.

From the gravity and magnetic maps, geological features, mainly of structural character, can be interpreted, such as: buried valleys, faulted and folded structures, basalt plateaus with windows, large dikes, etc. The airborne electromagnetic map yields more information about stratigraphy. The electromagnetic type of survey can also be used to map the electric resistivity of rocks lying at relatively shallow depths. Buried channels and spatial variations in water quality (seawater encroachment) can be mapped in this way since the electric resistivity of rocks is determined by the grain structure (matrix) of rocks as well as by the chemical composition of water filling the pores.

The airborne geophysical exploration can be considered as a geomorphological study only to a certain extent. Its greater emphasis belongs to structural or stratigraphical geology. The methods of airborne geophysics together with air photography were discussed here because they belong methodically to one big group of methods of exploration called remote sensing. The term remote means that the registering or measuring instruments are far from the object of sensing. The farthest air photos are produced by satellites equipped with suitable sensors, e.g., the ERTS program. It is necessary to emphasize at this point, that although the methods of remote sensing give very valuable help to the geomorphological studies, they cannot substitute for field work in geomorphology. The interpretation of the maps prepared by remote sensing exploration cannot be executed without using as much field data as possible.

Hydrography. The survey of hydrographic systems belongs to the reconnaissance methods of hydrogeological exploration. The drainage patterns, perennial and intermittent streams, watershed divides, springs and lakes, are characteristic features of topography. Patterns of valleys, or drainage patterns, depend on the distribution of bedrock, the attitude of strata, and the arrangement of surfaces of weakness (such as joints or faults). Consequently, hydrography may be used to assist in the interpretation of geologic structures during the geomorphological studies (Fig. 1-33).

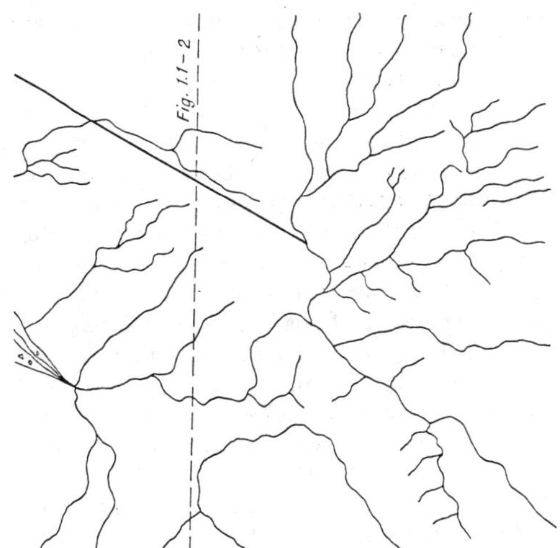

Fig. 1-33. Angular drainage pattern controlled by fault-system. This map shows the eastward continuation of the area represented in Fig. 1-2.

The Hydrogeological Mapping and Types of Maps are:

The complete hydrogeological exploration, i.e., fieldwork and interpretation together, includes the following studies: surface studies (geomorphology, hydrography, hydrochemistry, air reconnaissance, and geological field work) and subsurface studies (surface geophysics, lithology of boreholes, geophysical logs, technical logs, geological logs, stratigraphical and structural interpretation of geophysics). The result of the exploration, the hydrogeological information for a given area, must involve:

- delineation of lithological units, boundaries, and contacts;
- determination of size, shape, and position of aquifers;
- determination of hydrostratigraphical units;
- determination of petrophysical properties (porosity, permeability) of aquifers.

The final form of the hydrogeological information and the evaluation of data are presented by various illustrations such as maps, sections, and logs.

The logs show the distribution of rocks characterized by petrographical, lithological, or other similar parameters in a vertical line, e.g., the distribution of sand and clays in a well or the distribution of porosity along the drilling. Cross sections are sections in the vertical plane constructed from data obtained in the studies of outcrops, excavations using data of surface geophysics, and mostly from borehole data. The hydrogeological cross sections are constructed generally by correlating the well-logs between pairs of wells. The geological sections of formations are generally far from being accurate in the information which they convey about the thickness and depth of individual layers. When the sources of information, generally the data obtained from boreholes and from soundings performed from the surface, are especially abundant and reliable, the formations may be better depicted in structure contour maps than by the cross section maps. A structure contour map depicts the configuration of a surface, determined by hydrogeological parameters, in a horizontal plane by lines of equal elevation generally referred to the mean sea level as the datum. In preparing a structure contour map of a formation, the surface chosen is that of the top or bottom of some layer which is easily recognized and is persistent over large areas. This surface, which separates layers of different hydrogeological (petrographical, lithological, physical, etc.) characteristics, is called the key horizon.

A structure map contoured on some key horizon lying within the lithosphere is known as a subsurface map. If structure maps are prepared for the upper and lower boundaries of a formation, then by subtracting the elevations, the map of the thickness of the formation, called an isopach map, can be constructed. Recalling the meaning of lithological facies, the lithofacies maps furnish important information about rocks in the study area. These maps may indicate the facies in terms of rock types (such as conglomerate, sandstone, evaporite, etc.) or they may be prepared on the basis of varying proportions of clastic, e.g., sand, and non-clastic, e.g., anhydrite, granite, constituents of the rocks, expressed in percentage. To display the areal distribution of petrophysical properties, petrophysical maps are constructed which are very similar to the lithofacies map but are contoured in terms of porosity or permeability. Hydrogeological field studies of non-geological character can be represented in the form of contour maps. These special maps are contoured in terms of physical parameters, e.g., electric resistivity, magnetic intensity, etc. The special maps can be transformed into hydrogeological maps using suitable methods of interpretation. Some of the special maps are so closely related to hydrogeological maps, e.g., to lithofacies maps, that they can be considered as quasi-geological maps. For example, the ground resistivity maps, prepared on a sedimentary area, may display the attitude of the substratum of the aquifer formation, and in coastal plains shows the position of the seawater intrusion directly.

In the preparation of hydrogeological maps patterns, lines, and symbols are used to indicate rocks, formations, and structures, depending upon the desires of the investigator or topographer. It is advisable to use the standard forms and symbols published by UNESCO in the "International Legend for Hydrogeological Maps." In any case, the legend is the key to the meaning of colors, patterns, and other symbols and therefore should be obtrusive on each map.

The results of a hydrogeological exploration are presented, in very comprehensive form, as the maps attached to the exploration. They should be chosen carefully in order to fulfill the type and objectives of the exploration. Among the many kind of hydrogeological maps and cross sections listed, there are illustrations which cannot be disregarded irrespective of the objectives of the survey. These are the following accompanying maps and cross sections:

- topographic map (or air photo);
- geomorphological and hydrographical map;
- stratigraphical maps and cross sections;
- structural maps and cross sections;
- contour map of substratum (in sedimentary areas);
- quasi-geological maps (remote sensing, geophysics);
- petrographic maps or cross sections.

These maps together with the hydrological and hydrochemical maps will aid the assessment and evaluation of the hydrological potential of a given area represented and illustrated finally by the groundwater probability and groundwater availability maps.

2-6 Occurrence and Types of Waters Stored in Various Layers

The rocks, as lithological units, were discussed in the earlier parts of this section mainly with respect to their solid framework (matrix). In this subsection, the water, filling the pores and interstices, will be considered.

Subsurface waters can be classified in many different ways, taking their various properties into consideration as the basis of the classification, e.g., temperature, chemical composition, character of movement, origin, etc. From the hydrological point of view, the two most important classifications are those which can be formed according to: the character of the water-bearing rocks, and their connection with the meteorological and hydrological processes on and above the surface. According to the classification based on these two characteristics, the soil-moisture zone and the groundwater zone can be distinguished.

Soil moisture is the water content of the pores of soil above the water table, retained in various forms against gravity. The most important forces resulting from the development of a soil-moisture zone are adhesion and capillarity. Adhesion creates a thin film of water covering the walls of the pores (adhesive water) while capillarity fills some of the pores completely with water (capillary water). In fractured and fissured solid rocks the same type of water is called moisture content of rocks. The soil-moisture zone will be investigated in detail in Part II.

For the complete clarification of definitions used, it is necessary to analyze the term water table. According to the generally accepted meaning, it is the surface of unconfined groundwater bodies dividing the groundwater and the soil-moisture zones (Fig. 1-34). Another characteristic feature of the water table is that the pressure of water included in the pores is equal to that of the atmosphere. It is necessary to note that the water table is not a real surface, only a fictitious level because here there is no rapid change in the saturation of pores. Thus, the level of the water table is indicated only by the change in the sign of the water pressure, being positive in the groundwater zone and negative above the water table. Having established the correct definition of water table, it is relatively easy to define the term groundwater. Then, considering the connection of groundwater with the hydrological and meteorological processes on and above the surface, two further groups may be distinguished, shallow and deep groundwater.

Groundwater is that part of the subsurface water which is stored (and/or moves) in the interstices of the lithosphere below the water table, and thus its total pressure is higher than atmospheric. This pressure may be hydrostatic or hydraulic, i.e., influenced by the movement of water, but its surplus value is always positive. In the groundwater zone all the pores and other types of interstices are completely filled with water, or with other fluids, and in exceptional cases, with gaseous media under pressure.

Shallow groundwater is stored near the surface, below the water table and about the first, largely extended, continuous

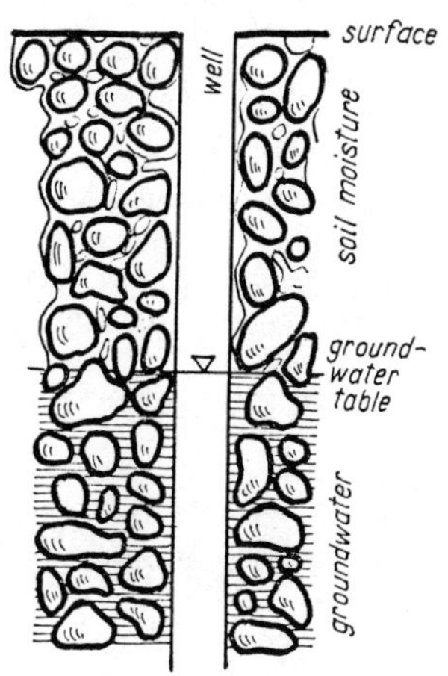

Fig. 1-34. The observation of water table separating groundwater from the unsaturated zone.

impervious formation. It is directly influenced by meteorological and hydrological events (recharged by precipitation, drained by evaporation and transpiration, having interactions with surface waters).

Deep groundwater can be found in aquifers lying below continuous, impervious beds which hinder the direct contact between surface and groundwaters over a very large area, as well as the influence of atmospheric processes on groundwater. It has no direct recharge from precipitation or that from surface waters, and is naturally drained only by the shallow groundwater.

It is very difficult to clarify the distinction between the two types of groundwater. Shallow groundwater may be partly covered by impervious strata, but if the covering layers are not continuous, or large enough, the effects of the meteorological and hydrological events can be clearly observed in the regime of this groundwater. On the other hand, it is also possible that the aquifer covered by impervious formations over a large area may be raised near the surface far from the place of investigation. Thus the same groundwater body may be regarded deep within the investigated area, but becomes shallow at another location. It is not possible, therefore, to draw a sharp line between the two types. There are several transition forms, and their distinction can be based only on the investigation of the dominating factors influencing the regime of groundwater. The means of these investigations are the observed groundwater data.

Subsurface waters can be classified according to their origin as it was made in case of sedimentary rocks. Three types of subsurface water are generally distinguished, according to their origin:

- internal or juvenile water derived from the interior of the earth as a new resource;

- connate or fossil water, a type of external water, which originates from atmospheric or surface water, and may be trapped in rocks at the time when the constituent material was depsited;

- absorbed water, also a type of external water adsorbed into the interstices of the layers some time after deposition, even quite recently.

The amount of juvenile water is negligible from the point of view of water resources development. Coming from very great depths, its occurrence in deep groundwater is more probable than in the shallow one. Because the upward movement of juvenile water generally follows faults, larger amounts may be expected in solid rocks, where it sometimes rises into shallow position. The fossil water, in most cases, preserves its original material characteristics, e.g., chemical composition, form, when the water-bearing layers were formed or when it infiltrated into the latter during their development. Changes may occur as a result of water exchange between aquifers, migration, and diffusion. For this reason, a higher ratio of the fossil water may be expected in layers where the movement of water is hindered considerably. Thus the highest amount is probably contained in cohesive, loose, clastic sediments. The adsorbed water indicates the recent influence of meteorological and hydrogeological processes. It is quite evident that soil moisture is entirely, and shallow groundwater is almost completely, composed of this type of water.

Two two kinds of classification explained above were formed irrespective of the formation containing the water. Considering the lithological situation, e.g., the cross section of a sedimentary basin given in Fig. 1-35, the deep groundwater may be divided into several subgroups, such as:

- closed groundwater, when the aquifer is completely surrounded by impervious layers;

- artesian water, which has recharge area where the water table lies higher than the surface of the investigated area thus producing water level rising above this surface in wells draining the artesian layer;

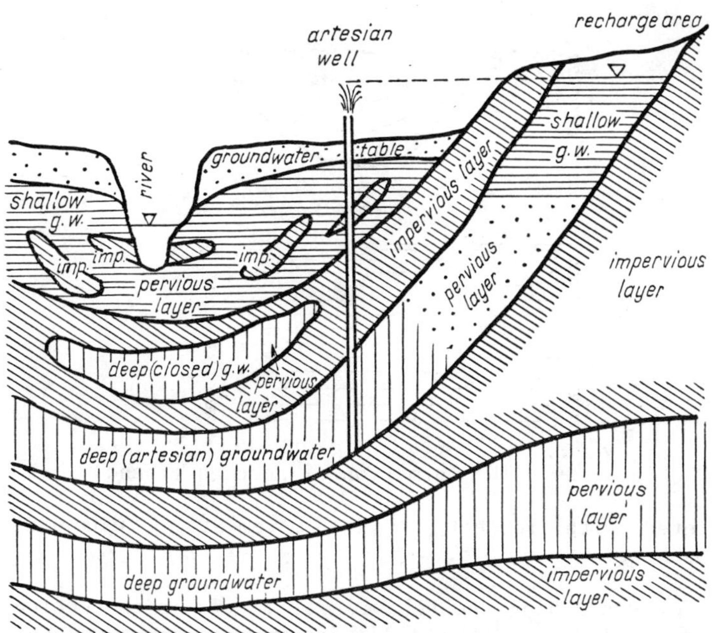

Fig. 1-35. Section representing the various forms of deep groundwater.

- confined water, when the water is contained in an aquifer covered by an impervious formation;

- unconfined water where the impervious covering layer is missing.

An exceptional subgroup should be mentioned here, and this is the water called perched groundwater. When a saturated zone develops above an impermeable lens in such a manner that the zone of soil moisture continues below the impermeable layer (Fig. 1-36), this is perched (or entrapped) groundwater. It can be seen from the foregoing, that such expressions as confined, closed, or artesian aquifer have special meanings and they do not characterize the hydrological behavior of the water-bearing layers completely. It is therefore advisable to use the adjectives of shallow and deep for distinguishing the groundwater directly influenced by meteorological and hydrological phenomena from that recharged or drained only through another groundwater space.

In Subsection 2-4 the classification of rocks was given mainly according to their lithological properties (development and structure). In constructing a full system of classification sufficient for characterizing the rocks hydrologically, the classification proposed in Subsection 2-4 must be

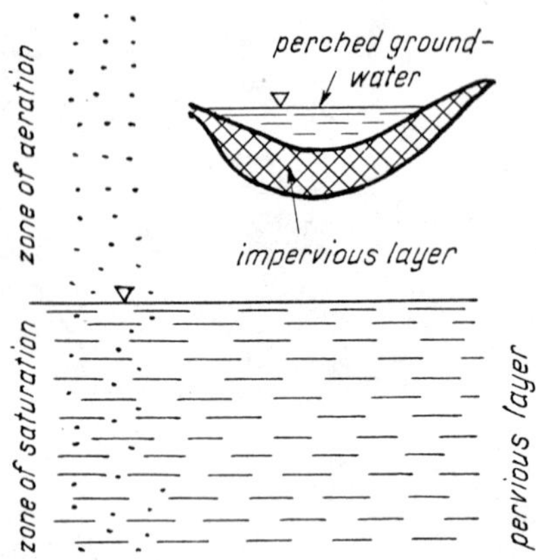

Fig. 1-36. The formation of perched groundwater.

completed with respect to the movement of water in the lithosphere.

For describing the behavior of a rock in connection with the flow of water, the following four terms are generally used: Aquifers are permeable geological formations having interconnected interstices which permit an appreciable quantity of water to move through them under ordinary field conditions. Aquifer, groundwater reservoir, water-bearing layer, and permeable rock are commonly used synonyms. Aquicludes are impermeable strata which may contain a great quantity of water (in some cases more than aquifers, e.g., in clay). Some aquicludes do not transmit water at all, and others only a very small quantity. Aquifuges are also impermeable formations but neither contain nor transmit water. These rocks have no interconnected pores or fissures and therefore cannot either absorb water or allow it to pass through. Aquicludes and aquifuges are opposite types of impermeable lithological formations. Aquitards represent a transition form between aquifers and aquicludes. This type of layer contains interconnected pores, but the water-conveying capacity of the channels composed of the pores is relatively small compared to that in aquifers. However, if the direction of flow is perpendicular to the large bordering surface of such a layer, the amount of water conveyed by seepage is not negligible, due to the great extent of the area. Aquitards are often called semi-pervious or leaky formations.

Considering the lithological types of rocks listed in Subsection 2-4, from the point of view of water movement, similarities may be found among different groups thus allowing the combination of types into two main groups, mentioned in Subsection 2-4, i.e., loose clastic sediments, and solid rocks. The most important difference in rocks involved with water movement occurs between the network composed of the pores surrounded by the individual grains of loose clastic sediments, and that built up by fissures and fractures. This is the reason why the groups mentioned above were selected as the basis of the hydrological characterization. Layers within the first group are constructed with small grains. The pores between the grains form an interconnected network of near uniformly sized channels. In the interior of a solid rock a similar structure is rather rare (sandstones, basaltic lavas). The known hydraulic equations of seepage, e.g., Darcy's law, are only valid in these cases. In the rocks belonging to the second group, water can move primarily along the laminated structure (shale, slate) or fissured and fractured zones (volcanic and metamorphic rocks). There are permeable solid rocks which can transmit water through dissolved openings and conduits which sometimes have become large caverns. These are karst rocks (limestone, dolomite). In these two cases, movement of water must be investigated by accounting for the special conditions, e.g., two-dimensional flow in slates or closed conduit flow in karst channels. The karst rocks, because of their large permeability and importance in connection with the exploration and exploitation of water, will be divided into a separate subgroup among the solid rocks.

According to the two main aspects, whose details were discussed previously, the general classification of groundwater can be summarized in Table 1-2.

Table 1-2. General classification of groundwater.

Location \ Formation Water Is In	Loose Clastic Sediments	Solid Rocks		
	gravel,sand, silt,clay	Karst (limestone, dolomite)		Non-karst
Above the water table	Soil moisture	Moisture content of rocks		
Below the water table	Shallow groundwater	Shallow karst water	Water content of rocks	In shallow position
	Deep groundwater	Deep karst water		In deep position

SECTION 3

CHARACTERIZATION OF THE SOLID MATRIX OF POROUS MEDIA

One of the basic principles of groundwater hydrology states that the study of subsurface waters should always include the investigation of their movement. In general, a special chapter on hydro- and aero-mechanics, dealing with flow through porous media and called seepage hydraulics, differs considerably from the other parts of hydrodynamics. In general, the calculation of hydraulic parameters of a flow system (e.g., distribution of velocity and pressure, flow rate, etc.) needs the solution of the movement equations while simultaneously considering the boundary conditions prevailing along the surfaces enveloping the moving fluid. Investigating a movement that develops in a relatively large, continuous water body, especially if the bordering surfaces have regular forms (e.g., pipes, prismatic open channels), the differential equations can be solved in most cases either analytically or by applying numerical methods. When a seepage field is investigated, the flow developing through the network of randomly interconnected pores or interstices has to be analyzed and the boundary conditions cannot be taken into account at the walls of the irregular channels. Attempts should be made, therefore, to substitute the water conveying network with a regular geometrical model having the same hydraulic resistivity as the original system. The structure of the pores, or that of the fractures and fissures forming the water transporting system, should be investigated first in order to determine the form and the main sizes of the models being suitable to simulate the natural network of interstices, which depends on the type and structure of the solid matrix of the various porous media. This analysis is the subject matter of this section.

The kinematic analysis of water movement always has statistical character, since it is based on the investigation of the motion of fictitious macroscopic water particles instead of the determination of the actual movement of microscopic water molecules. The size of the particles (so called the representative elementary unit) should be so large that the statistical average of the random movement of molecules within such a unit should be equal to zero. In seepage hydraulics the continuum approach must be applied not only for the description of the behavior of the flowing fluid, but also for the characterization of the flow space composed of the randomly changing structure of interconnected pores or fractures. It is necessary, therefore, to determine those smallest units of the solid matrix within which the structural characteristics can be approximated by the average of the relevant parameters. This is the reason why the general interpretation of porosity, the parameter most commonly used to describe the internal structure of the solid matrix, will be given as the first point of this section, in the context of a continuum approach.

The structure of water transporting channels and both the number and size of the interstices depend on the type of the layer. As it was already explained, a basic distinction should be made between loose clastic sediments and solid rocks. Thus the characterization of the solid matrix formed by the two different types of layers will also be dealt with. In loose clastic sediments the network conveying the water is composed of the pores surrounded by the grains. The pores are interconnected almost continuously and distributed at random in the flow space. The resistivity of the sediments depends mostly on the size and shape of the channels formed by the pores and may be expressed as a function with the following parameters: the size and shape of grains, the ratio of grains having different sizes (grain size distribution), and porosity. It is necessary, therefore, that the determination of these physical soil data, and the relationships existing between the parameters, be summarized. Since the adhesion between the solid matrix and the water may also have a considerable role among the retarding forces, and this force is affected by the mineralogical and chemical composition of the grains, the analysis of these characteristics provides the investigators with important information as well. In the case of solid rocks the various joints, fissures, and fractures are the water transporting elements. Their structure mainly depends on the type of the rock in question (intrusive rocks, extrusive volcanic rocks, metamorphic rocks, non-carbonate indurated sediments and carbonate rocks). The hydraulic properties of these formations and the relationships between these parameters and the types of rocks will be discussed in Part V. Only the general interpretation of porosity in solid rocks will be summarized here.

There are cases when the absolute size of openings and the statistical distribution of this geometrical parameter influences the character of flow considerably. Therefore, the water transporting capacity of the channels composed of the interstices cannot be determined by only considering the average size of the pores or joints. The pore-size distribution of the solid matrix must be analyzed in these cases. The statistical models derived in this way give supplementary information and they should be combined with the geometrical models. The statistical investigation of the openings having different diameters or widths will be separately introduced for the loose clastic sediments and for the solid rocks.

Finally, in this section the special structure of the cultivated soil is discussed. This relatively thin but very important layer is composed of clods which are aggregated from grains and the latter surround randomly interconnected pores (similar to the loose clastic sediments). Large openings run between the clods, where the behavior of water differs fundamentally from that in the pores.

This fact requires the special analysis of the processes developing in the cultivated zone. The investigation of the double system of openings provides the basis for the physical and hydrological studies where the influence of this layer must be considered.

3-1 General Description of the Porosity of the Solid Matrix (continuum approach)

According to the definition accepted most frequently, porosity is the ratio of the volume of pores (V_p) within the solid matrix of a given sample related to the total volume of the latter (V_t). Naturally the pore volume can be calculated as the difference of the bulk volume and that of the solid matrix (V_s):

$$n = \frac{V_p}{V_t} = \frac{V_t - V_s}{V_t} = 1 - \frac{V_s}{V_t} \quad . \qquad (1\text{-}1)$$

The same properties of the solid matrix may be expressed as the rate of pore volume to the volume of the solid matrix. This parameter is the void ratio, which can be interrelated with porosity as well (Fig. 1-37):

$$e = \frac{V_p}{V_s} = \frac{n}{1-n} \quad . \qquad (1\text{-}2)$$

An important aspect in connection with the determination of porosity is the requirement that the interstices, being part of the interconnected water-conveying network, should be distinguished from those not taking part in water transport. The pores may be inactive either because they are completely separated from the interconnected channels or some pores may compose the dead ends of the network, or the opening may be so small that the fluids contained in them are rendered essentially static by the close proximity of the force fields at the surface of the solid matrix (De Wiest et al., 1969). Considering the differences between the numerical values of the pore volume used as the basis of the calcultion of porosity, different terms may be distinguished, such as: total porosity, when the volume of all pores are related to that of

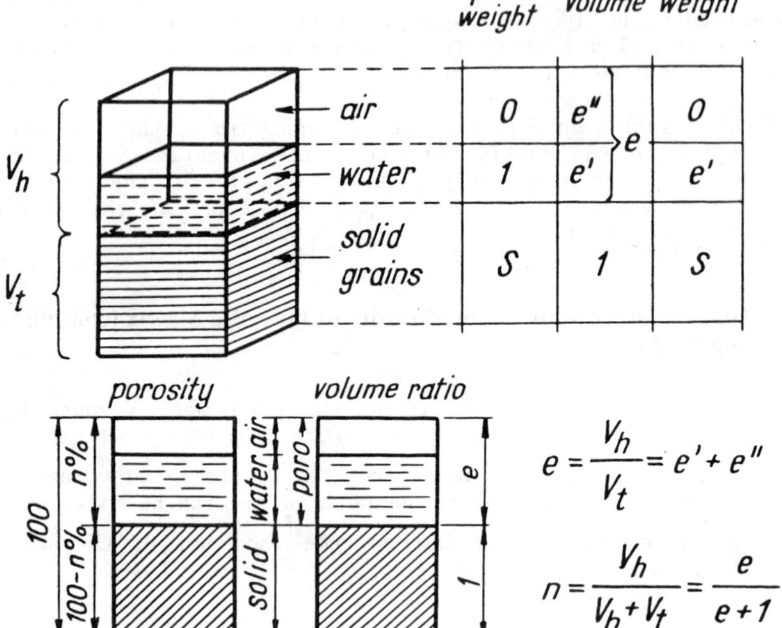

Fig. 1-37. Symbols used in connection with the determination of porosity and void ratio.

the sample; effective porosity considering only the pores available for the transition of fluid in the porous medium; and isolated porosity due to separated and inactive interstices.

It is necessary to note also, that there are other interpretations of porosity different from that given previously as the ratio of the pore volume and the bulk volume of the sample (Fig. 1-38).

Analyzing the porosity of a sample along a line, the lengths of the pores are related to the total length of the sample in the investigated direction. The calculated ratio is the parameter called linear porosity:

$$n_L = \frac{\sum_{i=1}^{k} \ell_i}{\Delta x} \quad . \tag{1-3}$$

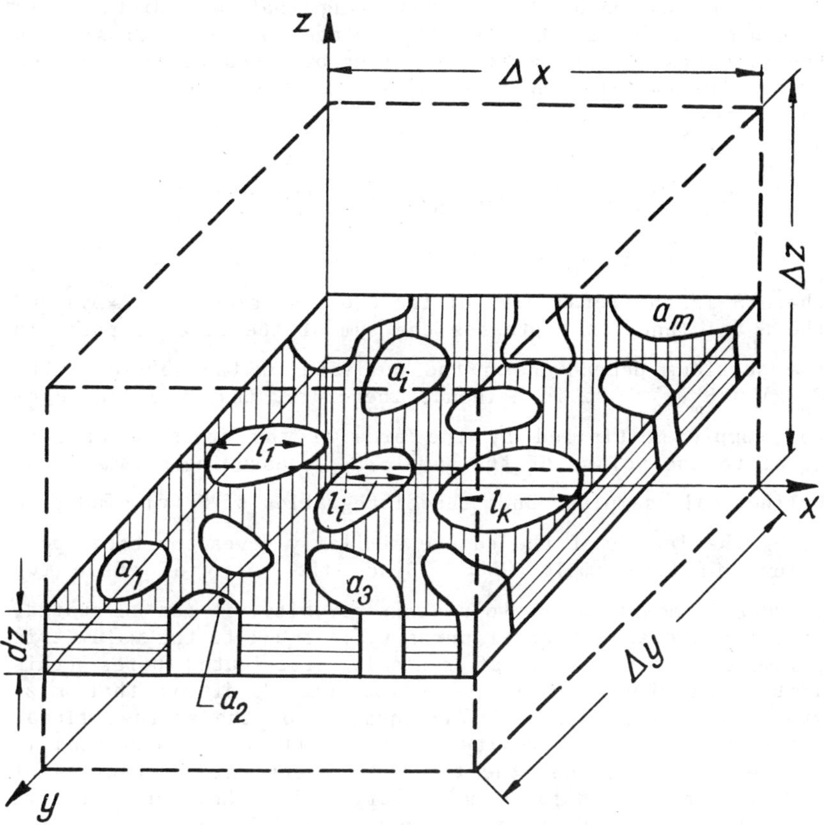

Fig. 1-38. Interpretation of linear, areal, and volumetric porosities.

Similarly the sum of the areas of pores in a given section may be related to the total area of the cross section. Areal porosity is thus computed in this way:

$$n_A = \frac{\sum_{i=1}^{m} a_i}{\Delta x \Delta y} \quad . \tag{1-4}$$

There are special relationships existing between the values of linear, areal, and volumetric porosity, on the basis of which they can be substituted by each other. The average areal porosity can be calculated as the mean value of the parameters determined for several sections, parallel to each other. The areal porosities measured in various sections have to be weighted according to the length belonging to the section in question, if the distances between the sections are not equal to each other. Supposing that the distance is elementarily small (dz) and the change of areal porosity in the direction of the z-axis is given by a continuous function [m(z)] the calculation of the average value can be expressed in the form of an integral:

$$n_A = \frac{1}{\Delta z} \int_o^{\Delta z} m(z)\, dz = \frac{1}{S\Delta z} \int_o^{\Delta z} S_p(z)\, dz = \frac{V_p}{V_t} = n \;; \tag{1-5}$$

where S is the total area of the cross section (s = ΔxΔy) and the $S_p(z)$ function expresses the sum of the area of pores in the sections depending on the position of the section, thus $S_p(z) = S_m(z)$. It is evident, therefore, that the length of the sample multiplied by the area of the cross section is equal to the volume of the sample (V_t), assuming prismatic or cylindrical samples, and the integration of $S_p(z)$ function along the length of the sample similarly gives the total pore volume in the sample (V_p). Since the ratio of these two volumes is equal to the volumetric porosity, it can be stated, that the average of areal porosity is equal to the volumetric parameter in the case of randomly distributed pores (this latter hypothesis originates from the condition that m(z) function be continuous). The equality of the average linear porosity and areal porosity can be similarly proved, thus it can be supposed, that the three different interpretations of porosity calculated for samples larger than the representative elementary volume provide the same numerical value.

The first condition of the previous derivation was the random distribution of pores. In the case of a regular structure of fissures, when the m(z) function is not continuous, some complications are caused by the difference between total and effective porosity. The orthogonal system of equidistant gaps between cubes may be used as an example (Fig. 1-39) for investigating the arising problems.

Two different values of linear porosity can be calculated along lines parallel to one of the main axes of the structure, depending on the position of the line in question (it may cross the cubes or run inside a gap):

$$n_{1L} = \frac{b}{a+b} \;; \qquad n_{2L} = 1 \;. \tag{1-6}$$

Similarly two parameters characterize the areal porosity as well:

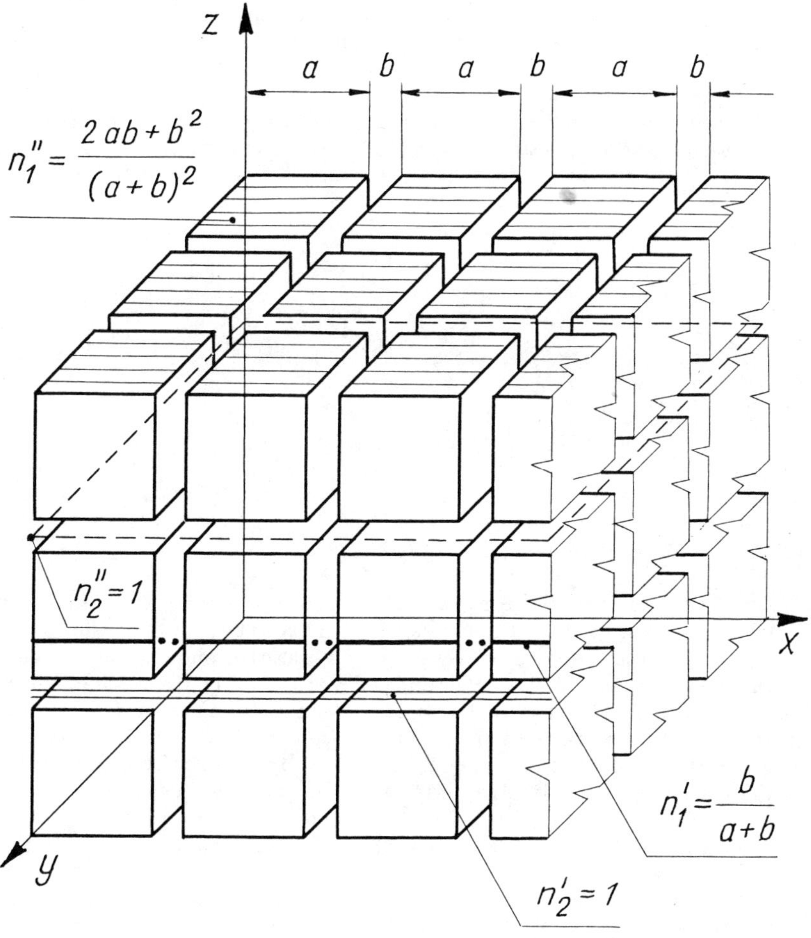

Fig. 1-39. Development of linear, areal, and volumetric porosity in a structure of interstices composed of equidistant gaps between cubes.

$$n_{1A} = \frac{(a+b)^2 - a^2}{(a+b)^2} = \frac{2ab + b^2}{(a+b)^2} \; ; \; n_{2A} = 1 \; ; \qquad (1\text{-}7)$$

while total volumetric porosity is determined by the following equation:

$$n = \frac{(a+b)^3 - a^3}{(a+b)^3} = \frac{3ab^2 + 3a^2b+b^3}{(a+b)^3} \; . \qquad (1\text{-}8)$$

Comparing the parameters not equal to unity, an apparent contradiction occurs because these values are not equal, although equality would be expected for randomly distributed pores. At the same time, it is evident that the areal porosity expressed by Eq. (1-1) supposes a flow perpendicular to one of the main planes of the structure, fitted to two axes of the coordinate system. In this case, however, a considerable part of the pores (those gaps perpendicular to the direction of flow) do not convey water, because there is no pressure difference along them. After decreasing the pore volume with the volume in the inactive gaps, effective porosity can be calculated

$$n_{eff} = \frac{[(a+b)^2 - a^2](a+b)}{(a+b)^3} = \frac{2\ ab+b^2}{(a+b)^2} = n_{1A} \quad . \quad (1\text{-}9)$$

It can be proved in this way, that the equality of areal and volumetric porosity is valid also in the case of regular network of fissures, but effective porosity has to be determined for such systems, only considering interstices taking part in water transport.

A second assumption was also mentioned as the condition for the validity of the concept of porosity, i.e., the size, area, or volume of the investigated sample should be larger, than the representative elementary volume. It is necessary to consider that the parameters calculated in this way are only statistical averages for a unit area or volume and do not represent the actual extreme values. The calculated porosity may have considerable scattering, depending on the position of the investigated unit within the seepage field, if the size of the analyzed sample is smaller than a given unit. The extreme values, occurring when infinitesimal volumes are used, are independent of the actual character of the porous medium. They are one or zero depending on whether the center of the investigated volume is in a pore or within the solid matrix. Thus, the total scattering is unity belonging to infinitesimally small volume. This range decreases gradually as the volume of the investigated sample increases and achieves a constant minimum at the representative elementary volume. The calculated porosity fluctuates randomly around the expected value when the volume is above this limit.

Naturally the size of the representative elementary unit is strongly influenced by the structure of the solid matrix of the porous medium in question. Porosity is a good example for the representation of this statement as well. In a loose clastic sediment, the limit above which porosity fluctuates randomly may be expected to be small, especially in sand and gravel. A slow change of the average porosity characterizes the layered formations after crossing an upper limit, indicating, that above a given size the investigated volume becomes nonhomogeneous (it also includes a layer having different properties than the basic one), and the average parameter is more and more influenced by the second layer. Figure 1-40 shows the interpretation of porosity as a function of the size of the investigated volume (Bear, 1972).

In fissured and fractured rocks, the distribution of the interstices is more uneven than that in loose clastic sediments. The representative elementary length is generally longer in such formations, and greatly varies with the structure of rock type. For demonstrating this statement the linear porosity of a dolomite covered with dense fine joints is compared to the same parameter of karst limestone. Four lines of equal length were investigated on a dolomite surface and the two extremes are represented in Fig. 1-41. Although the joints are relatively well distributed, the scattering of porosity, determined for various stretches of 10 cms, is still about 30-70 percent of the average parameter, while it is reduced below 10-20 percent if the investigated length is 25 cms. In karst limestone, where chemically dissolved openings as large as 50 cms can be found, the length of the acceptable representative elementary unit is 50-100 times longer than in dolomite (Fig. 1-41).

On the basis of a continuum approach, and considering the interpretation of the general seepage law determining the relationship between seepage velocity and hydraulic gradient, the definition of seepage can be clarified. The movement of any type of subsurface water is called seepage, if seepage velocity, as a characteristic parameter of the flow space, can be calculated and rendered to any point of the domain, as a function of the local hydraulic gradient. The applicability of this concept depends also on the ratio of the size of the seepage field related to the representative elementary unit determined by considering the scattering of porosity. In loose clastic sediments and solid rocks, having similar structure (e.g., sandstone), the flow may be called seepage almost without any restriction according to the extension of the flow domain. In the case of fissured and fractured rocks, the minimum size of the flow space may be determined, above which the aquifer can be substituted with continuous domain and thus the movement can be investigated by applying the seepage law, while the water transport through the tortuous, separate channels should be calculated if the size of the investigated layer is smaller than the limit.

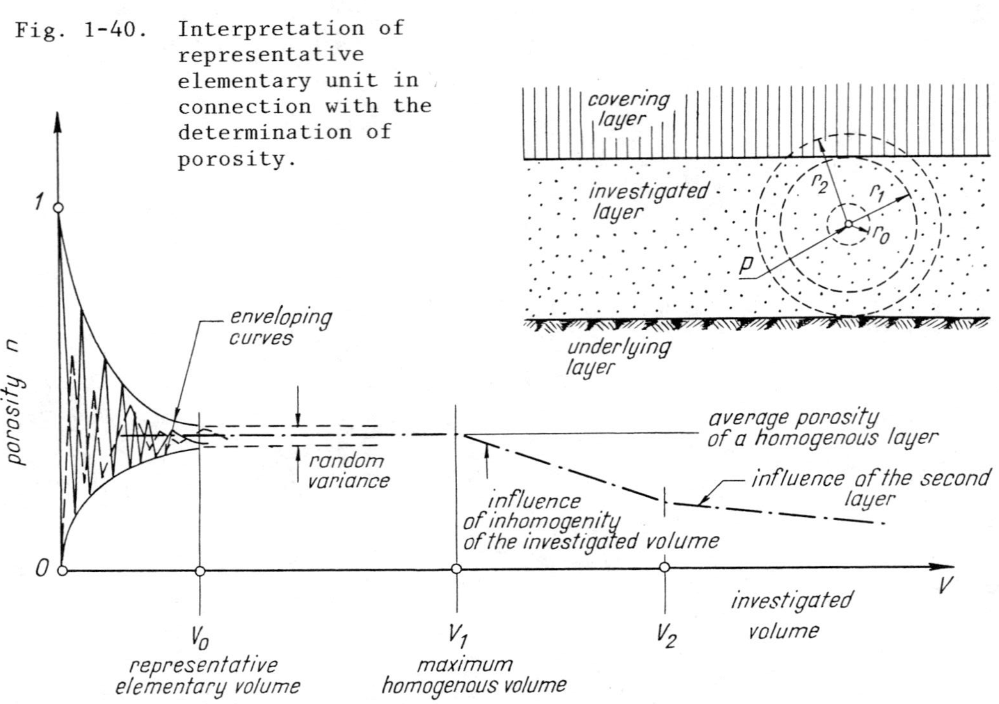

Fig. 1-40. Interpretation of representative elementary unit in connection with the determination of porosity.

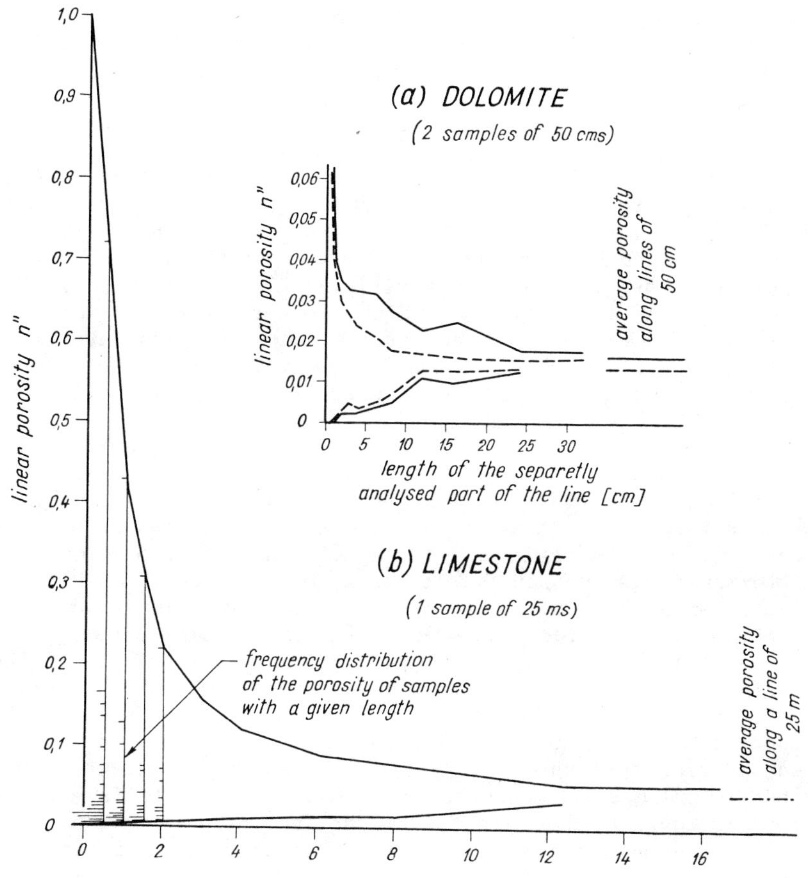

Fig. 1-41. Comparison of porosity distribution of limestone and dolomite.

3-2 Shape and Grain-Size Distribution of Particles Composing Loose Clastic Sediments

The grains and their main parameters (surface, volume, and others) can be characterized by one or more sizes of the particle only in the case of regular geometrical forms. In nature, the grains forming the various layers differ considerably from these regular forms, and especially from that of the sphere, which is generally used in soil physics to substitute the actual shape of grains because this is the simplest form, all the parameters of which can be determined knowing only the diameter of the sphere. This is the reason why the introduction of a shape coefficient is also necessary in addition to the diameter of the grains in order to characterize the difference between the sphere and the actual form.

There are two different ways to determine the diameter of a sphere equivalent to the actual grain. Both are related to the methods used in soil physics for measuring the size of grains. If the grain is larger than 0.1 mm, this measurement is made by using sieves. Thus the diameter of the equivalent sphere in this zone is equal to the size of holes on the sieve (which is the diameter for a circular hole, or the length of the side of a square) in which the grain still can fall through. In the case of finer grains, whose size is determined by hydrometric method, the equivalent diameter is the size of a sphere settling in quiescent water with the same velocity, as the actual grain in question.

In the first case, the diameter of the grain is consequently equal to the diameter of the sphere encircling the grain. In the lower zone, such clear relationship between the data of the grain and the substituting sphere cannot be determined unanimously. A further problem is caused in the latter case by the fact that the colloid grains may sink separately or joined together as a flake, according to the colloidical character of the suspension (morphology of the surface of grains, coagulation, peptization). The settling velocity of an aggregate composed of many grains is equal to that of a single grain having the same diameter as the flake. Considering this uncertainty, it is also acceptable to assume that a grain size measured by hydrometric method is approximately equal to the diameter of the encircling sphere, and to ensure the uniformity of the interpretation of grain size in this way.

There are several shape coefficients for characterizing the difference between the actual grain and the equivalent sphere. These are generally calculated from the three main lengths of the grain, perpendicular to each other. A further basic requirement is that the shape coefficient should characterize the difference mentioned above from an aspect relevant to the investigation in question. Thus there are shape coefficients used in geology to determine the origin of the grains and the length of their travel (Hagermann, 1936, 1938; Miháltz and Ungár, 1954; Szádeczky Kardos, 1933) and in hydraulics for the characterization of sediment transport and settling velocity (Heywood, 1938; Ivicsics, 1957; Stelczer, 1967). To investigate seepage, a shape coefficient should be chosen which suits the physical description of this process (Kovács, 1968a).

Friction is the most important force among those retarding the percolation of water. Apart from that, inertia and adhesion may be dominant. Two of these forces (friction and adhesion) are proportional to the contacting surface between the solid and liquid phases of the system. Gravity, accelerating the movement, is proportional to the volume of water. The latter can be substituted by the volume of solid grains using porosity as a transforming factor. Finally, the ratio of retarding and accelerating forces, which characterize the flow physically, can be approximated by the ratio of the surface of a grain (A) to its volume (V). The dimension of this ratio is $[L^{-1}]$, thus it can be expressed as a quotient of a dimensionless shape coefficient (α) and a characteristic diameter (d):

$$\frac{A}{V} = \frac{\alpha_d}{d} = \frac{\alpha_D}{D} \quad . \tag{1-10}$$

As it is shown by the equation, α is a function of the diameter chosen to characterize the grain. According to the soil physics determination, this diameter can be that of the encircling sphere (D). The corresponding value of the shape coefficient (α_D) for some particles having regular geometrical form is: sphere, $\alpha_D = 6$; cube, $\alpha_D = 10.4$; octahedron, $\alpha_D = 10.4$; and tetrahedron, $\alpha_D = 18$.

There are some crystals, the ratio of whose two main axes of the three (those determining its cross section), is constant and the third can vary considerably, thus the grains may have any form from the laminated plate to the shape of an elongated nail (pyramid, tetragonal, or hexagonal prism, circular cylinder, tetragonal or hexagonal pyramid). The shape coefficient of such grains can be expressed and represented as a function of the quotient calculated from the diameter of a circle encircling the main section of the grain (d_1) and the length of the grain (ℓ) (Fig. 1-42).

The dotted lines show the values of α_{d_1} (the shape coefficient calculated not from the diameter of the encircling sphere, but from d_1. The comparison of the graphs reveals that there is practically no difference between the two types of shape coefficients (i.e., α_D and α_{d_1}) in the zone of the disc-shaped grains, but the two values differ considerably in the case of nail-shaped crystals. Another conclusion, which can be drawn from the figure, is that the shape coefficients of prismatic forms having a square and oblong cross section, respectively, may also differ only in the zone of elongated

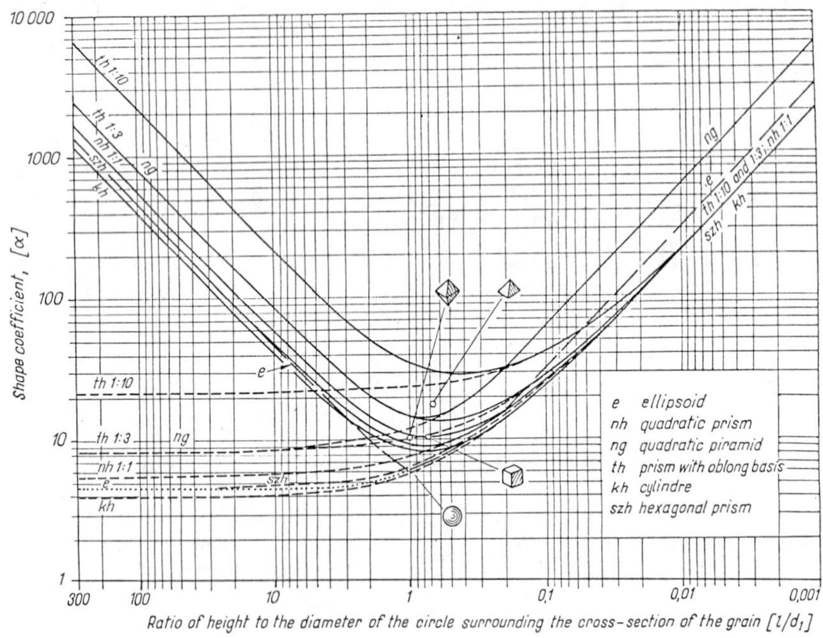

Fig. 1-42. Shape coefficient of grains as a function of the ratio of their length to the diameter of the circle encircling their cross section.

grains, and only in the case when the ratio of the sides of the oblong is significant (e.g., 1:10). The shape coefficient depends on the mineralogical character of the grains, while the latter may be related to the size of grains.

In gravels, the dominating material, of which is quartz, are generally stubby and the lengths of their main axes do not differ considerably, the quotient ℓ/d_1 is nearly unity. Their form can be approximated by a sphere, or a cube, depending on the measure of abrasion occurred during their transport from the place of origin to the location of sedimentation. Thus their shape coefficient is generally about α_D = 7-11. The case of gravels originating from laminated rocks (sandstone, slate) is an exception, the shape coefficient of such particles may be as high as α_D = 20.

Quartz is the main component of sand as well. The shape of sand grains is very similar to that of gravels. In this group, abrasion depends mostly on whether the investigated sediment is a fluvial, or a wind-borne deposit. In the first case, the shape coefficient is α_D = 9-11, while the quartz grains of a wind-borne sand can be characterized by a coefficient of about α_D = 7-9. In sands, and especially among their fine grains, a considerable amount of mica can also be found. These are characteristically laminated minerals, thus their shape coefficient is much higher than that of quartz (α_D = 20-50).

The clay minerals form a quite separate group due to their shape. Their diameter is smaller than 2μ, and they are generally very thin plates (illites, kaolinite, montmorillonite), but some special types can show the form of a nail, which is really a tube rolled from a thin plate (halloyisite). These minerals have the highest shape coefficient (Albert, 1967; Kuhn, 1963) (Table 1-3).

Table 1-3. Shape coefficients of clay minerals.

	ℓ[μ]	d_1[μ]	ℓ/d_1	α_D
Kaolinite	0.15-0.40	1.4-0.40	0.10-0.030	30-70
Illite	0.04-0.03	0.5-0.15	0.20-0.050	20-60
Montmorillonite	$\leq$ 0.01	0.7-0.30	0.03-0.015	70-100
Na montmorillonite	$\leq$ 0.001		0.003-0.0015	700-1000
Nail shaped grains (e.g., halloysite, attapulgite)	2.0-0.5	0.25-0.04	10-30	40-100

According to this summary, the shape coefficient of clay minerals varies generally between 30 and 100. Na-montmorillonite is an exceptional case whose lattices are separated to independent, very thin lamellae and thus the internal surface of the crystals becomes active.

The mean value of the shape coefficient for various grains of different sizes, origin, and mineralogical character can be determined by statistical evaluation of numerous measurements (Fig. 1-43). Such studies furnish suitable data for the approximation of the actual shape coefficient which is applicable in practice without any further detailed investigation.

All parameters discussed in the preceding paragraphs are related only to a single grain. In nature, however, there is no layer built up from particles with identical size and

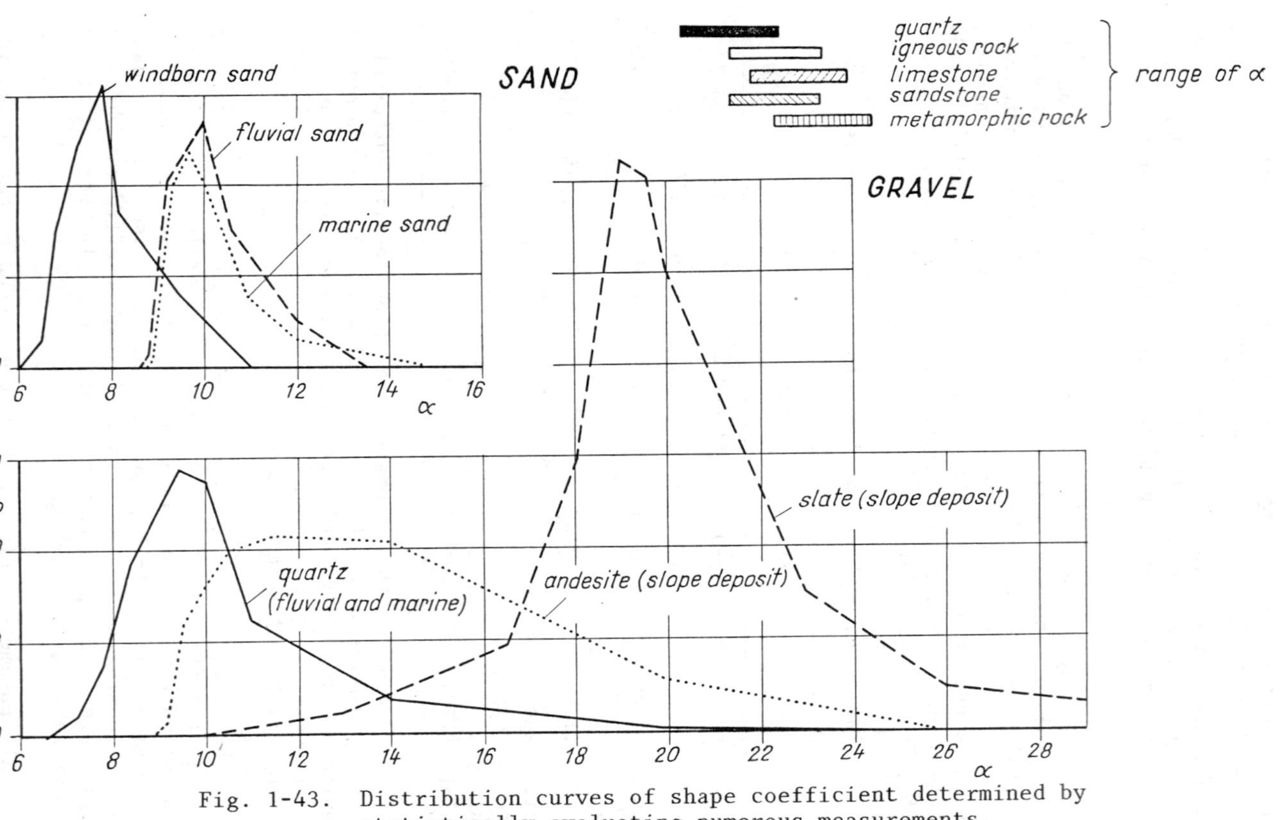

Fig. 1-43. Distribution curves of shape coefficient determined by statistically evaluating numerous measurements.

shape. It is therefore necessary to find methods by which the layer can be characterized as a mixture of various grains.

For comparing the amount of grains having various sizes, the so-called grain-size distribution curve is generally used, the definition of which is well known in soil physics (Fig. 1-44) (Jaki, 1944; Kézdi, 1961, 1962). In manuals, some comparison can also be found showing the different names used to indicate the various grain sizes in different countries or organizations (Bear, 1972). There have been many attempts made to find only one parameter by which the sample can be described to identify it numerically, and to characterize its most important physical behavior. Such parameters are the various characteristic diameters (e.g., maximum diameter D_{max}; diameter corresponding to the 50 percent value of the distribution curve d_{50}; diameter of grains present in the sample with maximum weight, i.e., the inflexion of the distribution curve D_m; etc.).

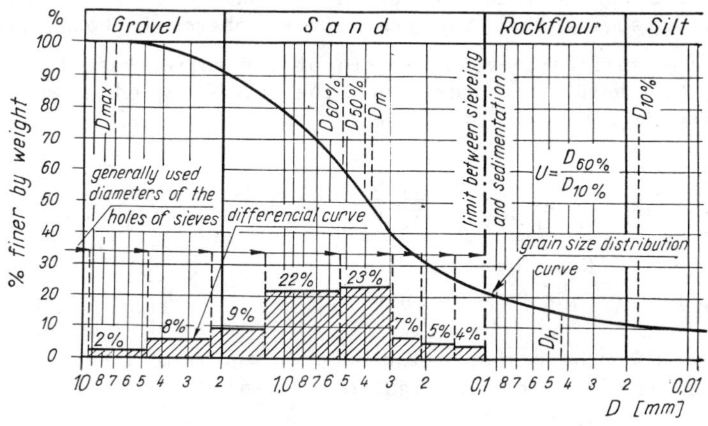

Fig. 1-44. The ratio of the grains having different diameter in the sample characterized by the grain-size distribution curve.

Hazen's design diameter, which pertains to the 10 percent value of the distribution curve (D_{10}), has an important role in seepage hydraulics (Hazen, 1895). Previously, its use was generally accepted, considering that both the size of pores and the internal surface of the sample are influenced mostly by the fine grains, thus the physical character of seepage can be related to a value measuring the amount of fine grains in the sample. This parameter is usually supplemented by the coefficient of uniformity which is the ratio of the grain diameter corresponding to the 60 percent value of the distribution curve (D_{60}) to Hazen's diameter:

$$U = \frac{D_{60}}{D_{10}} . \qquad (1\text{-}11)$$

On the basis of recent investigations, the use of Kozeny's effective diameter (D_h) can be unanimously proposed as a parameter which can correctly describe the heterodisperse sample in seepage hydraulics (Kozeny, 1953). According to the original definition, this is the diameter of a sphere, whose homodisperse sample (a sample of particles having the same diameter) has the same surface-to-volume ratio as the investigated sample of heterodisperse spheres. The determination of the effective diameter can be given as follows (Fig. 1-45). The distance between the maximum and minimum diameter is divided into "n" intervals. The average diameter in the i-th interval is D_i, and the weight of grains in this interval compared to the total weight of the sample is represented by ΔS_i. Assuming that the grains are spheres in both systems, and the surface-to-volume ratio of the solid phase is identical for both the model and the investigated sample, the following relationship can be given:

$$\frac{N D_h^2 \Pi}{N D_h^3 \Pi/6} = \frac{\Sigma \frac{G_i}{\gamma_s} \frac{6}{D_i}}{\Sigma \frac{G_i}{\gamma_s}} ; \qquad D_h = \frac{1}{\Sigma \frac{\Delta S_i}{D_i}} ; \qquad (1\text{-}12)$$

where N is the number of spheres in the homodisperse sample, and γ_s is the specific weight of the solid grains.

A similar relationship can be determined in the case of grains with shapes differing from a sphere using the shape coefficient described previously. It has to be assumed that the shape coefficient of grains can be characterized approximately with an average value (α_i) in each interval. Following the same steps made to derive Eq. 1-12 the quotient of the effective diameter and the average shape coefficient can be calculated as

$$\frac{D_h}{\alpha} = \frac{1}{\Sigma \Delta S_i \frac{\alpha_i}{D_i}} . \qquad (1\text{-}13)$$

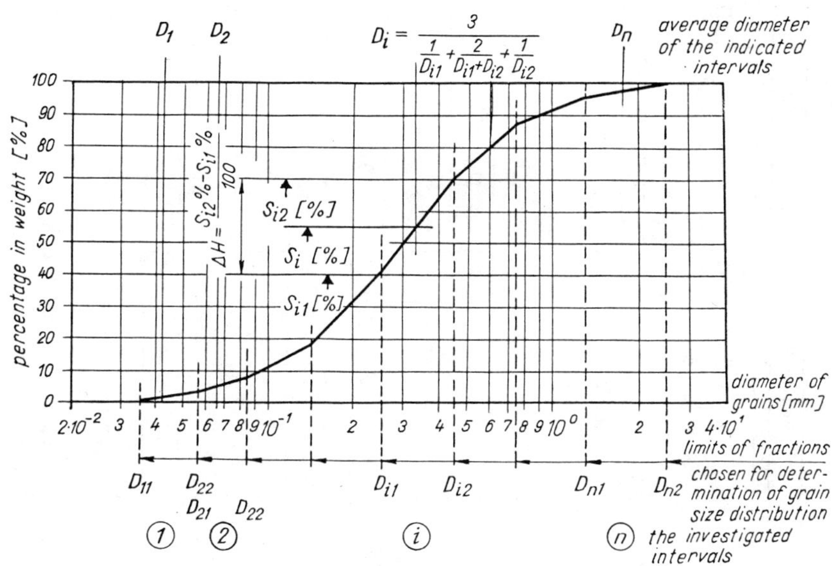

Fig. 1-45. Symbols used for the calculation of effective diameter.

The determination of the shape coefficient for every fraction requires mineralogical and microscopic investigation. For the practical application of Eq. 1-13, however, some informative data can be given on the basis of statistical evaluation of numerous samples. An example is given in Fig. 1-46, which indicates the probable shape coefficient of grains according to their origin (alluvial or aeolian), mineralogical composition (quartz, feldspar, mica, clay mineral), and size as a headline of the forms generally used to represent grain-size distribution.

Average of fluvial deposit		8 - 11	9 - 12	10 - 15	12 - 17	15 - 50
clay minerals	montm.					70 - 100
	Na montm.					700 - 1000
	illite					20 - 60
	kaolinite					30 - 70
	pins					40 - 100
mices			20 - 50	20 - 50	20 - 50	20 - 50
metamorph rock fragments	fluvial dep.	11 - 15				
	slope dep.	12 - 20				
Igneous rock fragments	fluvial dep.	9,5 - 11,5				
	slope dep.	11 - 13				
sand stone	fluvial dep.	9,5 - 11,5				
	slope dep.	11 - 13				
limestone	fluvial dep.	10 - 12	10 - 12	10 - 12	10 - 12	10 - 12
	slope dep.	11,5 - 13,5	12 - 14			
quartz	fluvial dep.	8 - 10	9 - 11	9 - 11	9 - 11	9 - 11
	wind born dep.		7 - 9	7 - 9	7 - 9	
	slope dep.	9,5 - 11,5	10 - 12			
		Gravel	Sand	Rockflour	Silt	Clay

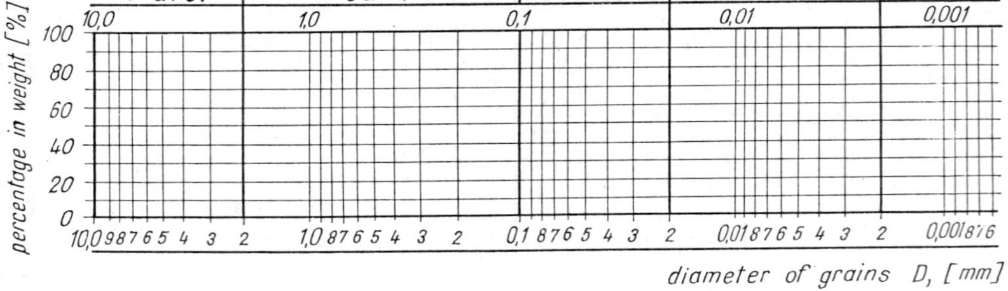

10,0 1,0 0,1 0,01 0,001

percentage in weight [%]

100 80 60 40 20 0

10,0 9 8 7 6 5 4 3 2 1,0 8 7 6 5 4 3 2 0,1 8 7 6 5 4 3 2 0,01 8 7 6 5 4 3 2 0,001 8 7 6

diameter of grains D, [mm]

Fig. 1-46. The average shape coefficients represented in the headline of forms used for plotting grain-size distribution curves.

The problem of aggregation of very fine grains has already been mentioned in connection with the determination of grain size. This process disturbs the lower section of the distribution curve measured by hydrometric method. The flakes settle with a velocity corresponding to their diameter, and not to that of the individual grains. A larger weight is measured in intervals including the diameter of flakes and this amount is missing at the lower section of the curve (Szilvágyi, 1966, 1967).

Figure 1-47 represents some examples showing the deformation of the distribution curve. The first sample is a bentonite with no grains above the limit of 0.1 mm. All but one curve were determined by hydrometric method, and the last by computing the number of grains smaller than 2μ using mineralogical methods (electromicroscope, x-ray analysis, differential thermoanalysis). The first settling was executed in distilled water, and the others by adding various chemicals to the suspension. The settling in distilled water measures the number of grains finer than a given diameter, e.g., D = 2μ, and not bound to flakes in a natural condition. This is the lowest curve in the example in question. The mineralogical method determines the total number of fine grains independently of aggregation (fully peptized conditions), and naturally the corresponding curve gives the highest value at a given diameter. The other curves are between these two, showing the various grades of peptization created by the chemicals used. In the second example, where the sample contains mostly kaolinite among the clay minerals, the effect of the chemicals is somewhat different. Some of these chemicals do not increase the grade of dispersity, but on the contrary, promote coagulation. Thus the curve determined in distilled water is not the lowest in this case. Finally, the third sample was an illite. Its distribution curve was not appreciably altered by the chemicals used.

After these examples, a question can be easily raised: From which distribution curve should the effective diameter be determined to investigate the seepage, when the sample contains clay minerals? In answering this question, it must be known that there are two kinds of bindings between the grains. The first is loose, the grains bound in this way can be easily separated either by applying chemicals, or mechanically. There are, however, irreversible bindings as well. In the latter case the aggregated grains behave, from the point of view of seepage, like only one solid grain. The task of the investigation is, therefore, to measure the amount of individual and loosely bound grains separately from that of irreversibly aggregated ones. The ratio of mobile, or active (individual or loosely aggregated) clay minerals to those bound irreversibly to aggregates (immobile or inactive) is called the

The problem of aggregation of very fine grains has already been mentioned in connection with the determination of grain size. This process disturbs the lower section of the distribution curve measured by hydrometric method. The flakes settle with a velocity corresponding to their diameter, and not to that of the individual grains. A larger weight is measured in intervals including the diameter of flakes and this amount is missing at the lower section of the curve (Szilvágyi, 1966, 1967).

Figure 1-47 represents some examples showing the deformation of the distribution curve. The first sample is a bentonite with no grains above the limit of 0.1 mm. All but one curve were determined by hydrometric method, and the last by computing the rate of grains smaller than 2μ using mineralogical methods (electromicroscope, x-ray analysis, differential thermoanalysis). The first settling was executed in distilled water, and the others by adding various chemicals to the suspension. The settling in distilled water measures the rate of grains finer than a given diameter, e.g., D = 2μ, and not bound to flakes in a natural condition. This is the lowest curve in the example in question. The mineralogical method determines the total rate of fine grains independently of aggregation (fully peptized conditions), and naturally the corresponding curve gives the highest value at a given diameter. The other curves are between these two, showing the various grades of peptization created by the chemicals used. In the second example, where the sample contains mostly kaolinite among the clay minerals, the effect of the chemicals is somewhat different. Some of these chemicals do not increase the grade of dispersity, but on the contrary, promote coagulation. Thus the curve determined in distilled water is not the lowest in this case. Finally, the third sample was an illite. Its distribution curve was not appreciably altered by the chemicals used.

After these examples, a question can be easily raised: From which distribution curve should the effective diameter be determined to investigate the seepage, when the sample contains clay minerals? In answering this question, it must be known that there are two kinds of bindings between the grains. The first is loose, the grains bound in this way can be easily separated either by applying chemicals, or mechanically. There are, however, irreversible bindings as well. In the latter case the aggregated grains behave, from the point of view of seepage, like only one solid grain. The task of the investigation is, therefore, to measure the amount of individual and loosely bound grains separately from that of irreversibly aggregated ones. The ratio of mobile, or active (individual or loosely aggregated) clay minerals to those bound irreversibly to aggregates (immobile or inactive) is called the

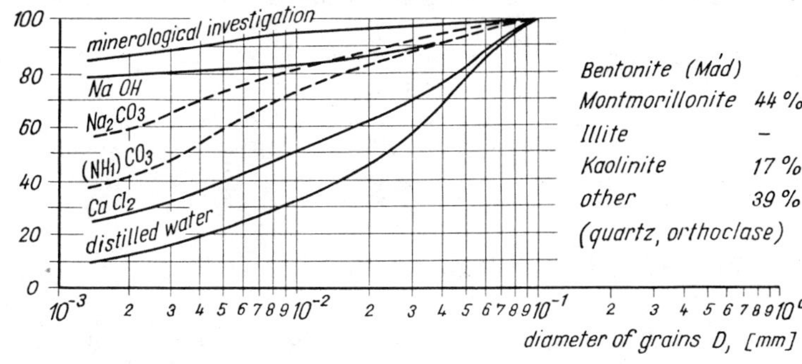

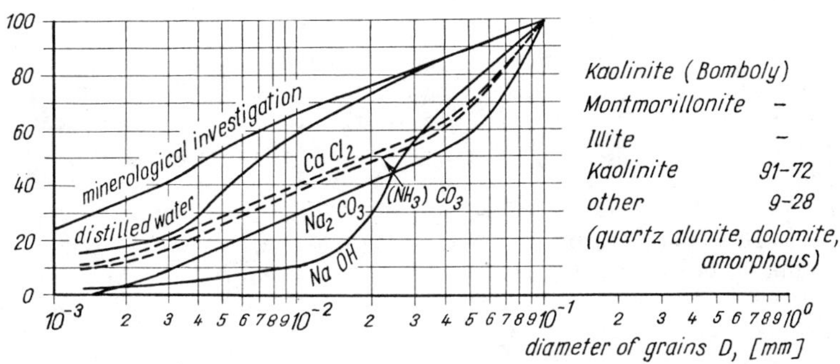

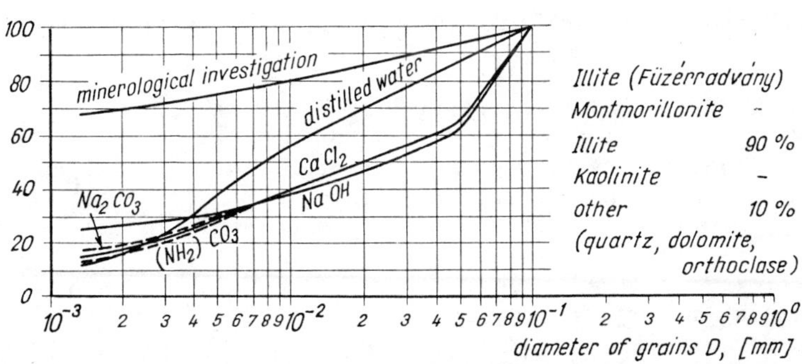

Fig. 1-47. Deformation of grain-size distribution curves caused by the presence of colloid particles.

morphological condition of colloid grains. This value has to be determined, and characterized by the distribution curve.

Strong drying, frost, or chemicals considerably changing the structure of the crystals can alter the morphological condition. This is the reason why treatments causing such effects must be avoided during the preparation of the sample, e.g., drying.

3-3 Mineralogical Characterization of Grains

The mineralogical character of grains has a close correlation with their size. Naturally it depends also on the type of rocks from which the particles of the sediments have originated. Grains larger in diameter than 2 mm (gravels) are mostly quartz and feldspars. There are, however, many rock fragments in this fraction composed of several minerals. The amount of calcite may also be high at some places. In sand (2 mm > D > 0.1 mm) the grains are composed mostly of only one mineral. Among the minerals, quartz is the most common, the amount of feldspars and calcite is less significant. The presence of mica may also be characteristic in this fraction. The next size interval (0.1 mm > D > 20μ) is called mo or rock-flour in soil physics. The mineral composition of this fraction does not greatly differ from that of sand, except perhaps the amount of mica is higher. The size of silt particles is 20μ > D > 2μ. The considerable change of the physical behavior of such samples, i.e., plasticity, is basically caused by clay minerals, which may be already present in this fraction in the vicinity of its lower size limit. Finally, the character of clays (D < 2μ) is mostly determined by clay minerals and micas, although other minerals (quartz, feldspar, carbonate) may remain dominating in amount. To illustrate the relationship between the size of grains and the mineralogical composition within a heterodisperse sample, Fig. 1-48 represents the distribution of various minerals according to the diameter of the particles (Kézdi, 1962).

The interaction between water and grains is mostly influenced by the surface-to-volume ratio of the latter. This ratio in a homodisperse sample of spheres is inversely proportional to their diameter. In the case of spheres of 2μ diameter (clay) this value is 10,000 times greater than that of a sample of particles having a diameter of 20 mm (gravel). The surface related to the volume is further increased in the zone of small grains by the fact that the laminated forms are more frequent here than in the zone of gravel and sand, as it is clearly indicated by the higher probable shape coefficient of the small particles. Finally, the mineralogical structure of the minerals of small diameters (mica and clay minerals) may

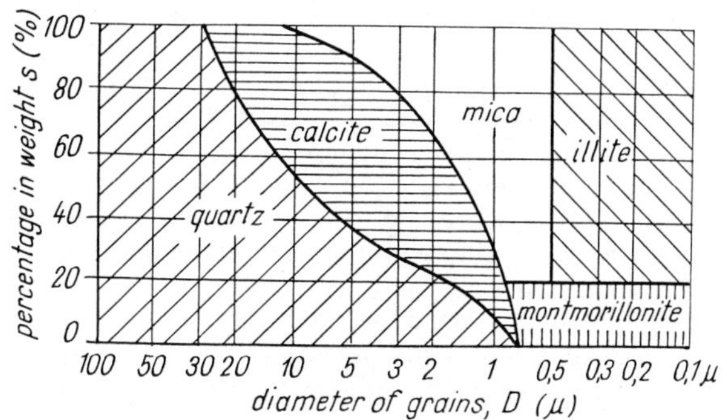

Fig. 1-48. Mineralogical composition of a sample as a function of the diameter of grains.

also increase the internal surface of the samples. At the same time, however, the aggregation of these small particles can cause contrary effects, as it was mentioned in the previous subsections. Before investigating the determination of the effective diameter of samples containing colloid particles, a short summary of the structure of clay minerals is given.

Clay minerals mostly originate from the decomposition of feldspars of eruptive rocks. Generally three main groups of these minerals are distinguished. The main representatives of these groups are kaolinite, illite and montmorillonite. A common character of clay minerals is the laminated structure of their crystals. One type of the lamellae is composed of the plain net of one silicium and four oxygen atoms bound to each other in the form of tetrahedrons, while the other lamellae is a net of octahedrons consisting of one aluminium and six oxygen atoms, or hydroxil radicals (Fig. 1-49) (Baver, 1948).

In the case of kaolinites $[Al_4Si_4\ O_{10}\ (OH)_8]$ the peaks of tetrahedra in the silica net are turned in the direction of the alumina net (the number of silica net and that of alumina net is the same) and thus the oxygen atoms here are built into the octahedra with their free charges. The charges of the silica net are, therefore, neutralized by the Al^{3+} and OH^- ions of the other lamellae and the whole system of kaolinite has free charges only at the edges of the crystals, where ions can be absorbed from the water surrounding the particles.

In the crystals of montmorillonite [n(Ca,Mg)O·Al_2O_3 4 SO_2·H_2O + H_2O] all lamellae of aluminiumhydroxide are surrounded by two silica nets (the aluminium can be substituted by iron or manganese in the octahedron net). According to the probable structure of the crystal, only every second tetrahedron of the silica net is joined to the alumina net, the peak of the others are turned outside, thus the crystal grid has free negative charges on the surface of each bunch of lamellae composed of three nets (one alumina and two silica nets). Ions can be, therefore, absorbed at these internal surfaces of the crystals, and the type of the bound ions influences the behavior of the montmorillonite. Water molecules can also be inserted between the crystal grids and absorbed by the free charges, causing the swelling of montmorillonite.

The third type of clay minerals is the illite (hydromica). Its structure, and also its behavior in connection with water, and thus its role in seepage, can be characterized as a transition between kaolinite and montmorillonite.

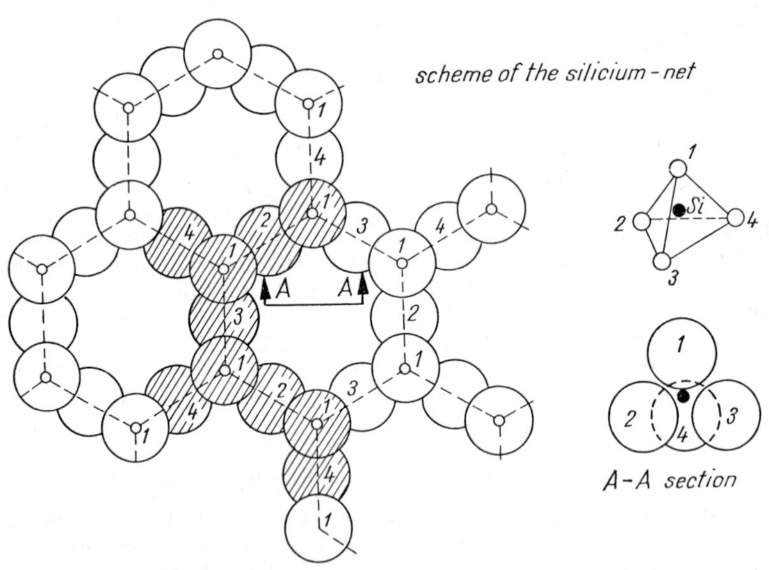

Fig. 1-49. The probable crystal structure of kaolinite and montmorillonite.

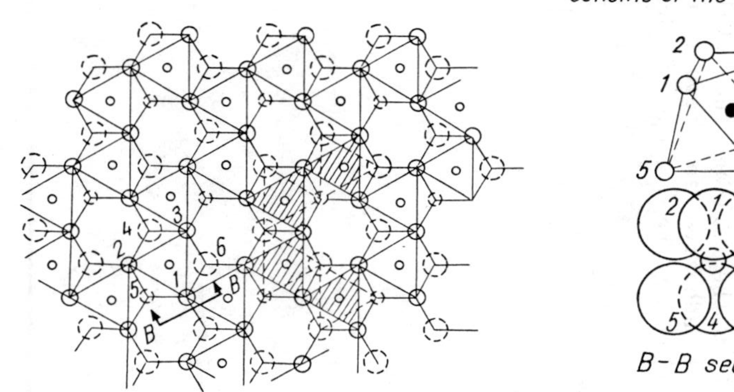

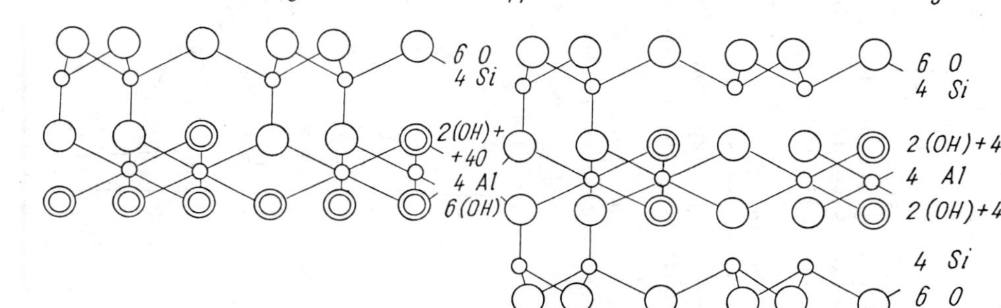

Fig. 1-49. (continued)

Summarizing the foregoing, it can be stated that clay minerals create an electrostatic field due to the free charges of the crystals. Thus they can absorb ions from the surrounding solution. The binding of these ions, however, is not permanent, and the so-called ion exchange can therefore alter the chemical character of the minerals as well. It is not yet clear whether the interaction between water and grains is also modified by the ion exchange or not. According to the investigations related to the adhesive forces, the tension is inversely proportional to the sixth power of the distance measured from the wall of the grains, and both the character of this relationship and the numerical parameters of the latter are constant, they are not influenced by the absorbed ions. The most acceptable hypothesis is, therefore, to assume that the adhesive force binding the water to the grains is independent of the mineralogical and chemical character of the particles (as it is generally stated in the literature concerning the so-called Van der Waals force). The interaction between grains in the suspension, however, is influenced by the adhesive and repulsive forces (Kovács, 1968b). Thus the relationship between the distance from the wall and the total energy, which is the resultant of adhesive and repulsive forces, will be considerably modified by the change of the repulsive force (Fig. 1-50).

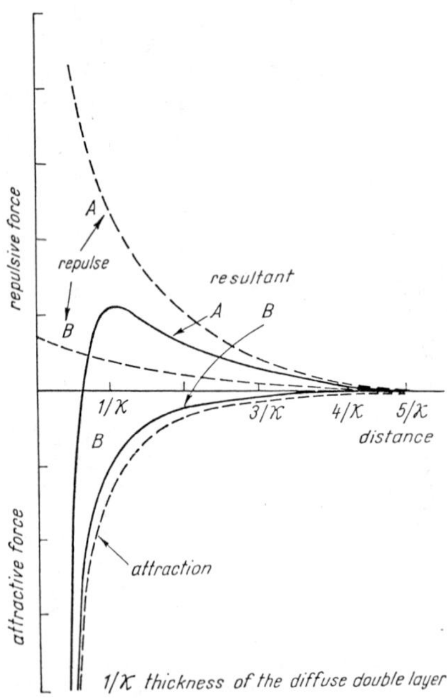

Fig. 1-50. Distribution of total electrostatic energy around a grain.

The result of this explanation is, that all effects causing an apparent change in the relationship of water and grains (ion exchange, salt content of water, etc.) directly alter only the repulsive forces, and thus the interaction between the grains. The grains will be, therefore, aggregated, or existing bindings will be disintegrated. The treatment of the sample may modify the morphology of the grains, the rate of independent, loosely aggregated, or irreversibly bound grains, and thus alter the effective diameter which depends on the active surface of grains related to their volume. Since the interaction between water and grains, accepting the constancy of the adhesive force, is influenced only by the active surface of grains, the change of this parameter modifies the hydraulic character of the sample (water content, hydraulic conductivity, etc.).

This way of thinking leads back to the problem mentioned at the end of the last section: which is the correct method to determine the effective grain diameter, the grain-size distribution curve, or some of its characteristic points (e.g., the ratio of the weight of grains smaller than 2μ to the total weight of the sample), when colloid particles are present in the sample. If the actual morphological character, the ratio of irreversible aggregates to the total amount of colloid grains, can be taken into consideration by the granulometric curve, the separate investigation of the influence of mineralogical and chemical character is not necessary, because the effective diameter expresses all the existing effects.

It has already been mentioned in connection with Fig. 1-47 that the mineralogical methods measure the total amount of fine grains, and, therefore, the clay content determined in this way is always greater than the actual amount of active particles, or it is equal to the latter, if there are no grains bound together by irreversible aggregation in the sample. Clay content represented by the symbol $S_{2\mu}$ means the weight of grains smaller than 2μ are related to the total weight of the sample. Meanwhile, the clay content measured by hydrometric method can be only smaller than, or equal to the amount of active grains smaller than 2μ. In the suspension, some loose aggregation may occur apart from the irreversible bindings, thus a part of the active particles will be measured among the larger grains. It is necessary, therefore, to find methods which are suitable for the direct determination of the actual active clay content.

Previous attempts use the plasticity of the sample to characterize its clay content. It is obvious that the plastic behavior is caused by the clay content, thus a relationship must exist between the two phenomena. At the same time the parameters characterizing plasticity, e.g., those used in soil physics--the limit of plasticity w_p, and the liquid limit w_L, can be measured easily and the methods of these measurements are well known, even standardized all over the world.

A large amount of measured data can be found in the literature which give the corresponding values of plasticity (w_p and w_L) and clay content ($S_{2\mu}$) of the same sample. Some of these data are listed in Table 1-4. Data obtained by various measuring methods greatly differ from each other, and can be divided into two main groups. The first group includes data measured by mineralogical methods, i.e., electron microscope, x-ray, DTA, and the second group is formed by those determined by hydrometric methods. The diversity of the results is further increased by the use of different chemical treatments of the soil sample before or during settling.

There are also numerous articles investigating the direct relationship between plasticity and clay content, and suggesting methods to describe it (Herczog et al., 1966; Seed et al., 1964; Szilvágyi, 1966, 1967), the results of which show great differences, depending mostly on the character of the measurements used for the determination of the clay content.

One article (Dumbleton and West, 1966) has to be dealt with in detail because the results achieved and explained in it have provided the basis for the further investigations (Kovács, 1971). The first result in their paper is represented by four curves which characterize w_L vs. $S_{2\mu}$, and w_p vs. $S_{2\mu}$ relationships for montmorillonite and kaolinite, respectively. The clay content of pure montmorillonite and kaolinite samples determined by mineralogical methods was found to be nearly 100 percent. The limit of plasticity and the liquid limit of samples with smaller clay content have been measured using mixtures of these pure samples of clay minerals with quartz sand and silt in various predetermined ratios.

The second part of the paper presents the measured parameters of numerous natural clays. The clay contents of the latter have been determined by hydrometric method. It is very interesting to note that the curves constructed from these data do not show such marked differences between the type of clay minerals and those based on the results of artificial samples, e.g., the curve of natural kaolinites more closely resembles the curve of artificial montmorillonite, than the curve representing artificial kaolinite samples (Fig. 1-51).

Table 1-4. Comparison of clay content values determined by various methods.

A. Materials corresponding to the proposed method

		Characteristics of Plasticity		Clay Content Determined By		Active Clay Content	
		w_L	w_P	mineralogical methods	sedimentation	from w_L	from w_P
1	2	3	4	5	6	7	8
Montmorillonite	Own measurement	0.87	0.35	0.70	0.27	0.67	0.66
	Szilvágyi	0.94	0.35	0.76	0.20	0.72	0.66
	Dumbleton-West	1.58	0.58	1.00	--	1.00	1.00
Illite	Own measurement	0.33	0.19	0.20	0.07	0.25	0.28
	Szilvágyi	0.43	0.23	0.40	0.14	0.35	0.39
	Craft	1.03	0.39	0.88	--	0.79	0.74
Kaolinite	Dumbleton-West	0.52	0.25	--	0.41	0.42	0.44
	Dumbleton-West	0.82	0.42	0.96	--	0.64	0.79
	Craft	0.59	0.30	0.88	--	0.47	0.55
Clay	Schmertmann	0.38	0.19	0.53	--	0.30	0.29
	Dumbleton-West	0.77	0.35	--	0.63	0.60	0.65
Silt	Own measurement	0.33	0.19	0.20	0.07	0.25	0.28
	Thompson	0.54	0.24	--	0.34	0.43	0.41
Keuper marl	Dumbleton-West	0.35	0.18	0.54	0.28	0.27	0.26
	Dumbleton-West	0.49	0.25	0.70	0.32	0.40	0.43

B. Materials the behavior of which is different from that described by the proposed method

1	2	3	4	5	6	7	8
Na Montmorillonite	Own measurement	1.43	0.14	0.34	--	1.00	0.11
	Szilvágyi	1.16	0.27	--	0.13	0.84	0.49
Halloysite	Craft	0.55	0.32	0.75	--	0.44	0.60
	Dumbleton-West	1.03	0.66	--	0.79	0.78	1.00
Bauxite	Dumbleton-West	0.60	0.34	--	0.54	0.49	0.63
Clays the liquid limit of which increases after the preparation of sample (probably amorphous colloids)	Dumbleton-West	0.87	0.48	--	0.87	0.67	0.89
	Dumbleton-West	0.78	0.38	--	0.85	0.61	0.74

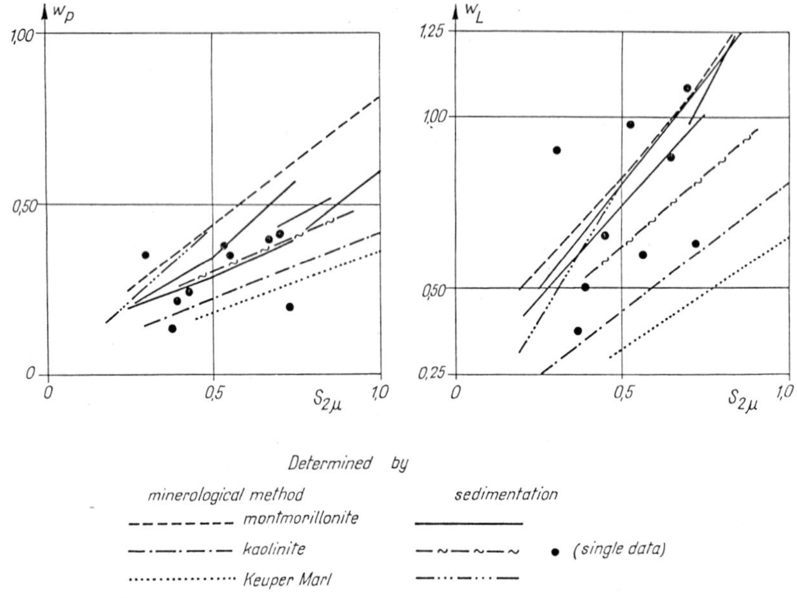

Fig. 1-51. Relationship between clay content and the parameters of plasticity (after Dumbleton and West).

A further important result of the paper, quoted, is the series of measurements, which shows that the w_p vs. $S_{2\mu}$ and w_L vs. $S_{2\mu}$ curves change according to the methods used to determine the clay content. In their investigations, the authors have used various samples of Keuper marl and the results are summarized in a figure which is reproduced here as Fig. 1-52. The data obtained by mineralogical measurements indicate a close relationship. The scattering of points representing the data measured by hydrometric method is greater, and the clay content corresponding to a given plasticity according to these data is considerably smaller than that obtained by using mineralogical data.

When data measured in different countries and by various investigators are plotted on one figure (w_L vs. $S_{2\mu}$ relationship in Fig. 1-53 and w_p vs. $S_{2\mu}$ relationship in Fig. 1-54),

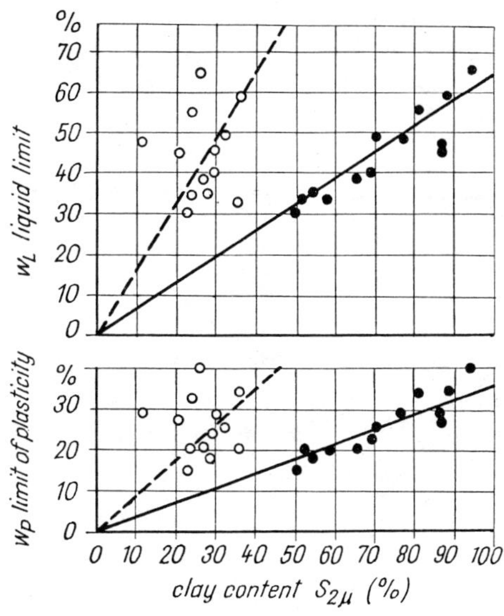

Fig. 1-52. Modification of relationships between clay content and the parameters of plasticity caused by the methods used to determine clay content (after Dumbleton and West).

the separation of the points can be considered to bear direct relation to the methods used to determine the clay content, and not to the various types of clay minerals. This statement is verified by Fig. 1-53(a) and 1-54(a), which represent almost all data listed in Table 1-2. Only a few data had to be omitted, their result being very different from that of the majority of the samples. The different behavior of these samples, however, could always be explained by the presence of special clay minerals (Na montmorillonite, halloysite, or amorphous colloids).

The left-hand side of each of these figures, constructed from the remaining more than 500 pieces of data, is covered with points measured by hydrometric method, while on the right-hand side are found the data determined by mineralogical methods. There is quite a well-defined border between these two regions in each case. For characterizing these borders the following curves expressed by mathematical equations can be fitted:

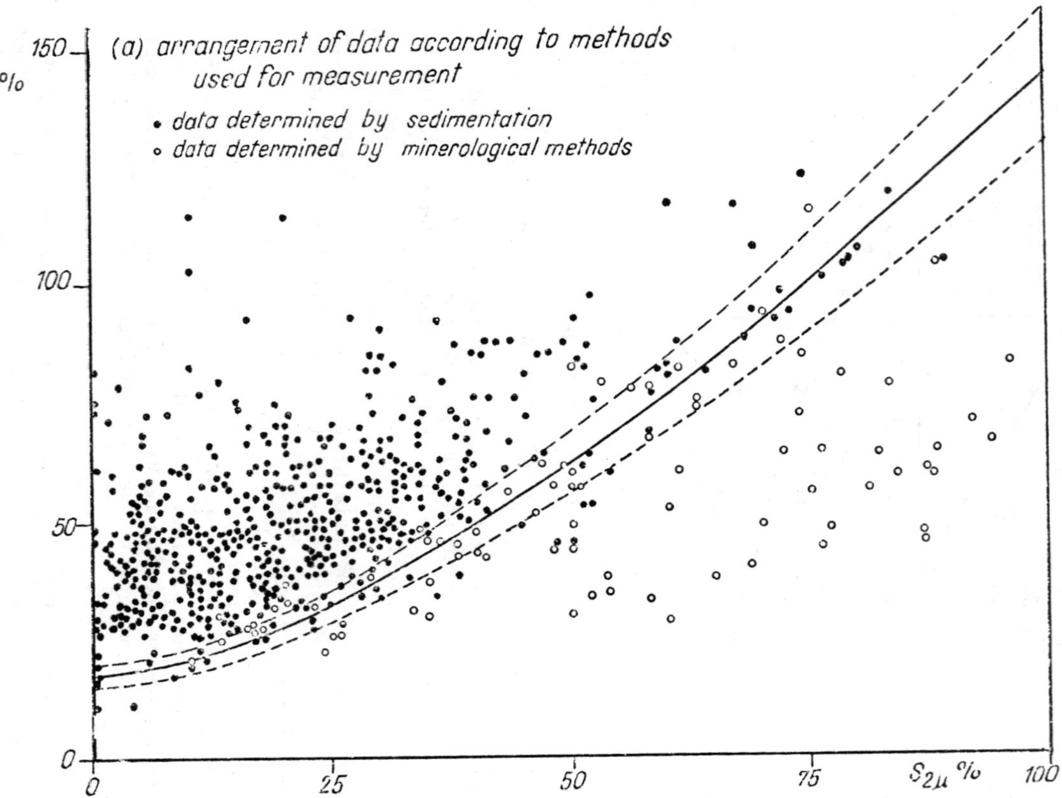

Fig. 1-53(a). Relationship between clay content and liquid limit.

(b) arrangement of data determined by mineralogical methods according to the type of clay minerals
montmorillonite Ca
Na
illite
(illite montmorillonite)
kaolinite
(halloysite)
mixed or undetermined samples
Keuper marl
w_L %
150
100
50
0
0
25
50
75
100
$S_{2\mu}$ %
montmorillonite
illite
kaolinite and others

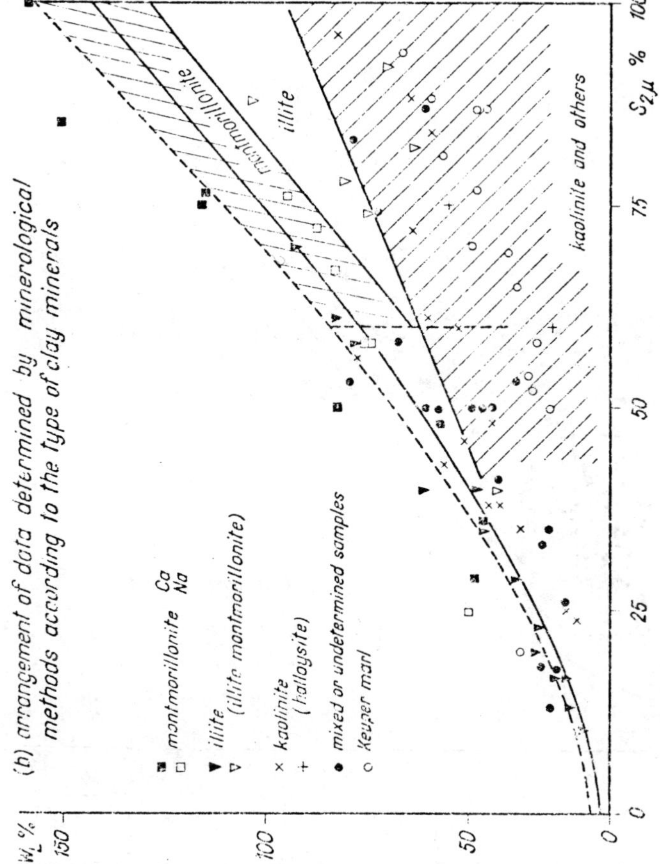

Fig. 1-53(b).

(c) arrangement of data determined by sedimentation according to methods used for peptization
w_L %
100
75
50
25
10
distilled water
Na CO3
English standard
English standard
Na CO3
without peptization
0
25
50
75
100
$S_{2\mu}$ %

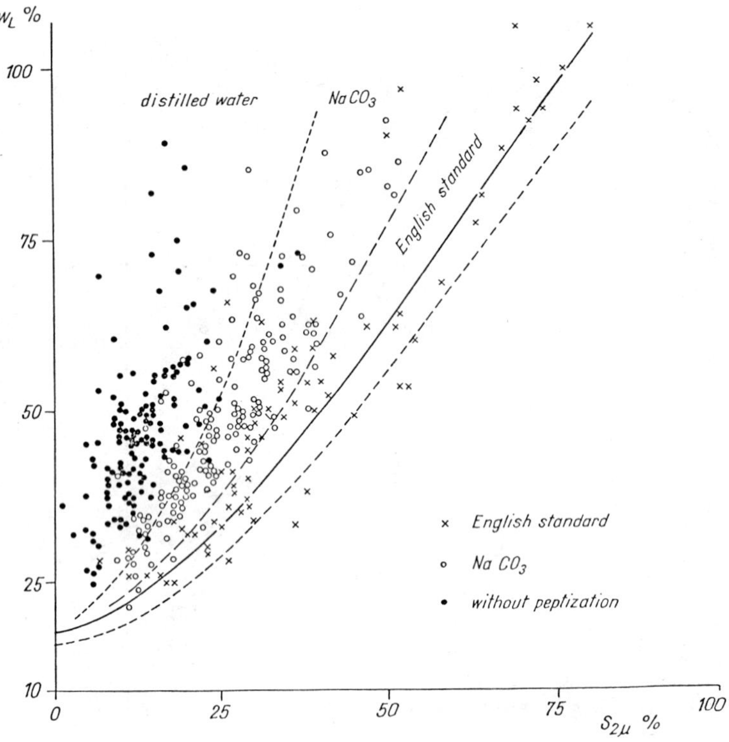

Fig. 1-53(c).

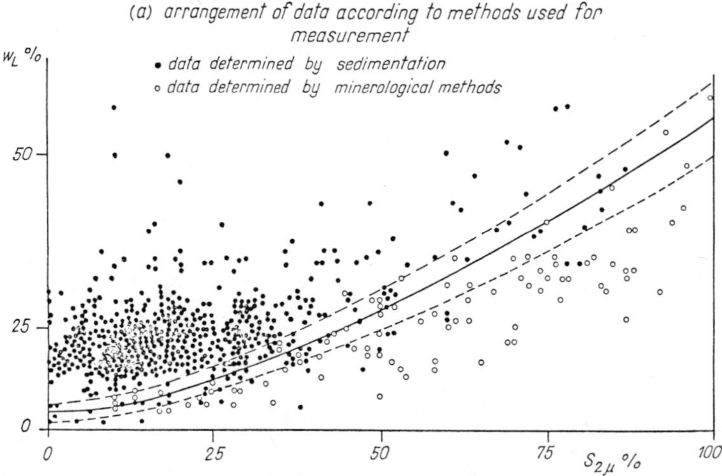

Fig. 1-54(a). Relationship between clay content and the limit of plasticity.

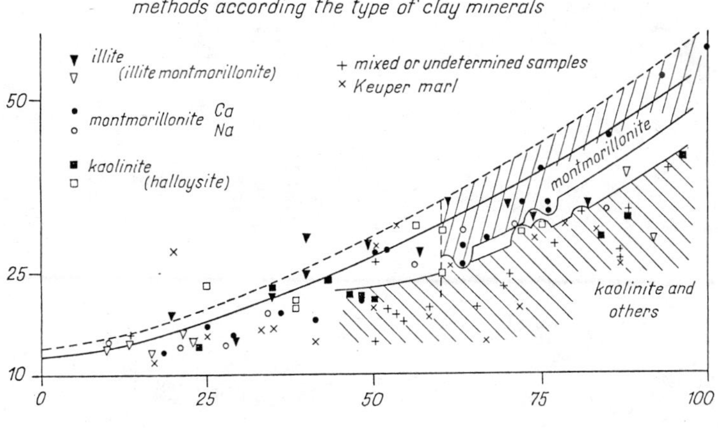

Fig. 1-54(b).

(c) arrangement of data determined by sedimentation according to methods used for peptization

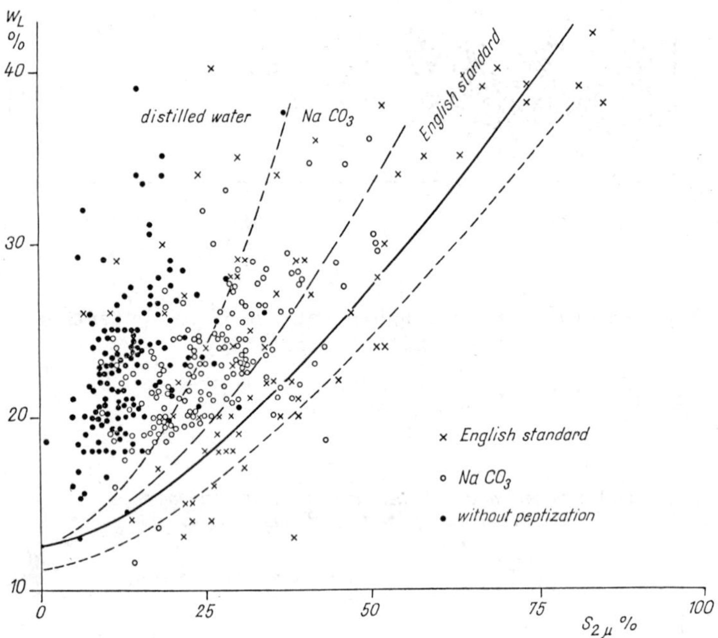

Fig. 1-54(c).

$$w_p = 0.125 + 0.42\ S_{2\mu}^{3/2} \quad ; \qquad (1\text{-}14)$$

$$w_L = 0.175 + 1.26\ S_{2\mu}^{3/2} \quad . \qquad (1\text{-}15)$$

Only a few points from each group do not lie on their own particular side of the border, and the distance of these points from the given curves is generally smaller than the probable error of the original measurements (in the figures the differences of ±10 percent are indicated).

The existence of a border between the two different groups of data is clearly demonstrated by the figures, and its significance cannot be ignored. The first question concerning the evaluation of the figure is, what do the lines characterized by Eqs. 1-17 and 1-18 mean? To answer this question it is necessary to recall that the total amount of clay content can be divided into three groups:

- individual grains (the total surface is active);

- grains united by loose bindings (almost the entire surface of each grain is active);

- grains bound by irreversible aggregation (the inner surface of these aggregates is not active).

It was already mentioned that the mineralogical methods measure the total amount of fine grains, and therefore, the clay content determined in this way is always greater than the actual amount of active grains or is equal to the latter if there are no grains bound together by irreversible aggregation in the sample. Meanwhile, the clay content measured by hydrometric methods can only be smaller than or equal to the amount of active grains smaller than 2μ, depending on the coagulation in the suspension.

The right-hand sides of the figures include data greater than (or equal to) the active clay content, and the left-hand sides, those which are smaller than (or at the limit, equal to) it. Thus it is evident that the borderline indicates the relationship between the really active clay content (including the first two groups of fine grains, but excluding the grains bound by irreversible aggregation), and the parameter of plasticity in question. Consequently, it can be stated that Eqs. 1-17 and 1-18 give the correct point of the grain-size distribution curve at the diameter of 2μ, which is needed for the determination of the effective diameter. This value ($S_{2\mu}$) is a function of the limit of plasticity and the liquid limit, respectively.

A second question has to be considered as well in connection with the present discussion: Do the various types of clay minerals have any influence on the proposed relation curves? Contrary to the previous investigations the answer is negative and requires further explanation. The points in the figures are not separated according to the clay minerals. The various points representing kaolinite, montmorillonite, illite, or other minerals can be found situated along the border with almost the same probability, which indicates that the borderline can stand for all types of clay minerals (except for Na-montmorillonite, halloysite, and amorphous colloids).

On the other hand, some separation can be observed, on the right-hand side of the figures, among data determined by mineralogical methods (see Fig. 1-53(b) and 1-54(b)). In general the points for kaolinites, or other small grains not being members of the three main clay mineral groups, e.g., Keuper marl, are dispersed farthest away from the borderline, while points for montmorillonite are located along it.

Recalling the energy curves shown in Fig. 1-50 some explanation of the mentioned difference can also be given. The energy curve of montmorillonite is similar to type A in the figure, while that of kaolinite to type B. In the first case the repulsive force is higher than the adhesive one along a considerable distance, and their difference increases when the distance decreases. The sign of the resultant changes only in the very close vicinity of the wall. The number of the irreversible aggregations is therefore very rare. The energy curve of type B shows, however, higher adhesive force at each point than the repulsion. A considerable amount of kaolinites, represented by this type of curve, can therefore be easily aggregated, and thus the total clay content is always higher than the active one.

Effects of various chemicals used to disintegrate the flakes in the suspension can also be investigated (Fig. 1-53(c) and 1-54(c)). It can be seen that the use of sodium-hexametaphosphate (English standard) is the most reliable method. However, in general, the hydrometric method includes a very high uncertainty.

As a result of the foregoing, it can be stated that the relationship between plasticity and the active clay content is not influenced by the type of clay minerals as long as Na-montmorillonite, halloysite, and amorphous colloids are excluded from the investigation. The probable ratio between active and total clay content depends on the fact whether montmorillonite, illite, kaolinite, or other minerals are dominant in the sample. As a next step it is necessary to check the proposed relationship and to give a method for determining the presence of disturbing clay minerals in the investigated sample.

The simplest check is to calculate the active clay content, using different parameters of plasticity, and to compare the values obtained. When the two values of clay content, determined from w_p and w_L, respectively, are close to each other, the proposed method is applicable, while a difference higher than a given limit, e.g., 10-15 percent, refers to the presence of clay minerals which excludes the application of the derived relationships. Some results of this type of comparison are given in Table 1-4.

Another way of the check mentioned is to use the Casagrande's A line. It is well known that the equation of this line gives a linear relationship between the index of plasticity, i.e., $I_p = w_L - w_p$, and the liquid limit:

$$I_p = w_L - w_p = 0.73\ (w_L - 0.2) \quad . \tag{1-16}$$

A similar relation curve can be derived from Eqs. 1-14 and 1-15:

$$I_p = w_L - w_p = 0.05 + 0.84\ S_2^{3/2} = 2/3\ (w_L - 0.1) \quad . \tag{1-17}$$

To evaluate the difference between Eqs. 1-16 and 1-17, the possible scattering of the points was determined by assuming that there may be an error of ±10 percent and ±15 percent, respectively, in the measurements of both liquid limit and the limit of plasticity. 84.2 percent of the points representing the samples listed in Table 1-2 are within the 10 percent error zone, uniformly covering this field: 32 points are located above this zone and 49 below it, from which 26 and 35, respectively, are within the zone of an error of ±15 percent (Fig. 1-55). This figure was supplemented by the results of some previous investigations as well (Rétháti, 1968; Herczog et al., 1966; Dumbleton and West, 1966).

The comparison shows the reliability of Eqs. 1-14 and 1-15. It is also reasonable to use Eq. 1-17 instead of Casagrande's equation (Eq. 1-16) in practice. Not only the accuracy of Eqs. 1-14 and 1-15 is supported by the existence of such a linear relationship, but the explanation of these equations serves, at the same time, as a physical background of Casagrande's A line, which was, until now, absent. The linear relationship of I_p and w_L is based on the fact that both values (more precisely w_p and w_L) depend on the active

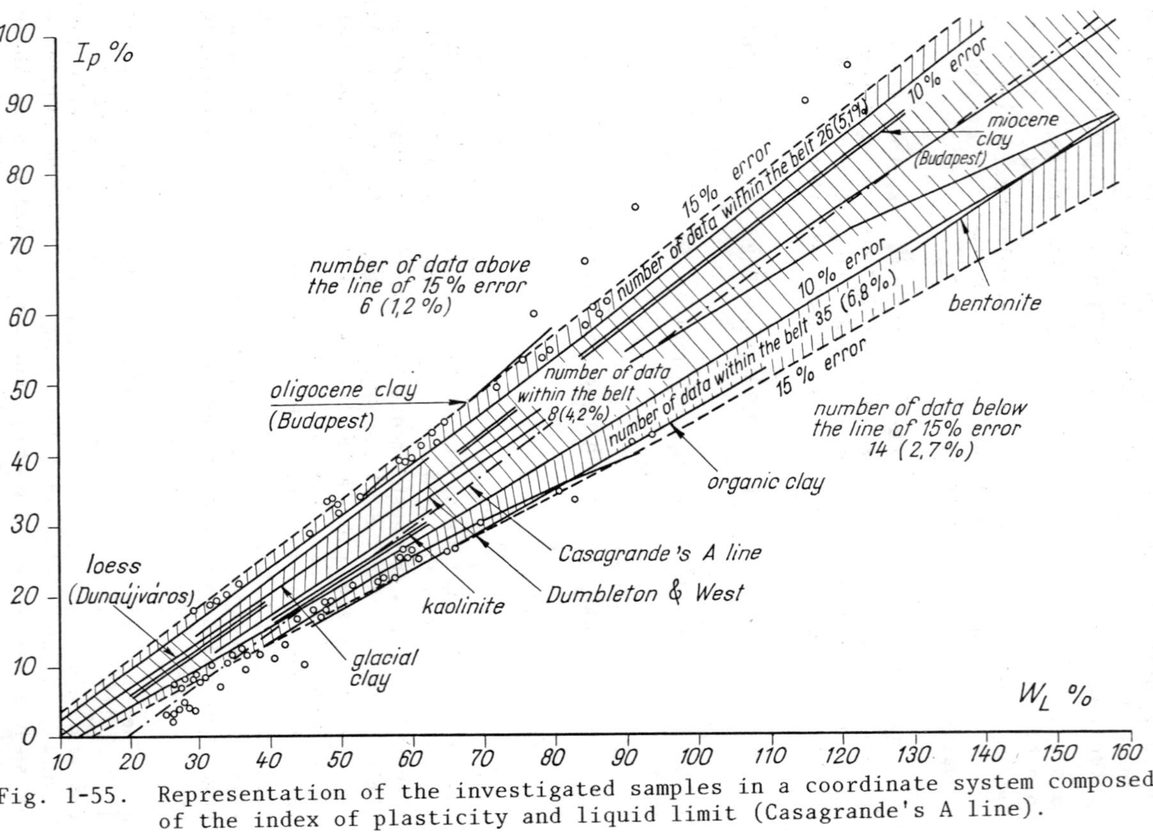

Fig. 1-55. Representation of the investigated samples in a coordinate system composed of the index of plasticity and liquid limit (Casagrande's A line).

clay content of the samples, and this common factor creates the close relationship between the parameters of plasticity. Excluding some special minerals (Na-montmorillonite, halloysite, amorphous colloids), this relationship is not influenced by the type of clay minerals. Knowing the measured w_p and w_L values of a sample, the latter can be represented in the I_p vs. w_L system with a point. Equations 1-14 and 1-15 can be used to calculate the active clay content of the sample only if this point is within the zone determined by the possible error of the measurements.

Considering the results of the investigation, the grain-size distribution curves, the lower stretches of which were either measured by hydrometric method or determined by mineralogical analysis, can be corrected so that they should represent the actual morphological character of the colloid particles. Following, is the method of this correction.

At first the distribution of grain diameters of particles larger than 0.1 mm should be determined by mechanical sieving. The hydrometric method is generally used for measuring the lower stretch of the distribution curve of the sample, from which the grains larger than 0.1 mm were already separated. It is advisable to execute this analysis both with distilled water and by adding sodiumhexametaphosphate to the suspension (English standard). The difference between the curves determined in these two ways already indicates the morphological character of the colloid grains. If the ordinates of the two curves belonging to D = 2μ diameter differ considerably, the sample contains a large amount of active colloidal particles. It is expected that the active clay content is equal to, or near the value determined by the analysis of the chemically treated suspension.

When there is a possibility for detailed investigation of the mineralogical composition of a sample (x-ray, electron microscope, DTA), information can be obtained not only on its total colloidal content, but also on the ratio of different minerals (montmorillonite, kaolinite, illite, amorphous colloids) in the fraction of the sample smaller than 2μ. This type of analysis is relatively expensive and therefore it is not generally applied in the every day routine work.

After measuring the two plastic properties of the sample (liquid limit and limit of plasticity) the active clay content (the ratio of the weight of colloidal particles having smaller diameter than 2μ, and not bound irreversibly in aggregates, to the total weight of the sample) has to be determined from Eqs. 1-14 and 1-15. The average of the two calculated $S_{2\mu}$ values can be accepted as the most probable active clay content, if the two values calculated in this way to not considerably differ, or the point representing the measured w_L and w_p fits the corrected Casagrande's A line in Fig. 1-55 well. This value indicates the height of the grain-size distribution curve belonging to the diameter of 2μ, thus the curve may be corrected to fit to this point.

The corrected distribution curve can be the basis of the determination of the effective diameter. It is hoped that this diameter correctly expresses the morphological character of the colloidal particles, since it was calculated from the grain-size distribution corrected by considering the active clay content of the sample. According to the hypothesis explained previously, the mineralogical composition of the grains modifies the behavior of the sample, related to water only through influencing the development of irreversible or loose aggregates. It is supposed, therefore, that the effective diameter calculated in this way is suitable for the characterization of the resistance to flow through the pores of the sample, without any further mineralogical investigations, if the colloidal particles behave normally.

It was already mentioned, that there are some minerals, the presence of which considerably changes the interaction between water and grains in the sample. The most important minerals causing such "non-normal" behavior are Na-montmorillonite, halloysite, and the amorphous colloids. The disturbing effects of these minerals are indicated by the large difference of the two values of $S_{2\mu}$ calculated from Eqs. 1-14 and 1-15, respectively, or by the fact that the point representing the measured w_L and w_p parameters of the sample does not fit to the Casagrande's A line. In this case the detailed mineralogical investigation of the sample is advisable.

Examples of the correction of grain-size distribution curve according to the calculated active clay content are shown in Fig. 1-56. The distribution curves of two samples are represented in the figure. The samples (Ca bentonite and clay) were analyzed both mineralogically (curve 1) and by hydrometric method, the latter with distilled water (curve 2), and after treating the suspension with $NaCO_3$ (curve 3). Thus, originally three curves were determined for each sample. The correction was executed on the basis of the calculated active clay content (curve 4), and the effective diameters were then determined. The ratio of the effective diameters determined from the corrected distribution curve, and that calculated on the basis of the analysis of the chemically treated suspension were in the range of 1:300-1:200.

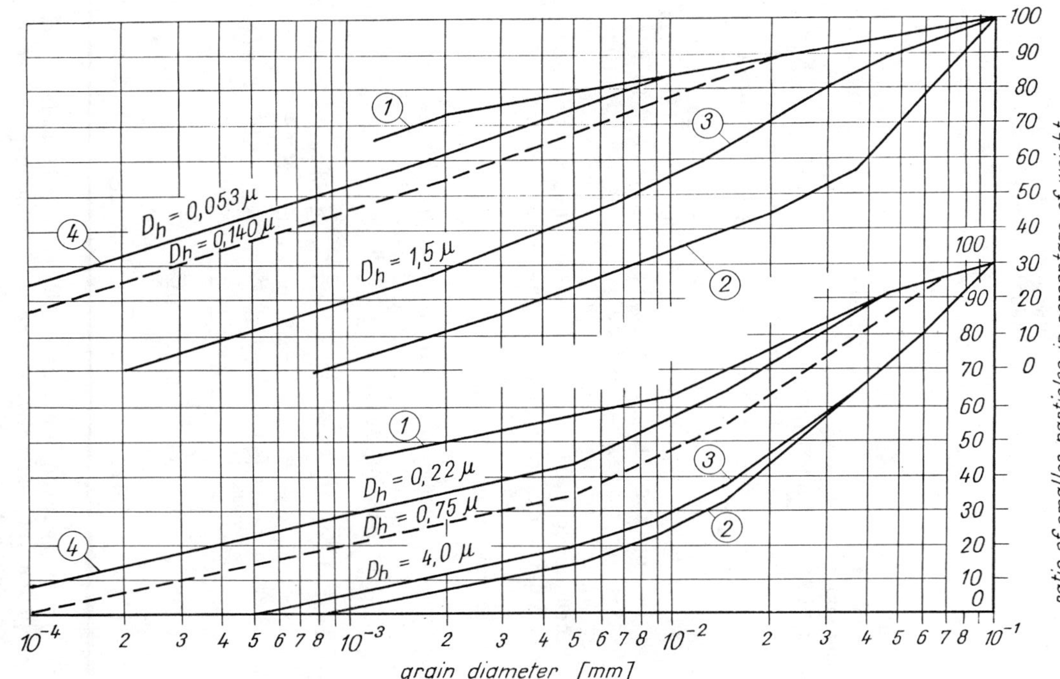

Fig. 1-56. Change of grain-size distribution curves as a result of drying the sample.

The examples can also be used to demonstrate the change occurring as the result of any modification of the morphological character of the colloidal particles. For this purpose, the influence of the strong and repeated drying was investigated. The water films surrounding the grains are decreaed by drying, thus the distance between the grains decreases as well, and this process increases the number of irreversible aggregates in the sample. The consequence of drying is therefore the lowering of plasticity and the increase of hydraulic conductivity. The latter parameter can be measured to control the correctness of the distribution curves constructed on the basis of the method explained previously.

The numerical results of the investigation are summarized in Table 1-5, and the probable grain-size distribution curves are represented by dotted lines. The decrease of the parameter, used for the characterization of plasticity, clearly indicates the increase of the number of the irreversibly bound particles. The new distribution curve shows the increase of the effective diameters, which is the expected result of aggregation. In the case of the first sample the effective diameter is 2.65 times greater than that of the original sample, while for the clay this ratio is 3.16:1. This result can be checked by comparing hydraulic conductivities measured before and after drying. As it will be proved further on, hydraulic conductivity is linearly proportional to the square of the effective diameter, thus the expected ratio of the increase of the former parameter is 7:1 and 10:1, respectively, while the measured quotients are 7.1 and 9.1.

Table 1-5. The change of maximum molecular water retention capacity and hydraulic conductivity as the results of drying.

The Investigated Material	In Natural Condition		After Strong and Repeated Drying	
	w_{mol}[%]	K[m/sec]	w_{mol}[%]	K[m/sec]
Ca bentonite	37.5	1.3×10^{-9}	29.0	9.3×10^{-9}
Clay	20.2	4.5×10^{-8}	16.9	4.1×10^{-7}

This investigation proves both the reliability of the method proposed to determine the relationship between the active clay content and the parameters characterizing plasticity, as well as the influence of the change of the number of irreversible aggregates on the active clay content.

3-4 Porosity of Clastic Sediments and its Relationship with Parameters of Soil Physics

The porosity of loose clastic sediments may vary within a very large interval. The porosity of the loosest structure of a homodisperse sample of spheres (Fig. 1-57), when the centers of the spheres form a hexahedron in space ($\alpha = 90°$), and each sphere contacts six other particles is n = 0.476. In the case of the most compact sample the angle of the same configuration is $\alpha = 60°$, each sphere contacts twelve others, and the porosity is n = 0.259. In the general case, when the angle mentioned previously lies between the two limits (α = 90°-60°), the porosity can be calculated from the following equation (Slichter, 1899):

$$n = 1 - \frac{\pi}{6\ (1 - \cos\alpha)\ \sqrt{1 + 2\cos\alpha}} \quad . \qquad (1\text{-}18)$$

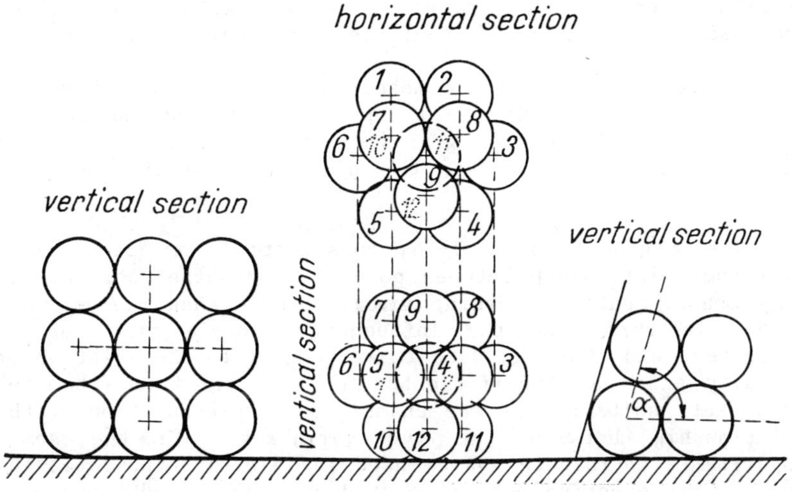

Fig. 1-57. Position of grains in a homodisperse sample of spheres having various porosity.

The upper limit of porosity, however, is not absolutely strict. A stable structure of spheres can be constructed where the particles form arches, and thus porosity is higher than the theoretical upper limit: n = 0.6-0.7 (Jáki, 1944).

The variability of porosity is even greater in the case of natural layers. It is influenced by the origin of the layer, the size and shape of the grains, the grain-size distribution, and the pressure affecting the layer as well.

Fluvial sediments (sand and gravel) are generally well compacted, while sand of lacustrine origin is usually deposited in less compact layers. The porosity of cohesive materials (clay) is influenced by chemicals dissolved in the water. When these chemicals increase coagulation, the settled flakes build up a loose layer, while a more compact layer is formed from suspensions in which the development of flakes is hindered.

The relationship between porosity and the size of particles can be explained by the fact that the number of contacts between grains related to the unit weight of the sample increases, when the size of particles decreases. Thus the resistance against the movement of a grain, attempting to have a more stable position, is also increased, the layer of fine grains is generally less compacted. This effect is strengthened by the shape of the grains as well. Among fine grains there are more laminated particles, and therefore, their surface-to-volume ratio is higher, this fact further increases the number of contacts, and helps form arches.

In Table 1-6 there are measured data collected from the literature representing the porosity of homodisperse samples (or samples built-up from grains having very small difference in size) as a function of the diameter, and the shape of grains.

The graphs in Fig. 1-58, constructed from these data, show the relationship between porosity and shape coefficient. Only measurements concerning grains larger than 0.2 mm were used. In this zone the influence of grain size can be neglected, and thus the data corresponding to the relation of porosity and shape coefficient can be regarded as a homogeneous set, suitable for the unambiguous representation of the relationship between the two variables. In the semi-logarithmic coordinate system used for plotting the data, the curves, representing both compacted and loose conditions, can be approximated by a parabola of the second order (see Eq. 1-20).

From the same series of data, a set can also be chosen representing grains with approximately the same shape. According to the graphs constructed from these data (Fig. 1-59), the porosity of homodisperse samples does not depend on the diameter of grains, if the latter is greater than 0.2 mm. Below this limit, porosity increases with decreasing diameter. It was also shown that some measurements (Hamilton and Menard, 1956; Früchtbauer and Reineck, 1958) contain data related to samples differing from each other not only in particle size,

Table 1-6. Data characterizing the relationship between grain size, shape coefficient, and porosity of homodisperse samples.

Homodisperse Samples Consist of Grains			The Porosity of the Sample		References
Material	Diameter (mm)	Shape Coefficient	In Loose	In Compacted	
			Condition		
1	2	3	4	5	6
lead balls	1.50	6.0	0.424	0.389	Fraser, 1935
sulphur balls	1.50	6.0	0.441	0.382	
sea sand	1.50		0.430	0.350	
sand from seashore	1.50	10^x	0.466	0.385	
dune sand	1.50		0.449	0.393	
ground calcite	1.50		0.545	0.427	
ground quartz	1.50	20^x	0.539	0.440	
ground rock-salt	1.50		0.520	0.435	
ground mica	1.50	90^x	0.924	0.873	
lead balls	2.42	6.0		0.369	Westmann and Hugill, 1930
	2.03	6.0		0.369	
	1.40	6.0		0.370	
round beans	3.43	10^x		0.375	
poppy seeds	0.51	8^x		0.398	
washed round sand	2.24	8^x		0.377	
	0.36	8^x		0.382	
	0.12	8^x		0.386	
	0.045	8^x		0.425	
sand	0.096			0.386	King, 1899
	0.152			0.368	
	0.483			0.329	
	0.611			0.352	
sea sand	0.015		0.657		Hamilton and Menard, 1956
	0.030		0.610		
	0.040		0.720		
	0.050		0.512-0.610		
	0.135		0.475		
	0.190		0.467		
	0.300		0.410		
sea sand	0.010		0.775		Früchtbauer and Reineck,
	0.030		0.750		
	0.040		0.615		
	0.060		0.630		
	0.070		0.505		
	0.080		0.504		
	0.090		0.575		
	0.110		0.490		
	0.120		0.425		
	0.130		0.425		
	0.140		0.445		
	0.150		0.430		
	0.160		0.445		
	0.170		0.420		
	0.190		0.410		
	0.200		0.435		
	0.210		0.435		
	0.220		0.405		
	0.230		0.450		
sand	1.385			0.375	Stakman, 1966a
	0.693			0.369	
	0.500			0.375	
	0.457			0.375	
	0.353			0.373	
	0.250	8^x		0.384	
	0.177			0.396	
	0.125			0.418	
	0.089			0.443	
	0.061			0.442	
aluminum discs	35.4	39.0	0.608	0.501	Own measurements
	18.5	23.1	0.553	0.467	
	14.3	20.4	0.570	0.403	
	10.4	23.9	0.586	0.410	
aluminum pins	10.0	35.6	0.586	0.500	

^{x}estimated values

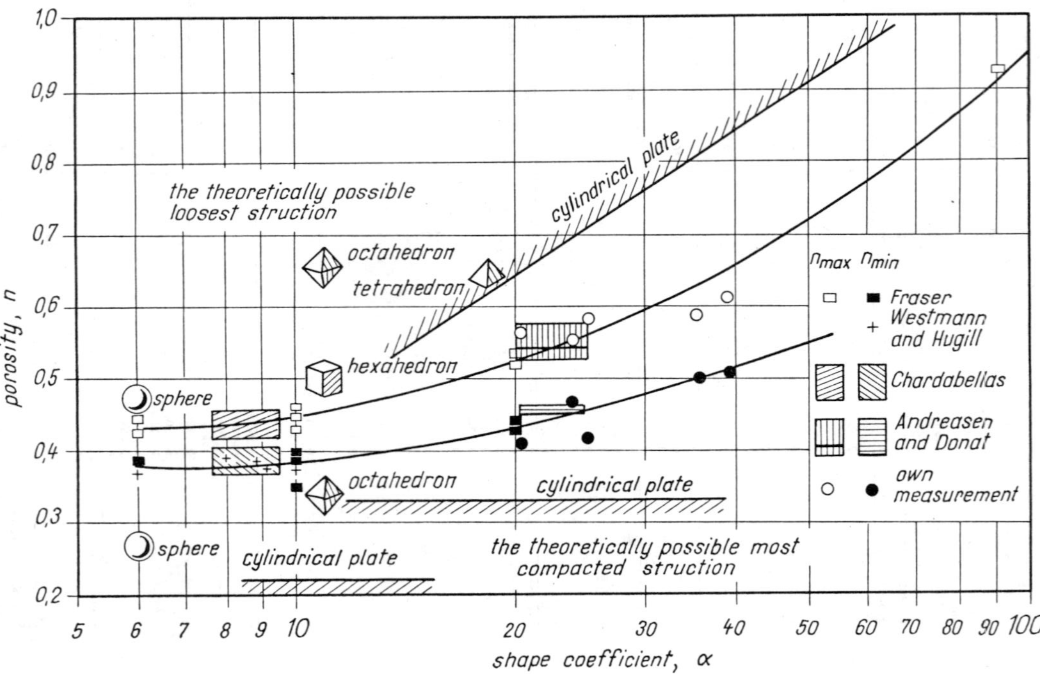

Fig. 1-58. Relationship between porosity and shape coefficient.

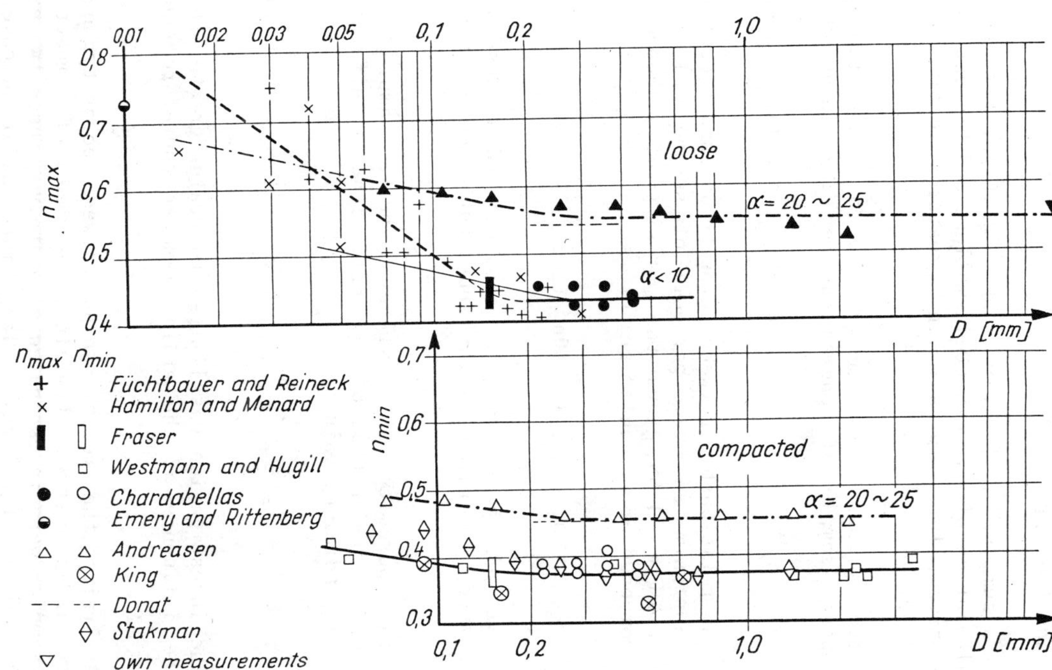

Fig. 1-59. Relationship between porosity and the diameter of grains.

with standard curves, as it was proposed by Chardabellas. Fig. 1-61, constructed from data listed in Table 1-7, shows that porosity is closely related to the coefficient of uniformity. Although there is some scattering of the points, the relationship can be well approximated by an exponential equation in both cases representing compacted and loose conditions (see Eq. 1-21).

As a summary of the results shown in Figs. 1-58 and 1-61, the calculation of the expected smallest and highest porosity can be given, for samples having greater effective diameter than 0.2 mm, as follows:

- the probable extreme values of porosity for a homodisperse sample of spheres can be taken in practice as

$$n_{o\ min} = 0.38; \quad \text{and} \quad n_{o\ max} = 0.43; \tag{1-19}$$

- the relationship between porosity and shape coefficient can be approximated by a parabola of second order in the system in Fig. 1-58, thus the porosity of a homodisperse sample composed of irregular grains is

$$n_1 = n_o \left[1 + 10n_o^3 \left(\log \frac{\alpha}{6}\right)^2\right] \quad , \tag{1-20}$$

which gives the porosity of a compacted and a loose sample, respectively, according to the substituted value of n_o; and finally

- the porosity of a heterodisperse sample can be determined as a function of the coefficient of uniformity and the corresponding n_1 value:

$$n = \frac{2}{3} n_1 + \frac{1}{3} n_1 \exp\left(- \frac{U-1}{2}\right) \quad . \tag{1-21}$$

The investigation was limited to grains greater than 0.2 mm, the influence of the grain diameter therefore could be neglected.

Analyzing the relationships between porosity and the geometrical parameters of grains, the cases of compacted and loose samples were investigated separately. These two extreme values indicate a relatively large interval which covers the probable porosity of the natural layers. It was already mentioned that the pressure acting on the layer also affects porosity. It can be added that porosity is influenced not only by the pressure at present, but by all those pressures which were acting since geological ages as well, due to the limited elasticity of the layers.

but in the shape of grains as well. The reliability of this statement is proved not only by the origin of the samples (natural samples from the seabed were investigated, where the grains may be separated according to both size and shape), but also by the comparison with data of grains with known shape coefficient. This is the reason why the rapid change of porosity, represented with a dotted line in the figure, cannot be accepted, although some authors have drawn this conclusion from the mentioned observations (Chardabellas, 1964; Engelhardt, 1960).

There is a further well-known relationship between porosity and grain-size distribution. Porosity is generally higher in a homodisperse sample, than that in a layer composed of grains of various sizes. In the latter case the very fine grains can fill the pores of the larger particles, decreasing the volume of the pores in this way, without changing the total volume of the sample. King's experiments visualize this effect very well, representing the change of porosity of a mixture of two homodisperse samples, as the ratio of the two samples in the mixture was modified (Fig. 1-60) (King, 1899).

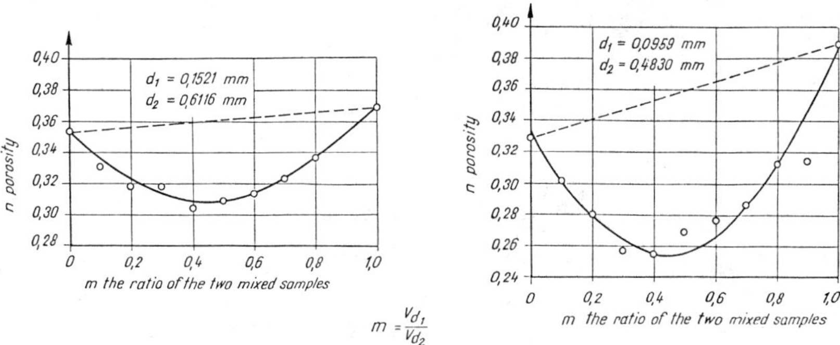

Fig. 1-60. The porosity of a mixture of two homodisperse samples and the modification of porosity as a function of the ratio of the weights of the two samples.

Chardabellas (1964) has made very detailed analyses, investigating how porosity is influenced by grain-size distribution. He has found the fact that the shape of the distribution curve, the location of its maximum or maxima, affect the porosity in both compacted and loose conditions. In practice, however, it is more preferable to give an equation which can describe the existing relationship numerically, as a function of only one, or a few, known parameter(s) with acceptable accuracy, than to use a classification based on the comparison

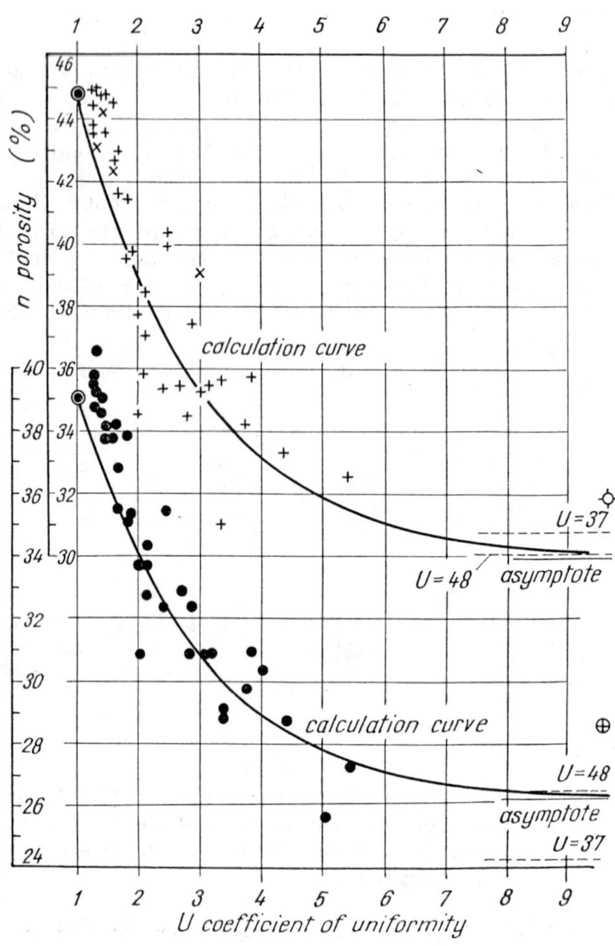

Fig. 1-61. Relationship between porosity and the coefficient of uniformity.

Table 1-7. Data characterizing the porosity of heterodisperse samples.

The Investigated Sample					The Porosity of the Sample				
Material	Effective Diameter (mm)	Estimated Shape Coefficient		Coefficient of Uniformity	In Loose Condition		In Compacted Condition		References
					Measured	Corrected for α=10	Measured	Corrected for α=10	
1	2	3	4	5	6	7	8	9	10
Heterodisperse sample of sand	0.455	10.9		2.10	0.384		0.343		Chardabellas, 1964
	0.455	9.7		2.00	0.377		0.336		
	0.575	10.4		3.16	0.354		0.308		
	0.552	9.6		3.03	0.352		0.308		
	0.378	9.1	10	1.81	(0.390)	0.395	(0.348)	0.351	
	0.387	9.0	10	1.85	(0.391)	0.397	(0.349)	0.353	
	0.739	8.3	10	5.40	(0.318)	0.325	(0.266)	0.271	
	0.162	10.4		1.66	0.429		0.368		
	0.162	10.3		1.66	0.416		0.355		
	0.355	9.2	10	2.09	(0.354)	0.358	(0.324)	0.327	
	0.355	9.1	10	2.09	(0.367)	0.370	(0.333)	0.336	
	0.478	9.6		3.35	0.310		0.290		
	0.404	9.6		2.45	0.399		0.354		
	0.404	9.2	10	2.45	(0.398)	0.403	(0.352)	0.355	
	0.466	10.0		2.67	0.354		0.328		
	0.430	9.1	10	2.39	(0.349)	0.353	(0.321)	0.323	
	0.460	8.7	10	3.83	(0.350)	0.357	(0.304)	0.309	
	0.454	9.9		3.38	0.356		0.287		
	0.183	10.4		1.81	0.414		0.378		
	0.802	9.2	10	4.38	(0.329)	0.333	(0.284)	0.286	
	0.810	8.3	10	3.75	(0.334)	0.342	(0.292)	0.297	
	0.306	10.6		2.80	0.344		0.308		
	0.261	10.3		2.00	0.345		0.308		
	0.211	10.7		2.89	0.374		0.324		

Table 1-7. (continued)

The Investigated Sample					The Porosity of the Sample				
Material	Effective Diameter (mm)	Estimated Shape Coefficient		Coefficient of Uniformity	In Loose Condition		In Compacted Condition		References
					Measured	Corrected for α=10	Measured	Corrected for α=10	
1	2	3	4	5	6	7	8	9	10
Mixture of	0.246	8.3	10	1.27	(0.435)	0.445	(0.386)	0.392	Chardabellas,
homodisperse	0.246	8.0	10	1.27	(0.426)	0.437	(0.398)	0.405	1964
samples of sand	0.320	7.5	10	1.26	(0.429)	0.444	(0.388)	0.397	
	0.320	7.5	10	1.26	(0.435)	0.449	(0.386)	0.394	
	0.401	8.0	10	1.26	(0.425)	0.436	(0.381)	0.387	
	0.401	7.6	10	1.26	(0.429)	0.443	(0.386)	0.394	
	0.277	7.6	10	1.41	(0.433)	0.447	(0.328)	0.390	
	0.277	7.0	10	1.41	(0.431)	0.447	(0.377)	0.385	
	0.349	7.3	10	1.45	(0.421)	0.435	(0.371)	0.381	
	0.349	7.0	10	1.45	(0.431)	0.447	(0.368)	0.377	
	0.300	6.9	10	1.59	(0.411)	0.426	(0.368)	0.377	
	0.295	7.3	10	1.59	(0.429)	0.444	(0.372)	0.382	
Mixture of homodisperse	0.185			5.05			0.255		King, 1899
samples of sand	0.288			4.00			0.303		
Heterodisperse	0.016			3.0	0.390				Wesseling
sample of sand	0.028			1.3	0.430				and Wit, 1966
Glass balls	0.173	6.0	10	1.45	(0.426)	0.442			Sine and
	0.136	6.0	10	1.59	(0.406)	0.422			Bantz, 1966
Sandy gravel		10		37.0	0.307		0.240		Own
		10		48.0	0.300		0.263		measurements
		10		10.4	0.318		0.285		

The relationship between porosity, or more precisely void ratio, and presently acting pressure can be expressed in terms of the depth of the investigated layer below the surface, the latter being proportional to the pressure. The equation generally used to represent this relationship is based on the hypothesis assuming linear correlation between the modulus of compressibility and pressure (Terzaghi, 1943):

$$e = e_o - c \log \frac{\sigma_z + \sigma_o}{\sigma_o} ; \qquad (1\text{-}22)$$

where σ_z is the actual effective normal stress at a depth of z; and σ_o is the actual effective normal stress at a starting level, near the surface, where the porosity of the layer is characterized by e_o (loose condition).

It is necessary to consider that Eq. 1-22 gives only a rough estimation because the elasticity of layers is very limited, only a partial expansion can develop after decreasing the pressure. Thus the compression is not an unambiguous function of the present load (i.e., the depth of the layer in question below the surface at the time of investigation), but porosity has preserved the influence of previous pressures as well.

Another possible cause of the difference between actual and theoretically calculated porosity is the slow development of consolidation. The total stress caused by the load on the layer can be divided into two parts: one taken over by the solid skeleton of grains (effective stress), and another balanced by water pressure (neutral stress). The latter does not cause any compression, thus porosity may be a function of the effective stress only. The rate of the effective and neutral stresses changes in time, as the surplus water is drained from the system. The process depends on the size of loads and on the permeability of layers. The actual porosity is therefore a function of all of these factors.

Finally it can be stated that the probable zone of the expected porosity can be determined on the basis of Eqs. 1-19, 1-20 and 1-21, knowing the geometrical parameters of the grains in the investigated layer. The likely location of the actual value within this interval can also be estimated by using Eq. 1-22. The reliability of the latter, however, is very low because of the effects of many unknown factors.

3-5 Pore-Size Distribution in Loose Clastic Sediments

As it was already explained in the introduction of this section, one of the basic principles of the characterization of water movement is the continuum approach. The properties of the solid matrix are described with the average values determined for the representative elementary units. The main sizes of the geometrical models being suitable to substitute the complicated channels composed of the interstices are generally determined from these average parameters. There are, however, cases when the absolute size of the water transporting conduits influences the character of the movement considerably and therefore the actual size of the openings should be considered.

The variance of pore size in loose clatic sediments is generally small and does not cause notable changes in flow conditions. This effect is therefore negligible when investigating seepage through saturated layers. The process of capillary saturation is however a function of pore diameter, thus the pore-size distribution should be taken into account in the investigation of both the moisture retention capacity of the soil-moisture zone and the water transport through the layer lying between the soil surface and the water table.

In a homogeneous porous medium, having pores evenly distributed and randomly interconnected, the pore-size distribution (the change of the probable number of the pores having the same diameter expressed as the function of their size) can be determined by applying statistical methods. Only the layers below the cultivated zone can be regarded as such a homogeneous system, the statistical analysis can be applied for the characterization of the physical processes in these layers. The cultivated zone is intersected by large openings (plowing, root channels) disturbing here the homogeneity of pore-size distribution. The investigation of the topmost layer of the soil profile therefore needs special attention.

The analysis of the cross-sectional areas of all pores, measured on various samples, has proved that the number of pores characterized by the same ratio of the individual areas (f) to the average value (f_o) can be well approximated by a Γ-distribution function of the random variable $x = f/f_o$. The general form of the function can be simplified by considering some special character of the pore-size distribution, i.e.:

- the area of the possible smallest pore tends to zero ($x_o = 0$);

- the random variable is the ratio of the area of the pores related to their mean value, the mean of this variable is, therefore, equal to unity (m = 1);

- considering the previous conditions, the other two parameters of the function have to be equal to one another ($k/\lambda = m$; $m = 1$; $k = \lambda = \lambda_o$).

Thus the final forms of the probability distribution functions are as follows:

$$f(x) = \frac{\lambda^k (x - x_o)^{k-1}}{\Gamma_{\lambda_o}(k)} \exp\,[-\lambda(x - x_o)] =$$

$$= \frac{\lambda_o\, x^{(\lambda_o - 1)}}{\Gamma(\lambda_o)} \exp\,(-\lambda_o x); \text{ and} \qquad (1\text{-}23)$$

$$F(x) = \int_o^x f(x)\, dx \quad .$$

In this special case, direct relationships exist between the λ and k parameters of the Γ-function and the m and σ parameters of the normal distribution ($\sqrt{k} = m/\sigma$, $\sqrt{k}/\lambda = \sigma$ and $k/\lambda = m$). Thus not only measurements of four different samples, with 7, 6, 8, and 4 different porosity values, respectively, were available for further analysis, but also data of a previous publication (which summarized the mean value and the variance of the pore-size distribution measured on seven different samples) was analyzed (Murota and Sato, 1969). The statistical analysis of these measurements and their graphical representation in λ_o vs. n coordinate system (Fig. 1-62), have shown that the mean value of λ_o is 1.14 with a standard deviation of $\sigma_\lambda = 0.42$. The application of the relationship:

$$\lambda_o = 2.19 - 3.75\, n \pm 0.34 \; ; \qquad (1\text{-}24)$$

determined by using regression analysis, can eliminate only 20 percent of uncertainties. The queuing of the points belonging to the same sample, but measured with different porosity, is parallel to the average line described by Eq. 1-24 in most cases, indicating that the remaining variance is mostly caused by some soil physics parameters not considered in the investigation.

Since the discrepancy between the λ_o mean and unit, is small and the probable λ_o value in the practically important range of porosity $(0.3 < n < 0.4)$ is also near unity, the function can be reduced to an exponential function, considering $\lambda_o = 1$:

$$f(x) = \exp(-x) \ ; \quad \text{and} \quad F(x) = 1 - \exp(-x) \quad . \qquad (1\text{-}25)$$

The results summarized by Eq. 1-25 were supported by the analysis of only 32 samples. It was, therefore, necessary to test the general applicability of the conclusions by comparing them with the results of other similar investigations.

Réthási (1960) has stated, on the basis of the investigation of the capillary behavior of soils, that the vertical soil-moisture distribution can be well approximated by some of the distribution functions used in statistical analysis for the characterization of the probability of various random events. This statement is based theoretically on the fact that the numerical value of capillary height is inversely proportional to the pore size. The latter is a random variable, the probable number of pores having a given size (related to the total number of pores within a cross section) can be described, therefore, by a distribution function. Réthási has proposed the use of the lognormal function for the characterization of pore-size distribution. According to his reasoning, there exists a close relationship between the distributions of the grain size and the pore size. There are, however, many investigations testifying that grain-size distribution may be well approximated by the normal distribution function, if the grain size is represented on an axis having logarithmic scale, thus the lognormal function is also acceptable for the characterization of pore-size distribution.

Although this result is not identical with our position, they are in proper harmony with each other. Both functions have a lower limit. This limit is zero in the case of a lognormal function having two parameters, and it was found in our investigation that a Γ-function with $x_o = 0$ parameter also gives a good approximation. Both lognormal and Γ-functions are asymmetrical, the median being smaller than the mean. The form of the curves describing the functions is determined by arbitrarily chosen parameters, thus the similarity of the curve can be ensured by selecting suitable numerical values for the parameters. Since the form of the lognormal function (its asymmetry) is more definitive than that of the Γ-distribution, the use of the latter is advisable because its fitting to the empirical distribution can be easily solved.

In the framework of the other investigation (Kézdi, 1968) used for comparison, the distribution of the volumes of the actual pores was determined by direct measurements for eight samples (four materials all in loose and compacted conditions). The published data had to be transformed, therefore, to achieve such characterization of the pore-size distribution which is comparable to Eq. 1-25. The relative volume of the pores was determined at first by relating the absolute values to their mean, and this dimensionless parameter was represented on the horizontal axis. Raising these ordinates to the 2/3 power, the distribution of the relative area of pores is obtained, if it can be assumed, that the shape coefficients of the various pores do not differ considerably. The correctness of this initial hypothesis was checked by using the grain-size distribution curves determined from diameters and volumes, respectively, for particles of the same sample.

After having the transformations determined, the numerical values of the cross-sectional areas having 10, 20, ..., 80, 90 percent probabilities were calculated for the eight samples. The next step was the determination of the λ_o parameters of the $\exp(-\lambda_o x)$ functions giving the best (least squares) fit to the points. The results are:

λ_o Values for the Samples Having

	loose	compacted
	condition	
Danube-gravel	0.917	0.644
pit gravel	1.001	0.609
crushed limestone 1.	0.964	0.441
crushed limestone 2.	0.585	0.585

These values, also represented in Fig. 1-62, prove the correctness of the concept outlined. The discrepancy between the mean of λ_o and unity is even smaller, if the set of the 32 data originally investigated is supplemented by Kézdi's measurements ($\lambda_{o\ mean}$ = 1.06). This result justifies the applicability of the previously made supposition, to simplify the statistical model, i.e., the pore-size distribution may be approximated by using the exponential function. Equation 1-25 may be used, therefore, for the general characterization of the distribution of the numbers of pores having different cross-sectional areas and therefore, this model will be applied further in the various investigations requiring the determination of pore-size distribution.

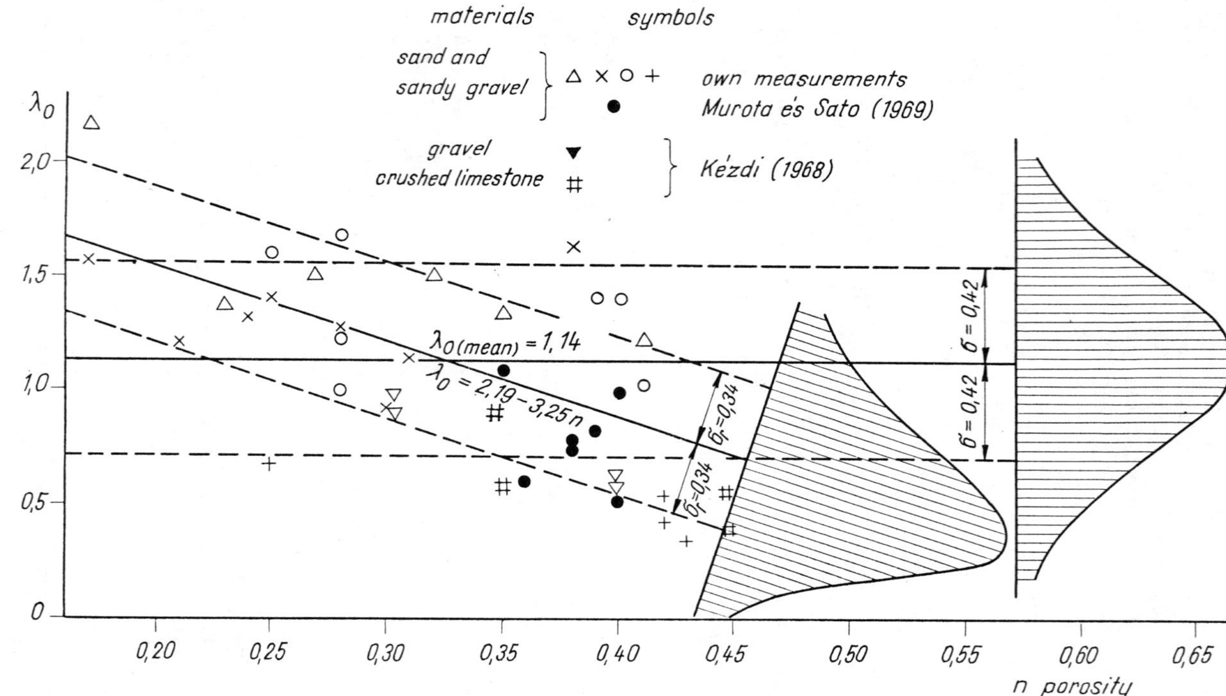

Fig. 1-62. The λ_o parameter of Γ-function describing the pore-size distribution represented as the function of porosity.

Dividing the entire range of f (area of pores) into m number of intervals, each having a length of Δf, the number of pores within a unit section of the sample, and belonging to the i-th interval, is indicated by N_i (Fig. 1-63). The ratio of this value related to the total number of pores within the same unit area (N) gives the empirical distribution of the specific number of pores (ν_i). It can be proved that the area covered by the distribution curve (either by the empirical one or by its continuous approximation) should be equal to the Δf interval chosen as the basis of the investigtion. If the Δf interval is small enough, areal porosity (n_A) can be approximated by multiplying the total number of pores by the sum of the $\nu_i f_i$ products (where f_i is the mean value within the i-th Δf interval):

$$N\left(\sum_{i=1}^{m} f_i \nu_i\right)' = n_A \; ; \quad \text{and} \quad \sum_{i=1}^{m} f_i \nu_i \Delta f = \frac{n_A}{N} \Delta f \; ; \tag{1-26}$$

or using the $(f;\Delta f)$ continuous function to approximate the empirical distribution of $_i$:

$$\int_0^\infty f \, \nu(f;\Delta f) \, df = \frac{n_A}{N} \Delta f = f_o \Delta f \; ; \tag{1-27}$$

since areal porosity divided by the number of pores within a unit area gives the average pore size (f_o).

Applying dimensionless variables [$x = f/f_o$; $\Delta x = \Delta f/f_o$; and $dx = df/f_o$], and transforming the $\nu(f;\Delta f)$-function accordingly into the form of $\xi(x;\Delta x)$, the previously given conditions can be expressed by

$$\int_0^\infty \xi(x;\Delta x) \, dx = \Delta x \; ; \quad \text{and} \quad \int_0^\infty x\xi(x;\Delta x) = \Delta x \; . \tag{1-28}$$

Considering Eqs. 1-25 and 1-28, the $\xi(x;\Delta x)$-function can be approximated with the following equation:

$$\xi(x;\Delta x) = \Delta x \exp(-x) \; . \tag{1-29}$$

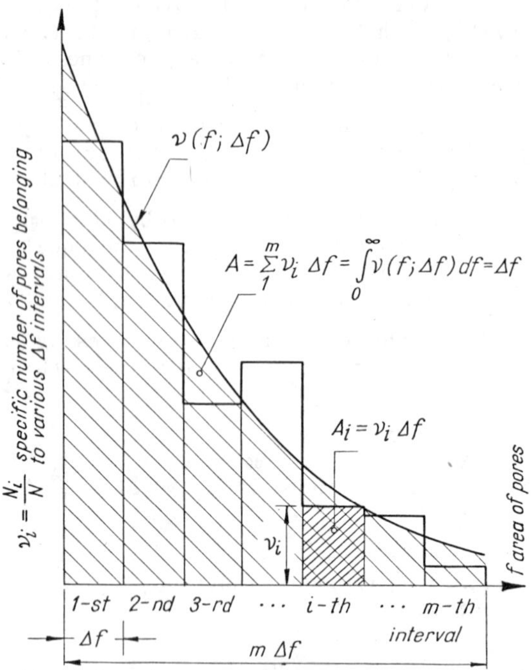

Fig. 1-63. Sketch for the determination of the continuous function characterizing pore-size distribution.

3-6 Special Structure of the Cultivated Layer

It is a general opinion that the physical investigation of seepage has to exclude the analysis of the phenomena occurring in the root zone because physiological (biological and biochemical) processes are predominant there, whereas below the layer affected by roots, the water movement is solely a physical process (van Bavel, 1966). Some investigations proved that the channels of earthworms could also modify the water movement through the top soil (Peterson and Dixon, 1971). Apart from these influenced of biological origin, the development of nonhomogeneous pore-size distribution also disturbs the processes developing near the soil surface and causes deviations from the character of flow through the lower homogeneous part of the soil profile. This special structure is created partly by the roots and the aggregation of grains influenced by the organic content of soil, and partly by plowing.

The basic difference between the physical behavior of the cultivated zone and that of the remaining part of the soil-moisture zone is caused by the fact that the special structure of the former is composed of aggregates, while in greater depth, the layer is generally built-up from evenly distributed individual grains and pores. The development of the secondary porosity created by the large channels between the clods is represented by Fig. 1-64.

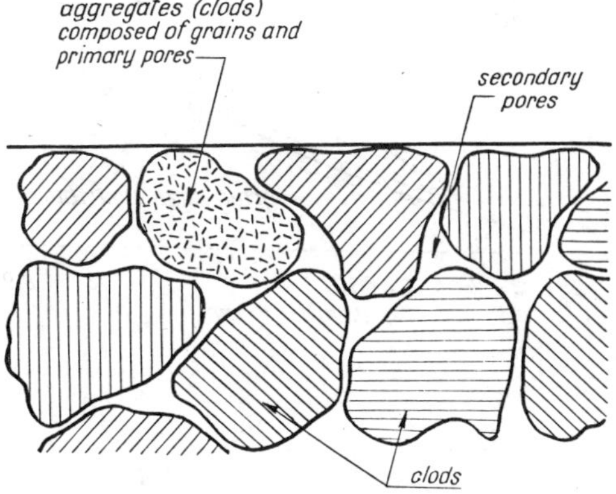

Fig. 1-64. Primary and secondary porosity in cultivated zone.

The comparison of grain-, aggregate- and pore-size distributions of the different soil layers (horizons) can be used to demonstrate and characterize the structural differences caused by cultivated. In the case of the example represented in Fig. 1-65 a chernozem soil developed on a loess terrace was investigated (Vucic, 1966). The almost identical granular composition of the layer was indicated by the grain-size distribution, thus, their similar physical behavior could be expected. There was only a slight difference between the lowest layer and the others, showing that the former was somehow finer, with this fact probably caused by the irreversible aggregation of fine grains in the upper part of the profile where the physicochemical properties of the particles had been influenced in a greater extent by drying and wetting or freezing and melting. The distribution of the aggregate size was determined only for the A-horizon (where the aggregates could be well distinguished) by sieving the samples in both wet and dry conditions. The wet sieving did not show any difference between the plowed layer and that under the plowing depth, while according to dry sieving the aggregates of the plowed layer were considerably coarser, indicating that the large clods created by plowing are less stable.

The most important information was provided by the pore-size distribution curves. The probable maximum and minimum pore sizes were calculated from the characteristics of soil physics of the samples. It was found that the diameter of the primary pores is probably smaller than 3μ and their total volume is about 30-32 percent of the bulk volume

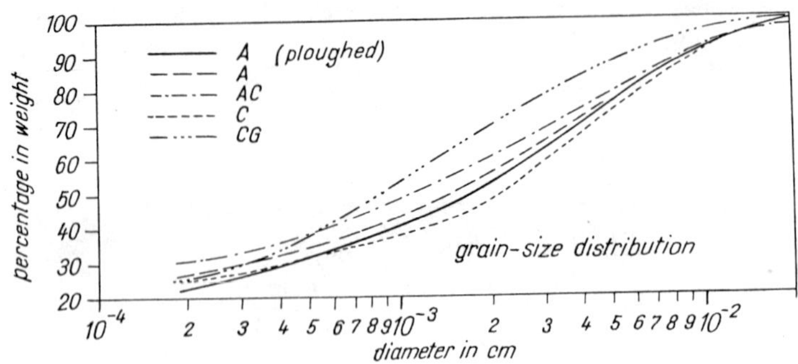

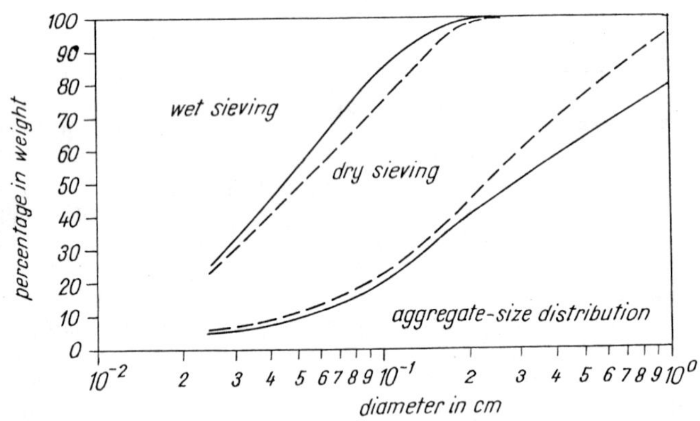

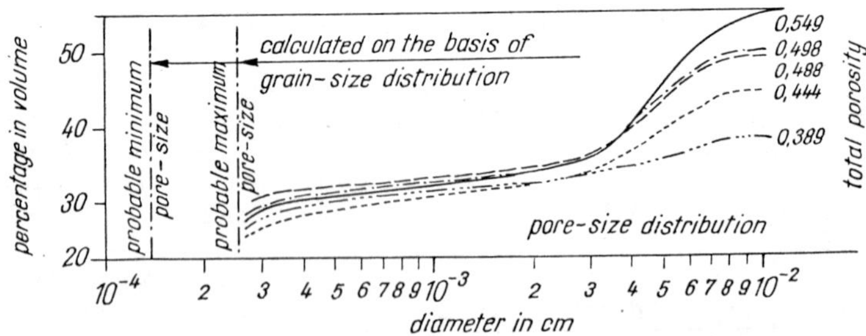

Fig. 1-65. Grain-size, aggregate-size, and pore-size distribution measured at different horizons of the cultivated zone.

(primary porosity n_1 = 0.30 - 0.32), which was in good agreement with the measured data. The measured total porosity of the plowed layer is $n \sim 0.55$, that of A- and AC-horizons is $n \sim 0.5$, at the C-horizon $n \sim 0.45$, while in the basic layer (CG) $n \sim 0.39$. Thus, secondary porosity ($n_2 = n - n_1$) decreases with the depth. Plowing raises this value to $n_2 \sim 0.24$ from $n_2 \sim 0.18$-0.19, with this value characteristic of the top soil. At the C-horizon, a secondary porosity of $n_2 \sim 0.16$ was found. In the basic layer, the secondary porosity is only a few percent, and its development is probably due to the deeply penetrating roots.

According to the continuum approach, porosity (more precisely its change) can be characterized by selecting a suitable representative elementary volume, then most of the hydraulic processes can be described in a macroscopic way by applying this parameter, even in the case of a porous medium having special structure. The relationship between tension and moisture content is, however, an exception because its development is basically influenced by the structure of the pores as well. The process has to be investigated, therefore, on a microscopic level, and the macroscopic approach is not applicable. The channels composing the secondary pores are generally large, and capillary menisci cannot develop in them, thus atmospheric pressure prevails everywhere around the aggregates. This is the reason why all clods have to be regarded as a separated system in which the relationship between tension and moisture content develops independently of the others, influenced only by the boundary conditions of the clod investigated. As it was already mentioned, the boundary condition along the large secondary channels is characterized by atmospheric pressure. Only a small fraction of the surface is contacted with other aggregates, where different boundary conditions prevail, governed by the moisture content and tension condition of the neighboring clods.

The aggregation of the particles occurs usually only in the very fine grained (cohesive) soils. Thus the tension belonging to saturated condition (height of the closed capillary zone characterizing the material of clods) considerably surpasses the size of clods and, therefore, the complete saturation of the primary pores is expected if the boundary condition is described with $p = p_o$ on the surface of the clod (even if there is a small suction caused by other clods contacting the investigated one). Only a few pores contain entrapped air, which slightly decreases the water content of the aggregates. From this condition the moisture content of the structured top layer (W_{s1}) can be calculated supposing that suction on the surfaces of the clods is smaller than the air entry pressure (which is approximately equal to the height of the closed capillary zone):

$$W_{s1} = \frac{V_{p1}}{V_t} = n_1 \; ; \quad s_{s1} = \frac{W_{s1}}{n} = \frac{n_1}{n_1 + n_2} \; ; \qquad (1\text{-}30)$$

where V_{p1} is the volume of pores within the clods, V_t is the total volume of the sample including both clods and secondary pores and s_{s1} is the rate of saturation belonging to water content W_{s1}.

The secondary pores also contain water, if the moisture content is higher than this limit value, in which condition the disaggregation of the clods is caused and the structure of the soil is destroyed. The moisture content of the aggregates is constant until the suction of their surfaces is smaller than a given limiting value, characterizing the porous medium of the clods. The suction determining the boundary condition can be raised above the given limit either by evaporation (around clods near the surface), or by the roots encircling the aggregates and even penetrating into them. Since the maximum suction created by the crops usually cultivated is about 16 atm (pF = 4.2), water will be available for the plants if the water content is higher in the clod than that belonging to the maximum suction (pF = 4.2). Thus the lower limit of moisture content (W_{s2}) below which agricultural production requires artificial water recharge can also be calculated:

$$W_{s2} = \frac{s_{(4.2)} V_{p1}}{V_t} = s_{(4.2)} n_1 \; ;$$

$$S_{s2} = \frac{W_{s2}}{n} = s_{(4.2)} \frac{n_1}{n_1 + n_2} \; ; \qquad (1\text{-}31)$$

where $s_{(4.2)}$ is the rate of saturation of a homogeneous sample from the material of the clods under the influence of suction described by the pF = 4.2 value. Using Eqs. 1-30 and 1-31 as well as Fig. 1-65, the probable characteristic values of the porosities of different horizons of the investigated chernomzom soil were summarized in Table 1-8, together with the range of moisture content within which the saturation has to be kept in arable land. The last line of Table 1-8, which was calculated presuming that the basic layer has no special structure (secondary porosity, large channels), but the total porosity (n = 0.38) is evenly distributed in the layer, shows that in this very fine material, agricultural production can

Table 1-8. Porosity values and representative elementary lengths of four different carbonate rocks (after Balásházy and J. Kovács, 1975).

Homodisperse Samples Consist of Grains			The Porosity of the Sample		References
Material	Diameter (mm)	Shape Coefficient	In Loose	In Compacted	
			Condition		
1	2	3	4	5	6
lead balls	1.50	6.0	0.424	0.389	Fraser,
sulphur balls	1.50	6.0	0.441	0.382	1935
sea sand	1.50		0.430	0.350	
sand from seashore	1.50	10^x	0.466	0.385	
dune sand	1.50		0.449	0.393	
ground calcite	1.50		0.545	0.427	
ground quartz	1.50	20^x	0.539	0.440	
ground rock-salt	1.50		0.520	0.435	
ground mica	1.50	90^x	0.924	0.873	
lead balls .	2.42	6.0		0.369	Westmann
	2.03	6.0		0.369	and Hugill,
	1.40	6.0		0.370	1930
round beans	3.43	10^x		0.375	
poppy seeds	0.51	8^x		0.398	
washed round sand	2.24	8^x		0.377	
	0.36	8^x		0.382	
	0.12	8^x		0.386	
	0.045	8^x		0.425	

hardly achieve good results without secondary porosity because total saturation will not occur until a pF = 1.8-2.0 is reached, thus plants cannot be provided with the necessary air.

3-7 Porosity of Solid Rocks and the Statistical Distribution of the Size of Openings

Fissures and fractures occurring in hard solid rocks are of different form and nature, depending on the type of rocks. The interstices were formed either during the process of lithification or as the results of geological forces and tectonic movements. Some of the pores remain more or less unchanged after their development, others are enlarged by the process of weathering (near the surface). In soluble layers (especially in carbonate sediments) the increase of the size of openings and porosity is also expected due to the solution and erosion of the rock. This very rough summary of the processes influencing the development of the interstices already indicates that the important hydraulic parameters, i.e., porosity, hydraulic conductivity and the yield of wells penetrating the fissured and fractured formations, are different and vary widely depending on the type of rock in which the network of interstices is formed. The detailed analysis of the relationships between the parameters characterizing the hydraulic properties and the material of these formations will be discussed in Section 5, which deals with the hydrology of solid-rock terrains. Only some general information is summarized here in connection with the porosity of hard rocks.

The combined character of the water transporting channels should always be considered when the porosity of solid rocks is investigated, because the variance of the size of openings is large due to the different actions having formed them. There are some formations having considerable primary porosity, pores formed between the grains composing the solid matrix of the layer. The role of this type of pore is the most important in indurated, non-carbonate sediments (e.g., sandstones). Primary porosity may range from 1 percent to 35 percent in these formations depending on the grain-size distribution and the packing of particles, as well as on the material and degree of their cementation. The volume of pores located between the grains is generally low in carbonate rocks and they are poorly interconnected, thus primary porosity hardly influences the hydraulic behavior of these aquifers. The young coarse-grained limestones are exceptions, which may have high primary porosity of 20-50 percent (chalks and tufts). The primary pores of igneous and metamorphic rocks are negligible in most cases.

The structures composed of primary and secondary pores, respectively, are basically different. The amount of inactive primary pores is higher than that of fissures and fractures in not taking part in water transport. The size of joints usually decreases with increasing depth due to the increasing load. This is the reason why the parameters describing the hydraulic properties of the aquifer (storage capacity, hydraulic conductivity) are approximated, in most cases, by monotonically decreasing functions (generally exponential functions) of the depth below the surface, if these characteristics are fundamentally determined by secondary porosity. On the contrary, the vertical distribution of primary pores depends mostly on the layering of the formation.

The structure of the porosity of solid rocks is further complicated by the fact that the secondary pores are caused by different actions and therefore, their system is not homogeneous. The dense network of generally very narrow joints, formed during the process of lithification or later on by and erosion of the rock. This very rough summary of the processes influencing the development of the interstices already indicates that the important hydraulic parameters, i.e., porosity, hydraulic conductivity and the yield of wells penetrating the fissured and fractured formations, are different and vary widely depend on the type of rock in which the network of interstices is formed. The detailed analysis of the relationships between the parameters characterizing the hydraulic properties and the material of these formations will be discussed in Section 5, which deals with the hydrology of solid-rock terrains. Only some general information is summarized here in connection with the porosity of hard rocks.

The combined character of the water transporting channels should always be considered when the porosity of solid rocks is investigated, because the variance of the size of openings is large due to the different actions having formed them. There are some formations having considerable primary porosity, pores formed between the grains composing the solid matrix of the layer. The role of this type of pore is the most important in indurated, non-carbonate sediments (e.g., sandstones). Primary porosity may range from 1 percent to 35 percent in these formations depending on the grain-size distribution and the packing of particles, as well as on the material and degree of their cementation. The volume of pores located between the grains is generally low in carbonate rocks and they are poorly interconnected, thus primary porosity hardly influences the hydraulic behavior of these aquifers. The young coarse-grained limestones are exceptions, which may have high primary porosity of 20-50 percent (chalks and tufts). The primary pores of igneous and metamorphic rocks are negligible in most cases.

The structures composed of primary and secondary pores, respectively, are basically different. The rate of inactive primary pores is higher than that of fissures and fractures in not taking part in water transport. The size of joints usually decreases with increasing depth due to the increasing load. This is the reason why the parameters describing the hydraulic properties of the aquifer (storage capacity, hydraulic conductivity) are approximated, in most cases, by monotonically decreasing functions (generally exponential functions) of the depth below the surface, if these characteristics are fundamentally determined by secondary porosity. On the contrary, the vertical distribution of primary pores depends mostly on the layering of the formation.

The structure of the porosity of solid rocks is further complicated by the fact that the secondary pores are caused by different actions and therefore, their system is not homogeneous. The dense network of generally very narrow joints, formed during the process of lithification or later on by tectonic movements, can be found together with large openings formed in faulting zones and those enlarged by the solution of the materials of rocks. This is the reason why the size of the water transporting channels in hard rocks varies within a considerably large range, resulting in the nonhomogeneity of flow conditions in the seepage fields formed by such aquifers.

The double, or even multiple, system of interstices causes the special hydrological behavior of hard rocks (especially sandstones, dolomites, and limestones). The hydraulic conductivity of these formations is basically determined by the degree of fissuring and fracturing, primarily influenced by the large openings. The storage capacity of these aquifers basically depends however on the amount of narrow interstices (in sandstones on primary porosity, in other rocks on the density of tectonic joints), because their volume generally surpasses that of large fractures, the latter being relatively low in number. It follows, from this structure, that the wells penetrating hard rock have a relatively high initial yield, by draining the network of large openings, but the yield characteristic for a long period of operation mostly depends on the density and the volume of small pores (on primary porosity in the case of sandstones, on the amount of narrow joints in carbonate rocks).

The same role of the double system of porosity may be observed in natural conditions as well, viz., infiltration and water transport is rapid through large fractures causing the high yield of springs after rains, while the second stretch of the recession curve (the graph showing the decrease of yield in time after a rainy period) is flat, indicating the relatively low permeability and high storage capacity of the solid matrix. This is the reason why the influence of fractures of different size is generally characterized by the recession curves of springs. In most cases, the numerical interpretation is based on a very simple mathematical model which describes the depletion of a cylindrical reservoir having constant cross section (its area is F). The water flows out through a pipe, which is filled with porous material. The length and the area of the cross section of the pipe are indicated with ℓ and f, respectively and the hydraulic conductivity of the material filling the pipe with K (Fig. 1-66a). At time t, the amount of the outflowing water is equal to the change of storage:

$$F\,dh = Q dt = fK\frac{h}{\ell}\,dt \; ;$$

$$t = \frac{F\ell}{fK}\,\ell n\, h + C \; ;$$

$$h = h_o \exp\,[-a(t - t_o)] \; ; \qquad (1\text{-}32)$$

$$Q = Q_o \exp[-a(t - t_o)] \; ;$$

where the initial discharge $Q_o = h_o\ Kf/\ell$ at time t_o depends on the initial height of the water level (h_o) at the same time and on the hydraulic transmissivity of the outlet (Kf/ℓ); the constant $a = fK/\ell F$ is inversely proportional to the storage capacity of the system (in the model, to the horizontal area of the reservoir), and the coefficient of proportionality is the reciprocal value of the parameter characterizing transmissivity.

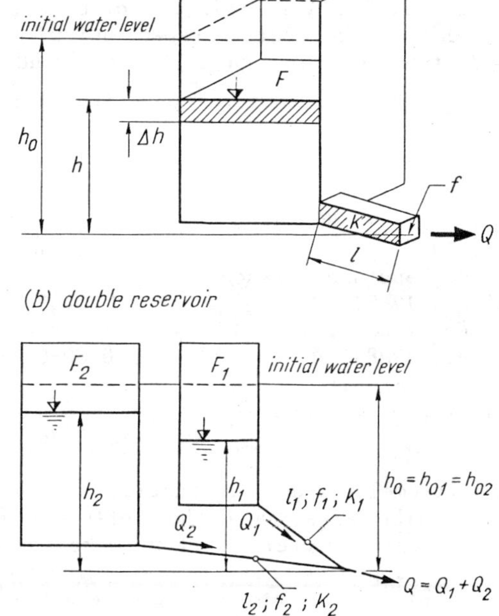

Fig. 1-66. Models for the derivation of the equation of recession curves.

If two reservoirs are drained through the same outlet (Fig. 1-66b), their flow rates have to be added to get the instantaneous discharge at time t:

$$Q = Q_{o1} \exp[-a_1(t - t_o)] + Q_{o2} \exp[-a_2(t - t_o)] \quad . \tag{1-33}$$

The comparison of the mathematical model with observed yields of karst springs proves the correctness of such approximation about the recession curves, even the constants can be calculated from the measured data (Schoeller, 1965; Korkasiewicz and Paloc, 1965). The example shown in Fig. 1-67 was constructed, and the numerical parameters of Eq. 1-33 indicated in the figure, were calculated on the basis of data of a karst spring in the Bükk mountain (northern Hungary) (Maucha, 1972). The ratio of the yield belonging to time t_o and the constant a, of Eq. 1-32, is proportional to the storage capacity of the system. This ratio can be calculated for the two components of the recession curve and, supposing that the initial water level was equal in both reservoirs ($h_{o1} = h_{o2}$), the quotient of these parameters is approximately equal to the amount of the porosities, originating from narrow and large fractures. There are some data published in the literature showing the ratio of the volumes of narrow and large pores calculated in this way or by using similar mathematical models. The data listed below give the porosity calculated from narrow openings (n_n) and that from wide fractures (n_w) related to total porosity (n):

	$\frac{n_n}{n}$	$\frac{n_w}{n}$
Nagytohonya Spring (Hungary) (Maucha, 1972)	0.71	0.29
Fountaine de Vaucluse (France) (Schoeller, 1965)	0.69	0.31
Spring of Kef (Tunisia) (Schoeller, 1948)	0.80-0.75	0.20-0.25
Spring of Loz Herault (Schoeller see Brown et al., 1972)	0.76	0.24

The extremely wide fluctuation of porosity, within a relatively small area, and the influence of the width of channels on the character of flow, make the consideration of the continuum approach and the determination of the size of the representative elementary unit more important here than in loose clastic sediments, and requires the investigation of the probability distribution of the varying sizes of openings. An example will be demonstrated here investigating the linear porosity on the fresh surfaces of four different carbonate rocks of Triassic age (Balásházy and J. Kovács, 1975):

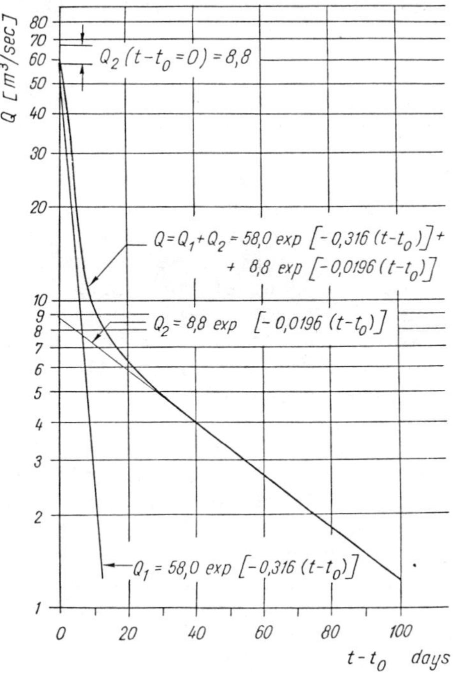

Fig. 1-67. Recession curve of Nagytohonya spring (Bükk Mountain, northern Hungary) (after Mauha, 1972).

- Middle Triassic (Karnanien) dolomite (Obarok);

- the same rock type in a fractured zone along a cross fault with well observable karst forms (chemically enlarged openings) (Szár);

- Upper Triassic (Norian) bedded limestone (Csolnok);

- Upper Triassic (Dachstein) karst limestone (Dorog).

At every place, three directions (one vertical and two horizontal, approximately perpendicular to each other) were fixed on the walls of mines. The scattering of linear porosity, determined for independent stretches having equal lengths, decreased gradually as the investigated length was increased. The enveloping curves, enclosing the points representing the data in a coordinate system having the investigated length on the horizontal axis and the determined linear porosity on the vertical one, tend to a horizontal line, indicating the average linear porosity.

A limit of the investigated length can be determined, above which the scattering of n_L is smaller than 10 percent of its average and the enveloping curves become almost horizontal. The length belonging to this scattering (λ) can be regarded as the representative elementary length characterizing the rock in the investigated direction.

Ten pairs of enveloping curves, determined for ten independent stretches at one investigated place for one given direction, are represented in Fig. 1-68 (Obarok, vertical direction). The comparison of these curves proves that they are very similar to each other and that both the average linear porosity and the representative elementary length have only random variation, thus these parameters can be regarded as probability variables characterizing the rock at the given place and in the investigated direction. The curves enveloping all the calculated data, grouped according to the measured directions, are represented in Figs. 1-69, 1-70, 1-71 and 1-72 for the four investigated rocks. The data were evaluated statistically, as well, if there were more than ten independently determined average linear porosity values in a direction. The empirical distribution curves are also represented in the figures in these cases. The distribution could always be well approximated by the normal distribution function. The results of the investigation (average linear porosity and representative elementary length) are listed in Table 1-8.

Considering the relationships existing between linear, areal, and volumetric porosity (see Eqs. 1-6, 1-7 and 1-9) all the parameters can be calculated if the linear porosity values are known in three orthogonal directions (n_{Lx}; n_{Ly}; n_{Lz}). Areal porosities in planes normal to the x-axis (n_{Ax}), the y-axis (n_{Ay}) and the z-axis (n_{Az}) respectively are:

$$n_{Ax} = 1 - (1-n_{Ly})(1-n_{Lz}) + n_{Ly} n_{Lz} ;$$

$$n_{Ay} = 1 - (1-n_{Lz})(1-n_{Lx}) + n_{Lz} n_{Lx} ; \qquad (1\text{-}34)$$

$$n_{Az} = 1 - (1-n_{Lx})(1-n_{Ly}) + n_{Lx} n_{Ly} .$$

The volumetric porosity (n) can be similarly calculated:

$$n = 1 - (1-n_{Lx})(1-n_{Ly})(1-n_{Lz}) + n_{Lx} n_{Ly} n_{Lz} \qquad (1\text{-}35)$$

The areal and volumetric porosity values determined on the basis of the investigation discussed here, as an example, are also listed in Table 1-8.

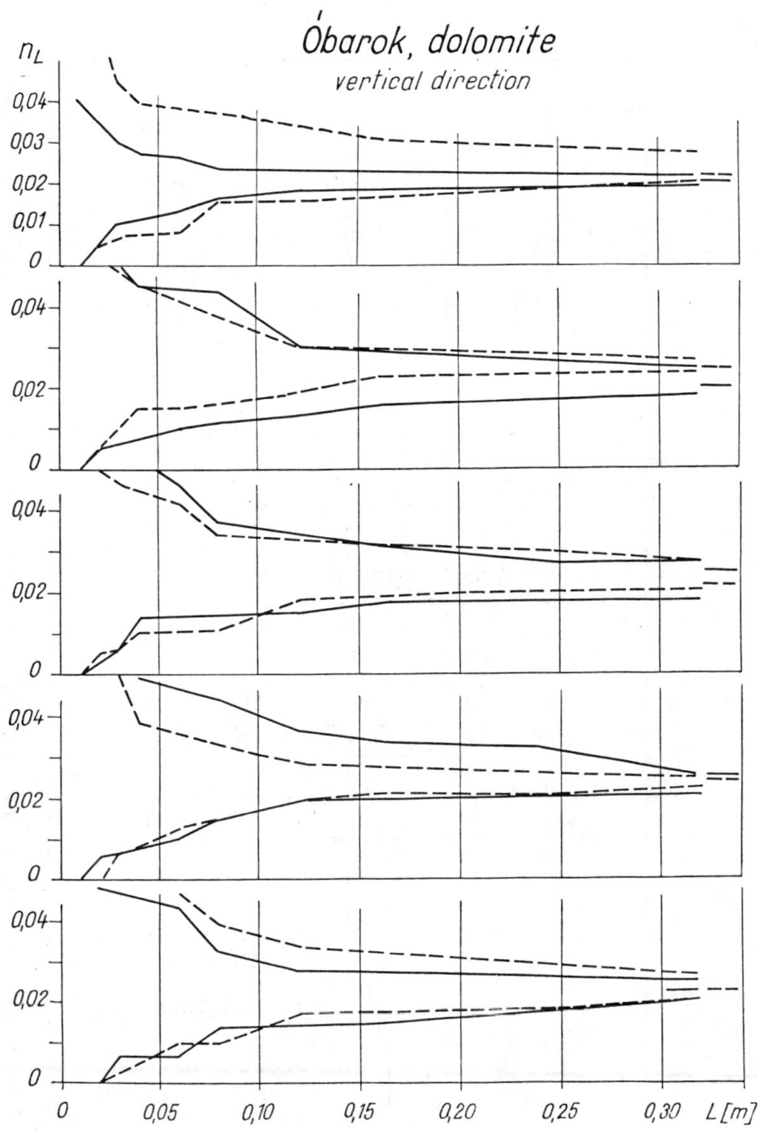

Fig. 1-68. Comparison of curves enveloping the linear porosity values calculated along independent lines having the same direction (after Balásházy and J. Kovács, 1975).

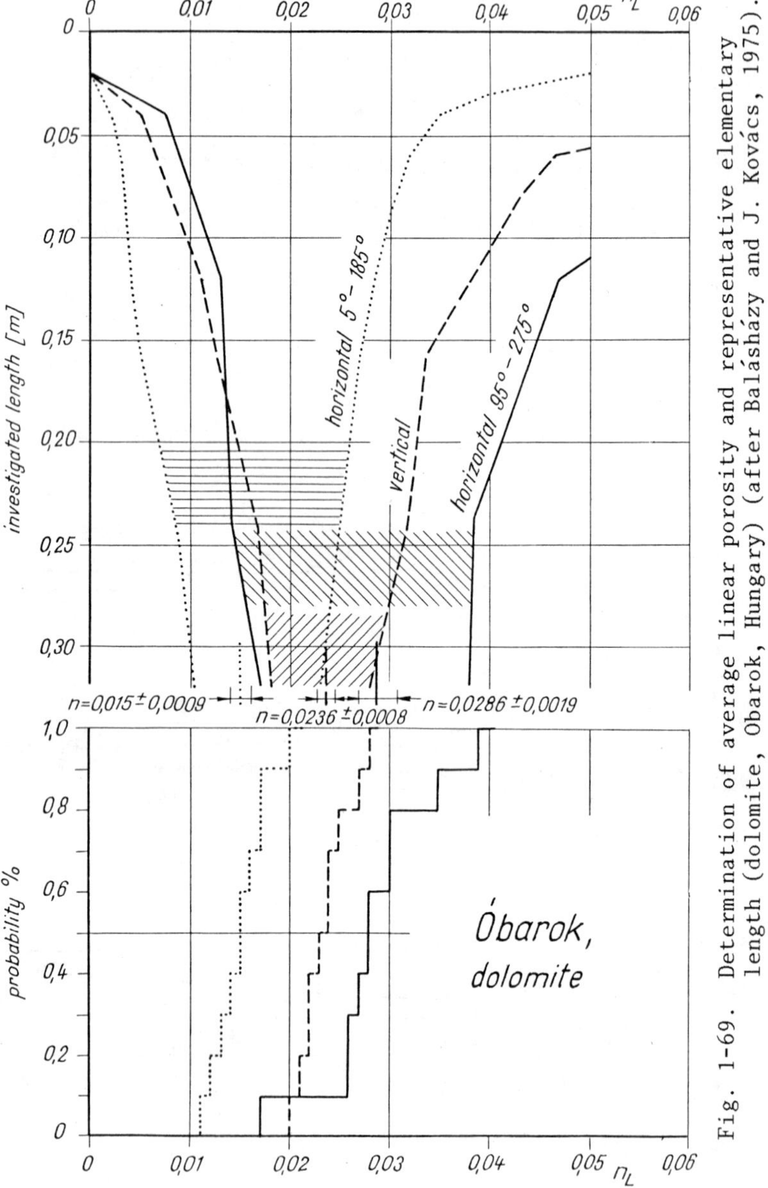

Fig. 1-69. Determination of average linear porosity and representative elementary length (dolomite, Obarok, Hungary) (after Balásházy and J. Kovács, 1975).

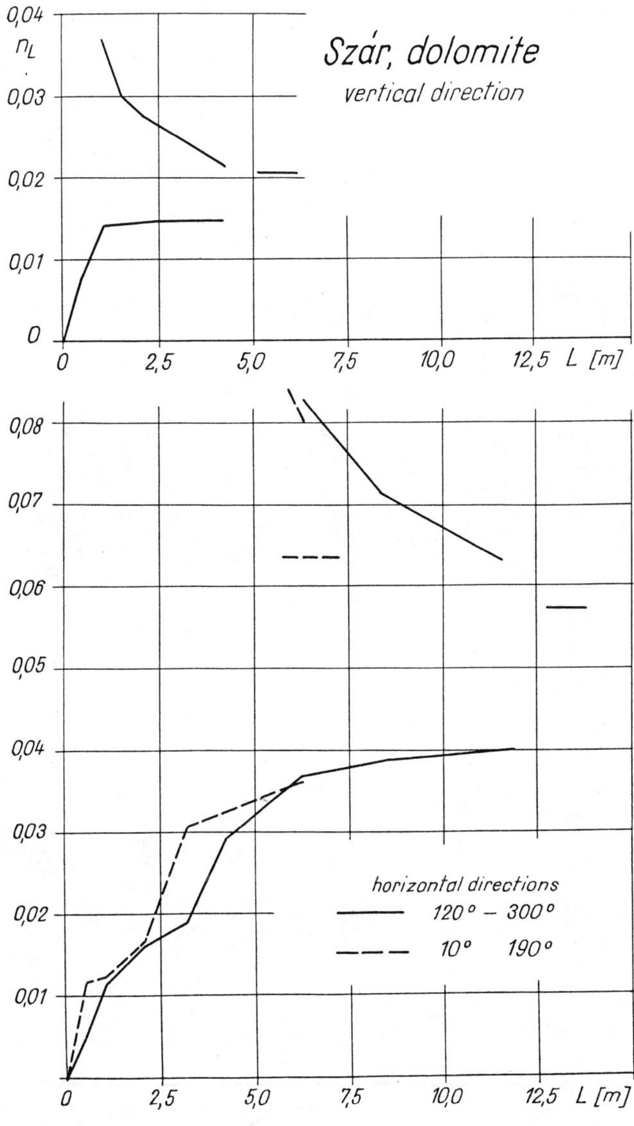

Fig. 1-70. Determination of average linear porosity and representative elementary length (dolomite, Szár, Hungary) (after Balásházy and J. Kovács, 1975).

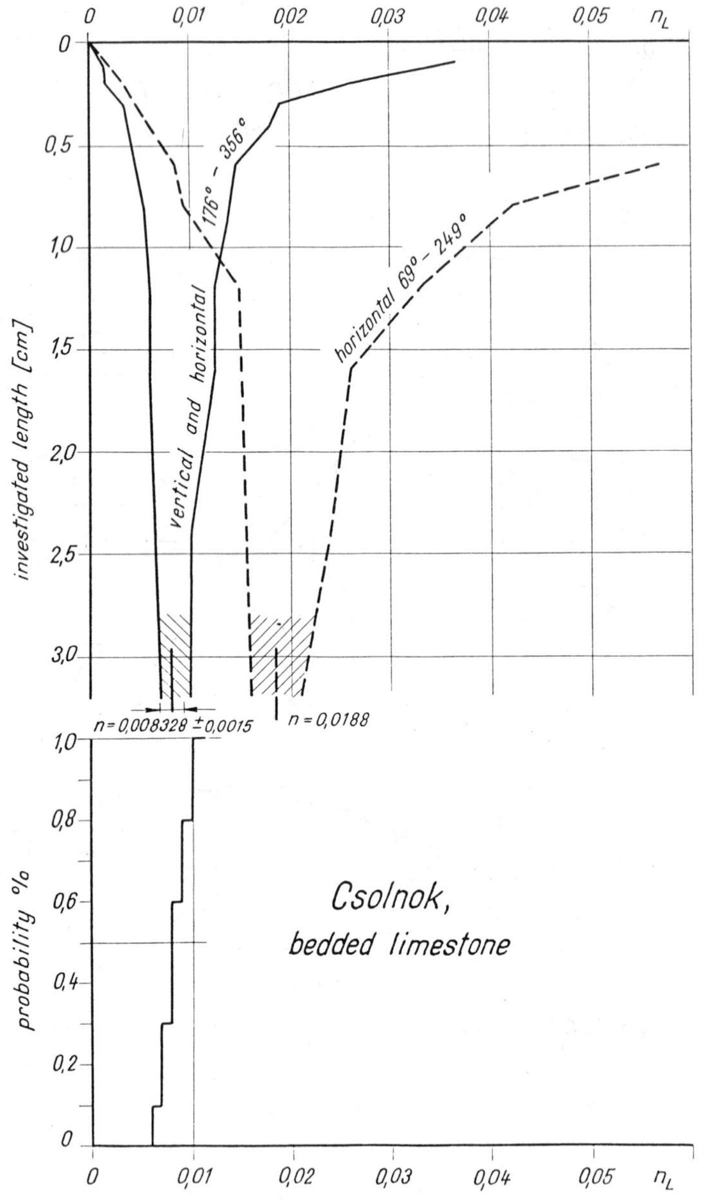

Fig. 1-71. Determination of average linear porosity and representative elementary length (bedded limestone, Csolnok, Hungary) (after Balásházy and J. Kovács, 1975).

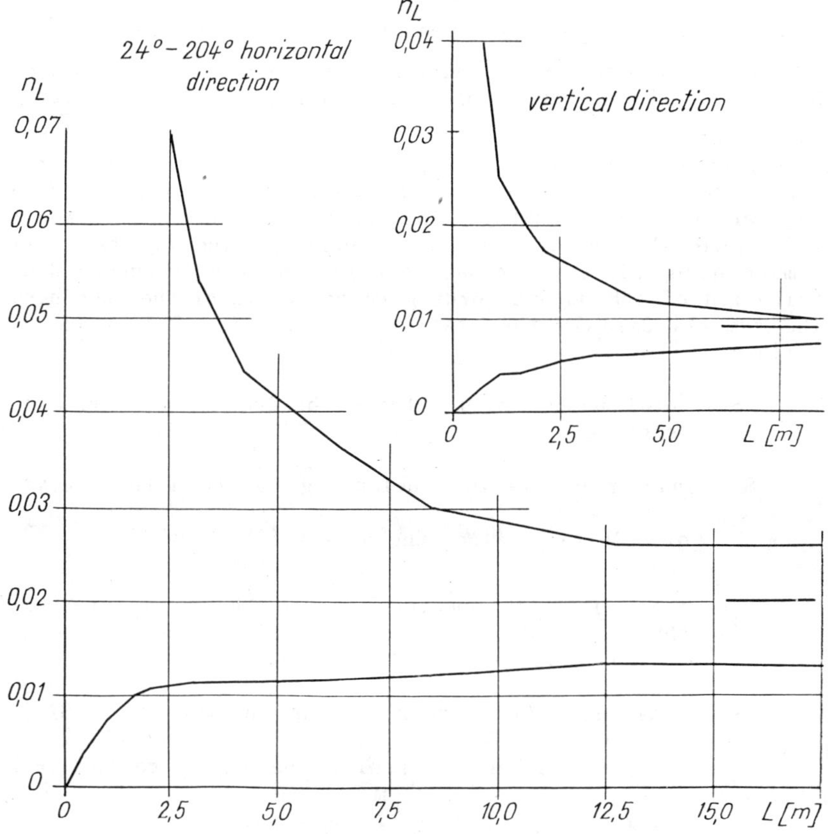

Fig. 1-72. Determination of average linear porosity and representative elementary length (karstic limestone, Dorog, Hungary) (after Balásházy and J. Kovács, 1975).

As it was already mentioned, the investigation should be continued by the statistical analysis of the frequency distribution of both the width and spacing of fractures. In the research carried out by Balásházy and J. Kovács, the parameters measured for studying the probability distribution of the openings having different widths were as follows:

- the number of the openings (S) within a given length (L);

- their width (b);

- their distance from the origin of the fixed line (the latter data were used only for the determination of the representative elementary length).

The total range of the measured widths ($0 < b < b_{max}$) was divided into m number of uniform intervals (Δb, where $m\Delta b > b_{max}$), and the number of openings belonging to each interval was counted. The total number of openings, and the number belonging to each Δb interval, were divided with the measured length to get their specific values (average number of openings within a unit length). The specific number of the intervals were also related to the specific value of the total number of openings, in order to determine the frequency distribution of the data according to the width of the openings. The symbols used in the statistical analysis of data are as follows:

S = total number of fractures observed along a line of length L;

S_i = number of openings belonging to the i-th interval of a Δb range along the b axis ($\sum_{i=1}^{m} S_i = S$);

b_i = mean width of openings within the i-th interval, thus the limits of the interval are determined by $b_i \pm \Delta b/2$ values;

s = total specific number of the openings ($s = S/L$);

s_i = specific number of openings belonging to the i-th interval ($s_i = S_i/L$; $\sum_{i=1}^{m} s_i = s$);

σ_i = frequency distribution value of the specific number of openings belonging to the i-th interval ($\sigma_i = s_i/s$; $1 > \sigma_i > 0$; $\sum_{i=1}^{m} \sigma_i = 1$);

$\sigma(b; \Delta b)$ is a mathematical expression to approximate the empirical frequency distribution curve, which depends on the random variable of the width and on the length of Δb-interval chosen arbitrarily (Fig. 1-73).

Apart from the upper and lower limit of the σ-value mentioned previously, there are two further conditions to determine the mathematical approximation of the empirical frequency distribution.

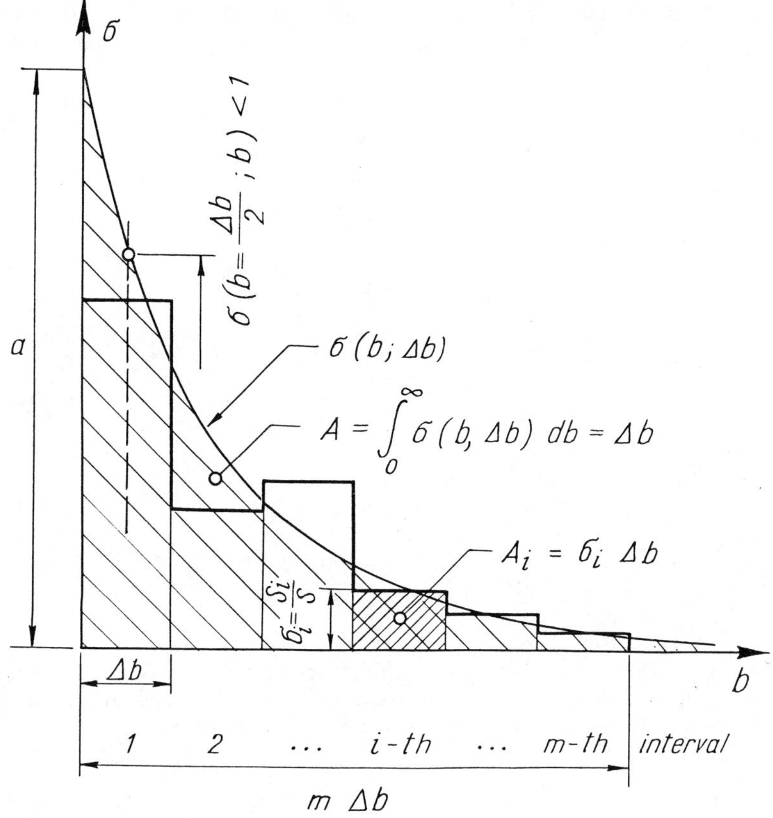

Fig. 1-73. Theoretical sketch to represent the frequency distribution of the widths of openings.

The area covered by the empirical frequency distribution is equal to the Δb interval, and the area enclosed by the horizontal axis and the approximating curve should be equal to this value,

$$A_i = \sigma_i \Delta b \; ;$$

$$A = \sum_{i=1}^{m} A_i = \sum_{i=1}^{m} \sigma_i \Delta b = \Delta b \sum_{i=1}^{m} \sigma_i = \Delta b \; ; \qquad (1\text{-}36)$$

$$A = \int_0^{\infty} \sigma(b,\Delta b)\, db = \Delta b$$

The linear porosity can be expressed as the sum of the measured width of openings related to the total length, and if the Δb-interval is small enough, the sum can be substituted with the product of the frequency distribution values and the mean widths in each interval. The relationship derived in this way should be valid also for the approximating curve:

$$n_L = \frac{\sum_{k=1}^{S} b_k}{L} ;$$

$$n_L = \frac{\sum_{i=1}^{m} S_i b_i}{L} = \sum_{i=1}^{m} s\,b = s \sum_{i=1}^{m} \sigma\,b ;$$

$$n_L \Delta b = s \sum_{i=1}^{m} \sigma_i b_i \Delta b ; \qquad (1\text{-}37)$$

$$n_L \Delta b = s \int_0^{\infty} \sigma(b,\Delta b)\; b\; db .$$

After plotting the actual measured data (see Figs. 1-74, 1-75, 1-76 and 1-77), and considering the listed conditions, the approximation with an exponential distribution function seems to be very suitable as first an estimation:

$$\sigma(b,\Delta b) = a \exp(-cb) ;$$

$$a \int_0^{\infty} \exp(-cb)\; db = -\frac{a}{c}\,[\exp(-cb)]_0^{\infty} = \frac{a}{c} = \Delta b ;$$

$$\frac{n_L \Delta b}{s} = a \int_0^{\infty} b \exp(-cb)\; db = -a[(\frac{1}{c^2} + \frac{b}{c}) \exp(-cb)]_0^{\infty} = \frac{a}{c^2} ; \qquad (1\text{-}38)$$

$$c = \frac{s}{n_L} ; \quad a = \Delta b \frac{s}{n_L} ;$$

$$\sigma(b,\Delta b) = \Delta b \frac{s}{n_L} \exp(-\frac{s}{n_L} b) .$$

Instead of the usual statistical investigation of fitting the form of the empirical frequency distributions of $b\sigma$, $b^2\sigma$ and $b^3\sigma$, the empirical data were compared to theoretical

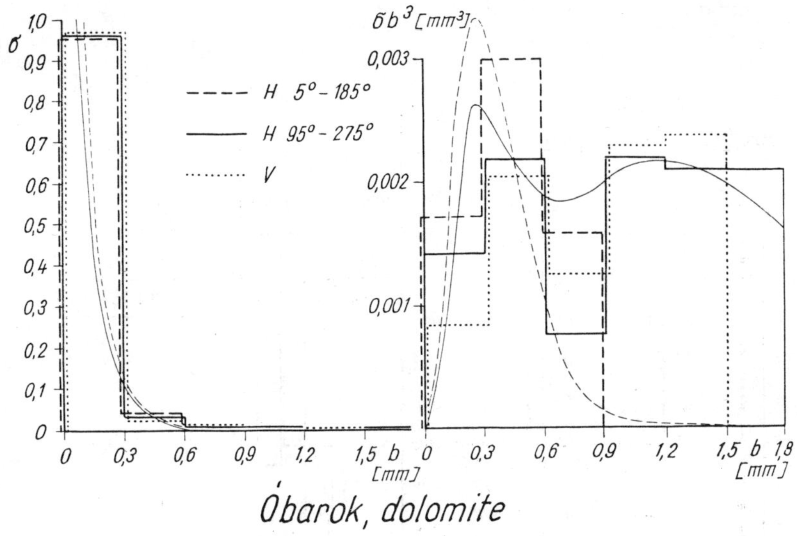

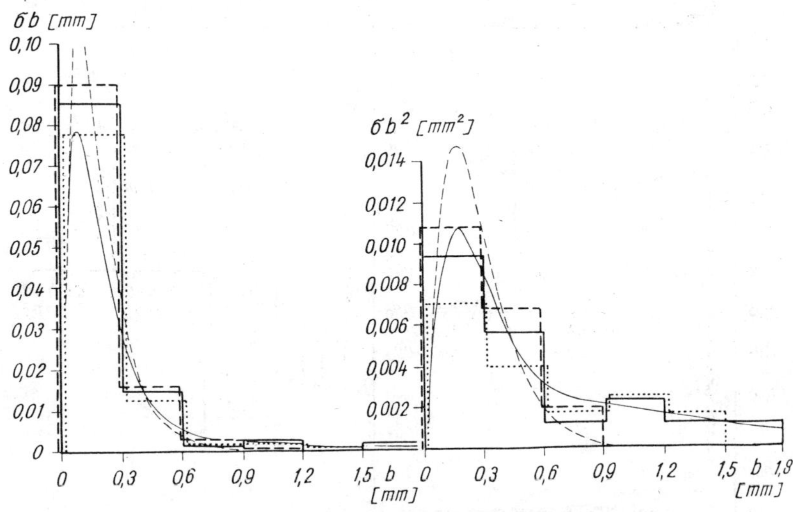

Fig. 1-74. Frequency distribution of the widths of openings in Middle Triassic dolomite (Obarok, Hungary) (after Balásházy and J. Kovács, 1975).

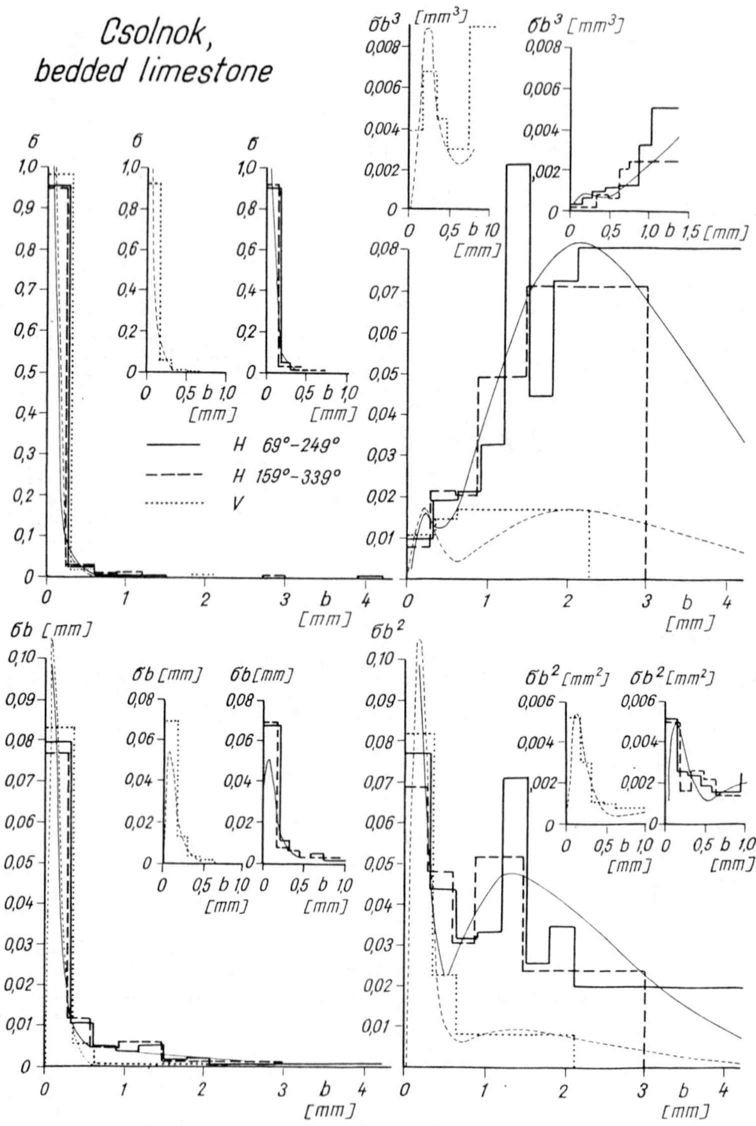

Fig. 1-75. Frequency distribution of the widths of openings in Upper Triassic bedded limestone (Csolnok, Hungary) (after Balásházy and J. Kovács, 1975).

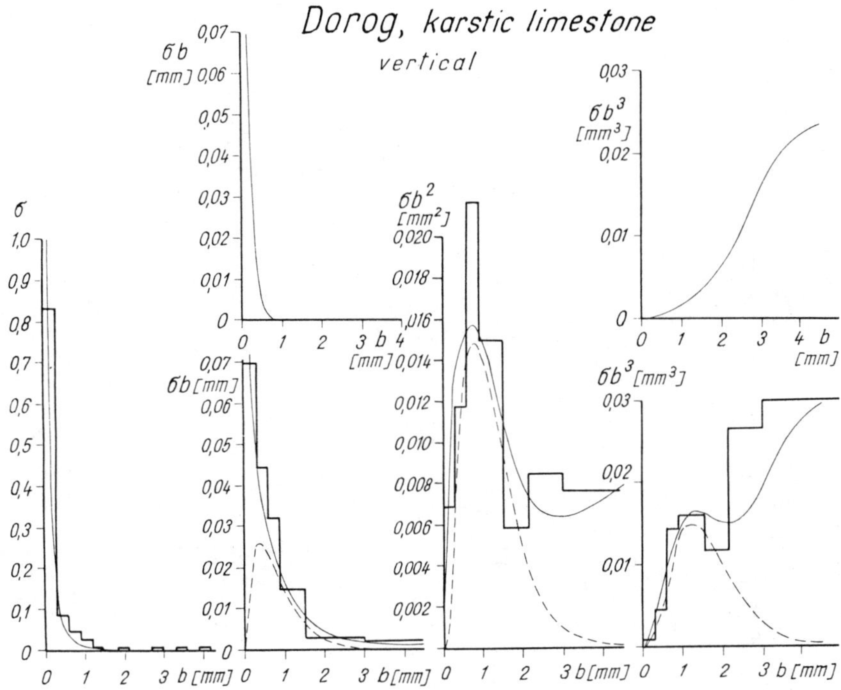

Fig. 1-76. Frequency distribution of the widths of openings in Upper Triassic karstic limestone (Dorog, vertical direction, Hungary) (after Balásházy and J. Kovács, 1975).

curves determined by the proposed approximation (Eq. 1-38). The application of this method was chosen because the products of the higher powers of the width will be used in further calculations, and it was necessary to also see the reliability of the listed values. It is evident that each of the products has only one local maximum, thus, if the value is approximated by Eq. 1-38, i.e., $[b\sigma]_{max}$ if $b = 1/c$; $[b^2\sigma]_{max}$ if $b = 2/c$; and $[b^3\sigma]_{max}$ if $b = 3/c$. The measured values show, however, two or sometimes three local maxima, indicating that the approximation is not acceptable. Also considering the previously listed conditions, it was proposed to use a series of exponential functions composed of r-members instead of the simple exponential frequency distribution, namely

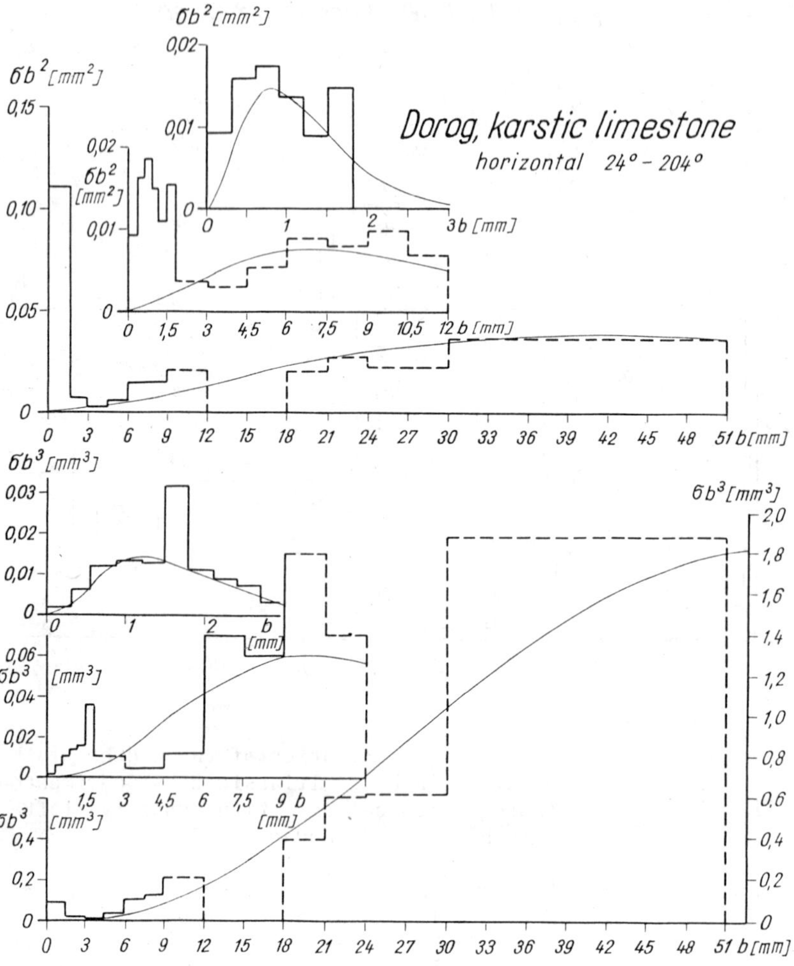

Fig. 1-77. Frequency distribution of the widths of openings in Middle Triassic karstic dolomite (Dorog, horizontal direction, Hungary) (after Balásházy and J. Kovács, 1975).

$$\sigma(b,\Delta b) = \sum_{j=1}^{r} \sigma_j = \sum_{j=1}^{r} a_j \exp(-c_j b) \quad . \qquad (1\text{-}39)$$

From Eqs. 1-36 and 1-37, also considering the upper limit of σ, the following conditions can be determined for the constants (a_j and c_j) in Eq. 1-39:

$$\Delta b = \sum_{j=1}^{r} \frac{a_j}{c_j} ;$$

$$\frac{n_L \Delta b}{s} = \sum_{j=1}^{r} \frac{a_j}{c_j^2} ; \qquad (1\text{-}40)$$

$$\sum_{j=1}^{r} a_j \exp(-c_j \frac{\Delta b}{2}) \leq 1 .$$

Some physical explanation can also be found to support the use of more than one exponential function for describing the frequency distributions of the widths of openings. There are more geologic processes taking part in the creation of the joints and governing the development of their size (lithification and initial tectonic stresses in the rocks, exfoliation and weathering, later tectonic movements, chemical and mechanical enlargement of fractures). It may be supposed, therefore, that the total number of openings observed at present reflects the influences of these actions in a combined form. It is reasonable to divide the whole system into sets of openings, each described with an exponential distribution function, getting a series composed of as many members as the number of the dominating factors in the development of the recent conditions.

Equation 1-40 does not contain sufficient information to calculate the constants in closed form if more than one exponential function is used. Their values can be estimated, however, after the construction of the empirical distribution function by using the trial and error method (Fig. 1-74). The accuracy of the method can be increased by applying smaller Δb-intervals in the zone of very narrow joints (Fig. 1-75). The interrelation between the position of the local maxima of $b\sigma$-, $b^2\sigma$- and $b^3\sigma$-curves and the c-parameters assists to determine the numerical values of the latter (Fig. 1-76). If the range of the measured widths is large, it is advisable to use a relatively high Δb-value for the determination of the smallest c-parameter, and repeat the investigation with the residual part of the frequency distribution, i.e., $\sigma_{exper.} - a_r \exp(-c_r b)$, by using a smaller interval (Fig. 1-77). In this case, it is necessary to consider that the a-parameter is linearly proportional to Δb ($a'/a = \Delta b'/\Delta b$). After the determination of the first estimation of the parameters in this way, the values can be checked and corrected to satisfy the conditions listed in Eq. 1-40 (Table 1-9).

It is quite evident that the parameters depend on the material of the rock and on the processes governing the development of the openings. It is interesting to note, however, that the grouping of the c-values can be observed in the investigated cases. A member of the series with a parameter c_1 being around 11 was characteristic for all places. This set of openings might be created by the initial tectonic stresses. The second member is also generally characteristic with a parameter c_2 = 1.5-3.5 due perhaps to weathering and exfoliation. Smaller c-values were determined only for karst layers and strongly faulted zones. The number of the investigated cases, however, is very low. It would therefore be premature to draw further conclusions from the data, but the investigation opens a promising technique in this subject.

Table 1-9. Parameters "a" and "c" characterizing the frequency distribution of the width of fractures (after Balásházy and J. Kovács, 1975).

Type of Rock	Direction	Specific Number of Openings [db/mm]	Graphically Estimated Parameters							
			a				c			
			1	2	3	4	1	2	3	4
Middle Triassic	x 5°-185°	0.1347	3.30	-	-	-	11.0	-	-	-
dolomite	y 95°-275°	0.2361	2.60	$2.7x10^{-2}$	-	-	11.0	2.5	-	-
(Obarok)	z vertical	0.1929	2.60	$2.7x10^{-2}$	-	-	11.0	2.5	-	-
Middle Triassic	x 120°-300°	0.1075	3.50	$1.6x10^{-2}$	$1.0x10^{-3}$	$1.6x10^{-5}$	12.0	1.2	0.20	0.01
dolomite in strongly faulted zone	y 10°-190°	0.2651	4.30	$4.0x10^{-3}$	-	$4.0x10^{-6}$	15.0	1.2	-	0.01
(Szár)	z vertical	0.2198	4.00	$4.5x10^{-3}$	-	-	13.5	1.2	-	-
Upper Triassic	x 69°-249°	0.1145	4.10	$1.8x10^{-2}$	-	-	14.0	1.4	-	-
bedded limestone	y 159°-339°	0.0751	4.10	$1.8x10^{-2}$	-	-	14.0	1.4	-	-
(Csolnok)	z vertical	0.0956	4.10	$8.0x10^{-3}$	-	-	14.0	1.4	-	-
Upper Triassic	x 24°-204°	0.0552	2.65	$1.5x10^{-1}$	$1.2x10^{-3}$	$1.8x10^{-4}$	11.5	2.5	0.30	0.06
karstic limestone	y non-measurable	-	-	-	-	-	-	-	-	-
(Dorog)	z vertical	0.0388	2.80	$1.7x10^{-1}$	$1.6x10^{-3}$	-	11.0	2.5	0.30	-

Table 1-9. (continued)

		Linear Porosity n_L [%]	$\frac{n_L \Delta b}{s}$	Parameters Corrected by Considering the Condition to be Satisfied							
				a				c			
				1	2	3	4	1	2	3	4
Obarok	x	1.46	0.0325	3.00	9.4×10^{-2}	-	-	11.0	3.5	-	-
	y	2.51	0.0320	3.08	3.1×10^{-2}	-	-	10.7	2.5	-	-
	z	1.88	0.0292	3.34	3.0×10^{-2}	-	-	11.7	2.5	-	-
Szár	x	8.02	0.2238	3.00	1.6×10^{-3}	1.0×10^{-3}	1.6×10^{-5}	10.5	1.2	0.20	0.01
	y	5.86	0.0663	3.76	4.0×10^{-3}	-	4.0×10^{-6}	12.7	1.2	-	0.01
	z	1.88	0.0263	3.80	4.5×10^{-3}	-	-	12.8	1.2	-	-
Csolnok	x	1.55	0.0406	3.15	2.0×10^{-2}	-	-	11.0	1.2	-	-
	y	0.80	0.0320	3.75	1.4×10^{-2}	-	-	13.0	1.2	-	-
	z	1.05	0.0329	3.20	8.0×10^{-3}	-	-	10.9	1.2	-	-
Dorog	x	1.95	0.1060	2.75	1.4×10^{-1}	1.2×10^{-3}	1.8×10^{-4}	11.8	2.5	0.30	0.06
	y	-	-	-	-	-	-	-	-	-	-
	z	0.66	0.0510	2.82	1.7×10^{-1}	1.7×10^{-7}	-	11.8	2.9	0.38	-

SECTION 4

DYNAMIC MODELING OF POROUS MEDIA

In any case when the goal is to describe a special type of movement, to discover the affecting physical processes together with their interrelation, and to determine the mechanically correct characterization of the motion, the detailed investigation of the acting forces is the first necessary step. The creation and maintenance of movement requires the action of an accelerating force, or that of a system of such forces. The retarding forces act against the former ones, and they generally increase with the velocity of the motion, thus there is a limit where the two groups of forces reach a state of equilibrium and a steady movement develops.

When investigating the dynamics of seepage, the first task is to analyze the acting forces. It is necessary to determine their physical properties, and to find the most suitable way to incorporate them into a conceptual model. There are many cases when the dynamic laws, established in theoretical physics, are not convenient for direct application because the model would be complicated, thus hindering the mathematical solution. The acceptable approximations have to be determined in these cases, which ensure both the necessary simplicity and the accuracy in practice.

One important way of approximation is to distinguish the dominant and the negligible forces. Forming the ratio of the considered and neglected forces, parameters can be established, the given numerical values of which indicate the limits of the validity zones of the various approximations applied by neglecting certain actions. Similar dimensionless quotients can be derived as the ratios of the considered accelerating and retarding forces. These parameters provide the theoretical form of the basic movement equations which have to be verified later by experiments.

The other important step of the theoretical investigation is to find a physical (geometrical) model of the analyzed system. When selecting this model, two contradictory aspects act against each other. The boundaries and the structure of the model should be as simple as possible to ensure the mathematical solution of the relevant movement equations. On the other hand, it is necessary to include the largest possible number of parameters from the actual system into the model, because each parameter may describe a special feature of the prototype.

Following the determination of the acting forces and the geometrical model, the combination of the results provides a conceptual model which is suitable to theoretically analyze the investigated process. The basic relationships may in this way be determined between the interrelated variables, and the negligible effects can also be selected.

In many cases, when the whole water transporting network can be approximated by an evenly distributed system of geometrically simplified conduits, all having the same sizes, the combination of the geometrical and dynamic models does not cause any problem. There are, however, special types of porous media, or special conditions of a given medium, whose investigation does not permit the application of this simplified model. The statistical model of pore-size distribution discussed in the previous section, in detail, must also be incorporated in such an investigation.

4-1 Dominating Forces Influencing the Flow Through the Interstices of a Solid Matrix

There are three main forces acting on the water in porous medium, when seepage does not yet develop and the water is in static condition. These forces, forming the dynamic balance in both the groundwater and soil-moisture zone, are: gravity, adhesion, and capillarity. When flow develops through the solid matrix, there are two further forces which act against the flow, i.e., friction and inertia.

Gravity is a mass force having potential and caused by the attraction of the earth. Its magnitude (G) is equal to the product of the acceleration due to gravity (g) and the moving mass (M), and it is directed vertically downwards:

$$G = Mg = V \rho g ; \qquad (1\text{-}41)$$

where V is the volume of the investigated mass and ρ is its density.

A force has potential when there exists a potential function, a single-valued scalar function of the space coordinate system, whose gradient is equal to the force in question. In a gravitational field, choosing one axis of the coordinate system vertically, the vector of acceleration has no component normal to this axis. The derivative of the potential function, in the vertical direction, has to be proportional to the acceleration due to gravity. Consequently the gravitational potential (U) can be calculated as a product of mass, gravitational acceleration, and the vertical distance (z) measured above a horizontal reference level chosen arbitrarily:

$$U = Mzg \quad . \tag{1-42}$$

The water pressure prevailing within a water body being in static condition is equal to the negative gravitational potential acting on a unit volume of water, supposing that the reference level is chosen at the water table where the excess pressure is equal to zero:

$$- p = h \rho g = h \gamma_v \quad ; \tag{1-43}$$

where γ_v is the specific weight of water, the h-axis is directed upwards and the so-called suction head (h) is measured above the water table (if h is directed downwards from the surface having zero excess pressure, it is called pressure head and Eq. 1-43 gives the positive pressure in the water body).

Internal friction is a retarding force in viscous fluids hindering the relative displacement of two water particles moving along two neighboring paths. This force (S) is directed against the flow and its size is proportional to the surface (A) parallel to the flow line and perpendicular to the direction of the velocity gradient (Fig. 1-78). The friction acting on a surface of unit size (specific value) is the shearing stress (τ), which is the product of dynamic viscosity (μ) and the change of velocity in a direction normal to the investigated area (velocity gradient):

$$S = A\tau \quad ; \quad \tau = \mu \frac{dv}{dn} \quad . \tag{1-44}$$

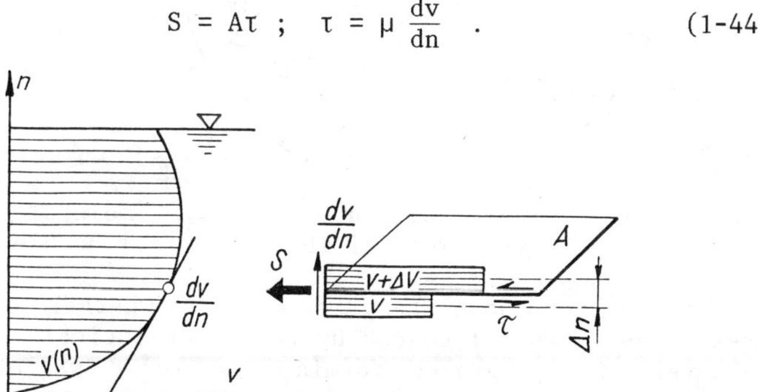

Fig. 1-78. Sketch for interpretation of viscosity and shearing stress.

The non-Newtonian behavior of water can be considered also in Eqs. 1-44 by supposing that the shearing stress is composed of two members, one (τ) prevailing also in static conditions and being the function of the distance (δ) measured from the interface between the solid matrix and the water, the other depending on viscosity and velocity gradient:

$$\tau = \tau_o(\delta) + \mu \frac{dv}{dn} \quad . \tag{1-45}$$

The development of adhesion between the water and the wall of grains can be explained by the dipole character of the water molecules. The protons and electrons, having positive and negative charges, respectively, are located asymmetrically within the molecule (Fig. 1-79). The solid wall has electrostatic charges related to the water, which polarize the dipole water molecules along the wall. The molecules are fully oriented in the first layer. They are bound with their opposite charge to the wall and their charge, which is the same as that of the wall, is oriented towards the interior of the water body (Fig. 1-80). The effect of polarization can also be observed within the water body inside this double layer. The number of the oriented molecules, however, is inversely proportional to the distance measured from the wall (Erdey-Gruz and Schay, 1954; Kézdi, 1962, 1974).

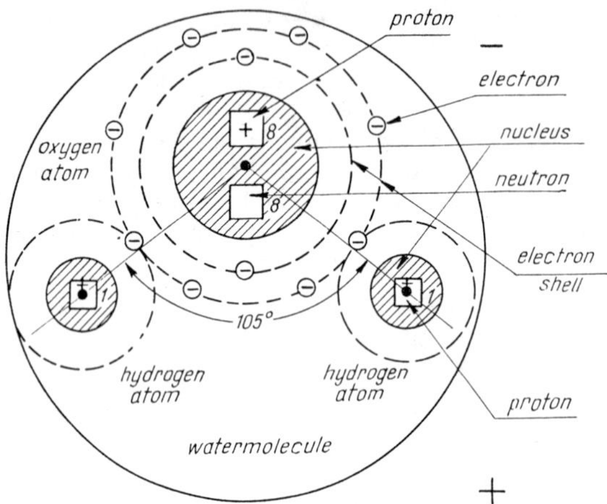

Fig. 1-79. Dipole structure of a water molecule and the development of the electrical double layer.

It is not absolutely clear yet whether or not the adhesive force is influenced by the mineralogical and chemical character of the grains forming the solid skeleton of the seepage field. On the basis of previous investigations concerning the interaction between water and grains, it is very likely that this relationship is determined by Van der Waals force (Fig. 1-81), the effect of which can be characterized by a tension (suction) (-p) inversely proportional to the sixth power of the distance from the wall (δ), independently of the mineralogical and chemical character of the latter.

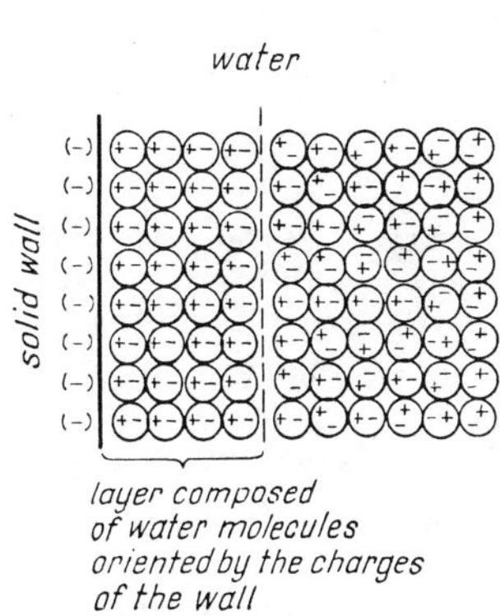

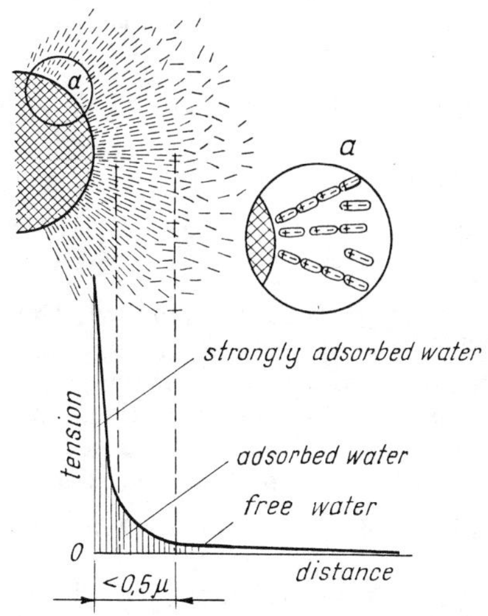

Fig. 1-80. Polarization of water molecules due to an electrostatically charged wall.

Fig. 1-81. Van der Waals force (tension) versus distance from a molecular wall.

$$- p = \frac{c}{\delta^6} \quad . \tag{1-46}$$

The fourth force dominantly influencing the flow between grains is capillarity, which occurs on the contact surface of water and air. The simplest explanation of this force can be given taking into consideration the attraction between water molecules. In the interior of the water body the attractive forces act from every direction with equal probability, thus all molecules are in a balanced condition. A molecule on the water surface is attracted only by molecules below the surface. Thus, in an infinitesimally thin layer along the contacting surface of two different media there is free energy which has to be balanced by pressure differences existing between the two sides of this layer (within the two contacting media).

To characterize the amount of the free energy, its specific value is used, which is equal to the amount of the surface energy $F^{(A)}$, and also the surface area (A). Relating this quotient to an infinitely small surface element, the interfacial tension, σ_{ik}, is achieved which is a material constant for any pair of media i and k depending only on absolute temperature (T):

$$\sigma_{ik} = \frac{\partial_F^{(A)}}{\partial A}\ T \ . \qquad (1\text{-}47)$$

The special value of the interfacial tension, developing on the surface of a fluid covered with its own vapor, is called surface tension (σ). According to Eq. 1-47 its dimension can be calculted as the ratio of energy (work) over the surface, which is equal to force over distance, thus surface tension can be interpreted as a force acting within the surface along a line having a length of unity $[FL^{-1}]$.

Let us investigate the contact point of three immiscible fluids (or gaseous media). In a section normal to the three contact surfaces, the balanced condition of the acting interfacial tensions requires that the vector triangle composed of the three tensions be closed (Fig. 1-82). From this condition, the angles (α_{ik}) of the contact surfaces, measured from an arbitrarily chosen axis, can be calculated as a function of the numerical values of interfacial tensions, or, if the axis is equal to the tangent of one of the contacting surfaces, the so-called contact angles (θ_{ik}) can be determined (Németh, 1963; Bear, 1972):

$$\sigma_{12} \cos\alpha_{12} + \sigma_{23} \cos\alpha_{23} + \sigma_{13} \cos\alpha_{13} = 0 \ ;$$
$$\sigma_{12} + \sigma_{23} \cos\theta_{23} + \sigma_{13} \cos\theta_{13} = 0 \ . \qquad (1\text{-}48)$$

This condition indicates at the same time, whether or not the development of the balanced condition is possible. It is evident, that if $\sigma_{12} > (\sigma_{23} + \sigma_{13})$ there is no possibility to fulfill Eq. 1-48, and therefore, equilibrium cannot be achieved at the contact of these media.

Equation 1-48 can be simplified if one of the contacting materials is solid. The molecular forces do not modify its form, thus the interfacial tensions between the solid and the two liquid (or gaseous) media are fitted to a straight line (Fig. 1-83):

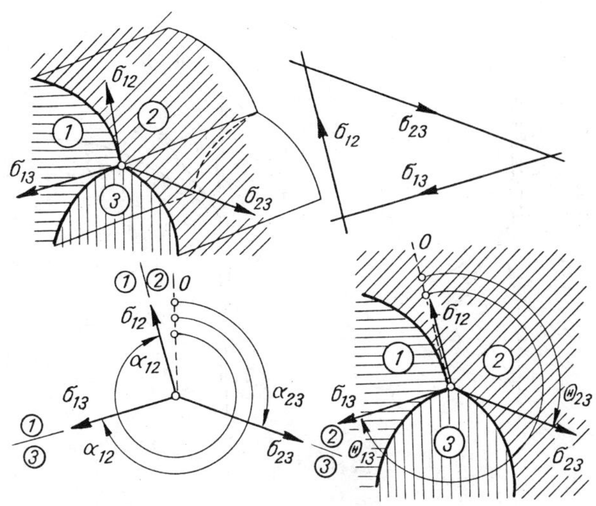

Fig. 1-82. Development of the balance of surface tensions at the contact point of three fluids.

$$\sigma_{12}\cos\theta + \sigma_{13} - \sigma_{23} = 0\ ;$$
$$\cos = \frac{\sigma_{23} - \sigma_{13}}{\sigma_{12}}\ . \qquad (1\text{-}49)$$

Knowing the parameters characterizing the surface tension of a fluid and the contact angle, which is enclosed by the surface of the fluid and the solid wall (σ and θ), the pressure difference between the two sides of the surface (capillary pressure p_c) can be determined:

$$p_c = p_1 - p_2 = \sigma\left(\frac{1}{R_1} + \frac{1}{R_2}\right)\ ; \qquad (1\text{-}50)$$

where R_1 and R_2 are the radii of the meniscus in its two main sections.

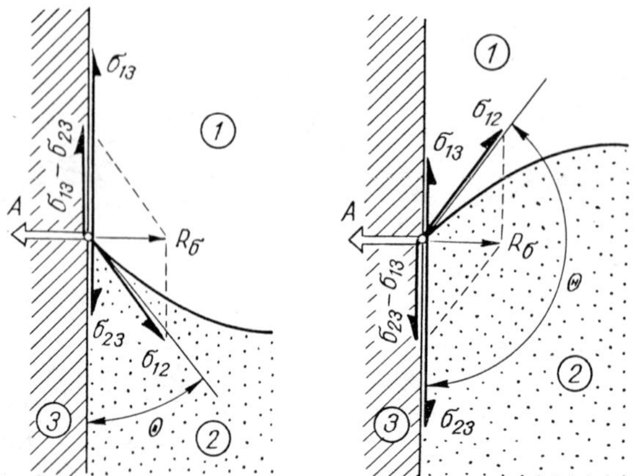

Fig. 1-83. Development of the balance of surface tensions at the border of a solid medium.

The consequence of the capillary pressure not equal to zero is that the position of the meniscus in the capillary tube or slit will be different from the level of the contacting water body having a large surface. Choosing this latter level as datum, where the excess pressure is zero, the average height of the meniscus (capillary rise) above this level can be calculated from capillary pressure:

$$h_c = \frac{p_c}{\rho g} ;$$

or in a circular tube

$$h_c = \frac{4\sigma}{\gamma} \frac{\cos\theta}{d} . \qquad (1\text{-}51)$$

Investigating the contact of water, air, and quartz (as fluid, gaseous, and solid media), it was found that the contact angle is near zero, although this value changes with the condition of the solid surface to a great extent. Substituting $\cos\theta = 1$ and the value of the surface tension belonging to the most generally occurring temperature (i.e., 15-20°C) into Eq. 1-51 a simple approximation for the capillary rise of water in a glass tube of diameter d can be used in practice:

$$h_c \text{ [cm]} = \frac{0.30 \text{ [cm}^2\text{]}}{d \text{ [cm]}} \quad . \tag{1-52}$$

Finally, the fifth dominating force, inertia, which acts against the movement of water, is also a body force similar to gravity, consequently it is the product of mass and acceleration. In this case the latter is the acceleration of the motion which is retarded by inertia:

$$T = - Ma = - M \frac{dv}{dt} \quad . \tag{1-53}$$

4-2 Dimensionless Numbers Characterizing the Validity Zones of Seepage

The seepage can be classified according to the accelerating forces, and further subgroups can be determined considering the acting retarding forces. According to the accelerating forces, the seepage developing in saturated porous medium should be distinguished from the water transport through an unsaturated solid matrix. In the first case, there is only one dominating force accelerating the movement, gravity. More precisely, all the other forces taking part in the maintenance of flow, e.g., pressure to the overlying layers, pressure differences due to the change of the specific weight of fluid, can be expressed as a supplementary term of gravity and therefore it is not necessary to investigate them separately. When seepage develops through unsaturated pores, adhesion and capillarity accelerate the movement as well.

Considering the complicated character of unsaturated seepage, its analysis is dealt with separately from the gravitational flow through the saturated interstices of porous media. Thus, one accelerating force (gravity) and three retarding ones (friction, inertia, and adhesion) have to be taken into account when the purpose is the dynamic classification of seepage. The various groups are characterized by the ratio of the retarding forces, having considerable role in the case of the groups in question related to gravity which accelerates the flow. The quotients, formed in this way, have the sum of the forces in the numerator and one force in the denominator, they are, therefore, dimensionless numbers. Considering the retarding forces which may act simultaneously, the following combinations of dimensionless numbers can be constructed (Kovács, 1966; Spranck, 1932; Mosonyi and Kovács, 1952):

Laminar zone

$$\frac{S}{G} = \frac{A\ \eta\ \frac{dv}{dn}}{V\ \rho\ g} = \frac{v\nu}{\ell^2 g} = MK\ ;$$

Turbulent zone

$$\frac{T}{G} = \frac{V\ \rho\ \frac{dv}{dt}}{V\ \rho\ g} = \frac{v^2}{\ell g} = Fr\ ; \tag{1-55}$$

Transition zone

$$\frac{T+S}{G} = \frac{V\ \rho\ \frac{dv}{dt} + A\ \eta\ \frac{dv}{dn}}{V\ \rho\ g} =$$

$$= \frac{1}{\ell g}\ (v^2 + \frac{v\nu}{\ell}) = Ko_4\ ; \tag{1-56}$$

Microseepage

$$\frac{E+S}{G} = \frac{I_o V\ \rho\ g + A\ \mu\ \frac{dv}{dn}}{V\ \rho\ g} =$$

$$= I_o + \frac{v\nu}{\ell^2 g} = Ko_3 \tag{1-57}$$

where A, V, ℓ and v are characteristic area, volume, length and velocity, respectively; ρ, μ, and ν are the physical parameters characterizing the percolating fluid (density, dynamic viscosity, and kinematic viscosity); g is the acceleration due to gravity; dv/dt and dv/dn are the changes of velocity in time and in space perpendicularly to the direction of flow, I_o is the threshold gradient of the system; MK, Fr, Ko_4 and Ko_3 are the dimensionless numbers.

Similar dimensionless numbers can be constructed from the neglected and considered retarding forces. It is quite evident that such a quotient indicates the ratio of the two forces in question. The dimensionless numbers can therefore be used to numerically characterize the limits of validity zones of the various types of seepage.

The limit between turbulent and transition zones is characterized, evidently, by the fact that friction can be neglected as compared to inertia, while the lower limit of the transition zone (the upper limit of laminar zone) can be marked where friction becomes dominant relative to inertia.

Thus both limits can be expressed as a ratio of inertia and friction:

$$\frac{T}{S} = \frac{V\ \rho\ \frac{dv}{dt}}{A\ \mu\ \frac{dv}{dn}} = \frac{\ell v}{\nu} = Re \quad . \qquad (1\text{-}58)$$

The lower limit of the laminar zone can be similarly expressed by a given numerical value of the ratio of adhesion to friction:

$$\frac{E}{S} = \frac{I_o V\ \rho\ g}{A\ \eta\ \frac{dv}{dn}} = I_o \frac{\ell^2 g}{v\ \nu} = Ko_2 = \frac{I_o}{MK} \quad . \qquad (1\text{-}59)$$

Finally, there is a further possible combination of the forces in question, i.e., the ratio of adhesive force to gravity, $E/G = I_o = Ko_1$. This dimensionless number characterizes a static condition below microseepage, where friction is zero and not a type of movement.

Some of the listed dimensionless numbers are well known from general hydraulics as the bases of the various modeling laws, such as the Froude number from Eq. 1-55, the Reynolds number in Eq. 1-56 or the Mosonyi-Kovács number (Eq. 1-54), which was derived previously to determine the model law valid for the characterization of seepage models, and for calculating the transforming factors between the corresponding measurements of the prototype and the model, respectively (Mosonyi and Kovács, 1952, 1956).

The structure of the other dimensionless numbers is the same as, or very similar to, those limited earlier. The only significant difference occurs in the numerator of Ko_3 and Ko_4 where the sum of two forces can be found instead of one force. This is the reason why these numbers cannot be used as model laws, because dynamic similarity can be only ensured in the case of two dominating forces, when the identify of the ratio of these two forces can be set-up in the prototype and the model. When there are three dominating forces, i.e., two retarding forces in the numerator and the accelerating gravity in the denominator, the identity of the dimensionless number for the two systems only ensures the same ratio of the retarding and accelerating effects, the ratio of the two retarding forces may be, however, quite different in the model from that in the prototype.

The dimensionless numbers characterizing the various zones of seepage, Eqs. 1-54 through 1-57, can also be used to determine the general forms of the movement equations valid for each zone. It is necessary to take into consideration that seepage is not a free movement. Only a part of the potential energy (increased with pressure energy) accelerates the movement. The actual accelerating force can be characterized numerically by the product of gravity and hydraulic gradient:

$$G = IMg = IV \rho g \quad . \tag{1-60}$$

The substitution of this value in the equations in question, instead of gravity, is in accordance with the theory of similitude as well. The total mechanical similarity includes the geometrical similarity apart from the dynamic one. The former is ensured, when the ratio of two corresponding lengths in the two systems is constant. Thus the gradient determined as a quotient of two lengths (pressure expressed in equivalent water column and the length of seepage) must be the same in the prototype and the model as well.

Mechanical similarity in Darcy's zone means the identity of both Mosonyi-Kovács number and hydraulic gradient in the two systems. Thus the general form of the movement equations can be derived as follows:

$$MK = \frac{1}{I}\frac{v\nu}{\ell^2 g} \;; \quad v = MK\frac{\ell^2 g}{\nu} \;\; \text{and} \;\; I = KI \;;$$
$$\left(\text{if} \;\; K = MK\frac{\ell^2 g}{\nu}\right) \quad . \tag{1-61}$$

Similarly, a well-known formula can be derived for the turbulent zone (Chezy's equation):

$$Fr = \frac{1}{I}\frac{v^2}{\ell g} \;; \quad v = \sqrt{Fr\ g}\,\sqrt{\ell I} = \sqrt{\frac{2g}{\lambda}}\,\sqrt{\ell I} \;;$$
$$\left(\text{if} \;\; \lambda = \frac{2}{Fr}\right) \quad . \tag{1-62}$$

In the transition zone between the two former zones, the general form is also the same as that proposed empirically to describe this type of movement (Forchheimer, 1924):

$$Ko_4 = \frac{1}{I\ell g}\left(v^2 + \frac{v\nu}{\ell}\right) \;; \quad I = av^2 + bv \;;$$
$$\left(\text{if} \;\; a = \frac{1}{Ko_4\ \ell g} \;; \;\; \text{and} \;\; b = \frac{\nu}{Ko_4\ \ell^2 g}\right) \quad . \tag{1-63}$$

Finally, the general form of the movement equation in the zone of microseepage is:

$$Ko_3 = \frac{I_o}{I} + \frac{\ell v}{I\ell^2 g} \; ; \quad v = \frac{\ell^2 g}{\nu} = (Ko_3 \; I - I_o) = c_1 \; I - c_2 \; I_o \; ; \tag{1-64}$$

$$(c_1 = \frac{\ell^2 g}{\nu} Ko_3 \; ; \quad \text{and} \quad c_2 = \frac{\ell^2 g}{\nu} = \frac{\ell^2 g}{\nu} = K \frac{1}{MK}) \; .$$

The correctness of such application of dimensionless numbers calculated as quotients of retarding and accelerating forces is also proved by Karádi's measurements (Karádi, 1963). The points representing his data were plotted in a coordinate system with the logarithm of Reynolds number on the horizontal axis and 2/MK on the vertical one. The fact that the points follow a horizontal line in the laminar zone verifies the constancy of the MK value, and also that of Darcy's hydraulic conductivity.

It has already been mentioned that the ratio of considered and neglected retarding forces is a suitable parameter to calculate the limits of the validity zones of each movement equation, because this quotient measures the magnitude of the two forces in question related to one another. It is possible, therefore, to determine such fixed numerical values of these dimensionless numbers which indicate where one retarding force becomes negligible as compared to the considered one. For this investigation it is necessary to define the characteristic length and velocity from which the dimensionless number in question should be calculated.

The Reynolds number is proportional to the ratio of inertia to friction. If this number is smaller than a given limit, inertia is also small and friction is high compared to one another. The validity of Darcy's law can probably be accepted below this limit. When this parameter is above another limit (which is higher than the previous one), the inertia in the numerator is dominant compared to friction and the latter is negligible. This is the turbulent zone. Between these two limits there is the zone of the transition condition, where both inertia and friction have to be considered.

According to the general practice in seepage hydraulics, the Reynolds number is calculated using the effective diameter as the characteristic length, and substituting the Darcy's seepage velocity. The latter is the discharge of flow divided by the total cross section including the area of both pores and solid grains. Consequently, the Reynolds number generally used in seepage hydraulics can be calculated from the following formula:

$$Re_s = \frac{D_h v}{\nu} \quad . \tag{1-65}$$

The authors presenting their experimental results give the upper limit of Darcy's zone as 2-5, expressed by the parameter calculated from Eq. 1-65 (Lindquist, 1933; Kozeny, 1953; Karádi and V. Nagy, 1960; Karádi, 1963). According to the listed papers, the lower limit of the turbulent zone is 100-200, similarly measured by Re_s. There are, however, some explanations mentioned in the same articles, indicating that this value gives only the lower limit of the second transition zone, and the real turbulent zone starts at a higher limit.

Some of the soil physics parameters (porosity, shape coefficient) are excluded from the Reynolds number calculated by Eq. 1-65. It is more correct to use the Reynolds number determined for the model pipe instead of Re_s. In this parameter (Re_p) the characteristic length is the average pipe diameter, and the mean velocity in the pores (or in the model pipes) is substituted as characteristic velocity. Thus the relationship between the two Reynolds numbers can be determined as

$$Re_p = \frac{d_o v_{eff}}{\nu} = \frac{4}{1-n} \frac{D_h}{\alpha} \frac{v}{\nu} = \frac{4}{1-n} \frac{Re_s}{\alpha} \; ;$$

because (1-66)

$$d_o = \frac{4}{\alpha} \frac{n}{1-n} D_h \; ; \quad \text{and} \quad v_{eff} = \frac{v}{n} \; .$$

Using this parameter, the following numerical limits can be assigned to various zones

$Re_p < 10$, laminar (Darcy's) zone;

$10 < Re_p < 100$, first transition (Lindquist's) zone;

$100 < Re_p < 1000$, second transition zone;

$1000 < Re_p$, turbulent (Froude's) zone.

The purpose of all the other proposals concerning the calculation of the Reynolds number is similarly to select the characteristic length and velocity, so that the information content of the parameter should be raised by including the soil physics data other than the effective diameter.

The velocity value generally used is either the seepage velocity (v) (Ward, 1964; Harleman et al., 1963; Perez Franco, 1973; Collins, 1961), or the effective mean value in the pores

($v_{eff} = v/n$) (Massanari, 1967; Kovács, 1969a, 1969b; Chauveteau and Thirriot, 1967; Thirriot, 1969; and Thirriot and Habib, 1970). Considering the possible difference between volumetric and areal porosity, the n value was introduced sometimes with a power lower than unity as it was explained by Thirriot (1969) that the power may be between 1 and 2/3, and Zampalione (1969) used a characteristic velocity $v_{eff} = vn^{-2/3}$ in the Reynolds number.

There is an even larger variety of parameters applied for the characteristic length in the calculation of the numerical value of the Reynolds number. There is a general attempt to find a length directly proportional to the resistance of the solid matrix. The most natural way is to use the square root of intrinsic permeability for this purpose ($\sqrt{k}$) (Ward, 1964; Harleman et al., 1963; Perez Franco, 1973; Valentin, 1970; Massanari, 1967). Some authors have related this parameter to the square root of porosity ($\sqrt{k/n}$) (Collins, 1961). Thirriot (1969) has derived a characteristic length assuming that the resistance of the porous medium is equivalent with that of a capillary tube having this length as diameter. According to his result this length (d^*) is equal to the product of $\sqrt{k/n}$ and a constant

$$d^* = \sqrt{32 \frac{k}{n}} \quad . \tag{1-67}$$

Substituting the value of intrinsic permeability according to the modified Kozeny-Carman equation, it is easy to prove that the d_o average pore diameter, used in Eq. 1-65 as the characteristic length, is similarly proportional to $\sqrt{k/n}$:

$$k = \frac{1}{5} \frac{n^3}{(1-n)^2} \left(\frac{D_h}{\alpha}\right)^2 \; ; \quad \sqrt{\frac{k}{n}} = \frac{1}{5} \frac{n}{1-n} \frac{D_h}{\alpha} \; ;$$

$$d_o = 4 \frac{n}{1-n} \frac{D_h}{\alpha} = \sqrt{80} \sqrt{\frac{k}{n}} \quad . \tag{1-68}$$

The characteristic length used by Zampaglione (1969) is derived also from the total resistance of the sample. The most important deviation from the other methods is the consideration of the difference of areal and volumetric values of porosity. The square of the length in question (Δ^2) is defined as a factor of proportionality between the mean pore velocity and the hydraulic gradient

$$v'_{eff} = CI \ ; \quad C = \frac{g}{\nu}\Delta^2 \ ; \quad \text{and because}$$

$$v'_{eff} = \frac{v}{n^{2/3}} \ ; \quad \text{therefore}$$

$$\Delta = \sqrt{\frac{\nu}{g}\,\frac{v'_{eff}}{I}} = \sqrt{\frac{\nu}{g}\,\frac{1}{n^{2/3}}\,\frac{v}{I}} = \sqrt{\frac{k}{n^{2/3}}} \ . \qquad (1\text{-}69)$$

A grouping according to the two applied parameters gives an excellent survey of the various proposals analyzed in the foregoing paragraphs (Table 1-10). The summarization clearly shows that the proposed dimensionless numbers differ only in the consideration of the effect of porosity (in the power used of this parameter), because the use of various numerical constants does not change the character of the Reynolds number. On the basis of the comparison, the transforming factors between the various Reynolds numbers can be also determined together with the ratio of these numbers to the classical seepage Reynolds number (Re_S) (Table 1-11).

After comparing the different Reynolds numbers, Valentin (1970) has already proposed that the use of intrinsic permeability should be generally accepted for the calculation of the characteristic length. The only remaining problem is the way of consideration of porosity and the choice of the applied constant. At the first glance it seems that this problem is only a question of decision, because knowing the various transforming factors of the numerical characteristics of the limits between the various types of seepage, what the application of Reynolds number services for can be easily transferred from one to the other system (i.e., the limits fixed previously in the form of given Re_p values). Aiming, however, at the establishment of the general seepage law that describes the relationship between the seepage velocity and the hydraulic gradient without any restriction of the validity zone, the power of porosity has to be chosen so that the points representing the measured resistances in a coordinate system, based on the applied form of the Reynolds numbers, should queue along a curve. Some guidance concerning the constant parameter to be used can be gained from the constant of the modified Kozeny-Carman equation proposed for the calculation of the Darcy's hydraulic conductivity. Considering these objectives it was found, that the multiplication of intrinsic permeability by porosity is more suitable than the previous proposals, and the application of a constant of 5 is advisable:

$$Re_K = 5\,\frac{v\sqrt{nk}}{\nu} \ . \qquad (1\text{-}70)$$

Table 1-10. Grouping of the various Reynolds numbers proposed for the characterization of seepage.

characteristic velocity	seepage velocity	effective velocity	reduced effective velocity	proposed Reynolds number expressed as a function of porosity and intrinsic permeability
characteristic length	v	$v_{eff} = \frac{v}{n}$	$v_{eff}' = \frac{v}{n^{2/3}}$	
square-root of intrinsic permeability $\sqrt{k}$	Harleman et al. (1963) Ward (1964) Perez (1973)	Massanari (1967)		$Re_W = \frac{v}{\nu} \sqrt{k}$ $Re_M = \frac{v}{\nu} \sqrt{\frac{k}{n^2}}$
square-root of the ratio of intrinsic permeability and porosity $\sqrt{\frac{k}{n}}$	Collins (1961)			$Re_C = \frac{v}{\nu} \sqrt{\frac{k}{n}}$
equivalent tube diameter d or d_o		Thirriot (1969) Kovács (1969)		$Re_T = \sqrt{32} \frac{v}{\nu} \sqrt{\frac{k}{n^3}}$ $Re_P = \sqrt{80} \frac{v}{\nu} \sqrt{\frac{k}{n^3}}$
parameter derived from the total resistance of the sample Δ			Zampalione (1969)	$Re_Z = \frac{v}{\nu} \sqrt{\frac{k}{n^2}}$
square-root of the product of intrinsic permeability and porosity $\sqrt{nk}$	Kovács			$Re_K = 5 \frac{v}{\nu} \sqrt{nk}$

Table 1-11. Transforming factors between the various Reynolds numbers.

	Re_W	Re_C	$Re_M = Re_Z$	Re_T	Re_P	Re_K	Re_S (using the modified Kozeny-Carman equation for the calculation of permeability)
Re_W	1	$Re_W = \sqrt{n}\ Re_C$	$Re_W = n\ Re_M$	$Re_W = \frac{n^{3/2}}{\sqrt{32}} Re_T$	$Re_W = \frac{n^{3/2}}{\sqrt{80}} Re_P$	$Re_W = \frac{1}{5} \frac{1}{\sqrt{n}} Re_K$	$Re_W = \frac{1}{\sqrt{5}} \frac{n^{3/2}}{1-n} \frac{1}{\alpha} Re_S$
Re_C	$Re_C = \frac{1}{\sqrt{n}} Re_W$	1	$Re_C = \sqrt{n}\ Re_M$	$Re_C = \frac{n}{\sqrt{32}} Re_T$	$Re_C = \frac{n}{\sqrt{80}} Re_P$	$Re_C = \frac{1}{5} \frac{1}{n} Re_C$	$Re_C = \frac{1}{\sqrt{5}} \frac{n}{1-n} \frac{1}{\alpha} Re_S$
$Re_M = Re_Z$	$Re_M = \frac{1}{n} Re_W$	$Re_M = \frac{1}{\sqrt{n}} Re_C$	1	$Re_M = \sqrt{\frac{n}{32}} Re_T$	$Re_M = \sqrt{\frac{n}{80}} Re_P$	$Re_M = \frac{1}{5} \frac{1}{n^{3/2}} Re_K$	$Re_M = \sqrt{\frac{n}{5}} \frac{1}{1-n} \frac{1}{\alpha} Re_S$
Re_T	$Re_T = \frac{\sqrt{32}}{n^{3/2}} Re_W$	$Re_T = \frac{\sqrt{32}}{n} Re_C$	$Re_T = \sqrt{\frac{32}{n}} Re_M$	1	$Re_T = \sqrt{\frac{2}{5}} Re_P$	$Re_T = \frac{\sqrt{32}}{5} \frac{1}{n^2} Re_K$	$Re_T = \sqrt{\frac{32}{5}} \frac{1}{1-n} \frac{1}{\alpha} Re_S$
Re_P	$Re_P = \frac{\sqrt{80}}{n^{3/2}} Re_W$	$Re_P = \frac{\sqrt{80}}{n} Re_C$	$Re_P = \sqrt{\frac{80}{n}} Re_M$	$Re_P = \sqrt{\frac{5}{2}} Re_T$	1	$Re_P = \sqrt{\frac{16}{5}} \frac{1}{n^2} Re_K$	$Re_P = \frac{4}{1-n} \frac{1}{\alpha} Re_S$
Re_K	$Re_K = 5 \sqrt{n}\ Re_W$	$Re_K = 5\ n\ Re_C$	$Re_K = 5\ n^{3/2}\ Re_M$	$Re_K = \frac{5}{\sqrt{32}} n^2\ Re_T$	$Re_K = \sqrt{\frac{5}{16}} n^2\ Re_P$	1	$Re_K = \sqrt{5} \frac{n^2}{1-n} \frac{1}{\alpha} Re_S$

Tables 1-10 and 1-11 are supplemented with this finally proposed parameter as well.

In the zone of microseepage, where seepage velocity is smaller than that pertaining to the laminar flow, the water movement is influenced also by two retarding forces (i.e., adhesion and friction) and, therefore, this zone can be regarded as transitional as well, which develops between laminar movement and the static condition. Similarly to the use of the Reynolds number, the limit between the laminar zone and microseepage can be indicated by a given numerical value of the dimensionless number, proportional to the ratio of adhesion and friction (Ko_2 in Eq. 1-59). Because of the lack of sufficient measured data, this limit can be determined only on the basis of theoretical consideration.

Equation 1-61 indicates the constancy of the MK' value in the Darcy's zone. Substituting the average pipe diameter in Eq. 1-61 as the characteristic length and the hydraulic conductivity in the form of the modified Kozeny-Carman equation, this constant can be calculated by

$$MK'_p = \frac{n}{80} \quad . \tag{1-71}$$

The dimensionless number MK' can be determined also from the grain diameter instead of the diameter of the model pipe:

$$MK'_s = \frac{1}{5\alpha^2} \frac{n^3}{(1-n)^2} \quad . \tag{1-72}$$

Similarly to the various forms of the Reynolds number, the square root of intrinsic permeability can also be substituted as a characteristic length. In this case the dimensionless parameter is equal to unity, or using $5\sqrt{nk}$ value it is inversely proportional to porosity

$$MK'_w = 1 \; ; \quad \text{or} \quad MK'_K = \frac{1}{25n} \quad . \tag{1-73}$$

Assuming that porosity is n = 0.40, MK'_s is approximately 10^{-3} for a sample of spheres (α = 6), which this value is in good agreement with Karádi's data. Combining Eqs. 1-59 and 1-71 the dimensionless number giving the limit in question can be expressed in Darcy's zone as a product of the ratio of threshold and actual gradient, and a constant depending only on porosity:

$$Ko_{2p} = \frac{I_o}{MK_p} = \frac{I_o}{MK'_p I} = \frac{I_o}{I} \cdot \frac{80}{n} \qquad (1\text{-}74)$$

The same parameter, if $\sqrt{k}$ or $5\sqrt{nk}$ are substituted for the characteristic length, is

$$Ko_{2W} = \frac{I_o}{I} ; \quad \text{or} \quad Ko_{2K} = 5n \frac{I_o}{I} . \qquad (1\text{-}75)$$

This type of relationship must be valid not only within the laminar zone, but also at its lower limit, thus the numerical value in question can also be determined by a constant value of the ratio between the threshold gradient and the actual one:

$$I_h = 12\ I_o \qquad (1\text{-}76)$$

where I_h is the actual gradient at the limit between the laminar zone and microseepage, and the constant of 12 is based on the detailed investigations of microseepage.

Comparing this result with the results of investigations concerning the value of the threshold gradient, it can be seen that this limit is about $I_h = 0.003$ in sand ($D_h > 0.01$ cm), while in clay ($D_h \sim 2\mu$) the gradient limit is higher, about $I_h = 0.5$, although the effect of adhesion is still very small here, smaller than the probable error in the measurement of the gradient.

The validity zones of the various types of seepage can be graphically represented as the function of either velocity or hydraulic gradient (Fig. 1-84).

The first graph shows the laminar zone and the zones having higher gradient. Here the Reynolds number is linearly proportional to velocity. Its given numerical values indicate the limits of the borders between the subsequent zones. The Reynolds number, indicates the ratio of inertia to friction, but the ratio of the two retarding forces can be also judged by this parameter. The Mosonyi-Kovács number, according to its definition, is independent of velocity in the laminar zone. Its value decreases gradually with increasing velocity in the transition and turbulent zones. The I(v) function is represented in two ways. The full line shows the relationship considering dynamically correct relationship in each zone, while the dotted line supposes the validity of the linear function within the whole range of movement. The difference

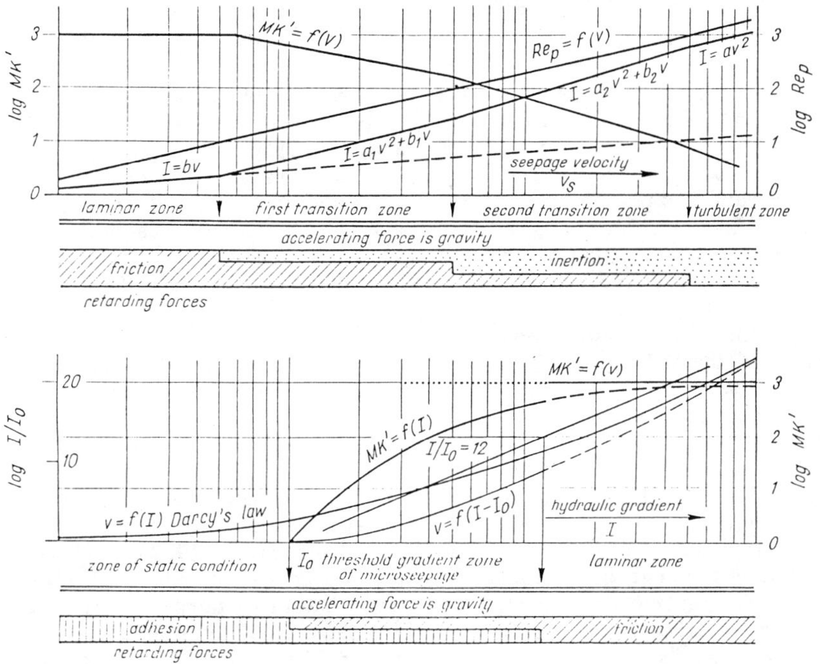

Fig. 1-84. Graphical representation of the validity zones of the various types of seepage.

between the two curves indicates the error, which may be committed by the unjustified application of Darcy's law.

The second graph summarizes the validity zones of the types of seepage having small velocity (including the laminar zone as well). Here the v(I) function is represented similarly as in the first part, but now the gradient is chosen as an independent varible (measured on the horizontal axis). The linear relationship also gives a higher velocity than the actual one in this range of seepage. From the dimensionless parameters, the Mosonyi-Kovács number is represented as a function of hydraulic gradient. It is nearly constant in the laminar zone, and gradually decreases below the lower limit of the latter. It becomes zero at the threshold gradient.

4-3 Capillary Tube Model to Characterize the Channels Composed of the Pores between Grains

There are several previously made attempts to substitute the irregularly contacted channels formed by the pores of loose clastic sediments with a regular physical model, in which the relationship between the geometrical and hydraulic variables can be determined theoretically. Naturally, the models provide only the most probable form of the movement equations, carefully executed experiments are necessary to prove the reliability of the model to examine its validity zone and the accuracy of the equations achieved in this way, and to determine the numerical constants of the formulae.

The best known models are those composed of capillary tubes (Scheidegger, 1953, 1960). The most simple form of these models is a bundle of straight pipes. In other cases the tortuosity of the capillary tubes is also considered. According to another proposal, straight tubes composed of small stretches of different diameters are very suitable to simulate the flow through the water conveying channels of the sediments. Such bundles can be composed of tubes having only two different diameters (Wyllie and Spangler, 1952), or many different stretches, ensuring only the constancy of the areas free for water transport in a cross section perpendicular to the main direction of flow (Wyllie and Gardner, 1958). Bear and Bachmat's model (1967) is a generalized form of the capillary tube models because it is composed of a spatial network of interconnected, random passages of varying length, cross section, and orientation, and of junctions, where the channels (minimum three) meet (Bear, 1972).

Another way usually applied for determining the resistivity of the porous medium is to consider the analogy between the drag force acting on a sphere settling in water and the resistance against flow in the interconnected pores (Goldstein, 1938; Ward, 1964). Rumer gives a detailed summary of the results achieved by applying this type of model to characterize the relationship between seepage velocity and hydraulic gradient in the various zones of seepage (De Wiest et al., 1969). Irmay's proposal can also be included in this group of models (Bear et al., 1967). In this theory, the resistance of spheres located in steady flow along a line perpendiculr to the flow direction is determined by averaging the solution of Navier-Stokes equations, also considering the inertial terms in them.

Although Rumer has stated, in the previously noted book (De Wiest et al., 1969), that the capillary tube model is not suitable to characterize non-laminar flow, it can be proved that the difficulties listed by him can be avoided by a conveniently chosen form of pipes. On the other hand, from practical point of view, it is desirable to select the simplest model, from which, however, the largest possible amount of information can be gained. It is this reason why the bundle of capillary tubes, directed in the main direction of flow, will be used subsequently as a physical model of the water conveying channels. For ensuring the reliable simulation of both non-laminar flow and capillary effect, it is necessary to apply tubes with varying diameters. As it will be seen, the simplest of these types of models (that also proposed by Wyllie and Spangler with two different diameters) is adequate for this purpose, and therefore, the numerical determination of the geometrical parameters of this model will be summarized here.

On the basis of the movement equation, valid for the investigated condition of flow, and determined for a model pipe (e.g., in the case of laminar flow, when the dominating forces are gravity and friction, this equation is the well-known Poiseuille's formula), the relationship between the hydrodynamic parameters of seepage can be determined, or the previously established empirical formulae can be controlled (in the case of laminar movement mentioned before as an example, the validity of Darcy's equation can be determined, and its factor, i.e., hydraulic conducitivity, can be calculated).

Assuming that the parameter D_h/α is known, e.g., it was calculated from Eq. 1-53, two conditions can be found to determine the average diameter of the model pipe (d_o), and the number of pipes crossing the unit area of the section, perpendicular to the main direction of flow (N). (It is necessary to note here that the surface-to-volume ratio proposed by Carman (1959), to characterize the water transporting system, is equal to D_h/α as it is proved by Eq. 1-50.) According to the first condition, the surface of the pipe wall related to its inner volume should be equal to the ratio of the grain surface (A) in the sample to the pore volume (V_p):

$$\frac{\ell\, d_o\, \Pi}{\ell\, d_o^2\, \Pi/4} = \frac{4}{d_o} = \frac{A}{V_p} = \frac{1-n}{n}\frac{A}{V_t} = \frac{1-n}{n}\frac{\alpha}{D_h} \quad ; \qquad (1\text{-}77)$$

consequently

$$d_o = \frac{4n}{1-n}\frac{D_h}{\alpha} \quad . \qquad (1\text{-}78)$$

The second condition states that the cross section of flow should be identical in both the original and the model systems. It follows from this condition, that the cross section of one model pipe multiplied by the number of pipes crossing a unit area should be equal to areal porosity, and therefore, the number of pipes can be calculated by

$$N = \frac{4n_A}{d_o^2 \Pi} = \frac{4n}{d_o^2 \Pi} , \qquad (1\text{-}79)$$

because according to Eq. 1-45, areal porosity is equal to effective volumetric porosity. Without repeating the derivation of the basic equation, only its final result is quoted here, according to which the mean velocity (v_{mean}) in a tube of diameter d_o can be calculated from

$$v_{mean} = \frac{1}{32} \frac{\gamma}{\mu} d_o^2 I = \frac{1}{32} \frac{g}{\nu} d_o^2 I \quad ; \qquad (1\text{-}80)$$

where the quotient of specific weight and dynamic viscosity ($\frac{\gamma}{\mu}$) (which is equal to the ratio of the acceleration due to gravity related to kinematic viscosity ($\frac{g}{\nu}$)) represents the character of the flowing fluid, and I is the hydraulic gradient.

The flux q, i.e., flow rate through a unit area perpendicular to the flow direction, can be calculated as the mean velocity of the model tube (v_{mean}), the area of the tube ($A = d_o^2 \frac{\Pi}{4}$), and the number of tubes crossing the unit area (N). Considering the dependency of d_o and N on soil-physics parameters (Eqs. 1-78 and 1-79), the final result is

$$q = \frac{1}{32} \frac{g}{\nu} n\, d_o^2 I = \frac{1}{2} \frac{g}{\nu} \frac{n^3}{(1-n)^2} \left(\frac{D_h}{\alpha}\right)^2 I \quad . \qquad (1\text{-}81)$$

This value can be compared to the results of some very carefully executed experiments (Zunker, 1930; Lindquist, 1933; Carman, 1956). The actual hydraulic conductivity is 2.5 times smaller than that recalculated from the resistance of the system of model tubes. The constant proposed by Carman to be used in Kozeny's equation (1/5 instead of 1/2 in Eq. 1-81) similarly means a multiplicator of 0.4 compared to the theoretical value calculated for straight tubes. The difference may be caused by three possible factors:

- the cross section of the actual channels is not circular;

- the channels in the network are longer than the length of the sample, and the tubes do not perpendicularly cross the cross section directed at a right angle to the main flow direction;

- the cross-sectional area of the channels is not constant, but they are composed of diffusors and confusors.

The discharge of laminar flow through pipes having a cross section different from the circle (triangle, ellipse, etc.) was calculated (Forchheimer, 1924; Engelhardt, 1960), and compared to that in a circular pipe. It was shown, by this comparison, that the effect of the pipe shape can be expressed by a multiplying factor from 1.2 to 0.8, and thus the difference in question cannot be explained by this cause.

The two phenomena summarized as the second possible cause are generally included in one term, tortuosity. As it was explained by Bear (1972), this effect can be correlated to the amount of the average length of channels (ℓ') and the length of the sample in the flow direction (ℓ), but this relationship is not linear, as it is supposed by several authors. The linear relationship considers only the difference in length, but the change of the area free for water transport has to be taken into account as well, and this influence may be also related to the amount of lengths. Thus it can be supposed that the parameter of tortuosity (T) is proportional or equal to the square of the ratio of the length of the sample to the average length of the channels

$$T = \left(\frac{\ell}{\ell'}\right)^2 \quad . \tag{1-82}$$

Many authors have proposed numerical values for considering tortuosity. Thus Irmay's parameter (0.4) (Bear et al., 1967), as multiplying factor, eliminates the difference mentioned before (hydraulic conductivity of the actual sample is 2.5 times smaller than that of a bundle of straight capillary tubes).

The tortuosity can be also investigated theoretically, if the average length of the water conveying channels can be determined. Bear has pointed out that two different averages can be formed depending on the aspects used as the basis of the calculation. It could be supposed that the average angle between the flow direction and the axes of the channels was $\pi/4$, if only the pore geometry were considered. This hypothesis results in a parameter $T = 0.5$. It is necessary, however, to investigate the kinematic differences between the various directions, which requirement necessitates the determination of the mean velocities of the channels as well (weighting the channels according to their mean velocities).

It is evident that channel velocity is higher in tubes parallel to the flow direction than in those directed at different angles, and therefore, the average position of the tubes, considering kinematic aspects, approaches a smaller angle with the flow direction, than the geometrical average. It was already mentioned that even inactive channels may be expected, thus the overall result can be characterized by a considerably higher tortuosity factor than 0.5.

Considering a value of T = 0.5 with the influence of the non-circular tubes, these two effects could explain the observed difference. Because of the kinematic character of the channels, the numerical results, however, can probably cover only half of the expected value, and therefore, the third cause has to be considered as well. It would be possible to take into account all three effects separately, and to summarize the numerical parameters finally. Effort is made, however, to apply only a very simple model and to give the total explanation of experimental observations, as well as to determine the numerical factors on the basis of this model. It is the writers' opinion, therefore, that the best method to eliminate the difference mentioned previously is the use of model pipes constructed from short stretches of various diameters. The tubes are parallel to the main flow direction, the degree of the inactive channels, perpendicular to this direction, may be considered by decreasing total porosity with a factor smaller than unity. The tubes are located such that the area free for water transport is constant. This model is suitable not only for the verification of the mentioned difference, characteristic of laminar movement, but the numerical comparison indicates good agreement between measured and calculated data in the case of non-laminar flow conditions as well.

According to Lindquist's investigation, the ratio of the largest and narrowest cross-sectional area of the channels in a homodisperse sample of spheres can be as high as 10. This value, however, is a very extreme one, which is valid only at a few points of the channel. To determine a model pipe having two different stretches, with a diameter of d_1 and d_2 respectively, the average ratio of the area of large and narrow sections in the channel should be investigated. A pipe was chosen as a final model (Fig. 1-85), the volume of which is the same as that of the pipe with constant diameter d_o, but constructed from short stretches of d_1 and d_2 diameters ($d_1 < d_o < d_2$) in such a way that half of the length has the smaller, and the other half the larger diameter. The discharge of this pipe was calculated and plotted against the ratio of the two cross-sectional areas ($d_1^2 : d_2^2$) (Fig. 1-86). The same graphs were determined not only for circular pipes, but for other shapes as well. It is shown by the result of this investigation, that the discharge becomes equal to that of the pipe with constant diameter multiplied by 0.4, the actual discharge of the original seepage, according to the experiments represented in the figure, when the ratio of the two areas is 1:3-1:3.5. On the basis of the foregoing, the following equations can be given to calculate the characteristic diameters of the pipe:

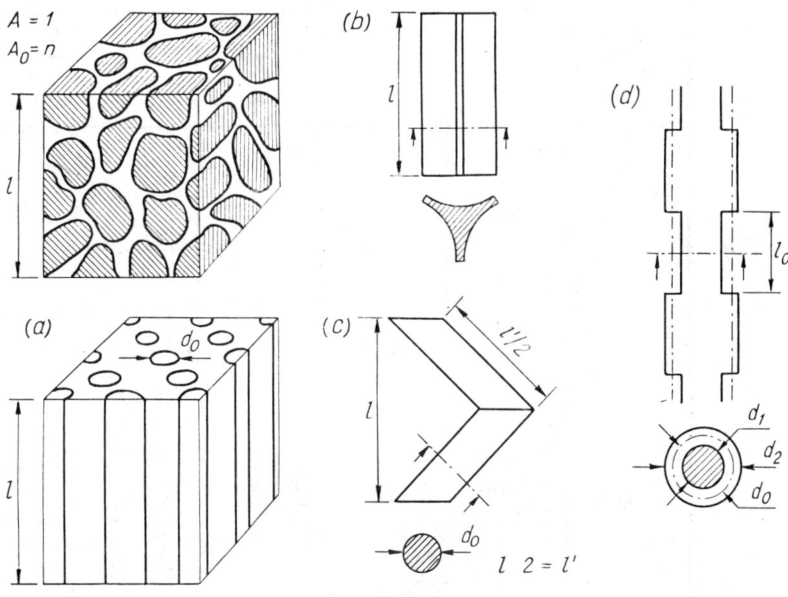

Fig. 1-85. Physical models to substitute the system of irregular channels composed of pores between grains.

$$d_2 = 1.87\ d_1\ ;\ d_1 = \frac{d_o}{1.5}\ ;\ d_2 = 1.25\ d_o\ ; \qquad (1\text{-}83)$$

while the length of each stretch can be approximated, taking into consideration the actual character of the channels by the following formula:

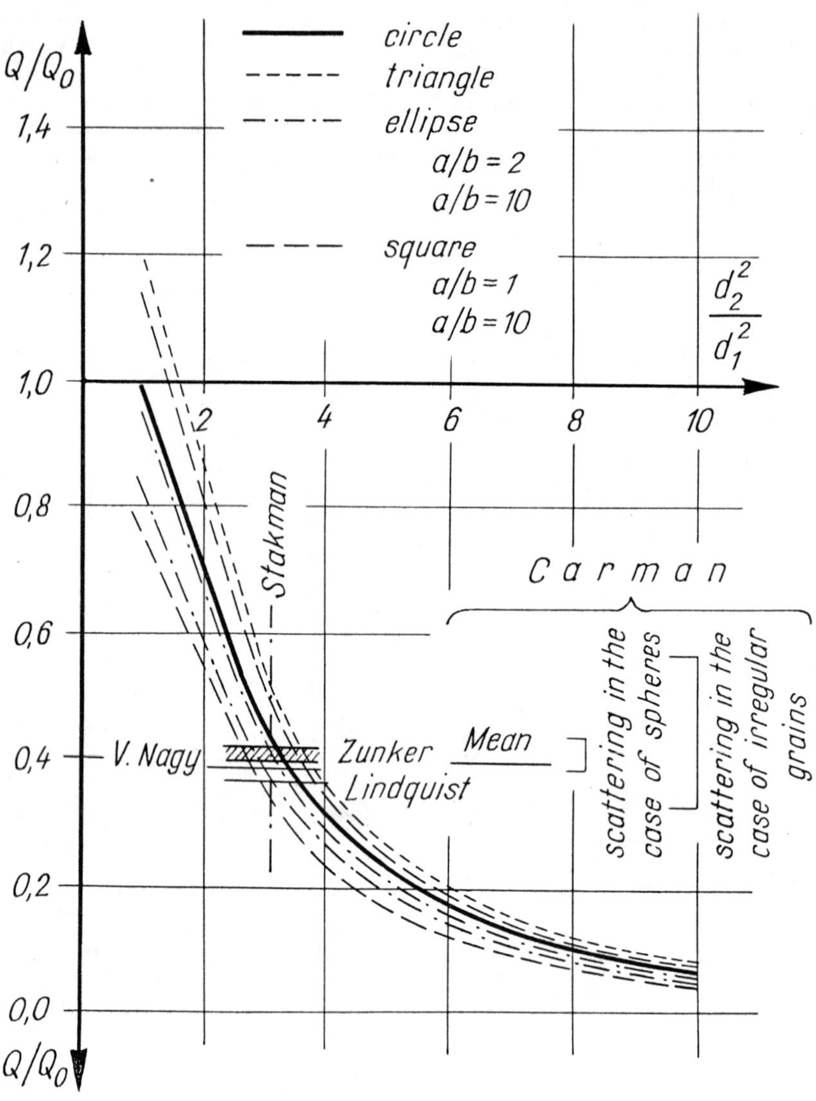

Fig. 1-86. The flow rate of a pipe composed of two types of short stretches with different diameters as a function of the ratio of the areas of the two different cross sections.

$$d_o \leq \ell_o \leq 1.5\ d_o \tag{1-84}$$

The mean velocity in this tube (v'_{mean}) and the flux through a unit area of porous medium simulated with such tubes (q') (the latter being equal to Darcy's (v_s) seepage velocity) can be calculted from the following equations

$$v'_{mean} = \frac{1}{80} \frac{g}{\nu} d_o^2 I \quad ;$$

$$q' = v_s = \frac{1}{80} \frac{g}{\nu} n\, d_o^2 I = \frac{1}{5} \frac{g}{\nu} \frac{n^3}{(1-n)^2} \frac{D_h^2}{\alpha^2} I \quad . \tag{1-85}$$

Although this model was developed, and has served primarily to describe the flow of water through the pores, d_1 and d_2 were found to be numerically equal to the actual smallest and largest pore diameters, as determined by Stakman with air-bubbling experiments. Thus the diameters calculated in this way can also be used as parameters characterizing the expected actual size of pores in a given sample, when studying the water movement through fine grains, or the capillary rise in the pores.

For the characterization of seepage through unsaturated porous medium, the capillary tube model can be similarly applied. It has to be considered, however, that one part of the pores are completely saturated in this zone, while the walls of the larger pores are covered only by water films and therefore the suitable model to simulate the flow occurs along the wall through an annular space. It is necessary to combine the geometrical model of the capillary tubes with two different dynamic models. The first supposes a ring-shaped cross section available for water transport, while the second one assumes that some pores are completely saturated and therefore the models derived to characterize saturated seepage are applicable (Fig. 1-87). A further consequence of the two different types of flow is that the openings cannot be simulated with tubes of similar diameter since saturated seepage occurs through the narrow pores and the annular flow space can develop only in the large ones. It is therefore necessary to determine the probability distribution of the opening sizes, and to model the actual condition by applying tubes with various diameters following the same distribution.

Investigating the water movement through the annual flow space attached to the wall of the tube (Fig. 1-88), a special type of Poiseuille's equation has to be applied considering

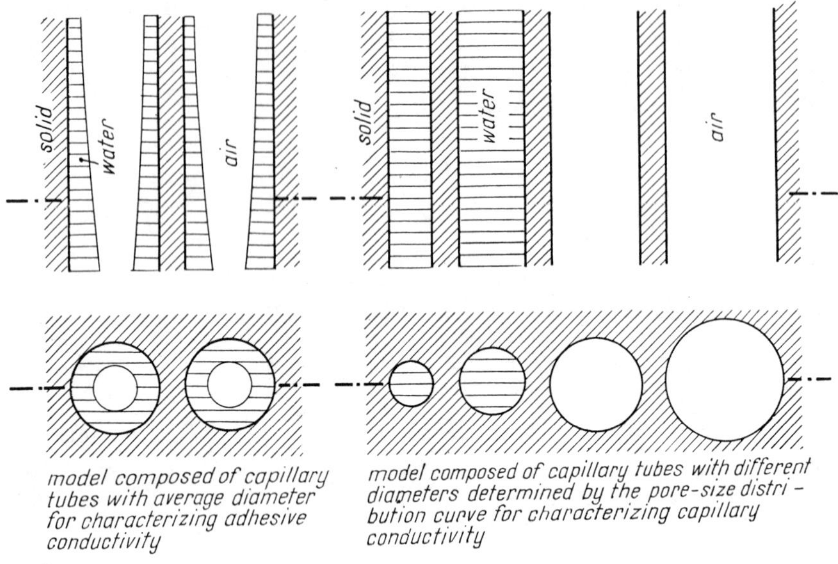

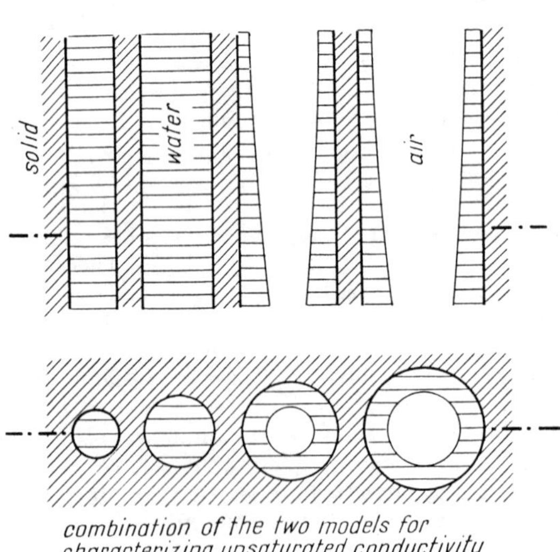

Fig. 1-87. Tube model to simulate the flow of adhesive water film, through saturated pores, by capillarity, and in a combined system respectively.

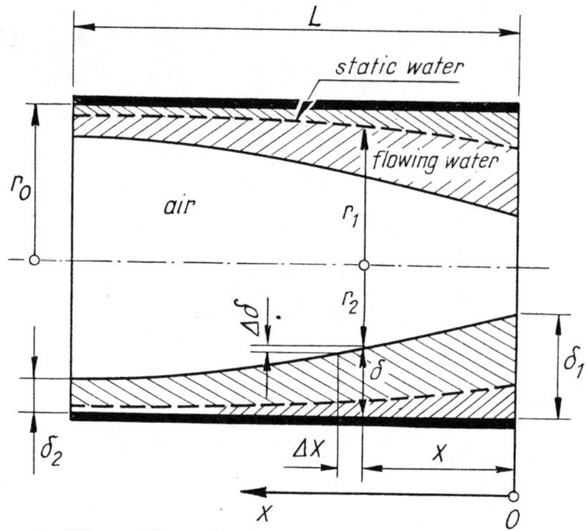

Fig. 1-88. Sketch for the calculation of flow of adhesive water films.

only the limited cross section between r_o and r_2 radii (the latter is the radius of the internal space filled with air):

$$\frac{I\gamma}{2\mu} \frac{r^2 - r_2^2}{r} = - \frac{dv}{dr} \quad . \tag{1-86}$$

After solving the differential equation and substituting the ratio between r_2 and r_o with the amount of saturation (s)

$$s = \frac{W}{n} = 1 - \left(\frac{r_2}{r_o}\right)^2 \quad ; \tag{1-87}$$

the adhesive hydraulic conductivity (K_a) related to the parameter belonging to saturated condition (K_s) can be given in the following form:

$$\kappa_a = \frac{K_a}{K_s} = s^2 - 2s(1-s) - 2(1-s)^2 \ln(1-s)$$

$$= \frac{2}{3} s^3 \cdot \left[1 + 3! \sum_{n=1}^{\infty} \frac{n}{(n+3)!} s^n \right] . \tag{1-88}$$

(This type of hydraulic conductivity is called "adhesive," because the annular flow space is created by the adhesion between the wall and water.)

Since the velocity is very small, it can be assumed that some layers of water molecules are inactive in the close vicinity of the wall, the distance is indicated by r_1 radius. Investigating this type of flow, the static shearing stress (τ_o) has to also be considered in Poiseuille's equation:

$$\frac{I\gamma}{2\mu} \frac{r^2 - r_o^2}{r} = \frac{\tau_o}{\mu} - \frac{dv}{dr} \quad . \tag{1-89}$$

The r_1/r_o quotient can also be interconnected with a characteristic degree of saturation belonging to the dry condition of the sample (s_o):

$$s_o = 1 - \left(\frac{r_1}{r_o}\right)^2 \quad . \tag{1-90}$$

The final result, expressed as the ratio of the relevant hydraulic conductivity values, is:

$$\kappa_a = \frac{K_a}{K_s} = (1-s_o)^2 \left[\left(\frac{s - s_o}{1 - s_o}\right)^2 - 2\,\frac{(1 - s)\,(s - s_o)}{(1 - s_o)^2} - 2\left(\frac{1-s}{1-s_o}\right)^2 \ell n\,\frac{(1-s)}{(1-s_o)}\right] . \tag{1-91}$$

Equations 1-88 and 1-91 are usually approximated by a simple potential function in practice:

$$\kappa_a = \left(\frac{s-s_o}{1-s_o}\right)^m \quad ;$$

or assuming

$$s_o = 0 \;;\quad \kappa_a = s^m \quad ; \tag{1-92}$$

where the m parameter may be between 3 and 4 (m=3; Irmay, 1954; m=3.5; Averyanov, 1949a, 1949b; Boreli and Vachaud, 1966; m=4; Johnson and Kunkel, 1963). Even the variation of

s_o between zero and 0.1, which is its reasonable range, does not considerably modify the final result and, therefore, the use of Eq. 1-92 gives a sufficient approximation ($s_o = 0$ and $m = 3$).

4-4 Application of Hele-Shaw Model to Stimulate the Network of Openings in Solid Rocks

The geometrical models proposed to characterize the flow through the openings of fissured and fractured rocks can be divided into two large groups:

- network of interconnected pipes with constant or varying diameters;

- one or more series of slits between parallel plates.

The first model is very similar to the network of the mains and distributing pipes of an urban water supply scheme. Generally, the water transporting channels are supposed to have circular cross section. The joints of the stretches may be distributed regularly in the space, e.g., forming an orthogonal network, or randomly. The diameter of the pipes may be constant between the joints, or it may be variable. Using constant diameter between the joints, there are two further possibilities, either all the stretches are modeled from pipes of the same diameter, or their sizes are different. The change of the diameter may be either random, or it can be determined according to some known behavior of the solid matrix, e.g., to simulate the anisotropic character of permeability, larger diameters are used in one direction. Some possible variations are represented in Fig. 1-89.

The dynamics of flow in a closed conduit is a thoroughly investigated process in hydromechanics. Relatively simple solutions are known for networks composed of circular pipes having constant diameters between the joints. This is the reason why this simplified form is generally applied to simulate the karst openings through carbonate rocks (Őllós, 1963). There is difficulty, however, hindering the application of these types of models, i.e., too many parameters have to be determined for the complete description of the models and these data cannot be derived from easily measurable physical characteristics of the aquifer rocks. Even using the most simple model, constant pipe diameter at every stretch and regular distribution of the joints, the pipe diameter, the length of stretches, the density of the joints in various directions all have to be estimated. The data listed depend on the position and the distribution of the large water conveying channels and they cannot be interrelated with such

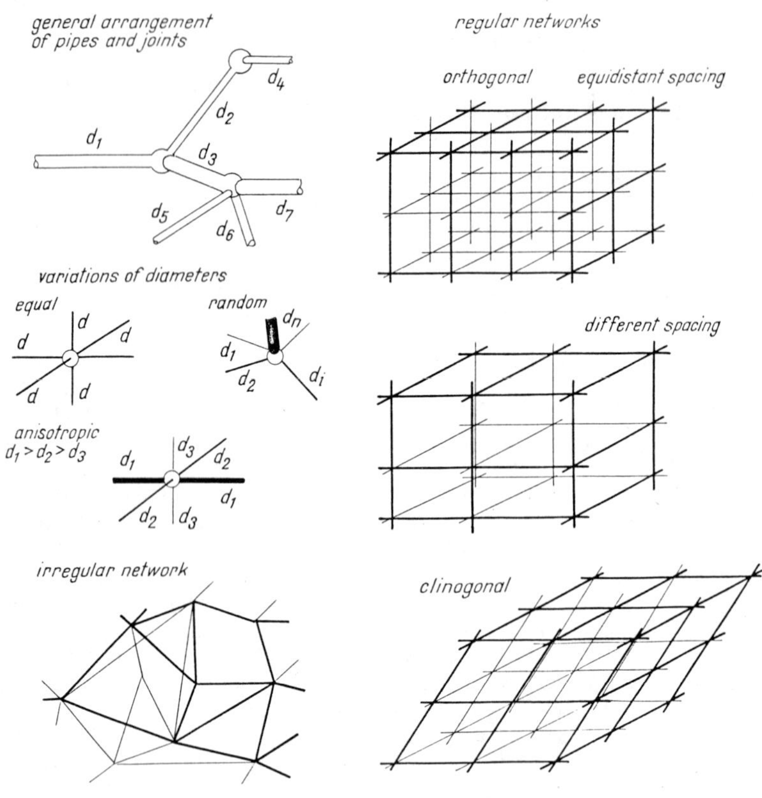

Fig. 1-89. Sketches showing various networks of pipes simulating the interconnected channels of carbonate rocks.

simple physical parameters as porosity or statistical distribution of openings. For this reason, the models composed of pipes are generally applied only to solve a given problem in a well-defined, and not too large, part of the carbonate rock terrain. In this case, the hydraulic parameters developing under the influence of known boundary conditions (e.g., the discharge, the position of the water table or piezometric surface, and its change in time) can be measured, and the best fitting network of pipes can be determined by a trial and error method. Thus, the network of pipes is proposed basically for small scale modeling of a special process occurring in karst zones of carbonate rocks (Mijatović, 1970) and not to generally describe the hydraulic conductivity of a large mass of the fissured and fractured aquifer.

Another reason can also be mentioned to support the previously given statement. In the entire rock mass, there are generally only a few large openings, while it is divided by thousands and thousands of fine joints, smaller or larger fractures, especially in the fault zones. Thus, the total permeability of the rock is more strongly influenced by the latter type of openings, than by karst channels. The role of dissolved channels becomes dominant mostly in the vicinity of draining structures where the resistivity of the aquifer is basically determined by the character of individual water transporting elements (karst channels or strongly fractured zones). The joints, fractures, and faults can be, however, better simulated by narrow slits, than pipes.

It is the general opinion, therefore, that the hydraulic conductivity of a large mass of carbonate rocks, or any other types of fissured and fractured aquifers, can be described by using a geometrical model composed of the network of narrow slits running through the solid mass in different directions and intersecting each other (Lomidze, 1951; Louis, 1970). This model may be composed of either regularly or randomly distributed slits, and the width of the slits may be either constant or varying. The anisotropy of hydraulic conductivity may be simulated either by choosing different spacing of the slits or by applying different widths in various directions (Fig. 1-90).

The dynamic characterization of water movement in a narrow slit bordered by two smooth walls parallel to one another is well-known, if the flow is laminar (gravity and friction are the dominant forces). Let us investigate the flow in a vertical slit (Fig. 1-91(a)). Because of the small width of the slit, the component of the velocity vector $v(v_x, v_y, v_z)$ normal to the wall (v_y) is zero, while it follows from the laminar condition of flow that the change of velocity in x and z directions compared to that in y direction is negligible. Since gravity is the only one accelerating force, the total potential acting on the unit mass of the fluid can be composed of two terms: potential and pressure energies.

Considering these conditions, the amount of water transported during a unit time in a section of the slit having unit length normal to the flow direction can be determined by

$$q = - \frac{1}{12} \frac{g}{\nu} b^3 I \quad ; \qquad (1\text{-}93)$$

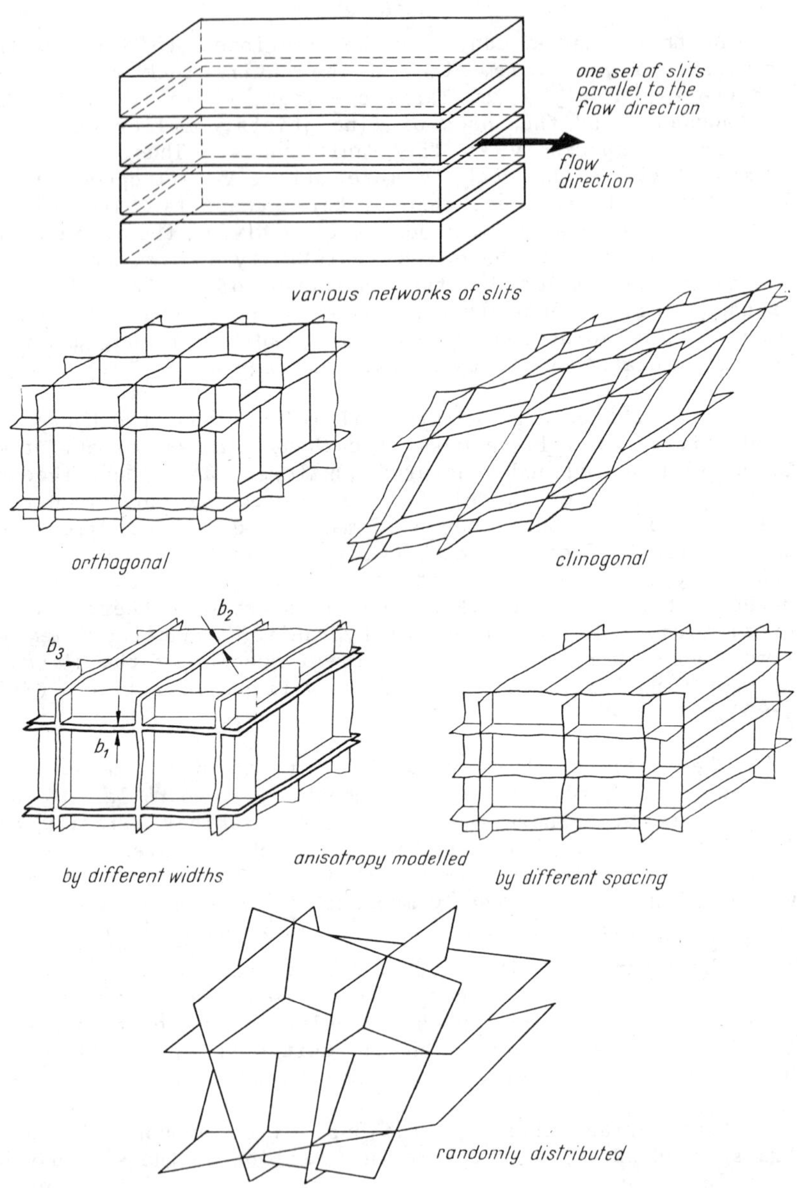

Fig. 1-90. Sketches showing various networks of slits simulating the interconnected joints of fractured rocks.

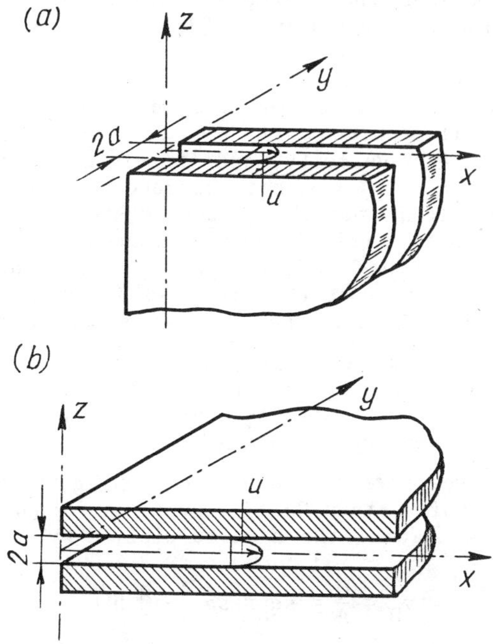

Fig. 1-91(a,b). Interpretation of symbols used for the derivation of Hele-Shaw equation.

where b is the width of the slit, ν is the kinematic viscosity of water, and I is the hydraulic gradient along the slit. The mean velocity of the fluid is equal to the ratio of discharge and area

$$v_{eff} = \frac{1}{12} \frac{g}{\nu} b^2 I \quad ; \tag{1-94}$$

and, using the analogy of Darcy's law, the hydraulic conductivity of a slit can be calculated

$$K_1 = \frac{1}{12} \frac{g}{\nu} b^2 \quad . \tag{1-95}$$

Investigating the flow in a horizontal slit (Fig. 1-91(b)) the modification of velocity distribution can be expected because the gravity force acts in a right angle to the walls. In practice, however, the gravity term of potential (z) is very small compared to pressure head (p/γ), and therefore its influence can be neglected. Thus, all equations (Eqs. 1-93, 1-94, and 1-95) derived for vertical slits can be applied for all fractures independently of their direction.

It is necessary to note here that there is a basic difference between the parameter determined by Eq. 1-95 and Darcy's hydraulic conductivity. The latter multiplied by the hydraulic gradient gives the seepage velocity (or flux) which is equal to the ratio of the flow rate and the total area normal to the flow direction and thus smaller than the mean velocity in the pores. On the contrary, Eq. 1-94 gives directly the mean velocity, and therefore the hydraulic conductivity of the slit does not characterize the rock, but only one opening. It is possible, however, to also derive Darcy's parameter from the relationships given previously, if it is assumed that s number of slits having b width and parallel to the flow direction cross the unit area (normal to the flow) of the rock (Fig. 1-92). Because s is the number of slits counted within a unit length, its dimension is $[L^{-1}]$. The total flow rate through this area is the product of the number of the slits and the water conveying capacity of one opening. This flow rate is equal to the flux (v_D seepage velocity), because the area investigated perpendicular to the flow has an extension of unity, thus Darcy's hydraulic conductivity can be calculated as that of the slit multiplied by linear porosity:

$$v_D = \frac{Q}{A} = sq = sb \frac{1}{12} \frac{g}{\nu} b^2 I \quad ; \tag{1-96}$$

where

$$n_L = sb \quad ;$$

because sb is the total width of the slits, and it has to be related to the total investigated length to get the linear porosity, but in the present case this length is equal to unity, the product sb directly yields the value of linear porosity.

There are many conditions arbitrarily chosen to derive Eq. 1-96 for the characterization of the conductivity of a fractured medium, and they hinder the direct application of this model when the purpose is the description of the behavior of natural rocks.

The most important hypotheses are:

- the walls of the openings are parallel to one another;

- the openings have smooth walls and are not intersected by other fractures;

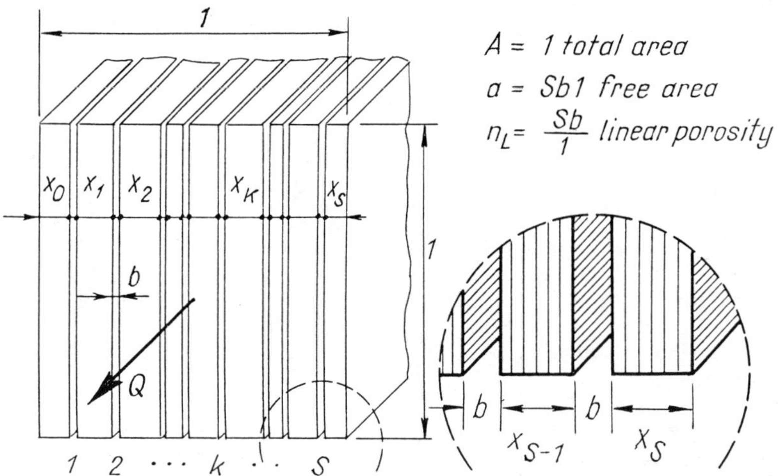

Fig. 1-92. Interpretation of Darcy's hydraulic conductivity in fractured rocks.

- the network of oepnings is composed of joints having equal width;

- the flow is laminar.

The influence of the changing width of slits can be taken into account by substituting a suitable average value (Whitherspoon, 1974). The roughness of the wall and the edges developing at the intersections of the joints increases the resistivity of the system, thus the actual velocity is smaller than the theoretical one. Experimental measurements have shown, however, that this influence is generally not very strong and it can be accounted for by reducing the velocity by 20-25 percent. It may be proposed that this effect be taken into account by increasing the numerical constant in the denominator of Eq. 1-96 from 12 to 15:

$$v = n_L \frac{1}{15} \frac{g}{\nu} b^2 I \quad ;$$

and (1-97)

$$K = n_L \frac{1}{15} \frac{g}{\nu} b^2 \quad .$$

There are some publications in the literature directed at determination of the average width of the uniform slits, and their number, within a given mass. These investigations are generally based on the statistical analysis of the probable distribution of both the widths of openings and the distances between the fractures, e.g., Király, 1973; Snow, 1970; Romm,

1966. These methods include, however, some uncertainties in connection with the determination of the hydraulically equivalent slits. It is advisable, therefore, that the statistical model explained in the previous section (see Eq. 1-79) be combined with the geometrical and dynamic model of the slit with parallel walls. This combined conceptual model is suitable to simulate not only the geometrical differences of the fractures but also the change of flow conditions caused by different widths of the slits, thus the non-laminar character of seepage can be considered in this way as well. The derivation of the conceptual model and its application to determine the hydraulic conductivity of fissured and fractured zone will be discussed in Section 5.

The last hypothesis listed as the precondition of the application of Eq. 1-96 includes two assumptions:

- the network of openings is composed of one system of straight fractures running parallel to each other;

- they are also parallel to the flow direction, and thus normal to the investigated cross section which is perpendicular to the flow direction.

A generally applicable geometrical model can be established only if the arbitrary position of the slits may be also considered. For this reason, let us suppose that two measuring lines (x and y direction) are fixed on the front of the rock perpendicularly to the flow direction (z). If the network of fractures were composed of two systems of slits having equal width of b, and if the slits were parallel to the flow direction (and thus perpendicular to the measuring plane on the front of the rock), and if the two systems were perpendicular to one another as well as to one of the two measuring lines, the flux through a unit area (which is equal to the seepage velocity) and the hydraulic conductivity in the z direction could be calculated by considering Eqs. 1-97, namely by:

$$(v_{max})_z = I\,(n_{Lx} + n_{Ly})\,\frac{1}{15}\,\frac{g}{\nu}\,b^2 \quad ;$$

$$(K_{max})_z = \frac{v_s}{I} = (n_{Lx} + n_{Ly})\,\frac{1}{15}\,\frac{g}{\nu}\,b^2 \quad . \qquad (1\text{-}98)$$

The index "max" indicates that both seepage velocity and hydraulic conductivity decrease if the slits decline from the position described previously.

Investigating the water transport through a slit having a general position, two types of declination have to be characterized. The slit may remain parallel to the flow direction, and thus perpendicular to the measuring plane (but closing a β angle not equal to $\pi/2$ with one of the measuring línes and an angle of $\pi/2-\beta$ with the other one). The other declination is measured by the α-angle enclosed by the axis of the slit and the flow direction. The interpretation of the two angles is shown in Fig. 1-93. Since the position of the slit is a random event both angles β and α can be regarded as probability variables, and therefore, neither the flux nor the hydraulic conductivity can be characterized by one given numerical value, but the range of their probable value has to be determined.

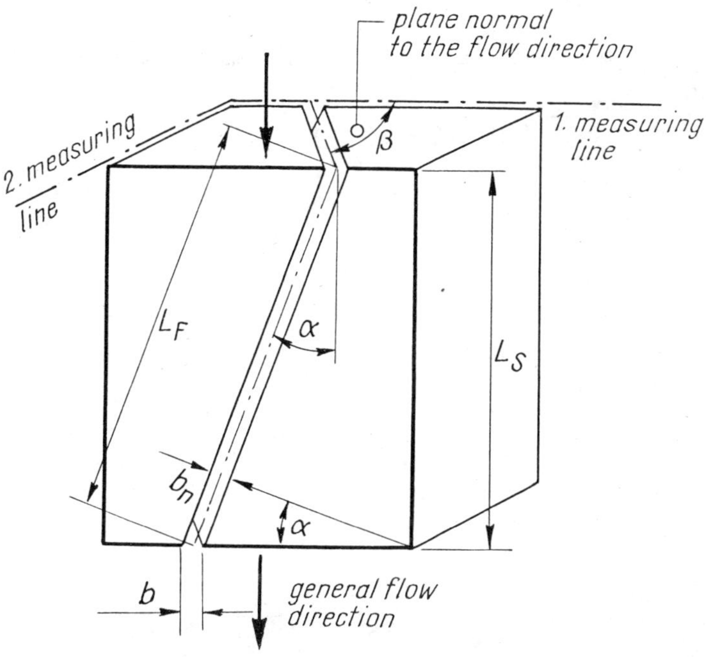

Fig. 1-93. Characterization of the position of a slit related to the main direction of flow.

It was already mentioned that Eqs. 1-98 give the upper limits of the possible scattering of parameters. Considering a pure geometrical concept, it can be stated, that the most probable value of both angles $\beta_{med} = \alpha_{med} = \pi/4$. Expressing the change of the normal width and the length of the slit as functions of the two angles, the influence of both declinations on the decrease of water transport can be determined mathematically. After having executed both reductions and having substituted the most probable value of the angles, it was found that the median (most probable) flux and hydraulic conductivity could be calculated by dividing the maximum parameters by 16. Considering the asymmetry of the frequency distribution of the flux having random character, it is generally approximated by using the lognormal distribution, the lower limit of the zone of confidence can be assumed to be one eighth of the median. Slightly modifying Eq. 1-98 to take into account the different average width of openings in the direction of the two measuring lines (b_x and b_y) the probable zone of the hydraulic conductivity of a fractured system, the network of interstices of which are composed of slits having random position, can be given in the following form:

$$K_{min} = \frac{1}{128} K_{max} < K_{med} = \frac{1}{16} K_{max} < K_{max} \quad ; \qquad (1\text{-}99)$$

where $K_{max} = \frac{1}{15} \frac{g}{\nu} (n_{Lx} b_x^2 + n_{Ly} b_y^2)$.

REFERENCES FOR PART I

Albert, J., 1967: The Materials of Brick and Their Use in Rough Ceramics (in Hungarian), Budapest.

Albertson, M. L., 1952: Effect of Shape on the Fall Velocity of Gravel Particles (in English), 5th Hydraulic Conference, Iowa.

Anderson, A. H. M., 1950: Adhesive Forces between Grains within Their Pores in Clastic Sediments (in German), Kolloid-Zeitung.

Arcan, L., 1966: The Determination of Clay Component of Soil by Infrared Spectrophotogrammetry (in French), Symposium on Water in Unsaturated Zone, Wageningen.

Averanov, S. V., 1949a: Relationship of Permeability of Soil with Air Content (in Russian), D.A.N., No. 2;
1949b: Approximative Evaluation of the Role of Seepage in the Capillary Fringe (in Russian), D.A.N., No. 3.

Bachmat, Y., 1965: Basic Transport Coefficients as Aquifer Characteristics (in English), IASH Symposium on Hydrology of Fractured Rocks, Dubrovnik.

Balásházy, L. and Kovács, J., 1975: Determination of Hydraulic Conductivity of Triassic Carbonate Rocks by Statistical Analysis of Silts (Manuscript in Hungarian), Budapest.

Barneman, R. R., 1975: Problems Associated with Development of Groundwater in Igneous and Metamorphic Rocks, Groundwater, Vol. 11, No. 5.

Baver, L. D., 1948: Soil Physics (in English), London, New York.

Bear, J., 1972: Dynamics of Fluids in Porous Media (in English), New York, London, Amsterdam.

Bear, J., Zaslavsky, D., and Irmay, S., 1968: Physical Principles of Water Percolation and Seepage (in English), Paris.

Bear, J. and Bachmat, Y., 1967: A Generalized Theory on Hydrodynamic Dispersion in Porous Media (in English), IASH Symposium on Artificial Recharge and Management of Aquifers, Haifa.

Bidlo, G., Kleb, B., and Török, M., 1967: Analysis of Aggregates of Concrete of Hydraulic Structures (in Hungarian), Épitoanyag, No. 11.

Bingham, E. C., 1922: Fluidity and Plasticity (in English), London.

Bondarenko, N. F., 1966: Nature of Seepage-Anomaly of Fluids (in Russian), Symposium on Seepage and Well Hydraulics, Budapest.
1973: Physics of the Movement of Groundwater (in Russian), Leningrad.

Bondarenko, N. F. and Nerpin, S., 1966: Influence of Viscous-Plastic Properties of Water on its Equilibrium and Transfer in Unsaturated Zone (in English), Symposium on Water in the Unsaturated Zone, Wageningen;
1967: Shearing Stress of Fluids and its Consideration in the Investigation of Surface Phenomena (in Russian), Publication of the Institute of Physicochemistry of Academy of Sciences of USSR.

Boreli, M. and Vachaud, G., 1966: Certain Problems of Infiltration in Unsaturated Porous Media (in English), Saopstenja Institute za Vodoprivredu Jaroslav Cerni, No. 39.

Boreli, M. and Vachaud, G., 1966: On the Determination of the Residual Water-Content and on the Variation of Relative Permeability of the Unsaturated Soil (in French), C. R. Academy of Sciences, Paris, Series A., Tom. 263.

Böcker, T., 1973: Theoretical Model for Karstic Rocks (in English), Karszt-és Barlang-Kutatás, Budapest, Vol. VII.

Brown, R. H., Konoplyantsev, A. A., Ineson, J., and Kovalovsky, V. S., 1972: Groundwater Studies Chapter 9, Analytical and Investigational Techniques for Fissured and Fractured Rocks by M. Schoaler (in English), UNESCO Studies and Reports in Hydrology, Paris.

Carman, P. C., 1956: Flow of Gases through Porous Media (in English), London.

Charadabellas, P., 1964: Standardization of k-Value Determination by Field Tests through Standardizing the Grain-Size Distribution Curve of Water-Bearing Clastic Sediments (in German), Mitteilung des Institutes für Wasserwirtschaft, No. 20.

Chauveteau-Thirriot, C., 1967: Regime of Flow in Porous Media and the Limit of Darcy's Law (in French), La Houille Blanche, No. 2.

Childs, E. C. and Tzimas, E., 1971: Darcy's Law at Small Potential Gradients (in English), Soil Science, No. 3.

Collins, R. E., 1961: Flow of Fluids through Porous Materials (in English), New York.

Craft, J., 1967: The Influence of Soil Mineralogical Composition on Cement Stabilization (in English), Geotechnique, No. 2.

Darab, K., Rédliné, and Reményiné, 1967: Development of Methods for Determination of Clay Minerals (in Hungarian), MTA Talajtani és Agrokémiai Intézet, Budapest (Manuscript).

Davis, S. N. and De Wiest, R. J. M., 1966: Hydrogeology, p. 261.

De Wiest, J. M., 1969: Flow Through Porous Media (in English), New York, London.

Dolgov, S. I., 1948: Studies Related to the Determination of Soil Moisture and its Availability for Plants (in Russian), Moscow, Leningrad.

Donat, J., 1929: Study on the Permeability of Sands (in German), Wasserkraft und Wasserwirtschaft, No. 17.

Dubleton, M. J. and West, G., 1966: Some Factors Affecting the Relationship between Clay-Minerals in Soils and Their Plasticity (in English), Clay Minerals, No. 3.

Elrick, D. E., 1966: The Mineralogic Characterization of Soils (in English), Symposium on Water in the Unsaturated Zone, Wageningen.

Emery, K. O. and Rittenberg, S. C., 1952: Early Diagenesis of California Basin Sediments in Relation to Origin of Oil (in English), Bulletin of American Association for Petrol-Geology.

Engelhardt, W., 1960: Porosity of Sediments (in German) Berlin, Göttingen, Heidelberg.

Eötvös, L., 1886: On the Relationship between Surface Tension and Chemical Properties of Various Fluids (in Hungarian), Matematikai és Természettudományi Értekezések, Vol. IV.

Erdey-Gruz, T. and Schay, A., 1954: Theoretical Physicochemistry (in Hungarian), Budapest.

Fekete, Z., 1950: Comparative Analysis of Parameters Describing the Water Content of Soils (in Hungarian), Hidrológiai Közlöny, No. 7-8.

Forkasiewicz, J. and Paloc, H., 1965: The Process of Recession of "La Toux de la Vis" - Preliminary Study (in French), IASH Symposium on Hydrology of Fractured Rocks, Dubrovnik.

Forchheimer, Ph., 1886: On the Yield of Wells and Drains (in German), Hannover Zeitschrift des Architecks and Ingenieurs;
1901: Water Movement through Soil (in German), Zeitschrift des Verbondes der deutschen Ingenieurs;
1924: Hydraulics (in Germany), Leipzig, Berlin.

Fraser, H. J., 1935: Experimental Study of the Porosity and Permeability of Clastic Sediments (in English), Journal of Geology.

Füchtbauer, H. and Reineck, H., 1958: Eclogae geologiae Helv. (in German).

Goldstein, S., 1938: Modern Developments in Fluid Dynamics (in English), Vol. II, London, New York.

Hagerman, T. H., 1938: About the Relation between the Distribution Field of the Relative Width of the Particles and the Genesis of the Sediment (in English), Geologiska Foreningens Förhandl, No. 3.

Hamilton, E. L. and Menard, H. W., 1956: Density and Porosity of Sea Floor Surface Sediments of San Diego, California (in English), Bulletin of American Association for Petrol-Geology.

Harleman, D. R. F., Mehlhorn, P. F., and Rumer, R. R., 1964: Dispersion-Permeability Correlation in Porous Media (in English), Proceedings of ASCE, HY2.

Hazen, A., 1895: The Filtration of Public Water Supplies (in English), New York.

Herczog, H., Kovács, G., Nagy, L., Papfalvi, F., and Öllős, G., 1966: Applied Soil Mechanics in Hydraulic Engineering (in Hungarian), Budapest.

Heywood, H., 1938: Measurement of the Fineness of Powdered Materials (in English), Proceedings of Institute of Mechanical Engineers.

Irmay, S., 1954: On the Hydraulic Conductivity of Unsaturated Soil (in English), T.A.G.U., No. 1.

Ivicsics, L., 1957: Hydraulic Design of Sedimentary Basins and Determination of Velocity of Sedimentation (in Hungarian) Vizügyi Közlemények, No. 3.

Jáki, J., 1944: Soil Mechanics (in Hungarian), Budapest.

Johnson, A. I., 1962: Methods of Measuring Soil-Moisture in the Field (in English), Geological Survey Water-Supply Paper, No. 1619-U.

Johnson, A. I. and Kunkel, F., 1963: Some Research Related to Ground-Water Recharge (Progress Report from the USGS) (in English), Conference on Groundwater Recharge and Groundwater Basin Management, University of California, Berkeley.

Juhász, J., 1958: Investigation of Seepage (in Hungarian), Hidrológiai Közlöny, No. 1;
1966: Dynamic Characterization of Seepage (in Hungarian), Symposium on Seepage and Well Hydraulics, Budapest.

Juhász, J., 1966: Calculation of Static Groundwater Resources Available as a Result of Consolidation (in Hungarian), Symposium on Seepage and Well Hydraulics, Budapest;
1967: Hydrogeology (in Hungarian), Budapest.

Karádi, G. and Török, L., 1955: Applicability of Seepage Law (in Hungarian), Hidrológiai Közlöny, No. 5-6.

Karádi, G. and V. Nagy, I., 1960: Investigation of Validity of Darcy's Law (in Hungarian), Conference on Hydraulics, Budapest.

Karádi, G., 1963: Hydraulics of Linear Drainage Systems (in Hungarian), Khartoum-Budapest (Thesis: Manuscript).

Kézdi, Á., 1962: Soil Mechanics (in Hungarian), Budapest;
1963: Neutral Stress and Pressure of Flow (in Hungarian), Vizügyi Közlemények, No. 1;
1968: Distribution of Grains and Pores According to Their Volume (in English), 3rd Budapest Conference of SMFE, Budapest;
1974: Handbook of Soil Mechanics (Volume I. Soil Physics) (in English), Amsterdam.

King, F. H., 1899: A Study on Porosity and Grain Relationship in Experimental Sands (in English), U.S. Geological Survey.

Király, L., 1973: Explanatory Notes for the Hydrogeological Map of Neuchatel Country (in French), Bulletin de la Société Neuchatoise des Sciences Naturelles, Tom. 96.

Kleb, B., 1968: Sediment-Geological Investigation of Pliocen Formations Southwards from Mecsek Mountains (in Hungarian), Földtani Közlöny, No. 2.

Kovács, G., 1957: Theory of Microseepage (in Hungarian), Hidrológiai Közlöny, No. 3;
1958: Theoretical Investigation into Microseepage (in English), Acta Technica Academiae Scienciarum Hungaricae, Tom. XXI, No. 1-2;
1966: Dynamic Investigation of Seepage by Invariant Numbers (in English), Symposium on Seepage and Well Hydraulics, Budapest;
1968: Characterization of Shape of Grains in Seepage Hydraulics (in Hungarian), Épités-és Közlekedéstudományi Közlemények, No. 1-2;
1968: Characterization of the Molecular Forces Influencing Seepage with the Help of the pF Curve (in English), Agrokémia és Talajtan (Supplement).

Kovács, G., 1969: General Characterization of Different Types of Seepage (in English), 13th Congress of IAHR, Kyoto;
1969: Relationship between Velocity of Seepage and Hydraulic Gradient in the Zone of High Velocity (in English), 13th Congress of IAHR, Kyoto;
1969: Seepage Law of Microseepage (in English), 13th Congress of IAHR, Kyoto;
1971: Seepage through Saturated and Unsaturated Layer (in English), Bulletin of IAHS, No. 2;
1971: Seepage through Unsaturated Porous Media (in English), 14th Congress of IAHR, Paris;
1971: Relationship of Plasticity and Clay-Content of Soils (in Hungarian), Agrokémia és Talajtan, No. 1-2;
1976: Combination of Geometrical, Dynamic and Statistical Models to Determine Hydraulic Conductivity (in English), IAHR Symposium on Stochastic Hydraulics, Lund;
1976: Statistical Model of the Pore-Size Distribution of Loose Clastic Sediments (in English), IAHR Symposium on Flow through Porous Media, Kiev.

Kozeny, J., 1953: Hydraulics (in German), Wien.

Kreyoig, L., 1963: Heat and Water Balance of Soils (in Hungarian), Budapest.

Kriván, P., 1957: Sediment-Geological Evaluation of Hagerman's Grain-Shape Method (in Hungarian), Földtani Közlöny, No. 3.

Kuhn, A., 1963: Handbook of Colloid Chemistry (in Hungarian), Budapest.

Kuron, H., 1932: Zeitschrift für Pflanzen (in German), Berlin.

Kutilek, M., 1965: Influence of Interface on Filtration of Water in Soil (in French), Science du Sol (Prague), No. 1.

Kutilek, M. and Salingerova, J., 1966: Flow of Water in Clay Minerals as Influenced by Absorbed Quinolimium and Pyrodinium (in English), Soil Science, No. 5.

Lahermo, P., 1973: The Groundwater of Central and West Lapland, Geol. Survey of Finland, Bulletin 262, p. 14.

Lamarck, J. B., 1964: Hydrogeology, Urbana, p. 152.

Landau, L. D. and Lifschits, E. M., 1974: Textbook of Theoretical Physics (in German), Berlin.

Levich, V. G., 1958: Physicochemical Hydrodynamics (in Hungarian), Budapest.

Lebediev, A. F., 1963: Soil-Moisture and Groundwater (in Russian), Izdatielstvo Academii Naukoj SSSR, Moscow.

Lindquist, E., 1933: On the Flow of Water through Porous Soil (in English), 1st Congress of ICOLD, Stockholm.

Lo, K. Y., 1961: Secondary Composition of Clays (in English), Proceedings of ASCE, SM. 6.

Lomidze, T. M., 1951: Percolation in Fissured Rocks (in Russian), Moscow.

Louis, C., 1970: Three-Dimensional Flow in Fractured Rocks (in French), Comité Francois de Mécanique des Roches, Paris.

Madge, E. W., 1934: The Viscosities of Liquids and Their Vapor Pressures (in English), Physics, No. 2.

Mados, L., 1939: Studies Related to Irrigation and Water Management in the Area of the Irrigation Scheme of Tiszafüred (in Hungarian), Öntözési Közlemények;
1941: Application of Soil Science in Connection with Drainage (in Hungarian), Öntözési Közlemények.

Massarani, G., 1967: Critics of the Validity of Darcy's Law in Heterogeneous Porous Media (in French), CRAS.

Mattson, S., 1932: Soil Science (in English), Baltimore.

Maucha, L., 1972: Investigation of Natural Processes Influencing Yield of Springs (in Hungarian), VITUKI, Budapest (Manuscript).

Melentev, V. A., 1960: Hydraulic Dams from Gravels and Sands (in Russian), Moscow.

Miháltz, J. and Ungár, T., 1954: Separation of Alluvial and Wind-borne Sands (in Hungarian), Földtani Közlöny, No. 1-2.

Mijatovic, B. F., 1970: Hydraulic Mechanism of Karst Aquifers in Deep-Lying Coastal Collectors (in English), Bulletin of Institute for Geological and Geophysical Research, Beograd, Series B, No. 7;
1975: Rational Exploitation of Karst Waters, Hydrogeology of Karstic Terrains, p. 131.

Mitscherlich, E. A., 1932: Soil Science for Agriculture and Forestry (in German), Berlin.

Mosonyi, E. and Kovács, G., 1952: Model Law by Joint Consideration of Gravity and Friction (in Hungarian), Hidrológiai Közlöny, No. 7-8;
1956: Model Law of Filtration (in French), IAHR Congress, Dijon.

Murota, A. and Sato, K., 1969: Statistical Determination of Permeability by the Pore-Size Distribution in Porous Media (in English), 13th IAHR Congress, Kyoto.

Németh, E., 1963: Hydromechanics (in Hungarian), Budapest.

Öllos, G., 1963: Hydraulic Investigtion of Flow Rate in Fractured Rocks by Small Scale Model (in French), IASH Bulletin, No. 2.

Perez Franco, D. F., 1973: Non-Darcy Flow of Groundwater Towards Wells and Trenches, Particularly within the Turbulent Range of Seepage (in English), Technical University of Budapest (Manuscript).

Seed-Woodward-Lungen, 1964: Clay-Mineralogical Aspects of the Atterberg's Limits (in English), Proceedings of ASCE, SM. 4.

Shevyakov, L. D. and Mankovsky, G. I., 1963: Experiences Gained by De-watering Mining Sites of Productive Minerals Having Complicated Hydrogeological Conditions (in Russian), Moscow.

Sine, L. and Bentz, A., 1966: Differences between Experimental and Theoretical Values of Capillary Diffusivity (in French), Symposium on Water in the Unsaturated Zone, Wageningen.

Snow, D. I., 1970: The Frequency and Apertures of Fractures in Rock (in English), International Journal of Rock Mechanics and Mining Science, No. 1.

Spronck, R., 1932: Hydrodynamic Similarity and the Investigation of Models (in French), Annales des Travaux Publiques de Belgique.

Stakman, W. P., 1966: The Relation between Particle Size, Pore Size and Hydraulic Conductivity of Sand Separate (in English), Symposium on Water in the Unsaturated Zone, Wageningen;
1966: Determination of Pore Size by the Air Bubbling Pressure Method (in English), Symposium on Water in the Unsaturated Zone, Wageningen.

Stelczer, K., 1967: Abrasion of Bed-Load Material (in Hungarian), Vizügyi Közlemények.

Swartz, C. A., 1931: The Variation on the Surface Tension of Gas-Saturated Petroleum with Pressure of Saturation (in English), Physics, Vol. 1.

Szádeczky Kardoss, E., 1933: Determination of the Coefficient of Roundness (in German), Centralblatt für Mineralogie.

Szilvágyi, I., 1966: Experiments to Characterize the Rheological Behavior of Clays (in Hungarian), Földtani Kutatás, No. 4;
1967: The Stability Problems of Clayey Valley Slopes in North Hungary (in English), 9th Congress of ICOLD, Istanbul.

Terzaghi, K., 1943: Theoretical Soil Mechanics (in English), New York, London.

Peterson, A. E. and Dixon, R. M., 1971: Water Movement in Large Soil Pores: Validity and Utility of the Channel System Concept (in English), Research Report, University of Wisconsin, No. 75.

Powell, J. W., 1885: U.S. Geological Survey, Fifth Annual Report, p. 17.

Rats, M. V. and Chervashov, S. N., 1965: Statistical Aspect of the Problem on the Permeability of the Jointy Rocks (in English), IASH Symposium on Hydrology of Fractured Rocks, Dubrovnik.

Reiner, M., 1952: Theory of Reology (in French), Paris.

Réthá́ti, L., 1960: Engineering Aspects of the Capillarity of Soils (in Hungarian), Vizügyi Közlemények, No. 1.

Rodier, A. A., 1952: Soil Moisture (in Russian), Izdatielstvo Academii Naukoj SSSR, Moscow.

Rose, D. A., 1966: Water Transport in Soils by Evaporation and Infiltration (in English), Symposium on Water in the Unsaturated Zone, Wageningen.

Scheidegger, A. E., 1953: Theoretical Models of Porous Matter (in English), Producers Monthly, No. 10, 17;
1960: The Physics of Flow through Porous Media (in English), Toronto.

Schlichter, C. S., 1899: Annual Report of the U.S. Geological Survey (in English).

Schmertmann, 1964: Comparison of One and Two-Specium CFS Tests (in English), Proceedings of ASCE, SM. 4.

Schmieder, A., Willems, T., Szilágyi, G., and Keserü, Zs., 1970: Investigation of Laws Governing the Movement of Water and Rocks in Aquifers (in Hungarian), Bányászati Kutató Intézet, Budapest (Closing Report, Manuscript), No. 13-2/69 K.

Schoeller, M., 1948: Hydrogeological Regime of Eocene Limestone in the Synclinal of Dyr-El-Kef (Tunisie) (in French), Bulletin de Society Geological de France, Tom. 18;
1962: Groundwater (in French), Paris;
1965: Hydrodynamics in the Karst (in French), IASH Symposium on Hydrology of Fractured Rocks, Dubrovnik.

Thirriot, C., 1969: Hydrodynamics of Flow in Porous Media (in French), 13th IAHR Congress, Kyoto.

Thirriot, C. and Habib, J., 1970: Experimental Study of Flow with Low Reynolds Number in Clays (in French), Sciences Agronomiques Rennes.

Thompson, 1966: Lime Reactivity of Illinois Soils (in English), Proceedings of ASCE, SM. 5.

Vachaud, G., 1966: General Verification of Darcy's Law and Determination of Capillary Conductivity (in French), Symposium on Water in the Unsaturated Zone, Wageningen; 1966: Study on Soil-Moisture Distribution after Stopping Horizontal Infiltration (in French), Symposium on Water in the Unsaturated Zone, Wageningen.

Vageler, P., 1935: Cations and Water-Content of Mineralogical Soils (in German), Berlin.

Valentin, F., 1970: Non-linear Resistance of Porous Media (in German), Mitteilungen, Institut für Hydraulik und Gewässerkunde, München, No. 6.

Várallyai, G., 1942: Experiences on Irrigation at Márialiget (in Hungarian), Öntözésügyi Közlemények, No. 2.

Ward, J., 1964: Turbulent Flow in Porous Media (in English), Journal of ASCE, Hydraulics Div., No. 0, 5-1.

Wentworth, Ch., 1922: A Scale of Grade and Class Terms for Clastic Sediments (in English), Journal of Geology.

Wesseling, J. and Wit, K., 1966: An Infiltration Method for the Determination of the Capillary Conductivity of Undisturbed Soil Cores (in English), Symposium on Water in the Unsaturated Zone, Wageningen.

Wilson, C. R. and Witherspoon, P. A., 1974: Steady-State Flow in Rigid Networks of Fractures (in English), Water Resources Research, No. 2.

Westmann, A. E. R. and Hugill, H. R., 1930: The Packing of Particles (in English), Journal of American Geran-Society.

Williams,, N. M., 1965: A Method of Indicating Pebble Shape with One Parameter (in English), Journal of Sedimentary Petrology, No. 4.

Wyllie, M. R. J. and Spangler, M. B., 1952: Application of Electrical Resistivity Measurements to Problems of Fluid Flow in Porous Media (in English), Bulletin of American Association for Petrol-Geology, No. 36.

Wyllie, M. R. J. and Gardner, G. H. F., 1958: The Generalized Kozeny-Carman Equation, II. (in English), World Oil Prod. Sect., No. 4.

Wind, G. P., 1966: Capillary Conductivity Data Estimated by a Simple Method (in English), Symposium on Water in the Unsaturated Zone, Wageningen.

Zampaglione, D., 1969: Turbulent Seepage Flow in Filters (in English), 13th IAHR Congress, Kyoto.

Zingg, Th., 1935-1936: Characterization of Clastic Sediments According to Grain-Size Distribution (in German), Schweizerische Mineralogische und Petrographische Mitteilungen.

Zunker, F., 1930: Behavior of Soils Related to Water (in German), Berlin.

PART II

HYDROLOGICAL INVESTIGATION OF THE SOIL-MOISTURE ZONE

by

G. Kovács

SECTION 5

GENERAL CHARACTERIZATION OF THE SOIL-MOISTURE ZONE

Soil moisture is the water content of the pores of soil above the water table retained in various forms against gravity. The most important forces resulting from the development of soil moisture are adhesion and capillarity. Adhesion creates a thin film of water covering the wall of the pores (adhesive water) while capillarity fills some of the pores completely with water (or all the pores, depending on the height of the considered point above the water table) (capillary water) (Fig. 2-1). In fissured and fractured solid rocks, the same type of water is called the moisture content of rocks.

For the complete clarification of this definition, it is necessary to analyze the term "water table" as well. According to the generally accepted meaning, it is the surface of unconfined groundwater bodies dividing the groundwater and the soil-moisture zones. Another characteristic feature of the water table is that the pressure of the water in the pores is equal to that of the atmosphere. It is necessary to note that the water table is not a real surface, only a fictitious level, because there is no rapid change in the saturation of pores here. Both the groundwater zone and the lower part of the capillary zone are completely saturated. The decrease of water content related to porosity starts gradually only at the upper reach of the closed capillary zone, and it is continued in the open capillary zone. Thus the level of the water table is only indicated by the changes in the sign of the water pressure, it is positive in the groundwater zone and negative (suction, tension) above the water table. It should be noted here that water pressure is always expressed in the form of excess pressure (p), i.e., the difference of the total pressure (p_t) and the atmospheric (p_o) value:

$$p = p_t - p_o \quad . \tag{2-1}$$

The change of the sign of the excess pressure is naturally a gradual quantitative modification, and no qualitative difference occurs at the boundary; thus the most important common character of any type of soil moisture (and of the moisture content of rocks) is the negative pressure governing the behavior and the movement of water in this zone.

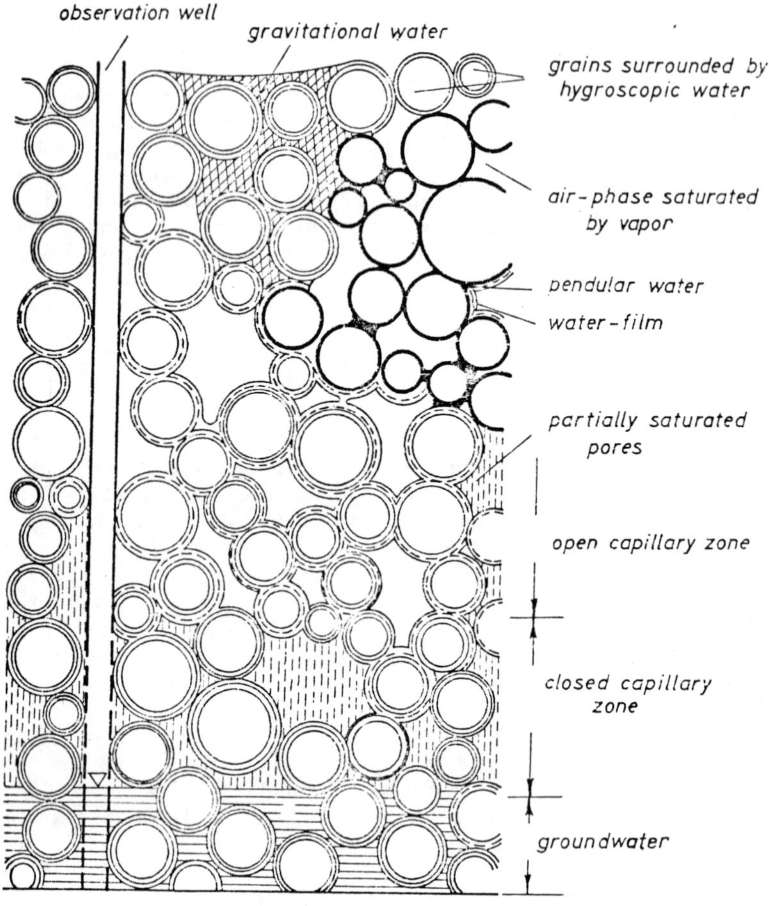

Fig. 2-1. The form of occurrence of water in the soil-moisture zone.

5-1 Components of the Water Balance above the Water Table

Although the thickness of the soil-moisture zone is relatively negligible, if it is compared to the entire thickness of the water storing crest, its hydrological importance is extremely large. The amount of water stored here in an easily accessible position, and the water conveying ability of this layer influence the ratio of actual and potential evapotranspiration (the latter is a function of climatic and surface conditions only). Surface runoff and infiltration depend on the actual condition of the soil-moisture zone (on its storage capacity and permeability influenced by the instantaneous moisture content of the soil). Thus the behavior of this zone affects the water regime of both surface and groundwaters.

To make a further explanation easier, it is useful to group and classify the various processes influencing the water movement through and acting in the soil-moisture zone. For this grouping, a model sketched in Fig. 2-2 is used. This model includes not only the soil-moisture zone, but also the zone of plants covering it, as well as a part of the gravitational groundwater space which is the lower boundary of the zone presently investigated. The path of water either in or through the soil-moisture zone can be easily followed in the sketch.

Regarding the model as a closed system, the water volumes moving through the boundaries of the separated column are the inputs and the outputs of the system. These effects, which can be regarded as the boundary conditions of the model, are indicated by arrows in the figure. The investigated prism is divided into three main subsystems: zone of air, soil-moisture zone and groundwater zone. Two further subsystems may also be distinguished, the soil surface and the active root zone (the latter being part of the soil-moisture zone). The soil surface has a special role in the system, it being the border between the air and soil-moisture zones, although the roots are crossing it. At the same time, a part of precipitation reaching the surface is turned back into the zone of air by evaporation, and surface runoff represents a special input or output here. This is the reason why the surface has to be regarded as an independent subsystem. The separation of the root zone as a fifth subsystem is justified because its structure differs from that of the remaining part of the soil-moisture zone (see Subsection 3-6).

Apart from inputs and outputs, a system is characterized by the operation within the system itself. The storage within the zones of plants and soil moisture, as well as the fluctuation of the water table (which represents the storage in the groundwater zone) are indicated in the figure as such internal operations. Some characteristic values governing the operation of the system are also represented, i.e., soil-moisture retention curve, actual moisture distribution, etc.

The investigated column is in contact at its upper boundary with the atmosphere. The input through this surface

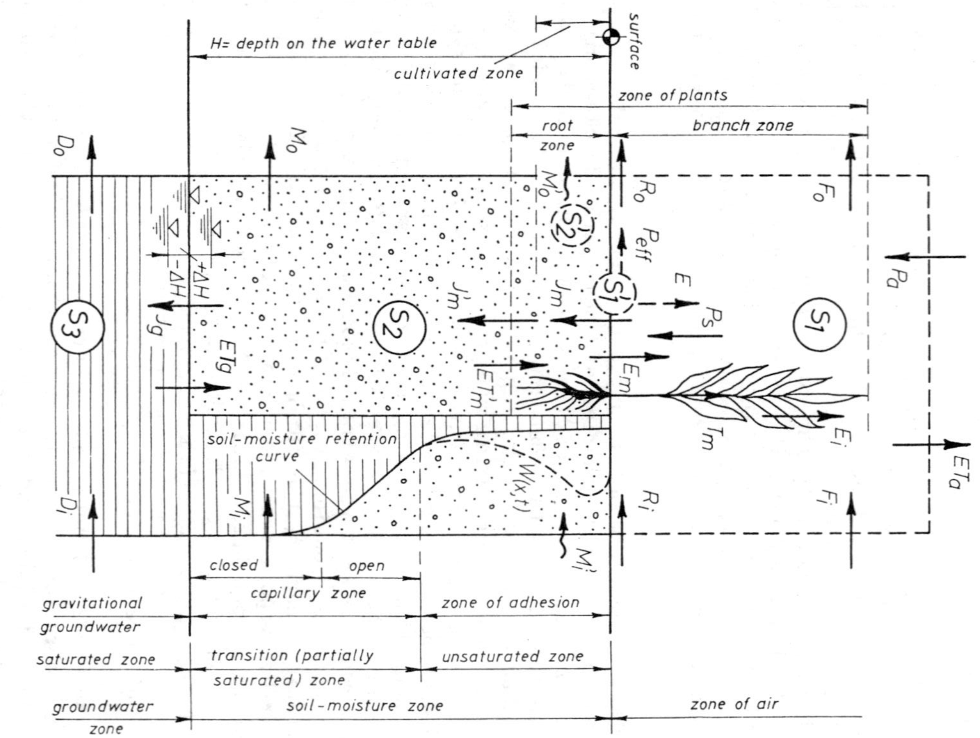

Fig. 2-2. Sketch showing the factors influencing the water balance in the soil-moisture zone.

is precipitation (either liquid or solid), while the outputs are combined under the term of actual evapotranspiration.

Along the vertical faces, the model is bordered by the same zones, which are included into the separated prism as well, i.e., zone of air, the lower part of which is the zone of vegetal stands above the surface, soil-moisture zone between the surface and the water table, and finally gravitational groundwater space below the water table. The lower horizontal boundary of the column is similar to the lower part of the vertical face, it is contacted by the groundwater zone.

By separating the model from its surroundings, a discontinuity is caused, along the listed boundaries, which has to be replaced by inputs and outputs acting on and originating from the system. Such replacing factors are: horizontal vapor movement in the zone of air, horizontal moisture movement in the soil-moisture zone, surface runoff, and groundwater flow.

The further groups of the acting processes can be distinguished according to the subsystems previously mentioned. The storage in the zone of vegetal stands and the evaporation of the water stored here may be included into the investigation of interception. The storage on and the direct evaporation from the surface is closely related to the former group, although these processes are already belonging to the next subsystem (to the surface itself), which is characterized by surface runoff as one of its inputs (inflow on the surfaces to the investigated space), and also as the output of this subsystem (outflow on the surface). The water exchange between the air and soil-moisture zones is the resultant of infiltration through the surface, as well as evaporation and transpiration from the soil moisture. The two draining processes mentioned (i.e., evaporation and transpiration) are only a part of actual evapotranspiration because the latter includes the direct evaporation from the surface of both plants and soil. The next subsystem is the soil-moisture zone, where the storage process is governed by direct evaporation from soil moisture, transpiration of plants from the root zone, and the water retaining capacity of the unsaturated soil, which can be characterized by the soil-moisture retention capacity compared to the actual moisture content of the layer. As it was already mentioned, the structure of the top layer (root zone, or cultivated zone) has to be dealt with separately because its structure differs basically from that of the remaining part of the soil-moisture zone. Thus the soil-moisture retention curve cannot be applied here, viz., the secondary porosity caused by the large channels of roots and plowing maintains the atmospheric pressure in deeper levels as well. The final group of phenomena creating a link between the subsystems is the water exchange of groundwater and soil moisture (accretion). Vertical recharge and drainage of groundwater are the components of this group, and their result can be characterized with the change of the amount of stored groundwater indicated by the fluctuation of the water table.

Surveying the processes listed, those selected which are most closely related to the hydrological investigation of the soil-moisture zone are: infiltration through the surface, evapotranspiration from soil moisture, storage in the unsaturated layer, and water exchange between soil moisture and groundwater.

5-2 Water Content and Parameters Characterizing the Special Condition of Unsaturated Samples

The pores of loose clastic sediments, between the grains, are filled with water either completely (saturated system) or partially (unsaturated system). It is necessary, therefore, to define numerical parameters for the quantitative characterization of the stored amount of water and that of the amount of saturation of the layer.

The amount of water stored in a unit volume of the sample is called specific water content or moisture content. According to its definition, water content is the weight or the volume of water related to the weight of the dried sample or to the total sample volume. To standardize the measurement, the sample should be kept at a temperature of 105°C until there is no change observed in its weight. The water content, expressed as the ratio of weights, can be calculated from the equation:

$$w = \frac{G_v - G_o}{G_o} , \qquad (2\text{-}2)$$

where G_v is the weight of the sample in its original condition (natural or representing a special given condition) and G_o is the weight of the same sample after drying. To calculate the water content in percentage of volume, the specific weight of the dry sample must be known:

$$W = \frac{G_v - G_o}{\gamma_v} \frac{G_o}{\gamma_o} = \frac{G_v - G_o}{G_o} \gamma_o = w\gamma_o , \qquad (2\text{-}3)$$

where γ_v is the specific weight of water, which is unity using centimeters and grams to measure length and weight, and $\gamma_o = \gamma_s(1-n)$, if the specific weight of the solid grains is indicated by γs.

This method of measurement requires a sample taken from the layer, which process disturbs the surface, and prevents the repetition of both the investigation at the same point once again, and the continuous observation of the change of soil moisture. There are equipments to measure soil moisture, which eliminate the disadvantages mentioned. The various measuring methods will be discussed in Section 5-2.

The water content determined by using Eqs. 2-2 and 2-3 represents the amount of water present in the sample at the moment of investigation. This value cannot be regarded, therefore, as a soil physics parameter. It gives only information on the instantaneous condition of the sample. When determining the water content related to a given condition of the sample, this value becomes a parameter suitable for the classification of soils. Such parameters are, e.g., the various values of moisture content used to characterize the cohesive samples in soil physics such as: liquid limit (w_L; or W_L) and the limit of plasticity (w_p; or W_p), or the index of plasticity which is the difference between the former two values expressed as the ratio of weights ($I_p = w_L - w_p$).

Although these parameters belong to arbitrarily chosen conditions (liquid limit to the water content when the sample closes the gap in Cassagrande's apparatus, and the limit of plasticity when a string of 3 mm can be rolled from the sample), they are useful means to compare various cohesive soils.

The maximum molecular water capacity (w_{mol}, or W_{mol}) is also an arbitrary parameter, viz., the water content of the sample (if the diameter of the sample is 50 mm, its thickness is 2 mm and it is loaded by a pressure of 64.5 kg/cm) after the development of consolidation. This way of measuring, however, is only an approximation to standardize the parameter. According to the original concept, maximum molecular water capacity is the remaining water content of a soil column after draining it by gravity (water retention capacity), and therefore,it characterizes a general physical property of the soil (Lebedyev, 1963).

It is necessary to note that the symbols applied here to indicate the two different types of moisture content, i.e., the amount of water stored in the sample and expressed as the ratio of weights, or that calculated as the volume of the stored water related to the total volume of the sample, are different from those generally used in the literature. It is the author's opinion that the same character of both parameters has to be emphasized by choosing the same letter for their representation. At the same time, the necessary distinction can be ensured by using the small and capital case of this letter similar to the indication of hydraulic conductivity and matrix porosity. It was decided, therefore, that the symbol of gravimetric moisture content should be "w" while the volumetric moisture content should be represented by "W".

The pores below the water table are completely saturated ($n = W$). Above this surface (where $0 \leq W \leq n$) the water content indicates the degree of saturation. Two samples having the same water content are differently filled with water, when the porosity of one of them is high and that of the other is low. This is the reason why the coefficient of saturation (or the amount of saturation) is also used to characterize the condition of the sample. This parameter is the quotient of the volumetric moisture content and porosity:

$$s = \frac{W}{n} \quad , \tag{2-4}$$

the value of which is unity in a saturated sample ($s_s = 1$), while in a completely dry sample it is zero. The total range of saturation lies, therefore, between zero and unity ($0 < s < 1$). The practical lower limit, however, differs from the theoretical value because a completely dry sample does not occur in nature. If the minimum natural water content (the volumetric moisture content of a so-called air-dry sample) is W_o, the possible range of the variation of the coefficient of saturation is $s_o < s < 1$, where the lower limit is $s_o = W_o/n$.

One can easily argue that the knowledge of the ratio of saturation is still insufficient information to judge the expected behavior of the soil because the change of saturation in a sand does not involve considerable modification, as may be the case in a clay soil. Measuring water contents belonging to a given state of the sample gives guidance to completely understanding the effects of the actual or instantaneous moisture content on the investigated soil. Knowing these specific values of water content of the soil in question, and comparing the actual amount of soil moisture to them, the behavior of the soil can be estimated.

When selecting the specific parameters, the most important requirement is that the state described by them should be characteristic, and easily repeatable. From this aspect, the parameters of plasticity used in soil physics are well determined values, although they belong to states of the sample chosen arbitrarily, as it was already explained. At the same time, the index of consistency, which is the ratio of two differences (that between liquid limit and instantaneous water content related to the difference of w_L and w_p):

$$K_i = \frac{w_L - w}{w_L - w_p} , \qquad (2\text{-}5)$$

is a good example how to compare the actual water content to the selected limit values.

Although the specific soil-moisture values generally used in soil science as parameters belong to more natural conditions (describing the real physical character of the sample), than the soil-physics parameters, their reproduction causes, however, some difficulties because the conditions described are not determined precisely enough.

The parameters generally applied in soil science are as follows:

- field capacity (w_{fc}, or W_{fc}) is the water content retained against gravity in the sample, thus the value is approximately equal to the difference of the total porosity and specific yield (sometimes the latter is called gravitational pores as well);

- wilting point (w_{wp}, or W_{wp}) is the amount of water, which cannot be used or removed by the suctions of the roots of plants;

- usable (disposable) water (w_d, or W_d) is the difference between the two parameters mentioned above.

Field capacity is a very important parameter, but it is influenced by numerous undetermined factors (e.g., temperature, relative humidity of air, etc., among which the position of the investigated point related to the actual water table is perhaps the most important, as it will be demonstrated later) and therefore, it can hardly be reproduced, or physically interpreted. The same statement can refer to the wilting point as well because the water not available for the plants depends on the suction of the roots which differs from plant to plant, and changes also according to their growing stage.

The most stable and repeatable parameter among the specific moisture-content values used in soil science are the various types of hygroscopic moisture content (or hygroscopicity) which measures the water content retained by the sample against the suction of sulphuric acid of different concentrations, in a closed space, and at a given temperature. Thus, Mitsherlich's hygroscopicity (w_{HY}, or W_{HY}) is determined with 10 percent concentration of sulphuric acid, while Kuron's hygroscopicity (w_{hy}, or W_{hy}) is measured with sulphuric acid of 50 percent concentration (Mitsherlich, 1932; Kuron, 1932). Because of the high stability of these parameters (supposing that constant temperature, e.g., 20°C is ensured for all the measurements), it is reasonable to accept their general use in the future and to regard the others only as rough approximations, which can be estimated as the functions of hygroscopicity using linear relationships.

According to Mados' investigations, the characteristic values of moisture content mentioned previously can be approximated as the functions of Kuron's hygroscopic moisture content by the following equations (Mados, 1939, 1941):

$$w_{fc} = 4w_{hy} + 0.12 \quad ;$$

$$w_{wp} = 4w_{hy} + 0.02 \quad ; \qquad (2\text{-}6)$$

$$w_d = w_{fc} - w_{wp} = 0.10$$

According to these relationships, the usable water content is 10 percent by weight, independently of the type of soil. Excluding very extreme cases, coarse sand and very cohesive clay, this average value can be accepted as a rough approximation.

Similar relationships were determined between Mitsherlich's hygroscopic moisture content and the various water contents used as parameters in soil physics and soil science (Juhász, 1967):

$$w_o = 0.5w_{HY} \quad ; \qquad w_{mol} = 1.7w_{HY} + 0.04 \quad ;$$

$$w_{wp} = 1.44w_{HY} + 0.02 \quad ; \qquad w_L = 2.0w_{HY} + 0.12 \quad ; \qquad (2\text{-}7)$$

$$w_{fc} = 1.7w_{HY} + 0.14 \quad ; \qquad w_p = 1.7w_{HY} + 0.065 \quad .$$

On the basis of Fekete's (1950) measurements, a further relationship can be established to connect Mitsherlich's and

Kuron's hygroscopicities and to construct a transition between Mados' and Juhász' equations in this way:

$$w_{hy} = 0.45\ w_{HY} \quad . \qquad (2\text{-}8)$$

Comparing Mados' and Juhász' equations, their predictions correspond closely. These linear relationships of course cannot substitute direct measurements. They give, however, very useful information on the connection between the parameters characterizing the various states or conditions of soils. Thus it can be recognized that the limit of plasticity and maximum molecular moisture content are very similar parameters, and liquid limit and field capacity are also almost identical characteristics.

5-3 Interpretation of Field Capacity and Gravitational Porosity in a Soil Profile

As it was already mentioned, the water retention capacity of a soil prism depends on the position of the investigated sample. It is necessary to know, therefore, whether the sample to be characterized is separated from its surroundings, or the prism is situated somewhere in a soil profile, as a continuous part. In the latter case, its elevation above the water table also influences the moisture retention capacity. The reliability of this statement can be easily understood. In the case of a sample separated from its surroundings, the effect of gravity is expressed by the weight of the water included in the pores. If the soil prism (the water retention capacity of which is investigated) is a part of a continuous soil profile, the sample is fitted into a space of gravitational potential, the datum (reference level) of which is the water table, where the excess pressure is zero (the gravitational potential is zero at this level). Everywhere in this space the effect of gravity should be expressed referring to this datum, thus above the reference level gravity causes suction proportional to the height of the investigated point above the water table (see Eq. 1-43). Physically, this process can be explained by imagining that the continuous chain of water films compose a closed system in which the pressure on a water particle (in this case negative pressure, i.e., suction because the particle is above the water table) is proportional to the weight of a straight water column between the particle and the reference level, and independent of the form of the container (the form of the chain composed of water films). This is the reason why the point value of field capacity will be larger near the water table, where the suction caused by gravity is smaller, than at a higher point of the profile, and why water retention capacity, measured in a separated sample, can be regarded as a meaningless parameter when investigating the field condition of a complete soil profile.

There is another parameter, generally used in the investigation of the soil-moisture zone, which becomes questionable by accepting the interpretation of field capacity given previously, and determining the latter as a function of the elevation above the water table. This parameter is the free or gravitational porosity (specific yield), generally indicated with the symbol n_s. Classically defined, it is a part of the total porosity where, under the given conditions that water is not bound to the grains by adhesion or by the influence of capillarity, it moves, therefore, freely. Consequently, it was proposed previously that gravitational porosity should be calculated as the difference of total porosity and field capacity, the latter expressed as a volumetric ratio. Accepting the concept of field capacity being the function of the position of the investigated point, it should be stated that the gravitational porosity can also be expressed only as a function of elevation above the water table.

In all types of soil profile a line could be determined which divides the total porosity into two parts: one is field capacity and the other one is gravitational porosity. It is evident that field capacity is relatively larger near the water table and its value decreases going upwards in the profile, while gravitational porosity changes inversely. It is also obvious that the sum of the two parameters at a given elevation is equal to total porosity and constant if porosity does not change in the profile (Fig. 2-3).

There is a further condition, which has to be considered as well, when determining the position of the line dividing the porosity of the soil profile into two parts. If the part covered by field capacity is filled with water and the remaining part is occupied by air, the soil moisture is in a dynamic equilibrium in the profile. There is no water movement in this case, only if an external force (e.g., evaporation from or infiltration to the top soil, change of the position of water table, etc.) induces it. Dynamic equilibrium can develop only when the acting internal forces are balancing each other.

The condition of the development of the dynamic balance can be expressed by equating the effect of gravity with the forces keeping the water in a raised position above the water table against gravity. It was already stated in Eq. 1-43 that the effect of gravity can be expressed as the product of the specific weight of water (γ_v) and the height of the investigated point above the water table (h). The forces retaining

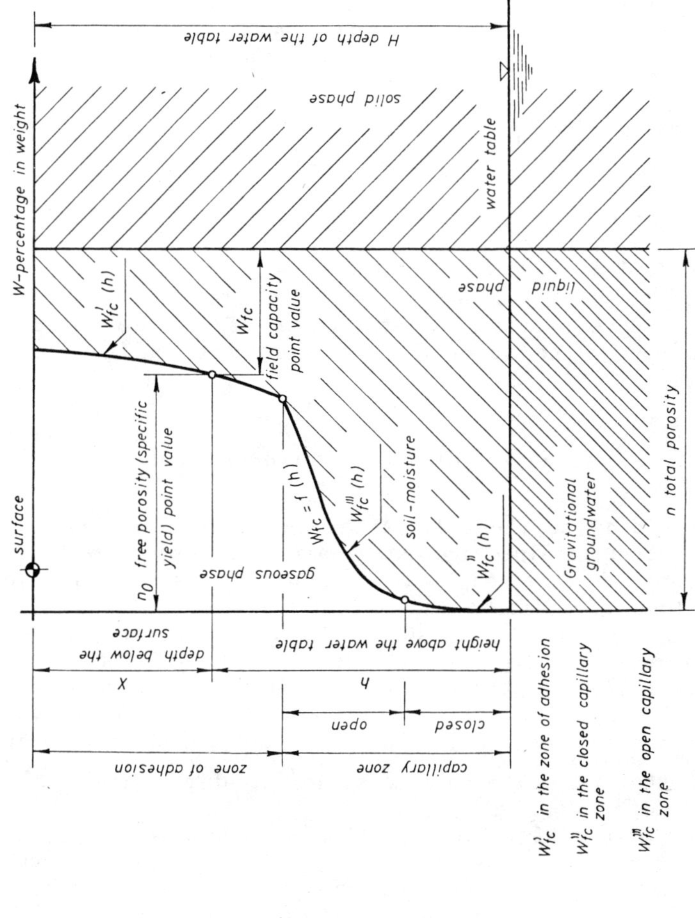

Fig. 2-3. Interpretation of field capacity and free porosity in the soil profile above the water table.

the water against gravity in a position raised above the reference level create a suction (-p negative pressure) on the surface of the water films or within the capillarly filled pores.

The total (Buckingham) potential can be expressed as the sum of gravitational and suction potential, or the difference of the elevation and the suction head ($\psi = -p/\gamma_v$) can be also used instead of the potentials for the dynamic characterization of the profile. The required dynamic balance means that the vertical gradient, the change of the energy head, has to be zero:

$$I = -\frac{d(h-\psi)}{dx} = \frac{d\psi}{dx} + 1 = \frac{d\psi}{dW}\frac{dW}{dx} + 1 = 0 \quad , \qquad (2\text{-}9)$$

where it is considered that $h = H - x$.

It follows from Eq. 2-9 that the suction head has to be equal to the height measured from the water table, and there exists a given vertical distribution of the soil moisture, when there is no vertical water movement in the profile:

$$h = \psi \quad ; \quad \text{and}$$
$$W_o = f(h) = f(\psi) = f(H-x) \quad . \qquad (2\text{-}10)$$

According to the definition, the field capacity is the water content retained against gravity, the development of the condition characterized by this parameter assumes, therefore, the balanced condition of gravity and the internal forces acting against it. This special vertical moisture distribution expressed by Eq. 2-10 should be applied for the characterization of the water retention capacity of the soil profile instead of the point values of field capacity generally used in practice ($W_{fc} = W_o(h)$). The curve representing Eq. 2-10 is called, therefore, the soil-moisture retention curve.

The main active forces differ, however, in the upper and the lower part of the soil profile, extending upwards from the water table to the lower boundary of the cultivated zone. This is the reason why the soil-moisture zone is not a homogeneous field, but the amount and type of saturation belonging to the balanced condition are also different near the soil surface and in the vicinity of the water table. The soil-moisture zone has to be divided, therefore, into subzones distinguished by considering either the dynamic character of the acting forces or the amount of saturation, as it is indicated in Fig. 2-2. It is necessary to note here that the amount of saturation considered as the basis of this classification develops in the soil profile only if there is no special external action (infiltration, evapotranspiration, change of the position of the water table) influencing the equilibrium of the internal forces (gravity, adhesion, and capillarity, the dynamic interpretation of which was given in Part I). The instantaneous vertical distribution of the moisture content in a soil profile may considerably differ, therefore, from that belonging to the balanced condition, but any external action is followed by the redistribution of soil moisture, which process always tends to re-establish the internal balance and the vertical distribution of the moisture content characterizing this condition.

It was already explained that within a given distance from the phreatic surface (water table), or if a separated sample is investigated below a given limit of suction prevailing in the sample, the main force acting against gravity is capillarity which fills a part of the pores completely with water (pores having smaller size than the diameter of the capillary tube in which the capillary rise is equal to the height of the investigated point above the water table or to the suction head applied in the sample). In the remaining larger pores and in all the pores if a part of the soil profile having high position above the water table is investigated, water films develop on the walls of grains due to adhesion (see Fig. 2-1).

Below the probable minimum capillary height (h_{cmin}, the numerical determination of which will be given in Section 2-3 together with that of the other characteristic values of capillary rise) all pores are completely saturated by capillarity (closed capillary zone). As an exception, the amount of saturation may be lower than unity if air bubbles are entrapped in the pores due to the relatively rapid wetting of the sample, but the influence of this phenomenon can be neglected in theoretical investigations because this condition occurs only temporarily.

The development of incomplete saturation in the closed capillary zone and even in the gravitational groundwater zone is demonstrated by Fig. 2-4, which includes two photos. The saturated zones are white in the pictures, while the gray spots indicate the remaining air bubbles in the sample, the darkness being proportional to the air content. When the first picture was taken the water table was at the bottom of the column and a closed capillary zone of about 8 cm was observed. The water table was raised with a velocity of 1 m/hour up to 25 cm. It is clearly indicated by the second picture that the soil remained unsaturated between 9 and 25 cm (Koenig's verbal information, Laboratorium voor Groundmechanica, Delft). It can be stated, therefore, that there is no difference from the point of view of saturation between

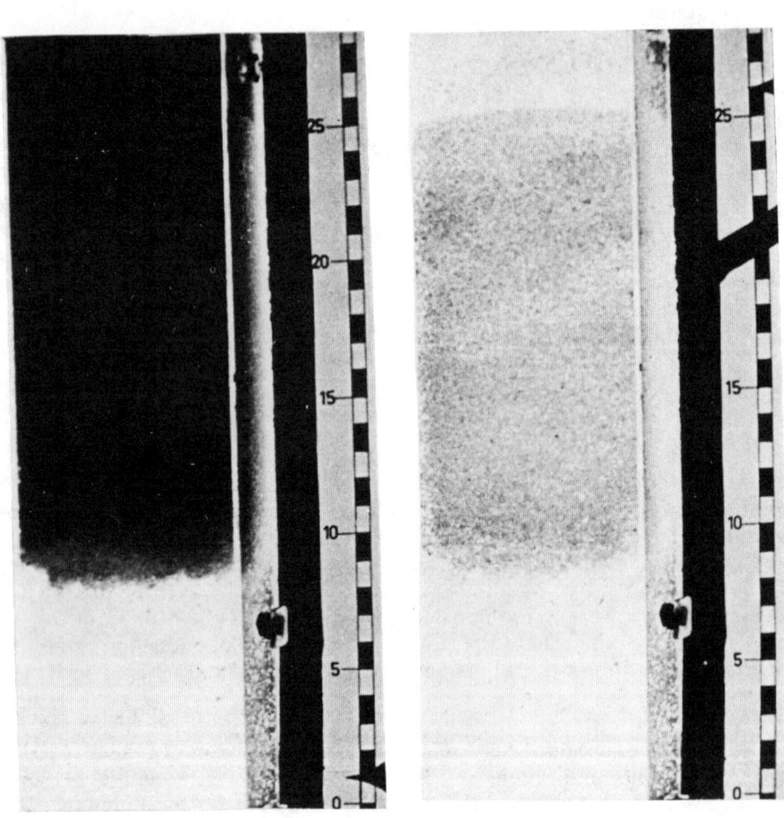

Fig. 2-4. Partial saturation of pores due to rapid rise of the water table.

groundwater zone and the lowest part of the soil-moisture zone. The two zones differ, however, dynamically because the excess pressure has positive sign below the water table, while suction (negative pressure) prevails above it, as it was explained earlier. This is the reason why the entire capillary zone is regarded as a transition form between saturated and unsaturated conditions (partially saturated zone), although its lower part, the closed capillary zone, is completely (or almost completely) saturated, and thus the hydraulic behavior of the solid matrix, its water transporting capacity, is the same here as that in the groundwater zone.

Going upwards in the soil profile, the next subzone is the open capillary zone between the probable maximum and minimum capillary rise ($h_{cmax} > h > h_{cmin}$). The ratio of the completely saturated pores to the total amount of pores decreases gradually here. Water transport develops through the entire section of the smaller pores which are completely saturated, while in the larger pores water movement can develop only through the water films. The two different types of water retention (films and saturated pores) indicate that the influence of both adhesion and capillarity governs the development of the balanced condition in this zone. The height of its upper border (h_{cmax}) is determined at the elevation below which the weight of capillarity is higher than that of adhesion, thus the dynamic classification may combine the entire part of the soil profile between the water table and the probable highest capillary rise as the capillary zone. The amount of saturation changes from complete saturation to the development of water films almost in all pores within this belt, which is regarded, therefore, from the aspect of saturation as a transition zone (or partially saturated zone as it was explained in the previous paragraph).

Finally, above the probable maximum capillary height, practically all pores contain air, and water occurs only in the form of water films. These films are bound to the wall of the solid matrix by adhesion, the zone is called, therefore, the zone of adhesion. Considering the amount of saturation, only this subzone is really unsaturated, although this adjective is used in many cases to indicate the entire zone of soil moisture. Because of the lack of completely saturated pores, water movement can develop only through the chains of the water films forming annular areas within the section of pores.

SECTION 6

EQUIPMENT AND METHODS TO DETERMINE THE PARAMETERS CHARACTERIZING THE HYDROLOGICAL PROCESSES IN THE SOIL-MOISTURE ZONE

6-1 Laboratory Statements

Some aspects of practical importance to hydrological processes developing in the soil-moisture zone, though already mentioned, are repeated here to emphasize and support the statements. Evaporation and transpiration are governed by these processes. The development of surface runoff also depends on the behavior of the layer above the water table. The most important part of the natural recharge of groundwater also percolates through the soil-moisture zone, thus the properties and the instantaneous state of the latter also have basic influence on the groundwater regime. The water consumed by the plants originates from soil moisture and the decreasing amount of the latter is partly recharged by the negative accretion of groundwater. It is this reason why the productivity of agriculture depends also on the storage and seepage occurring below the surface. The migration of salts in the soil is also closely related to the percolation of water because soil moisture is the carrier of the dissolved salts. Thus the accumulation and leaching of salts, and through these processes, the development of soils (their structure, the ratio of exchangeable cations, etc.) depend also on the dynamics of moisture content in the soil profile. Since the properties of the arable layer have a fundamental role among the factors determining agricultural productivity of a given area, the investigation of the soil-moisture conditions related to the salt balance in the soil profile has also high practical importance.

As a consequence of the interrelations listed in the previous paragraph the detailed investigation and the adequate knowledge of the physical processes governing the storage and the movement of water in the unsaturated zone are inevitable, not only for the complete understanding of the hydrological phenomena, but also for better planning as well as efficient operation of water management and agricultural projects. The development of both the theoretical analysis and the practical observation of these processes have been considerably accelerated in the last decades to achieve the goals mentioned.

There are many obstacles, however, hindering the hydrological investigation of the soil-moisture zone. The following problems may be mentioned as the most important consequences of the high number and the versatile character of the processes acting.

The system of the pores transporting and storing the water are determined by the texture and the structure of soils. The texture (grain size, mineralogical character) depends on the parent materials from which soils have developed as well as on the transport, settling, and decomposition of grains. The structure is also influenced by the origin of the layer, but apart from this factor, human activities (cultivation, application of fertilizers), the exchangeable ions associated with the minerals, the roots of plants penetrating into the soil, and even fauna (worms, bugs, insects) have important roles in its development as well. The chemical character may change the dynamic interaction between the grains and water. The suction exerted by the roots is an important action in the maintenance of flow. Although the list of the factors is far from being complete, it clearly demonstrates the diversity of processes governing the movement of soil moisture, and shows that the correct characterization of this zone requires the consideration of physical, chemical and biological parameters.

The fact that the investigated layer is near the surface and thus the water stored and moving here is under the direct influence of many acting atmospheric processes, causes the rapid change of both the stored amount and the flow of soil moisture. It is necessary, therefore, to investigate the time-variant character of the processes and not just the static conditions developing as the result of the forces forming a dynamically balanced system.

The interactions between groundwater and soil moisture are important components of the water balance in the topmost layer. When the correct interpretation of field capacity or water retention capacity was discussed in Part I, the influence of the depth of water table on the dynamically balanced condition of soil moisture was already exhibited, and the change of the prevailing forces, depending on the position of the investigated point related to the water table, was explained. Considering this dependence, it can be stated that neither the investigation of the parameters in the close vicinity of a point, nor the analysis of a limited depth, give sufficient information, but always the whole soil profile, extending from the surface down to the water table, should be characterized.

Almost all the parameters (soil structure and texture, type and growing phase of plants, cultivation) are characterized by high local variance, and the whole system of the occurring processes exhibit, therefore, strong nonhomogeneity. The nonhomogeneous character is further increased by the micro and macro morphology of the land surface. Precipitation is collected in local depressions, increasing the infiltration there. The position of the water table does not follow the relief of the terrain, thus the depth of the former and the influence of the groundwater on the soil-moisture zone also vary from place to place.

The high number of the factors acting complicate the theoretical derivation of mathematical models. Notwithstanding the difficulties, important theoretical results have been already achieved, the detailed discussion of which is given in the following sections. A general characteristic of the models is that the application of many rough approximations was necessary for their derivation, in both cases, either the description of the dynamically balanced state was their purpose or they were concerned with the approximation of the flow conditions. This is the reason why the relationships usually contain empirical parameters, the determination of which require the execution of field measurements. The nonhomogeneity of the layer and that of the developing processes can be explored only by observing the properties of soils and the movement of flow at as many points as feasible. This need emphasizes once again the importance of measurements. Thus, the principle generally accepted in each field of hydrology, according to which any analysis has to be based on a large number of measured and observed data, is valid in the hydrological investigation of soil moisture perhaps to the greatest extent. The topic of this chapter, dealing with the measuring methods of soil-moisture parameters has, therefore, very high practical importance.

When discussing the measuring methods, the first requirement is to survey the parameters, the determination of which is inevitable for understanding the development of the water balance in the soil-moisture zone.

Although the characterization of the texture and structure of the solid matrix has primary importance, the methods serving the description of these properties are not discussed here because they are supposed to be known from soil physics and soil science. Among the numerous parameters (grain size and the shape of grains; distribution of grain, pore, and aggregate sizes; porosity; mineralogical composition) three characteristics are generally used later in the following sections dealing with theoretical investigations, the detailed interpretation of which can be found in Part I, together with the methods suitable to measure or calculate them. These parameters are:

- D_h effective grain diameter (Kozeny, 1954);

- α shape coefficient (Kovács, 1968);

- n porosity (Engelhart, 1960; Kézdi, 1972; Kovács, 1977).

The description of special structures in the cultivated zone requires more detailed investigation, as it was also explained in Part I. Since the purpose of this book is the physical analysis of the hydrological processes occurring below the surface, and since the biological and chemical factors have a dominant role in the cultivated zone, the survey of these special problems is also excluded from the further discussion.

The determination of chemical and biological parameters are not included in this part either because the knowledge of the necessary methods described in soil science, soil chemistry, and agronomy is also assumed as a precondition of the investigation of physical hydrology of the soil-moisture zone. Another reason for this restriction is that these parameters, especially the biological ones, become important mostly in the cultivated zone, the behavior of which is dealt with here only in a rough approximating manner.

The measurement of the instantaneous moisture content (or water content) in the soil profile provides the most important information among the remaining, strictly physical parameters. It was already explained in connection with the definition of water content that two different parameters are generally used to describe this characteristic:

- the weight of water stored in a sample, whose solid matrix has a dry weight of unity (w gravimetric moisture content, or water content related to weight);

- the volume of water included in a unit volume of the investigated solid matrix (W volumetric moisture content, or water content related to volume).

These two parameters are applied in a complimentary way and they are generally supplemented with the determination of the amount of saturation (s) which is the ratio of volumetric water content to porosity. The results of the measurements of water content can be expressed, therefore, either in the ratio of weights or that of volumes, but the parameters necessary to calculate the interrelations between w, W and s, i.e., density, specific weight, and porosity of the solid matrix, have to also be determined in each case.

The other parameter important in the physical characterization of soil moisture is the tension prevailing at the investigated point in the liquid phase of the unsaturated or partially saturated medium. It is known that the water table is not a real surface; its position is defined only by the change of sign of water pressure. Thus, the distinction between groundwater and soil moisture is based on the knowledge of suction. Above the water table, the gradient maintaining the movement of water is composed of two members: change in gravitational potential and that in suction. This is the reason why one point value of suction does not give sufficient information, but the determination of the vertical distribution of suction in the whole profile should be the aim of an investigation.

When investigating the change in storage, the specific yield of the layer has to be known. According to its definition, specific yield (n_s) is the volume of water drained by gravity from a completely saturated sample, related to the bulk volume of the latter. It is supposed that this parameter is equal to the amount of water released or stored in a column of unit horizontal cross section, if the position of the water table changes with a height of unity. There are many uncertainties influencing the determination of this parameter. It is advisable, therefore, to look for simple methods suitable to calculate this parameter from soil physics data instead of the application of more complicated measurements. Since the reliability of the simple estimation is not any worse than that of measuring methods, a mathematical model will be proposed for the determination of specific yield. Some theoretical concepts (soil moisture retention curve derived on the basis of dynamic conditions, hysteresis influencing the development of the balanced moisture distribution) are applied for the determination of the model, although the detailed explanation of the relationships used here as starting conditions will only be given in the following parts.

The characterization of the water balance in the soil-moisture zone requires the observation of its components separately. Lysimeters are the equipment most commonly used for this purpose. It is necessary to recognize, however, that the data measured in a lysimeter generally represent the combined actions of a few factors. The grouping of the factors depends on the structure and operation of the lysimeter and on the method of observation. This is the reason why the data themselves are influenced by the measuring methods and they are comparable only if the observations were executed under identical conditions.

There are some components of the water balance which can be directly measured, or calculated, on the basis of more simple observations than those performed by lysimeters. The infiltration through the soil surface may be mentioned among them which characterizes the positive component (input) of the upper boundary condition of the system.

Considering the aspects explained in this introduction, the topics to be discussed in detailed form can be summarized as follows:

- equipment and methods suitable to measure moisture content;

- determination of the tension prevailing in the liquid phase of the unsaturated and partially saturated zone of the soil profile;

- methods to measure specific yield or to estimate it numerically;

- application of lysimeters to characterize the water balance of the soil-moisture zone;

- methods to investigate infiltration;

- areal characterization of the distribution of soil moisture.

6-2 Methods for Measuring the Moisture Content

The determination of the amount of soil moiture is one of the most difficult measurements required in the field of hydrology. Not only point values have to be measured, but a series of point's should be observed to facilitate the characterization of the vertical distribution of water content. The change of the parameter in time also provides important information, the continuous recording of the data is desirable or at least the regular repetition of measurements should be ensured. In special cases, e.g., when the observation is intended to be used for automatic control of irrigation, the possibility of telemetering is also necessary. Naturally the reliability, the simplicity, and the low operational cost are also requirements.

The versatile character of the requirements listed may be mentioned as one of the main reasons, apart from the basic difficulty of soil-moisture measurements, why there are many different methods developed in the last decades to carry out field measurements. After evaluating the advantages of the various methods and the obstacles hindering their application, it is not yet feasible to select one that is superior to the others, but the choice has to be always made, according to the character of the problem, the solution of which is the purpose of the collection of data.

Gravimetric method. The oldest method generally applied to measure the water content of soil is the gravimetric method. There are references in the literature indicating that it was already applied in the mid-19th century (Whitney, 1894; Russell, 1950). The measurement is executed by following exactly the definition of water content. A sample is taken from the soil profile, and after measuring its weight corresponding to the instantaneous condition of wetness (which should be characterized), it is placed into an oven. The sample is kept there, at a temperature of 105°C, until there is no decrease observed in its weight. Measuring the weight of the solid matrix of the dried sample, the gravimetric water content can be calculated as the difference between the wet and dry weights related to the dry weight (see Eq. 2-2).

If the volume of the sample and that of the solid matrix are also determined, the average specific weight of the grains, as well as the porosity and the bulk density, of the sample can be calculated. The supplementary measurements provide us with the transforming factors necessary to determine the volumetric water content and the degree of saturation.

When investigating field conditions, samples have to be taken from the layers. Generally, light drilling equipments (sampling augers) are used for sampling. Care must be taken to achieve homogeneous samples free from roots, organic matter, and stones. Seldom are all these conditions met. The technique and equipment used for the collection of samples should ensure that the samples do not lose or gain moisture during sampling and transportation. Special problems arise when samples are taken from a dry layer after penetrating a wet one. A further requirement is that the natural porosity should not be disturbed or the modification should at least be negligible. Sampling tubes (core barrels) are generally applied to satisfy the listed requirements to the greatest possible extent. Naturally all the difficulties associated with core sampling have to be overcome (coarse-grained dry samples may slide out the end of the core barrel as it is withdrawn; it is difficult to drive the core barrel in dry, hard, cohesive layers, etc.).

The characterization of the vertical moisture distribution can be solved by using the gravimetric method if a long continuous core is taken from the borehole. It is divided into short samples and the water content of the separated stretches are determined individually. There are special sampling equipment and core barrels developed for collecting such long cores (Johnson, 1962; Hvorslev, 1949).

One of the disadvantages of the gravimetric method is the time and effort required for the determination of the water content. The collection of the samples, especially from depths greater than a few decimeters, is time consuming. The drying and weighing of many samples also mean a great burden for the laboratory.

The most difficult problem, however, is caused by the fact that the determination of water content, in this way, requires samples taken from the layer. This procedure modifies the surface conditions of the investigated area and disturbs the natural processes, e.g., infiltration increases through the bore hole opened to take the sample. The measurement cannot be repeated at the same point, thus the continuous recording and even the regular repetition of observations are excluded.

The gravimetric method is considered to be the most accurate way of soil-moisture measuring. This is the reason why it is generally used to check the reliability of other methods and to construct calibration curves needed for their application, although the method itself is hardly used nowadays for field measurements.

Measuring electrical resistivity. This method also has a long tradition. Its first application was reported at the end of the last century (Whitney et al., 1897). The method is based on the relationship existing between the water content of the soil and the electrical resistivity of the latter. The moisture content is determined, therefore, in an indirect way by measuring another physical property of the medium and using a previously determined relationship (calibration curve) to transform the measured values into the desired parameters.

The equipment is composed of two electrodes. The current developing through the circuit, formed by the porous medium between the electrodes, is measured. The electrical resistivity, changing with water content of the medium, is calculated from the known values of voltage and current (or resistivity is read directly on the measuring device), and this parameter is converted to a moisture-content value by using calibration curves or tables. Considering the required safety, low voltage is generally used to maintain the current. The application of direct current is not advisable because the polarization developing on the poles may disturb the measurements (Vilkner, 1959; Sprigade, 1969; Linder, 1970a). The frequency of the alternating current may be different, its value is generally determined according to the internal structure of the measuring equipment.

At an early stage, the electrodes were buried directly into the soil at a given distance from each other. The system was, however, very sensitive to changes in temperature. The results were influenced to a great extent by the texture, structure, and electrolyte content of the soil as well. Considering these influencing factors, it was necessary that the equipment should have been calibrated on the spot. The changes occurring in the chemical composition of soil moisture influenced by infiltration and evaporation could not be taken into account even by calibration; the process caused, therefore, high uncertainties. The transitional resistivity between the soil and the electrodes also lowered the reliability of the method (Crony et al., 1951; Bushmann, 1956; Vilkner, 1959; Vetterlein, 1961a).

The electrodes were covered by a porous block to eliminate the source of errors previously listed. Plaster of paris was used at the beginning (gypsum blocks). The moisture content of the porous block depends on the water amount stored in the soil contacted by the block because a balance develops between the tensions prevailing at the two opposite sides of the contacting surface. Thus, the moisture content of the block determined by measuring electrical resistance inside the block can be used to observe the moisture content of the soil.

The change in the electrolyte content of soil moisture is smaller inside the block, than in the soil, where water moves freely. It was found that the influence of the chemical composition of the water hardly influenced the results measured with gypsum blocks (Höschele, 1957; Schendel, 1962; Lindner, 1970a). The effect of the change in temperature is not negligible even in the case of gypsum block either. It is estimated that 2-3 percent of the gravimetric water content is induced by one degree centigrade (Vilkner, 1959; Vetterlein, 1961b). It is necessary, therefore, to measure the temperature near the block and correct the resistivity measurements by considering its influence. A formula suitable for such correction was published by Lindner (1970a). Another solution may be the application of special devices to compensate for the effect of the change in temperature (Bouvoucos, 1949; Rentschler, 1956; Hübener, 1970).

Since the moisture content of the soil is measured by double transmissions (i.e., the water content to be measured is supplemented by that of the block through the balanced condition of tensions and the electrical resistivity of the block, depending on whether its moisture content is measured directly) and one of them (i.e., the development of balanced tension) needs considerable time to transfer the impulse from the soil into the block, the change in soil moisture is followed by the measurements only with a non-negligible time lag. The reaction time can be decreased by changing the porosity of the block (Höschele, 1957). A similar result can be achieved by adding nylon fibers (Bonyoucos, 1949; Kurosaki,

1954) or fiberglass (Colman, 1950; Youker and Dreibelbis, 1951) to the material of the blocks. It was found, however, that the influence of the electrolyte concentration increased with decreasing reaction time (Colman and Hendrix, 1949).

The accuracy claimed by the inventors of the various blocks is at best one percent by weight. The measuring range of the blocks is limited because the relationship between resistance and moisture content is not linear, the change in resistance belonging to a unit change in moisture content is very small above a given limit. This limit was found to be 10 percent in gravimetric water content in the case of silty sand while in sand it was still lower (about 6 percent) (Lindner, 1979b). The advantage of the method is that the observations can be directly transformed into electrical signals, thus they can continuously record and the telemetering of the signals can be easily solved as well. More blocks have to be placed above each other when the task is the determination of moisture distribution in the profile.

Method based on the measurement of a dielectric constant. When a porous medium is placed between the armatures of a condenser, the dielectric constant of the latter changes depending on the moisture content of the medium. This principle is also used for measuring the water content of soils (Borguis, 1942; Marel, 1959; Rame and Whinner, 1960). It was found, however, that the dielectric adsorption disturbed the measurements (Sprigade, 1969). Calibration curves have to be determined for the application of such equipment. The measuring error was found to be as high as 7 percent, but the results were considerably worse under field conditions than those achieved in laboratories.

According to recent investigations (Gálfi, 1976) the low accuracy is caused by the fact that the total resistivity of the system is composed of two parts, i.e.:

- reactance (reactive or capacitive resistivity) which is the inverse value of capacity;

- ohmic resistivity due to the current between the armatures of the condenser (capacitor).

The measured dielectric constant can be accepted only as a relible parameter of the system if the capacitive resistivity surpasses with some magnitude the ohmic value. To meet this condition, the frequency of the applied alternating current has to be very high (e.g., in the case of dry soil, 10 MHz frequency is required for the measurements to lower the ohmic resistance below one percentage of reactance, and this value increases rapidly with the increasing moisture content). At the same time, the sensitivity of the system is considerably lowered in the important range, from a practical point of view, by the application of high frequency.

The insulation of the armatures with a very thin layer may be proposed to avoid the development of current or more precisely to decrease it. It was found that the ohmic resistivity became negligible when the armatures were covered by a shellac layer of 0.1 mm. Relatively high sensitivity was achieved by such equipment between W = 0.005 and W = 0.10 volumetric water content by using 1 KHz frequency. The lower limit of the measuring range can be extended by applying a lower frequency, while a higher range of water content became measurable if higher frequency is used.

The introduction of further variables (thickness and dielectric constant of the insulating layer) makes the system more complicated. Theoretically, however, the problem is completely solved, only further development is needed for the practical solution of the devices and for their calibration. The mathematical models, derived theoretically and describing the process developing in the sensor, provide assistance for the calibration. According to the control measurements executed in the laboratory, the accuracy achieved with condensers surpasses the accuracy of the most up-to-date sensors based on the determination of ohmic resistance, especially if the measurements are repeated by applying two or more different frequency values. A great advantage of the use of condensers contrary to the gypsum blocks is that the effects of temperature and dissolved salts do not considerably influence the dielectric constant, at least within the probable range of the scattering of these parameters. The continuous recording and telemetering can be similarly solved as it was mentioned in connection with resistivity methods. The application of a series of equipment is necessary to characterize the vertical moisture distribution. It is expected that the measurement of a dielectric constant will develop to be the best method to observe the water content in the form of electric signals, if the field control under execution now shows similarly good results as the laboratory calibrations.

Measuring of wave propagation. It was found that the propagation velocity of both electromagnetic waves (Westhuizen, 1967) and acoustic waves (Golerik et al., 1969) is proportional to the moisture content of the soil. The advantages of these methods in measuring soil moisture are:

- the measuring field is large, and thus average information can be gained for a great volume of the investigated layer;

- the measuring range is also large because the relationship between moisture content and velocity is almost linear.

The application of the methods is hindered by the fact that the density of the whole layer, penetrated by the waves between the transmitter and the sensor, has to be known. The high cost of the equipment and their complicated structure may be mentioned as further obstacles of the wide application of these methods, which is not yet fully developed. The continuation of their improvement indicate a promising way to achieve new, practical means of observation.

Heat-diffusion method. One type of equipment is based on the principle that the heat conductivity of the soil changes with its moisture content. An electrically heated source is placed in the soil and the rise of temperature caused by the heater is recorded. Wet soil will conduct the heat rapidly away from the source and thus a smaller rise in temperature will be observable than in dry soil.

Three types of the equipment have been developed:

- both the heater and the sensor are imbedded into a porous block;

- the electrical elements are in direct contact with the soil;

- a modified type of the latter, where the heat source and the thermometer are separated by a portion of the soil to be tested (thermal-conductivity cell).

Johnson (1962) has given a detailed summary of research works performed with such equipment. He has also published a very complete bibliography of the publications dealing with this topic. According to his evaluation, the heat-diffusion cells do not give satisfactory results in the range of high moisture content. The shrinkage of soils and the change of moisture content caused by the heat in the vicinity of the source may disturb the observation. The thermal-conductivity cell has been the most satisfactory among the three types, but needs further development.

Absorption of infrared rays. The absorption line of water, in the infrared range of the spectrum, is well-defined. The comparison of results gained by using different wavelength facilitates the quantitative determination of the water content (Becker, 1969). The advantages of the method are:

- it has a high degree of accuracy;

- it is independent of temperature and the components of soil.

The method is costly and is not suited for the description of vertical moisture distribution. Using airborne pictures, large areas can be characterized, thus it is the best method to investigate territorial nonhomogeneity. The evaluation can be considerably simplified by combining areal photos and field measurements because the qualitative evaluation of the pictures gives sufficient results, in this case, to determine the change of moisture content between the points of field observation.

Radioactive methods for borehole logging in general. The radiation emitted by unstable isotopes during their radioactive decay consists of alpha particles, positive and negative beta particles, and gamma photons. Among these radiations, only the measurement of gamma rays is applied in borehole geophysics. Neutrons produced by an artificial source are also used for well logging.

The various types of nuclear logging are:

- natural gamma logs that record the amount of gamma radiation emitted by the different rocks under natural conditions;

- in the case of gamma-gamma logging, a source emitting gamma rays is applied and the intensity of radiation reflected by or penetrating through the medium to be measured is recorded;

- neutron logging is based on the measurements of the change in the energy level of the neutron emitted with high energy content by an artificially introduced source. Neutron-gamma log counts the gamma photons produced by neutron reactions. Other probes indicate the number of reflected neutrons, the energy of which was reduced to a given level by collisions. Thus the neutron-epithermal neutron tool gives the number of neutrons reflected, and having energy between 0.1 and 100 electron volts, while the neutron-thermal neutron probes count those having lower energy level than 0.024 electron volt (thermalized neutrons).

All types of nuclear logging can be executed in the form of point measurements or by using probes moving in the borehole with continuous speed. In the latter case, the time constant of the sensor (the length of the period within which the sensor averages the number of pulses following each other in random intervals) has to be harmonized with the speed of the probe.

Any probe applied in radioactive methods has to be calibrated individually, and the probes have to always be used in boreholes having similar geometry and casing because any change of the systems may alter the output. The calibration has to be frequently checked because the equipment may have a tendency to drift.

There is always some radioactive hazard when nuclear probes are used or transported. There are regulations, therefore, in every country ordering that only qualified personnel (properly trained and licensed) may use such equipment. Other precautions are also specified, generally in connection with the handling of nuclear probes.

An excellent summary and evaluation was published by Keys and Maclary (1971) about the application of nuclear borehole geophysics to water resources investigation. It is not necessary, therefore, to deal with this topic in general. Only two methods which can be successfully applied to determine the water content in the soil-moisture zone will be discussed in detail. These are:

- gamma-gamma logging;

- neutron probe counting the reflected thermal neutrons.

Gamma-gamma logging. As it was explained in the general description of nuclear probes, the equipment consists of a source emitting gamma radiation and a sensor recording the intensity of the photons reflected by or penetrating through the material to be tested. If the source and the sensor are combined in one probe separated only by a shield impermeable to the radiation (generally called a spacer and constructed, e.g., from lead), the reflected rays are measured and one borehole is adequate for the operation. When applying the other technique, two boreholes have to be drilled near to each other for field measurements. The source is moving in one hole and the detector in the adjacent one. The synchronous movement of the two devices has to be ensured. The system measures the rays penetrating through the soil between the two boreholes, and thus the volume of material investigated is increased and the influence of the geometry of the hole is lowered.

Gamma radiation is absorbed and scattered by the materials through which it travels. The energy loss is approximately proportional to the electron density of the material penetrated. This relationship is slightly influenced by the chemical nature of the medium, e.g., in the case of salt and gypsum some correction has to be made, but it has no practical importance when investigating the soil-moisture zone. Since the proportionality between electron density and bulk density of a material can also be assumed, bulk density and porosity of the solid matrix can be determined from the output of the probe, if the average specific weight of the grains and that of the material filling the pores are known.

For certain source-to-detector spacings, and over a limited range of density, a linear relationship is obtained when bulk density is plotted against the logarithm of the count rate. This fact makes the determination of the calibration curves easier. The measuring error in bulk density of a gamma-gamma probe, combining the gamma source and the sensor, is estimated to be 0.03-0.04 g/cm^3. The error in porosity calculated from the measured values of reflected gamma radiation depends on the accuracy of the other data used for transforming the data, i.e., average specific weight of grains and that of the material filling up the pores. This is the reason why porosity values are more reliable if the pores are saturated by a homogeneous medium (by water below the water table or by air in completely dry layer).

Considering the character of the process measured by gamma-gamma logging, it can be stated that the method can be used for measuring moisture content only in an indirect way. The parameter determined by the probe is bulk density. For this reason, only the change in bulk density caused by the water content can be observed and the change of moisture content calculated (or if bulk density of the dry solid matrix is known the absolute values of the water content can be estimated from the difference of measured bulk density and a known basic value).

The indirect way of the determination of moisture content influences the accuracy of the measurements to a great extent. The change in bulk density caused by the change of water content is relatively small compared to the total bulk density of the solid matrix. It was reported that an accuracy of 0.01 cm^3/cm^3 was achieved in the absolute value of volumetric moisture content under laboratory conditions (Vachaud et al., 1970). An Americium-241 source and a detector moved separately but synchronously along two opposite sides of a vertical column composed of fine sand to be tested. The diameter of the column was 5.0 cm. Two investigations were carried-out, producing a measured bulk density of ρ_d = 1.43 g/cm^3 and ρ_d = 1.26 g/cm^3, respectively. The counting period was 40 seconds at one point and the average number of pulses per second was calculated (N-cps). The graphical representation of the measured data (Fig. 2-5) indicates,

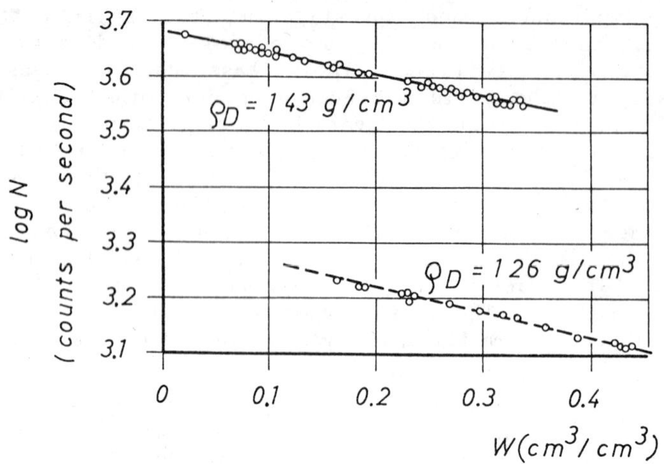

Fig. 2-5. Relationship between counts and water content measured by gamma-gamma probe.

however, a low sensitivity. The change in the number of counts is only 900 and 500 cps respectively, while the change in volumetric moisture content is 0.3 cm^3/cm^3. The accuracy is worse under field conditions and the use of the combined probe further lowers the reliability. Thus, the application of the method cannot be advised for field measurements. It has, however, an important role giving supplementary information needed for the application of neutron probes, the results of which have to be corrected according to the bulk density (or porosity) of the medium.

Thermal neutron logging. The basic principle of the method was already explained. The source emitting the high energy content neutrons and the detector counting the reflected thermal neutrons are generally combined in one probe separated only by a lead shield. The decrease of the energy level is relatively the highest if the colliding particles have similar masses. Thus, the slowdown is caused primarily by collisions of neutrons with hydrogen atoms, both particles having a mass of unity. The output of the probe can be approximated, therefore, as a function of the hydrogen content of the medium penetrated. Most of the hydrogen atoms are present in loose clastic sediments in the form of water, consequently the method is primarily suitable to determine the water content of the soil.

The vertical distribution of moisture content can be determined by drilling and casing boreholes in which the probe can be moved vertically to measure the water content from the surface down to the water table. Considering the fluctuation of the water table, the hole has to penetrate into the groundwater space as well. The access tube should be sealed to exclude the inflow of water below the water table, and to ensure, in this way, the same measuring conditions in the tube.

The access tube stabilizes the place of measurements as well, and thus the repetition of observations at regular intervals does not cause any difficulty. The tube may disturb, however, the accuracy and the sensitivity of the method. Detailed research was carried out, therefore, to determine the most suitable casing of the boreholes used for neutron logging (Teasdale and Johnson, 1970). The influence of the material and that of the diameter of the access tubes was tested (Fig. 2-6). The aluminum tube was found to be superior to plastic and sensitivity increased with a decrease in diameter. Air space remained free between the layer drilled and the access tube and caused uncertainties and lowered the sensitivity of the measurements. Gravel packing placed behind the casing improved these conditions, but the best result was achieved by pressing the tube into the hole, ensuring tight fitting (Fig. 2-7). The comparison of the volumetric moisture-content values determined by gravimetric method and neutron probes using casing packed with gravel shows the reliability of the neutron method (Fig. 2-8). Results, even better than those represented in the figure, can be expected if the access tube fits tightly in the hole. The conclusions of the research indicate that:

- thin aluminum tubes should be used for casing having a diameter as small as possible (the diameter depends on the size of the probe);

- the access tube should be pressed into the hole to achieve tight fitting, or the space between the wall of the borehole and casing should be packed, if the tight fitting cannot be ensured by pressing because of the great depth of the hole;

- the neutron probes should be calibrated in tubing of identical composition, diameter, and wall thickness as that used in the field.

The moisture probes utilize a relatively small source and the spacing between the source and the detector is small. The result of this arrangement is that the count rate increases with higher moisture content. There are, however, materials disturbing the proportionality between moisture content and counting rate. Since the probes measure the amount of hydrogen atoms, hydrogen present in forms other than water may

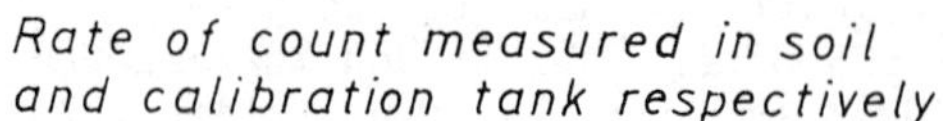

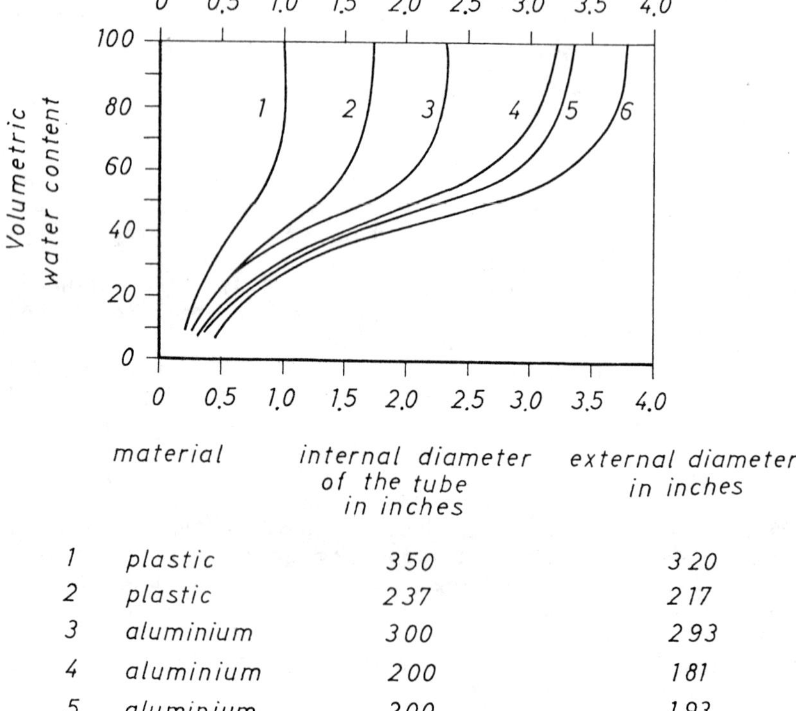

	material	internal diameter of the tube in inches	external diameter in inches
1	plastic	350	3 20
2	plastic	237	217
3	aluminium	300	293
4	aluminium	200	181
5	aluminium	200	193
6	aluminium	1 63	155

Fig. 2-6. Influence of the material and the diameter of the protective tube on the results of neutron logging.

cause uncertainties. Organic matters (roots), hydrocarbons, and chemically bound water may result in anomalous values.

Most of the observations are executed in the form of point measurements, but continuous logging is also applicable. Considering the relative small depth of the access tubes, the point measurements do not require too great an effort. It is necessary, however, to determine the optimum spacing between the measuring points necessary to achieve reliable characterization of the vertical moisture distribution. Kristensen's research (1973) gives guidance in this problem. In the table published by him, the moisture contents measured at 10 cm intervals below a barley crop, at four different points in time, are listed (Table 2-1). The water content values summarized for the upper layer of one meter are indicated in the

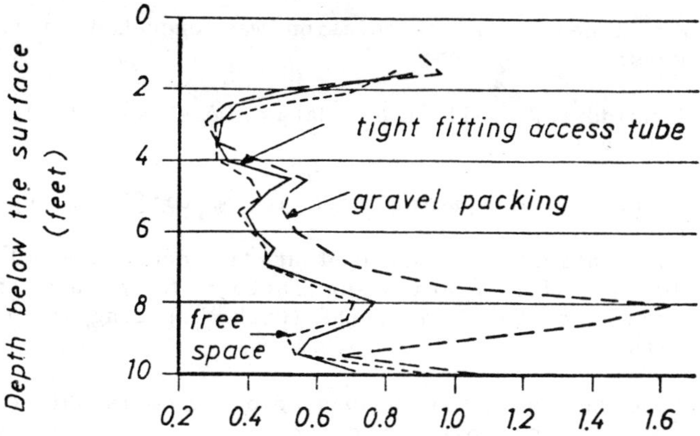

Fig. 2-7. Change of counts depending on the development of casing.

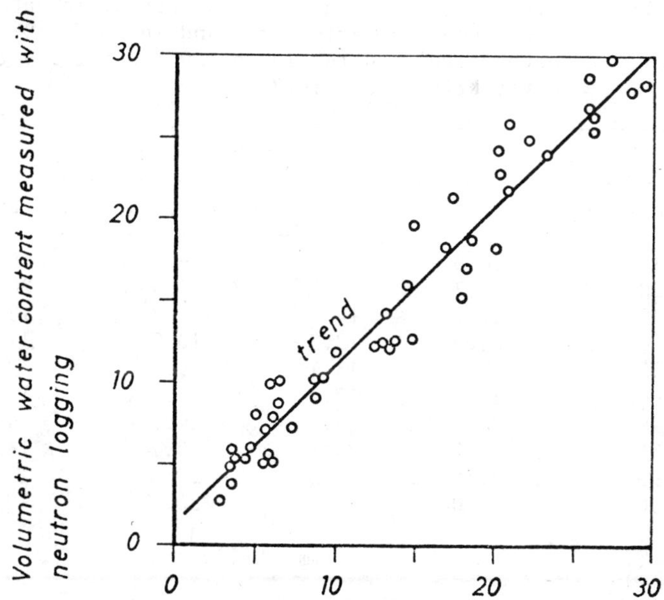

Fig. 2-8. Relationship between water content values determined by neutron probe and gravimetric method, respectively.

lowest three lines. The calculation was executed in three different ways:

1. Considering all the data measured at 10 cm intervals;

2. Using 20 cm spacing from 10 cm to 90 cm depths;

3. Calculating data measured at the depths of 10 and 100 cm with a weighting factor of 1/2 and those from 20 cm to 80 cm depths (with a spacing of 20 cm) with 1.

The comparison has shown that a spacing of 20 cm is sufficient in most cases if the moisture distribution in a relatively homogeneous soil is measured. A regular decrease can be observed only at the top of the profile due to the fact that the measuring radius, starting from the topmost point, extends above the surface. This error can be decreased by covering the surface with a sheet, thus reflecting the thermal neutrons (e.g., plastic).

Table 2-1. Moisture content in percentage at different times (1969) under a barley crop and in mm for 0-100 cm depth calculated for 10 and 20 cm soil layers (after Kristensen, 1973).

	Percent by Volume			
Depth (cm)	April 18	June 23	August 11	August 18
10	20.6	16.9	9.1	16.7
20	26.7	17.5	13.4	20.0
30	27.6	14.9	11.7	15.4
40	25.9	12.3	10.0	11.3
50	25.0	13.1	9.5	9.9
60	24.7	17.0	11.9	11.9
70	24.8	20.6	14.7	14.8
80	26.9	25.8	18.7	18.1
90	27.6	27.5	20.0	18.9
100	28.0	28.7	20.3	19.6
		mm (0-100 cm)		
I	254	188	134	155
II	252	186	130	151
III	257	191	137	159

The volume of influence, or radius of investigation, is a function of both the spacing between the source and the detector, and the type of material within the volume in question.

The radius decreases with an increase in the amount of saturation because of the reduced range of neutrons prior to capture. The radius of the sphere from which about 90 percent of the responses originate is estimated to vary between 60 and 15 cm (in dry and wet sand, respectively).

The zone of influence and the calibration curve also change with the bulk density (porosity) of the layer being tested. It is necessary, therefore, to know the porosity of the medium around the access tube, and check this property from time to time. A gamma-gamma log may be used with a good result. Where a regularly observed network was developed to record the change of moisture content at more points, it was advisable to supplement the set of neutron probes with a gamma-gamma probe having the same size. Thus, the same access tubes can be used to control the changes in porosity.

Almost all the requirements listed in the introduction in connection with the measuring methods (accuracy, reliability, regular repetition, determination of vertical moisture content, consideration of secondary effects) can be met by applying neutron probes, only continuous recording and telemetering cause problems. The automatic operation together with tele-recording are already solved, technically, but the equipment is too costly to be installed in each measuring tube. Leaving the probe in the field is dangerous too, because of the radioactive hazards.

Comparison of various methods. As a conclusion of the evaluation of the various methods developed for measuring soil moisture, their advantages and the aspects hindering their application are listed in Table 2-2. The classical gravimetric method is used nowadays only for calibration of other equipments. Among the electrically operated equipments, that measuring electrical resistivity is the most commonly used. Although the accuracy of this method is not quite satisfactory, because secondary effects may disturb the meaurements, there are tasks (continuous recording, telemetering), the fulfillment of which can be best achieved by applying improved gypsum blocks. The other electrical methods and those based on wave propagation need further development. It is likely that heat diffusion, as a basic principle of measurement, can be excluded from practical use. Neutron probes counting the reflected thermal neutrons are the best measuring devices. This method meets almost all requirements needed to be satisfied by moisture measurements. The other types of nuclear logging, i.e., gamma-gamma, gives satisfactory accuracy only under laboratory conditions. The gamma-gamma probe can be used in the field only as an equipment supplementing the neutron probe to check the bulk density of the medium around the access tubes from time to time.

Table 2-2. Comparison of the various methods applicable to measure moisture content.

Method	Advantages	Disadvantages	Remarks
gravimetric method	high accuracy	sample has to be taken, therefore: continuous recording excluded; measurement cannot be repeated; conditions of infiltration are disturbed; time consuming and laborious.	advisable to use only for calibration of other methods
electrical resistivity	electrical output makes recording, automation, and telemetry possible; simple instruments.	results depend on temperature and chemical composition of of water; measuring range is limited.	generally applied for systems; the limited measuring range and the secondary effects should be considered.
measuring dielectric constant	electrical output makes recording, automation, and telemetry possible; simple instruments; independent of temperature and chemical composition of water.	considerable measuring errors due to low sensitivity; limited measuring range.	promising possibility to develop a simple measuring device, the signal of which can be continuously recorded automatically and by telemetering.
measuring wave propagation	measuring field is large; measuring range is not limited.	density of the field has to be known; complicated structure and high cost of equipment.	further development is required before practical application
heat-diffusion method		results are not satisfactory in the range of high moisture content; heat causes the shrinkage of soil and the change of moisture content, disturbing the observations.	practical application is not advisable

Table 2-2. (continued)

absorption of infrared rays	high accuracy; independent of secondary effects.	high costs; quantitative evaluation is complicated.	applicable for characterization of large areas by using airborne photos and accepting qualitative evaluation (combination with field records)
gamma-gamma logging	high accuracy; unlimited measuring range.	low sensitivity; radioactive hazards; high costs.	applicable for the determination of soil density as supplementation of neutron probes; independent application is advisable only for laboratory measurements.
neutron probe	high accuracy; good for repetition of measurements; provides the complete vertical distribution in the profile.	high costs; radioactive hazards.	presently the best field method, but automation and telemetering is not solved

6-3 Tensiometers and Their Application to Characterize Moisture Flux

The equipment generally applied to measure the suction prevailing in the liquid phase, at various points, in the soil-moisture zone is composed of a head contacting the soil and a measuring device. The head is a porous point or cup (usually ceramic) which is connected through a tube to a manometer. The most frequently applied types of the latter are the Bourdon-tube vacuum gauge, and the mercury manometer. The tensiometer may also be attached to a pressure recorder (Richards and Gardner, 1936).

The basic principle of the operation of tensiometers is the development of a balanced condition between the tension developing in the water contained by the porous sensor of the equipment and that prevailing in the soil moisture contained by the soil connected by the sensor. When the moisture content of the soil is lowered, its tension increases and water is drained from the porous cup, thus increasing the suction in it. The wetting of soil initiates an opposite process, the suction is lowered in the sensor. The change of suction in the measuring head is indicated or registered by the manometer.

The method of operation of tensiometers determines the limit of their operation as well. The development of an absolute vacuum has to be avoided, the suction being measured should always be lower than the atmospheric pressure. Considering also some uncertainties, the upper measuring limit of tensiometers can be characterized with a suction, the absolute value of which is 0.9 atm, i.e., about 700 mm Hg.

Anomalies in measurements are caused by air leaking through the porous head of the sensor and disturbing the pressure conditions registered by the manometer. It is necessary, therefore, that the porous material applied in the sensor should have a high air-entry pressure. This is the reason why ceramics with evenly distributed small pores are generally used for this purpose. Since the possibility of leakage of air cannot be completely excluded, the equipment has to be checked from time to time and the air has to be exhausted from them if necessary. The application of media having fine pores as a contacting sensor extends the reaction time of the tensiometer. The time lag between the change of moisture content in the soil and the development of the balanced tension in the porous cap may be considerable, which has to be considered when making readings on the measuring device.

The size of the sensor also influences the records. A large head contacts not only the surface of water films but penetrates at some points into the adsorbed water shell as well, where higher suction prevails. The equipment averages the suctions along the contacting surface, thus the measurement does not provide a real point value, but characterizes a larger surface. Attempts are made, therefore, to decrease the size of the sensors, e.g., development of porous points, and to lower the uncertainties in this way. The development of small cavities between the soil and the sensor, disrupting the continuity of the system, is also more likely in the case of large porous caps. The application of small sensors is advantageous from this aspect as well. Great attention should be paid, in any case, to avoid the existence of such cavities when installing tensiometers.

The tensiometers are mentioned, in most publications, as one of the equipment suitable to determine the moisture content of soils. The relationship between water content and the tension prevailing in the liquid phase of the soil has to be known if tensiometers are intended to be used for this purpose. It will be demonstrated later in the framework of theoretical investigations that there is no single-valued relationship between the two variables, i.e., suction and moisture content. Thus, the reliability of moisture values determined in this way is very low and the method cannot be proposed, therefore, to be used as one of the alternatives to measure moisture content for practice.

In the author's opinion, the practical users of data (especially in agriculture) do not really need the knowledge of moisture content but that of tension. The start of wilting does not depend on the amount of water stored in the pores, but the deterioration of plants commences when the tension of soil moisture surpasses the highest suction that the roots are able to exert. Tensiometers provide us with more important information than the various methods measuring moisture content. Automatically governed irrigation systems have to be operated by using tensiometer signals instead of water content values telemetered by the observation network.

The continuous recording or the regularly repeated measurements of tension also characterize the dynamics of soil moisture. The records can be compared with meteorological data (air temperature; moisture content of air; wind velocity; precipitation), the influence of the latter on soil-moisture conditions can also be demonstrated (Fig. 2-9). Using more tensiometers at the same depth, the scattering caused partly by measuring errors and partly by the nonhomogeneity of the layer can also be analyzed. In the case represented in Fig. 2-9, four tensiometers were installed within an area of

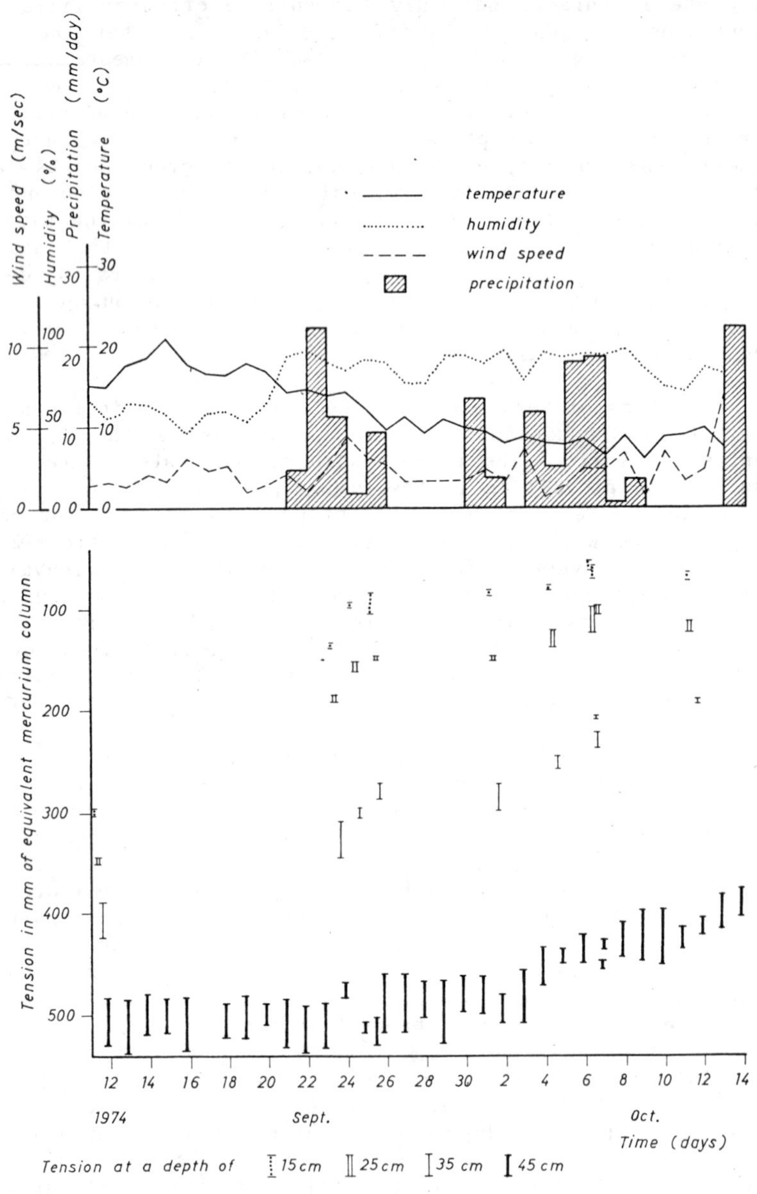

Fig. 2-9. The change in time of tension measured at different depths and its relationship with meteorological parameters.

$6x10^5 m^2$ at a depth of 45 cm. The observed scattering was small, testifying that the tension conditions can be well characterized on the basis of relatively few measuring points.

Most of the requirements mentioned in connection with the equipment measuring moisture content (accuracy and reliability; continuous recording; telemetering) can be already met in the case of tensiometers. For the characterization of tension distribution in a vertical profile, more tensiometers have to be installed above each other. On the basis of automatic records of such a data set, the changes in soil-moisture conditions can be well recognized, as it is demonstrated in Fig. 2-10, constructed by Vachaud et al. (1974) using the data recorded at the representative station in Grenoble. The detailed technical description of a newly developed measuring system recording the tensions automatically is also given in the cited publication.

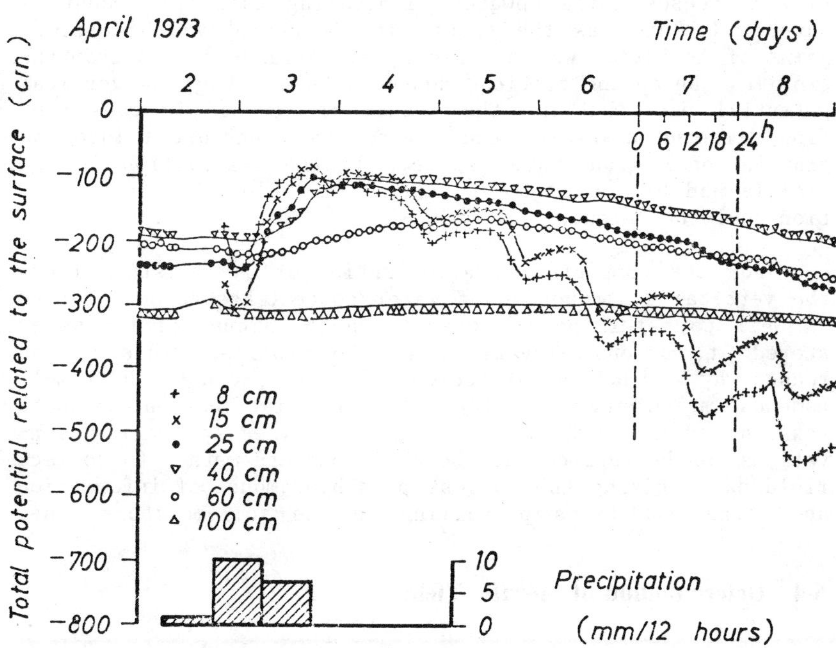

Fig. 2-10. The change in time of total potential at different depths.

It is known that there are two dominating forces initiating and maintaining the moisture flux: i.e., gravity and tension. The total potential (more precisely the potential head expressed as an equivalent water column) is equal to the difference between the elevation of the investigated point above the water table (potential energy) and the suction head (the suction related to the specific weight of water). The gradient maintaining the flow can be calculated by differentiating total potential in the direction of the movement.

Recording the tension values along a profile, the vertical distribution of total potential can be easily constructed (Fig. 2-11). The figure is partly copied from the publication of Vachaud et al. (1974) and the whole concept of the determination of evaporation and accretion is also discussed on the basis of their research. It has to be noted that the surface is used as a reference level in the figure instead of the water table. This change does not modify, however, the character of the relationship, only the numerical values of potential are changed. Where the potential versus depth curve has a maximum point, the vertical gradient is zero, flow does not develop through this horizontal section. The gradient is directed upwards above this depth (the potential decreases going upwards) indicating that the amount of water calculated as the change in storage above the maximum point of potential was drained by evapotranspiration from the profile. Below the critical depth, determined by the vertical potential distribution, the change in storage is caused by flow directed downward, thus, water is transported into the gravitational groundwater space. If the observation in the profile had been extended to the water table, positive accretion could be calculted in this way.

For the complete characterization of the moisture flux, the vertical distribution of water content has to be recorded as well to determine the changes in the amount of the water stored at various elevations in the profile. This is the reason why Vachaud and his co-workers have proposed the simultaneous measurement of the vertical distribution of both tension and moisture content. This combined system of observations can be regarded as the most complete method to collect field data, giving the largest possible amount of information about the conditions prevailing in the soil-moisture zone.

6-4 Determination of Specific Yield

As it was already mentioned, specific yield characterizes the storage capacity of unconfined aquifers. Considering the definition of specific yield, this parameter can be calculated as the difference between porosity and the volumetric water content retained in the sample against gravity (W_o) (the latter used to be approximated by the Lebedyev's maximum molecular water capacity (W_{mol}), or field capacity (W_{fc}) as well):

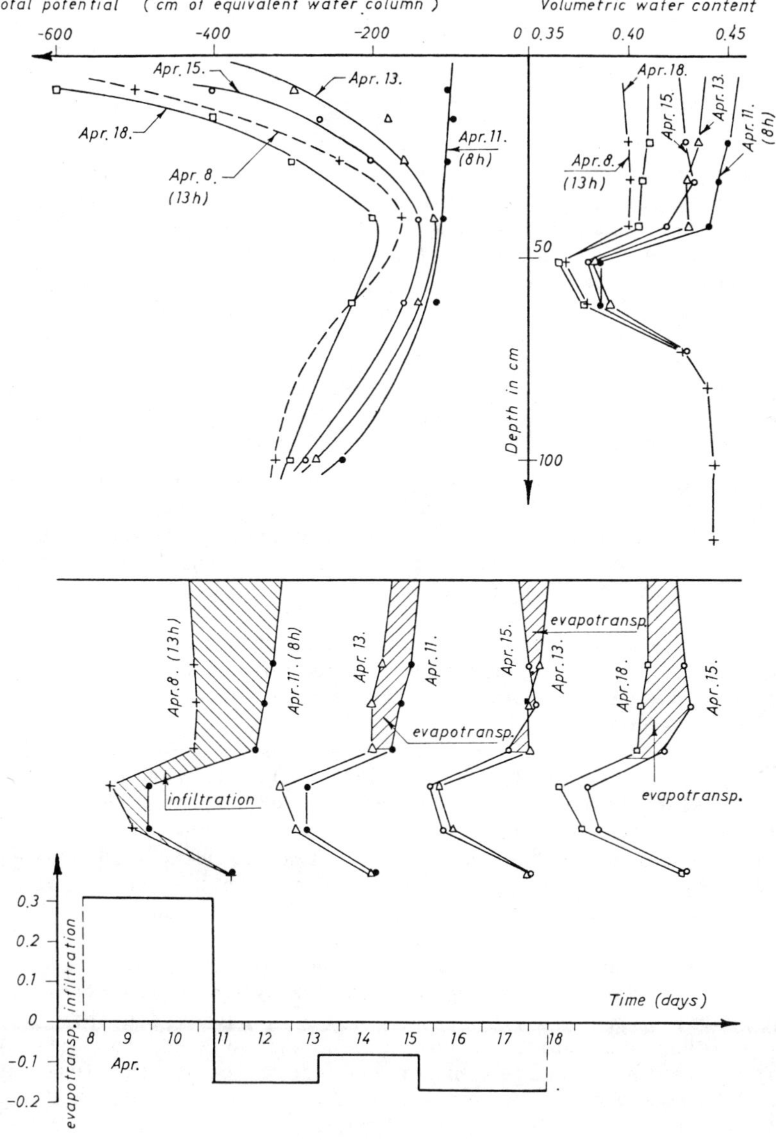

Fig. 2-11. Calculation of evapotranspiration by combining the measured values of tension and water content.

$$n_s = n - W_o \quad . \tag{2-11}$$

The difficulties arising in connection with the application of this concept are caused by the uncomplete physical interpretation of the water content in question, and by the uncertainties in connection with its measuring methods.

The most commonly applied method for determining both specific yield and W_o (so called water retention capacity of a sample) is in a drainage column. A column composed of the material to be investigated is completely saturated at first, and the volume of water drained from the sample is measured afterwards, determining the water content retained by the sample as well. Schoeller (1962) has collected numerous data from the literature representing the results of such measurements. This information together with some newly published materials are summarized in Table 2-3. Representing the data in the form of graphs (Fig. 2-12), the tendency of the relationship between the water retention capacity and grain diameter is clearly indicated (the former slightly increases with decreasing diameter until D_h = 0.02 cm and below this limit the increase is more rapid), although the scattering of the points is very large. The data, especially those measured by Hazen, also indicate the influence of porosity (or uniformity, which is closely related to porosity), but the uncertainties hinder the numerical evaluation of this relationship.

The most probable reason causing the large discrepancy between the measured data is the development of the soil-moisture retention curve in the drained sample. The U.S. Geological Survey in cooperation with the California Department of Water Resources has executed a series of experiments (Johnson et al., 1963; Prill et al., 1965), the results of which, also listed in Table 2-3, prove the possible large differences between the parameters determined in different ways from the same observations. The soil columns were drained either by gravity or by suction (in the latter case a semipermeable membrane was placed below the column, the water level was maintained at a given elevation, and it was contacted with the experimental column through a tube). The results are compared in Fig. 2-13, proving the validity of the previously given statement, that field capacity (in this case water retention capacity, which is equal to field capacity in this concept) is a function of the elevation of the investigated point above the water table. The development of the soil-moisture retention curve is also well demonstrated in the figure.

The time dependent character of specific yield was also investigated in the framework of the USGS experiments. It was

Table 2-3. Specific yield of various samples determined by column drainage.

Diameter Given by the Authors (cm)	Effective Diameter Calculated or Estimated D_h (cm)	Porosity n	Coefficient of Uniformity U	Water Retention Capacity W_o	Specific Yield n_S	References	Comments
0.003	0.0051	0.44	2.3	0.19	0.25	Hazen (1892)	The characteristic diameter is given in the form of D_{10}
0.006	0.0102	0.42	2.3	0.16	0.26		
0.017	0.0272	0.42	2.0	0.11	0.31		
0.035	0.0875	0.325	7.8	0.095	0.23		
0.048	0.0864	0.40	2.4	0.080	0.32		
0.14	0.252	0.415	2.4	0.075	0.34		
0.5	0.750	0.45	1.8	0.070	0.37		
0.0284	0.020	0.413	near unity	0.051	0.362	Woolny (1885)	The experiments were executed with almost homodisperse samples
0.035	0.035	0.406		0.044	0.352		
0.069	0.070	0.391		0.041	0.350		
0.139	0.240	0.382		0.037	0.345		
0.0082	0.008	0.397	near unity	0.066	0.331	King (1898)	The experiments were executed with almost homodisperse samples; Directly published values for specific yield and water retention capacity
0.0118	0.012	0.406		0.050	0.356		
0.0155	0.015	0.408		0.041	0.367		
0.0185	0.018	0.401		0.035	0.366		
0.0473	0.050	0.389		0.034	0.355		
	0.008	0.397		0.180	0.208		Parameters recalculated from data published for the characterization of time dependency of the process
	0.012	0.406		0.148	0.258		
	0.015	0.408		0.114	0.294		
	0.018	0.401		0.080	0.330		
	0.050	0.389		0.069	0.320		
0.0139	0.014	0.404	near unity	0.051	0.355	Atterberg (1911)	The experiments were executed with almost homodisperse samples
0.0305	0.03	0.405		0.048	0.357		
0.0694	0.07	0.418		0.040	0.378		
0.139	0.14	0.404		0.028	0.376		
0.303	0.30	0.401		0.032	0.369		

Table 2-3. (continued)

0.0014	0.0015	0.422	near unity	0.282	0.140	Zunker	The experiments were executed
0.0047	0.0050	0.466		0.174	0.292	(1930)	with almost homodisperse
0.0067	0.0060	0.381		0.134	0.347		samples, although the smaller
0.022	0.022	0.372		0.052	0.320		porosity of more coarse
0.045	0.045	0.352		0.043	0.309		materials indicates that their
0.072	0.072	0.350		0.040	0.310		coefficient of uniformity was
0.112	0.112	0.355		0.038	0.317		above three
Given in the	0.0436	0.36 0.38		0.015 +	0.35 0.36	Johnson and	Glass beads
form of				0.031 ++	0.33 0.35	et al.	
grain-size				0.120 +++	0.24 0.26	(1963);	
distribution	0.0120	0.39 0.41		0.030 +	0.36 0.38	Prill et al.	
				0.043 ++	0.35 0.37	(1965)	
				0.280 +++	0.11 0.13		
	0.0016	0.40 0.42		0.062 ++	0.36 0.38		
				0.290 +++	0.11 0.13		
	0.0587	0.35		0.040 +	0.310		20. Del Monte sand
				0.063 ++	0.287		
				0.186 +++	0.164		
	0.0212	0.37		0.081 +	0.289		Fresno medium sand
				0.090 ++	0.280		
				0.142 +++	0.228		
	0.0114	0.40		0.075 +	0.325		Natural alluvial sand
				0.100 ++	0.300		(Holbrook)
				0.213 +++	0.187		

+ at the top of sample
++ around the top of the open capillary zone
+++ average for the column

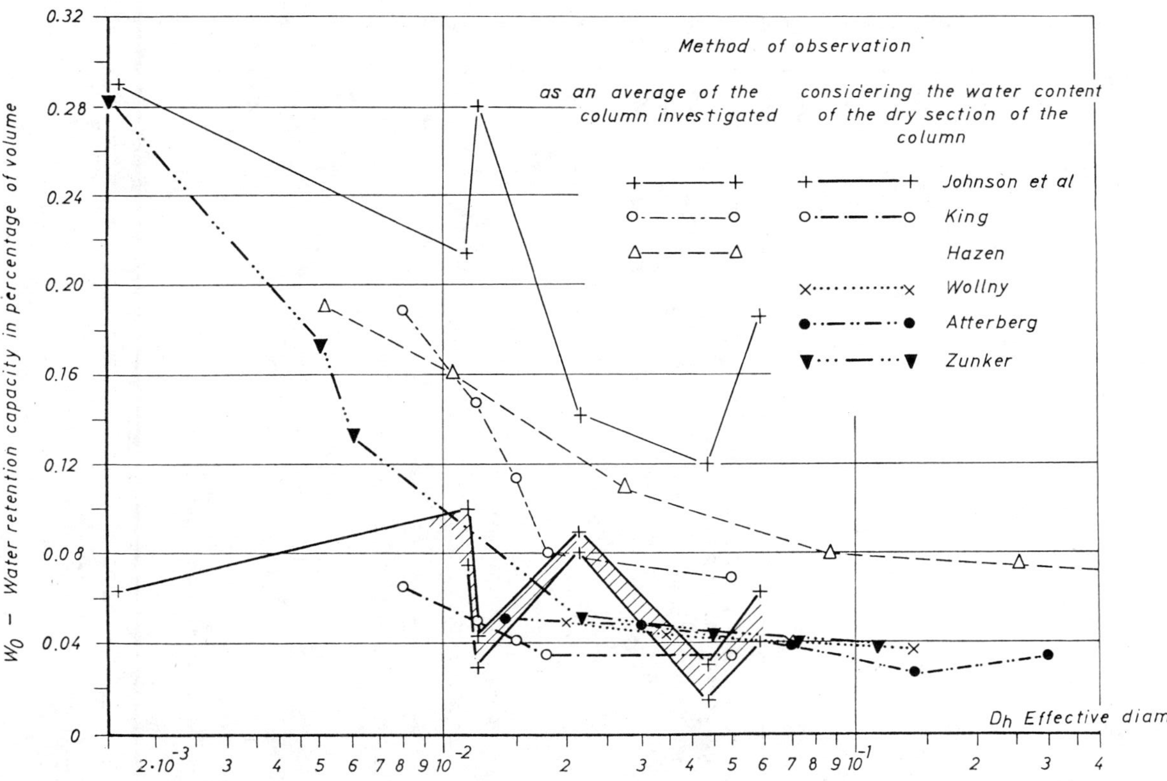

Fig. 2-12. Relationship between water retention capacity and effective grain diameter.

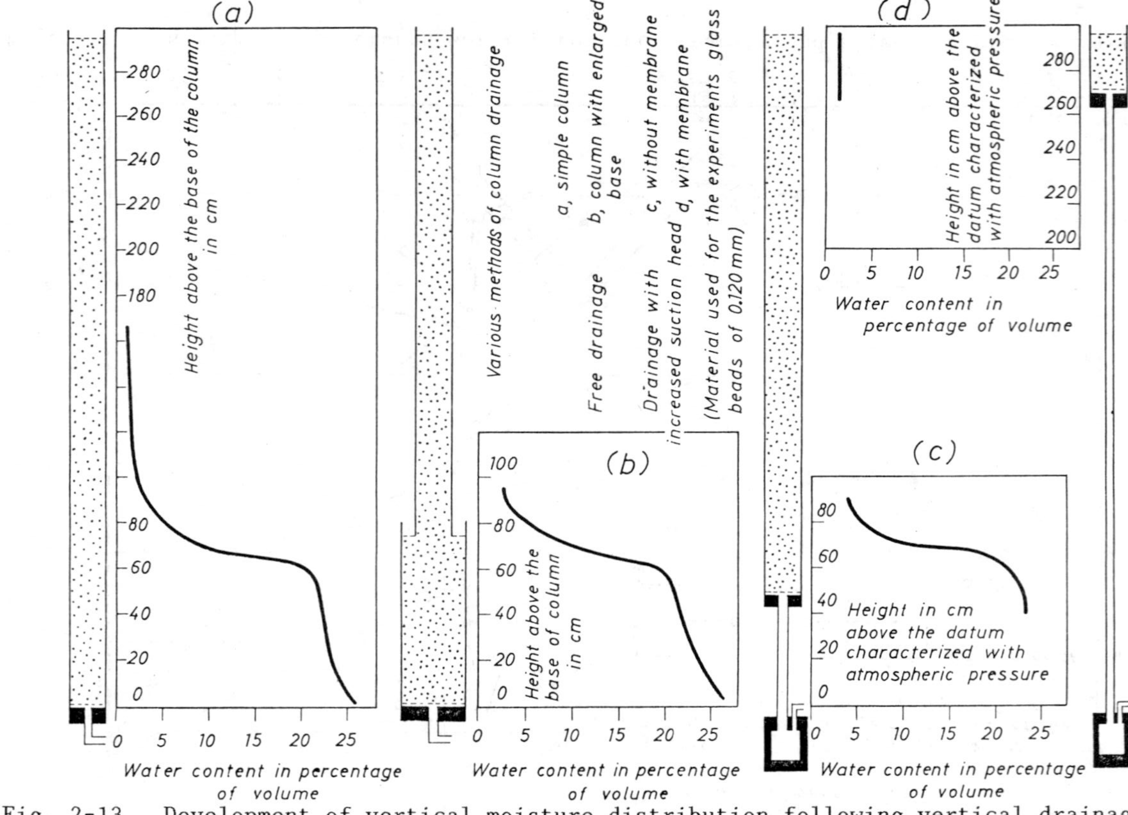

Fig. 2-13. Development of vertical moisture distribution following vertical drainage.

found that even 100 hours are not enough to reach the development of the complete dynamic equilibrium, which can be characterized by a hydrostatic tension distribution (the tension is equal everywhere to the product of the height above the water table and the specific weight of water). In Fig. 2-14, the change of tension in time is represented to show an example as it was measured in one of the experimental columns. It can be seen that in a very fine sand the dynamic equilibrium is achieved only in the lower 80 cm after 100 hours, while at the top of the column (around 150 cm above the water table) the hydraulic gradient remains at almost unity. At the same time, the saturation of the sample is very low at this elevation and, therefore, unsaturated conductivity is also very low as it is well-known and will be proved later. Thus, the water amount transported by this gradient is negligible as well. This supposition can also be proved by the graphs showing the relationship between the drained amount of water and the time elapsed from the beginning of the experiments (Fig. 2-15).

Among the previous investigations, King's measurements also gave data for representing the dependence of specific yield on time (Fig. 2-16). The length of these experiments was 2½ years, and water was released from the sample up to the end of this very long period. The result is in good agreement with the previous conclusions, showing that the process is very slow, but the amount drained at the end of the prolonged experiments is almost negligible. Another interesting experience can be gained from King's data as well: i.e., about 13-18 percent of the total water stored in the sample was lost during the long experiments, probably through evaporation (the sum of the drained water and the soil moisture retained at the end of the investigation in the sample is smaller than the total pore volume).

As a consequence of the numerous uncertainties, occurring in connection with the determination of both specific yield and water retention capacities, as it was listed previously, there are attempts in the literature to give average values as rough estimations for the parameters in question. Such proposals are summarized in Fig. 2-17 (Castany, 1967; Major, 1973; De Wiest, 1968).

Another way to avoid both the development of soil-moisture retention curve and the slow process of drainage is the use of a centrifuge for dewatering the sample. The parameter obtained in this way may numerically differ from water retention measured with column drainage, thus it is called centrifuge moisture equivalent. This parameter is also influenced by many factors. It was necessary, therefore, to standardize its measurement. The American Society for Testing Materials (1958) has defined the centrifuge moisture

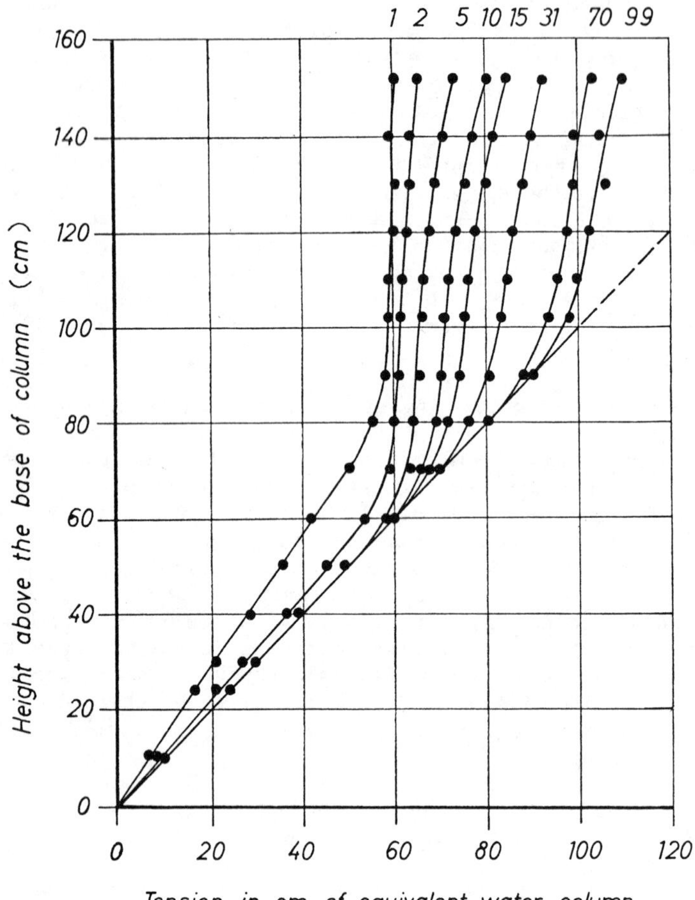

Fig. 2-14. Investigation of the development of vertical tension distribution in time.

equivalent as follows: "water content retained by a soil which has been first saturated with water, and then subjected to a force equal to one thousand times the force of gravity for one hour."

The detailed investigations performed by the USGS (Johnson et al., 1963) included the analysis of the moisture content measured by centrifuge as well. They have found that this parameter depends on the temperature, the length of

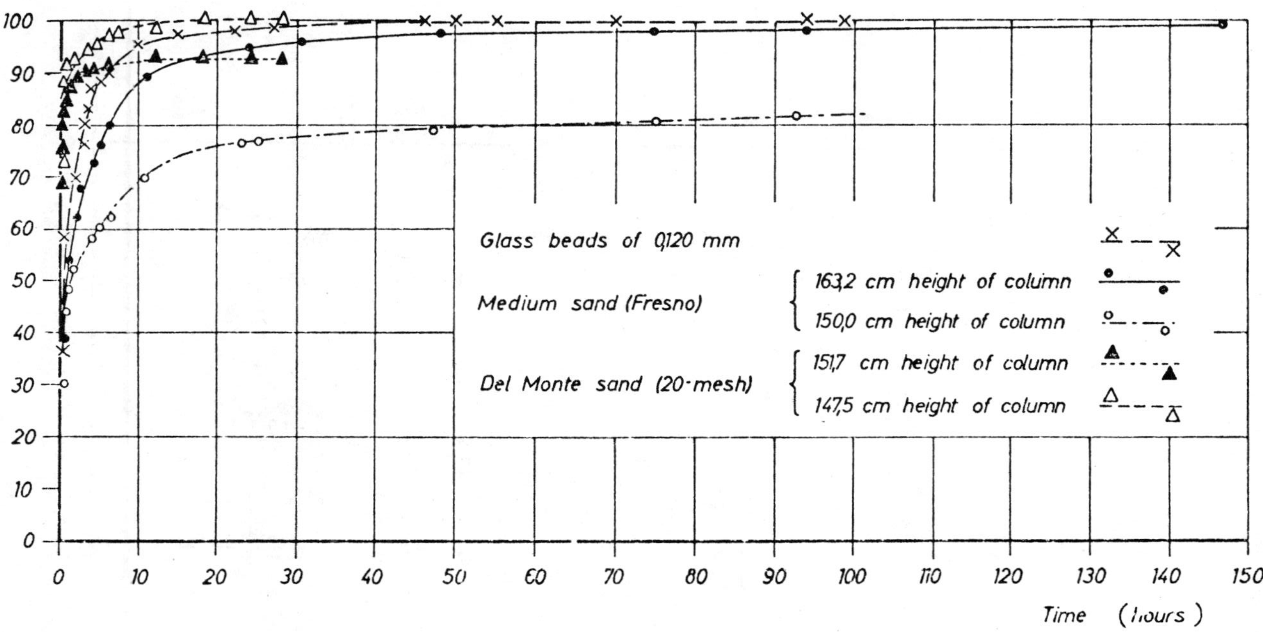

Fig. 2-15. The change in time of the flow rate measured by column drainage (1)

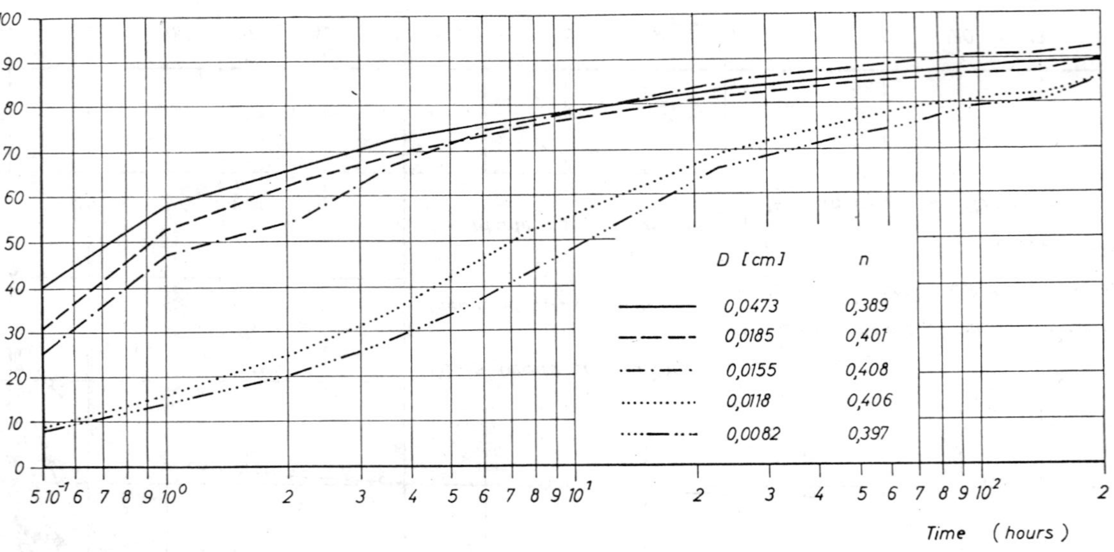

Fig. 2-16. The change in time of the flow rate measured by column drainage (2).

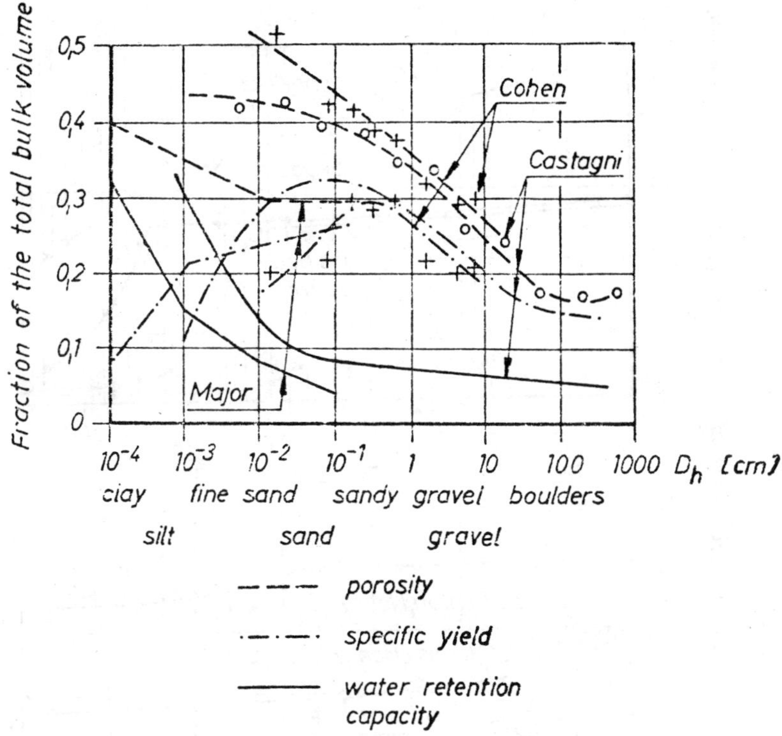

Fig. 2-17. Estimated values of porosity, specific yield, and water retention capacity depending on effective grain diameter.

period of centrifuge, the size and structure (porosity) of the sample, and the magnitude of the force acting (angular velocity). Because the force changes with the distance from the center of rotation, the moisture content also changes within the sample. This latter effect can be well demonstrated if the measured moisture content is represented as a function of the force acting (Fig. 2-18). The relationship can be characterized by considering that the tension created by the rotation (h_{CR}), expressed in centimeters of equivalent water column, depends on the angular velocity (ω) (T^{-1}), as well as on the distances of both the bottom of the sample (r_1) and the specified point (r_2) from the center of rotation measured in centimeters:

$$h_{CR} = \frac{\omega^2}{2g} (r_1^{\ 2} - r_2^{\ 2}) \quad . \tag{2-12}$$

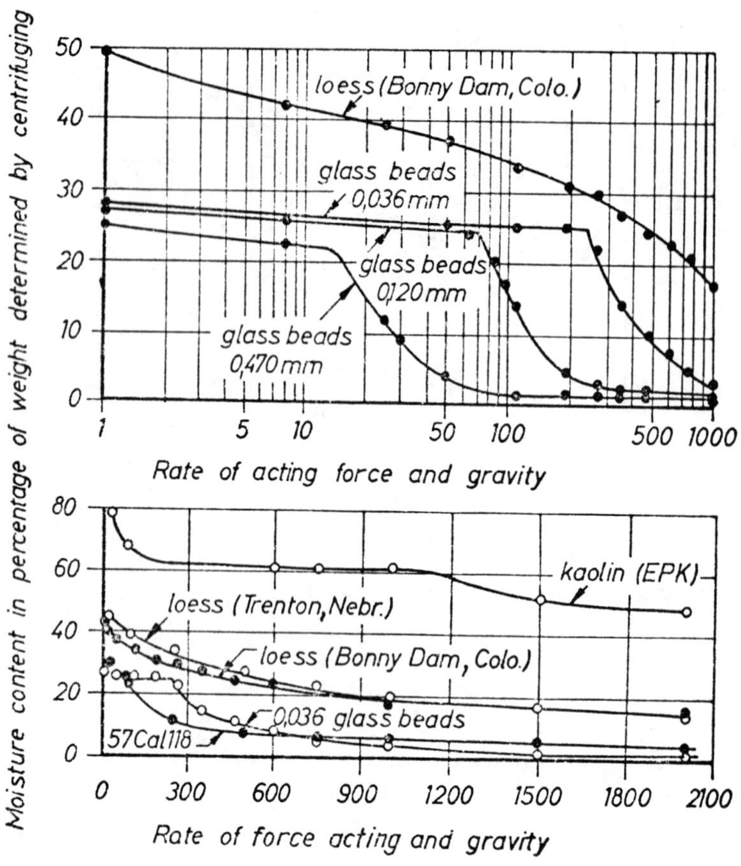

Fig. 2-18. Centrifuge moisture content as the function of the force acting.

In Table 2-4, the centrifuge moisture equivalent (the data measured with the standardized method) are listed for each sample investigated, while in the further columns the range of moisture content is given within which the parameter was found when one of the influencing factors was modified.

The data listed in Table 2-4 are also graphically represented in Fig. 2-19 similarly to water retention capacity. The relationship with grain diameter shows the same tendency, although the numerical values are different. The advantage of the use of a centrifuge is that the process is much faster than column drainage although the scattering of the measured points cannot considerably decreased. At the same time, the real physical meaning of specific yield is lost in this way.

Table 2-4. Centrifuge moisture equivalent and moisture content for various samples (after Johnson et al., 1963).

Material		D_h (cm)	U	Centrifuge Moisture Equivalent	Range of data scattering if some of the conditions are modified			
					Size of Sample	Structure	Temperature 30-50°C	Uncontrolled Temperature
Del Monte sand	60-mesh	0.0156	2.42	1.4	1.4 ~ 1.6		1.2 ~ 1.8	
	30-mesh	0.0300	1.40	0.4			0.3 ~ 0.6	
	20-mesh	0.0587		0.5	0.5 ~ 0.6		0.4 ~ 0.6	
Glass beads	0.47 mm	0.0436	1.11	1.2			0.9 ~ 1.3	
	0.12 mm	0.012	1.30	1.3	1.3 ~ 1.9		1.0 ~ 1.7	
	0.036 mm	0.0036	1.90	3.4			3.1 ~ 3.9	1.2
59 CAL 191-244		0.0212	4.0		1.9 ~ 2.2		1.6 ~ 2.0	
57 CAL 122		0.0026	5.43	4.0				
57 CAL 66		0.0045	2.66	4.2				1.4
57 CAL 110		0.0030	26.00	5.4				1.2
57 CAL 118		0.0022	8.00	6.0			6.0 ~ 7.0	3.0
59 CAL 258-261		0.0017	15.55	9.8		13.0 ~ 13.6		
59 CAL 251-252		0.0013	22.50	10.6		13.1 ~ 13.6		
59 CAL 253-254		0.00084	61.11	12.4		11.0 ~ 14.6		
59 CAL 137-138		0.00071	33.00	13.2		14.4 ~ 15.1		
57 CAL 72		0.00227	56.00	14.1				8.0
58 OKL 15		0.00057	26.66	16.5		18.0 ~ 21.1		
Loess (Bonny Dam Colo.)		0.00074	5.16	17.3	17.3			
57 CAL 106		0.00071	30.95	17.6				
Loess (Trenton, Nebr.)		0.00075	4.69	19.0			14.0 ~ 20.0	
59 CAL 248		0.00070	30.00	20.2		20.6		
58 OKL 42		0.00091	18.75	20.4		18.8 ~ 23.4		
57 CAL 155		0.00092	30.40	21.6				12.4
59 CAL 130-131		0.00038	15.00	23.1		21.9 ~ 25.9		
57 CAL 139		0.00149	16.66	24.8				25.5
57 CAL 152		0.00061	38.88	28.8				21.1
57 CAL 162		0.00029	66.66	30.0				20.2
57 CAL 173		0.00048	13.33	32.3			31.0 ~ 35.0	27.2
58 ARK 146		0.00014	4.16	37.2		29.4 ~ 33.6		
57 CAL 112		0.00017	6.43	6.43		40.9		34.6
57 CAL 123		0.00053	11.10	43.2				32.8
57 CAL 2		0.00020	8.33	46.4				41.4
57 CAL 205		0.00013	8.33	49.3				46.9
Kaolin (N.F.)		0.00029	3.36	49.6				
Kaolin (EPK)		0.00007	2.86	61.2			42.0 ~ 48.0	
Fuller's earth		0.00048	16.66	109.0			75.0 ~ 80.0	

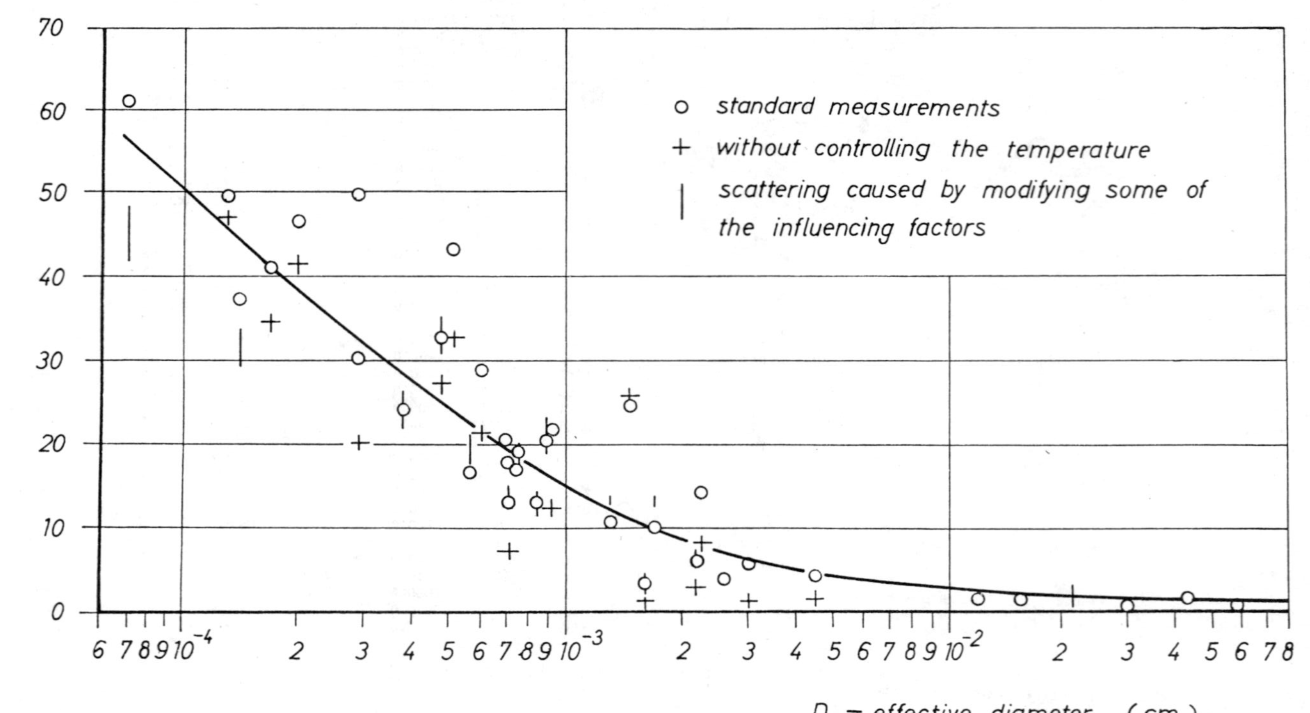

Fig. 2-19. Centrifuge moisture content depending on effective grain diameter.

To see all the difficulties encountered in connection with the determination of specific yield, it is necessary to recognize that most of the problems were caused by the insufficient physical interpretation of the term. In an unconfined system, capillary zones always develop above the water table, thus, the dynamic balance is characterized with the soil-moisture retention curve. The correct way for the determination of specific yield is, therefore, to calculate the difference of the amount of water stored above a reference level arbitrarily chosen below the original and the modified water table, assuming that the change in the elevation of the latter is unity according to the definition of specific yield. The redistribution of soil moisture following the change of the position of the water table needs some time, the length of which depends on the hydraulic conductivity of the layer. Thus, the comparison of the stored amounts has to be performed after the development of the dynamic equilibrium and the determination of specific yield can be reduced, therefore, to the calculation of the differences between the areas of the two soil-moisture retention curves (Fig. 2-20).

Let us indicate two positions of the water table with suffixes 1 and 2, respectively. It is quite indifferent which is the original condition because specific yield ought to have the same numerical value when either lowering or raising the water table. To fulfill the condition given by the definition, the distance between the two water tables is equal to unity. The stored amount of water in a column having a cross section of unit area can be easily calculated above the datum. Executing the calculation for both positions of the water table, and formig the difference of the two determined values, the physically correct specific yield is achieved:

$$n_s = \left[z_1 n + \int_o^{h_1} W_{fc}^{(1)}(z)\,dz \right] - \left[z_2 n + \int_o^{h_2} W_{fc}^{(2)}(z)\,dz \right]$$

$$= n - \left[\int_o^{h_2} W_{fc}^{(2)}(z)\,dz - \int_o^{h_1} W_{fc}^{(1)}(z)\,dz \right] \; ; \qquad (2\text{-}13)$$

$$(z_1 - z_2 = 1) \; .$$

Comparing Eq. 2-13 with Eq. 2-11, it can be easily seen that the difference of the two expressions to be integrated is equal to the water retention capacity.

Although theoretically this concept is correct, difficulties, however, are raised when it is applied in practice. The two most important ones are those caused by the

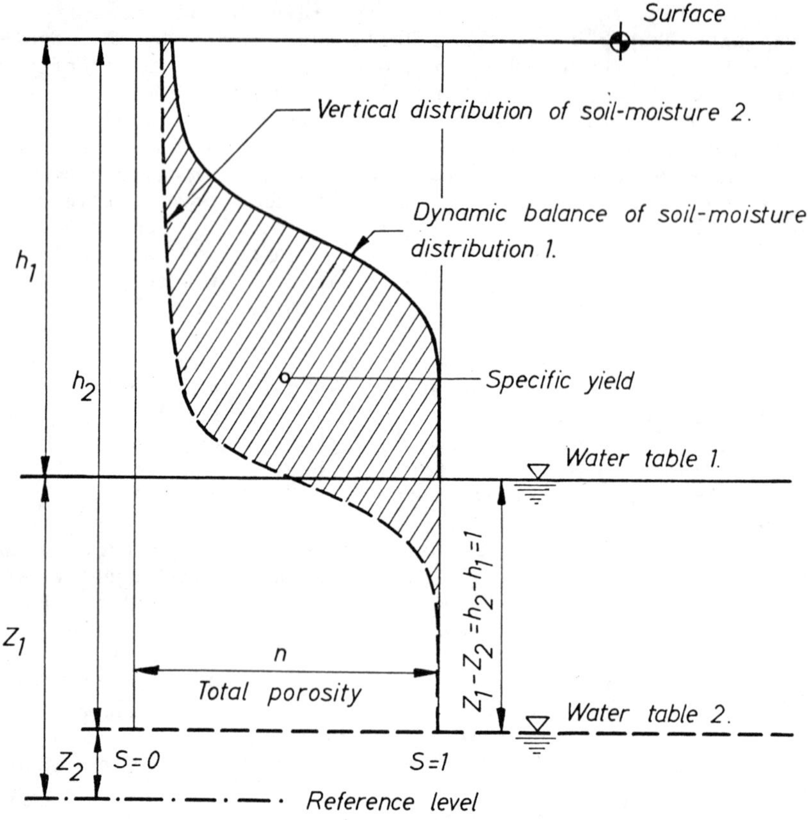

Fig. 2-20. Interpretation of specific yield considering the soil-moisture retention curve.

hysteresis of the soil-moisture retention curve and by the time lag between the modification of the water table and the final development of the dynamic equilibrium. Further problems are encountered when the unsaturated zone is layered, but these details will not be discussed here.

As it was already mentioned, the relationship between tension and water content is not unambiguous because of the phenomenon called hysteresis. Dynamic equilibrium develops at lower positions, if the soil-moisture zone is wetted from the direction of the water table, while the highest curve can be achieved if the zone was completely saturated previously and drained vertically later. Between these two extreme positions the balanced condition can develop at every elevation. In Fig. 2-21, four different cases are represented.

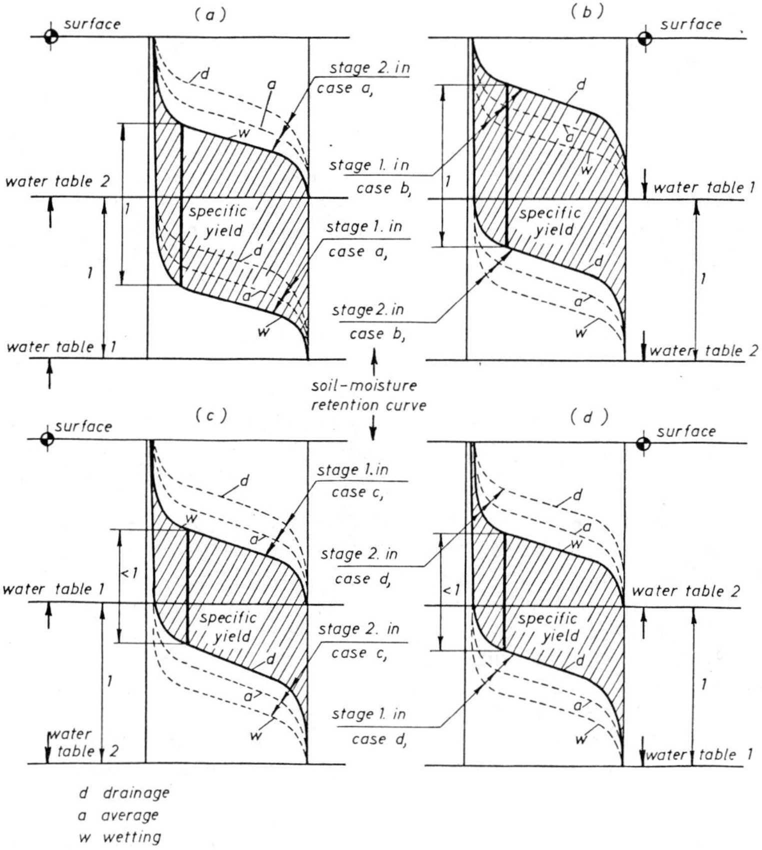

Fig. 2-21. Influence of the hysteresis of the retention curve on specific yield.

(a) A wetting process is followed with further rise of the water table, thus, both balanced conditions are characterized with the lower stretch of the hysteresis loop. The curves are parallel and their distance is unity, equal to the modification of the water table.

(b) The original condition was achieved by drainage, and the lowering of the water table is investigated. Both soil-moisture retention curves follow the upper limit, thus, their distance is equal to case (a).

(c) After wetting the soil-moisture zone, the water table is lowered once again. The lower limit of hysteresis belongs to the initial condition, and the upper to the final state. Thus, the difference between the two areas of the retention curve is smaller than that calculated from the average curves. Theoretically the difference can be negative as well, but in this case the upper limit is never achieved, the dynamic equilibrium develops at an intermediate position, and thus the lower limit of the calculated difference is zero.

(d) The case opposite to (c) is the rise of the water table in a previously drained system. Now the final state is characterized by low retention curve, and the initial condition with a high one. The result is also the decrease of the difference of the two areas, and its lower limit can be similarly zero.

Further special cases can also be expected (e.g., wetting of the soil-moisture zone previously and infiltration from the surface afterwards, raising the water table) when the distance between the initial and final retention curves are greater than unity, its upper limit being the sum of the change of water table and the thickness of the hysteresis loop.

It would be very difficult to investigate the processes having prevailed previously, in each case, separately to determine the difference between the actual and average specific yield. Only one possible way can be proposed: to accept the specific yield calculated from the average soil-moisture retention curves, and to neglect the influence of hysteresis. Realizing the uncertainty caused in this way, some further approximations can also be applied to derive a formula amenable to mathematical treatment, and suitable for describing the relationship between water retention capacity and soil-physics parameters.

In the following section, where the soil-moisture retention curve will be analyzed, it will be proved, that in the adhesive zone, the relationship between tension and moisture content can be well approximated with a hyperbola of sixth order (see Eq. 2-18). At this very steep stretch, the curves belonging to the initial and the final stage are essentially the same (Fig. 2-22). Accepting the use of the average curve (without hysteresis) for the characterization of both conditions (before and after the modification of the water table), it can be seen that the change of the areas covered by the retention curves can be calculated as the free gravitational porosity at the adhesive zone multiplied by the change in the deviation of the water table. The latter being unity, if the purpose is the determination of specific yield, or the water retention capacity, the only remaining problem is to select the numerical value suitable for the characterization of both water content and gravitational porosity at high tension.

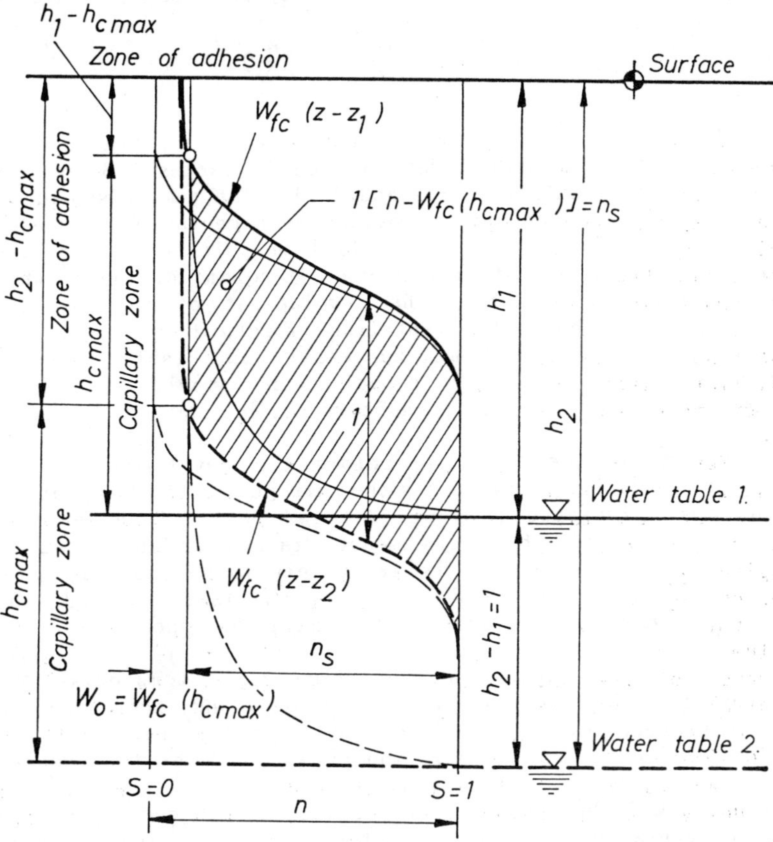

Fig. 2-22. Sketch for the calculation of specific yield and water retention capacity.

It can be proposed that the parameter considered in the calculation should be the specific amount of adhesive soil moisture which belongs to the highest capillary rise. The water retention capacity can be calculated, therefore, by substituting the probable highest capillary rise into the equation describing adhesive moisture content (the derivation of the equations quoted here can be found in Section 7):

$$W_o = 2.5 \times 10^{-3}\ \frac{1-n}{(h_{cmax})^{1/6}}\ \left(\frac{\alpha}{D_h}\right)^{2/3} = 3.6 \times 10^{-3} (1-n)^{5/6} n^{1/6} \left(\frac{\alpha}{D_h}\right)^{\frac{1}{2}} ;$$

$$h_{cmax} = \frac{0.45}{4}\ \frac{1-n}{n}\ \frac{\alpha}{D_h} ;$$

(2-14)

or the same parameter expressed with the amount of saturation

$$s_o = 3.6 \times 10^{-3} \left(\frac{1-n}{n}\right)^{5/6} \left(\frac{\alpha}{D_h}\right)^{1/2} . \qquad (2\text{-}15)$$

In Fig. 2-23, the relationship given by Eq. 2-14 is compared to the measured data listed in Tables 2-3 and 2-4, and represented in Figs. 2-12 and 2-19. The comparison proves that the proposed equation provides an acceptable and reasonable compromise between the various measurements. It gives the lower limit of data measured with column drainage (which is acceptable considering the development of the capillary zone) but runs above the centrifuge moisture equivalent. It ensures, at the same time, an easily applicable method to calculate water retention capacity and specific yield, as functions of soil physics parameters (n, D_h, α).

Naturally the time lag before the development of the dynamic equilibrium disturbs the direct application of the parameter calculated from Eq. 2-14 as well. In a permeable layer, the ratio of water amount drained or stored at the beginning of the process is relatively large and only a few percent of specific yield occurs in the latter period, when the water has to be transferred through the upper part of the column where the unsaturated conductivity is very small because of low saturation. In fine grained material, the hydraulic conductivity is relatively small and, therefore, the whole process is considerably prolonged. Figures 2-15 and 2-16 give sufficient information on the size of the expected time lag, although it is also mentioned in the literature that the delay will be longer if the balance is achieved by wetting the unsaturated zone, than in the case of drainage (Smith, 1933).

6-5 Lysimeters and the Use of Their Data

According to the most simple and very general definition, lysimeters are soil columns separated from their surroundings. They can be regarded, therefore, to be the physical realizations of a theoretical model, represented in Fig. 2-2. Their purpose is to measure the various components of the water balance (or some groups of the components), indicated in the figure. To fulfill this task, the instrument (which consists basically of a soil column and its container) is equipped with different kinds of devices to observe and to regulate the development of the various processes.

There are many possible ways to construct the required closed space in the soil. The review of the literature dealing with lysimeters is omitted here. The readers are referred

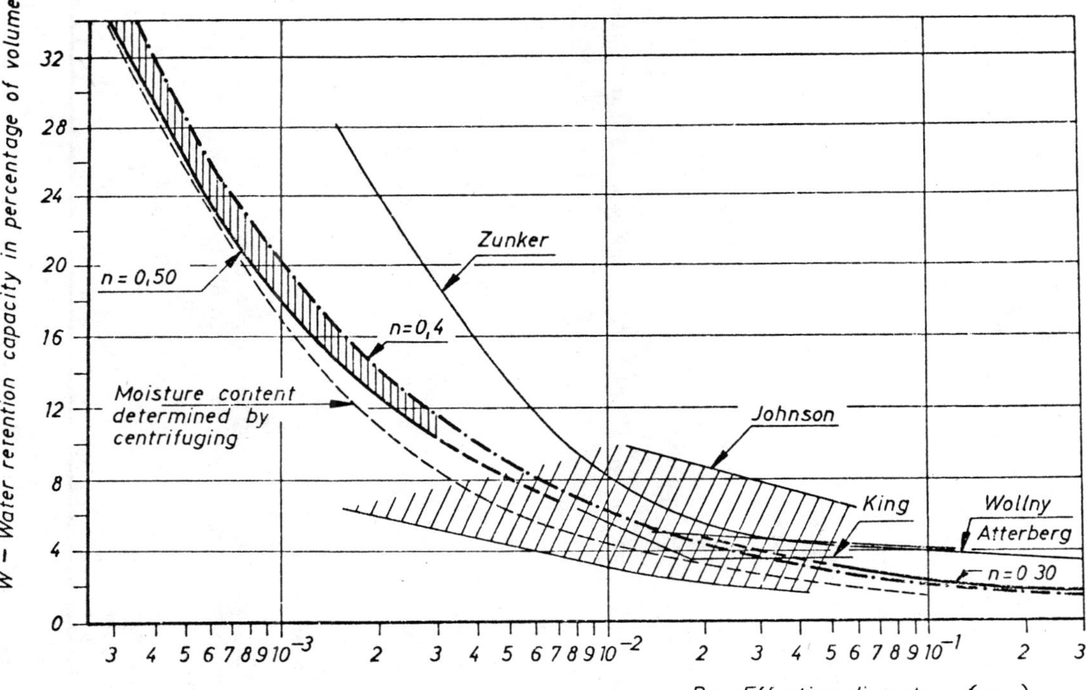

Fig. 2-23. Comparison of measured and calculated water retention capacity values depending on effective grain diameter.

to Harrold and Dreibelbis publication (1967) where a detailed list and evaluation of the previous works can be found. The lysimeters may differ not only in their size (depth, surface, area), but also in their construction and operation. This latter group of the influencing factors includes such aspects as:

- type of the surface (bare or covered with vegetation);

- character of the cultivated zone (depth, structure, root system);

- type of soil (homogeneous or layered, cohesive or coarse-grained);

- regulation of moisture content (position of the water table, maintenance of a given tension at the lower boundary);

- recharge and discharge of water (natural or artificial, in the second case recharge through either the surface or within the groundwater space);

- measuring techniques of the components of water balance, etc.

The constructional and operational aspects influence not only the reliability and accuracy of the measurements, but at the same time, they determine which components or which groups of them are measurable by the lysimeter. This is the reason why the data observed by different equipment are not comparable, and any observation is applicable in practice only if information is available on the measuring conditions.

It is not possible to give preference to one type of construction or operation over the others because the possible solutions observe the phenomena to be measured in different combinations, and therefore, the purposes served by the lysimeter determine which group of the components of the water balance should be measured, and thus, which type of the usually applied constructions should be chosen. It is necessary, however, to survey the possible solutions from the point of view of the various aspects of application to give some guidance for the determination of the suitable construction and operation.

The size of the investigated soil column. The form of the horizontal section of the investigated soil column does not influence the development of the hydrological phenomena. It can be determined, therefore, according to the requirements of the applied construction (circular, square, rectangular). The size of this area is, however, an important parameter. It would be desirable to have as large a lysimeter as possible because the disturbing effects increase, relatively, as the horizontal size decreases. These effects are:

- the nonhomogeneity of the surface (covering vegetation) and the soil (especially the cultivated zone) which requires that the horizontal area should be many times larger than the representative elementary unit of the surface;

- the discontinuity between the surface of the lysimeter and that of the surrounding terrain which can modify the water balance of the investigated column (e.g., oasis effect);

- the concentrated infiltration along the wall of the container, the influence of which is proportional to the ratio between the perimeter and the area of the horizontal section.

On the other hand, the horizontal extension of the lysimeter is limited by the rapidly increasing costs and is influenced by the structure and the material of the container. If the change of the stored water is determined by measuring the weight, the type of the weighing balance can also give an upper limit in the form of the measurable weight.

It is advisable that the horizontal surface of a bare lysimeter should not be smaller than 0.5 m^2, while this limit is higher in the case of lysimeters covered by vegetation. It can be proposed, as a rough guidance in this case, that selecting 1/30-1/50 of the total area as a representative elementary unit, the parameters describing the conditions of the plants and the cultivated zone should be constants (having only random variation) independent of the location of the unit area within the surface of the lysimeter.

The other main geometrical parameter of lysimeters is their depth. The most important factor influencing this size is the character of the zones to be modeled. Thus, the depth of a lysimeter with vegetation should be greater than the natural root zone of the applied plants. If the task is the investigation of the influence of the water table, the depth of the container is determined by thc position and the fluctuation of the water table at the simulated area.

In the case of weighing lysimeters, the upper limit of the load determines the total weight, which influences both the horizontal area and the depth. At the same time, the accuracy of weighing is generally given as the percentage of the total load. It is necessary in each case to fix the smallest change of stored water which should be measurable by the equipment. This parameter is expressed as a water column over the unit horizontal area, its weight can be related, therefore, to the weight of a part of the soil column also having a cross section of unity. The ratio should be smaller than the accuracy of the measurement which gives the upper limit of the applicable depth.

The purpose of the lysimeter determines its minimum depth. The construction and material of the container has to be designed according to this requirement. A further condition is that the container should be absolutely waterproof, excluding any water exchange between the internal and external spaces. Its rim has to be raised above the surface, because in other cases, capillary flow can develop through the border. This requirement causes, however, further problems: surface runoff can be ensured only by applying special arrangements; the upper part of the container has to be constructed from heat-insulating material, because in other cases, the lysimeter is heated by solar radiation.

Location of the lysimeter. Theoretically, the location of the investigated soil column does not influence the development of the water balance because it is separated from the surroundings. There is, however, one zone where the communication is not excluded, i.e., the zone of air. The problem is increased by the facts that the mobility of water is the greatest in the vapor phase, and the measurement of vapor flux is not yet solved. It is advisable, therefore, to ensure the same condition above the lysimeter, as that prevailing in the simulated area. To achieve this goal, the container is generally lowered below the surface maintaining the same level of the soil surface inside the lysimeter, as that of the surrounding terrain. The type, the rate of growing, and the cultivation of plants covering the investigated column should be the same as those outside the container and the discontinuity caused by the structure of the latter should be minimized.

Examples can be mentioned to demonstrate the necessity of having continuous and homogeneous cover within and around the lysimeter. Makkink (1957) has found that the vegetative gap between the lysimeter and the surrounding grass caused by the structure of the container having a width of about 25 cm increased the incoming radiation in the lysimeter grass by 10 to 20 percent. Because of the same reason, the rim of the lysimeters at the Coshocton station (Ohio, USA) was rebuilt to decrease the vegetation gap and it is now considered to be nearly representative. To demonstrate the influence of the condition of the plants, the ratio of the measured daily actual evapotranspiration and the calculated potential value (characterized by the evaporation of lakes determined by using the U.S. Weather Bureau formula, LE) was used by Mustonen and

McGuinness (1968), and their figure is repeated here, showing the decrease of the actual parameter by one-half at the time of haycut (Fig. 2-24).

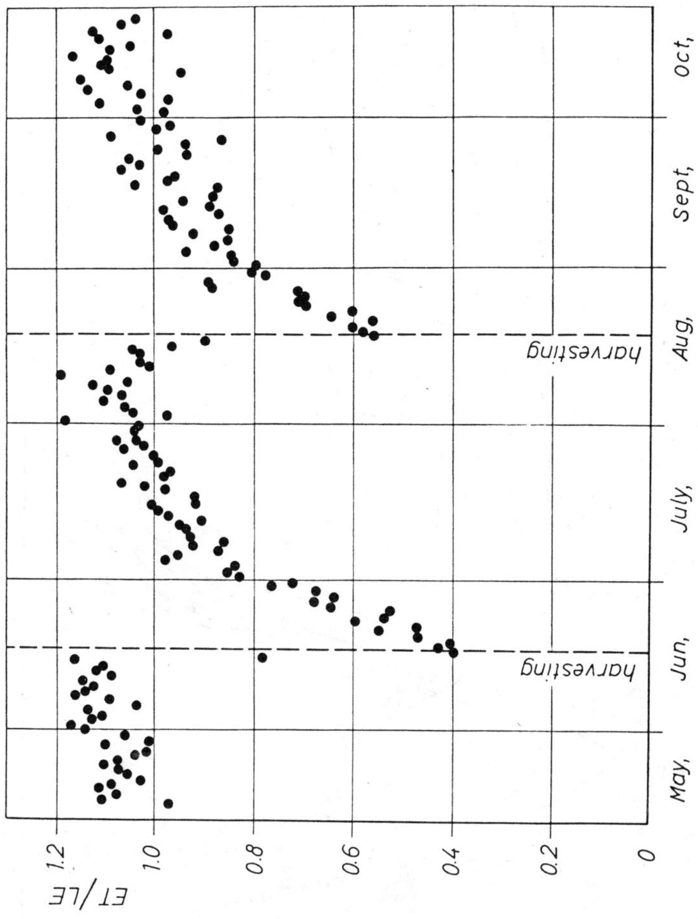

Fig. 2-24. Change of the ratio of actual evapotranspiration to lake evaporation, depending on the development of plants.

The structure of the various zones. The factors most strongly influencing the development of the water balance in the soil column are those originating from and governed by the structure of the zones modeled in the lysimeter. The basic differences caused by the various structures, or the missing of some of the zones, are demonstrated in Fig. 2-25 where the

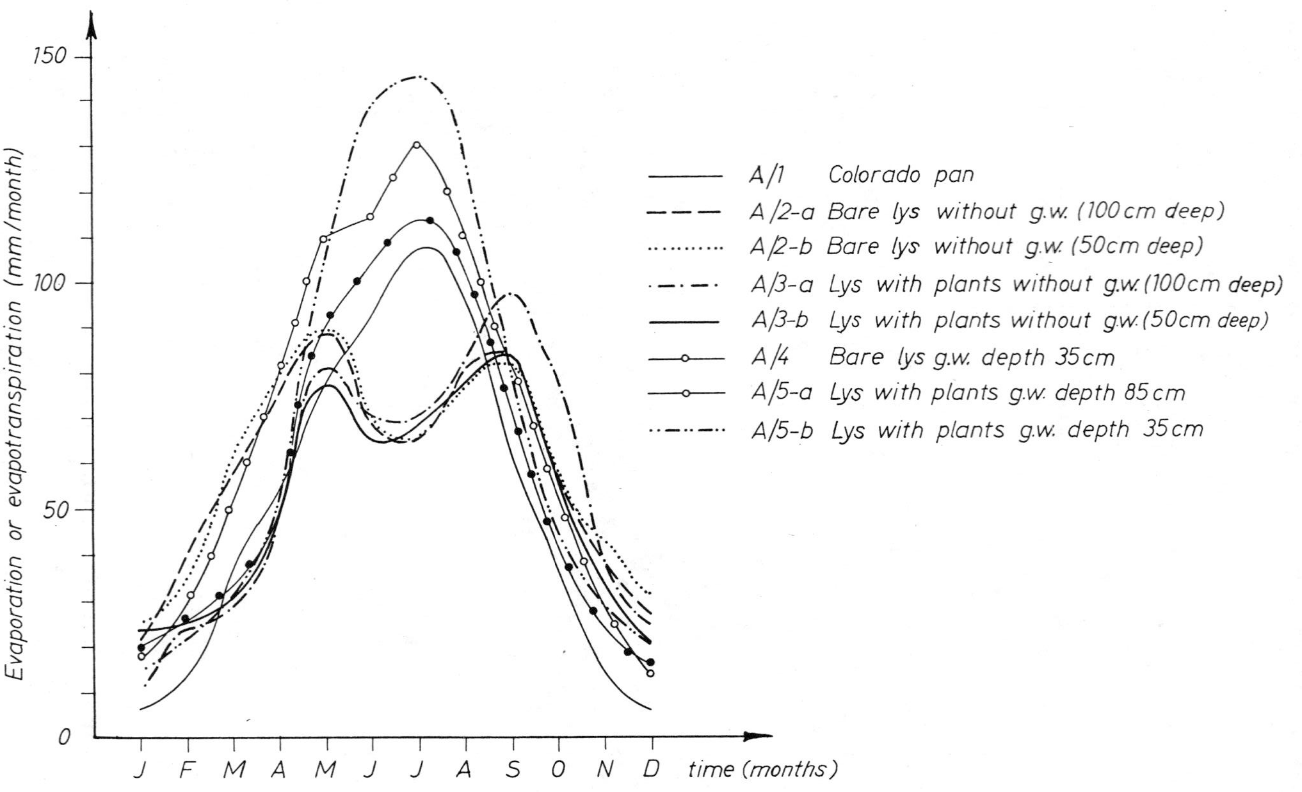

Fig. 2-25. Comparison of the data measured by lysimeters operating in different ways.

data recorded in different types of lysimeters at the Montpellier research station (France) are compared on the basis of personal information received from Y. Cormary. Analyzing the various zones separately, the main aspects to be considered in the evaluation of lysimeter data can be summarized.

The greatest difference which may occur in the zone of air is that existing between a bare surface and a lysimeter covered by vegetation (see A/2 and A/3 or A/4 and A/5 curves of Fig. 2-25). In the first case there is no interception ($S_1 = 0$), thus, the input of the second subsystem is equal to the precipitation measured by rain gages ($P_a = P_s$). There is no cultivated (root) zone ($S_2 = 0$; $I_m = I_m$) and the soil moisture is drained only by evaporation ($ET_a = E + E_m$; $T_m = 0$; $ET_m = E_m$). (The symbols applied here are the same indicated in Fig. 2-6.)

Although the character of the processes acting does not change considerably if different plants are used in the lysimeters, the numerical discrepancy may be even larger between the recorded data, depending on the type, the growing phase, and the applied cultivation of the plants. The influence of the type of plants is demonstrated by Fig. 2-26 where the data recorded at two stations of the Hungarian lysimeter network (Gödöllo and Nagyhegyes) are compared (Balogh, 1973). (The operation of the lysimeters is identical, the sole difference is that of the covering plants.) The effects of cultivation were already shown (hay cutting in Fig. 2-24). The forested area is the most suitable one to characterize the changes of the elements of the water balance with the growing phase of the plants. The construction of lysimeters in forest is, however, very difficult and expensive because of the large size needed. The water balance data of a forested area, recalculated from groundwater data for an 18-year long period is used, therefore, to give a numerical value showing the influence of the growing phase (Fig. 2-27) (Major, 1975). The examples clearly indicate that the lysimeter data should always be supplemented with information concerning the type and condition of plants covering the surface.

The information needed to characterize the second subsystem include the data describing the condition of the surface (slope, roughness, bare or covered with vegetation, it is loosened or smoothed time to time, etc.). Most of these parameters influence the internal operation of the subsystem (surface storage S_1) and determine the ratio of direct evaporation (E), specific surface runoff or effective precipitation (P_{eff}), and infiltration (I_m) in this way.

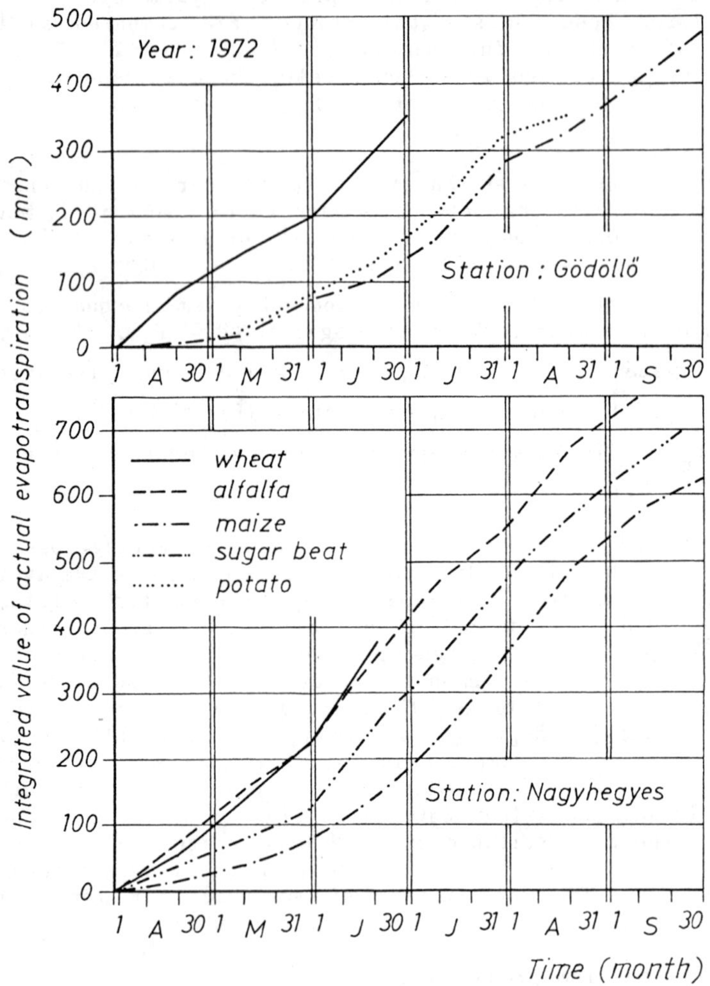

Fig. 2-26. Actual evapotranspiration changing in time and measured by lysimeters covered by different plants.

Apart from the precipitation reaching the surface (P_s) the other important input of the subsystem is the incoming surface flow (R_i). The free development of this process as well as that of the runoff (R_o) is hindered by the rim of the lysimeter. The outflow has to be ensured in every lysimeter without any time lag because the inundation of the surface creates a condition basically different from the natural situation and disturbs the development of the water

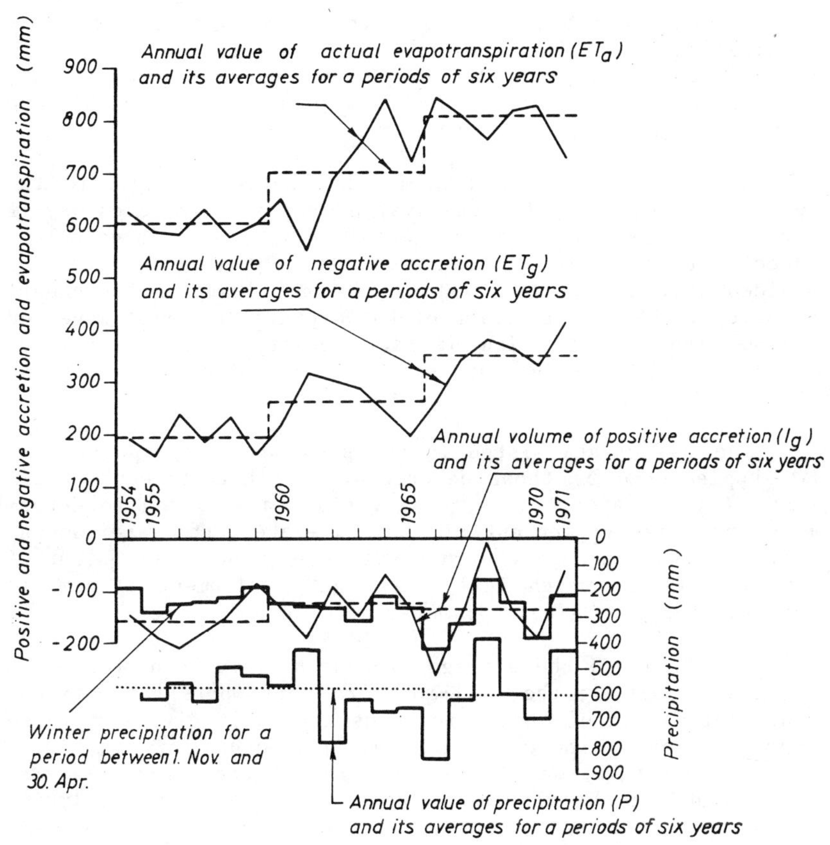

Fig. 2-27. Increase of annual evapotranspiration of a forested area due to the growth of trees.

balance elements. The lysimeter simulates only the upmost part of the catchment if the surface inflow is excluded because on the lower part of the catchment, infiltration is governed by local precipitation and surface flow, in each case, when the former is smaller than the possible upper limit of infiltration. It is advisable, however, to exclude the incoming surface runoff because the boundary conditions are, in this way, simplified, and to apply special devices for the simulation of R_i. In any case, the measurement of both R_i and R_o is an indispensible requirement, and the records are integrated parts of data collected by the lysimeter.

The most important parameters characterizing the behavior of the soil-moisture zone are those describing its soil-physical properties (the number and the sequence of the layers as well as porosity, effective diameter, shape-coefficient, grain-size and pore-size distributions, mineralogical and chemical composition for each layer, etc.).

Within the third subsystem, the characterization of the special structure of the cultivated zone is also necessary if such a zone exists in the lysimeter, e.g., there is no cultivated zone in bare lysimeters. Thus, porosity has to be divided into two parts, distinguishing primary and secondary porosity. The determination of the depth of the root zone and the vertical distribution of root density is also required. Data concerning the type and growing phase of plants inform on the probable suction exerted by the vegetation.

The basic data listed in the previous paragraph have to be supplemented by those describing the instantaneous condition of this layer. Thus, observations are needed to determine the moisture content of the aggregates, the water amount stored in the secondary structural pores, the salt content of the soil water (which influences the development of osmotic forces), the biological parameters of the soil, etc.

The flow and the storage process have to be described in the lower part of the soil-moisture zone where the physical characterization of these processes generally give acceptable approximations. There are many methods known for the hydrodynamic investigation of the flow through unsaturated layers. The results of the flow are measured, however, in the lysimeters as inputs and outputs, therefore, the storage and its change are the most important data to be determined for the water balance equation. The instantaneous storage is the difference of the actual vertical moisture distribution and the water retention curve. The latter can be determined either from soil-physical parameters or from direct measurements, while the observation of the instantaneous condition needs the application of special measuring methods to record the moisture distribution regularly, e.g., location of a measuring tube inside the lysimeter for a neutron probe.

The change of the stored water can be measured either by weighing the lysimeter or recording the water amount recharged into and drained from the equipment and calculating the difference of the two values. Weighing measures all the storages

(S_1, S'_1, S'_2, S_2 and S_3) in a combined form while the number of the jointly observed reservoirs depends on the position of the recording points of inputs and outputs if the mass balance method is used. Efforts should be made to determine the changes of storage in each zone separately. It is absolutely necessary, therefore, to have regular soil-moisture measurements within the lysimeter.

As it was already mentioned previously, the soil-moisture distribution curve is a parameter characterizing the soil-moisture zone of the lysimeter. Its position depends, however, on the depth where the pressure of the water in the pores is equal to the atmospheric value (water table). The correct simulation of the natural conditions can be ensured, therefore, only by those lysimeters in which a water table is maintained and even the fluctuation of the latter is regulated according to the depth of the water table observed at the simulated area.

The sensitivity of weighing lysimeters depends on the total weight of a vertical column with unit area, thus, the required accuracy excludes almost completely the modeling of a water table condition, since the latter requires the investigation of a relatively high column. The error caused by the aeration of the lower boundary can be well demonstrated by comparing the total moisture content in a 137 cm (54 in.) thick layer within the lysimeter to that in the surrounding area at the Coshocton research station (Ohio, USA) (Harrold and Dreibelbis, 1967) (Fig. 2-28). It can be seen from the figure that the stored amount was practically equal in the two profiles if this value was above a given limit. When the soil strated to dry out, the process was more rapid in the lysimeter, where the profile was aerated from both sides, and the moisture content decreased gradually below the natural value.

There is one solution, however, to eliminate this error, if other aspects exclude the maintenance of the water table. A given tension or moisture content has to be ensured at the lower boundary of the column by draining and moistening the pores here from a container where the water is under the required suction. The system can even be automated, measuring the tension outside the lysimeter at a given depth and regulating the suction of the water supplying system (having its distribution devices at the same depth as the tensiometers) accordingly to model the effect of the water table fluctuation.

The final conclusions of the analysis of the most important zone of lysimeters can be summarized as follows:

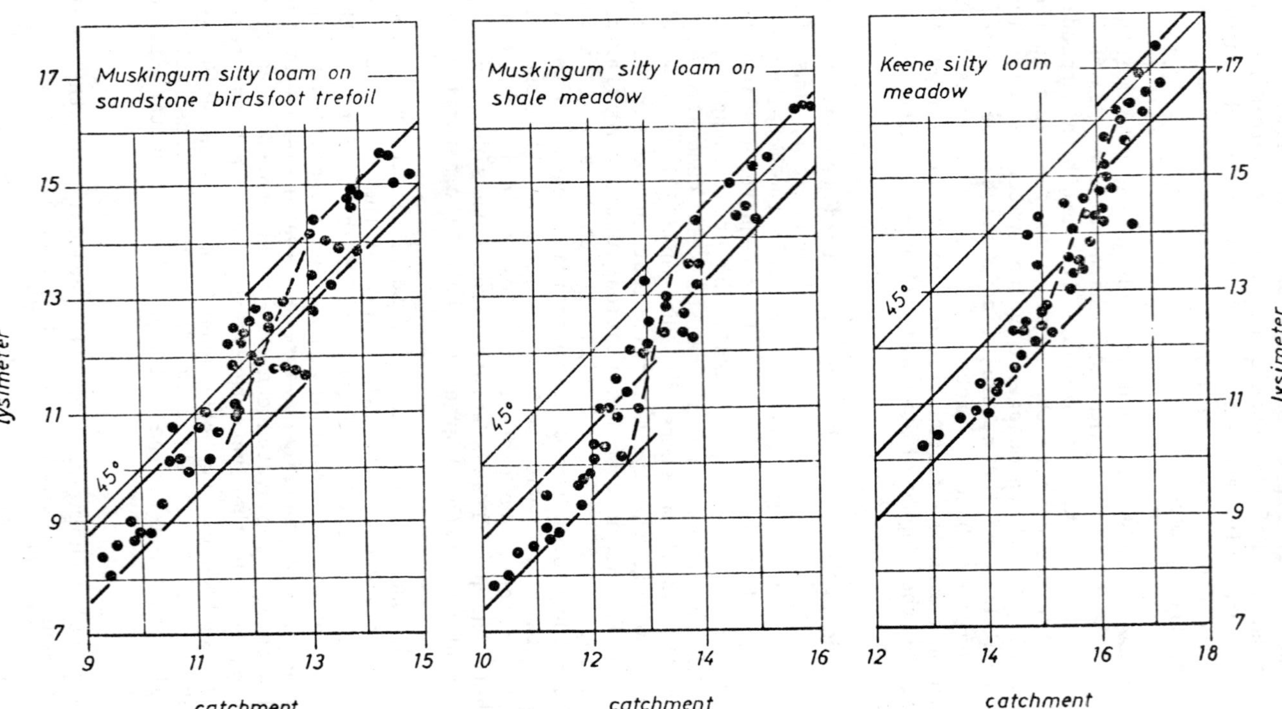

Fig. 2-28. Decrease of moisture content of lysimeters without water table due to drying at the lower boundary of the soil column.

- the existence or the lack of the water table inside the lysimeter causes a basic difference in the water balance (see A/2 and A/4 or A/3 and A/5 curves in Fig. 2-25), the equipment has to be divided into separate groups according to these two types of operation;

- to eliminate the error caused by the aeration of the profile from both sides, the moisture content has to be regulated at the lower boundary of the lysimeter if there is no water table maintained in the column;

- the basic value of the storage capacity of the soil-moisture zone can be characterized with the retention curve and, therefore, it is an important parameter of lysimeters;

- the regular observation of the vertical moisture distribution is absolutely necessary in all lysimeters, to facilitate the separate characterization of the storage in the various zones, thus, these records form an integral part of the lysimeter data.

The lowest zone of the investigated column is the gravitational groundwater space which may even be left out in some cases if suitable arrangements are made to simulate its influence. For the characterization of its composition and structure, the parameters generally used in seepage hydraulics may be applied (effective diameter, porosity, shape coefficient). Because the system is unconfined, the specific yield has to be determined as well. For recording the instantaneous condition, the position of the water table has to be observed.

When maintaining a water table condition in the lysimeter, the level of the groundwater has to be controlled. The possible ways of this operation are:

- to drain or recharge the groundwater with a given amount of water (which can be constant or changing according to a seasonal program) and to leave the water table to develop freely as the result of the natural and artificial effects, only recording the position of the water table;

- to keep the water table at an a priori determined level (which can be similarly constant or fluctuating according to seasons) and to ensure (and to record) the recharge and discharge necessary for the maintenance of this level.

The most perfect solution would be to regulate the water table within the lysimeter similar to how it develops in the surroundings and to measure the recharge and drainage necessary for this operation. The technical solution of this construction raises, however, difficulties and, therefore, either the maintenance of constant water level or the application of constant recharge (or drainage) are the generally accepted methods for operation. The maintenance of a constant water table, or recharging (draining) the lysimeter with constant amount of water, creates somehow unnatural conditions different from the groundwater regime in the surrounding layers. It is advisable, therefore, to construct a group of lysimeters, the members of which differ from each other only in the numerical value (depth of the water level maintained; the amount recharged or drained) characterizing the operation. In this way the suitable model of the surroundings can be selected for each period.

Water balance of lysimeters characterized by the directly measured and calculated data. On the basis of the previous explanations, four basic groups of lysimeters can be distinguished:

- bare lysimeter without groundwater (LA);

- bare lysimeter with groundwater (LB);

- lysimeter covered by vegetation without groundwater (LC);

- lysimeter covered by vegetation with groundwater (LD).

The measured and calculated data in the various groups are summarized in Table 2-5. The numbers used as subscripts of the symbols indicate the measuring methods usually applied:

1 - weighing lysimeter without soil-moisture measurement;

2 - weighing with soil-moisture measurement;

3 - weighing supplemented with the measurement of the drained or recharged amount of water, but without soil-moisture measurement;

4 - weighing and measuring both the drained and recharged amount of water and the soil moisture;

5 - measuring the recharge and discharge but not the soil moisture;

51- the version of type 5, when the control of the groundwater is ensured by recharging to or draining from it a given amount of water (constant or varying according to an a priori determined program);

Table 2-5. Possibility of evaluation of lysimeter data collected by various equipment.

Type of Lys.	Directly Measured Parameters	Parameters Calculated from Measured Values	Remarks	Type of Lys.	Directly Measured Parameters	Parameters Calculated from Measured Values	Remarks
I.	II.	III.	IV.	I.	II.	III.	IV.
	$P_T=P_s$	$\Delta S_2=\Delta G/\gamma A$			P_T	$\Delta S_1+\Delta S_2=\Delta G/\gamma A-\Delta P1$	4.
LA_1	$\Delta R=R_i-R_o$	$E+I_s=P_s-\Delta R$	1.		$\Delta R=R_i-R_o$	$E+\Delta S_1+E_1+I_s=P_T-\Delta R$	1.
	$\Delta G=G(t_2)-G(t_1)$	$E_s=\Delta S_2$	2.	LC_1	$\Delta G=G(t_2)-G(t_1)$	$ET_A=E+E_I+E_s+T_s=\Delta G/\gamma A-\Delta P1$	2.
					P1		
					P_T		
	$P_T=P_s$	$\Delta S_2=\Delta G/\gamma A$			$\Delta R=R_i-R_o$	$\Delta S_2=\int_o^M [W(x,t_2)-W(x,t_1)]dx$	
	$\Delta R=R_i-R_o$	$E+I_s=P_s-\Delta R$	1.		$\Delta G=G(t_2)-G(t_1)$	$\Delta S_1=\Delta G/\gamma A-\Delta S_2-\Delta P1$	4.
LA_2	$\Delta G=G(t_2)-G(t_1)$	$E_s=\Delta S_2$	2.	LC_2	$\Delta P1$	$P_s=P_T-\Delta S_1-E_I$	1.
	$W(x,t)$	$\Delta S_2=\int_o^M [W(x,t_2)-W(x,t_1)]dx$	3.		$W(x,t)$	$E+E_I+I_s=P_T-\Delta R-\Delta S_1$	1.
						$E_s+T_s=\Delta S_2$	2.
						$ET_A=E+E_I+E_s+T_s=\Delta G/\gamma A-\Delta P1$	2.
	$P_T=P_s$	$\Delta S_2=\Delta G/\gamma A$			P_T	$\Delta S_1+\Delta S_2=\Delta G/\gamma A-\Delta P1$	4.
	$\Delta R=R_i-R_o$	$E+I_s=P_s-\Delta R$	1.		$\Delta R=R_i-R_o$	$E+\Delta S_1+E_1+I_s=P_T-\Delta R$	1.
LA_3	$\Delta G=G(t_2)-G(t_1)$	$I_s=\Delta S_2-(-D)$	1.	LC_3		$I_s=\Delta S_2-(-D)$	1.
	(-D)	$E=P_s-\Delta R-\Delta S_2+(-D)$	1.		$\Delta G=G(t_2)-G(t_1)$	$ET_A=E+E_I+E_s+T_s=\Delta G/\gamma A-\Delta P1$	2.
		$E_s=\Delta S_2$	2.		$\Delta P1$		
					(-D)		

Table 2-5. (continued)

					P_T	$\Delta S_2=\int_o^M [W(x,t_2)-W(x,t_1)]dx$	
	$P_T=P_s$	$\Delta S_2=\Delta G/\gamma A$			$\Delta R=R_i-R_o$		
	$\Delta R=R_i-R_o$	$E+I_s=P_s-\Delta R$	1.		$\Delta G=G(t_2)-G(t_1)$	$\Delta S_1=\Delta G/\gamma A-\Delta S_2-\Delta P1$	4.
	$\Delta G=G(t_2)-G(t_1)$	$I_s=\Delta S_2-(-D)$	1.		$\Delta P1$	$P_s=P_T-\Delta S_1-E_I$	1.
LA_4	$(-D)$	$E=P_s-\Delta R-\Delta S_2+(-D)$	1.	LC_4	$(-D)$	$I_s=\Delta S_2-(-D)$	1.
	$W(x,t)$	$E_s=\Delta S_2$	2.		$W(x,t)$	$E+E_I=P_T-\Delta R-\Delta S_1-I_s$	1.
		$\Delta S_2=\int_o^M [W(x,t_2)-W(x,t_1)]dx$	3.			$E_s+T_s=\Delta S_2$	2.
						$ET_A=E+E_I+E_s+T_s=\Delta G/\gamma A-\Delta P1$	2.
	$P_T=P_s$	$E+I_s=P_s-\Delta R$	1.		P_T	$E+\Delta S_1+E_1+I_s=P_T-\Delta R$	1.
	$\Delta R=R_i-R_o$	$\Delta S_2=I_s+(-D)=P_s-\Delta R-E+(-D)$			$\Delta R=R_i-R_o$	$\Delta S_2=I_s+(-D)=P_T-\Delta R-E-\Delta S_1-E_1$	
	$(-D)$				$(-D)$		
LA_5	$V(t)=S_{2max}-S_2(t)$	$\Delta S_2=S_2(t_2)-S_2(t_1)=V(t_1)-V(t_2)$		LC_5	$V(t)=S_{2max}-S_2(t)$	$\Delta S_2=S_2(t_2)-S_2(t_1)=V(t_1)-V(t_2)$	
	periodically saturating the soil column until maximum water capacity	$E_s=\Delta S_2$ only approximation	2.		periodically saturating the soil column until maximum water capacity	$E_s+T_s=\Delta S_2$ only approximation	2.

Table 2-5. (continued)

	$P_T=P_s$	$\Delta S_2=\int_o^M [W(x,t_2)-W(x,t_1)]dx$			P_T	$\Delta S_2=\int_o^M [W(x,t_2)-W(x,t_1)]dx$	
	$R=R_i-R_o$				$\Delta R=R_i-R_o$		
LA_6	(-D)	$E+I_s=P_s-\Delta R$	1.	LC_6	(-D)	$I_s=\Delta S_2-(-D)$	1.
	W(x,t)	$I_s=\Delta S_2-(-D)$	1.		W(x,t)	$E+E_I+\Delta S_1=P_1-\Delta R-I_s$	1.
		$E=P_s-\Delta R-\Delta S_2+(-D)$	1.			$E_s+T_s=\Delta S_2$	2.
		$E_s=\Delta S_2$	2.				
	$P_T=P_s$	$\Delta S_3=n\cdot\Delta H=I_g-E_g$			P_T	$\Delta S_3=n\cdot\Delta H=I_g-E_g$	
	$\Delta R=R_i-R_o$	$\Delta S_2=\Delta G/\gamma A-\Delta S_3$			$\Delta R=R_i-R_o$	$\Delta S_1+\Delta S_2=\Delta G/\gamma A-\Delta S_3-\Delta P1$	4.
LB_1	$\Delta G=G(t_2)-G(t_1)$	$E+I_s=P_s-\Delta R$	1.	LD_1	$\Delta G=G(t_2)-G(t_1)$	$E+I_s+\Delta S_1=P_T-\Delta R$	1.
	$\Delta H=-[H(t_2)-H(t_1)]$	$I_s=\Delta G/\gamma A$	1.		$\Delta H=-[H(t_2)-H(t_1)]$	$I_s=\Delta G/\gamma A$	1.
		$E=P_s-\Delta R-\Delta G/\gamma A$	1.		$\Delta P1$	$\Delta S_1+E=P_T-\Delta R-\Delta G/\gamma A$	1.
		$E_s=\Delta G/\gamma A$	2.			$ET_A=E_s+T_s+E+E_I=\Delta G/\gamma A-\Delta P1$	2.

Table 2-5. (continued)

	$P_T=P_s$	$\Delta S_3=n\cdot\Delta H-I_g-E_g$			P_T	$\Delta S_3=n\cdot\Delta H=I_g-E_g$	
	$\Delta R=R_i-R_o$	$\Delta S_2=\Delta G/\gamma A-\Delta S_3$			$\Delta R=R_i-R_o$	$\Delta S_2=\int_o^M [W(x,t_2)-W(x,t_1)]dx$	
	$\Delta G=G(t_2)-G(t_1)$	$E+I_s=P_s-\Delta R$	1.		$\Delta G=G(t_2)-G(t_1)$		
LB_2	$\Delta H=-[H(t_2)-H(t_1)]$	$I_s=\Delta G/\gamma A$	1.	LD_2	$\Delta H=-[H(t_2)-H(t_1)]$	$\Delta S_1=\Delta G/\gamma A-\Delta S_3-\Delta S_2-\Delta P1$	4.
	$W(x,t)$	$E=P_s-\Delta R-\Delta G/\gamma A$	1.		$W(x,t)$	$P_s=P_T-\Delta S_1-E_I$	1.
		$E_s=\Delta G/\gamma A$	2.		$\Delta P1$	$E+E_I+I_s+\Delta S_1=P_T-\Delta R$	1.
						$I_g=\Delta G/\gamma A-\Delta P1$	1.
		$\Delta S_2=\int_o^M [W(x,t_2)-W(x,t_1)]dx$	3.			$E+E_I=P_T-\Delta R-\Delta S_1-\Delta G/\gamma A-\Delta P1$	1.
						$ET_A=E_s+T_s+E+E_I=\Delta G/\gamma A-\Delta P1$	2.
	$P_T=P_s$	$\Delta S_3=n\cdot\Delta H$			P_T	$\Delta S_3=n\cdot\Delta H$	
	$\Delta R=R_i-R_o$	$I_g-ET_g=\Delta S_3-D$			$\Delta R=R_i-R_o$	$I_g-ET_g=\Delta S_3-D$	
	$\Delta G=G(t_2)-G(t_1)$	$\Delta S_2=\Delta G/\gamma A-\Delta S_3$			$\Delta G=G(t_2)-G(t_1)$	$\Delta S_1+\Delta S_2=\Delta G/\gamma A-\Delta S_3-\Delta P1$	4.
	$\Delta H=-[H(t_2)-H(t_1)]$	$E+I_s=P_s-\Delta R$	1.		$\Delta H=-[H(t_2)-H(t_1)]$	$E+E_I+E_s+\Delta S_1=P_T-\Delta R$	1.
	D	$I_s=\Delta G/\gamma A-D$	1.		$\Delta P1$	$\Delta S_1+I_s=\Delta G/\gamma A-D-\Delta P1$	1.
LB_3		$E=P_s-\Delta R-\Delta G/\gamma A+D$	1.	LD_3	D	$E+E_I=P_T-\Delta R-\Delta G/\gamma A+D+\Delta P1$	1.
		$E_s=\Delta G/\gamma A-D$	2.			$ET_A=E_s+T_s+E+E_I=\Delta G/\gamma A-D-\Delta P1$	2.
	in the special case of maintenance of a given water table, groundwater drainage and recharge are measured separately	$I_g=(-D)$ $E_g=(+D)$			in the special case of maintenance of a given water table, groundwater drainage and recharge are measured separately	$I_g=(-D)$ $ET_g=(+D)$	

Table 2-5. (continued)

	$P_T=P_s$	$\Delta S_3=n\cdot\Delta H$			P_T	$\Delta S_3=n\cdot\Delta H$	
	$\Delta R=R_i-R_o$	$I_g-E_g=\Delta S_3-D$			$\Delta R=R_i-R_o$	$I_g-ET_g=\Delta S_3-D$	
	$\Delta G=G(t_2)-G(t_1)$	$\Delta S_2=\Delta G/\gamma A-\Delta S_3$			$\Delta G=G(t_2)-G(t_1)$	$\Delta S_2=\int_o^M [W(x,t_2)-W(x,t_1)]dx$	
		$E+I_s=P_s-\Delta R$	1.		$\Delta H=-[H(t_2)-H(t_1)]$		
	$\Delta H=-[H(t_2)-H(t_1)]$	$I_s=\Delta G/\gamma A-D$	1.		$\Delta P1$	$\Delta S_1=\Delta G/\gamma A-\Delta S_3-\Delta S_2-\Delta P1$	4.
LB_4	D	$E=P_s-\Delta R-\Delta G/\gamma A+D$	1.		D	$P_s=P_T-\Delta S_1$	
	$W(x,t)$	$E_s=\Delta G/\gamma A-D$	2.	LD_4	$W(x,t)$	$E+E_I+I_s+\Delta S_1=P_T-\Delta R$	1.
						$\Delta S_1+I_s=\Delta G/\gamma A-D-\Delta P1$	1.
						$E+E_I=P_T-\Delta R-\Delta G/\gamma A+D+\Delta P1$	1.
		$\Delta S_2=\int_o^M [W(x,t_2)-W(x,t_1)]dx$	3.			$ET_A=E_s+T_s+E+E_I=\Delta G/\gamma A-D-\Delta P1$	2.
	in the special case of maintenance of a given water table, groundwater drainage and recharge is measured separately	$I_g=(-D)$ $E_g=(+D)$			in the special case of maintenance of a given water table, groundwater drainage and recharge is measured separately	$I_g=(-D)$ $ET_g=(+D)$	

Table 2-5. (continued)

	$P_T=P_s$	$\Delta S_3=n\cdot\Delta H$			P_T	$\Delta S_3=n\cdot\Delta H$	
	$\Delta R=R_i-R_o$	$I_g-E_g=\Delta S_3-D$			$\Delta R=R_i-R_o$	$I_g-ET_g=\Delta S_3-D$	
LB_{51}	$\Delta H=-[H(t_2)-H(t_1)]$	$E+I_s=P_s-\Delta R$	1.	LD_{51}	$\Delta H=-[H(t_2)-H(t_1)]$	$E+E_I+I_s+\Delta S_1=P_T-R$	1.
	D	$\Delta S_2=I_s-\Delta S_3+D-P_s-\Delta R-E-\Delta S_3+D$	1.		D	$\Delta S_2=I_s-\Delta S_3+D$	1.
						$\Delta S_1+\Delta S_2=P_T-\Delta R-E-E_I-\Delta S_3+D$	1.
	$P_T=P_s$	$\Delta S_3=n\cdot\Delta H$			P_T	$\Delta S_3=n\cdot\Delta H$	
	$\Delta R=R_i-R_o$	$I_g-ET_g=\Delta S_3-D$			$\Delta R=R_i-R_o$	$I_g-ET_g=\Delta S_3'-D$	
	$\Delta H=-[H(t_2)-H(t_1)]$	$I_g=(-D)$			$\Delta H=-[H(t_2)-H(t_1)]$	$I_g=(-D)$	
LB_{52}	D	$ET_g=(+D)$		LD_{52}	D	$ET_g=(+D)$	
		$E+I_s=P_s-\Delta R$	1.			$E+E_I+I_s+\Delta S_1=P_T-\Delta R$	1.
		$\Delta S_2=I_s-\Delta S_3+D=P_s-\Delta R-E-\Delta S_3+D$	1.			$\Delta S_2=I_s-\Delta S_3+D$	1.
						$\Delta S_1+\Delta S_2=P_T-\Delta R-E-E_I\Delta S_3+D$	1.
	$P_T=P_s$	$\Delta S_3=n\cdot\Delta H$			P_T	$\Delta S_3=n\cdot\Delta H$	
	$\Delta R=R_i-R_o$	$I_g-ET_g=\Delta S_3-D$			$\Delta R=R_i-R_o$	$I_g-ET_g=\Delta S_3-D$	
	$\Delta H=-[H(t_2)-H(t_1)]$	$\Delta S_2=\int_o^M [W(x,t_2)-W(x,t_1)]dx$			$\Delta H=-[H(t_2)-H(t_1)]$	$\Delta S_2=\int_o^M [W(x,t_2)-W(x,t_1)]dx$	
LB_{61}	D	$E+I_s=P_s-\Delta R$	1.	LD_{61}	D	$E+E_I+I_s+\Delta S_1=P_T-\Delta R$	1.
	$W(x,t)$	$I_s=\Delta S_2+\Delta S_3-D$	1.		$W(x,t)$	$I_s=\Delta S_2+\Delta S_3-D$	1.
		$E=P_s-\Delta R-\Delta S_2-\Delta S_3+D$	1.			$E+E_I+\Delta S_1=P_T-\Delta R-\Delta S_2-\Delta S_3+D$	1.
		$E_s=\Delta S_2+\Delta S_3-D$	2.			$P_s=P_T-E-E_I-\Delta S_1$	1.
						$E_s+T_s=\Delta S_2+\Delta S_3-D$	2.

Table 2-5. (continued)

	$P_T=P_s$	$\Delta S_3=n\cdot\Delta H$			P_T	$\Delta S_3=n\cdot\Delta H$	
	$\Delta R=R_i-R_o$	$I_g-ET_g=\Delta S_3-D$			$\Delta R=R_i-R_o$	$I_g-ET_g=\Delta S_3-D$	
	$\Delta H=-[H(t_2)-H(t_1)]$	$I_g=(-D)$			$\Delta H=-[H(t_2)-H(t_1)]$	$I_g=(-D)$	
	D	$ET_g=(+D)$			D	$ET_g=(+D)$	
LB_{62}	$W(x,t)$	$\Delta S_2=\int_o^M [W(x,t_2)-W(x,t_1)]dx$		LD_{62}	$W(x,t)$	$\Delta S_2=\int_o^M [W(x,t_2)-W(x,t_1)]dx$	
		$E+I_s=P_s-\Delta R$	1.			$E+E_I+I_s+\Delta S_1=P_T-\Delta R$	1.
		$I_s=\Delta S_2+\Delta S_3-D$	1.			$I_s=\Delta S_2+\Delta S_3-D$	1.
		$E=P_s-\Delta R-\Delta S_2-\Delta S_3+D$	1.			$E+E_I+\Delta S_1=P_T-\Delta R-\Delta S_2-\Delta S_3+D$	1.
		$E_s=\Delta S_2+\Delta S_3-D$	2.			$P_s=P_T-E-E_I-\Delta S_1$	1.
						$E_s+T_s=\Delta S_2+\Delta S_3-D$	2.

52- the version of type 5, when the water table is controlled by keeping it at a constant depth, or regulating according to a given program;

6 - measuring the recharge, discharge, and soil moisture;

61- the version of type 6, when the control of the groundwater is ensured by recharging to or draining from it a given amount of water (constant or varying according to an a priori determined program);

62- the version of type 6, when the water table is controlled by keeping it at a constant depth, or regulating according to a given program.

Most of the symbols applied in the table are represented in Fig. 2-2. The supplementary terms are as follows:

A the horizontal area of the lysimeter (m^2);

γ specific weight of water (kg/dm^3);

M thickness of the investigated soil column in lysimeters without groundwater (mm);

P1 the increase of plants in weight expressed by equivalent water column (mm);

G(t) the weight of the lysimeter measured at time t of the investigation;

V(t) the free storage capacity of the unsaturated zone which can be determined as the amount of water necessary to completely saturate the pores at time t of the investigation.

The meaning of figures applied in the last column of the table (Remarks) is as follows:

1. In periods with precipitation

2. In dry periods (without precipitation)

3. Control

4. If the change of the weight of plants is not measured, the period between two measurements should be short and thus P1 can be neglected.

Evaluating the various groups of the elements of the water balance being measurable by different types of lysimeters, the following proposals can be made in connection with the development of standardized forms of lysimeters.

- The lysimeter should always be equipped with devices ensuring the regular observation of the vertical soil-moisture distribution within the investigated soil column.

- If there is no water table maintained in the lysimeter, the regulation of the moisture content (or tension) has to be solved at the lower boundary of the soil column.

- The data recorded in lysimeters, the surface of which is covered by vegetation, have to be supplemented with information concerning the conditions of the plants.

- Supposing that the purpose served by the data collection can be fulfilled by using lysimeters without groundwater, the application of the weighing type can be proposed (in the case of a bare lysimeter (LA_4), with plants (LC_4). In large networks, if more lysimeters are built as members of the network, (LA_6) or (LC_6) can also be applied considering the advantage of the cheaper construction.

- When the modeling of the water table is necessary, the use of the weighing lysimeter is almost excluded because the accuracy of weighing is proportional to the total weight, and thus the achievement of the required sensitivity is hindered by the high soil column. Thus the type LB_6 or LD_6 is the suitable construction which gives almost the same amount of information (except interception) as the weighing lysimeter.

- The technical solution of a lysimeter with continuously regulated water table is difficult and expensive. It is advisable, therefore, to operate the lysimeter either with constant recharge or discharge (types LB_{61} and LD_{61}) or with constant water table (types LB_{62} and LD_{62}). The complete investigation of the unsaturated zone requires the construction of more lysimeters as a set, each having different recharge (or discharge) in the case of types LB_{61} and LD_{61}, and different constant water table if types LB_{62} and LD_{62} are applied.

6-6 Methods of Characterizing Infiltration

The purpose of lysimeters is to measure the components of the water balance developing in the soil moisture zone separately from each other, or at least some combined effects of those members of the balance, the actions of which cannot be separated. The construction of a lysimeter fully equipped is expensive. Its operation and maintenance as well as the evaluation of the collected data require a well trained staff of considerable number. For this reason, the number of lysimeters is generally low everywhere compared to the area investigated. The example of the Hungarian lowland (which is one of the best equipped areas in the world) will be used to demonstrate the density of lysimeters. There exists a regularly observed lysimeter network composed of 12 stations within the area, the extension of which is about 50,000 km^2. The purpose of the network is the determination of the water consumption of various plants under different climatic conditions. There are, therefore, an average of 10 lysimeters at each station covered by different plants. Apart from the network, there are eight research stations, where lysimeters are applied to investigate various hydrological processes (groundwater recharge; relationship between the depth and accretion of groundwater; evaporation of aquatic vegetation; comparison of actual and potential evaporation; influence of irrigation and sewage disposal in agriculture, etc.). The total number of lysimeters within the research stations is about 130. Their structures are different depending on the actual purpose of the investigation and also on the time of their construction. Considering the number of the stations, one station per 2500 km^2 characterizes their density (the number of equipment related to the area, i.e., 1 lysimeter/200 km^2 is not a characteristic parameter, since their grouping is necessary to collect the information required at one place).

The average area of one lysimeter is generally between 0.5-1.0 m^2. The number of large lysimeters is low because the cost rises rapidly with their size. Since the reliability becomes worse and worse as the size decreases, the use of smaller equipment is not advisable. Supposing that 1 m^2 can be achieved as the optimum horizontal area of lysimeters, only a 1:2x10^8 ratio of the total area would be under observation by lysimeters if the relatively high density demonstrated by the Hungarian example were characteristic for an investigation. The low ratio draws the attention to the possible high sampling errors, and emphasizes the difficulties encountered when investigating hydrological processes occurring in the soil-moisture zone over large areas. The latter task can be solved only if more samples and cheaper methods are also applied, measuring at least one or two components of the water balance at as many points as feasible, apart from lysimeters giving the most complete set of information.

There is only one process directly influencing the storage and flow of water in the soil-moisture zone which is observable on the surface: i.e., infiltration. The quantitative determination of this parameter raises, however, similarly difficult technical and theoretical problems. Among the generally applied methods, the use of environmental isotopes and radioactive tracers is suitable to characterize the actual seepage velocity of the infiltrating water, thus, its amount can be calculated by combining these methods with the determination of the change of moisture content stored in the influenced zone. The other method frequently used to investigate the process of infiltration is the application of infiltrometers. This equipment gives information, however, only on the probable character of infiltration and not on the actual process. It is necessary, therefore, to analyze the possible evaluation of data collected in this way.

Use of radioactive tracers and environmental isotopes. Investigations performed by applying radioactive isotopes have proved that the water moves downward layer by layer, in the soil-moisture zone (Blume et al., 1967). Rain falling on the surface and infiltrating into the soil pushes the older water ahead and takes its place. This phenomenon allows a simple determination of seepage velocity, i.e., to artificially mark a water layer just below the surface by radioactive isotopes and observe its movement downwards after rain or after accelerating this movement by applying sprinkling on the surface.

Some authors have reported the use of radioactive tracers for this purpose (^{36}Cl, ^{60}Co, etc.). The adsorption of the tracers by soil may cause problems in such cases. The best results were achieved by metallic radioactive isotopes in the form of a chelatic compound, e.g., triton B about -2.5 percent (Bondarenko and Globus, 1967). The effect of the adsorption of anions was also observed by other authors (e.g., Frissel et al., 1973). There are many examples mentioned in the literature to demonstrate that the problems arising in connection with adsorption can be avoided by applying tritium as a tracer instead of other isotopes occurring only as dissolved materials in water (Blume et al., 1967; Frissel et al., 1973; Charreau and Jaquinot, 1967).

The propagation of the marked layer can be easily followed if the radiation of the tracer is strong enough, and detectable by a simple sensor in a borehole cased similarly as those used for nuclear logging. Using tritium as a tracer, samples have to be collected from different depths by drilling, and the water extracted from them have to be condensated and analyzed to determine the tritium content.

There are two other ways to use tritium as a natural environmental isotope for the investigation of infiltration.

The first is the tagging of the positions of the peaks which have occurred in the time series of atmospheric tritium content as the results of experimental explosions of nuclear bombs. The other is the comparison of the tritium content of recent precipitation and groundwater, respectively.

The first thermonuclear tests took place in 1954, and two eminent peaks of the tritium content are known in precipitation: the first between 1958 and 1959 and the second in 1963.

A complete core has to be taken from the soil profile for the investigation and it has to be divided into subsequent samples. The determination of the tritium content values in the soil moisture extracted from the samples is the following step, and the vertical distribution of tritium values can be constructed from the data. The position of the relative peaks can be related to the time of the extreme values in precipitation.

The absolute value of the tritium ratio measured in soil moisture also gives assistance to determine the corresponding peaks in the time series and in the profile. The possible mixing in the profile caused by diffusion and concentration due to evapotranspiration have to also be considered in this comparison apart from the decay of radiation.

It follows from the basic principles of the investigation that the movement of water infiltrated into the soil before 1954 cannot be tagged. The great change of the atmospheric tritium content in 1954 and the two eminent peaks mentioned are the most easily detectable points. There is also a lower limit of the age of water detectable in this way: i.e., the stopping of the nuclear explosions on the surface. Precipitation fallen since that time contained and still contains an almost constant amount of tritium having only a small seasonal fluctuation. It is not possible, therefore, to make any distinction between the younger soil-moisture layers by measuring their tritium contents.

Where the percolation of water from the surface to the groundwater does not require more than ten years because of the higher permeability or the smaller thickness of the soil-moisture zone, the determination of the tritium content of the groundwater and its comparison with the parameter of precipitation can be used for the characterization of infiltration, although this method does not give quantitative information, it is sufficient only to draw-up qualitative conclusions.

The seasonal fluctuation of the tritium content of precipitation has to be determined at first from regularly observed data. Where the change of the tritium value can be observed from month to month in the groundwater as well, a rapid infiltration and the direct influence of precipitation on soil moisture can be expected. The character of the yearly fluctuation of tritium content gives information even on the time lag between infiltration and groundwater accretion, in this way, facilitating the estimation of the velocity of infiltration.

The application of this method is demonstrated with an example represented in Fig. 2-29. The fluctuation of the tritium content observed in six observation wells is compared to that of precipitation. The wells are located in a closed groundwater basin in the northern part of Hungary and all drain the first unconfined water horizon. The graph of Well No. 14 runs together with precipitation. Wells No. 7 and No. 12 show a similar fluctuation, but their amplitude is smaller and a considerable time lag can also be observed between the peaks. In other cases, e.g., Well No. 4, there is no resemblance between the graphs of groundwater and precipitation, respectively. A common character of all the observation points is, however, that the average TU value is around or above 100 units indicating the relatively strong and fast influence of precipitation.

The wells having very similar graphs as that of precipitation indicate the strongest infiltration within the area. Such seasonal fluctuation as that of Well No. 7 (decreased amplitude and well pronounced time lag but similar form as that determined from precipitation data) characterizes the area covered by a less impermeable layer where the vertical infiltration is still the dominant process influencing the development of groundwater balance. The comparison of the graphs of Wells No. 7, No. 12 and No. 14 shows that the data can be used even to characterize the permeability of the covering layer. Where the fluctuation of the tritium content is irregular or almost constant, the mixing of various processes is expected. Horizontal flow collecting groundwaters with various tritium values also becomes dominant, apart from local infiltration.

Infiltrometers. The original purpose of infiltrometers was to have a simple method applicable under field conditions for the determination of hydraulic conductivity of layers lying just under the surface. The equipment is suitable, at the same time, to characterize the process of infiltration through the unsaturated layer.

The first form of this test was developed by Müntz and Lainé at the beginning of this century (Németh, 1942). After taking away the upper layer of soil being penetrated by roots and having special structure composed of aggregates and large channels between the clods (cultivated zone), the surface of the remaining part of the soil-moisture zone has to be cleaned

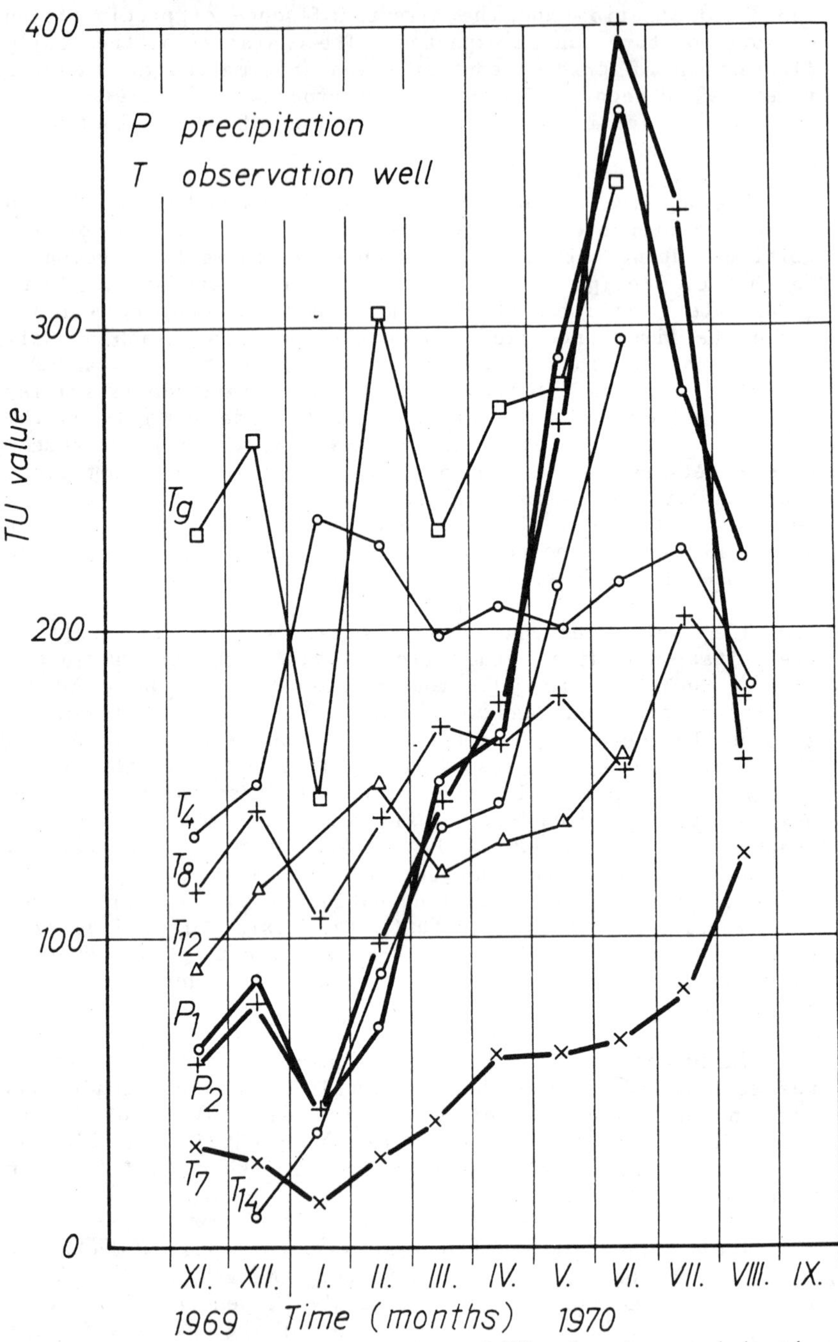

Fig. 2-29. Seasonal fluctuation of TU value in precipitation and groundwater.

and smoothed. A cylinder prepared with a thin metallic wall, having standardized diameter and height is stuck into the layer afterwards. There is a collar welded to the outer side of the wall around the cylinder. The distance between the lower sharp edge of the cylinder and the collar is generally 10 cm. It is in this way ensured that the infiltrometer be always pushed into the soil to a depth of 10 cm.

The internal space surrounded by the cylinder is filled with water. A continuous water supply maintains a constant water level here, at a given elevation (h_o is constant) above the surface, by recharging the amount infiltrated from the cylinder into the soil. The most simple and generally applied method of supplying the necessary water is the use of a Mariott bottle. The neck of the bottle is cut obliquely. Water can flow out of the turned-over bottle if the water level is below the upper part of the orifice. The Mariott bottle has to be placed, therefore, above the infiltrometer so that its orifice is at the required elevation of the water level. Thus, water is released from the bottle continuously and its amount is always equal to that necessary to maintain the water level at the h_o elevation because the water supply would be automatically stopped if the level were raised above the orifice. The wall of the bottle is generally calibrated, thus the amount of water recharged into the cylinder between two times t_1 and t_2 can be directly determined.

The volume released from the bottle (V) during a time interval, $\Delta t = t_2 - t_1$, divided by Δt gives the average flow rate of infiltration (Q) in the interval in question ($Q = V/\Delta t$). Choosing small time intervals, the Q value can be represented as a continuous function of time (Fig. 2-30). In other cases, the flow rate is divided by the horizontal area A of the cylinder and the flux (q, the flow rate through a unit area which is equal to seepage velocity v, $q = v = Q/A$) is plotted versus time.

When the purpose of the evaluation of the measured data is the determination of the hydraulic conductivity of the layer, a series of rough approximations have to be made. The most important hypothesis is that the development of piston flow is assumed (i.e., the infiltrating water percolates through a cylindrical flow space having the same diameter as the infiltrometer; the flow has no extension sidewards; there is a well-defined wetting front, above which the flow space is completely saturated the hydraulic conductivity is, therefore, constant here). Accepting this hypothesis, the seepage velocity is equal to the product of hydraulic conductivity (K) and hydraulic gradient (I), and the latter can be calculated from the geometrical parameters of the flow system. The total

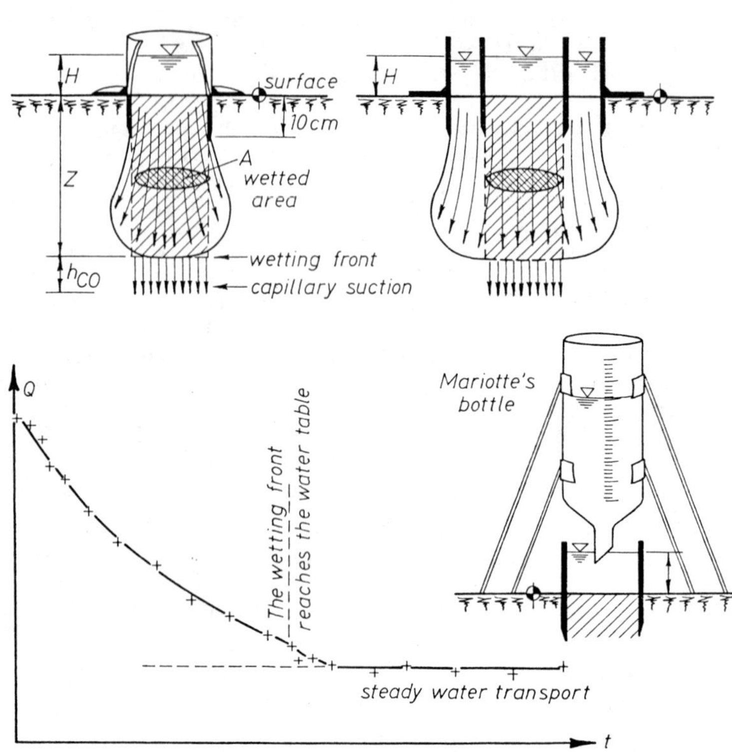

Fig. 2-30. Simple and double infiltrometer.

energy head prevailing between the entry face and the wetting front is composed of three members: i.e., h_o water column above the surface; z position of the wetting front related to the surface which is the reference level of the system (this z parameter is equal to the length of seepage as well); and h_{co} average suction head acting along the wetting front. The energy head divided by the length of seepage is equal to the hydraulic gradient, thus an equation can be derived to calculate the hydraulic conductivity characterizing the supposed saturated condition of the layer overlying the water table (Green-Ampt equation):

$$v = \frac{Q}{A} = KI = K \frac{h_o + z + h_{co}}{z} ;$$

$$v \rightarrow K, \text{ if } z \rightarrow \infty ; \text{ because } I = \frac{h_o + z + h_{co}}{z} \rightarrow 1 . \quad (2\text{-}16)$$

Contrary to the theoretical hypothesis, the infiltration is disturbed by the random character of the process of saturation along the wetting front. The front is not a well-defined plane, water moves faster in the large pores while some delay can be observed in the narrow ones. The suction is also a function of the pore size, changing randomly along the path. A partially saturated zone develops, therefore, at the border of the wetting instead of a sharp front. The effects of these complicated processes can be approximated only very roughly by considering the average suction head in the equation.

Both the development of the partially saturated zone and the effect of capillary suction is terminated when the wetting front reaches the water table, or more precisely the close capillary zone. This point in time is generally indicated by the break of the curve representing the infiltrating flow rate versus time relationship, thus, the development of this condition is clearly recognizable (Grishkan, 1966). The velocity of infiltration, as well as the flow rate, is constant after this time and Eq. 2-16 can be simplified by neglecting the suction head. In most cases, the depth of the ponding water above the surface can even be neglected compared to the distance between the surface and the water table, thus the unity of the hydraulic gradient can be assumed although the depth of infiltration is not infinite. The calculation of hydraulic conductivity is simpler if the flow rate independent of time can be observed and used as basic data for the investigation.

Observations in the first period of infiltration (when the gradient is higher than unity) can also be used to calculate hydraulic conductivity. The application of the complete form of Eq. 2-16 requires, however, the knowledge of the depth of infiltration and the average capillary suction. The instantaneous position of the wetting front can be determined with augers, but capillary suction has to be determined from soil-physics parameters. Applying a neutron probe to measure the depth of the wetting front, the thickness of the partially saturated zone between the undisturbed and saturated parts of the profile can also be determined. The thickness measured in this way may be assumed to be equal to the average suction head.

Error may be caused by the incomplete saturation of the layer above the wetting front. Air bubbles entrapped by the infiltrating water decrease the amount of saturation and since hydraulic conductivity is a function of this parameter, the conductivity determined by this method is generally smaller than that belonging to the completely saturated condition.

The most serious uncertainty in connection with the application of infiltrometers is caused by the fact that the infiltrating jet is not cylindrical, but the flow space extends sidewards. To decrease this disturbing effect, double cylinders are generally used, e.g., Nesterov's infiltrometer. In this case, two concentrical cylinders are stuck into the soil. The same water level is maintained within the annular space between the two cylinders as that acting inside the inner cylinder. The water infiltrating from the space enclosed by the two cylinders is recharged separately from the measuring space surrounded by the inner cylinder (where the water supply is solved in the same way as it was explained in connection with the simple infiltrometer). The amount infiltrating from the outer ring is not measured because its purpose is only to recharge the water consumed by the extension of the flow space sidewards, and to ensure, in this way, the approximately cylindrical form of the inner jet as well as its vertical direction. The evaluation of the observed data is the same as that applied when using a simple infiltrometer.

The relative decrease of the sidewards flow compared to the total amount of recharge can also be achieved by using large infiltration basins. Their further advantage is that a larger soil column is investigated and thus the average conditions are better represented by the measured data. The greater size of the equipment and the higher amount of water to be recharged make the operation more difficult which fact hinders the application of the method (Makaricheva, 1966; Szabó, 1954).

The data measured by infiltrometers can be used not only to determine hydraulic conductivity, but also to characterize the influence of the properties as well as the instantaneous conditions of the topmost layer of infiltration. Supposing that the infiltration versus time relationship can be approximated by a simple mathematical equation, e.g., the exponential form proposed by Horton, the constants of such formulae can be calculated from the corresponding flow rate and time values. Repeating the measurements, starting with different initial moisture contents, even the effect of the amount of saturation prevailing at the beginning of infiltration can be characterized. Knowing the infiltration potential of the soil-moisture zone belonging to a given condition of saturation, the amount of water expected to be infiltrated from a rain or during a period of irrigation can also be estimated.

6-7 Areal Characterization of Soil-Moisture Distribution

It was already mentioned that the parameters characterizing the moisture content of the soil can be determined only as point values or related to relatively small areas. At the same time, the great nonhomogeneity of the processes acting causes the considerable scattering of the observed values, which fact hinders the calculation of average parameters acceptable as representative values for large areas. There are many attempts, therefore, to characterize the territorial distribution of soil moisture within unified regions.

The most promising way to achieve the general description of moisture conditions over large areas is the use of airborne infrared photos, as it was mentioned among the methods suitable for soil-moisture measurements. Since this technique does not give information about the vertical moisture distribution in the soil profile, and since its application is much easier if only qualitative results are required (interpolation between points where the parameters are determined by using field measurements) and the quantitative evaluation of the photos is not necessary, it is advisable to combine air photos and data collected from a regularly observed network of soil moisture measuring points.

Both the establishment of such a network and the general analysis of the variability of soil-moisture parameters need the investigation of the representativeness of data measured at one point. The research used here to demonstrate the influence of the different variables on the dynamics of soil moisture was executed in the northwest part of Hungary as a preparatory investigation for the establishment of a network for soil-moisture observations. Two research fields were selected, one 25 km^2 and the other 15 km^2, and a great number of access tubes were installed for neutron probes. The tubes were located at different distances from each other. The soil and the vegetation were also variables. Thus, after collecting the data of the measurements repeated regularly during one year, the influence of distance, soil, and vegetation on soil moisture became calculable. A special condition characterizes both experimental areas: i.e., there is a covering layer of cohesive material (from clay to sandy silt) of about 2 m, and it is underlain by a coarse gravel. The water table fluctuates within the gravel. The covering layer has no capillary recharge from the groundwater and the analysis of the influence of the depth of the water table could be excluded, therefore, from the investigation.

The measured data collected under different conditions can be compared in two different ways:

- either the point values measured at the same time and in the same depth at two different observation profiles are related to each other;

- the moisture contents are calculated for given depth intervals and the simultaneous values are compared.

Figure 2-31 demonstrates the influence of the distance between the measuring points. The first pair of access tubes were located at a distance of 3 m, thus all the other conditions were the same. One tube of the second pair was in the middle of an agricultural plot and the other at its borders. The distance was a few tens of meters, but the effect of cultivation was slightly different. The distance between the members of the third pair was some hundreds of meters in different plots, but the soil and the vegetation were similar at both places. Finally, similar soil and plant conditions within the two experimental areas were compared by the fourth pair of access tubes. The distance between them was over 20 km, thus, some climatic change can also be expected.

The comparison shows that the scattering of the point values increases from 20 percent to 40 percent with increasing distance but the character of the relationship is not changed if the points are located within the same experimental field and the conditions (soil and vegetation) are the same (the relationship between the two measuring points is represented by a stragitht regression line having a slope of 45°). At greater distance (comparing two profiles located at the two experimental areas), the character of the data is also modified, although the scattering is not higher than in the previous cases. The use of the cumulative amount of soil moisture considerably decreases the scattering, due to the fact that the uncertainties caused by the local changes in soil material are in this way eliminated. The data observed within the same experimental area show close correlation while the comparison of two profiles from different areas indicates the different character of the latter.

Different soil conditions and plants were compared in the second step of the investigation (Fig. 2-32). As it is demonstrated by the figures, the point values have extremely large scattering if the soil or the vegetation is not the same at the two investigated points. The cumulated moisture content also indicates very weak correlation.

Since the airborne infrared photos determine the cumulative moisture content of the upper few dm of the profile, it can be stated that these photos combined with a soil survey and with maps showing the distribution of vegetation can be used for the characterization of the territorial distribution of soil moisture. Complete quantitative evaluation and the characterization of the vertical moisture distribution can be achieved, however, only if the set of systematically repeated photos is supplemented with data collected at the grid points of a regularly observed network where the soil moisture is measured between the surface and the water table by using field methods.

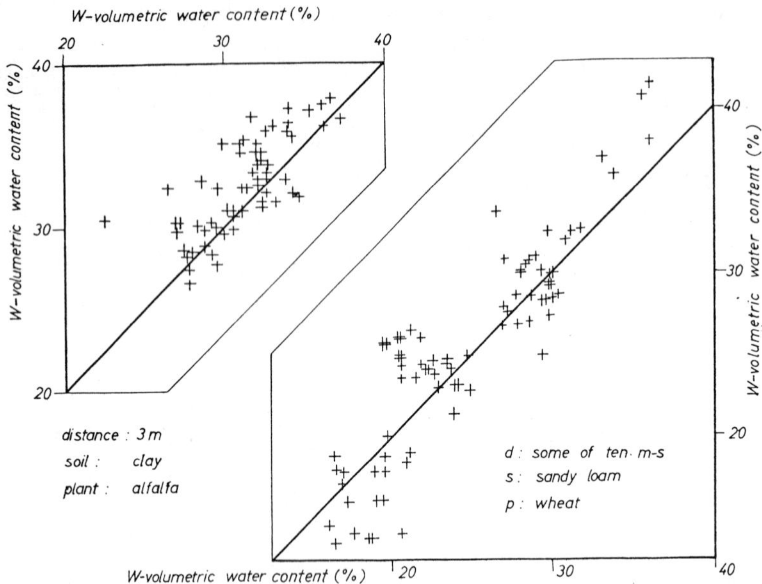

(a) Comparison of the point values of soil-moisture (soil and plant are the same at compared places)

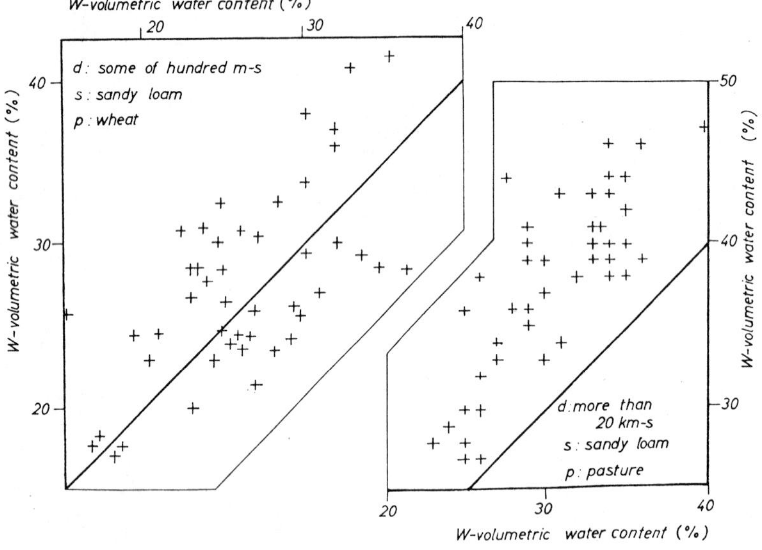

Fig. 2-31. Comparison of moisture-content values measured under similar conditions.
(a) point values

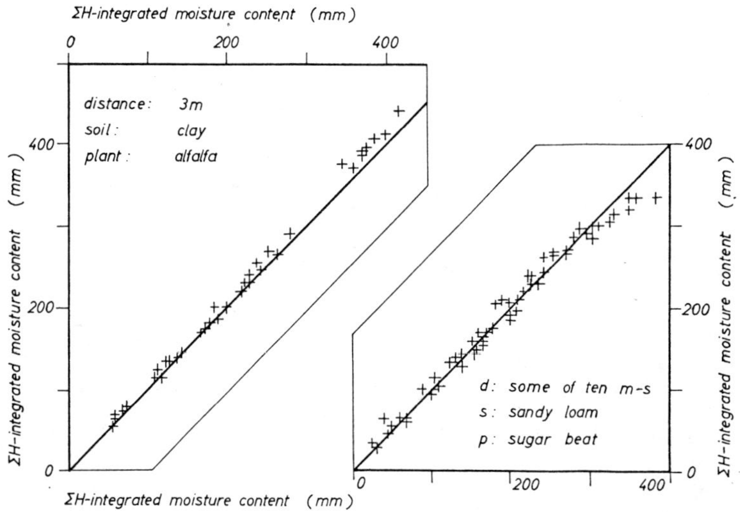

(b) Comparison of total moisture content integrated for given depth interval (soil and plant are the same at compared places)

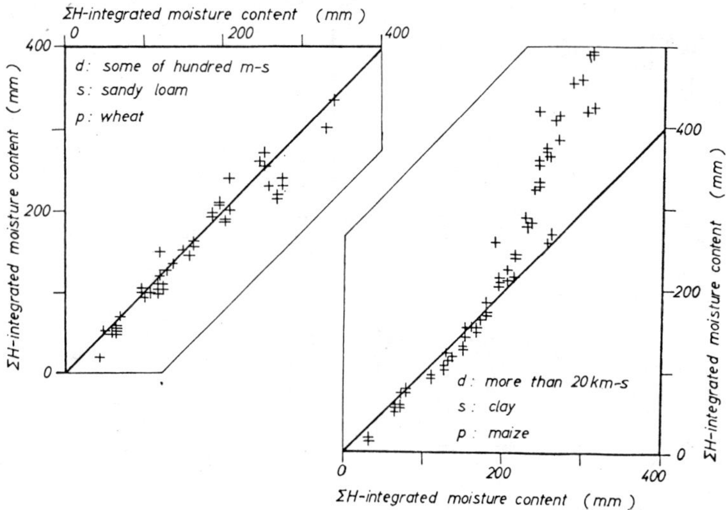

Fig. 2-31. Comparison of moisture-content values measured under similar conditions.
(b) values summarized for a given interval of depth

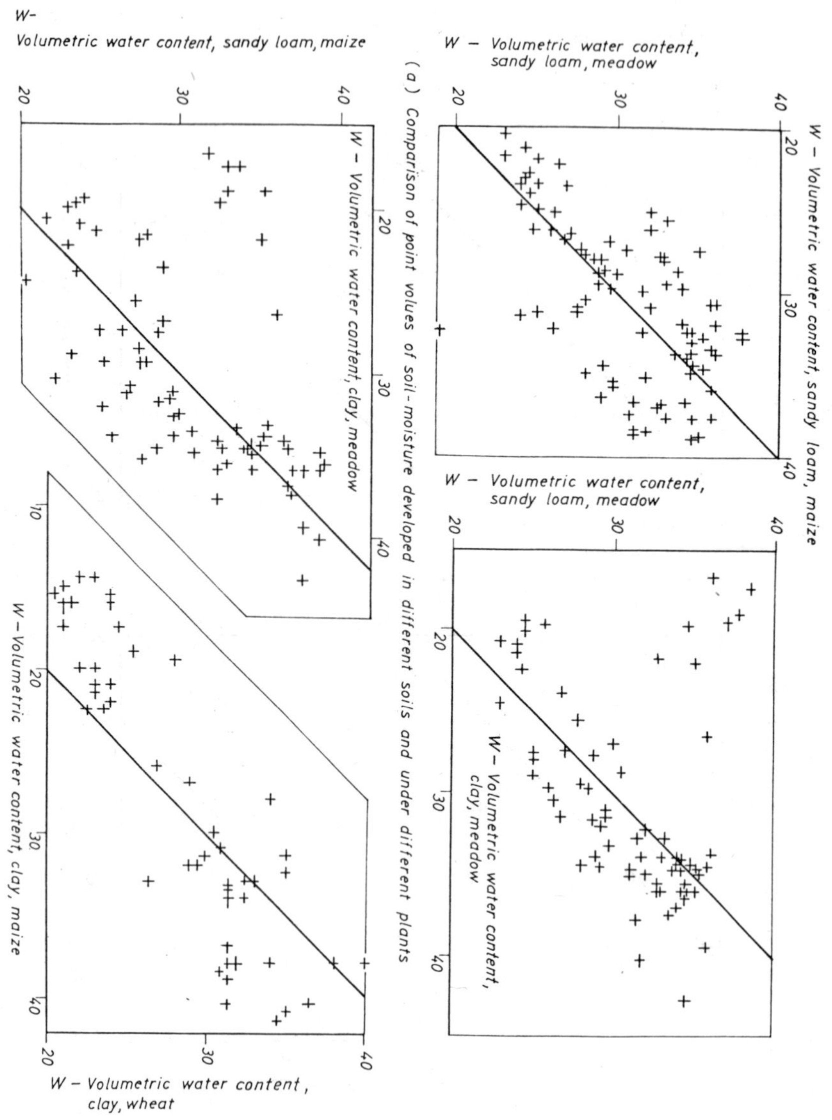

Fig. 2-32. Comparison of moisture-content values measured under different conditions.
(a) point values

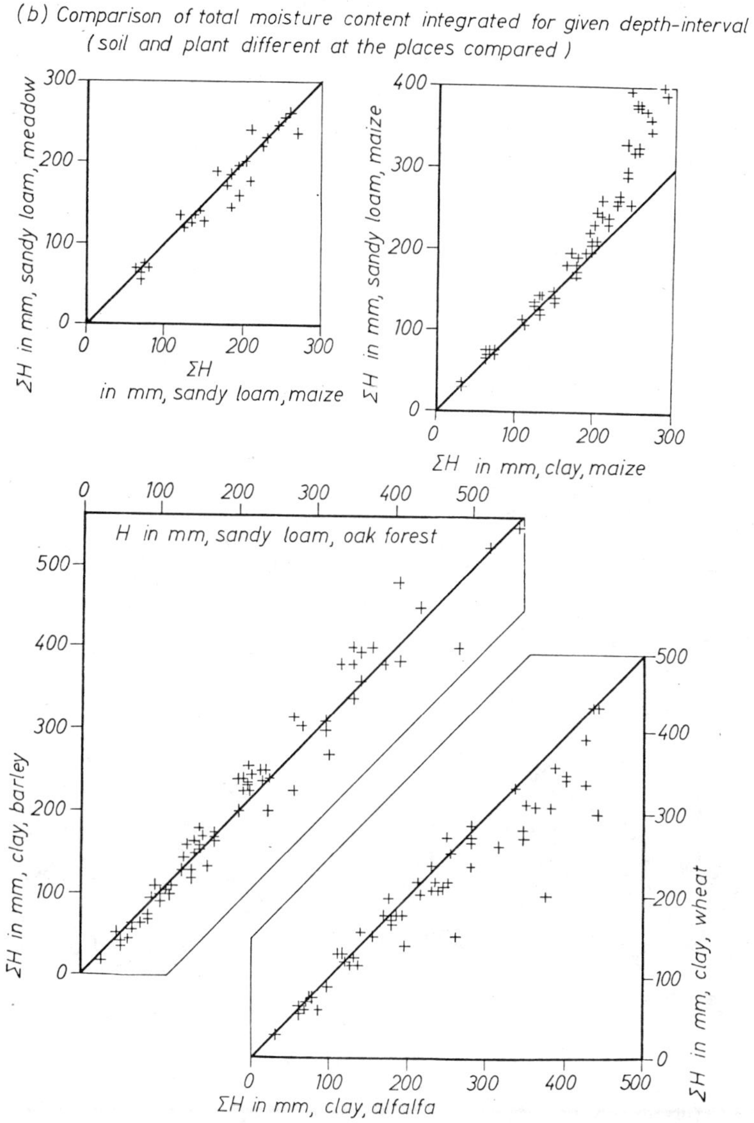

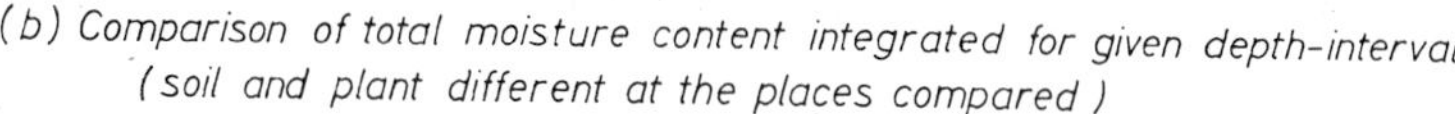

Fig. 2-32. Comparison of moisture-content values measured under different conditions.
(b) values summarized for a given interval of depth

SECTION 7

DETERMINATION OF SOIL-MOISTURE RETENTION CURVES

It was already mentioned that the water content does not give sufficient information on the behavior of the soil. It is necessary to compare these data to parameters belonging to special conditions of the sample. This restriction should be enlarged, indicating that point measurements, even compared to such specific characteristics as hygroscopic moisture content, liquid limit, or limit of plasticity, are insufficient to describe field conditions. The complete vertical distribution of moisture content has to always be determined in a profile and compared to the soil-moisture distribution belonging to the dynamic equilibrium (see Eq. 2-10). Only the differences between the actual distribution and the soil-moisture retention curve shows where there is water deficit or excess in the profile.

The only remaining problem is the numerical determination of the position of the line dividing the field capacity and the gravitational porosity, which line characterizes the state of equilibrium of soil moisture in a profile.

The definition of the retention curve immediately induces the idea that a pF curve could be used to determine the required matching curve since, according to its definition, a pF curve also represents the dynamic balance of the forces acting on the water included into the pores of the porous medium.

7-1 Construction of a pF Curve from Measured Data

The usually accepted way of determining a pF curve is to apply various suctions on the investigated sample, and to measure the soil-moisture values retained against the suctions in question prevailing in the pores. After plotting the points determined by the pairs of suction and water content data on a semi-log plot (suction, which is supposed to be equal to the tension on the surface of water films, or to a capillary height in the case of saturated pores, in the logarithm of the equivalent water columns and water content in arithmetic scale), the pF curve can be easily constructed (Fig. 2-33) (Wind, 1966).

There are different equipment proposed to create suction on the sample, the construction of which mostly depends on the range of suctions to be applied. The suction can be created

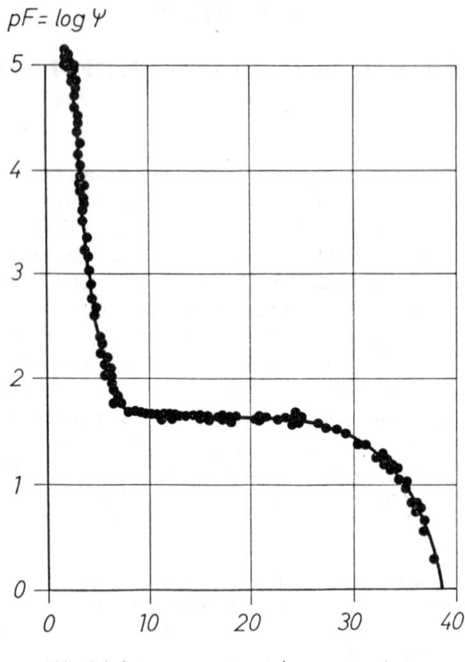

Fig. 2-33. pF curve determined on the basis of detailed measurements.

mechanically or by applying chemical methods. Two different basic principles are also applied for the construction of the mechanical equipment. The data to be measured are the moisture-content values remained in the sample after the development of the balanced condition within the water included into the pores. The amount of the retained water has to be related to the difference of the pressures prevailing in the water and in the air, respectively. The measurement can be executed, therefore, in two different ways, i.e., applying suction on the water or raising the pressure of the air in the pores. The variations of the applicable measuring methods are further increased by the fact that suction can be created either by contacting the bottom of the soil column to be investigated directly with a container where a constant free water level is maintained (column drainage), or by applying a separate tank (which is connected to the sample through a flexible pipe and a semi-permeable membrane) to regulate the suction. Thus the various methods used for the determination of the points of the pF curves can be summarized as follows (also indicating the range of the pF values, within which the method is generally applied):

Mechanical methods

applying suction:	column drainage	pF < 2.3
	hanging water column	pF < 3.0
applying pressure:	pressure plate	pF < 3.0
	pressure membrane	3.0 < pF < 4.2

Chemical method 4.4 < pF

Column drainage can be executed with the most simple equipment but its measuring range is very limited, not only because a high homogeneous soil column can hardly be prepared, but the long drainage time should also be considered. This is the reason why this method can be applied only for the determination of pF curves of sands. The data determined in the adhesion zone are uncertain even in these very permeable materials because of the low hydraulic conductivity of the dry material, and thus 100 hours are still not enough for the development of the dynamic equilibrium in the upper part of the column (Johnson et al., 1963; Prill et al., 1965). The determination of the amount of water remaining at various heights in the column after the development of the balanced condition raises technical problems. Earlier the container had been prepared as a series of rings which were taken apart to measure the water content by weighing. After the development of the isotope methods for measuring water content, a source of gamma rays and a sensor moving parallel on the two sides of the column are generally used to determine the vertical distribution of moisture content (Vachaud et al., 1970). A theoretical difficulty has to also be mentioned in connection with this method, i.e., the vertical distribution of the dynamically balanced soil moisture is different depending on whether this condition was achieved by wetting or by drying. This phenomenon is called hysteresis and will be discussed in more detailed form later.

The advantage of the hanging water column method is that a series of pairs of suction and moisture-content data can be determined using a relatively small sample by changing the position of the container (in which the free water level is maintained) related to the sample. The errors occurring from the slow development of the balanced condition in a soil column can be avoided, in this way, by applying a sample of 1 or 2 cm in thickness. The homogeneous packing of such a shallow column is also much more simple than that of a deep one. To cut an undisturbed sample of this small size is also relatively easier than the preparation of a long undisturbed column (Prill and Johnson, 1966). The most critical part of the equipment is the membrane, which transmits the suction created by the water table, maintained in low position from the connecting tube to the sample. Its material (ceramics, asbestos, cellophane, or other artificial membranes) has to be permeable for water, but the air-bubbling pressure should be higher than the highest suction applied because otherwise, the continuity of water between the water table and the sample would be interrupted by the air bubbles penetrating through the membrane and collected in the upper part of the connecting tube. The upper limit of the measuring range is determined, similarly, by cutting-off the water column, which occurs if the vertical distance between the water table and the sample is longer than the height of a water column equivalent to the atmospheric pressure.

The relative pressure difference between the water content of the sample and the air included into the pores can be modified not only by suction applied on the water but also by raising the external pressure around and inside the sample. High pressure equipment is generally used to mechanically determine the moisture content belonging to high pF values.

The sample is placed in a closed cell, inside which the required pressure is maintained by compressed air or N_2 gas. The vapor saturation of the applied gas has to be ensured, otherwise the gaseous material is also able to take-up a considerable amount of moisture from the sample, causing excess suction. The air or gas is, therefore, generally transported through a humidity control before introducing it into the measuring cell.

The sample has to be placed on a semi-permeable material, the air-bubbling pressure of which is higher than the pressure applied within the measuring cell, or chamber if an equipment with larger closed place is used suitable for instantaneous measuring of more than one sample. A considerable pressure difference develops through the semi-permeable plate or membrane, the upper surface of which prevails the maintained high pressure while its lower border is contacted with the free air having atmospheric pressure. A flow starts through the plate, the water for which is provided from the sample through its contact with the plate. The dynamic balance is indicated by the cessation of flow at the free surface of the plate.

The various equipment differ from each other mostly in the technical solution of the drainage of the samples. the ceramic plates can be applied for low pressure generally below 1 atm, which is equal to pF of 3.0 (pressure plate method). For higher pressure, special artificial membranes are used (pressure membrane method) which are suitable up to a pressure of about 15 atm (pF < 4.2). Leaking of air as well as inadequate contact between the plate and the sample may disturb the measurements, as it is testified by Prill and Johnson (1966). These equipment details have to be solved very carefully.

The application of the chemical method is based on the principle that there is always a balanced condition between the water content of the sample and the moisture content of the surrounding air. The vapor tension can be easily controlled in a vacuum exsiccator by filling it with a chemical solution with known concentration because the tension of the saturated vapor depends only on the material of the applied fluid and on temperature. The use of the chemical method requires, therefore, the maintenance of constant and known temperature. Sulfuric acid of various concentration is generally used to produce a well-defined vapor space, but other saturated salt solutions can also be applied. Some interrelated pF values and relative vapor contents achieved by using different chemicals are listed as follows (supposing in each case a constant temperature of 20°C):

Concentration of H_2SO_4 in %	10	20	30	40	50	60	70	80
Relative vapor content of air in %	97	87	75	56	35	17	5	2
pF	4.62	5.28	5.60	5.90	6.16	6.39	6.61	6.73

Saturated solution of

	K_2SO_4	KCL	$CaCL_2$-6HOH	LiC1.x HOH	$ZnC1_2$
Relative vapor content of air in %	98	85	32	15	10
pF	4.45	5.38	6.19	6.42	6.50

On the basis of this short summary of the various measuring techniques, the combined application of the hanging water column and the chemical method can be proposed for the determination of the points of pF curves. The use of high pressure cells becomes necessary only if the rapidly changing suction of the curves lies in the $3.0 < pF < 4.4$ zone.

The two curves (i.e., pF and soil-moisture retention) are identical if the osmotic forces (chemical-, electro-, or thermo-osmosis) are negligible because the pF curve expresses the relationship between water content and tension, the latter including the effects of all the forces. Neglecting osmosis, the remaining forces acting are gravity and adhesion supplemented with capillarity in the partially saturated transition zone. Gravity, adhesion, and capillarity are also the forces whose balanced condition (or more precisely the water content belonging to this condition) is characterized by the retention curve.

Differences occur only in the determination and the interpretation of the two curves. The pF curve describes the relationship of tension and water content of a sample determined by applying various tensions on a separated soil sample. Considering that the tension expressed in equivalent water column is the same as suction caused by gravity at an elevation above the water table equal to the water column mentioned (see Eq. 2-10), it is evident that the two curves have to be identical if the profile is homogeneous and composed of the same material (with the same porosity) as the sample investigated, supposing once again, that osmotic forces are negligible. This is the reason why the pF curves, constructed from data measured on a sample taken out of the profile, can also be used for the determination of the dynamically balanced vertical distribution.

There is a special problem, however, to be discussed in connection with the proposed application of pF curves, i.e., the change of the bulk volume of the investigated sample as a result of increasing moisture content.

A common feature of the soil-moisture retention curves is that they are composed of three clearly recognizable sections (Kovács, 1968) (Fig. 2-34a). The upper part of the curve is almost vertical (zone of adhesion), which is followed by a nearly horizontal section where water content increases rapidly with decreasing tension (open capillary zone). Finally, the soil-moisture curves have to be closed by a vertical line at the moisture content equal to porosity ($W = n$; $s = 1$), this value being the upper limit of water amount stored in the pores (closed capillary zone). There are only a few exceptions, in the case of very fine materials, when the first two stretches are substituted by a line curved monotonically with decreasing slope, as the water content increases (Fig. 2-34), but the vertical closing section below the minimum capillary rise intersects the horizontal axis at $W = n$ in this case as well.

Contrary to this concept based on the physical upper limit of the water-retention capacity (which states that the moisture content cannot be higher than that belonging to total saturation), most of the pF curves published in the literature show gradually increasing water content considerably surpassing the original porosity of the sample as the tension approaches the zero value. The sloping character of the closing stretch of the pF curve is especially characteristic for cohesive materials (clay, silt), but many similar curves are also published for fine sand as well. The theoretically correct vertical closing face was generally measured only in

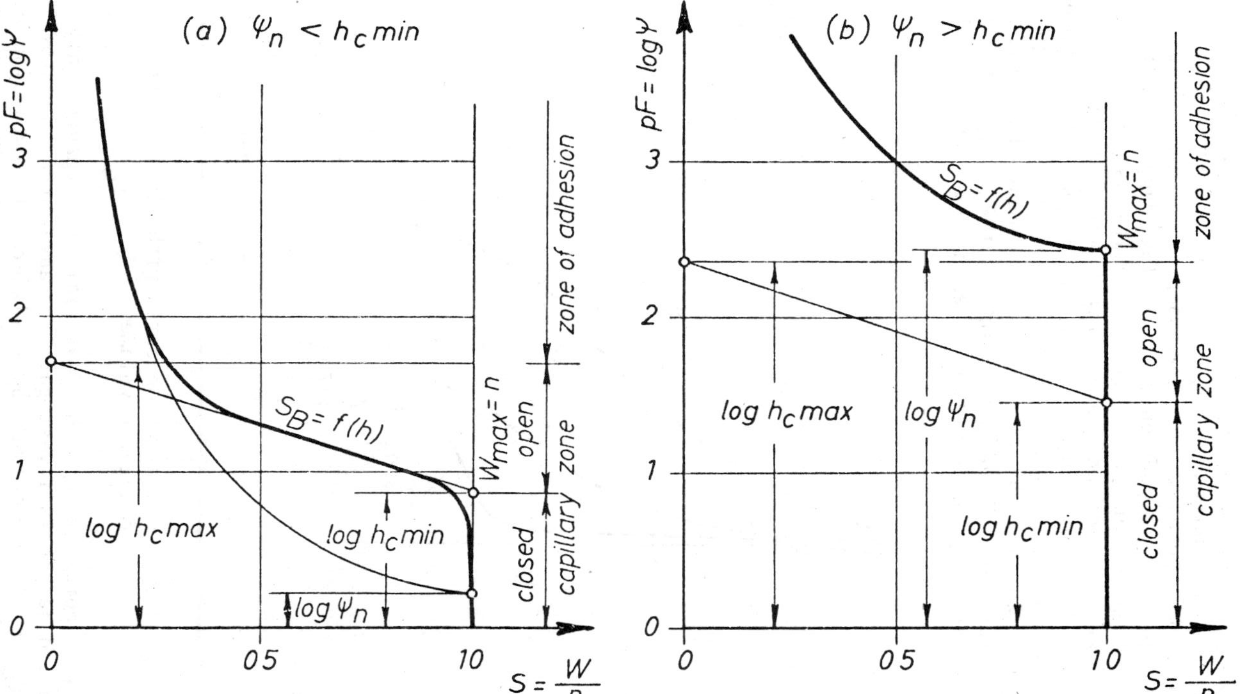

Fig. 2-34. Two characteristic forms of pF curves.

coarse grained sands. Instead of repeating these curves from other publications (Klute, 1952; Sudnitsin, 1966; Várallyai, 1975), three pF curves were carefully measured and represented in Fig. 2-35, together with the characteristic soil-physics parameters of the samples. It can be clearly seen that the total moisture content belonging to low tension is higher than the original porosity of the samples. This contradictory result can be explained only by the fact that the bulk volume of the samples, and thus their porosity as well, increased considerably as the moisture content was increased. In Fig. 2-36 the observed changes in volume of the three samples is also represented to prove the reliability of the previous statement (Zotter, 1975). These measurements have indicated, that in the case of cohesive samples, the change of porosity could cause some uncertainties even in the zone adhesion as well.

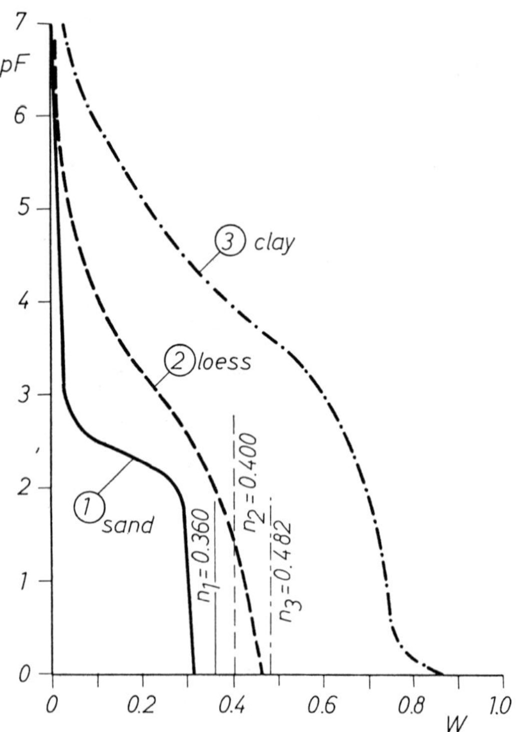

Fig. 2-35. pF curves measured without avoiding the change in volume.

The porosity of the material can change only if a separated sample is investigated because in a soil profile the weight of the overlying layer hinders the expansion of the medium. Thus, the sloping closing face can be observed only in the case of the pF curve and these measurements cannot be

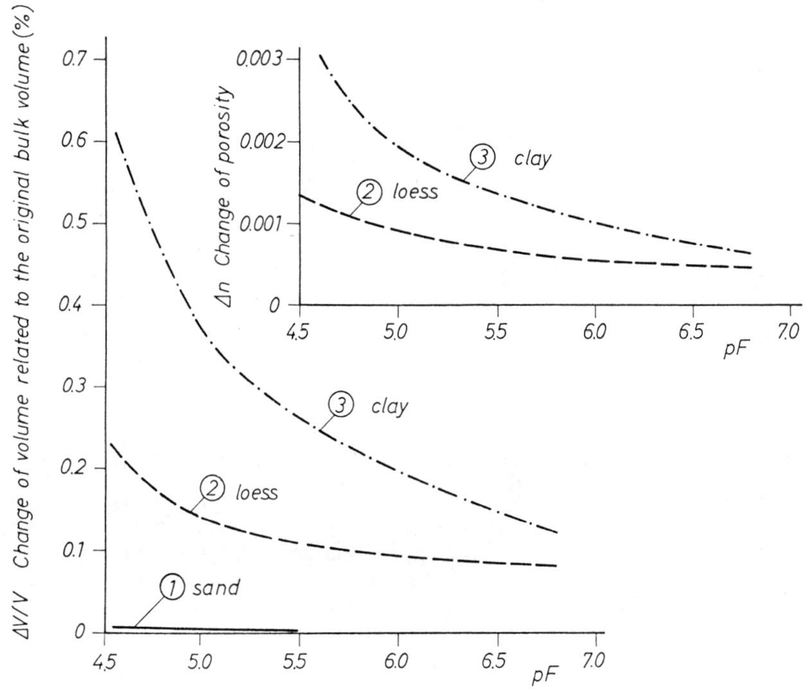

Fig. 2-36. Change in volume of samples due to water intake as a function of tension.

accepted for the characterization of the soil-moisture retention curves. It is also evident that pF values belonging to different porosities must not be used for the construction of a unified soil-moisture retention curve. The basis of this statement is the fact that the relationship between tension and water content includes porosity as an independent variable because both the size of pores (capillarity) and the relative internal surface (adhesion) change with the relative pore volume. It is advisable, therefore, either to ensure the constancy of pore volume of the sample when measuring pF values, or to correct the observed data on the basis of the measured change of pore volume in the open capillary zone and to close the soil-moisture retention curve with a vertical face in the closed capillary zone at $W = n$ values.

7-2 Calculation of Soil-Moisture Retention Curve from Soil-Physics Data

Accepting the capillary tube model influenced partly by adhesion and partly by capillarity (see Fig. 1-87) for the characterization of the partially saturated layer, and applying the statistical model to consider the probable pore-size distribution, it is evident that the soil-moisture retention curve has to be composed of two different curves because there are two molecular forces acting against gravity. If adhesion between the solid wall and water molecules were the sole retarding force, only the adhesive water content (W_a) would be kept at a given height in the form of water films. The attraction between the water molecules creates surface tension on the water-air interface, the result of which is the development of the curved menisci in closed conduits having relatively small diameters, and that of capillary suction which is able to raise the water in these conduits. Thus, the capillary effect can completely saturate some pores, but since the capillary rise is inversely proportional to the pore diameter, the vertical distribution of capillary water content (W_c) is closely related to the pore-size distribution of the medium (Réthāty, 1960). Considering that at a given height above the water table one part of the pores is completely saturated by capillarity (the ratio of these pores is measured by the amount of capillary saturation $s_c = W_c/n$) and water films develop in the remaining pores (the relative water content of these pores is measured by adhesive saturation $s_a = W_a/n$), the actual saturation of the solid matrix around the investigated point can be characterized by the combination of capillary and adhesive amounts of saturation (Fig. 2-37):

$$s = s_c + s_a(1-s_c) \qquad (2\text{-}17)$$

Considering that the effect of Van der Waals force (adhesion between the solid wall and water molecules) is generally characterized with a hyperbola of sixth order relating the suction head to the distance measured from the wall (see Eq. 1-85), and using the geometrical model of capillary tubes, a theoretical relationship can be derived between suction head and adhesive amount of saturation (Kovács, 1968). The reliability of the function was proved by comparing calculated and measured data. It was found at the same time that in the range of $4 \times 10^{-5} < D_h < 2 \times 10^{-2}$ the theoretical value can be well approximated by a simple empirical equation (Kovács and Péczely, 1975):

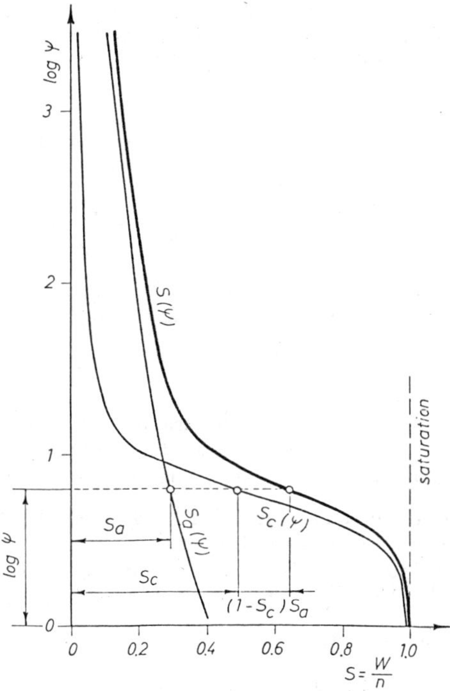

Fig. 2-37. Superposition of adhesive and capillary amounts of saturation.

$$s_a = \frac{1-n}{n} \frac{2.5x10^{-3}}{\psi^{1/6}} \left(\frac{\alpha}{D_h}\right)^{2/3} = 1.4 \times 10^{-2}\left(\frac{h_{co}}{\psi}\right)^{1/6}\left(\frac{1-n}{n}\right)^{1/3} h_{co}^{1/2} \; ; \tag{2-18}$$

where the relationship between the average capillary height and the soil-physics parameter is already considered according to Eq. 2-19. The parameter ψ, h_{co}, and D_h have to be substituted in cm, since the constants are not dimensionless values in Eq. 2-18.

There is, however, another molecular force influencing the development of the amount of saturation in the vicinity of the water table, i.e., capillarity. Its effect is generally characterized by the capillary rise (h_c) which is the suction head caused by capillarity. This parameter depends on the diameter of the capillary tube, and its numerical value can be calculated by applying Eq. 1-52. Accepting the capillary tube model to simulate the pores and using Eqs. 1-78 and 1-83 to express the average, maximum, and minimum pore diameters depending on the soil-physics parameters, the average as well as the probable lowest and highest capillary rise expected in a soil profile can also be given depending on the same parameters:

$$\text{average capillary height} \quad h_{co} = \frac{0.3}{d_o} = 0.075 \frac{1-n}{n} \frac{\alpha}{D_h} \quad ;$$

$$\text{minimum capillary height} \quad h_{cmin} = \frac{0.3}{d_2} = 0.06 \frac{1-n}{n} \frac{\alpha}{D_h} \quad ;$$

$$\text{maximum capillary height} \quad h_{cmax} = \frac{0.3}{d_1} = 0.11 \frac{1-n}{n} \frac{\alpha}{D_h} \tag{2-19}$$

After having calculated the amount of adhesive saturation from Eq. 2-22, the next necessary step is the determination of the ψ vs. s_c relationship, i.e., the amount of capillary saturation depending on suction head. The statistical model of pore-size distribution can be used to solve this problem. Knowing the probable number of the pores having smaller area than a given limit (f_c), the sum of their area can be easily calculated by considering Eq. 1-29.

At an elevation of h_c above the water table (which can also be expressed with equivalent suction head ($h_c = \psi$), all the tubes having smaller cross section than f_c are completely saturated by capillarity. It is supposed now, as a working hypothesis, that the larger pores remain filled with air. Actually, the wall of large pores are covered by water films, as a result of the effect of adhesion, and this effect will be taken into account by superimposing capillary and adhesive saturation according to Eq. 2-17. The amount of capillary saturation can be determined, therefore, by relating the sum of the area of pores having smaller section than f_c to the total free area in the cross section of the sample. Investigating a cross-sectional area of unity, using dimensionless parameters and considering Eq. 1-29, the amount of capillary saturation can be expressed depending on either the relative pore size (x), or the relative pore diameter ($d/d_o = \sqrt{x}$), or taking into account that the capillary rise is inversely proportional to the diameter; capillary saturation can even be determined as a function of the prevailing suction head (ψ) (or height above the water table) related to the average capillary rise of the sample (Fig. 2-38):

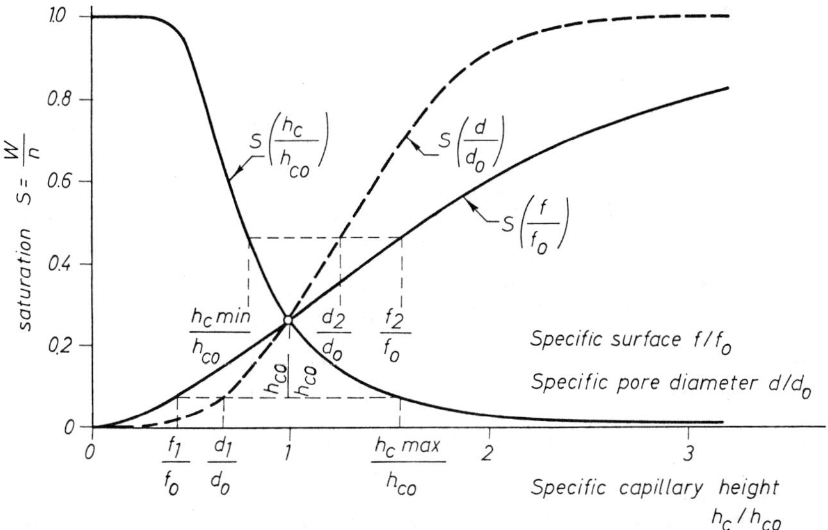

Fig. 2-38. Characterization of the amount of capillary saturation as a function of relative pore area, relative pore diameter, and relative suction head, respectively.

$$s_c = \frac{W_c}{n} = \frac{\int_0^{\ell_c} x\ \xi(x,\Delta x)\ dx}{\int_0^{\infty} x\ \xi(x,\Delta x)\ dx} = 1 - \left(\frac{fc}{fo} + 1\right) \exp\left(-\frac{fc}{fo}\right)$$

$$= 1 - \left[\left(\frac{d_c}{d_o}\right)^2 + 1\right] \exp\left[-\left(\frac{d_c}{d_o}\right)^2\right]$$

$$= 1 - \left[\left(\frac{h_{co}}{\psi}\right)^2 + 1\right] \exp\left[-\left(\frac{h_{co}}{\psi}\right)^2\right]; \tag{2-20}$$

where

$$x_c = \frac{fc}{fo} = \left(\frac{d_c}{d_o}\right)^2 = \left(\frac{h_{co}}{\psi}\right)^2$$

Considering the h_{co} vs. α/D_h relationship (see Eq. 2-19) and substituting Eqs. 2-18 and 2-20 into Eq. 2-17, the form of the soil-moisture retention curve is:

$$s = 1-\left[\left(\frac{h_{co}}{\psi}\right)^2+1\right]\exp\left[-\left(\frac{h_{co}}{\psi}\right)^2\right]\left[1-1,1.4x10^{-2}\left(\frac{h_{co}}{\psi}\right)^{1/6}\left(\frac{1-n}{n}\right)^{1/2} h_{co}^{1/2}\right]. \tag{2-21}$$

This generalized equation expresses the amount of saturation as a function of the relative suction head (ψ/h_{co}), but the relationship also contains a material constant $(1-n/n)\ h_{co}^{3/2}$, where the average capillary height has to be substituted in cm. Some corresponding values of the $M = (1-n/n)\ h_{co}^{3/2}$ parameter, introduced here, and the other soil-physics parameters used in general, are summarized in Table 2-6 to give information about the probable range of this material constant. Some limitations also have to be mentioned in connection with the application of the equation. It is valid only in the case of a homogeneous layer which does not have special structures (large openings, through which the atmospheric pressure can penetrate deeply between the clods), the water content of the cultivated zone and the root zone cannot be characterized, therefore, by this method. The hysteresis of the retention curve is not considered in the investigation, therefore, Eq. 2-21 represents only the average conditions, and the investigation of the influences caused by wetting or draining the profile needs special analyses.

7-3 Hysteresis of the Retention Curve

The basis of the previous derivation is a model composed of straight capillary tubes having different diameters where each tube is of constant diameter. The network composed of the pores is, however, a system of channels with changing cross section and not that of straight pipes with constant diameters. The capillary height is influenced, therefore, not only by the distribution of the pore sizes in a horizontal section but by the vertical change of the pore diameters as well. This is the reason why some pores are not saturated when the dry sample is wetted from the direction of the water table, while the same pores can retain capillary water when the process starts with the complete saturation of the sample and the dynamic equilibrium is achieved by drainage. This phenomenon is the hysteresis of the soil-moisture retention curve (the nearly horizontal section of the curve has a lower

Table 2-6. Interrelation between the M material constant and other soil-physics parameters.

		Compacted			Loose		
A) Pervious materials		U > 5	5 > U > 2	2 > U > 1	U > 5	5 > U > 2	2 > U > 1
sandy gravel	n	0.25	0.30	0.35	0.30	0.35	0.40
	D_h(cm)	0.81	0.53	0.36	0.53	0.36	0.25
	α	8	8	8	8	8	8
M = 10	h_{co}(cm)	2.64	3.07	3.54	3.07	3.54	4.06
	K_s(cm/sec)	5.5	4.8	4.1	4.8	4.1	3.5
	n	0.25	0.30	0.35	0.30	0.35	0.40
fine sand	D_h(cm)	0.047	0.031	0.021	0.031	0.021	0.015
	α	10	10	10	10	10	10
M = 1000	h_{co}(cm)	48.1	56.8	66.2	56.8	66.2	76.3
	K_s(cm/sec)	1.2×10^{-2}	1.0×10^{-2}	8.8×10^{-3}	1.0×10^{-2}	8.8×10^{-3}	7.6×10^{-3}
B) Semipervious materials							
	n	0.30	0.35	0.40	0.35	0.40	0.45
loess	D_h(cm)	7.3×10^{-3}	5.0×10^{-3}	3.8×10^{-3}	5.0×10^{-3}	3.8×10^{-3}	2.4×10^{-3}
	α	12	12	12	12	12	12
M = 10,000	h_{co}(cm)	264	307	354	307	354	406
	K_s(cm/sec)	4.8×10^{-4}	4.1×10^{-4}	3.5×10^{-4}	4.1×10^{-4}	3.5×10^{-4}	3.0×10^{-4}
	n	0.35	0.40	0.50	0.40	0.45	0.55
	D_h(cm)	1.4×10^{-3}	9.6×10^{-4}	5.0×10^{-4}	9.6×10^{-4}	6.8×10^{-4}	4.0×10^{-4}
loam	α	14	14	14	14	14	14
	h_{co}(cm)	1430	1650	2150	1650	1890	2500
M = 100,000	K_s(cm/sec)	1.9×10^{-5}	1.6×10^{-5}	1.2×10^{-5}	1.6×10^{-5}	1.4×10^{-5}	1.0×10^{-5}

Table 2-6. (continued)

C) Impervious materials							
	n	0.40	0.45	0.55	0.50	0.55	0.65
silt	D_h(cm)	5.2×10^{-4}	3.0×10^{-4}	1.9×10^{-4}	2.2×10^{-4}	1.9×10^{-4}	1.3×10^{-4}
	α	16	16	16	16	16	16
M = 250,000	h_{co}(cm)	3000	3500	4500	4000	4500	6000
	K_s(cm/sec)	4.8×10^{-6}	4.1×10^{-6}	2.9×10^{-6}	3.5×10^{-6}	2.9×10^{-6}	2.0×10^{-6}
Clay	n	0.50	0.55	0.60	0.60	0.65	0.70
	D_h(cm)	1.5×10^{-4}	1.1×10^{-4}	7.6×10^{-5}	7.6×10^{-5}	5.3×10^{-5}	3.7×10^{-5}
	α	20	20	20	20	20	20
M = 1,000,000	h_{co}(cm)	10.000	11.400	13.100	13.100	15.100	17.300
	K_s(cm/sec)	5.5×10^{-7}	4.7×10^{-7}	3.9×10^{-7}	3.1×10^{-7}	3.1×10^{-7}	2.6×10^{-7}

position if it is determined by wetting and a higher one in the case of drainage).

In the literature, the phenomenon of hysteresis is generally explained by considering two different physical causes (Bear; 1972).

The first possible reason may be the raindrop effect according to which the contact angle of the meniscus to the wall of the capillary tube (θ_1) is larger when the water advances in the tube than that in a static condition (θ_0) and the decrease of the angle can be observed ($\theta_2 < \theta_0$) in the case of a recessing water column (Fig. 2-39). The same difference of the contact angles can be observed between the front and the back side of a raindrop moving along a sloping glass plate (this similarity gave the name to the phenomenon). The water level may remain, according to the other explanation, in a higher position when a capillary tube with changing diameter is drained compared to the raising condition because the downwards movement is stopped when the meniscus reaches a small cross section (where capillarity can balance the weight of a relatively higher water column), while similarly the rise of the meniscus stops at a large diameter. This phenomenon is called the ink-bottle effect because in old ink bottles this effect was used to keep the level of ink near the mouth of the bottle.

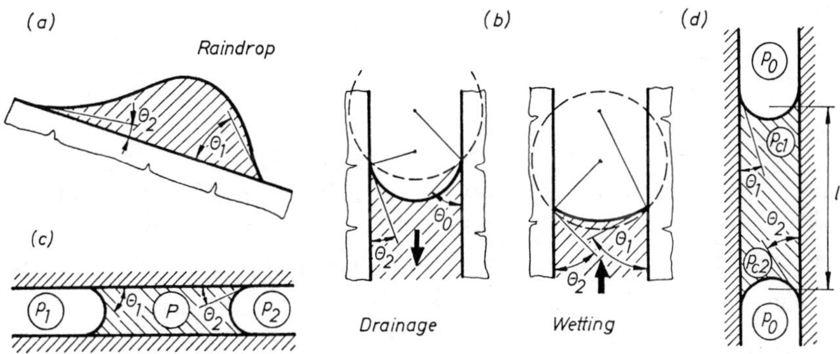

Fig. 2-39. Change of the wetting angle due to the movement of water and pressure differences in the contacting air space.

From the two possible causes, the raindrop effect cannot be accepted as an exaplanation of hysteresis. The soil-moisture retention curve represents the vertical moisture distribution in a static condition. On the contrary, the raindrop effect is a process occurring only in the case of flow in capillary tubes. Thus, a phenomenon characterizing the movement of water cannot be used as a cause to explain another phenomenon (the two possible positions of the retention curve) which describes the static condition of the moisture distribution.

This is the reason why the changing diameter of the channels will be used in the foregoing to explain the existence of hysteresis. It is necessary, therefore, to use a model composed of tubes, the longitudinal section of which is represented by two wavy lines having a minimum distance of d_1, a maximum one of d_2, and a wavelength of 2a (Figs. 2-40 and 2-41), instead of the straight capillary tubes mentioned previously. Even if the wavelength is constant, all the tubes are identical with each other as in Fig. 2-40, the required pore-size distribution in a horizontal cross section described previously can be easily ensured by placing the sections of the tubes with different diameters at the level of the water table. The same result can be achieved with the other model, where the amplitude, the fluctuation of the diameter of the tubes, is the same as in the previous case, but the wavelength varies from tube to tube (the parameter of a is also a random variable, described by a distribution function) (Fig. 2-41).

Investigating at first the model represented in Fig. 2-40, it can be seen that the maximum capillary height, belonging to diameter d_1, can never be achieved by wetting the tubes from the direction of the water table, if the difference between the maximum and minimum capillary rise is greater than the wavelength because in the zone of $h_{cmin} < h_c < (h_{cmin} + 2a)$ all the tubes have a section with diameter d_2 and, therefore, the rising water cannot surpass the level of the cross section with a diameter of d_2, the capillary rise belonging to this diameter being equal to h_{cmin}. The opposite process occurs in the system when it is saturated and the dynamic equilibrium is achieved by drainage. In this case the capillary level should be in the zone $h_{cmax} > h_c > (h_{cmax} - 2a)$ because all the tubes here have a section with diameter d_1, where the capillary tension can balance a water column of h_{cmax}.

In the case of the more realistic model composed of different wavy tubes, the wavelength of which is a random variable (Fig. 2-41), and investigating the wetting process of the sample, the water can be raised up to the level of h_{cmax}

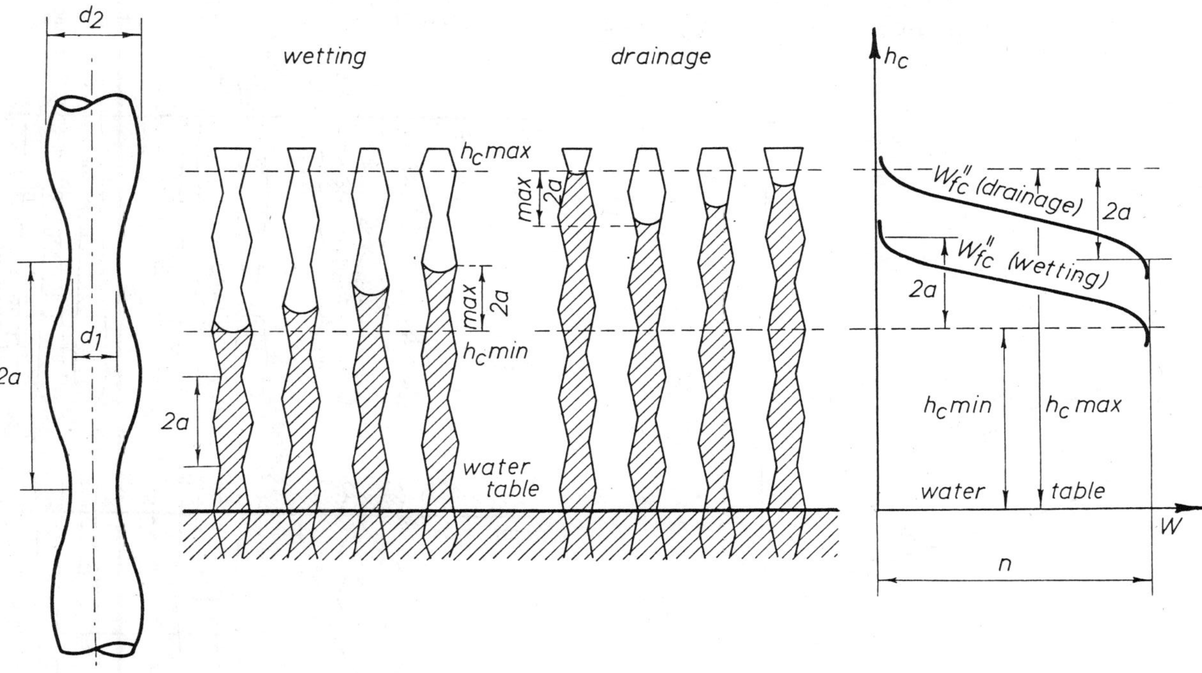

Fig. 2-40. Development of hysteresis in a model system composed of identical tubes with varying diameter.

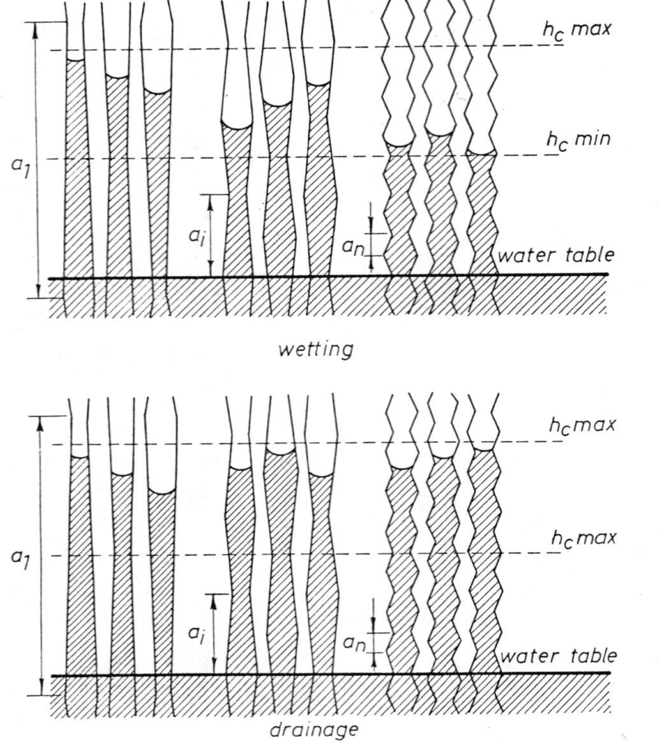

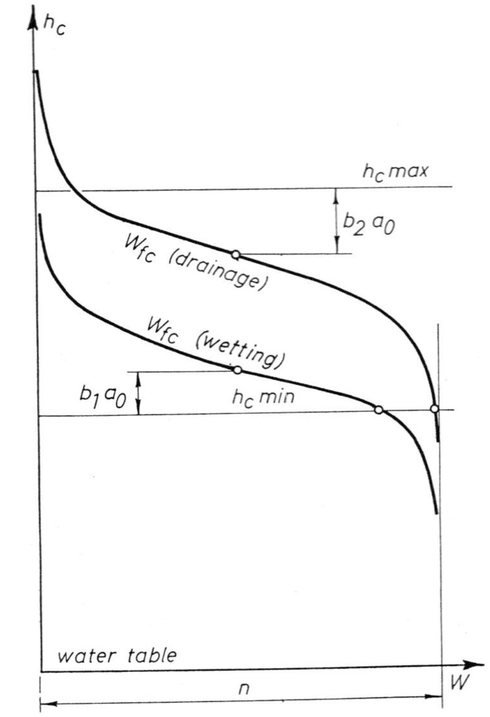

Fig. 2-41. Development of hysteresis in a model system composed of different tubes with varying diameter.

by capillarity in some of the tubes, the wavelength of which is greater than the difference of $(h_{c\ max} - h_{c\ min})$, and if the capillary rise is not hindered below the maximum level by a large stretch of the tube. In the tubes having shorter wavelengths, the capillary rise is limited at a level of $(h_{c\ min} + 2a)$ as it is described in the previous paragraph or (if the wavelengths are long enough but have large diameters below the highest capillary rise) at the elevation where the weight of the water column is balanced by capillarity belonging to the diameter at this level. In this case the average (most probable) capillary height is between the minimum capillary rise and the level characterized by the sum of the former and the most probable wavelength (2a).

Only the vertical shifting of the nearly horizontal stretch of the soil-moisture retention curve can be explained by the ink-bottle effect. The observations have shown, however, that the ψ vs. W relationship is not a single-valued function in the zone of adhesion either. One of the reasons of the different water contents at a given elevation above the water table is the considerable time lag between the time points of the change of position of the water table and the development of the dynamic equilibrium. Figure 2-42 shows the results of an experiment aimed at the observation of the process in time. The data indicate the extremely slow development of equilibrium in the adhesive zone when the column is drained, viz., in this case the upper part of the sample has very small hydraulic conductivity because of the low water content, and thus, the excess water percolates downwards with a very small velocity, generally the wetting is a faster process, but a considerable difference can be observed between the initial wetting (when the process starts with a completely dry sample) and rewetting (when the sample was previously drained and some soil moisture was retained in the pores). In the first case, the development of the dynamic equilibrium in the adhesive zone was perhaps greater time lag than that of drainage. Thus, in many cases, the difference between the water content observed at the same elevation in a wetted and drained column, respectively, can be explained by the insufficient length of observations.

The horizontal shifting of the vertical stretch of the retention curve is, however, also indicated by data determined directly from simultaneous measurements of tension and soil-moisture values (Fig. 2-43). In this case, the corresponding data are not influenced by the time-dependent development of dynamic equilibrium. The reason for hysteresis may be the water retained in the corners of the pores after draining, and the raindrop effect, the influence of which cannot be neglected in a moving system.

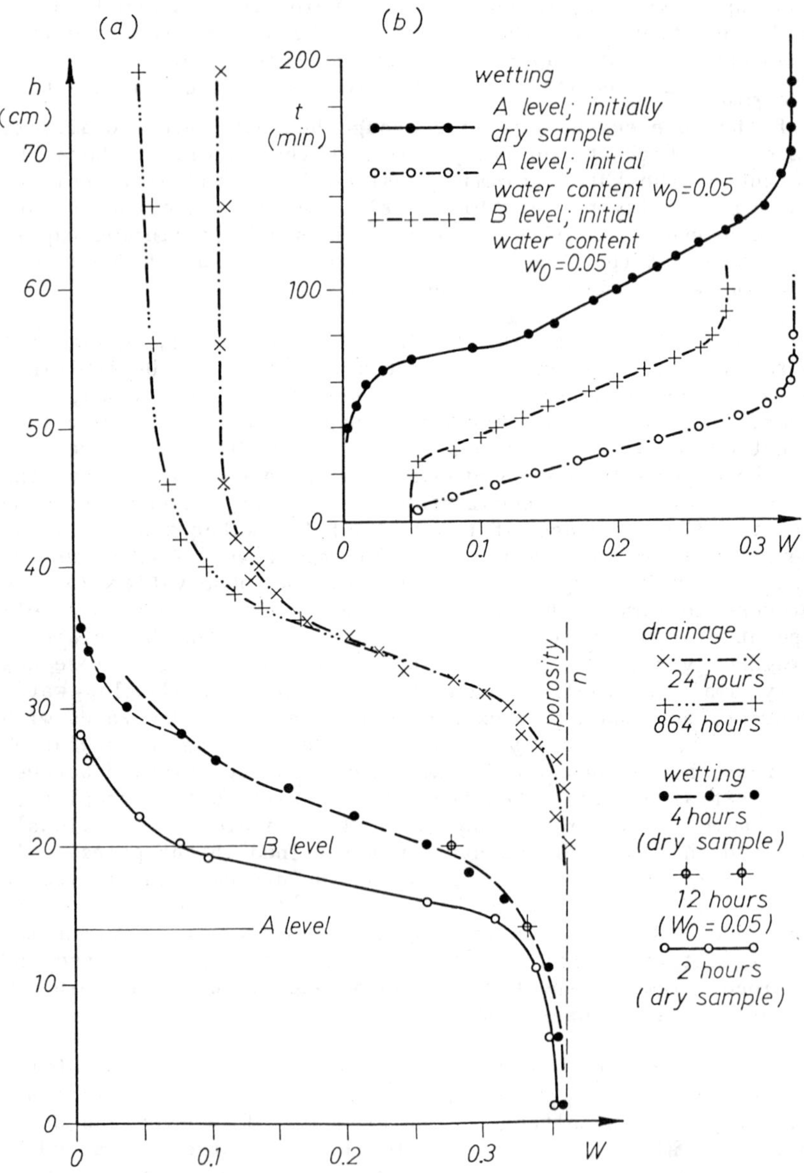

Fig. 2-42. Development in time of the processes of wetting and draining.

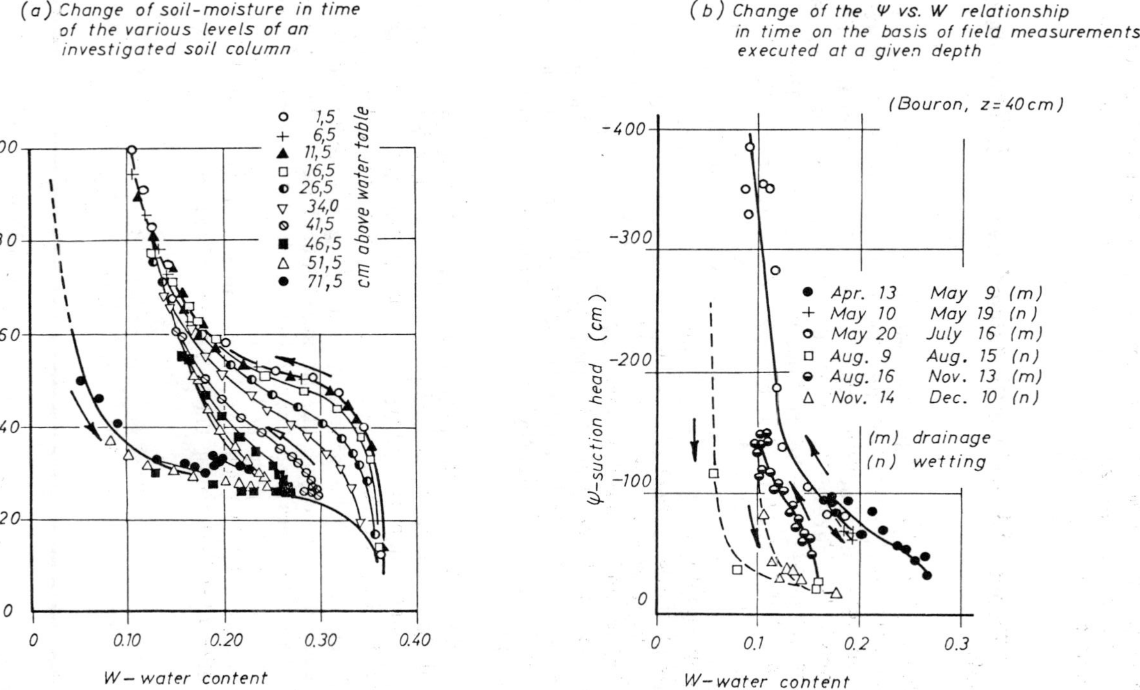

Fig. 2-43. Characterization of hysteresis on the basis of corresponding tension and moisture-content values measured in the field.

A further reason may also be mentioned for explaining different water-content values belonging to the same tension. The vertical transport of water, even that of a very small amount, may consume considerable energy, especially in cohesive materials. Numerical examples will be shown in the next section to characterize the expected change of the retention curve due to vertical flow.

7-4 Comparison of Calculated Retention Curves by Using Measured Data

The explanation of the phenomenon of hysteresis indicates many uncertainties influencing the form of the soil-moisture retention curve and causing its multi-valued character. It is not expected, therefore, that the history of wetting and draining processes can be followed theoretically. The random position of the retention curve can be characterized only by comparing the calculated average curve and measured data. Observations published by three different laboratories were used supplementing those by the author's measurements for this comparison (Figs. 2-44a and 2-44b, Johnson et al., 1963; Prill et al., 1965; Figs. 2-44c and 2-44d, Laliberte et al., 1966; Figs. 2-44e and 2-44f, Vachaud et al., 1975; Vauclin, 1975; Khanji, 1975; Vachaud et al., 1970; Vachaud and Thony, 1971; Figs. 2-44g and 2-44h, author's measurements). The measured data prove the reliability of the theoretical relationship (Fig. 2-37).

To characterize the probable zone of hysteresis, it is proposed that separate envelope curves be applied along the capillary and adhesive stretches, respectively. Considering the scattering of the data in the zone of adhesion, the envelope curves can be given by modifying the constant in Eq. 2-18. About 95 percent of the measurements have shown scattering in this zone smaller than ±30 percent. Thus, the following equations can be proposed to describe the positions of the curves enveloping the zone of confidence of data expected with 95 percent probability in the adhesive zone:

$$s_{a1} = 1.75 \times 10^{-3} \frac{1-n}{n} \left(\frac{\alpha}{D_h}\right)^{2/3} \frac{1}{\psi^{1/6}} \;;$$

$$s_{a2} = 3.25 \times 10^{-3} \frac{1-n}{n} \left(\frac{\alpha}{D_h}\right)^{2/3} \frac{1}{\psi^{1/6}} \,. \qquad (2\text{-}22)$$

According to the explanation given previously, the influence of wetting or draining can be characterized in the

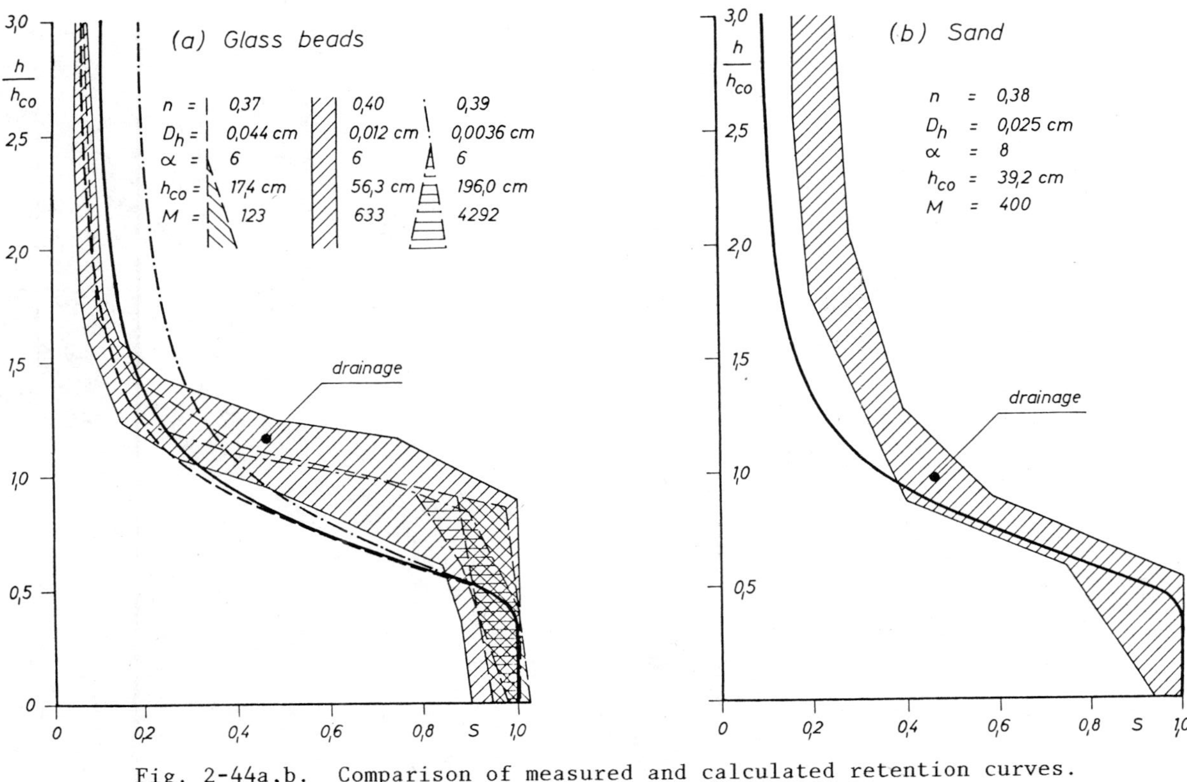

Fig. 2-44a,b. Comparison of measured and calculated retention curves.

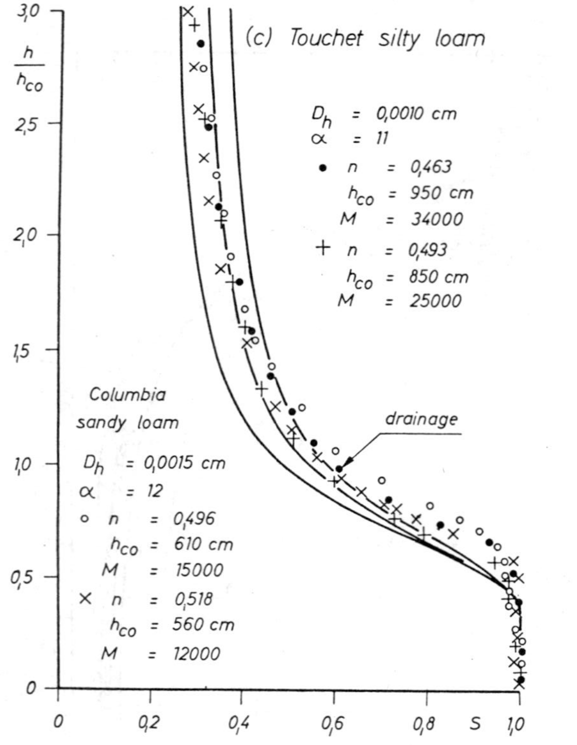

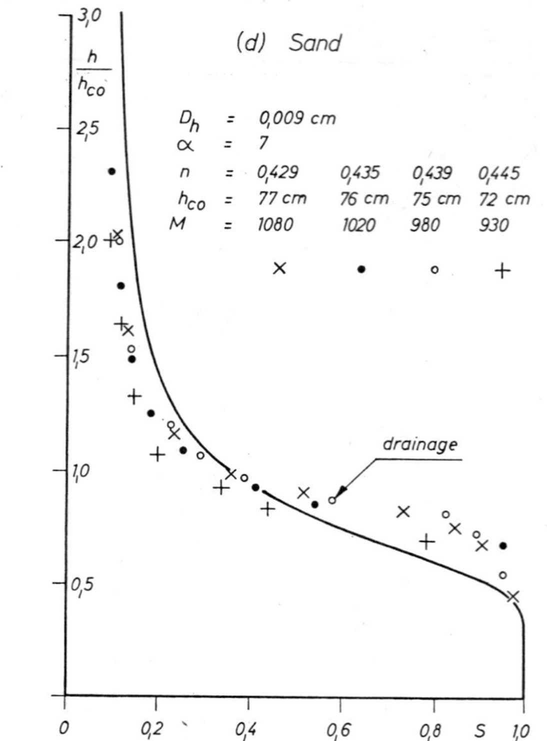

Fig. 2-44c,d. Comparison of measured and calculated retention curves.

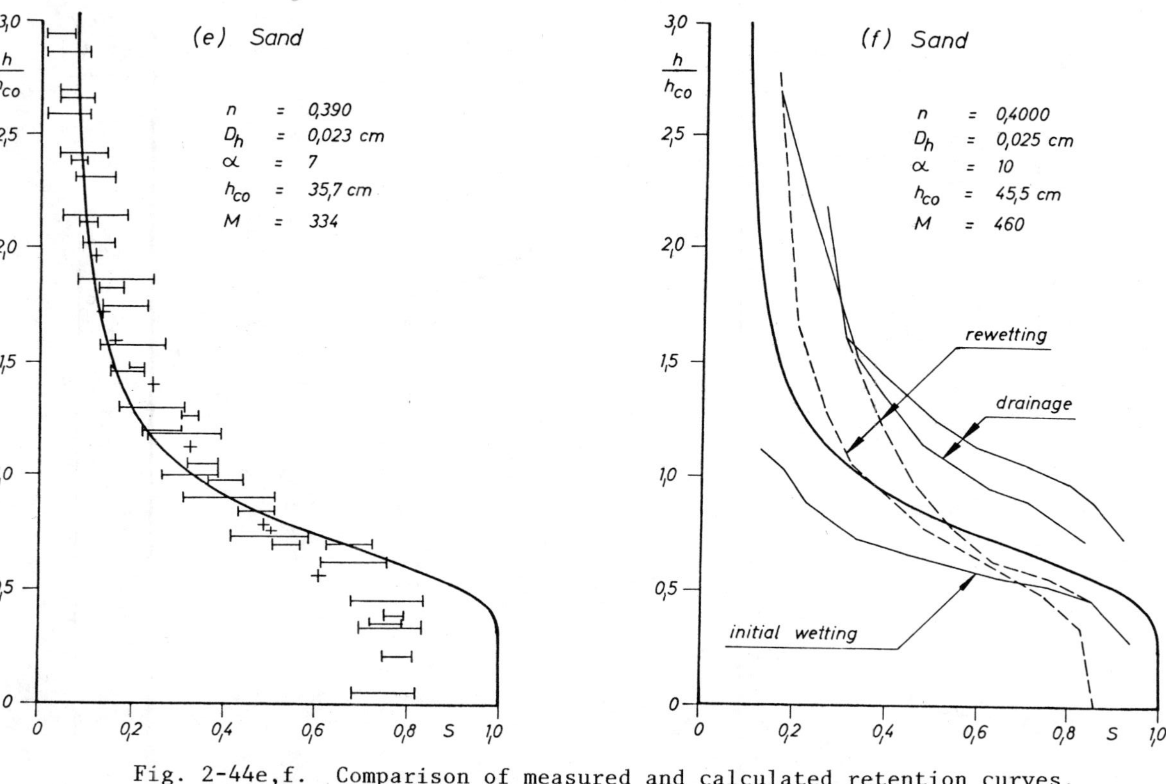

Fig. 2-44e,f. Comparison of measured and calculated retention curves.

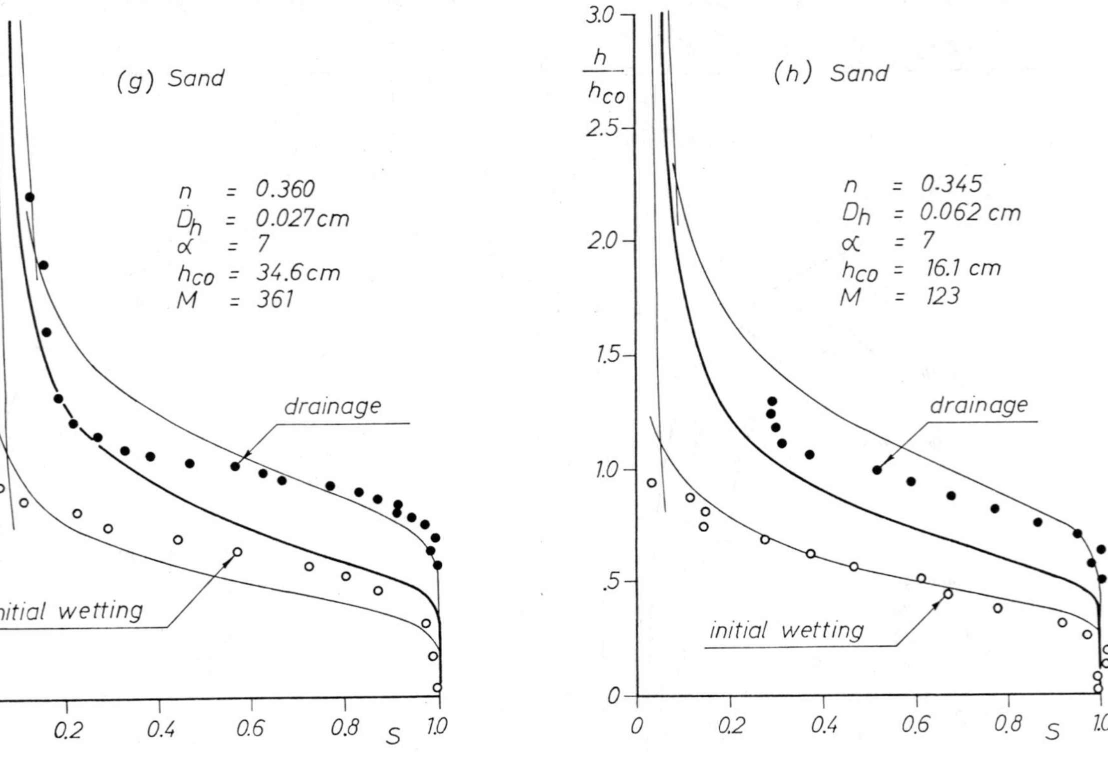

Fig. 2-44g,h. Comparison of measured and calculated retention curves.

capillary zone by shifting the average curve vertically with a height proportional to the wavelength of the changing diameter. This parameter is, however, also a random variable, the calculation of its most probable value requiring further statistical analysis. The investigation can be simplified if the positions of the retention curves, measured experimentally after wetting and draining the samples, respectively, are considered. Using this approximation, it was found that the envelope curves can be determined in the capillary zone similarly as it was done in the adhesive zone, viz., by using a factor to multiply the relative suction in the distribution function characterizing the amount of capillary saturation:

$$s_{c1} = 1 - \left[\left(0.7\,\frac{h_{co}}{\psi}\right)^2 + 1\right] \exp\left[-\left(0.7\,\frac{h_{co}}{\psi}\right)^2\right] ;$$
$$s_{c2} = 1 - \left[\left(1.5\,\frac{h_{co}}{\psi}\right)^2 + 1\right] \exp\left[-\left(1.5\,\frac{h_{co}}{\psi}\right)^2\right] . \qquad (2\text{-}23)$$

The curves calculated by Eqs. 2-22 and 2-23, which border the zone of confidence with 95 percent probability, are also represented in Figs. 2-44g and 2-44h. Their position proves that Eqs. 2-22 and 2-23 are suitable for the approximate characterization of the expected scattering caused by hysteresis.

SECTION 8

HYDROLOGICAL ANALYSIS OF THE PROCESSES OCCURRING IN THE SOIL-MOISTURE ZONE

The flow of water can develop either in liquid or in vapor phase through the soil-moisture zone, and water exchange may occur between the liquid and gaseous phases through the surface of water films in the form of evaporation and condensation. The water may be transported in both phases not only by convective flow (maintained by the forces acting), but also by diffusion caused by the uneven distribution of some intensive properties of either the liquid or the vapor (e.g., temperature, chemical concentration, etc.). Thus, the total mass flux is composed of four members, i.e., diffusion in liquid and gaseous phases and the convective transport of fluid and vapor (Klute, 1952).

To simplify the hydraulic investigation of the process, the diffusion is generally neglected, assuming isothermal conditions and that the differences of other intensive properties between the various points of the system are also negligible. Laboratory and field measurements have proved that the effect of diffusion is really smaller than uncertainties only if extreme conditions were maintained artificially, e.g., high temperature gradient by heating one side of the sample.

In connection with vapor flux, it was found that this component can be significant only if the moisture content of the medium is sufficiently low enough to provide a continuous gas phase open for vapor movement. In this case, however, the differences in vapor pressure at various points are relatively small, and therefore, the convective vapor flux cannot have a considerable role. It was already mentioned that measurable diffusive flux develops only under the influence of a high temperature gradient. In nature, such condition does not occur except in the upper few cm of the soil profile.

Considering these aspects, the investigation of seepage in the soil-moisture zone can be limited to the analysis of the convective flux of liquid water. The chains composed of water films and saturated pores may be regarded, therefore, as a closed system, through the border of which water exchange does not occur.

The movement of water through and within the upper zone of soil has very high practical importance. It determines the ratio of surface runoff, infiltration, and evaporation, and governs, in this way, the whole continental branch of the hydrological cycle. The productivity of agriculture depends on the sufficient water supply of plants and this process is also basically influenced by the storage and flow of soil moisture. The salt balance is also closely related to the movement of soil moisture, which fact raises the agricultural importance of the latter. This is the reason why the investigation of seepage developing above the water table has a very long tradition.

At the same time, the versatile character of the phenomenon, the high number of the factors acting, makes the physical description of the process difficult. It is not yet possible to propose a theoretically well based method, which can be easily applied in practice to characterize soil-moisture flow. Instead of such a complete solution, the development of the classical theories will be discussed at first. It will be followed by the critical analysis of Richard's equation which is generally applied recently for the calculation of the convective flux of soil moisture. The interpretation of the various soil-physics parameters used in this theory are summarized separately. The investigation of the seepage in the soil-moisture zone is assisted by the development of a theory to determine the hydraulic conductivity of unsaturated and partially saturated porous media. The conceptual model serving to calculate this important parameter is based on the combination of the dynamic, geometrical, and statistical models explained previously. Its reliability is proved by comparing the calculated results with measured data. Finally, examples are shown to demonstrate the practical application of the theoretical results.

8-1 Development of Methods to Describe Infiltration

The first investigations were limited to the analysis of infiltration through the surface, although this process only describes one of the boundary conditions of the whole system. This component of the system, being relatively easily observable and measurable, had numerous attempts made to approximate the flow rate infiltrating through the surface and decreasing monotonically in time. The proposed models express the flow rate as a function of the physical parameters and the instantaneous condition of the soil. These empirical formulae generally consider only a few limiting conditions (e.g., after a very long time having elapsed from the beginning of infiltration, the flow rate has to be equal to the flux transported by the saturated medium under the influence of a unit gradient) when selecting the mathematical form of the approximation. Horton's equation (Horton, 1939) can be mentioned as an example of the empirical formulae, which assumes that the time dependent infiltration through a unit area of the surface [f(t)] has to be divided into two parts: the first decreases with increasing time exponentially and the second is constant, characterizing the water conveyance of the saturated soil (Fig. 2-45):

$$f(t) = (a - f_o) \exp(-bt) + f_o \quad ; \qquad (2\text{-}24)$$

where f_o is the flux through the saturated medium (which is approximately equal to the saturated hydraulic conductivity if the water table is deep, because the final gradient tends to unity in this case), while the other constants (a and b) depend on the physical parameters and the instantaneous conditions of the soil.

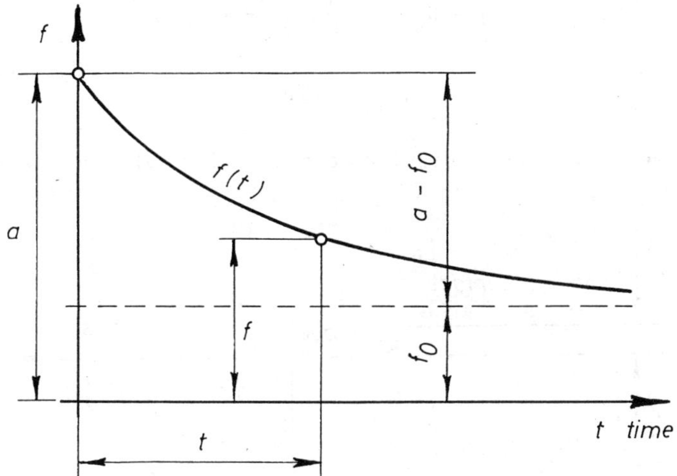

Fig. 2-45. Characterization of symbols used in Horton's infiltration model.

The next development of the theoretical analysis of infiltration was the supposition that there is a well-defined horizontal boundary (wetting front) between the upper and lower domains of the investigated vertical column. Above this front the medium is completely saturated from the water ponded on the surface, while below it the rate of saturation is not yet influenced, the original constant water content is characteristic there. The conditions considered in the mathematical description of this so-called piston flow are represented in Fig. 2-46. The hydraulic parameters (the flow rate and the propagation of the wetting front) can be theoretically determined by the integration of the basic differential equations derived from the kinematic concept of the laminar flow through a continuous field (Laplace's equation). The same result can be achieved, however, by substituting the gradient into

Darcy's equation. In this case the gradient is equal to the total pressure difference between the surface and the wetting front related to the length of the seepage. The numerator (the total pressure difference) is composed of three members: the depth of the ponded water (H), the length of the seepage (being a function of time) $[x_f(t)]$, and the average capillary suction at the wetting front (h_c). Thus, the flux of the piston flow (the infiltration through a unit area of the surface q) and the velocity of the propagation of the wetting front (v_{eff}) can be calculated from the following equations (Green and Ampt, 1911):

$$q = K_s \frac{H + h_c + x_f(t)}{x_f(t)} \quad ; \qquad v_{eff} = q/n \quad ; \tag{2-25}$$

where K_s is the hydraulic conductivity of the saturated medium.

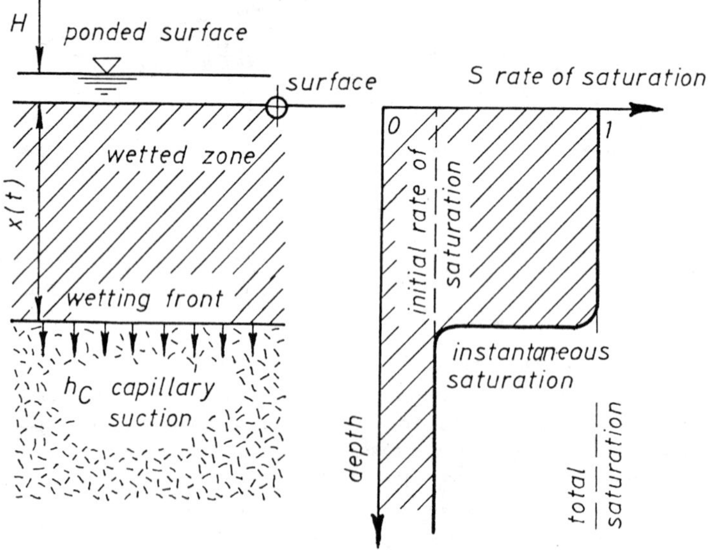

Fig. 2-46. Characterization of the process of piston flow.

Considering the continuity of the movement, a further condition can be determined: the total amount of water crossing the surface, from the beginning of infiltration until a time point t, should be equal to the storage of a prism having a base of unity and located between the surface and the wetting front. Thus the variables have to satisfy the following condition:

$$\int_0^t q \, dt = K_s \int_0^t \frac{H + h_c + x_f(t)}{x_f(t)} \, dt = n_s \, x_f(t) \quad ; \quad (2\text{-}26)$$

where n_s is the specific yield of the soil.

There are three parameters in Eqs. 2-25 characterizing the soil through which the infiltration takes place (K_s; n_s; h_c). Among them the hydraulic conductivity of the saturated medium and the specific yield are well-defined characteristics, although from those determined from a completely saturated sample, because the air bubbles entrapped above the wetting front may result in smaller K_s and n_s than those parameters belonging to the saturated condition. The capillary suction is even more uncertain than the other two characteristics. The actual wetting front is never a horizontal plane. The water runs ahead in the large pores and propagates more slowly in the small ones, while the capillary suction is larger in the narrow channels. This action of capillarity varying from point to point has to be taken into account, therefore, with an average value. If the suction is expressed by the probable equivalent height of a water column, the average capillary rise lies between the probable maximum and minimum capillary height.

In spite of the numerous uncertainties and the rough approximations applied to derive the mathematical model, the latter was proved to be quite reliable in practice if the soil-physics parameters were sufficiently well chosen. It is quite evident that parameters derived from the observation of a process and evaluated by assuming a given relationship between the measured variables will provide an acceptable description of the phenomenon, if they are substituted into the same equation which was used to determine them. The application of such mathematical models and soil-physics parameters always requires, however, the execution of local measurements and hinders the generalization of data determined elsewhere.

Another limitation arising in connection with the application of the Green-Ampt equation is caused by the fact that its form given as Eq. 2-25 describes the vertical movement of the water through the unsaturated zone only if the flow is directed downwards, the pores are completely saturated at the surface, and the infiltrating water is continuously replenished on the surface. Similar equations can be determined to characterize the upward movement of water from the water table up to the supposed average capillary height, or the lowering of the upper surface of the saturated zone to a height of h_c above the water. The condition of the application of this relationship is similarly the assumption of a sharp border between the dry and completely saturated domains. Thus, this approximation is not suitable to describe those types of movements which also occur in an unsaturated medium without requiring the complete saturation of any part of the profile (e.g., the effect of evaporation on groundwater and the redistribution of the soil moisture).

A further important development concerning the investigation of the water movement in unsaturated porous media was the combination of Buckingham's potential and Darcy's law (Richards, 1931). It was already explained that the total energy, expressed in units having the dimension of length, available above the water table to create and maintain seepage is composed of two terms: the height of the point related to the arbitrarily chosen reference level and the suction head. The hydraulic gradient vector can be calculated as the divergence of the total energy (the change of the energy head in the direction of movement):

$$I = \nabla(h-\psi) = \frac{d(h-\psi)}{ds} = \frac{\partial(h-\psi)}{\partial x} + \frac{\partial(h-\psi)}{\partial y} + \frac{\partial(h-\psi)}{\partial z} \quad . \tag{2-27}$$

Substituting this gradient into Darcy's equation, the flux (the flow rate through a unit area normal to the main direction of flow) can be obtained:

$$q = -KI = -K\left[\frac{\partial(h-\psi)}{\partial x} + \frac{\partial(h-\psi)}{\partial y} + \frac{\partial(h-\psi)}{\partial z}\right] \quad . \tag{2-28}$$

According to the equation of continuity, the local change of flux is equal to the change of the water content in time within a prism of unit volume;

$$\nabla q = -\frac{\partial W}{\partial t} \quad . \tag{2-29}$$

Considering that hydraulic conductivity is also a function of water content K(W), and using the relationship interconnecting suction head and water content [$\psi(W)$], all the terms can be given as functions of either volumetric moisture content (W) or suction (ψ). A differential equation can be determined by combining Eqs. 2-28 and 2-29 which gives a relationship between time, the space coordinates, and either W or ψ. Choosing the positive z-axis directed vertically upwards, the various forms of the differential equation are as follows:

$$\frac{\partial K(\psi)}{\partial z} - \frac{\partial}{\partial x}\left[K(\psi)\ \frac{\partial\psi}{\partial x}\right] - \frac{\partial}{\partial y}\left[K(\psi)\ \frac{\partial\psi}{\partial y}\right] - \frac{\partial}{\partial z}\left[K(\psi)\ \frac{\partial\psi}{\partial z}\right] = A(\psi)\ \frac{\partial\psi}{\partial t} \quad ;$$

where (2-30)

$$A(\psi) = \frac{\partial W}{\partial\psi}$$ (specific water capacity, Richards, 1931); or

$$\frac{\partial K(W)}{\partial z} + \frac{\partial}{\partial x}\left[D(W)\ \frac{\partial W}{\partial x}\right] + \frac{\partial}{\partial y}\left[D(W)\ \frac{\partial W}{\partial y}\right] + \frac{\partial}{\partial z}\left[D(W)\ \frac{\partial W}{\partial z}\right] = \frac{\partial W}{\partial t} \quad ;$$

where (2-30)

$$D(W) = K(W)\left[-\ \frac{\partial\psi}{\partial W}\right]$$ (soil-water diffusivity, Childs and Collis-George, 1948).

The great advantages of Eqs. 2-30 are that the seepage in any direction can be described by them and they do not include any restrictions concerning the rate of saturation of the medium. Thus it can be applied for the determination of the hydraulic parameters of infiltration, upwards movement of water which starts from the groundwater to replenish the moisture content evaporated from the unsaturated zone, as well as those of the horizontal seepage of soil moisture (or its flow in any direction) created by the tension differences between the various points of the medium. Difficulties are, however, caused by the fact that there is no unique relationship between the suction head and the water content because of the hysteresis of the soil-moisture retention curve. Supposing that the hydraulic conductivity of the unsaturated solid matrix is an unambiguous function of the water content, the $K(\psi)$ relationship and the specific water capacity (Eq. 2-30) are not single-valued functions in Richard's formula, while in the other form of the differential equation (Eq. 2-30) the diffusivity term has the same uncertainty.

Apart from the problems caused by hysteresis, the very complicated structure of the equation also hindered earlier practical application of the method. Those simplifications, which could ensure the analytic solution of the differential equation, would be very rough approximations physically and, therefore, reliable results can only be achieved by the numerical handling of the formula. This is the reason why this theory became widely used only when large computers were developed. One of the pioneers in preparing the various numerical methods was Philip (1955; 1957; 1966). Assuming that both D and K are single-valued functions of W, and considering the boundary conditions $W = W_1$ if $t = 0$, $z > 0$, and $W = W_2$ if $z = 0$, $t > 0$, he solved the differential equation simplified for the one-dimensional movement of infiltration:

$$\frac{\partial W}{\partial t} = \frac{\partial}{\partial z}\left(D \frac{\partial W}{\partial z}\right) + \frac{\partial K}{\partial z} \qquad (2\text{-}31)$$

by expanding it into a power series in $t^{1/2}$. One of the most important results of this analysis is the statement that, excluding the vicinity of the time point $t = 0$, the propagation of the wetting front (s_f) can be assumed to be proportional to the square root of the time

$$s_f(t) = \eta\, t^{1/2} \qquad (2\text{-}32)$$

In the last twenty years, the number of the proposed solutions was increased very rapidly. Some publications are listed here, as examples: Bouwer, 1964; Liakopulos, 1966; Boreli and Vachaud, 1966a; Vachaud, 1966; Kobayashi, 1966; Rubin, 1966; Whisler and Klute, 1966; Irmay, 1966; Watson, 1967; Whisler and Watson, 1969; Freeze, 1969; Whisler and Bouwer, 1970; Hornberger and Remson, 1970; Kastenak, 1971; Braester, 1973; Mein and Larson, 1973. The differences between the various methods are caused mostly by the different boundary conditions considered in the derivations and by the approximations applied to achieve a numerically amenable form of the differential equations. Theoretically they do not considerably differ from each other. This is the reason why these methods are not separately analyzed here in detail but reference is given only to Swartzengruber's excellent summary (De Wiest, 1969), where the application of the methods based on the explained theory (so-called diffusion theory) is represented in connection with different practical problems (one-dimensional horizontal flow; downwards vertical infiltration; vertical steady-state flow from the groundwater having a water table with constant position; change of moisture content caused by constant drainage through the surface; influence of the flow having a flow rate decreasing in time; vertical drainage of the soil moisture to the groundwater space). There are, however, several theoretical objections in connection with Eq. 2-30.

In recent publications, the influence of air compressed in the lower lying pores on the flow rate of infiltration is investigated (Youngs and Peck, 1964; Wilson and Luthin, 1963; Adrian and Franzini, 1966; Morel-Seytoux, 1973; Morel-Seytoux and Khanji, 1974; Brustkern and Morel-Seytoux, 1975; Vachaud et al., 1974). The basis of these analyses is the assumption that a part of the total available energy is consumed by the flow of air moving against the infiltrating water and only the remaining part is used for maintaining the seepage. This action can be observed in experimental columns if the lateral escape of the air is not ensured. The infiltrating water seals the surface and the pressure of the entrapped air gradually increased as the wetting front propagates downwards. Thus, the total potential between the free surface of the ponded water and the front decreases, retarding the seepage. A further consequence of the air compression is that after achieving a given pressure, the air starts to move upwards, escaping from the pores through the surface. Thus, the upper part of the profile can be only partially saturated because some fraction of the pores has to remain unsaturated to ensure the free path of air. The theory of the investigation of simultaneous air and water flow is still under development. It would be premature to give a detailed analysis of the problem. Referring to the publications listed previously only one example is presented here where the development of the saturation of two experimental soil column are compared: the mantle of the first was perforated to ensure the lateral escape of air, while in the second case the air could leave the pores only through the surface (Fig. 2-47, after Vachaud et al., 1974).

It was already explained that the root zone (or the cultivated zone) has to be excluded from the physical investigation of the soil-moisture zone not only because the structure of the pores is very uneven, but it has to be taken into account as well, that the biological processes strongly influence the water movement here and, therefore, the seepage cannot be investigated as a pure physical phenomenon. The remaining part of the partially saturated medium has to be divided into further zones according to the dynamic differences occurring at different elevations above the water table (adhesion or capillarity is the dominating force acting against gravity or the influence of both has to be taken into account).

Another objection, making the application of diffusion theory questionable, originates from the fact that the differences between the dynamically separated zones above the water table are not considered and all the parameters [$K(W)$; $\psi(W)$ and $D(W)$] are described with continuous functions in the whole soil profile. Since the approximation of the listed relationships with single-valued functions already includes some uncertainties, e.g., those caused by the hysteresis of the retention curve, this type of function seems to be generally acceptable although the separate determination of the relationships between the variables in the different zones is advisable in order to find the best continuous approximation.

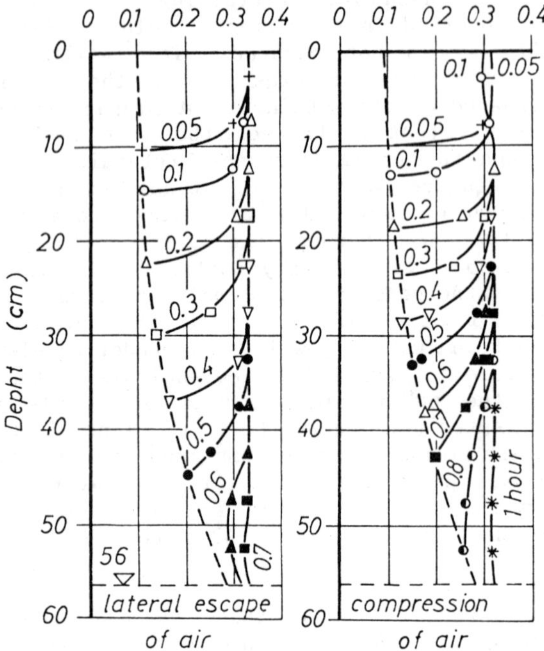

Fig. 2-47. Characterization of the influence of entrapped air on infiltration by using laboratory measurements.

There are cases, however, when the dynamic differences cannot be neglected. If the closed capillary zone reaches the surface, the hydraulic conductivity of the whole layer (where the water has lower pressure than the atmospheric) is equal to that belonging to the saturated condition. In the open capillary zone, some pores are completely saturated, and only thin water films develop in the others. Thus, flow starts directly through the continuous chains of fully saturated pores, reaching the surface without storage, while in the large openings, where the movement is maintained partially by tension differences, the process starts with thickening the films and water transport is always associated with storage. Finally, in the adhesive zone, the movement occurs through the water films and in the form of the development of capillary membranes at the narrow stretches of pores, from which the water propagates downwards by dropping when the weight of the overlying water breaks through the membranes. The various possible forms of the water movement is summarized in Fig. 2-48.

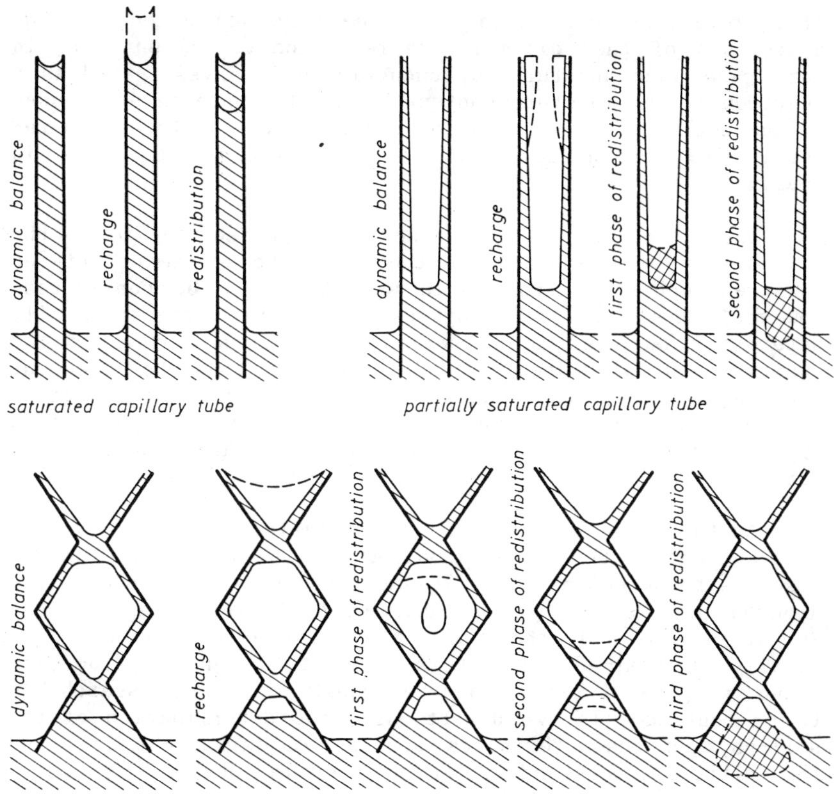

Fig. 2-48. Change of the character of infiltration in tubes saturated by capillarity, wetted by adhesion, and those having changing diameter, respectively.

The third objective, which may be raised in connection with the derivation of Eq. 2-30, is the fact that the static shearing stress is neglected. If the continuum approach is acceptable, the flow of water through the pores of the solid matrix can be described by applying the seepage law, only the nonlinear character of velocity vs. gradient relationship has to be taken into account by using the generalized form of Darcy's equation, which can be achieved by substituting a hydraulic conductivity depending on either velocity or gradient.

On the contrary to this principle, the unsaturated hydraulic conductivity is regarded, in most cases, as a single-valued function of saturation, neglecting its dependency on the local gradient. This is the reason why the measured data indicate considerable scattering not only in the

$K(\psi)$ relationship (which is generally explained by the hysteresis of the soil-moisture retention curve) but even in that case when unsaturated conductivity is investigated as a function of the moisture content K(W). The uncertainty caused by the influence of gradient on conductivity has to be, therefore, also considered when the reliability of Eq. 2-30 is analyzed.

The strongest reason generally applied to prove the reliability of the diffusion theory is the agreement between measured and theoretical values. It has to be considered, however, that the soil-physics parameters were generally determined on the same experimental columns for which the observed and calculated data are compared, thus the agreement between measured and calculated data testified the correctness of the whole system composed of the model and the parameters applied in it. This draws attention to the importance of the determination of the soil-physics parameters.

Another consequence of uncertainties in the physical description of the water movement in the unsaturated media also has to be emphasized. The models derived from the diffusion theory do not give a completely correct image of the actual process and, therefore, they can be applied only with locally determined parameters, and thus these parameters cannot be generalized as material constants. The research has to be continued in two directions: the determination of the soil-physics characteristics applied in existing models and the better understanding of the actual physical process.

8-2 Soil-Physics Parameters Used in Diffusion Theory

There are three soil-physics parameters which have to be known for the characterization of the water movement in unsaturated medium by applying the diffusion theory, i.e., soil-moisture retention curve (relationship between the suction head and the moisture content; (W)), unsaturated hydraulic conductivity (which can be expressed depending on either suction head or moisture content; $K(\psi)$ or K(W)), and soil-water diffusivity (the product of hydraulic conductivity and the derivative of the retention curve; D(W), see Eq. 2-30).

The physical interpretation of the soil-moisture retention curve was discussed earlier. It was proved there that the volumetric moisture content (W) or its value related to the total porosity of the solid skeleton (s the amount of saturation) can be expressed as a function of the suction head (ψ) prevailing in the sample, or depending on the height of the investigated point above the water table (h), if the expected soil-moisture distribution belonging to the dynamic balance has to be determined. One of the assumptions applied in the derivation of the equations describing this relationship is that the osmotic forces may be neglected because the cultivated zone is excluded from the investigation, thus the conditions can be assumed to be isothermal and the concentration of the dissolved salts in the soil moisture is relatively low. This is the reason why the two parameters can be regarded to be equal to one another ($\psi = h$). Since the double system of the forces (adhesion and capillarity) acting against gravity has to be taken into account, the rate of saturation is composed of two terms, one created by capillarity (s_c) and the other by adhesion (s_a). Thus, the final form of the equation combines s_c and s_a parameters (see Eq. 2-21).

Equation 2-21 can be regarded as an average characterization of the investigated relationship, not only because it contains many rough approximations, but the interrelation between suction head and moisture content is made uncertain by the hysteresis of the soil-moisture retention curve as well. Thus, the function describing the contact between the two variables is not a single-valued one, but it is also influenced by the history of wetting and drying of the profile. Analyzing the movement of the soil-moisture zone mathematically, it is not possible, however, to consider the difference caused by the present state and the previous conditions of the solid matrix and, therefore, only average relationships can be used to characterize the $\psi(W)$ function. The uncertainty of such approximations must not be forgotten, however, and the mathematical methods applied to solve the movement equations have to be selected so that the accuracy of the method and that of the basic data are in balance.

As it is indicated by Eq. 2-30, the physical parameters differing in the case of water movement through an unsaturated solid matrix from thoe of seepage under saturated condition can be expressed by the special behavior of two parameters depending on the moisture content, i.e., the hydraulic gradient is composed of two terms, one characterizing the potential energy and the other depending on the suction differences; the hydraulic conductivity is also a function of the water content. The determination of the total hydraulic gradient is based on the suction head vs. moisture-content relationship discussed already. The next physical process to be investigated is, therefore, the development of hydraulic conductivity in an unsaturated porous medium.

According to experimental measurements, the resistance of an unsaturated solid matrix is greater than that of the same medium in saturated condition. An easy explanation of this observation can be given considering that hydraulic conductivity is equal to the flux (flow rate through a unit area normal to the main direction of the flow) related to the gradient maintaining the movement. The actual cross section, through which the water movement develops within a unit area, is measured by the areal porosity of the sample, if the latter is saturated. Under unsaturated conditions, only a part of the pores is filled with water and conveys water. The actual cross section transporting the water decreases, therefore, as the saturation decreases. This simple geometrical reason testifies that both flux and hydraulic conductivity have to be smaller through the unsaturated porous medium than that in a saturated condition, even if the dynamic differences are not considered.

It was also proved by the experimental measurements that the two values of hydraulic conductivity, i.e., those belonging to the saturated (K_s) and unsaturated (K_u) conditions, respectively, are proportional to one another. The coefficient of proportionality can be given as the function of either saturation (depending on the degree of saturation) or that of the suction head (ψ) prevailing at the investigated point:

$$K_u = K_s \ f_1(s) \quad ;$$

or (2-33)

$$K_u = K_s \ f_2(\psi) \quad .$$

(Naturally the same proportionality is also valid for the ratio of intrinsic permeability values because the fluid is the same in the two systems.)

Since the resistance of the solid matrix is supposed to depend mostly on the amount of those pores which are filled with water, the general opinion is that the $K_u(W)$ function is applied to create contact between hydraulic conductivity and suction head and since this equation is multi-valued (because of the hysteresis of the soil-moisture retention curve), it is not possible to determine a unique relationship between the two variables in question. It is necessary, however, to mention that the $K_u(W)$ equation is not a single-valued function either because there are special dynamic processes (change of the capillary contact angle, influence of the actual gradient on adhesion) acting and not considered in the relationship.

The form most generally applied to describe the K_u vs. W relationship was already quoted (see Eq. 1-92), and is repeated here without detailed explanation:

$$K_u = K_s \left(\frac{s - s_o}{1 - s_o}\right) \qquad (2\text{-}34)$$

The relationship described by Eq. 2-34 is represented in Fig. 2-49. Other research workers have published their observations in the form of the actually measured data (Rose, 1966; Vachaud, 1966; Elrick, 1966; Peck, 1966; Elrick and Bowman, 1964). Some of these measurements are summarized in Fig. 2-49c).

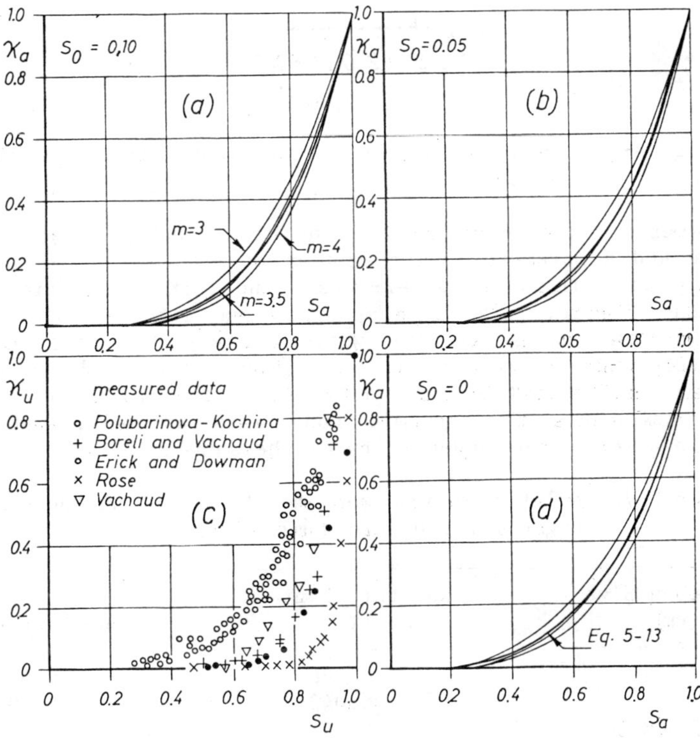

Fig. 2-49. Relationship between relative unsaturated conductivity and the amount of saturation on the basis of measured and calculated data.

The observations indicate the uncertainty of the data. One of the processes which may influence the scattering of conductivity values is the change of the capillary contact angle. Wladitchensky's (1966) experiments prove the reliability of this hypothesis. Figure 2-50 shows the change of the cosine of the contact angle resulted from the various

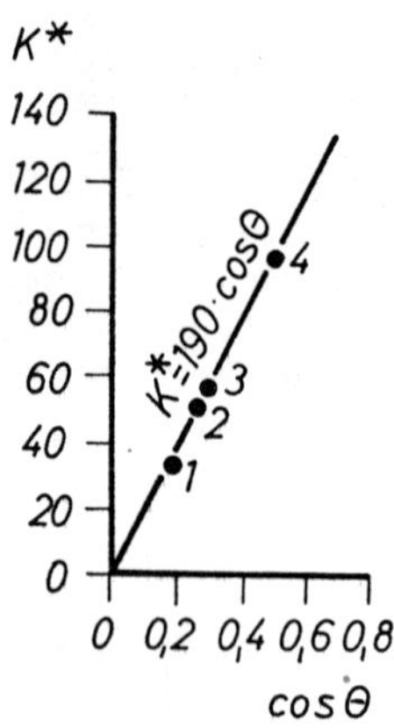

Fig. 2-50. Relationship between permeability and wetting angle.

treatments of a given sand. The relationship between this value and a parameter being proportional to hydraulic conductivity and called the water-uptake index (K*) is represented in the figure. The same publication also contains data proving the change of the wetting angle depending on the moisture content of sand samples (Table 2-7). The observed values testify that in unsaturated media the amount of saturation influences the dynamic conditions as well, and this change affects the conductivity of the sample.

Table 2-7. Relationship between moisture content and wetting angle for two sand samples.

Fraction Size (cm)	Moisture Content (%)	Cosine of the Wetting Angle
	0.036	0.66
	0.049	0.59
0.05-0.025	0.068	0.20
	0.09	0.36
	8.27	0.44
	11.97	1.00
	0.03	0.52
	0.08	0.40
0.025-0.01	0.13	0.17
	2.49	0.35
	5.19	0.59
	8.32	1.00

Another investigation (Vachaud et al., 1974) has drawn the attention to the errors which may occur in connection with the determination of the basic parameter, conductivity under saturated condition, to which all the measured capillary conductivity values are related. It was found that the complete saturation could hardly be achieved in the investigated unsaturated soil column because of entrapped air bubbles. The highest degree of saturation after the rapid wetting of the column was about $s_w = 0.8$ and, therefore, only the lowest part of the K_u/K_s vs. W curve could be determined (Fig. 2-51). The same result is shown by Fig. 2-4.

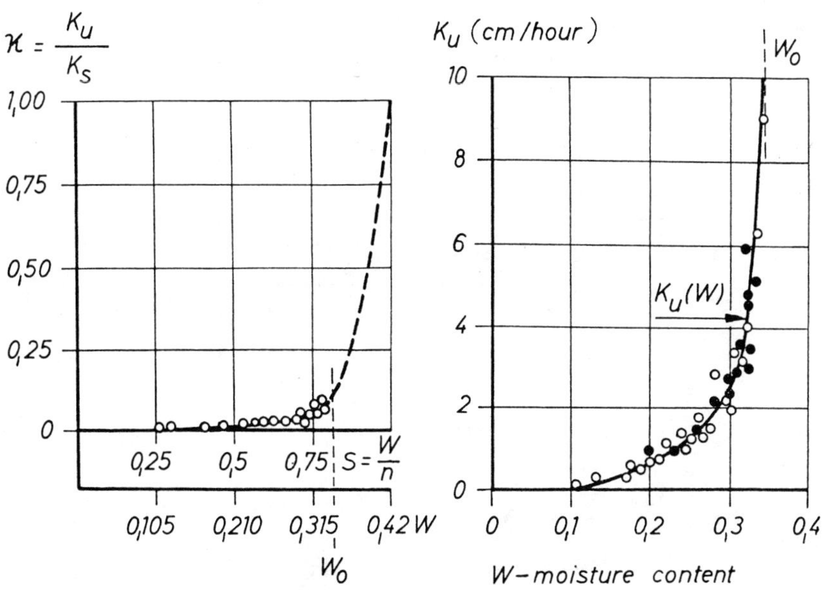

Fig. 2-51. Influence of the partial saturation on the evaluation of measured data of unsaturated conductivity.

The expression of hydraulic conductivity as a function of suction head is more uncertain than the characterization of the $K_u(W)$ function. Two examples are shown in Fig. 2-52. The data of the first was determined by Wind (1966) on samples including two different peats, one clay, and one sand (Fig. 2-52a). Wesseling and Wit (1966) have used disturbed and undisturbed samples of grained sediments (Fig. 2-52b). The second publication gives also a complete and clearly arranged survey of the various formulae proposed for the characterization of the $K_u(\psi)$ function:

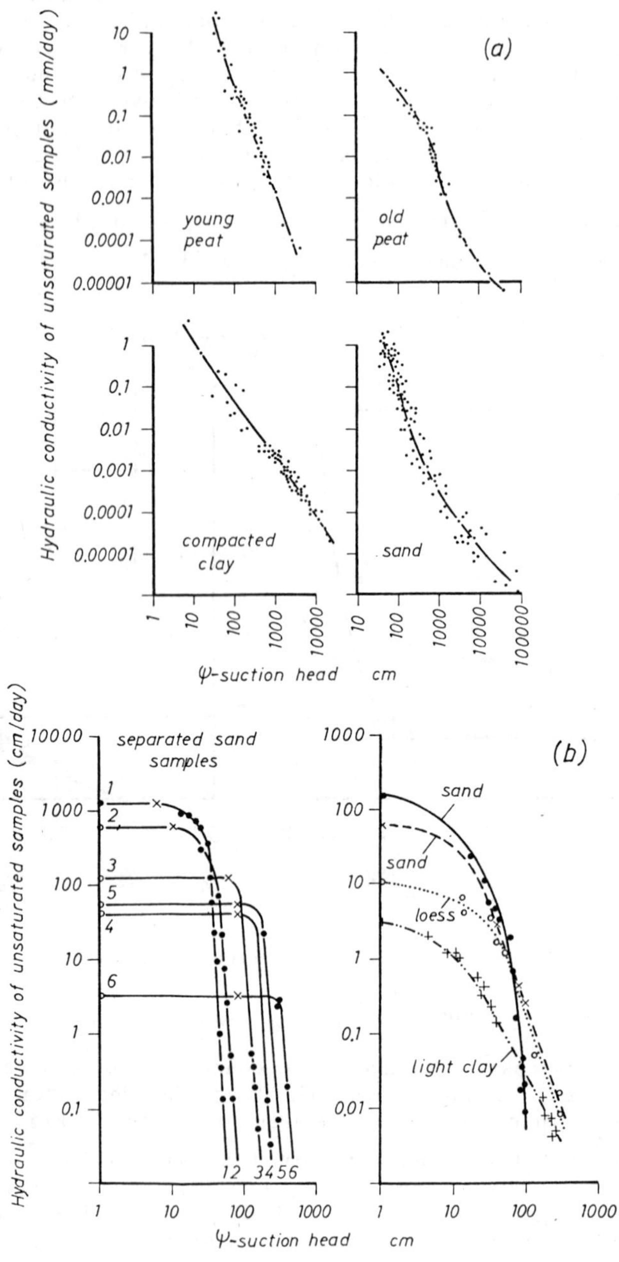

Fig. 2-52. Characterization of the relationship between unsaturated conductivity and suction head by using measured data.

$$K_u = a\,\psi^{-b} \quad ; \qquad \text{(Wesseling, 1957);}$$

$$K_u = a(\psi^b + c)^{-1} \quad ; \qquad \text{(Gardner, 1958);}$$

$$K_u = K \exp(-a\psi) \quad ; \quad \text{or} \qquad (2\text{-}35)$$

$$K_u = K \exp[-a(\psi - \psi_o)] \qquad \text{(Rijtema, 1965).}$$

Among the parameters applied in the diffusion theory, the soil-water diffusivity is the less reliable. It has no physical interpretation, its definition being only a formal mathematical operation giving the parameter as the product of hydraulic conductivity and the negative tangent of the soil-moisture retention curve (see Eq. 2-30). Even its name "diffusivity" does not refer to the physical process of diffusion. The parameter has only a similar place and character in Eq. 2-30 as the real coefficient of diffusivity in the general transport equations. This is the reason why the parameter was named soil-water diffusivity.

The uncertainty of this parameter is also shown by the publications giving generally the measured data in a direct graphical or tabulated form without trying to approximate them with some mathematical formula. Even the trends of the D vs. W curves fitted to the measured points are different according to the results of the various experiments. In some cases, a diffusivity monotonically increasing with increasing soil moisture was found (Fig. 2-53a; Boreli and Vachaud, 1966). Other observations showed inflection in the middle of the investigated range of soil moisture (Fig. 2-53b; Black et al., 1969), or the experiments indicated a local minimum of the curve (Fig. 2-53c; Rose, 1966).

Considering the fact that both the hydraulic conductivity and the soil-moisture retention curve are multi-valued functions of the water content and even the character of these relationships is different in the adhesive and capillary zones, it is quite obvious that soil-water diffusivity, being composed of the other two parameters, cannot be determined as a unique function of the moisture content. The better understanding of the role and the behavior of this special term has to be based, therefore, on the theoretical investigation of ψ vs. W and K vs. W relationships.

8-3 Determination of Unsaturated Hydraulic Conductivity

The influence of the two different molecular forces acting against gravity, i.e., adhesion and capillarity, also

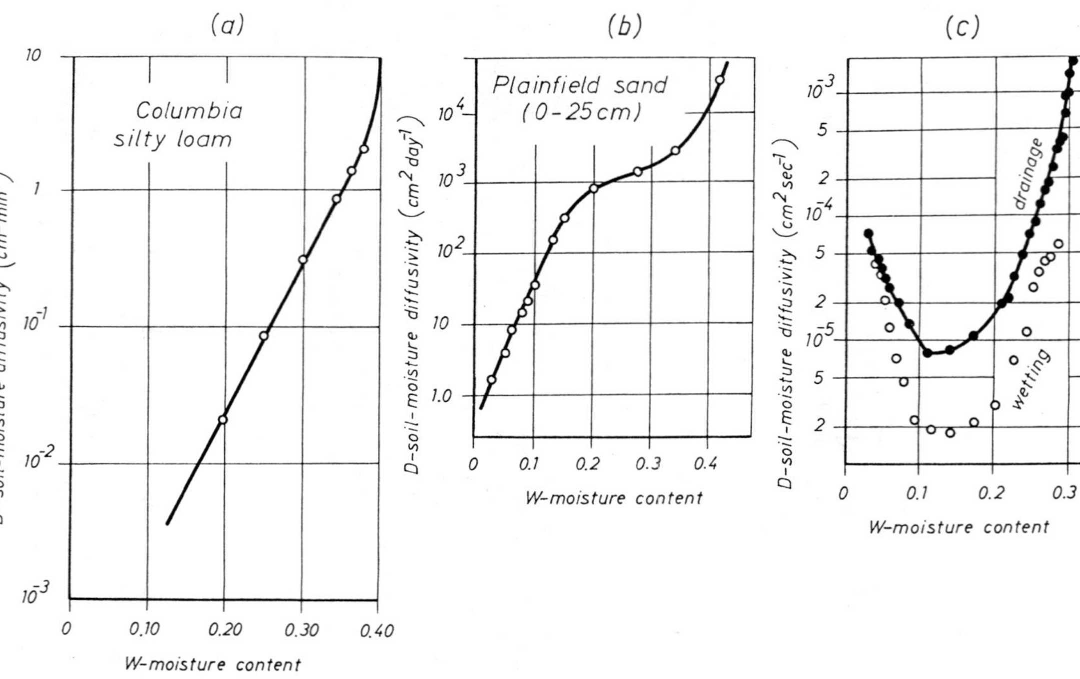

Fig. 2-53. Characterization of the relationship between moisture content and soil-water diffusivity by using different measurements.

has to be considered when the water transporting capacity of the unsaturated layer is investigated. A double system of models have to be established, one representing the narrow pores in which water is conveyed through the whole area of the channels, and the other characterizing the large pores having only annular water transporting section in the form of water films as it was explained earlier.

The dynamic model established to determine adhesive conductivity, K_a the transport through an unsaturated sample having soil moisture only in the form of water films retained by adhesion and not containing completely saturated pores, was derived by using the capillary tube model, supposing that the tubes convey water only through the annual water films covering the walls (Kovács, 1971a; 1971b). The result of this analysis is the detailed discussion which can be found in subsection 4-4 (Eqs. 1-88 and 1-91), and is only quoted here by repeating the final equation giving the relative adhesive conductivity χ_a which gives adhesive conductivity K_a related to saturated hydraulic conductivity K_s as a function of the amount of saturation:

$$\chi_a = \frac{K_a}{K_s} = \left(\frac{s-s_o}{1-s_o}\right)^2 - 2\,\frac{(1-s)(s-s_o)}{(1-s_o)^2} \;;\; -2\left(\frac{1-s}{1-s_o}\right)^2 \ln\frac{(1-s)}{(1-s_o)};$$

or, if $s_o = 0$ (2-36)

$$\chi_a = \frac{K_a}{K_s} = s^2-2s(1-s) - 2(1-s)^2 \ln(1-s)$$

$$= \frac{2}{3}\,s^3\left[1 + 3!\sum_{n=1}^{\infty}\frac{n!}{(n+3)!}\,s^n\right].$$

It was already mentioned in connection with the derivation of this function that the relationships generally proposed in the literature to calculate relative unsaturated hydraulic conductivity (see Eq. 2-34) can be regarded as the approximation of Eq. 2-36. It also has been proved earlier that a simplified form of Eq. 2-34 can be accepted to determine relative adhesive conductivity because neither the change of the power m nor that of the initial amount of saturation s_o does not considerably modify the final result if the changes occur within the reasonable ranges of the two parameters. It is this reason why a very simplified form of relative adhesive conductivity is used in further derivations (see also Eq. 1-92):

$$\frac{K_a}{K_s} = s^3 \quad . \qquad (2\text{-}37)$$

The calculation of capillary conductivity (water transport through the narrow pores completely filled with water by capillarity) requires once again the application of the statistical model. The flow rate through a capillary tube (q_i) is proportional to the cross section of the tube (f_i) and to the hydraulic gradient (I). The coefficient of proportionality is the hydraulic conductivity of the tube (K_i), which is the function of the second power of the diameter. It is linearly interrelated, therefore, to the area of the cross section f_i:

$$q_i = K_i f_i I \; ; \quad K_i = C_1 d_1^{\,2} = C_2 f_i \quad ;$$

therefore (2-38)

$$q_i = C_2 f_i^{\,2} I \quad .$$

The flux transported through the pores being smaller than f_c within a sample having a cross section of unity (q_c) can be calculated by summing up the product of q_i and the probable number of f_i pores until a limit f_c:

$$q_c = N \sum_{i=1}^{k} q_i \;{}_i = C_2 IN \sum_{i=1}^{k} f_i^{\,2} \;{}_i$$

$$= \frac{C_2 INf_o^{\,2}}{\Delta f} \sum_{i=1}^{k} (f_i/f_o)^2 \;{}_i \;\Delta f \; ; \quad \text{or} \qquad (2\text{-}39)$$

$$q_c = \frac{C_2 IN}{\Delta f} \int_0^{f_c} f^2 \quad (f,\Delta f)\; df = \frac{C_2 INf_o^{\,2}}{\Delta x} \int_0^{x_c} x^2 \; \xi(x,\Delta x)\; dx \; ;$$

where the upper limit k of the sum is the number of the Δf interval, the mean value of which is equal to f_c.

The same integral equation is also valid for saturated media, but, in this case, the water transporting capacity of each tube has to be taken into account and, therefore, the upper limit of the integral is infinity. Thus, the flux through a unit area of the saturated sample (q_s) is as follows:

$$q_s = \frac{C_2 INf_o^{\ 2}}{\Delta x} \int_o^\infty x^2 \ \xi(x,\Delta x) \ dx \quad . \qquad (2\text{-}40)$$

The flux is equal to seepage velocity, thus its value divided by the hydraulic gradient gives the hydraulic conductivity. Consequently the ratio of capillary conductivity (K_o) and saturated conductivity (K_s) can be calculated as the quotient of Eqs. 2-39 and 2-40, where the integrals can be solved by substituting the statistical model:

$$K_c = \frac{K_c}{K_s} = \frac{q_c}{q_s} = \frac{\int_o^{x_c} x^2 \ \exp(-x) \ dx}{\int_o^\infty x^2 \ \exp(-x) \ dx} = 1-(\frac{1}{2}\ x^2+x+1)\ \exp(-x) \quad .$$

(2-41)

To express the unsaturated conductivity as a function of the suction head, it has to be considered that the flux through a unit area of the unsaturated soil (q_u) is composed of two parts: viz., capillary flux (q_c) through the completely filled pores, and adhesive flux (q_a) developing through the water films in the remaining part of the pores and, therefore, the reduction of porosity (i.e., ($1-s_c$)) has to be considered when expressing the adhesive flux:

$$q_u = q_c + q_a = [K_c + (1-s_c)\ K_a]\ I \quad ;$$

and, therefore, (2-42)

$$K_u = \frac{q_u}{I} = K_c + (1-s_c)\ K_a \quad .$$

Combining Eqs. 2-37, 2-41 and 2-42, the final equation to calculate relative unsaturated conductivity is as follows:

$$\chi_u = \frac{K_u}{K_s} = 1 - \exp\left[-\left(\frac{h_{co}}{\psi}\right)^2\right] \ .$$

(2-43)

$$\cdot \left\{\frac{1}{2}\left(\frac{h_{co}}{\psi}\right)^4 + \left[1-2.75\text{x}10^{-6}\left(\frac{h_{co}}{\psi}\right)^{1/2}\left(\frac{1-n}{n}\right) h_{co}^{3/2}\right]\left[\left(\frac{h_{co}}{\psi}\right)^2 + 1\right]\right\} .$$

Some conclusions attained from the combined evaluation of the equations giving the average position of the soil-moisture retention curve (see Eq. 2-21) and the theoretical relationship proposed to calculate unsaturated conductivity (Eq. 2-43) are as follows.

- Both the amount of saturation and the relative unsaturated conductivity can be expressed depending on the relative suction. This dimensionless parameter is the average capillary height (h_{co}) related to the prevailing suction (ψ).

- Neither s nor K is a single-valued function of the h_{co}/ψ quotient, but the absolute value of a material constant has to be introduced in both relationships. According to the theoretical investigation, a combination of the average capillary height and porosity gives a suitable parameter to be applied for characterizing the material of the porous medium

$$\frac{1-n}{n} h_{co}^{3/2} \quad .$$

- Using this material constant as a parameter, the ψ vs. W or s relationship (soil-moisture retention curve; Fig. 2-54a) and the contact between unsaturated conductivity and suction head (Fig. 2-54b) can be represented in coordinate systems having dimensionless quantities measured along their axes. Using the corresponding values of s and K (Table 2-8), even the relationship between these two parameters can be graphically reconstructed and the curves depend on the absolute value of a parameter calculated from the average capillary height and porosity or from the basic soil-physics parameters (D_h grain diameter, α shape coefficient, and n porosity) by considering the dependence of the average capillary height on these values (see Eqs. 2-19) as a third variable (Fig. 2-54c).

- Comparing the theoretical values constructed in this way with measured values, the reliability of the results achieved by the combined application of the geometrical, dynamic, and statistical models can be proved. The good agreement between the calculated and measured values of the soil-moisture retention was already testified in Section 2-3 (see Fig. 2-45). Figure 2-54b can be compared to Fig. 2-52 to testify that the observed character of the K vs. ψ relationship is well reproduced by the theoretical curves. The influence of the properties of the investigated materials is also reflected correctly by the material constant used in Eqs. 2-21 and 2-34, i.e., the curves belonging to coarse materials have an almost vertical elongated closing section while those characterizing fine and cohesive samples are

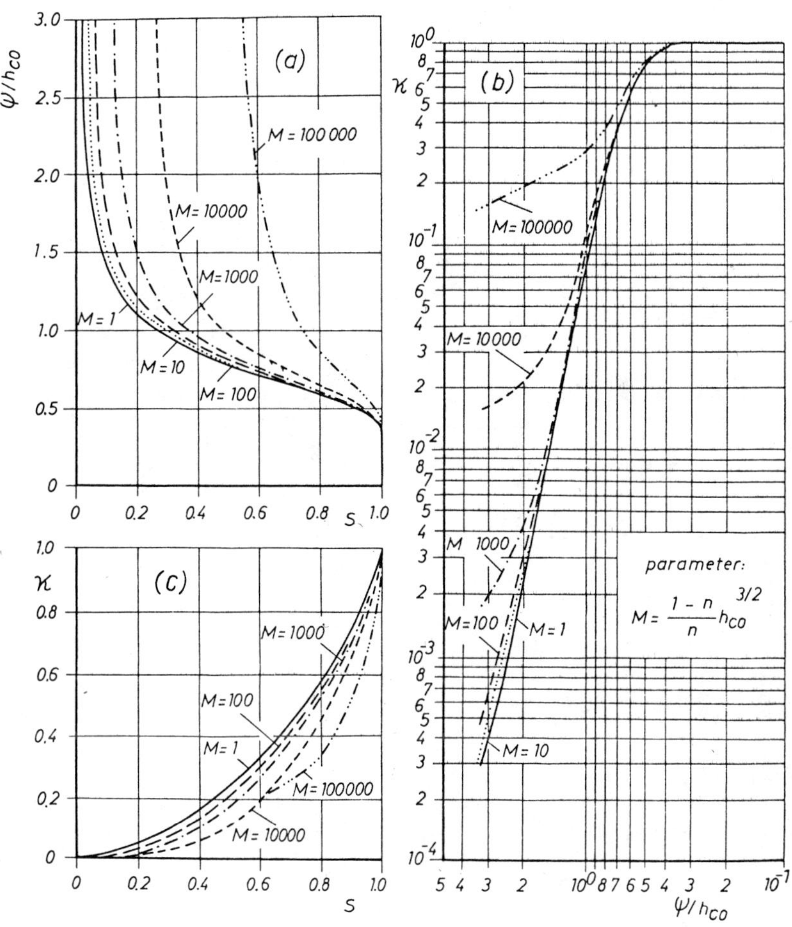

Fig. 2-54. $\psi/h_{co} = f_1(s)$; $K_u = f_2(s)$; and $K_u = f_3(\psi/h_{co})$ functions determined by theoretical analysis.

Table 2-8. Corresponding values of the amount of saturation, relative suction, and relative conductivity.

Relative Suction Head h_{co}/ψ	if the parameter characterizing the material $\frac{1-n}{n} h_{co}^{3/2}$: 1		10		100		1000		10,000		100,000	
	s	K	s	K	s	K	s	K	s	K	s	K
0.3	0.0155	3.00×10^{-4}	0.0287	3.14×10^{-4}	0.0572	4.49×10^{-4}	0.119	1.80×10^{-3}	0.251	1.53×10^{-2}	0.536	0.150
0.4	0.0239	1.10×10^{-3}	0.0376	1.12×10^{-3}	0.0670	1.27×10^{-3}	0.131	2.82×10^{-3}	0.268	1.83×10^{-2}	0.562	0.173
0.5	0.0392	2.70×10^{-3}	0.0532	2.72×10^{-3}	0.0834	2.89×10^{-3}	0.149	4.69×10^{-3}	0.289	2.16×10^{-2}	0.591	0.192
0.6	0.0632	5.80×10^{-3}	0.0774	5.82×10^{-3}	0.108	6.00×10^{-3}	0.173	7.82×10^{-3}	0.315	2.60×10^{-2}	0.619	0.208
0.7	0.0991	1.35×10^{-2}	0.113	1.35×10^{-2}	0.143	1.37×10^{-2}	0.208	1.56×10^{-2}	0.346	3.45×10^{-2}	0.647	0.224
0.8	0.147	2.70×10^{-2}	0.160	2.70×10^{-2}	0.189	2.70×10^{-2}	0.252	2.91×10^{-2}	0.388	4.83×10^{-2}	0.677	0.240
0.9	0.206	4.90×10^{-2}	0.219	4.90×10^{-2}	0.247	4.90×10^{-2}	0.306	5.11×10^{-2}	0.434	7.00×10^{-2}	0.711	0.259
1.0	0.254	8.00×10^{-2}	0.286	8.00×10^{-2}	0.312	8.00×10^{-2}	0.367	8.20×10^{-2}	0.486	0.100	0.742	0.282
1.2	0.430	0.176	0.444	0.176	0.461	0.176	0.505	0.178	0.601	0.193	0.808	0.350
1.4	0.589	0.312	0.596	0.312	0.612	0.312	0.645	0.313	0.716	0.326	0.869	0.448
1.6	0.729	0.472	0.734	0.472	0.744	0.772	0.767	0.473	0.815	0.482	0.919	0.568
1.8	0.837	0.628	0.840	0.628	0.846	0.628	0.860	0.629	0.889	0.634	0.953	0.689
2.0	0.910	0.761	0.910	0.761	0.915	0.761	0.923	0.762	0.939	0.765	0.975	0.797
2.2	0.955	0.861	0.955	0.861	0.957	0.861	0.961	0.861	0.970	0.863	0.988	0.880
2.4	0.979	0.926	0.979	0.926	0.979	0.926	0.982	0.926	0.986	0.927	0.995	0.935
2.6	0.991	0.967	0.991	0.967	0.991	0.967	0.991	0.967	0.994	0.967	0.998	0.970
2.8	0.997	0.984	0.997	0.984	0.997	0.984	0.997	0.984	0.997	0.984	0.999	0.986
3.0	0.999	0.994	0.999	0.994	0.999	0.994	0.999	0.994	0.999	0.994	0.999	0.994

curved upwards at the tailing part. Finally, it can be seen by comparing Figs. 2-41 and 2-54c that the set of calculated curves practically covers the zone of scattering of the measured data. The third variable applied as the parameter of the curves characterizes the material of the samples. Unfortunately, the soil-physics parameters characterizing the materials used for the measurements are not published in the relevant literature in detailed form. It can only be stated that the data belonging to finer materials are generally queuing along lower curves than those characterizing coarse sands. This general situation is also in good accordance with theoretical results.

8-4 Practical Application of the Theoretical Investigation of Soil-Moisture Movement

The general opinion concerning the solution of practical problems of the seepage of soil moisture is that there is no other presently applicable method but the use of diffusion theory. The various numerical methods assist the easy solution of the basic differential equation, and there are also proposals, in the papers quoted earlier, to consider the influence of the entrapped air in the pores. The basic problem in connection with the practical application of the existing models (independently of the fact whether they are neglecting the counterflow of air or not) is caused by the rough approximations applied to achieve the mathematical description of the water movement and by the uncertainties which occur in connection with limited knowledge related to the actual physical processes. These are the reasons why the models do not correctly reflect the percolation of soil moisture in each detail. Sufficiently accurate results can be achieved by using these models only if the soil-physics parameters included in the equations are determined from local measurements observing natural processes developing in the same surroundings, where that phenomenon develops, the hydraulic parameters have to be predicted. Thus, the relationships derived on the basis of the diffusion theory cannot be generalized. They are integrated systems together with the parameters which express the local conditions only in an indirect form.

The most important result of the recent theoretical investigations is the correct dynamic explanation of the forces acting against gravity in the soil-moisture zone and the establishment of relationships to describe the soil-moisture retention curve and the contact between hydraulic conductivity and suction head (or water content). The construction of a system of dynamic, geometrical, and statistical models (all considering the actual conditions of the porous medium till the possible greatest extent) was the basis of this analysis.

Apart from these results many important new aspects were recognized from the experiments, forming the basis of the investigations which have assisted the better understanding of the phenomena influencing the movement of the soil moisture. Some of these aspects are as follows.

- The compression of the entrapped air consumes a part of the available energy, and the counterflow of air hinders the development of the complete saturation in the upper part of the profile. The influence of air always has to be considered when infiltration is investigated.

- The large openings (secondary porosity) created by cultivation and by the action of roots form a special structure in the topmost layer of soil through which the atmospheric pressure can penetrate deeply between the clods. The concept of the soil-moisture retention curve cannot be applied, therefore, in this zone. Since the biological and chemical processes also have dominating influence on the moisture flux near the surface, the cultivated zone should be separated from the lower part of the soil profile, and the physical investigation of water movement has to be restricted to the remaining zone extending from the water table upwards to the lower border of cultivation. This boundary is a fictive horizontal plane below the roots, where the secondary structure does not disturb the porous material and where the influence of the upper lying layer has to be substituted in the form of boundary conditions.

- The mode of occurrence of water in the pores, whether the pores are completely saturated by capillarity or only water films cover the grains, influences not only the dynamics of the system, but the type of movement as well. The seepage starts through the continuous chains of pores completely filled with water without change in storage. The infiltrating water presses-out the corresponding amount from the capillary system at the bottom into the groundwater space while evapotranspiration is similarly followed by the rise of all water particles, replenishing the drained amount from the groundwater. In partially saturated pores, the development of the gradient is associated with the change of the thickness of water films, the water transport, therefore, goes together with change in storage. For the correct characterization of transient flow, it would be necessary to clarify the combination of the two different forms of movement and their ratio on the transport through a given system of partially saturated pores above the water table.

- In coarse-grained sediments, the transient seepage directed downwards is further complicated by the development of water membranes at the narrow stretches of the channels. This phenomenon may break the continuity of seepage and the storage process also becomes periodic.

The investigations carried-out until now were sufficient only to draw the attention to the problems listed previously. The observations, however, are not yet adequate to derive mathematical models considering these special aspects. There are proposals only to take into account the influence of air, which indicate the start of the development of more reliable methods, but further research works are required to achieve the better understanding of the actual physical processes and the suitable way of their simulation.

The limits created by the insufficient knowledge of the real character of movement, which hinders the generalization of the diffusion theory, cause similar uncertainties when our intention is to use the theoretically derived soil-moisture retention curve and hydraulic conductivity. Such a practical example has to be selected for the demonstration of the application of these parameters in which case the previously listed aspects do not influence the development of seepage.

The investigation of infiltration has to be excluded from the possible choices because:

- it is influenced by the compression of air;

- the water ponded on the surface directly penetrates into the large openings maintaining hydrostatic pressure there, and the boundary condition is, therefore, uncertain at the upper surface of the physically investigated zone;

- the downwards movement of water depends on the rate of flux developing through the fully saturated pores and the periodic storage may also disrupt the continuity of flow.

The storage process is made uncertain by the various possible forms of seepage in any direction other than that of infiltration. It is advisable, therefore, to restrict the application of the generalized parameters to the investigation of steady flow. The effect of evaporation and transpiration (the suction of the root system) can be characterized by a given value of suction acting evenly distributed along the interface between the cultivated zone and the lower part of the soil profile. This boundary condition may be supposed to be time independent if a long dry period is investigated after the water-content available in the upper part of the soil-moisture zone has been consumed. The single source supplying water in this period for evaporation is the seepage developing through the soil-moisture zone and recharged from the groundwater.

Since the flux of this seepage is relatively small, the change of the position of the water table in time is negligible. It can be supposed, therefore, that the lower boundary condition of the investigated seepage space is also constant, viz., the excess pressure is zero at the elevation of the phreatic surface, the depth of which is known and does not change in time. Thus, the investigation of a steady seepage gives acceptable approximation to determine the amount of water provided from the groundwater space for evapotranspiration. Supposing different values of the constant flux, a set of curves can be constructed showing the vertical soil-moisture distribution, the development of which is necessary to convey a continuously given amount of water from the water table to the lower boundary of the cultivated zone (naturally the relationship depends on the properties of the layer and on the vertical length of the investigated flow domain). Using these graphs, or their tabulated form, and regularly measuring the actual vertical distribution of the soil moisture, the time series of the component of actual evapotranspiration originating from the groundwater can be determined.

Considering the symbols represented in Fig. 2-55 and calculating the hydraulic gradient as the vertical derivative of the total energy head, the flux (flow rate through a unit area perpendicular to the main direction of flow) or the vertical seepage can be determined (assuming the positive direction upwards):

$$q = -K_u I = -K_u \left(1 - \frac{d\psi}{dh}\right) \quad ;$$

where,

$$I = \frac{d(h-\psi)}{dh} = 1 - \frac{d\psi}{dh}$$

is the gradient in the direction of the positive h-axis; and

$$h + \frac{p}{\gamma} = h-\psi \qquad (2\text{-}44)$$

is the energy head.

When $d\psi/dh < 1$ the flow is directed downwards (in the direction of the negative h-axis). This case is characterized in the ψ vs. h coordinate system by a curve (line 1) running

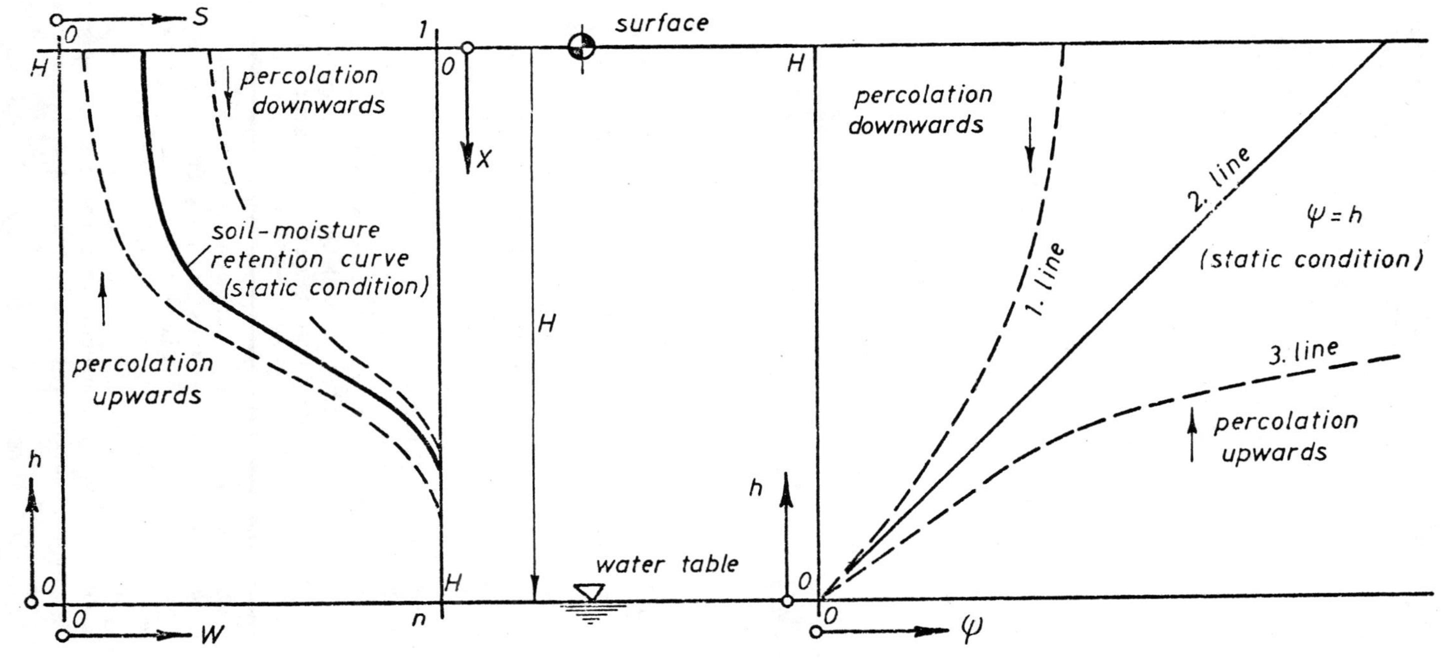

Fig. 2-55. Relationships between the parameters characterizing vertical flow in the soil-moisture zone.

above the straight line and approaches 45° with the horizontal (ψ) axis. The balanced condition is described by the $d\psi/dh = 1$ equation (line 2, being straight and halving the angle of the axes). Finally, $d\psi/dh > 1$ characterizes the seepage starting from the groundwater space, raising the water near the surface. In this case, the movement follows the direction of the positive h-axis and the ψ vs. h relationship is represented by line 3 which lies below the 45° line.

Investigating the steady flux of negative accretion, it can be assumed that the flow rate is a positive constant (+q). Using the symbol of relative unsaturated conductivity $K_u = K_u/K_s$, the final form of the differential equation to be integrated is as follows:

$$dh = \frac{K_u}{K_u + \frac{q}{K_s}} d\psi \;;$$

or using the dimensionless variables

$$d \frac{h}{h_{co}} = \frac{K_u}{K_u + \frac{q}{K_s}} d \frac{\psi}{h_{co}} \quad . \tag{2-45}$$

Substituting K_u according to Eq. 2-43, the relationship becomes very complicated, its direct integration is not possible. Similar problems may often arise when the theoretical equation of the soil-moisture retention curve is substituted into a movement equation to solve some practical problems. It is a great advantage of both relationships, however, that they determine the theoretically correct interrelations between the parameters in question and the characteristics of the soil. This contact can be empirically approximated by more simple mathematical formula without any physical simplifications. The structure of the equation used to approximate the theoretical relationship can be chosen according to the requirements of the given problem (in the present case, the approximation has to result in a new form of Eq. 2-45, which can be integrated). It is not at all necessary that the empirical formula should describe a continuous curve within the whole range of variables. In many cases it is sufficient to have good approximation only in a limited zone, because the occurrence of the extreme values of some variables is not probable in practice, or the entire domain of the theoretical relationship can be substituted by consecutive stretches of empirical curves. The uncertainties of the theoretical equations (hysteresis, errors resulted from the simulation of the actual channels by simplified models, etc.) have to be always considered when the fitting required between the original and approximate curves is determined.

In the present case, the K_u vs. ψ/h_{co} relationship has to be approximated by using the M value as a parameter. The requirement is that the empirical formula should ensure the direct integration of Eq. 2-45. The total probable range of both M and ψ/h_{co} can be divided into separated zones. According to the M parameter, three different materials are distinguished:

- impervious ($M > 220,000$);

- semi-pervious ($220,000 > M > 10,000$);

- pervious ($10,000 > M$).

Those materials regarded to be impervious have soil-moisture retention curves which are composed of two stretches (one hyperbola, of sixth order, characterizing the adhesive zone, and the vertical closing section), because adhesion surpasses the effect of capillarity in each pore (see Fig. 2-34). The amount of saturation is inversely proportional to the 1/6-th power of ψ/h_{co} (Eq. 2-18) above the intersection of the vertical axis and the retention curve, and the pores are completely saturated below the elevation of this point. In the adhesive zone, the relative unsaturated conductivity is supposed to be proportional to the third power of the amount of saturation (Eq. 2-37). Considering Eqs. 2-18 and 2-37, the K_u vs. ψ/h_{co} relationship can be easily determined if the latter variable is above a given limit. Below this $(\psi/h_{co})_1$ limit value, which can be calculated by substituting $K_u = 1$ into the K_u vs. ψ/h_{co} equation, the relative unsaturated conductivity is equal to unity:

$$K_u = 1 \ ; \qquad \text{if} \quad \psi/h_{co} < (\psi/h_{co})_1 \ ;$$

$$K_u = \frac{2.7\text{x}10^{-6}M}{\sqrt{\psi/h_{co}}} \ ; \qquad \text{if} \quad \psi/h_{co} > (\psi/h_{co})_1 \ ,$$

where,

$$(\psi/h_{co})_1 = 7.29\text{x}10^{-12}M^2 \ ; \text{ assuming } \ M > 220,000 \ . \tag{2-46}$$

In the case of pervious materials (gravel and sand), the influence of adhesion can be completely neglected. Two zones are distinguished only, the open and closed capillary zone. The K_u vs. ψ/h_{co} curve is divided further into two stretches in the closed capillary zone in order to consider that air bubbles may often decrease the amount of saturation in the upper part of the zone. Good approximation was achieved by assuming that the height of the closed capillary zone is constant $[(\chi/h_{co})_c = 0.5]$ having complete saturation below $\psi/h_{co} = 0.3$. Thus, the empirical approximations are determined for three stretches separately:

$$K_u = 1 \quad ; \qquad \text{if} \quad \psi/h_{co} < 0.3 \quad ;$$

$$K_u = 1-(6-Mx10^{-5})(\psi/h_{co}-0.3)^2 \quad ; \qquad \text{if} \quad 0.3 < \psi/h_{co} < 0.5 \quad ;$$

$$K_u = A \exp[-B(\psi/h_{co}-0.5] \quad ; \qquad \text{if} \quad 0.5 < \psi/h_{co} \quad ;$$

where $A = 1-0.04\ (6-Mx10^{-5})$; and

$B = 4.4-5x10^{-3}\ \sqrt{M}$;

assuming $M < 10,000$.

(2-47)

The characterization of the semi-pervious materials can be achieved by combining the systems represented by Eqs. 2-46 and 2-47. The relationships described by Eqs. 2-47 are used in the lower domain of ψ/h_{co} and Eq. 2-46 in the adhesive zone. The limit of the validity zones can be determined by expressing the equality of K_u calculated for the open capillary and adhesive zones respectively. The $(\psi/h_{co})_2$ limit can be calculated from this equation (after expanding some terms into series) as the function of the M parameter:

$$K_u = 1; \text{ if } \quad \psi/h_{co} < 0.3 \quad ;$$

$$K_u = 1-(6-Mx10^{-5})(\psi/h_{co}-0.3)^2; \text{ if } \quad 0.3 < \psi/h_{co} < 0.5 \quad ;$$

$$K_u = A \exp[-B(\psi/h_{co}-0.5)]; \text{ if } \quad 0.5 < \psi/h_{co} < (\psi/h_{co})_2 \quad ;$$

$$K_u = \frac{2.7x10^{-6}M}{\sqrt{\psi/h_{co}}} \quad ; \text{ if } \quad \psi/h_{co} > (\psi/h_{co})_2 \quad ;$$

where

$$(\psi/h_{co})_2 = 0.5-(7.3-\sqrt{M}x10^{-2})$$

$$+ \sqrt{(7.3-\sqrt{M}x10^{-2})^2 + 2.89-4 \ln Mx10^{-5}} \;;$$

assuming $10,000 < M < 220,000$. (2-48)

The correctness of the empirical formulae cannot be proved theoretically only by comparing these curves to points determined from Eq. 2-43 (Fig. 2-56). The discrepancy between the values calculated in two different ways is not larger than the uncertainties of the theoretical relationship. The approximations can be accepted, therefore, to calculate the steady flux of the negative accretion.

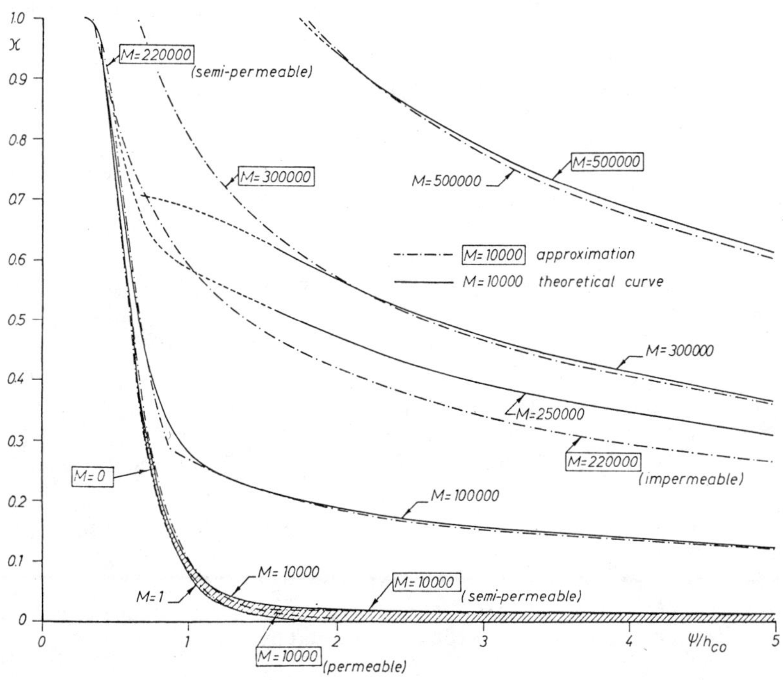

Fig. 2-56. Practical approximation of the theoretical $K_u = f(\psi/h_{co})$ function.

Substituting any of the K_u vs. ψ/h_{co} relationships into Eq. 2-45, the integration can be solved in closed form and the result provides the suction head ψ as a function of the height above the water table depending on the constant flow rate of accretion. Many practical problems can be answered on the basis of this relationship, e.g.:

- the estimation of the realistic range of the possible negative accretion depending on the material of the soil and the distance between the water table and the lower boundary of the cultivated zone;

- the actual flux (the component of evapotranspiration originating from groundwater), if the change of suction head along a given vertical stretch of the profile is known;

- the necessary depth of the water table to achieve a given flow rate of the vertical seepage, when a constant suction is maintained at the lower surface of the cultivated zone (the control of the water balance of the cultivated zone by regulating the water table).

Three different materials were chosen to demonstrate the application of the method:

- sand (D_h = 1.2×10^{-2} cm; α = 12; n = 0.30; U = 1.3;

  K_S = 1.2×10^{-3} cm/sec; h_{co} = 166 cm; M = 5000);

- loess or sandy loam (D_h = 2.48×10^{-3} cm; α = 16; U = 3.2;

  K_S = 4.55×10^{-5} cm/sec; h_{co} = 900 cm; M = 50,000);

- silt (D_h = 3.32×10^{-4} cm; α = 20; n = 0.45; U = 8.0;

  K_S = 7.35×10^{-7} cm/sec; h_{co} = 5500; M = 500,000).

Considering that the maximum potential evaporation hardly surpasses 10 mm/day, the $q < 100$ mm/day range of the constant flux was investigated. The data calculated in this way give sufficient information, therefore, on the processes, the occurrence of which is probable in nature.

The results were represented at first in h vs. s coordinate system (Fig. 2-57) which shows the vertical moisture-distribution and its change related to the soil-moisture retention curve depending on the water amount

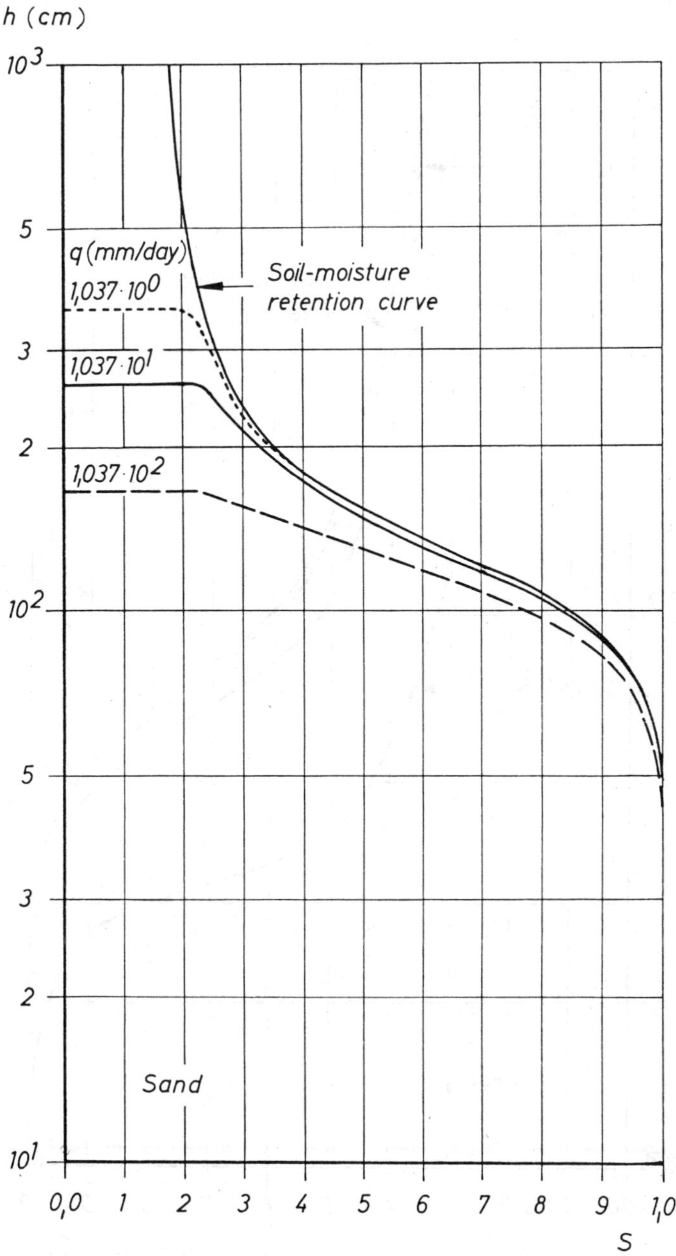

Fig. 2-57a. Change of the retention curve due to vertical seepage directed upwards depending on the flow rate of seepage.

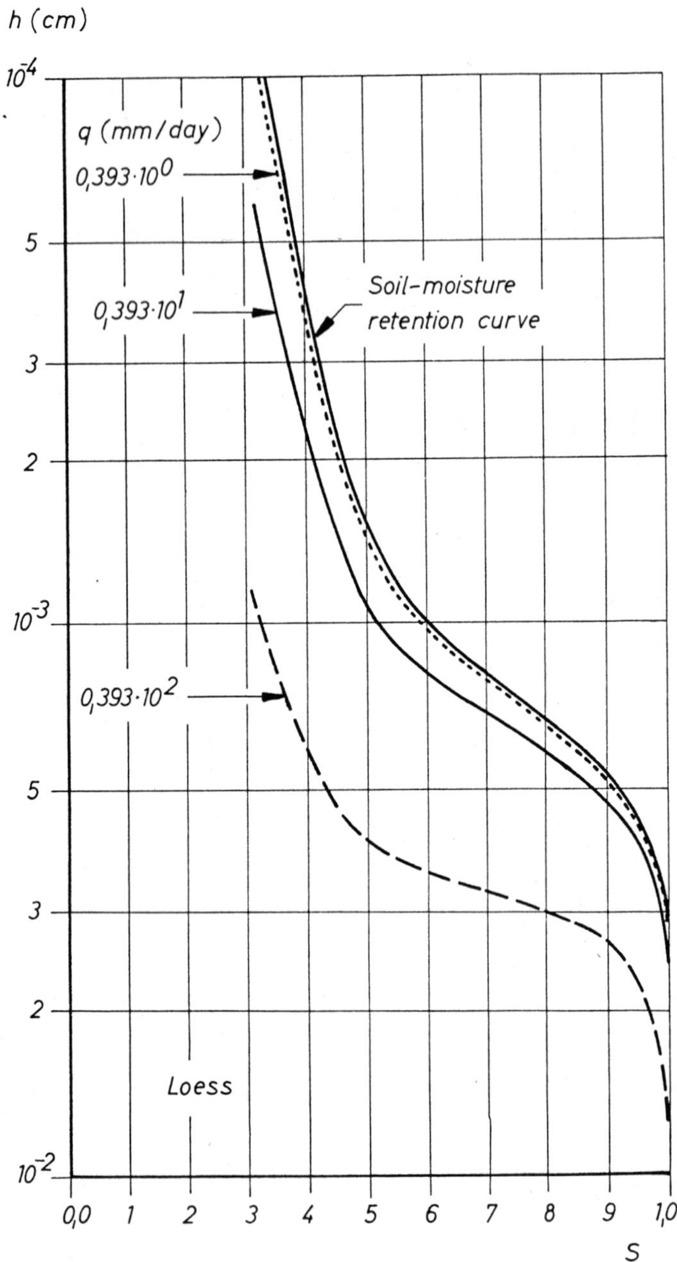

Fig. 2-57b. (continued)

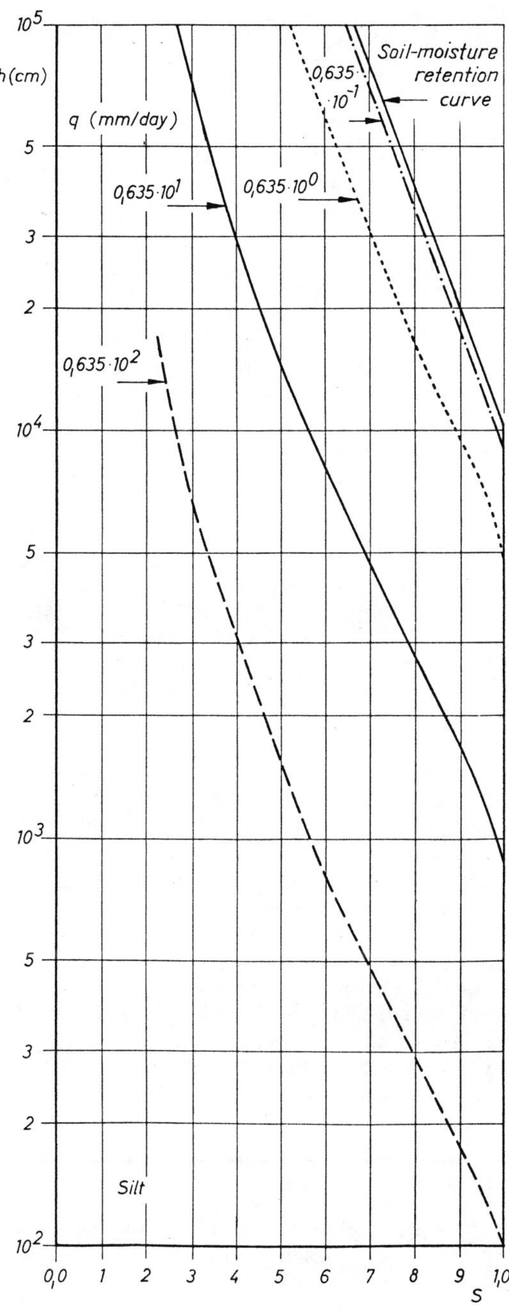

Fig. 2-57c. (continued)

transported in the profile. The conclusions drawn up from the figure are as follows.

- In sand, the vertical moisture distribution does not change if $q < 10$ mm/day. The lowering of the curve is smaller than the uncertainty of the retention curve, even in the case of $q = 10$ mm/day. This fact indicates that the influence of storage is negligible, the investigation of steady seepage instead of the actual transient process is, therefore, an acceptable approximation.

- The transport capacity of sand is limited, a given flow rate can be conveyed only if the water content is higher than a determined value, depending on the flux in question. An upper limit of the possible rise belonging to each flow rate can be calculated, therefore, which is indicated by the position of the horizontal closing section of the vertical moisture distribution.

- The range of flux (within which the change in storage is negligible) is smaller in loess ($q < 0.4$ mm/day). The evapotranspiration may cause, therefore, a considerable lowering of the curve, representing the vertical moisture distribution under dynamic conditions. The amount of water being available, in this way, has to be considered in the calculation of evapotranspiration.

- The change of the position of the moisture-distribution curve is even higher in the case of silt. The lowering becomes considerable if the flux surpasses the $q = 0.05$ mm/day value. The height of the completely saturated zone may be 100 m in this material under static condition, and the value is lowered below 10 m if the flow rate of vertical seepage is 6 mm/day.

- The almost parallel lowering of the distribution curve in loess and silt gives a reasonable explanation of the "hysteresis" observed in natural soil profiles. The vertical displacement is more than 10 m in silt, if $q > 0.05$ mm/day, and in loess if $q > 0.5$ mm/day. This fact indicates that a relatively small flux may cause a great change in the position of the curve. This is one of the reasons why the scattering of the measurements cannot be explained only as hysteresis, its considerable part may be caused by vertical seepage.

The next form of the evaluation of the data was the construction of h vs. ψ curves (Fig. 2-58). These graphs led to further conclusions.

- The suction belonging to a hydrostatic condition is hardly raised by seepage until a determined elevation in sand. At this limit (which depends on the flux), the curves have a

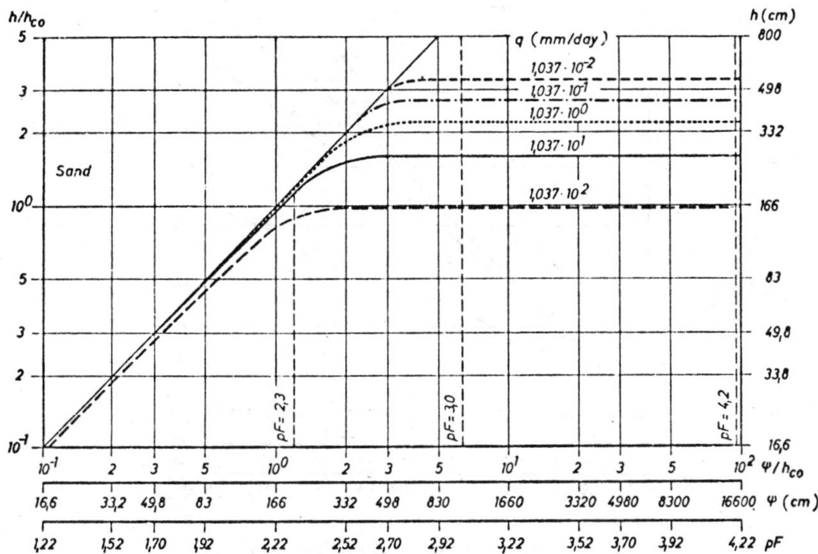

Fig. 2-58a. Change of the vertical distribution of tension due to vertical seepage directed upwards depending on the rate of seepage.

sharp turn and they become almost horizontal. This fact indicates that a given amount of water cannot be raised above a corresponding level in sand even if the suction is considerably increased.

- The character of the h vs. ψ curves of sand shows a great similarity to the relationship measured between the two variables in column drainage (see Fig. 2-14), but their position is reflected to the straight line describing the hydrostatic condition. The measurements observed the pressure distribution under the influence of vertical seepage directed downwards (the flow rate of which decreased in time). The similarity not only proves the correctness of the theoretical calculation, but also indicates that the method can be applied for the characterization of the vertical drainage of the profiles composed of sand. This conclusion is in good accordance with the previous results, i.e., the change in storage

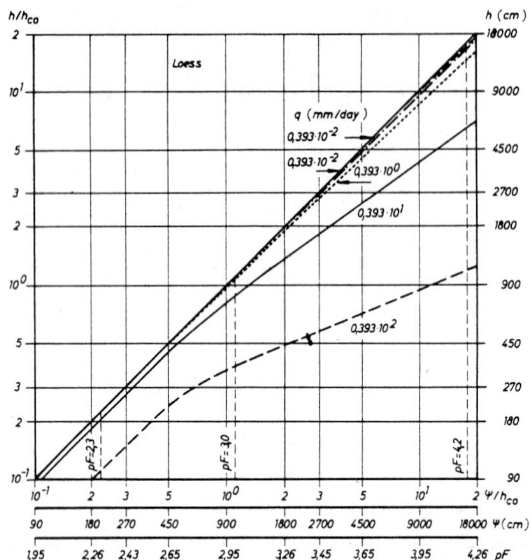

Fig. 2-58b. (continued)

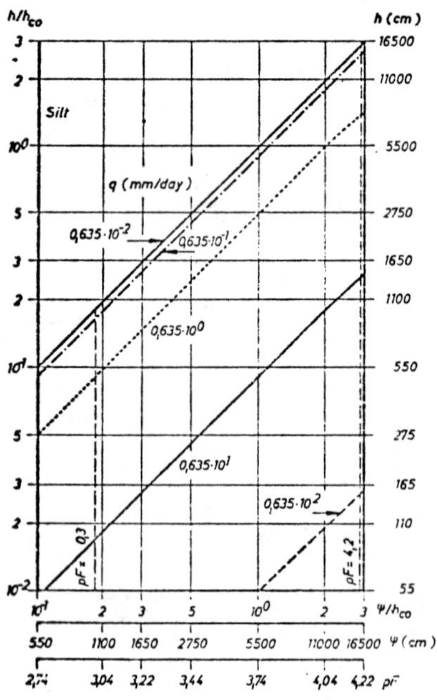

Fig. 2-58c. (continued)

and the adhesive water transport can be neglected in pervious materials.

- In loess and silt the elevation of a given flow rate is not limited, as it is in sand. The h vs. ψ curves are lowered almost parallel to each other within the investigated range of flow rate, if silt is investigated. In loess, the parallel displacement is characteristic only in the lower range of suction ($\psi < 500$ cm in the case of the investigated material), and a monotonic increase with a smaller amount (which amount decreases as the flux increases) can be observed in the zone of higher suctions.

A third version of the evaluation of the results is the investigation of the transported flow rate depending on the height of elevation and the applied suction (Fig. 2-59). The maximum suction of roots is estimated to be pF = 4.2 ($\psi = 1.6 \times 10^4$ cm water column). The cultivated zone is excluded from the investigation and, therefore, its influence is substituted in the form of an average boundary condition at the upper bordering plane of the analyzed part of the soil-moisture zone. It is evident, that this average value is considerably smaller than the suction of roots. The influence of three suction values was studied (pF = 4.2, $\psi = 1.6 \times 10^4$ cm; pF = 3.0, $\psi = 1 \times 10^3$ cm; pF = 2.3, $\psi = 2 \times 10^2$ cm) and it is assumed that a suction head of $\psi = 1 \times 10^3$ cm may be regarded as a realistic average value below the root zone. This parameter is used, therefore, when the role of negative accretion in evapotranspiration is evaluated. The consequences gained by this representation of the data are as follows.

- The maximum rise of any flow rate is equal to the applied suction head. This elevation is the horizontal asymptote of the h vs. q curve which is approached as q tends to zero.

- Accepting the pF = 3.0, $\psi = 1 \times 10^3$ cm value as a probable high suction at the lower horizontal boundary of the cultivated zone, the expected flow of the steady seepage depending on the elevation of this border above the water table can be summarized as follows:

- The listed numbers indicate that sandy layers ensure high water transport only near the water table. The flow rate through loess is the highest, if the path of seepage is longer than 1 m. At an elevation more than 4 m, even the amount transported by silt is higher than that in sand. The development of the local maximum of the relationship between flow rate and grain diameter can be explained by the fact that the water conveyance is influenced by conductivity and capillary

Table 2-9. Expected steady seepage flux from groundwater as a function of the distance (m) above the water table.

elevation of the lower boundary of the cultivated zone above the water table (m)	Material	1	2	3	4	5	6	7	8	9
flow rate of the steady negative accretion in mm/day, if the material is	sand	> 100	35.0	4.0	0.4	$4x10^{-2}$	$4x10^{-3}$	$< 1x10^{-3}$	$< 1x10^{-3}$	$< 1x10^{-3}$
	loess	> 100	> 100	47.0	26.0	15.0	9.0	4.5	2.2	1.0
	silt	7.7	2.0	1.3	0.9	0.6	0.35	0.22	0.13	$5x10^{-2}$

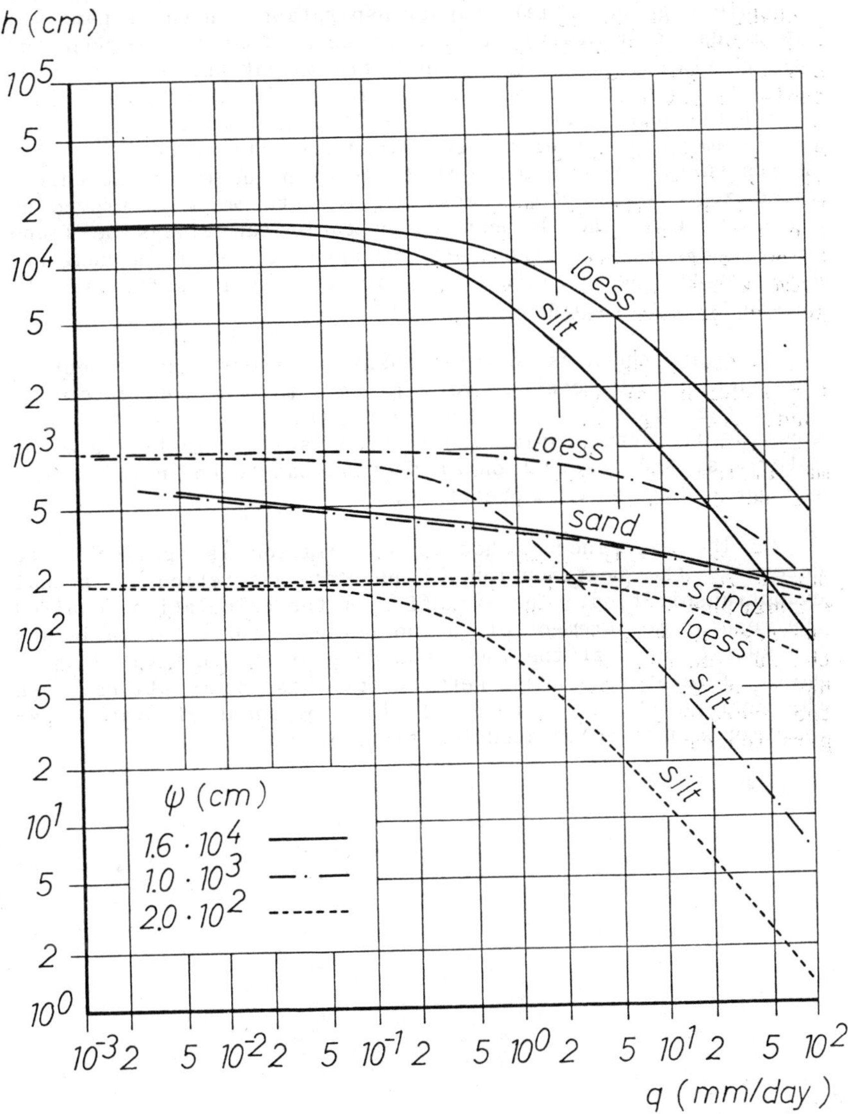

Fig. 2-59. Relationship between the height and steady flow rate of vertical seepage directed upwards depending on applied suction.

rise. The optimum of this combination is in the range of loess and loam. The small conductivity limits the water transport in silt, while the low capillary rise hinders it in sand, if the elevation is higher than 1 m.

- Comparing the calculated numerical values with the probable high potential evapotranspiration (which is namely a few mm/day in Hungary), it can be seen, that the groundwater has no significant role in evapotranspiration, if its water table is covered by fine sand having a thickness more than 3 m. This depth is about 4 m in silt. Through loess, considerable amount of water can be raised from the groundwater even if its surface is 8~9 m below the lower boundary of the cultivated zone. The correctness of this statement is proved by the fact that the deepest water table can always be found below loess layers (the average depth is 7~8 m in Hungary, from where the negative accretion is still sufficient to govern the water balance).

- Since the h vs. q relationship is almost independent of the suction (at least above a given limit of the latter) in sand, the regulation of the water table ensures the exact control of evapotranspiration in sandy soils. In fine grained materials, the observation of suction has to be included into the basic data of this operation.

- The explained method of calculation is applicable to determine the contribution of negative accretion to actual evapotranspiration. The execution of the calculation requires the regular measurement of the position of the water table and the suction head at the lower boundary of the cultivated zone. Having a relatively dense network for these observations, even the average characteristics of this component of evapotranspiration can be calculated for large areas.

REFERENCES FOR PART II

Adrian, D. D. and Franzini, J. B., 1966: Impedance to Infiltration by Pressure Built Up Ahead of the Wetting Front (in English), Journal of Geophysical Research.

ASTM, 1958: Procedures for Testing Soils (in English), American Society for Testing Materials, Philadelphia.

Atterberg, A., 1911: Plasticity of Clays (in German), Oversatt fr. svenska ur Kungl. Lautbruksakademiens Handlinger och Tidakon, No. 50.

Balogh, J., 1973: Determination of the Quantitative Norms and the Suitable Time Point of Irrigation for Design (in Hungarian), VITUKI Reports, No. 2443 (Manuscript).

Bear, J., 1972: Dynamics of Fluids in Porous Media (in English), New York, London, Amsterdam.

Bear, J. and Bachmat, Y., 1967: A Generalized Theory on Hydrodynamic Dispersion in Porous Media (in English), IASH Symposium on Artificial Recharge and Management of Aquifers, Haifa.

Bear, J., Zaslavsky, D., and Irmay, S., 1968: Physical Principles of Water Percolation and Seepage (in English), Paris.

Black, T. A., Gardner, W. R., and Thurtess, G. W., 1969: The Prediction of Evaporation, Drainage and Soil Water Storage for a Bare Soil (in English), Proceedings of American Soil Science Society, No. 5.

Blume, H. P., Zimmermann, U., and Muennich, K. O., 1967: Tritium Tagging of Soil Moisture; The Water Balance of Forest Soils (in English), Isotope and Radiation Techniques in Soil Physics and Irrigation Studies, IAEA, No. 315.

Bondarenko, N. F. and Globus, A. M., 1967: Radioactive Indicators for Up-to-Date Investigation of the Hydrophysical Problems of Soils (in Russian), Isotope and Radiation Techniques in Soil Physics and Irrigation Studies, IAEA, No. 291.

Boreli, M. and Vachaud, G., 1966: Certain Problems of Infiltration in Unsaturated Porous Media (in English), Saopstenja Institute za Vodoprivredu Jaroslav Cerni, No. 39.

Boreli, M. and Vachaud, G., 1966: On the Determination of the Residual Water Content and on the Variation of Relative Permeability of the Unsaturated Soil (in French), C. R. Academy of Sciences, Paris, Series A. Tom. 263.

Bonyoucos, G. I., 1949: Nylon Electrical Resistance Unit for Continuous Measurement of Soil Moisture in the Field (in English), Soil Science, Vol. 67;
1950: A Practical Soil Moisture Meter as a Scientific Guide to Irrigation Practices (in English), Agronomy Journal, Vol. 42;
1951: The Effect of Fertilizers on the Plaster of Paris Electrical Resistance Method for Measuring Soil Moisture in the Field (in English), Agronomy Journal, Vol. 43;
1954: New Type Electrode for Plaster of Paris Moisture Blocks (in English), Soil Science, Vol. 78.

Bouwer, H., 1964: Unsaturated Flow in Groundwater Hydraulics (in English), Proceedings of ASCE, HY5.

Braester, C., 1973: Moisture Variation at the Soil Surface and the Advance of the Wetting Front during Infiltration at Constant Flux (in English), Water Resources Research, No. 3.

Brustkern, R. L. and Morel-Seytoux, H. J., 1975: Description of Water and Air Movement during Infiltration (in English), Journal of Hydrology.

Buckingham, E., 1907: Studies in the Movement of Soil Moisture (in English), USDA Bureau of Soils Bulletin, No. 38.

Buschmann, H., 1956: Determination of Soil Moisture with V_2A Steel-Plexiglas-Electrode (in German), Zeitschrift für Pfalanzen, Vol. 72.

Castagny, G., 1967: Practical Investigation of Subsurface Waters (in French), Paris.

Charreau, C. and Jacquinot, J., 1967: Study on the Movement of Water through Sandy Soil in Senegal by Applying the Method of Tritium Content of Water (in French), Isotope and Radiation Techniques in Soil Physics and Irrigation Studies, IAEA, No. 301.

Childs, E. C. and Collis George, N. C., 1948: Soil Geometry and Soil-Water Equilibria (in English), Faraday Society, Discussion, No. 3.

Colman, E. A., 1946: The Place of Electrical Soil-Moisture Meters in Hydrologic Research (in English), Transactions of AGU, Vol. 27. 1950: Manual of Instructions for Use of the Fiberglass Soil-Moisture Instruments (in English), USDA, California Forest Range Experimental Station.

Colman, E. A. and Hendrix, T. M., 1949: The Fiberglass Electrical Soil-Moisture Instrument (in English), Soil Science, Vol. 67.

Crony, D., Coleman, J. D., and Currer, E. W. H., 1951: Soil-Moisture Measurement by Applying Electrical Resistivity (in English), British Journal of Applied Physics, No. 2.

De Wiest, J. M., 1969: Flow through Porous Media (in English), New York, London.

Dolgov, S. I., 1948: Investigation of the Movement and the Availability of Soil Moisture (in Russian), Moscow, Leningrad.

Elrick, D. E. and Bowman, D. H., 1964: Note on an Improved Apparatus for Soil-Moisture Flow Measurements (in English), Proceedings of American Soil Science Society, No. 3.

Elrick, D. E., 1966: The Microhydrologic Characterization of Soils (in English), IASH Symposium on Water in the Unsaturated Zone, Wageningen.

Engelhardt, W., 1960: Porosity of Sediments (in German), Berlin, Göttingen, Heidelberg.

Freeze, R. A., 1969: The Mechanism of Natural Groundwater Recharge and Discharge (in English), Water Resources Research, No. 1.

Frissel, M. J., Poelstra, P., Harmsen, K., and Bolt, G. H., 1973: Tracing Soil-Moisture Migration with ^{36}U, $^{60}C_o$ and Tritium (in English), IAEA/FAO Symposium on Isotopes and Radiation Techniques in Studies of Soil Physics, No. SM-176/13, Vienna.

Gálfi, J., 1976: Measuring Dielectric Constant with Plain Condensators and Determining the Relationship between Soil Moisture and Dielectric Constant (in Hungarian), VITUKI Scientific Reports, No. III.2.3.26, Budapest.

Gardner, W. R., 1958: Some Steady State Solution of the Unsaturated Moisture Flux Equation with Application to Evaporation from a Water Table (in English), Soil Science.

Golerik, L. V., Petrenko, A. A., and Nezametdinova, S. S., 1969: On the Problems of Relationship of the Density and Moisture Content of Soils with Their Accoustical Characteristics (in Russian), Isvestija Instituta Geotechniqui imeni be VEDENEVA, No. 88.

Green, W. H. and Ampt, G. A., 1911: Studies on Soil Physics. 1. Flow of Air and Water through Soils (in English), Journal of Agricultural Sciences, No. 4.

Grishkan, C. A., 1966: Hydraulic Conductivity of Soils as the Most Important Parameter of Seepage Calculations (in Russian), Symposium on Seepage and Well Hydraulics, Budapest.

Harrold, L. L. and Dreibelbis, F. R., 1967: Evaluation of Agricultural Hydrology by Monolith Lysimeters (in English), Technical Bulletin of USDA, No. 1367.

Hazen, A., 1889-1891: Experiments upon the Purification of Seepage and Water at Lawrence Experimental Station (in English), Health Publication of Massachusetts State Board, No. 1.

Hornberger, G. M. and Remson, J., 1973: A Moving Boundary Model of a One-Dimensional Saturated-Unsaturated Transient Porous Flow System (in English), Water Resources Research, No., 3.

Horton, R. E., 1939: Analysis of Runoff-Plot Experiments with Varying Infiltration Capacity (in English), Transactions of AGU.

Höschele, K., 1957: Investigation of the Method of Electrical Soil-Moisture Measurements (in German), University of Hohenheim, Doctor Thesis.

Hübener, J., 1969-1970: Determination of Soil Moisture with Gypsum-Block Method (in German), Ingenieurpraktikum, IML.

Hvorskev, M. J., 1949: Subsurface Exploration and Sampling of Soils for Civil Engineering Purposes (in English), American Society of Civil Engineers Research Project, U.S. Corps of Engineers, Waterways Experimental Station.

IAHS, 1974: Hydrological Investigation of the Unsaturated Zone; Second Circular (in English), Budapest.

Irmay, S., 1954: On the Hydraulic Conductivity of Unsaturated Soils (in English), Transactions of AGU, No. 1;
1966: Solution of the Nonlinear Diffusion Equation with a Gravity Term in Hydrology (in English), IASH Symposium on Water in the Unsaturated Zone, Wageningen.

Johnson, A. I., 1962: Methods of Measuring Soil Moisture in the Field (in English), Geological Survey Water-Supply Paper, No. 1619-U.

Johnson, A. I. and Kunkel, F., 1963: Some Research Related to Groundwater Recharge (Progress Report from the USGS) (in English), Conference on Groundwater Recharge and Groundwater Basin Management, University of California, Berkeley.

Johnson, A. I., Prill, R. C., and Morris, D. A., 1963: Specific Yield-Column Drainage and Centrifuge Moisture Equivalent (in English), Geological Survey Water-Supply Paper, No. 1662-A.

Juhász, J., 1967: Hydrogeology (in Hungarian), Budapest.

Kamensky, G. N., 1940: Differential Equations of the Non-Steady Groundwater Flow and Their Application to Calculate Backwater Effects (in Russian), Izvestija Akademii Nauk, USSR, No. 4;
1947: Investigation of Groundwaters (in Russian), Moscow.

Kastenak, F., 1971: Numerical Simulation Technique for Vertical Drainage from a Soil Column (in English), Journal of Hydrology;
1972: Soil Mechanics (in Hungarian), Budapest;
1974: Handbook of Soil Mechanics, Vol. I. Soil Physics (in English), Amsterdam, Budapest.

Keys, W. S. and Maclary, L. M., 1971: Application of Borehole Geophysics to Water-Resources Investigations (in English), USGS Techniques of Water-Resources Investigations, Book 2, Chapter 1, Washington.

Khanji, J. D., 1975: Investigation of the Recharge of Groundwater with Free Water Table by Infiltration (in French), University of Grenoble, Doctor Thesis.

King, F. H., 1897-1898: Principles and Conditions of the Movements of Groundwater (in English), 19th Annual Report of USGS.

Klute, A., 1952: Some Theoretical Aspects of the Flow of Water in Unsaturated Soils (in English), Proceedings of American Soil Sciences Society, Vol. 16, No. 2;
1967: The Movement in Unsaturated Soils (in English), MIT Summer Session on Groundwater Hydrology and Flow through Porous Media.

Kobayashi, H., 1966: A Theoretical Analysis and Numerical Solution of Unsaturated Flow in Soil (in English), IASH Symposium on Water in the Unsaturated Zone, Wageningen.

Kovács, G., 1968: Characterization of Shape of Grains in Seepage Hydraulics (in Hungarian), Épités-és Közlekedéstudományi Közlemények, No. 12;
1968: Characterization of the Molecular Forces Influencing Seepage with the Help of the pF Curve (in English), Agrokémia és Talajtan (Supplementum);
1969: General Characterization of Different Types of Seepage (in English), 13th IAHR Congress, Kyoto;
1969: Relationship between Velocity of Seepage and Hydraulic Gradient in the Zone of High Velocity (in English), 13th IAHR Congress, Kyoto;
1969: Seepage Law for Micro-seepage (in English), 13th IAHR Congress, Kyoto;
1971: Seepage through Unsaturated Porous Media (in English), 14th IAHR Congress, Paris;
1971: Seepage through Saturated and Unsaturated Layers (in English), IAHS Bulletin, No. 2;
1972: Seepage Hydraulics (in Hungarian), Budapest;
1975: The Use of Lysimeters in the Hydrological Investigation of the Unsaturated Zone (in English), (Report of HIUZ), IAHS Project Presented at the IAHS Symposium on Hydrological Characteristics of River Basins, Tokyo, IAHS Bulletin, No. 4;
1976: Combination of Geometrical, Dynamic and Statistical Models to Determine Hydraulic Conductivity (in English), IAHR Symposium on Stochastic Hydraulics, Lund.

Kovács, G., 1976: Statistical Model of the Pore-Size Distribution of Loose Clastic Sediments (in English), IAHR Symposium on Flow through Porous Media, Kiev.

Kovács, G. and Péczely, T., 1975: Water Retention Capacity of Soil Profiles (in Hungarian), Agrokémia és Talajtan, No. 1-2.

Kozeny, J., 1953: Hydraulics (in German), Vienna.

Kristensen, K. J., 1973: Depth Intervals and Topsoil Moisture Measurement with the Neutron Depth Probe (in English), Nordic Hydrology, No. 4.

Laliberte, G. E., Corey, A. T., and Brooks, R. H., 1966: Properties of Unsaturated Porous Media (in English), Hydrology Papers, Colorado State University, No. 17.

Liakopulos, A. G., 1966: Theoretical Approach to the Solution of the Infiltration Problem (in English), IASH Bulletin, No. 1.

Lindner, H., 1970: Comparative Investigation of Soil-Moisture Measurements (in German) (F/E-Work), Berlin;
1970: Construction and Checking of Soil-Like Gypsum-Dederon Measuringhead as Soil-Moisture Sensor for Operation of Automatized Sprinkling Irrigation System (in German) (F/E-Work), Berlin.

Major, P., 1972: Determination of Infiltration and Evaporation Considering Capillary Condition of the Covering Layer (in Hungarian), VITUKI Scientific Days' 72, Budapest;
1973: Investigation of Balance Parameters in Plains (in Hungarian), VITUKI Scientific Reports, No. 2152 (Manuscript);
1975: Investigation of the Process of the Accretion of the Groundwater in an Experimental Station (in French), IASH Symposium on the Hydrological Characteristics of River Basins, Tokyo.

Makaricheva, E. A., 1966: Field Determination of the Hydraulic Conductivity of Soils by the Method of Large Measuring Ring (in Russian), Moscow.

Makkink, G. P., 1957: Testing the Penman Formula by Means of Lysimeters (in English), Journal of Institute of Water Engineering, No. 11.

Marel, H. W., 1959: Rapid Determination of Soil Water by Dielectric Measurement of Dioxan Extract (in English), Soil Science.

Mein, R. G. and Larson, C. L., 1973: Modeling Infiltration during a Steady Rain (in English), Water Resources Research, No. 2.

Morel-Seytoux, H. J., 1973: On the Modified Theory of Infiltration 1st and 2nd Parts (in French), Cahiers of ORSTOM, No. 1 and 2.

Morel-Seytoux, H. J. and Khanji, J., 1974: Derivation of an Equation of Infiltration (in English), Water Resources Research, No. 4.

Mustonen, S. and McGuinness, J. L., 1968: Estimating Evapotranspiration in a Humid Region (in English), Technical Bulletin of USDA, No. 1389.

Németh, E., 1942: Water Problems in Modern Agriculture (in Hungarian), Budapest;
1963: Hydromechanics (in Hungarian), Budapest.

Peck, A. J., 1966: Diffusivity Determination by a New Outflow Method (in English), IASH Symposium on Water in the Unsaturated Zone, Wageningen.

Péczely, T. and Zotter, K., 1973: Summary of Interpretation and Measuring Methods of pF Curves (in Hungarian), VITUKI, Scientific Reports (Manuscript).

Philip, J. R., 1955: Numerical Solution of Equations of the Diffusion Type with Diffusivity Concentration-Dependent (in English), Transactions of Faraday Society, No. 51;
1957: Numerical Solution of Equations of the Diffusion Type with Diffusivity Concentration-Dependent; Part 2 (in English), Australian Journal of Physics, No. 10;
1966: A Linearization Technique for the Study of Infiltration (in English), IASH Symposium on Water in the Unsaturated Zone, Wageningen.

Prill, R. C., Johnson, A. I., and Morris, D. A., 1965: Specific Yield--Laboratory Experiments Showing the Effect of Time on Column Drainage (in English), Geological Survey Water-Supply Paper, No. 1662-B.

Rame and Whinnery, 1960: Fields and Waves in Modern Radio-Technique (in German), Berlin.

Rentschler, W., 1956: A Recording Instrument for Soil-Moisture Measurement with Gypsum-Block Method (in German), Netherland Journal of Agricultural Sciences, No. 4.

Réthàti, L., 1960: Engineering Aspects of the Capillarity of Soils (in Hungarian), Vizügyi Közlemények, No. 1.

Richards, L. A., 1931: Capillary Conduction of Liquids through Porous Media (in English), Physics.

Richards, L. A. and Gardner, W., 1936: Tensiometers for Measuring the Capillary Tension of Soil Water (in English), Agronomy Journal, Vol. 28.

Rijtema, P. E., 1965: An Analysis of Actual Evapotranspiration (in English), Wageningen, Doctor Thesis, VLO. 659.

Rode, A. A., 1952: Soil Moisture (in Russian), Moscow.

Rose, D. A., 1966: Water Transport in Soils by Evaporation and Infiltration (in English), IASH Symposium on Water in the Unsaturated Zone, Wageningen.

Rowe, 1955: Difference Approximation to Partial Derivative for Spacings in the Network (in English), Transactions of AGU, No. 36.

Royer, J. M. and Vachaud, G., 1975: Field Determination of Hysteresis in Soil Science Characteristics (in English), Proceedings of American Soil Science Society, No. 2.

Rubin, J., 1966: Numerical Analysis of Ponded Rainfall Infiltration (in English), IASH Symposium on Water in the Unsaturated Zone.

Russel, J. C., 1950: A Simplified Air-picnometer for Field Use (in English), Proceedings of American Soil Science Society, Vol. 14.

Schendel, U., 1962: Soil-Moisture Measurement with Tensiometers in Black Sandy Loam (in German), Zeitschrift für Kulturtechnik, No. 3.

Schoeller, H., 1962: Groundwaters (in French), Paris.

Smith, W. O., 1933: Minimum Capillary Rise in an Ideal Uniform Soil (in English), Physics.

Smith, W. O., Foote, P. D., and Busang, P. F., 1931: Capillary Rise in Sands of Uniform Spherical Grains (in English), Physics.

Sprigade, H., 1969: Mechanization and Automation of Large Irrigation Systems (in German), Berlin (F/E-Work).

Sudnitsen, L. I., 1966: Soil-Moisture Pressure in Some Climatic Zones (in English), IASH Symposium on Water in the Unsaturated Zone, Wageningen.

Szabó, L., 1954: Influence of Seepage Directed Sidewards on the Norms of Surface Irrigation (in Hungarian), Hidrológiai Közlöny, No. 7-8.

Teasdale, W. E. and Johnson, A. I., 1970: Evaluation of Installation Methods for Neutron-Meter Access Tubes (in English), USGS Professional Paper, No. 700-C.

Vachaud, G., 1966: Study on Redistribution after Stopping Horizontal Infiltration (in French), IASH Symposium on Water in the Unsaturated Zone, Wageningen;
1966: Verification of Generalized Darcy' Law and the Determination of Capillary Conductivity by Analyzing Horizontal Infiltration (in French), IASH Symposium on Water in the Unsaturated Zone, Wageningen.

Vachaud, G., Cisler, J., Thony, J. L., and De Backer, L., 1970: Utilization of Gamma Emission of Americium-241 for Measuring the Soil-Moisture Content of Unsaturated Soils (in French), Isotope Hydrology, IAEA.

Vachaud, G. and Thony, J. L., 1971: Hysteresis during Infiltration and Redistribution in a Soil Column of Different Initial Water Contents (in English), Water Resources Research, No. 1.

Vachaud, G., Gaudel, J. P., and Kuraz, V., 1974: Air and Water Flow during Ponded Infiltration in a Vertical Bounded Column of Soil (in English), Journal of Hydrology.

Vachaud, G., Vauclin, M., and Haverkamp, R., 1975: Towards a Comprehensive Simulation of Transient Water Table Flow Problems (in English), Seminar on Modeling and Simulation of Water Resources Systems.

van Bavel, C. H. M., 1966: Three-Phase Domain in Hydrology (in English), IASH Symposium on Water in the Unsaturated Zone, Wageningen.

Várallyai, G., 1974: Investigation of Water Movement in Unsaturated Soils (in Hungarian), Agrokémia és Talajtan, Vol. 23 and 3-4.

Vauclin, M., 1975: Experimental and Numerical Investigation of the Drainage of Groundwater with Free Water Table. The Influence of the Unsaturated Zone (in French), University of Grenoble, Doctor Thesis.

Vetterlein, E., 1961: Methodological Investigation of Continuous Soil-Moisture Measurements in Sandy Forest Soil (in German), Part 1: Measuring Electrical Conductivity; Part 2: Investigations with Gypsum Blocks, Zeitschrift für Pflanzen.

Vilkner, H., 1959: Soil-Moisture Measurement with Electrical Resistivity (in German), Zeitschrift für Meteorologie, No. 13.

Watson, K. K., 1967: Experimental and Numerical Study of Column Drainage (in English), Proceedings of ASCE, HY2.

Watson, K. K. and Whisler, F. D., 1969: Analysis of Infiltration into Draining Porous Media (in English), Proceedings of ASCE, IR4.

Wesseling, J., 1957: Some Aspects of Water Movement in Soils (in Dutch), Verst. Landb. Onderz., No. 5.

Wesseling, J. and Wit, K. E., 1966: An Infiltration Method for the Determination of the Capillary Conductivity of Undisturbed Soil Cores (in English), IASH Symposium on Water in the Unsaturated Zone, Wageningen.

Whisler, F. D. and Klute, A., 1966: Analysis of Infiltration into Stratified Soil Column (in English), IASH Symposium on Water in the Unsaturated Zone, Wageningen.

Whisler, F. D. and Bouwer, H., 1970: Comparison of Methods for Calculating Vertical Drainage and Infiltration for Soils (in English), Journal of Hydrology.

Whitney, M. A., 1894: Instructions for Taking Samples of Soil for Moisture Determination (in English), USDA Div. Soils, Circ. 2.

Whitney, M. A., Gardner, F. D., and Briggs, L. J., 1897: An Electrical Method of Determining the Moisture Content of Arable Soils (in English), USDA Div. Soils, Bull. 6.

Wind, G. P., 1966: Capillary Conductivity Data Estimated by a Simple Method (in English), IASH Symposium on Water in the Unsaturated Zone, Wageningen.

Wladitchensky, S. A., 1966: Moisture Content and Hydrophility as Related to the Water Capillary Rise in Soils (in English), IASH Symposium on Water in the Unsaturated Zone, Wageningen.

Wollny, E., 1885: Investigation of Water Retention Capacity of Soils (in German), Ferschungen auf dem Gebiete der Agrikultur Physik.

Youker, R. E. and Dreibelbis, F. R., 1951: An Improved Soil-Moisture Measuring Unit for Hydrologic Studies (in English), Transactions of AGU, Vol. 32.

Youngs, E. G., 1957: Moisture Profiles during Vertical Infiltration (in English), Soil Science.

Youngs, E. G. and Peck, A. J., 1964: Moisture Profile Development and Air Compression during Water Uptake by Bounded Porous Bodies (in English), Soil Science.

Zetter, K., 1975: An Instrument for Determining the Upper Stretch of the pF Curve (in Hungarian), VITUKI, Scientific Report (Manuscript).

Zunker, F., 1930: Behavior of Soils in Connection with Water (in German), Handbook of Soil Sciences, Vol. VI, Berlin.

PART III

HYDROLOGY OF SHALLOW GROUNDWATERS

by

G. Kovács

SECTION 9

GENERAL CHARACTERISTICS OF SHALLOW GROUNDWATERS

It was already explained in Part I, where the occurrence of water stored in various layers was discussed and the types of subsurface water were classified, that, the character of the water-bearing rocks and their contact with the meteorological and hydrological processes on and above the surface are the two most important aspects of grouping when the behavior of the water stored and moving below the surface is investigated from a hydrological point of view.

The consideration of the character of layers is needed to determine the physical laws describing the storage capacity of the porous medium and its resistivity to flow. It has been explained that some simplification can be accepted in the lithological classification of various rocks such as: loose or cemented clastic, mechanical, sediments; chemical and biological sediments; effusive, volcanic, intermediate and intrusive igneous rocks; and metamorphic rocks, because the interrelation between the porous medium and the water is mostly governed by the structure of the interconnected network of pores. The character of this structure depends on the fact whether the channels transporting and storing the water are composed of intergranular pores, or the network is constructed by fissures and fractures. Thus, two basic types of water bearing formations were distinguished: loose clastic sediments, and solid (hard) rocks.

Since the hydrological processes governing the water regime in hard-rock formations will be discussed in a separate part (Part V), the topic of both the present and the next parts is restricted to the hydrological investigation of groundwaters stored and moving in loose clastic sediments. The distinction between Part III and Part IV is made according to the contact of the water-bearing formations and the hydrological phenomena on the surface, which aspect was mentioned as the second basis of the hydrological classification of subsurface waters. According to this aspect, three horizons were distinguished considering the contact of the subsurface waters stored in loose clastic sediments with surface phenomena:

- the zone of aeration between the surface and the water table, where the soil moisture having negative pressure is stored;

- the zone of saturation below the water table where all the interstices are completely (or almost completely) filled with groundwater, which zone is further divided into two horizons: the zone of shallow groundwater, and the zone of deep groundwater.

The hydrological investigation of soil moisture was discussed in Part II. The topic of Part III is the analysis of hydrological processes governing the water regime of the shallow groundwater, while the same investigation will be continued in Part IV, concerning the deep groundwater.

In the introductory section of the present discussion, the difference between the shallow and deep groundwaters will be dealt with, at first giving a more detailed description of the groundwater zone and especially that of its upper part. This analysis is followed by the investigation of the processes governing the regime of the shallow groundwater. These hydrological processes are, however, very versatile because of the numerous influencing factors acting in the vicinity of the surface. The description of the water regime contains, therefore, some uncertainties, if its investigation is based only on hydrological observations and on the analysis of the water balance. The knowledge of the changes in other properties of groundwater (e.g., temperature, chemical constituents, isotope content) may assist the investigators to a great extent in better understanding and more precisely characterizing the interconnection of the shallow groundwaters with other water horizons (soil moisture, surface waters, and water stored in deep-lying layers), as well as numerically describing such processes as infiltration, evaporation and transpiration, percolation of water from surface waters into the water bearing-layers or vice versa, water exchange between various aquifers, etc. This is the reason why the general characteristics of the shallow groundwater are supplemented by some information on temperature, and chemical, and isotopic contents of the groundwater.

9-1 Distinction between Shallow and Deep Groundwaters

According to the definitions given in Part I and already recalled in the introduction, the crust of the earth may be divided into two parts when the position of the water stored in the interstices of the layers is investigated, i.e., the zone of aeration and the zone of saturation. The two zones are separated by the irregular surface of the water table which is the upper boundary of rocks completely saturated by water. It was also explained in Part II that this boundary is a fictive surface, the position of which can be observed only in wells and not in the soil profile, because the lower part of the soil-moisture zone is also completely or almost completely filled by water and, therefore, a sharp change does not occur in the condition of saturation at the water table. The main physical difference, on the basis of which distinction may be used between the two basic types of subsurface water stored in the zone of aeration and below the water table, respectively (i.e., soil moisture and groundwater) is the pressure prevailing on the water. Above the water table, the pressure is always less than the atmospheric pressure, and thus the soil moisture is characterized by suction if the pressure is related to the atmospheric value. Hydrostatic pressure conditions prevail in the gravitational groundwater zone, the pressure being equal to the atmospheric one (the excess pressure is equal to zero) at the water table, and it increases linearly with the depth. (It is necessary to note, that the pressure may be increased, in comparison with the hydrostatic value, by the weight of overlying formations (geostatic pressure) in confined systems, or decreased by discharging the layer, but in the gravitational groundwater pressure always remains higher than atmospheric.)

Some exception was also mentioned in Part I which disturbs this simplified pattern of the two zones, i.e., the perched groundwater (see Fig. 1-35). This type of completely saturated zone, storing water having positive excess pressure, develops within the zone of aeration above impervious lenses in such a manner that the soil moisture (water content of the pores under suction) is continued below it. Usually the extension of such lenses is, however, very small and the development of perched groundwater is only a periodic process in most cases (the water is collected there after heavy rains forming a water-table condition, but it is consumed by evapotranspiration in a relatively short time). These are the reasons why the influence of the development of such local disturbing effects on the regime of groundwater is negligible and its detailed analysis is omitted in the further discussion.

There are also considerable differences, occurring in various parts of large groundwater reservoirs, in the behavior of water. The water-bearing layer may be completely surrounded by impervious formations (closed groundwater) and the water exchange is excluded, or at least hindered, between the aquifer investigated and the other groundwater reservoirs located in its vicinity. At other places, the pressure of the water stored in a pervious layer, covered locally by impervious formations, is governed by the position of the water table developing in the same aquifer at some distance where the layer is raised near the surface and characterized as unconfined (artesian water) (see Fig. 1-36). The adjectives unconfined and confined describe the pressure conditions in the system. The former indicates the development of water-table conditions in the water-bearing layer, while the latter describes an aquifer covered by an impervious layer. The pressure is higher, therefore, at the upper boundary of confined aquifers than the atmospheric value. The large sedimentary formations are generally composed of subsequent series of aquifers and aquicludes (pervious and impervious layers) (aquifuges, i.e., a layer neither transmitting nor containing water, cannot be found among loose clastic sediments) and the interrelations between water and grains differ considerably according to the character of the layer as well, thus, the permeability of the porous matrix also has to be considered when the behavior of the groundwater is investigated.

Although the terms listed in the previous paragraph are well-known and generally applied in hydrogeology, some further distinction is required to emphasize the great difference, from a hydrological point of view, between the behavior of water being stored near the surface and that filling the pores of deep-lying aquifers. This need is caused by the fact that the character and the number of effects influencing the development of water balance differ considerably depending on the position of the aquifer investigated.

It is evident that the groundwater just below the water table has water exchange with the soil-moisture zone and thus, it is directly recharged by infiltrating precipitation and discharged by evaporation and transpiration. Under the influence of gravity, it moves in the direction of the slope of the water table, or, where the aquifer is confined, in that of the piezometric surface. As a result of this movement, there is a continuous connection between the groundwater and the various bodies of surface water (rivers and lakes) where the beds of the latter penetrate into the aquifer, storing the groundwater. The surface water can recharge or discharge the groundwater, depending upon whether its level is higher or lower than the water table. Due to the fluctuation of the water level in rivers and lakes, there are places where the groundwater is drained in one period and recharged in another by the same surface water body. A third type of water exchange may also influence the water regime of this topmost groundwater horizon, i.e., the seepage between the layer investigated and the lower lying aquifers, which depends on the pressure conditions and the hydraulic conductivity of the impervious formations separating the water-bearing layers.

In the opposite to the previous case, the deep aquifers have no contact with the hydrological events occurring at the surface and in the soil-moisture zone (infiltration, evaporation, transpiration). The beds of rivers and lakes hardly penetrate so deep that they may directly reach these water-bearing layers. This is the reason why the most important recharging and discharging factor of these systems is the water exchange between the neighboring aquifers, which effect is governed by the accumulation of pressure energy at various parts of the formation. Thus, the processes influencing the pressure conditions (development of hydrostatic and geostatic pressures and their changes due to the consolidation of the solid matrix or caused by the flow of water) should be analyzed first of all, when the task of the investigation is the characterization of the water balance in deep-lying aquifers.

Considering the differences listed in the previous paragraph, it was felt necessary to distinguish shallow and deep groundwaters. Definitions of the two types of groundwaters, separated in this way, are as follows:

- Shallow groundwater is stored near the surface, below the water table and above the first, largely extended continuous impervious formation. Its regime is influenced by meteorological and hydrological events through the soil-moisture zone by the recharging and discharging effects of surface waters and by the water exchange developing through the lower impervious bed bordering the aquifer of the shallow groundwater system.

- Deep groundwater can be found in aquifers lying below continuous impervious strata, which hinder the development of the influence of atmospheric processes on groundwater over a very large area. The surface waters hardly penetrate directly into such aquifers. The recharge and discharge of the deep groundwater systems depend, therefore, mostly on their contact with shallow groundwaters and on the water exchange between neighboring systems.

It was also explained in Part I that clear distinction cannot be made between the two types of groundwaters. Shallow groundwater may be partly covered by an impervious layer, but if the covering formation is not continuous, or its extension is not large enough, the groundwater below it can be directly recharged by precipitation and drained by evapotranspiration, thus the influences of the hydrological and meteorological processes occurring on and above the surface can be clearly observed in this regime of groundwater. On the other hand, it is also possible that a water-bearing layer covered by an impervious formation over a very large area and consequently containing deep groundwater may be raised near the surface far from the place of the investigated area (e.g., artesian aquifers), where a water-table condition develops in the aquifer in question and the water regime is influenced by precipitation, evapotranspiration, and by percolation from or to surface waters. Thus, the same groundwater which has deep character at the place of investigation, may become shallow groundwater at another point, as it is indicated by Fig. 1-36.

Considering all the aspects explained previously, it can be stated that there is no sharp borderline between the two types of groundwaters distinguished from a hydrological point of view. There are several transition forms and the distinction can be based only on the investigation of the dominating factors influencing the regime of groundwater. The means of these investigations are the observed groundwater data.

In spite of all the difficulties arising in connection with the determination of the character of the groundwater, it is advisable to separate the two hydrological types (i.e., shallow and deep groundwaters), because the methods of their investigation differ basically due to the different factors influencing the water regime. The first step is, therefore, to survey the acting effects. The observation as well as the evaluation and the hydrological analysis of the data may be planned and executed on the basis of this survey.

9-2 Processes Influencing the Regime of Shallow Groundwater

One of the fundamental aspects of the hydrological investigation of groundwater is the assumption of the balanced condition of the interconnected groundwater reservoirs. All groundwater catchments or any separated part of a drainage basin should be regarded as a storage system, and thus the equation of continuity can always be explained to describe this balance. This equation expresses that the difference between inflow and outflow, summarized for the investigated period, is equal to the change in the amount of water stored in the system. The application of this approach requires, however, the determination of the processes resulting in recharge and discharge as well as that of storage.

The very versatile system of the natural effects recharging and discharging the shallow groundwater was already analyzed in the previous section. The influencing factors listed there can be divided into three main groups, according to the direction of their actions, i.e.:

- water exchange between surface and groundwater (horizontal groundwater flow developing between rivers and the aquifers);

- processes affecting the groundwater space from the surface through the soil-moisture zone (surface effects acting vertically at the upper border of the shallow groundwater, including the infiltrating reaching the water table and the capillary process raising water from the groundwater to replenish the evaporated soil moisture);

- interaction between deep and shallow groundwaters (vertical percolation through the lower boundary of the shallow groundwater, which may either recharge or discharge the upper water horizon depending on the direction of flow).

The storage process has to be taken into account apart from the recharging and draining actions when the natural water balance of the shallow groundwater is investigated, and all these effects should be supplemented by considering the artificial factors when the purpose of the investigation is to characterize the human influences on the shallow groundwater.

The general description of the natural and artificial processes recharging and draining the shallow groundwater will be given here. The qualitative survey of the problems will assist to better understand the objectives of the hydrological investigation of the upmost horizon of the groundwater. After investigating the ways and means of data collection, as well as the numerical determination of the parameters describing the flow conditions in saturated porous media, we shall return to the hydrological analysis of these processes with the intent of their quantitative characterization.

Where a bed of surface water penetrates into a water-bearing layer, a continuous flow may exist, either to or from the groundwater, depending on the direction of the hydraulic gradient created by the difference between the levels of the two adjacent water bodies (surface- and groundwater), or between the level of the surface water and the piezometric head of the confined groundwater. It is also possible that the groundwater is recharged by a river in flood periods at the same stretch where it is drained when a low water level prevails in the river. This change of flow direction is caused by the fluctuation of water levels in rivers and lakes.

The most general geological structure of river valleys together with some hydrological (hydrogeological) characteristics are represented in a very simplified form in Fig. 3-1. The less permeable basic formations are usually covered by pervious alluvial sediments in the valleys. The river bed penetrates, in most cases, into the previous layers and, therefore, the water regime of the shallow groundwater is

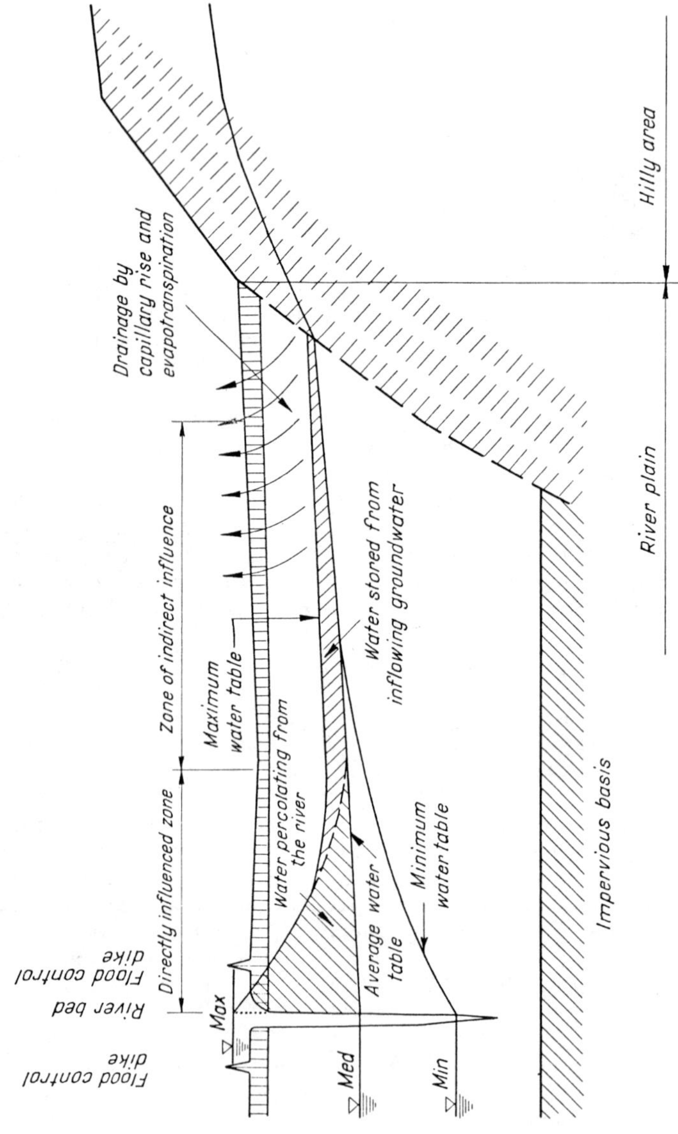

Fig. 3-1. General geological structure of river plains.

influenced by the surface water in the river valleys. At the outer edges of the river valleys the groundwater contacts the layers of the surrounding higher areas as well. Where these layers are not absolutely impervious, the water balance of the valleys is also affected by the groundwater developed under the adjacent areas.

It can be stated, however, that absolutely impervious layers hardly exist in nature, and when investigating over a long time period (studying natural processes, geological ages may be considered), the amount of water conveyed by groundwater flow developed between the investigated areas cannot be neglected. Thus, dynamic contact between the groundwater in the surrounding area and that of the river valleys must be always assumed to exist.

Based on the investigations of processes affecting the groundwater through its water table (which analyses will be given later in this text), it is evident that groundwater flow will develop everywhere from the higher areas to the deeper lying ones because the water table is at a relatively greater depth below the surface in hilly and mountainous areas than in the valleys and on the plains. Infiltration surpasses, therefore, the amount of water removed from the groundwater by capillarity to be evaporated or transpired. The excess water moves to the deep-lying areas following the slope of the shallow groundwater. This water transport is supplemented by the water conveyance of the regional flow systems of the deep groundwater as well (if the latter exists at all) because the river valleys are the natural discharging lines of such flow systems also.

Where the level of the river is deep below the terrain, the water table created by the drawdown effect of the river will develop at a considerable depth, and the infiltration will probably exceed the vertical removal of the groundwater even within the river plains. Thus, both groundwater flow coming from outside and excess infiltration on the plains are discharged by the river. In other cases, the groundwater flow is drained partly by evapotranspiration within the plains and partly by percolation to the river.

Where the level of the surface water has no fluctuation (or it is negligible as in the case of some lakes and reservoirs), groundwater supplies the surface water continuously. The amount of percolation to the surface water may, of course, vary in this case as well because of the seasonal fluctuation of both infiltration and evaporation which processes influence the flux of groundwater flow.

It was already mentioned that a considerable change in river stages may even modify the direction of flow through the bank from time to time. During flood periods, the water level is generally higher in the river than the water table. The river recharges the groundwater within a distance until the slope of the backwater curve is reversed during this period (zone of direct influence of river). At the same time, the groundwater flow is also stopped by the high river stage. The water arriving from the adjacent higher areas is stored by raising the water table and causing high evaporation on the river plains, exceeding the average value (zone of indirect influence of the river). When there is low water in the river, its level lies deeper than the average elevation of the water table. The drawdown curve developing due to the discharging effect gradually propagates, emptying at first the water from the pores of the riparian layers, where it has been stored during the flood. After a certain period has elapsed, the drawdown effect surpasses the border of the directly influenced zone, lowering the water table at a greater distance as well. Finally, a condition of equilibrium develops when groundwater flow coming from outside is balanced by the drainage to the river and the excess of evapotranspiration on the river plains.

It was assumed, so far, that the groundwater is drained by rivers because this is the general case in nature. There are areas, however, where the level of the surface water is always above the water table and, therefore, the groundwater is continuously recharged by the river, lake, or reservoir. Under natural conditions, such areas can be generally found on the alluvial cones of rivers, where the riverbed was raised by deposited sediments above the level of the surrounding terrain. The occurrence of recharged groundwater is more frequent in the vicinity of artificial surface waters (reservoirs, ponds).

Where conditions develop, the deepest part of the riparian plains is recharged from two directions: by percolation from the river and by groundwater flow arriving from higher areas. In such plains, the possible way of drainage should be investigated. There are cases when the sole discharging effect is the excess evapotranspiration from areas where the groundwater level is near the surface (or it sometimes reaches the surface, causing the development of swamps). In other cases, there is a deeper lying surface water (sometimes a lower stretch of the same river) which, apart from the high evapotranspiration, can drain the recharged part of the river plains (Fig. 3-2).

The balanced condition of the storage system composed of the interconnected aquifers was already mentioned as a basic principle of groundwater hydrology. When analyzing the water balance of the shallow groundwater, it can be seen that the water table generally exhibits a regular annual fluctuation

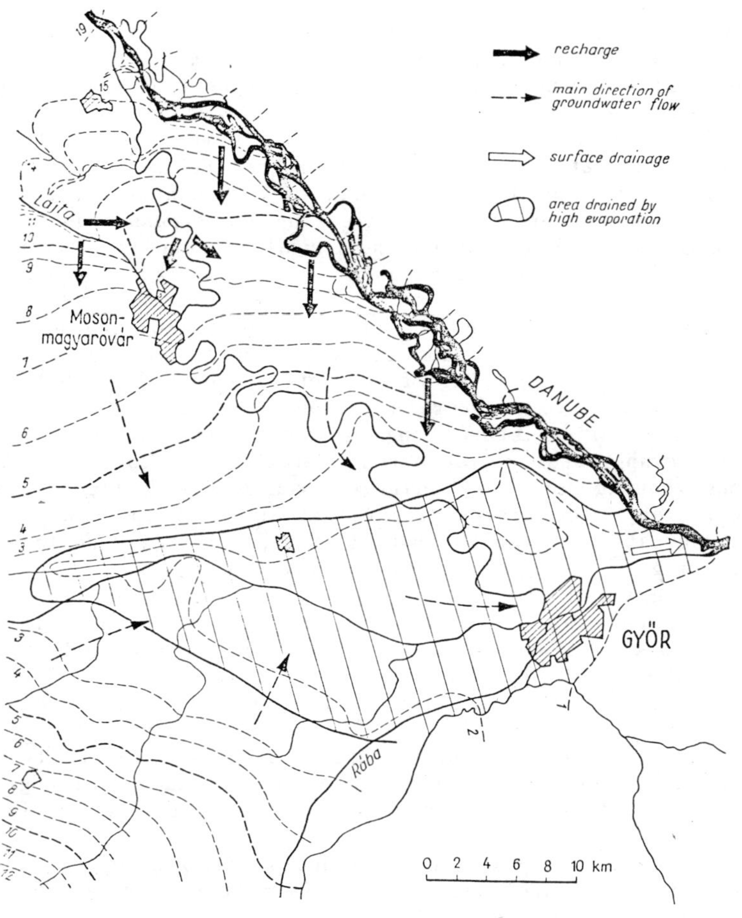

Fig. 3-2. Recharge and drainage of river plains in the vicinity of an alluvial cone.

and in most cases longer changes as well. Under natural conditions, a balanced state can, however, be expected if the period of the investigation is sufficiently long.

Some water from precipitation infiltrates through the land surface, the amount depending on the conditions of the surface (vegetation, cultivation, slope, etc.). The infiltrating amount has, therefore, seasonal, long-term, and local variation. Supposing average conditions characterized by mean values determined for a long period, the amount of water expected to infiltrate annually through the surface can be expressed by a fixed value of an equivalent water column (mm per year). A part of the infiltrated water is stored in the soil-moisture zone and evaporates from there. The amount reaching the water table as vertical input (positive accretion or recharge to the groundwater) is always smaller than that infiltrating through the surface. It is also evident that the deeper the water table the greater the difference between the infiltrating amount and the input of groundwater due to accretion. This monotonically decreasing relationship is valid to a given level below which the annual recharge is independent of the depth of the water table.

One part of water recharged to the shallow groundwater is drained by the rivers and lakes. There is, however, another considerable discharging effect acting evenly distributed over large areas, i.e., water is raised from the groundwater space into the soil-moisture zone by capillarity to replenish the water evaporated or transpired. This process causes a continuous output of the shallow groundwater called negative accretion. The discharging effect is equal to actual evapotranspiration if the water table reaches the land surface. The lowering of the water table decreases the actual evapotranspiration which reaches a constant value at a given depth of the groundwater. One part of evapotranspiration originates from soil moisture decreasing its stored amount and only the remaining part is replenished by negative accretion from groundwater. The numerical value of the latter is always smaller, therefore, than the actual evapotranspiration. The depth at which it becomes constant indicates the position of the water table below which the output of the shallow groundwater due to negative accretion is practically negligible.

The ratio of the two important discharging effects of the shallow groundwater (i.e., percolation to rivers and negative accretion) depends on the location of the investigated area. The drainage maintained by the percolation to surface waters may be the dominant factor within the river plains as it was previously explained. At the internal part of large sedimentary basins, far away from networks of water courses, the influence of horizontal flow may be negligible compared to the vertical discharging effect due to capillarity and evapotranspiration. One of the main tasks of the hydrological analysis of the groundwater data is to numerically characterize these components of the water balance and to describe the water regime of the shallow groundwater in this way.

The third natural factor, which may either recharge or discharge the shallow groundwater, is the water exchange through the impervious formation which forms the lower boundary of the aquifer containing the first water horizon. Since, according to the definition of the shallow groundwater, this impervious formation is large and continuous, the seepage developing between the separated aquifers generally transports smaller amounts of water than those conveyed by positive and negative accretion and horizontal groundwater flow. Thus, the consideration of this factor is more important for the investigation of the deep groundwater than for the analysis of the shallow one. It is quite evident that the direction and the flux of seepage depends on the gradient developing through the impervious or semipervious formation separating the aquifers investigated. It is also necessary to consider that the pressure difference between two neighboring water-bearing layers indicates only the possibility of movement. Actual seepage can develop only if the gradient surpasses the threshold gradient which is an important physical parameter of the porous media, the detailed analysis of which will be given later in this text.

The consideration of the human effects is relatively easier in most cases than that of the natural ones, since the amount of water either taken out from or recharged to the groundwater artificially is measurable, e.g., yields of wells and water works, seepage losses from canals, etc. There are, however, some exceptions when the artificial influence acts in a form distributed over large areas, e.g., recharge due to irrigation, the increase of capillary drainage caused by the modification of land use, etc. The human activity changes, in these cases, the hydrological processes developing in the soil-moisture zone and thus the boundary conditions prevailing along the water table as well. Since the positive and negative accretion is basically changed by these types of artificial effects, the influence of the latter can be taken into account only by reconsidering the conditions governing the water balance of the shallow groundwater.

Finally, the last factor to be considered in the balance equation is the change in storage during the period investigated. It is well-known that this change is always characterized by the change of the position of the water table and the piezometric level, respectively, depending on the fact whether an unconfined or a confined system is investigated. In both cases, the difference observed in the elevations of the level in question, at the beginning and at the end of the investigation, should be multiplied by a parameter characterizing the storage capacity of the layer to get the amount of water stored or released from storage during the period within an area having a horizontal extension of unity. The confined or unconfined character of the aquifer causes, however, some differences in the interpretation of storage capacity.

In general, the storage capacity of a porous medium depends on:

- the change of the ratio of the saturated and unsaturated thickness;

- the compressibility of both solid grains and water;

- the change in the porosity of the saturated layer.

The deformation of the grains, and thus, the compressibility of the solid matrix, is always negligible. In an unconfined system, only the change of saturation has a dominant role. The storage capacity may be, therefore, characterized by the specific yield (n_s), the detailed explanation of which was given in Part II, Subsection 6-4. On the contrary, the specific yield cannot be applied when a confined aquifer is investigated because the saturation of the water-bearing layer does not change as a result of raising or lowering the piezometric level. In this case, the storage coefficient (S) should be used for the characterization of the storage capacity of the system which expresses the influence of the compressibility of water and the change in the porosity of the solid matrix. Since the shallow groundwater is generally characterized by a water table condition, the knowledge of specific yield, already discussed, is sufficient for describing the storage process in this system. The interpretation of the storage coefficient will be dealt with in the next chapter in connection with the hydrological investigation of deep groundwater (Subsection 13-3).

9-3 Temperature, Chemical Composition, and Isotope Content of the Shallow Groundwater

The components of the water balance were analyzed in the previous section. It is necessary to remember, however, that the numerical characterization of the elements of the water balance is very uncertain and, therefore, the study of any properties of water other than its quantity provides the investigators with important information. There are three parameters generally used to get supplementary data, i.e., the temperature, the chemical composition, and the isotope content of the groundwater.

The temperature of the groundwater depends on many factors. In the case of shallow groundwaters being stored near the surface and having, therefore, close contact with the meteorological events, the air temperature is one of the most important influencing factors beside the terrestrial heat flux, which has a dominant role in forming the temperature of any type of groundwater. Since solar radiation also influences the development of groundwater temperature, the effects of the covering vegetation, the overlying soil, and the direction of slope are not negligible either, although this influence is not considerable compared to that of air temperature below a given depth. The heat conductivity of the layer storing the water and that of the neighboring formations, the temperature of the infiltrating precipitation as well as the groundwater flow may be mentioned as further affecting factors.

The validity of the statement emphasizing the strong influence of air temperature on the temperature of shallow groundwater is clearly demonstrated by Fig. 3-3 which represents the seasonal fluctuation of the two variables in question, observed at the same place (Léczfalvy, 1966). The water temperature was meaured at the water table in two observation wells. The depth of the groundwater was different at the two measuring points, thus the data indicate, at the same time, the effect of the thickness of the covering layer as well. It can be seen that the maximum and the minimum of the fluctuation in water temperature follow the corresponding values of air with a relatively short time lag which increases as the depth of the water table increases. The amplitude of the fluctuation is always smaller in water than in air and it decreases with an increase of water depth.

On the basis of regular observation of water temperature near the water table in such shallow groundwaters, where the flow is negligible, a graph can be constructed showing the interrelationship between the fluctuation of temperature and the depth of the water table (Fig. 3-4) (Léczfalvy, 1966). The characteristic depths indicated in the figure are only valid at the place where the basic data were observed, e.g., the depth of the daily fluctuation is about 80 cm in Hungary and the thickness of the neutral zone below which fluctuation does not occur at all, but the temperature increases according to the geothermic gradient, is about 15 m. Similar graphs can be determined, however, for any location, thus the relationship gives a guidance to check whether the shallow groundwater is in a static condition or not. Even the type of the influencing factor, e.g., inflow from surface water, can be estimated by considering the discrepancy of the actual temperature from the normal value.

The water moving along the atmospheric side of the hydrological cycle and reaching the continents is chemically relatively pure, and its character develops slowly as a result of the contact with the layers through which the water percolates. Thus, the composition of water, the type and the amount of its chemical constituents, also give information on

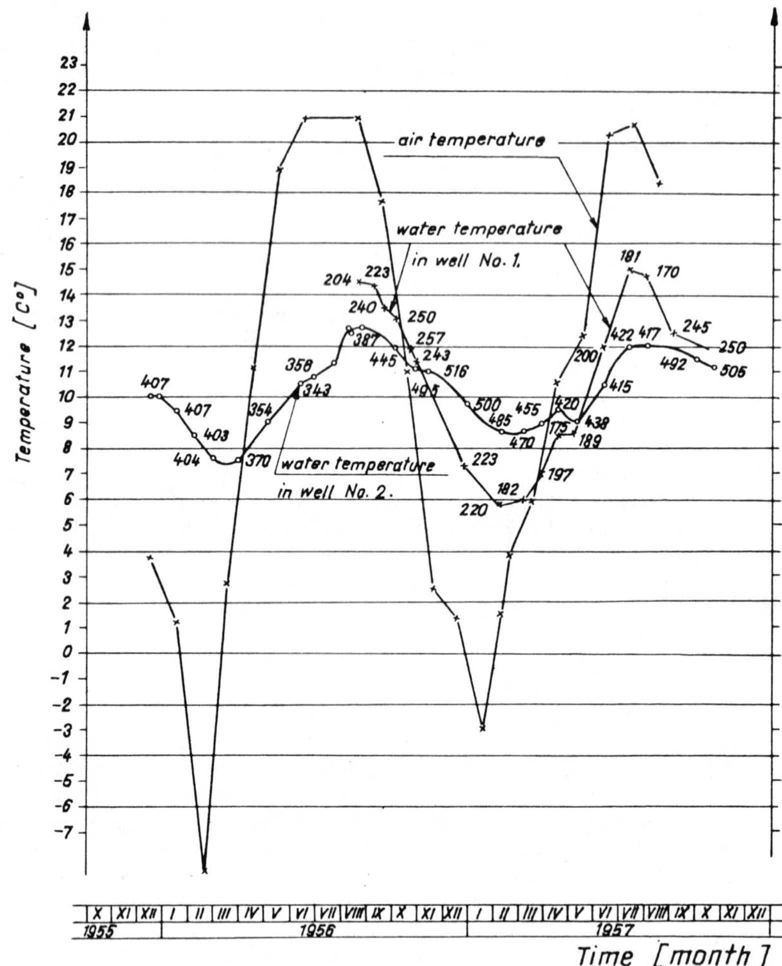

Fig. 3-3. Data series showing the change of air temperature and that of water temperature observed at two wells.

the history of the groundwater and on its path, since the infiltration of precipitation.

When the water reaches the land surface, it contains only a small amount of gases and some impurities absorbed from the air. The carbon dioxide is the most important among these constituents becuase it increases the solution capacity of water. The total salt content of water increases only to a small extent if the terrain is formed from igneous or metamorphic rocks because the minerals composing these rocks are hardly soluble. The water percolating through carbonate rocks becomes almost saturated by dissolving salts, mostly calcium

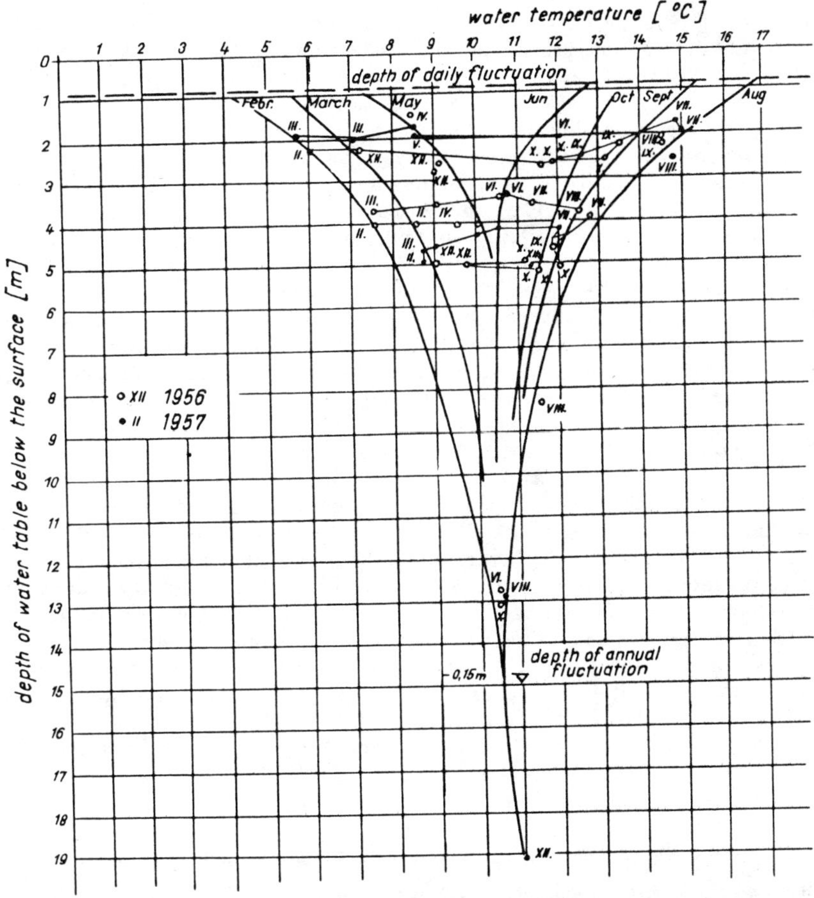

Fig. 3-4. Development of the temperature of groundwater at the water table as a function of the depth of the latter.

and magnesium carbonates. Similarly, high salt content characterizes the water stored in loose clastic sediments, especially when it percolates through the upper soil which is the case when the shallow groundwater is investigated because the fine grained sediments generally contain large amounts of soluble salts.

A great part of soil is composed of clay minerals which represent the final stage of the decomposition of various silicates. Besides the development of clay minerals, some cations become free during the decomposition. They are partly bound to the free charges of clay minerals while their other part is dissolved in the groundwater and soil moisture. There is a well-defined order of these exchangeable cations according to their adsorption capacity. The weakest binding characterizes sodium which is followed by potassium and the other alkaline metals. The second part of the series is composed of the alkaline earth metals, magnesium being the last, but one in the series which is closed by calcium having the strongest attraction to the crystalline grains. This is the reason why sodium is leached out first by the infiltrating water from the profile and calcium remains dominant in the leached soil. During the further development of the process, calcium ions reach the groundwater as well, and the free electrostatic charges of clay minerals are neutralized by hydrogen ions. Thus, the development of calcareous and acidic leached soils characterizes the recharged areas of the shallow groundwater. Cations washed into the groundwater increase the salt content of the latter and they are carried by groundwater flow to the drainage areas where the water is discharged by high evapotranspiration. The salts transported by the water remains in the soil after evaporation, causing salt accumulation. Thus, the drainage areas are characterized by high salt content, sodium being the dominant cation in the profile. The chemical analysis of the groundwater and the soil assists the investigators in determining the location of recharging and drainage areas (recharge is indicated by low and discharge by high salt content). The order of the cations in the soil even gives information on the direction of the dominant vertical movement of water. After a period when evapotranspiration exceeded infiltration, sodium is near the surface and calcium is at the deepest position. In the opposite case sodium can be found in the deeper layers below calcium (sodium is always at the front, up gradient, of the movement).

The chemical analysis is not limited to the determination of the concentration of cations, among which Na^{+}, K^{+}, Mg^{++}, and Ca^{++} are generally measured, but the measurement of the amount and the sum of dissolved salts as well as total hardness also give important information.

According to Chebotarev, the character of the dissolved anions also changes gradually as the water infiltrated through the surface propagates along the subsurface stretch of the hydrological cycle. Hydrocarbonates are the dominant salts at the beginning of the process. In the second phase, the occurrence of chlorides is characteristic, and in the third stage they become dominant beside HCO_3^-. The hydrocarbonates are subsequently supplemented by sulphates and finally, all the other anions are negligible compared to chlorides, e.g., in seawater. The shallow groundwater, having direct contact with the surface, should be a water of hydrocarbonate type. Connate water stored in marine sediments preserves the original character of the seawater, having a high rate of chlorides. The presence of chlorides in the shallow groundwater indicates, therefore, the water exchange with deep-lying layers. Sulphates in shallow groundwater originate, in most cases, from special minerals in the aeration zone, e.g., pyrite.

The hardness of water characterizes the total amount of calcium and magnesium salts dissolved in the water. Various systems are distinguished, depending on the numerical parameter describing the concentration of these salts (e.g., the hardness of water is one German grade, if the amount of calcium and magnesium salts dissolved in one liter of water is equivalent to 10 mg CaO; one German grade is equal to 1.79 French grade or 1.25 English grade). Water percolating through carbonate rocks generally has an almost constant characteristic value of hardness (between 22 and 25 German grades), thus this parameter is very suitable to trace the water exchange between shallow groundwater and karst water.

While the type of salts indicates the origin of water and the process of water exchange, the place of recharging and discharging effects is shown by the total concentration of salts. This parameter is usually low where the infiltration from precipitation or surface water is high, and evaporation raises the concentration under the drainage areas. Two examples are shown here to demonstrate the use of the amount of dissolved salts and total hardness for the qualitative characterization of water balance. Further examples will be given in Part IV, where the quantitative analysis of the balance of the chemical constituents is applied for investigating deep groundwater.

Figure 3-5 is a schematic E-W geological cross section of Bajos Submeridionales (Northern Plains) in Argentina lying on the right bank of the Parana River. The parameters showing the average salt concentration of groundwater at various places and at different depths are also indicated in the cross section. Where the pervious surface ensures higher infiltration of precipitation, and where surface water has direct contact with the water stored in sandy layers, the total amount of dissolved salts is lower than 500 mg/l. The same parameter is 1000-3000 mg/l in and below the impervious alluvial layers where the concentration is gradually raised by low infiltration and high evaporation. Near the surface of marine sediments, the water exchange results in a concentration between 5000-10,000 mg/l. The connate water preserved the high concentration of seawater (and the type of water is also NaCl) and it was even increased by evaporation. Thus, the average 10,000-20,000 mg/l concentration may reach extremely high values at some places (even 80,000 mg/l concentration was measured in tertiary marine clay).

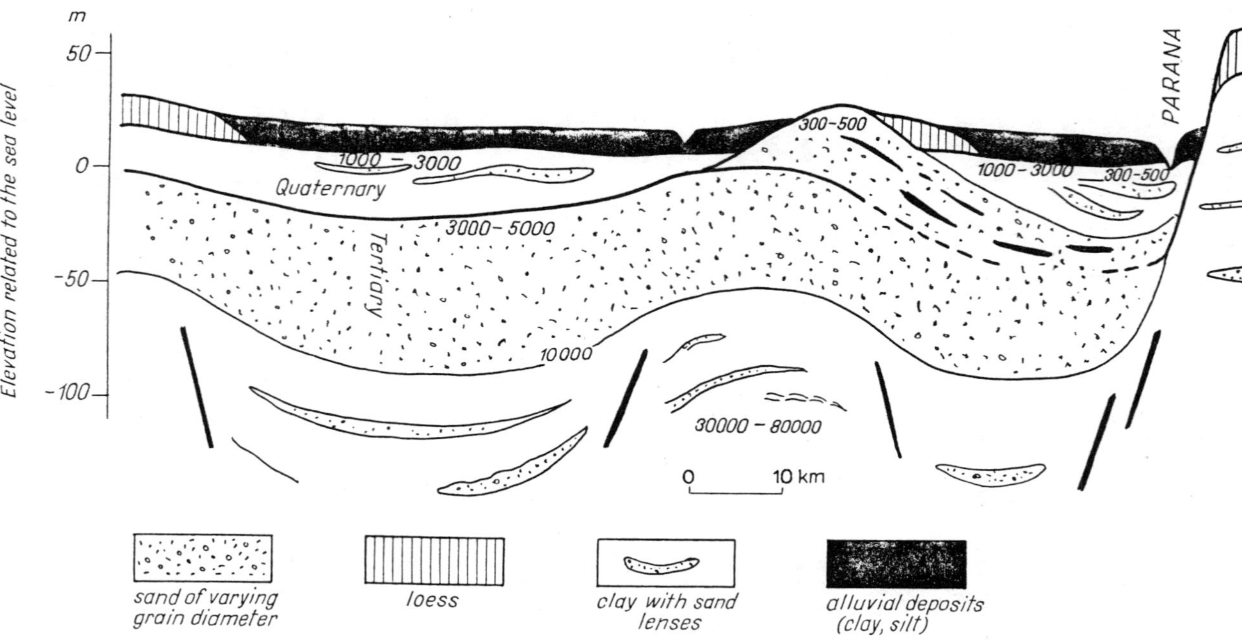

Fig. 3-5. General pattern of the development of total salt content in alluvial plains.

Figure 3-6 is a map of groundwater quality showing the hardness of the shallow groundwater within a part of the river plains of the Danube. There is a belt of a few hundreds of meters in width along the river where the hardness is equal to that of the Danube (8-11 German grades). This is the zone directly influenced by the river. The other characteristic value of hardness is about 25 grades which can be found in two narrow strips almost perpendicular to the bank. These lines continue the two main fault zones of the karst mountain forming the southern border of the plains. It is clearly indicated by the quality map that the karst water is drained by the river through the shallow groundwater and the discharged amount is transported in the two strips. At other points, the measured hardness was above 50 German grades, showing that the horizontal flow is not considerable within those parts of the plains, but evapotranspiration consumes most of the water arriving from the higher terrains at the edge of the area.

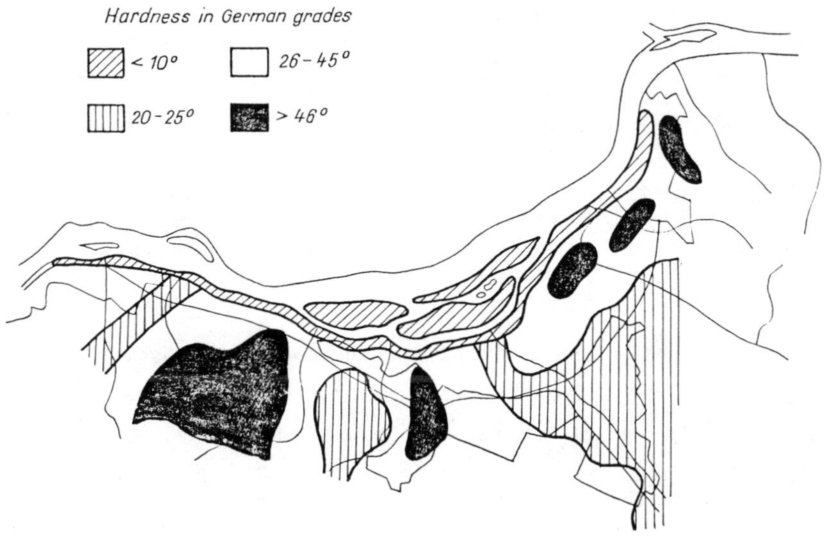

Fig. 3-6. Tracing of horizontal groundwater flow by considering the total hardness of the shallow groundwater.

The analysis of the isotopes of various constituents has opened a new and special way for the investigation of water quality, i.e., the determination of the stable and radioactie isotope content of water. The environmental isotopes are the special variants of the two components of water (hydrogen and oxygen) and also those of the atoms absorbed by the water, among which the carbon atoms are the most important. These variants differ from the normal forms of atoms by having higher atomic weight. The environmental isotopes investigated usually are:

	normal form	isotope
hydrogen	^{1}H	^{2}H or D (deuterium) and ^{3}H or T (tritium)
oxygen	^{16}O	^{18}O
carbon	^{12}C	^{13}C and ^{14}C

The D, ^{18}O, and ^{13}C are stable isotopes, while T and ^{14}C are radioactive ones; they are reduced to stable forms by emitting radiation. The amount of the number of stable isotopes related to that of normal atoms is a well-defined constant value characteristic for the material and for the conditions prevailing at the time of its origin. The number of the originating radioactive isotopes is also a fixed value having only small fluctuation depending on the conditions at that time, but their ratio to the total number of atoms decreases because of the radiation. The half-life, the period where the ratio of the radioactive isotope is reduced half its original value in the whole mass, is 12.26 years for tritium and 5,730 years for ^{14}C. The stable environmental isotopes can be used, therefore, to determine the origin of water, and the radioactive ones are suitable for determining the age of the water if the condition of its origin (the original number of the radioactive isotopes) is known or can be estimated.

The normal atoms and the stable isotopes of hydrogen and oxygen may form three different molecules: $H_2{}^{18}O$, $HD{}^{16}O$ or $H_2{}^{18}O$. The weight of the two types including D or ^{18}O is higher than that of the normal molecules. The heavy molecules behave differently than the light ones. Their evaporation is slower, the number of heavy molecules compared to the total number of the molecules increases in the remaining water and decreases in the vapor. Similar separation occurs during the process of vapor condensation, the precipitation originating from the vapor has a higher ratio than the remaining vapor. This is the reason why there exists a close relationship between the isotopic composition of precipitation and its temperature of formation. This dependency on temperature produces seasonal variation (summer precipitation has heavy isotopes in a higher number than winter precipitation) and similar changes are observable according to the latitude and altitude of the place investigated (the ratio decreases with increasing latitude and altitude).

The numerical characterization of both D and ^{18}O is generally expressed in terms of per mill difference (δ‰) with respect to the ratios of mean ocean water which is called SMOW (standard mean ocean water):

$$\delta‰ = \left(\frac{R_{sample}}{R_{SMOW}} - 1\right) 1000 \; ; \qquad (3\text{-}1)$$

where R is the isotope ratio D/H, or $^{18}O/^{16}O$, and the subscript indicates the sample investigated and SMOW, respectively.

It was found from the data observed that the variations of D and ^{18}O contents in precipitation are linearly correlated:

$$\delta\, D‰ = 8\, \delta\, ^{18}O‰ + 10 \; ; \qquad (3\text{-}2)$$

which line is indicated in Fig. 3-7.

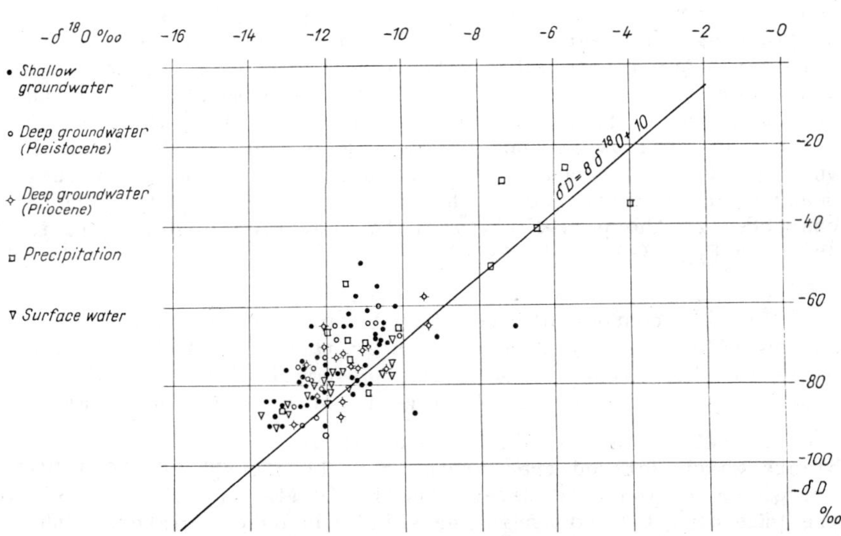

Fig. 3-7. Relationship between the ratios of deuterium and ^{18}O.

The isotopic composition of groundwater is related to that of precipitation in the recharge region of the aquifer at the time of infiltration, thus, the deuterium and oxygen-18 ratio can be used to trace the origin of groundwaters and the recharging areas. Figure 3-7 shows an example of the application of this method where the isotope content of precipitation, surface water and three groundwater horizons were compared indicating the common origin of the groundwaters and their probable recharge from the surface water. The latter has lower isotope content than the local precipitation which condition corresponds to the geographical situation because the catchment lies on higher altitude than the area investigated.

The application of the tritium isotope and its theoretical background was already discussed in Part II (see Fig. 2-29). The tritium concentration in water is generally expressed in tritium units (TU). The concentration is 1 TU when the $^{1}H/T$ ratio is equal to 10^{18}. The cosmic radiation establishes a concentration of sometimes 10 TU, having both geographical variation and seasonal fluctuation. This value was raised considerably as the result of thermonuclear testings, reaching a peak of about 10,000 TU. Its average decreased to about 100 TU after the tests on the earth surface were stopped (Fig. 3-8). Since the reliability of tritium measurements is some TU unit (background value), the tritium concentration cannot be used for determining the age of water originating from precipitation having fallen before 1953. This parameter is applicable, therefore, to trace the movement and water exchange first of all in the soil-moisture zone, as it was shown in the example presented in Part II. The tritium content of shallow groundwater mostly depends on the strength and speed of vertical infiltration. Thus, the TU values measured in the first water horizon of a basin can be used to characterize the permeability of the covering layer, as it is shown in Fig. 3-9 (Deák, 1976).

The ^{14}C content of water can be applied for dating water fallen before 1953 but being younger than 30,000 years. The basis of the investigation is the assumption that the carbon dioxide in soil water is of biogenic origin and the original ^{14}C concentration of water was equal to the ^{14}C content of plants which derived the carbon from the atmosphere. Considering the radioactive decay (the half-life of 5,730 years), the time of infiltration in years (T) can be calculated if the initial concentration (C_o) and the ^{14}C content measured in the sample (C) are known:

$$T = 8270 \ln \frac{C_o}{C} \quad . \qquad (3\text{-}3)$$

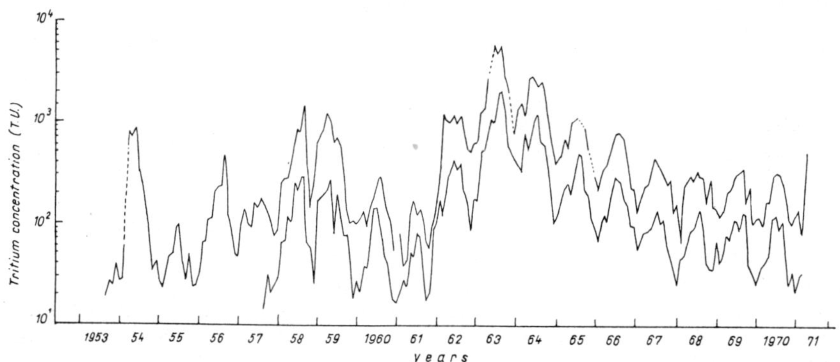

Fig. 3-8. Tritium concentration in Ottawa (Canada), Valentia (Ireland), and Vienna (Austria) from 1953 to 1970 (after Brown et al., 1972).

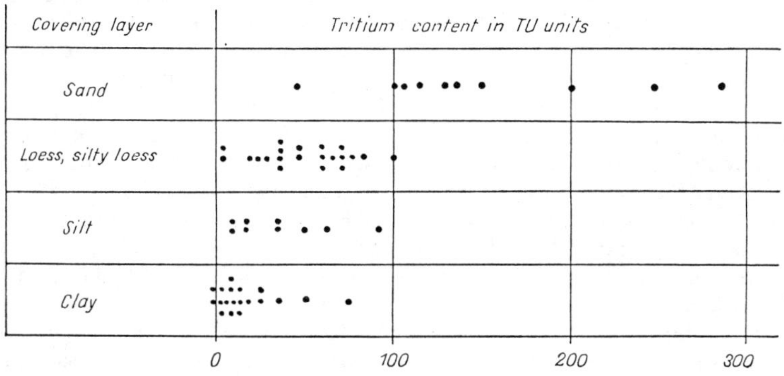

Covering layer | Tritium content in TU units
Sand
Loess, silty loess
Silt
Clay
0 100 200 300

Fig. 3-9. Relationship between the tritium content of the shallow groundwater and the texture of the overlying layers.

Some uncertainties are also caused by the fact that the carbon content of water may be of nonorganic origin (dissolved from carbonate rocks, absorbed from gaseous CO_2 originating from the interior of earth, e.g., postvolcanic activity). These processes increase the total carbon content without changing the ^{14}C concentration, thus they lower the amount of ^{14}C related to the total amount, indicating older water in this way. The ^{13}C concentration can be used to derive a corrective factor for eliminating this error. The $\delta\ ^{13}C$ value of plants is -25 ± 3‰, water having dissolved CO_2 of biogenic origin is characterized by a value of -17 ± 3‰ and the limestones have in general a $\delta\ ^{13}C$ value of 0 ± 2‰. ^{13}C is a stable isotope, thus the ^{13}C ratio measured in the water can be used to estimate the amount of carbon having biogenic and chemical origin, respectively.

The age of water determined by investigating the amount of carbon isotopes can be used to characterize the water exchange between shallow and deep groundwater as it is demonstrated by Fig. 3-10. The layered geological structure of one part of the sedimentary basin investigated is shown in the profile indicating the age of water estimated from ^{14}C data and corrected by considering ^{13}C. It can be seen that older waters are stored at the higher levels of the Pleistocene formation than at its lower part, indicating a slow but continuous seepage upwards. The shallow groundwater is very young. The postglacial sandy layer has a ^{14}C concentration between those values observed in the recent shallow groundwater and in the Upper Pleistocene layer, respectively. The probable process creating this pattern is the recharge of the aquifer in question from both sides, at the upper border of the layer there is an infiltration through the shallow groundwater and the Pleistocene water percolating upwards recharges the sandy aquifer through its lower bed. Water balance can be expected only if horizontal groundwater flow discharges the sandy layer and the excess water reaching it vertically is transported into the river crossing the area. After determining the most probable flow directions within the system, the influencing factors can be numerically characterized by calculating the various elements of the water balance. The knowledge of the amount of environmental isotopes may also give some assistance for this quantitative analysis by estimating the balance of the isotopes in the system.

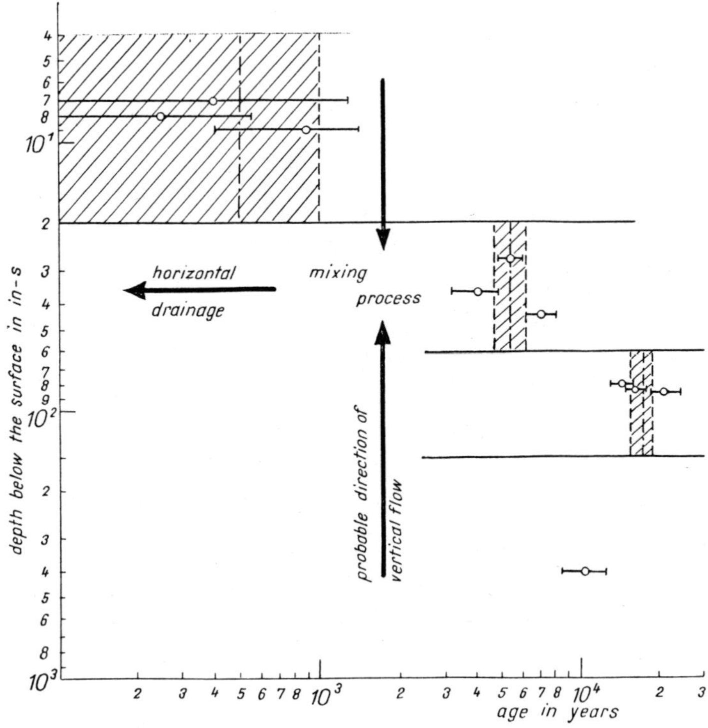

Fig. 3-10. Estimation of the system of groundwater flow on the basis of the investigation of carbon isotopes.

SECTION 10

OBSERVATIONS AND PROCESSING OF GROUNDWATER DATA

Similar to any other hydrological study, the observation of the water level provides the most important data for the investigation of shallow groundwater. It may even be stated that the other hydrological parameters characterizing the regime of the groundwater can hardly be observed regularly, which fact further raises the importance of the observation of the position of the water table.

In this case, however, wells must be constructed and used to record the water level instead of simple gauges. It is insufficient to observe data along a line, as it is done in the case of surface water, gauging the water level, e.g., along the bank of a river or along the shore of a lake, but observation wells should be situated all over the investigated area in order to determine and to map the position of the water table (or that of the piezometric surface in a confined system). The observations should be extended even in the third dimension to obtain the pressure conditions in each water horizon separately, when a multilayered structure is investigated.

Another basic difference between the observation of groundwater and surface waters is caused by the fact that the cross section of flow, e.g., a river bed, can be easily determined with sufficient accuracy in the latter case, while this geometrical parameter is not directly measurable in groundwater systems. For the determination of the boundaries of the groundwater flow, the geological structure of the pervious and impervious layers should be investigated within the formation. Several boreholes should be drilled, therefore, in the area and the samples removed should be carefully analyzed. A geological survey of the surface can also provide some assistance in this respect. Geophysical methods offer further information on the character of the layers penetrated by boreholes (geophysical logging). The knowledge concerning the position of some boundaries between various layers can even be extended, and the depth of some formations determined along measuring lines between boreholes by applying the surface methods of geophysics. The geological and geophysical surveys are, however, much more expensive than sounding, and the exact structure is only known at several points and along some lines, i.e., where there are boreholes or geophysical profiles. The probable geological structure of the entire system should be estimated from these data, thus the determination of the flow space can be regarded only as a rough approximation.

The inaccuracy of the determination of hydrological parameters is further increased by the fact that the homogeneity of layers must be assumed in almost every hydrological calculation: In contrary to this assumption, the properties of sediments (e.g., porosity, grain size, hydraulic conductivity, specific yield) vary from point to point and sometimes they even differ at the same point depending upon the direction of the investigated flow. It is therefore essential to examine as many samples as feasible from the different layers and to estimate the averages of the varying properties, namely their expected means and variances.

The velocity of flow in a riverbed can be measured at every point of a cross section with a current meter and the number of measurements required to determine the flow rate with sufficient accuracy can be decided upon. Current meters, however, cannot be used between the grains. Although there are methods to measure seepage velocity (e.g., placing a dye or some other tracer into a well and detecting the time of its arrival at another well by sampling the water at the second point or by using nuclear technique if radioactive tracer was applied) their accuracy is much smaller than that of the use of current meters due to numerous disturbing factors. Such processes influencing the measured value of velocity are, e.g., the diffusion and the adhesion of the material used as a tracer; the difference between the main direction of flow and the direction of the measuring line; the variation of velocity along a vertical line due to the nonhomogeneity of layers; etc.

There is a well-defined relationship between the velocity distribution and the shape of the bed in the case of surface waters. The discharge of a river can be calculated, therefore, from relatively few measurements. In the case of seepage, however, the distribution of velocity is influenced by the varying properties of the formation. Considering the large extension of the seepage space and the fact that the methods suitable to measure velocity are more expensive and less reliable, it is evident that the estimated flow rate of a groundwater system is less accurate than the discharge of rivers.

Sometimes the velocity of groundwater flow is calculated from the yields of a pumped well. It is necessary to consider that only some physical properties of the layers, e.g., hydraulic conductivity, specific yield, can be determined by this method, as it will be discussed in Subsection 11-3. The actual velocity of a natural system of groundwater flow cannot be characterized, however, in this way because a flow initiated and maintained artificially is introduced by pumping the well. The data observed in the vicinity of the well are primarily influenced by this effect and, therefore, they do not characterize the natural conditions, but can be regarded as the response of the system to human action.

Owing to the problems listed, the accurate observation of the water table and that of the piezometric surface is perhaps more important than the gauging of surface waters. All further data required to characterize the regime and the balance of groundwaters can only be theoretically deducted from water level records. This is the reason why the construction of observation wells, the collection of water level data, as well as the processing and evaluation of the latter will be discussed in greater detail in ths section.

10-1 Development of Observation Wells and Recording Water Levels

Practically all types of wells are suitable for the observation of groundwater because the water level in a well always develops at the elevation where the water table (characterized by zero excess pressure) lies around it in the unconfined layer drained by the well or at the height of the piezometric level showing the pressure head prevailing at the screen of the well penetrating into a confined aquifer (Fig. 3-11). Some discrepancy may occur due to the difference in the specific weight of water inside the well and that stored in the pores of the layer, respectively. This effect causes, however, considerable errors (surpassing the accuracy of the measurement of the depth of the water table) only if the depth of the well is more than several tens of meters. The influence of the temperature and compressibility of water and that of solid and gaseous media dissolved in the water can be neglected, when investigating the hydrology of shallow groundwaters. This is the reason why these problems will be discussed in the next chapter dealing with deep groundwaters.

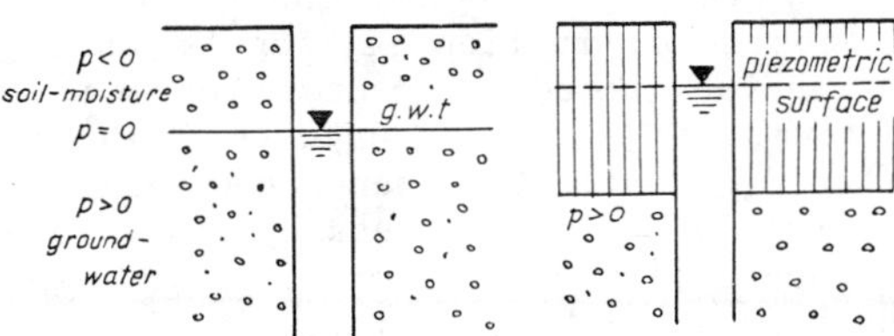

Fig. 3-11. Observation of the water table and the piezometric surface by wells.

Although any well can be used to observe the level of the water table or that of the piezometric surface, some restrictions should be considered when selecting one or more of the existing wells to use also for observation apart from their original goals, or when a new observation well is drilled and developed:

- the point selected for observation should be characteristic of a larger area;

- only one aquifer should be drained by the well;

- the use of the well (discharging water from the well, or recharging into it) may bias the readings;

- both too large and too small of a diameter may hinder the application of the well for observation;

- the well should be protected against silting.

The representativeness of observation wells may decrease their required number. In a large plain, the location of the wells should be selected by considering the geological structure, the depth of the water table, and even the type of the covering vegetation. This is the reason why only those existing wells, whose full geological description is available, are considered. This description includes the sequence of layers penetrated when the well was drilled or dug, together with their soilphysics parameters if possible; the depth, the thickness, and the texture (grain-size distribution) of the aquifer where the well was screened or left open; the pressure in the aquifer when it was opened and the change of the water level just after the development of the well; the structure applied for casing and screening the well; the yielding capacity of the well, the interrelated values of depression and yield and the result of pumping tests if any of such investigations were performed; records showing the use of the well since its development; any human activity in the vicinity of the well which might modify the regime of the shallow groundwater (pumping or recharging the groundwater by nearby wells, irrigation, construction of canals which may either drain or recharge the groundwater according to the position of their water level related to the water table). The same information should be collected when a new well is constructed solely for observation. It is necessary, therefore, to take a complete core when drilling an observation well and to perform all the possible soil-physics investigations to get the largest amount of information about the aquifer under observation and that of the overlying layers. It is also advisable to perform pumping tests in new wells whenever it is possible and to calculate the transmissibility (thickness multiplied by hydraulic conductivity) of the layers drained as well as to determine the yield vs. depression relationship for the wells in this way.

The influence of the surrounding vegetation is demonstrated here by showing it as an example. The well in question was drilled in a large meadow having nearly homogeneous vegetation. Near the well, however, there was a pine tree standing alone in the middle of the meadow. A regular daily fluctuation of a few cm was observed in the well although the nearby wells did not show similar behavior (Fig. 3-12). It was assumed that the evapotranspiration of the tree surpassed that of the surrounding grassland and created a local drawdown cone. The depression extends during daytime, when transpiration is higher, and it regresses during nights, and this fluctuation is recorded by the well. To prove the validity of the hypothesis, the canopy of the tree was completely covered and sealed by plastic sheets in this way avoiding the development of transpiration. The fluctuation of the water table was indeed stopped by this operation. Even the cloudiness of the sky (which decreases the evapotranspiration) reduces the fluctuation, as it is shown by the corresponding constant stretch of drawdown indicated in the figure.

All information collected in connection with the observation well should be arranged, evaluated, and preserved as a basic document of the well. The knowledge of the geological structure at the same time assists to meet the second requirement also, according to which no more than one aquifer should be drained by the well which is used for groundwater observation. Otherwise, when the well connects two or more aquifers, a movement may develop through the well, assuming that the piezometric heads are different in the interconnected layers. The flow transports water from the stratum having higher piezometric head to the other layers, where smaller pressure prevails. The water level measured in the well does not characterize the natural condition of any of the layers, but shows the pressure head developed at a given stretch (in the well) of the transport process (Fig. 3-13). A separate well should be drilled for each aquifer, if the observation of more than one water horizon is necessary at the same point. More precisely, a separate piezometric tube must be located for observing each water-bearing layer, but this tube may be placed into one borehole of larger diameter, if the reliable sealing of the aquifers can be solved by cementing the refilled part of the borehole at the elevations required.

It was also mentioned that a well is suitable for observation only if no water is removed from it or if it has been abandoned for a considerable period of time after having been used. When water is taken out of a well the water level is lowered. The two water levels, those within and around the well, will be equalized after awhile, following the withdrawal of water. The same process is initiated by recharging a well, but with an opposite sign. The length of time before the equalization is affected depends on the amount of water

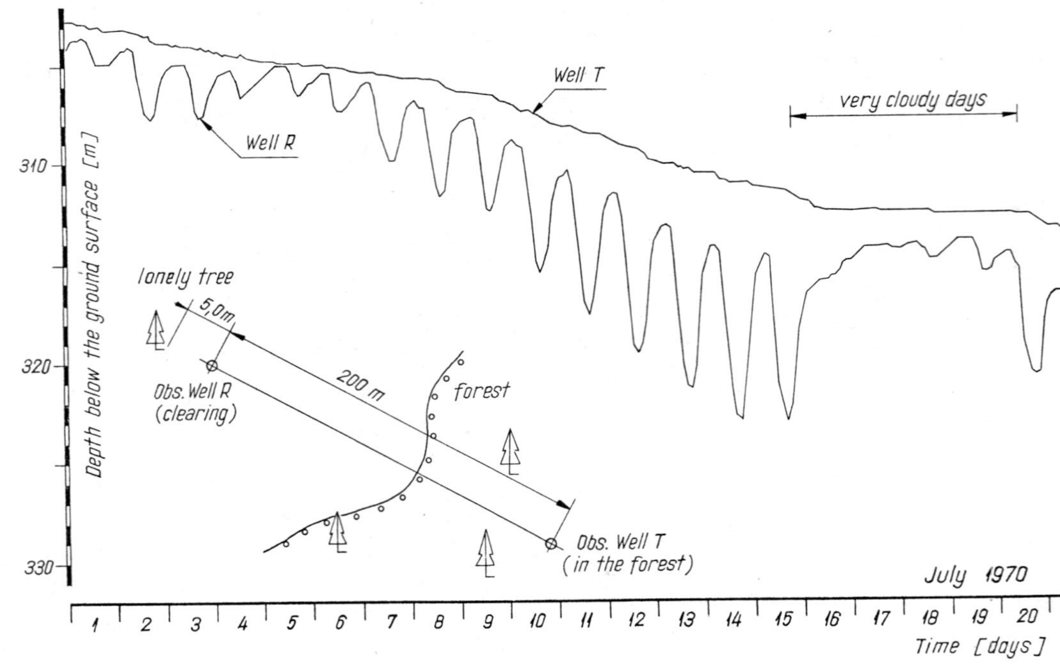

Fig. 3-12. Special behavior of groundwater drawdown influenced by the daily fluctuation of evaporation.

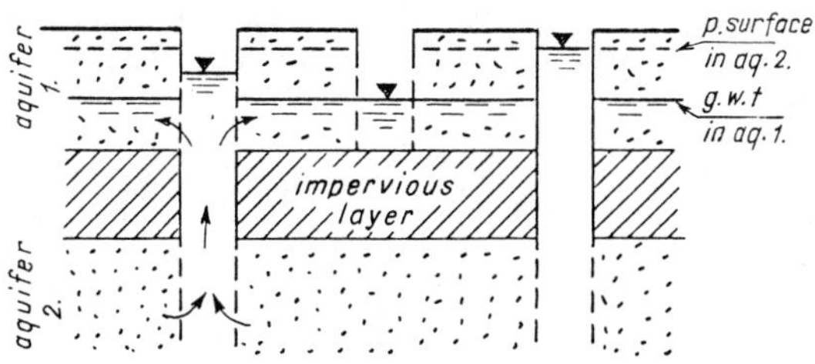

Fig. 3-13. Flow developing through observation wells interconnecting more than one aquifer.

removed from or recharged into the well, on the hydraulic conductivity and the specific yield (or storage coefficient) of the aquifer, and on the structure of the well.

The last aspect leads us to the next problem, i.e., the influence of the diameter of wells on observation. It is evident that the larger the diameter the longer the time necessary for the equalization of the internal and external water levels after having discharged or recharged water. At the same time, fluctuation occurs under natural conditions as well and the storage capacity of the internal space of the well may cause some time lag between the development of a given pressure in the layer and that of the corresponding water level in the well. The error caused by the non-simultaneous pressure and water level may be considerable only if the well is screened in a hardly permeable formation and the time lag can be reduced by selecting tubes of smaller diameter for casing. To avoid uncertainties, it is advisable to use as small a diameter as possible for observation wells, also considering some requirements determining the lower limit of the size of casing. The depth of the borehole influences the type of the equipment and the size of the tools used for drilling. The casing can be, however, smaller in diameter than the borehole, but the latter should be refilled and sealed carefully. Thus, the smallest internal size of casing is determined mainly by the tools which should be lowered into the well. The equipment used for gauging the water level should be mentioned at first. Its size can be minimized in both cases when the position of the water level is read manually or when it is recorded by automatic gauge. Water samples should be taken periodically from the well for its quality analysis and the well should be cleaned from time to time if silting occurs. The size of the tools used for these activities should be harmonized with the diameter of the well.

It was also mentioned that performing pumping tests in observation wells is advisable. The diameter of the suction tube or that of the driving pump may determine the smallest size of the tube applied for casing and screening of the observation wells.

Although the water movement through the screen is almost negligible, if the well is used only for observation the screening should be carefully prepared to ensure the reliability and the long life of the well, especially if pumping tests are also intended to be performed in the well. The degree of perforation may be lower than that of other wells because there is no interest to raise the yielding capacity of the well. The size of perforation, or that of the covering sieves, should protect the well against silting. It is also dangerous to select too small a diameter of sieve, especially if internal erosion may occur in the layer because the small grains may seal the well. The diameter of perforations of sieves has to be designed, therefore, according to the guidelines valid for normal wells (see Part IV). After finishing the construction of the well, and especially after executing pumping tests, it has to be cleaned carefully. The cleaning should be repeated periodically to remove the silt which entered the well since the last cleaning. To avoid any difficulty which may be caused by the silt accumulated in the well, a storage space should be formed continuing the casing below the screen. The bottom of the well should be sealed to hinder the development of connection between the well and other groundwater horizons. Since only some of the observation wells have to be pumped (it will be explained later that some grouping of wells is advisable, and it is sufficient if pumping tests are performed only in one well from each group), two standard designs of wells are represented in Fig. 3-14, one for pumped wells and one for those serving only for water level observation, which may be applied for the development of shallow tube wells for groundwater observation.

The most important part of data collected by the observation wells is composed of the water level readings. Information concerning groundwater flow and the accretion of the shallow groundwater can be derived from these data. Apart from the water level, some water quality parameters can be and have to be determined, for which purpose samples should be taken from the well.

The depth of the water level in the well can be measured manually, or automatic gauges can be installed to continuously record the position of the water table. In the first case, a measuring device is lowered into the well hanging on a chain or string. The measuring head gives a signal when it reaches the water, and the depth is read on the string. The more sensitive the head to signal the water, the more accurate the

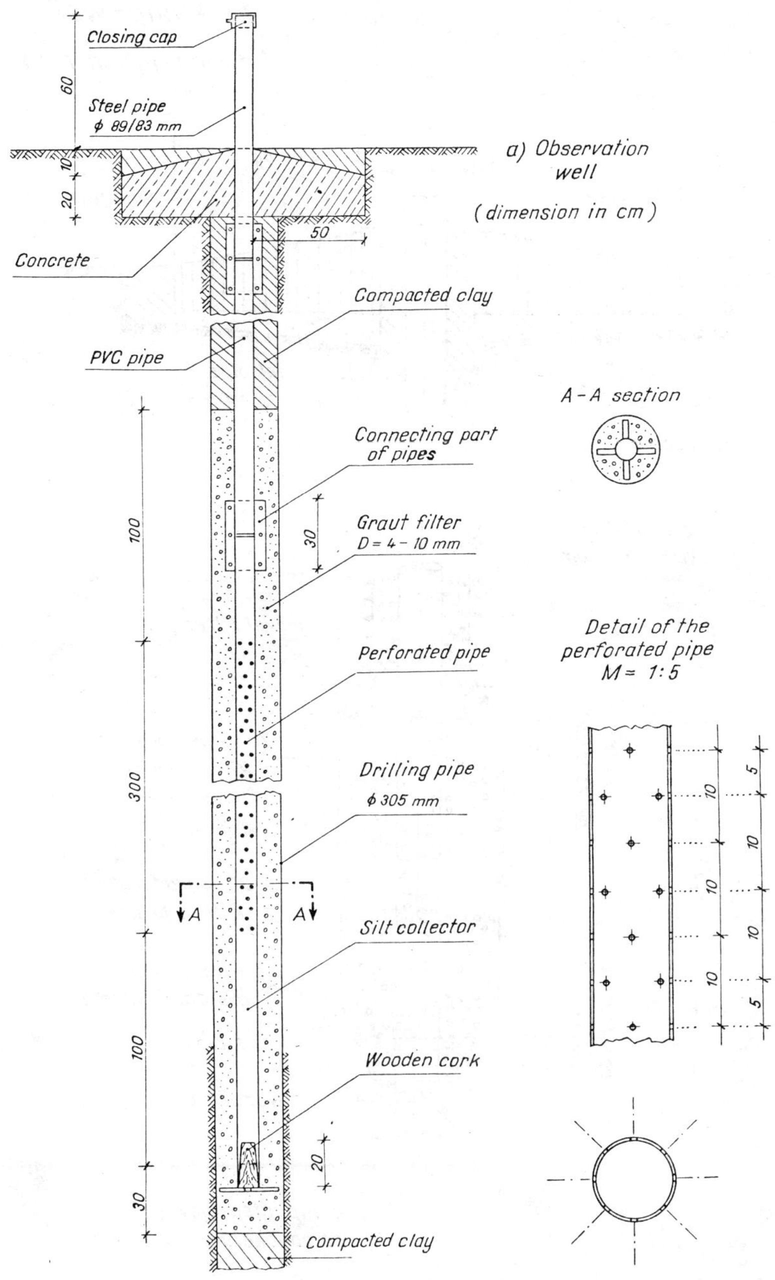

Fig. 3-14a. Standard designs of observation wells.

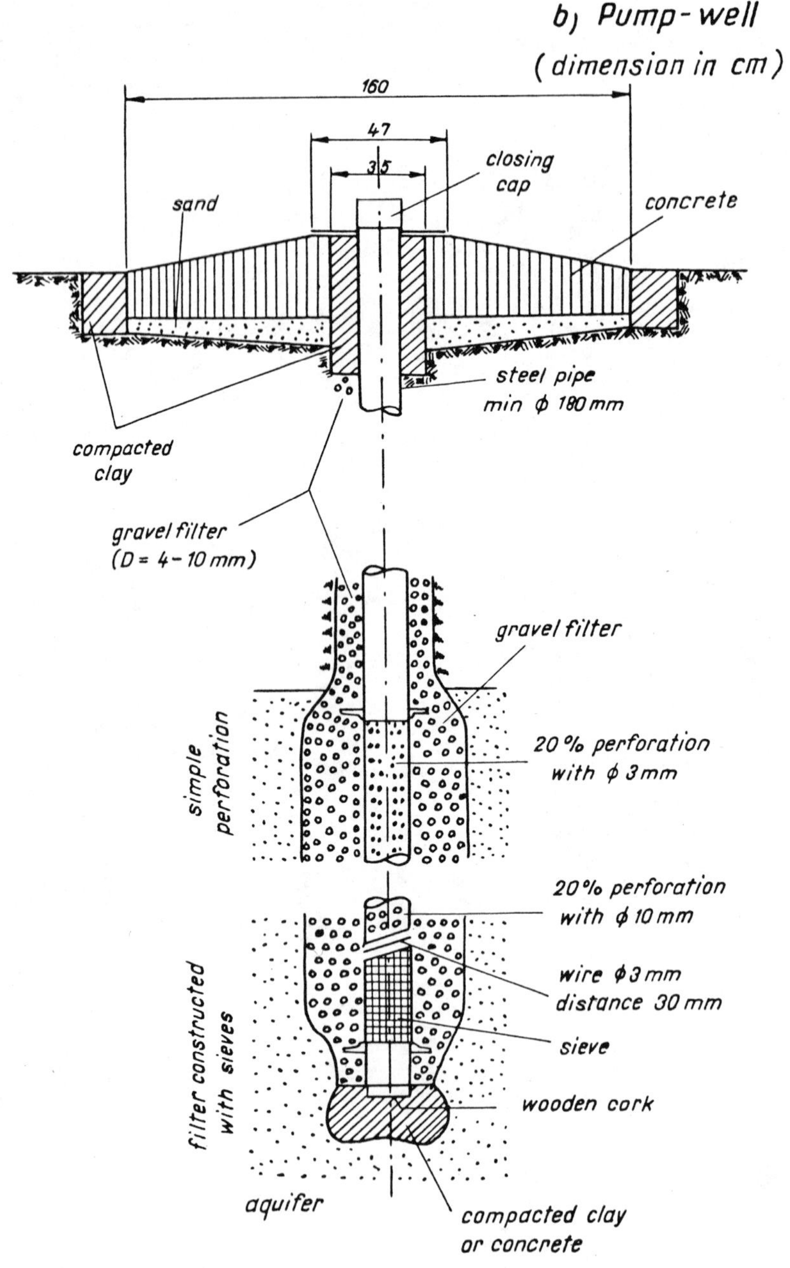

Fig. 3-14b. Standard designs of observation wells.

reading of the depth. The automatic gauge may operate mechanically (a small floating buoy follows the position of the water level, turning a drum on which the string of the floater is wound up) or electrically (separate resistances are built in a vertical column and the water creates connection between the units already submerged, or a floating core moves in an electromagnetic field and gives signals according to its position). The position of the floating buoy can be recorded locally by drawing a line on a slowly turning paper, governed by a timing clock in the case of using mechanical device, but the turn of the drum holding the string of the floater can be transformed also into electrical signals and the position of the water table can be telemetered or tele-recorded in this way. The signals are directly suitable for telemetering if electrical equipment is used to indicate the depth of the water table.

Since the water table does not fluctuate as rapidly as the level of surface waters, it is not necessary to carry-out daily readings of the wells. Weekly observations, or readings taken once in a three day period are usually quite sufficient (groundwater in coarse grained sediments near river banks can be mentioned as exceptions because the water table follows the fluctuation of the river in this case). Due to the same reason, the recording gauges can operate with very slowly turning time clocks or the signals of telemetered systems can be taken and transmitted in the same intervals as the readings in manually observed wells.

The change of the quality composition of the groundwater is even slower than the fluctuation of its water table. The sampling of water in each month is almost quite sufficient in every case and many times larger intervals can be also accepted. It is necessary, however, to consider that the processes influencing the regime of the shallow groundwater have seaonal fluctuation and the change of the chemical composition of the water follows the same rhythm. The minimum number of the determinations of water quality is, therefore, four samplings per year in three month intervals.

The depth of the water table is generally measured by relating it to a reference level indicated on the casing of the well (in most cases this level is the height of the rim of the casing). When the groundwater data are evaluated, either the depth of the water table below the surface is analyzed or the instantaneous position of the water table is characterized in maps using the absolute height of the observed level related to a common horizontal datum, e.g., height above sea level. It is this reason why the groundwater data can be given in three different forms:

- directly observed readings measured by relating the position of the water table to an arbitrarily chosen local reference level;

- the depth of the water table below the surface;

- the height of the water table above sea level (or any common horizontal datum).

To transform the data from one system into another, there is no difficulty if the height of the local reference level, e.g., the rim of the casing, is determined above the land surface being characteristic for the surroundings of the well and above the sea level, respectively. These two parameters, which are used as additive factors when transforming data from one system into another, are, therefore, important constants of the observation wells which should be determined by leveling immediately after having finished the construction of the well. It is also necessary to indicate, when publishing data, which system is applied, i.e., the direct readings or the position of the water table related to the land surface and to a common datum, respectively.

10-2 Primary Evaluation of Water Level Data

There are several methods for the primary evaluation of the water level data collected by using observation wells. The purpose of the analysis is always to draw-up the possible largest amount of information concerning the water balance of shallow groundwaters. The methods most frequently applied can be roughly divided into three main groups:

- time series data of one observation well are analyzed separately;

- the information originating from a given number of wells (e.g., wells being located within a groundwater basin or along a section defined by the main direction of groundwater flow; a specially arranged group of wells) is investigated jointly;

- the relationship between sets of data observed simultaneously at different points is investigated or similar connection is to be determined between data characterizing the position of the water table and some other hydrological parameters.

The simplest way to process data collected at one observation well is to plot them against time, i.e., the construction of groundwater hydrographs. Each reading, either daily or weekly, can be considered separately, contacting the points representing the observations with a continuous line. In some cases, if the fluctuation of the water table is very slow, the construction of a step-like graph from the monthly averages may be sufficient (Fig. 3-15). The rising stretches of the graph indicate the periods when the recharge of the groundwater exceeds its drainage (period of accumulation). When the losses caused by evapotranspiration and outflow are greater than the recharge, the quantity of the groundwater stored in the layer decreases and consequently the water table is lowered (period of depletion).

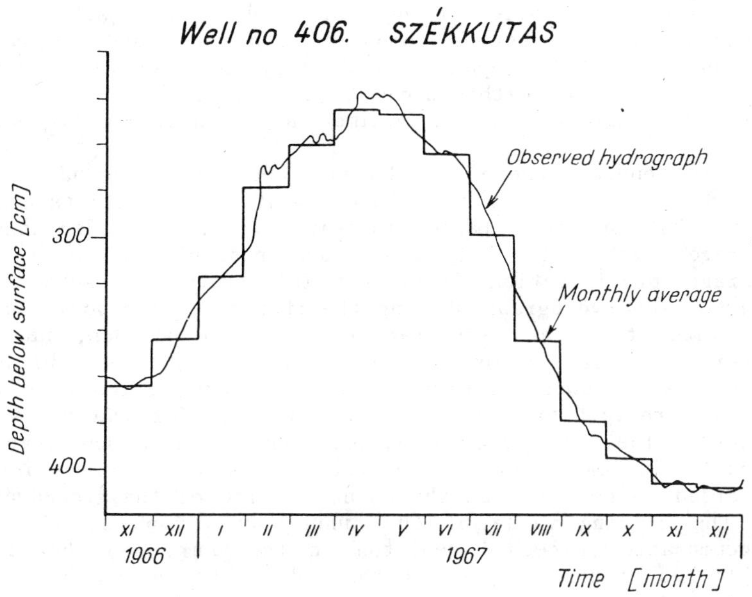

Fig. 3-15. Groundwater hydrograph.

It is necessary to note here that any of the three types of data (direct readings, the absolute values related to a common reference level, e.g., height above sea level, and the depth below the land surface) can be used to construct a hydrograph if one observation well is analyzed separately because the form of the graph is not changed by transferring data from one system to another, only the numerical scale along the vertical axis is modified as it is demonstrated in Fig. 3-15.

The water table has regular periodic fluctuations if the groundwater is influenced primarily by hydrometeorological processes, i.e., infiltration from precipitation, capillary discharge to replenish soil moisture evaporated or transpired. In these cases, the periods of accumulation and depletion follow each other in a regular order according to the sequence of dry and wet seasons as it is shown in Fig. 3-15.

Late autumn, winter, and early spring (low evaporation because of the low temperature, infiltration from precipitation and snow melting) is the period of accumulation in the temperate climatic zone, and the water stored during these times is depleted during the summer half-year (which includes late spring and early autumn as well) when evapotranspiration is high relative to precipitation. Thus, the length of the period of fluctuation is one year. Similar wavelength characterizes the fluctuation of the hydrograph in those subtropical areas where the year is composed of one dry and one rainy season. Near the equator, two rainy and two dry periods follow each other within a year. The hydrograph has, therefore, two maxima and two minima in one hydrological year.

In general, there is a balance between the recharge and the drainage of groundwater under natural conditions which means that the water table fluctuates around the multi-annual average (Fig. 3-16). It is of course possible that more than average precipitation falls and infiltrates in subsequent years. The hydrograph showing the changes of the position of the water table has a rising trend in such periods, but the water table will be lowered again after the wet years because they are followed in most cases by a dry period when the annual precipitation is below the average for the years of record. The wet, dry, or balanced character of the year is clearly indicated by the hydrograph as well. A year is regarded to be wet from the point of view of the groundwater if the closing value of the annual hydrograph is positive (accumulation exceeded depleton in the year, and the water table has higher position at the end of the period than that at its beginning). Similarly, the lowering trend of the annual change of the water table indicates the dryness of the year. When the rise of the water table during the period of accumulation is nearly equal to its lowering due to depletion in the same hydrological year, infiltration and drainage caused by capillarity and evapotranspiration balance each other.

When the trend of a groundwater hydrograph covering a long period is horizontal (as it is in Fig. 3-16), the balance of the groundwater is clearly shown. When the natural equilibrium is affected by human activities (excess infiltration from irrigation; continuous depletion by pumping; construction of a canal which either recharges or drains the groundwater depending on the position of its water level relative to water table, etc.), a sharp change can be observed on the hydrograph

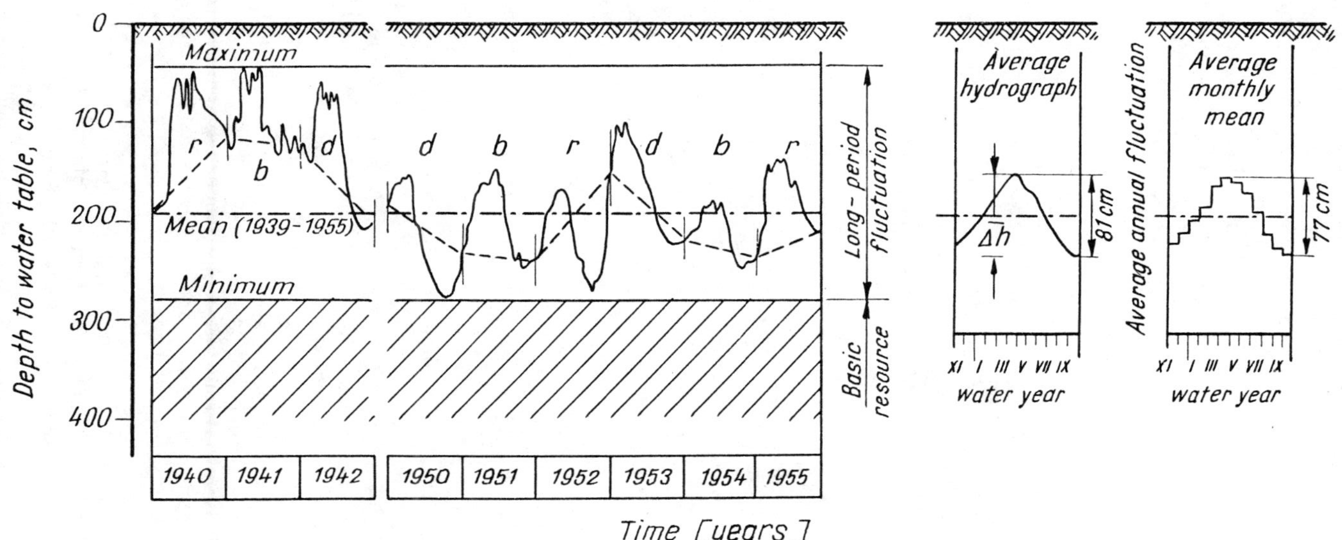

Fig. 3-16. Seasonal and long-term fluctuation of groundwater.

(Fig. 3-17). The multi-annual average and the long-term trend of the graph may also be considerably modified by artificial effects. To demonsrate the changes caused by irrigation, the hydrographs of three observation wells are compared in Fig. 3-18. One well is located outside the irrigation area, one at the edge of it, and the third at the outlet of the irrigated groundwater basin where the influence of the action was the strongest. It can be clearly seen that the excess infiltration due to irrigation not only raised the water table but shortened the length of the fluctuation of the hydrograph, creating a local maximum in summer which is the period of depletion under natural conditions.

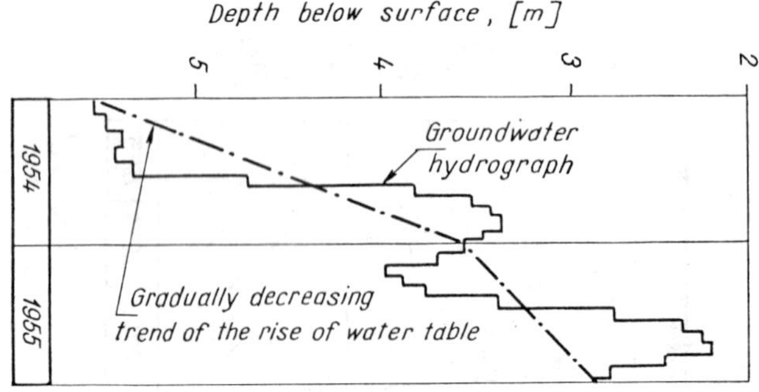

Fig. 3-17. Rapid rise of the water table influenced by irrigation.

Investigating a balance condition of groundwater which is indicated by the horizontal trend of the hydrograph, the multi-annual average of the water table can be calculated separately for each month or even the average of all readings made at the same calendar date of subsequent years can be determined. The hydrograph formed from these values is a closed one, i.e., the line at the end of the hydrological year is at the same level as it was at its beginning (see Fig. 3-16). The fluctuation shown by these graphs indicates the regular alterations of wet and dry seasons within a year and their effects on groundwater.

The method most commonly applied to joint analysis of the water level data collected by several observation wells is the construction of groundwater maps of a basin or at least for a part of the groundwater catchment. Since the data of various wells have to be homogeneous, the direct readings, which

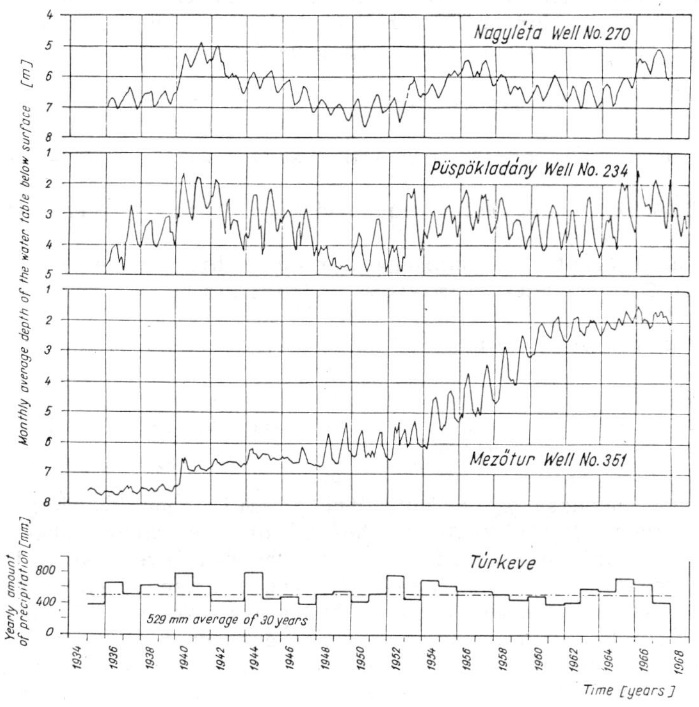

Fig. 3-18. Comparison of groundwater hydrographs to show the influence of human activity.

relate the depth of the water table to a reference level fixed on the casing of the well in question, cannot be used for mapping. When the isolines are constructed from data related to the land surface, they are interconnecting the points where the water table is at a given depth below surface. The maps, using the readings in a form which indicates heights above sea level, show the contour lines of the water table or that of the piezometric surface. The first type of map can be used to select the homogeneous areas within the basin (the depth of the water table being one of the most decisive factors influencing the development of the balance of the shallow groundwater) while the direction of the groundwater flow and the slope of the water table or the piezometric surface (the gradient maintaining seepage) can be determined on the basis of contour maps, thus they provide an important information for calculating the groundwater flow.

The variety of maps is further increased by the fact that they can be constructed either from data measured simultaneously at various wells or by considering characteristic values at each well determined statistically from the observations recorded during a given period. The parameters composing the second group of data usually are:

- the multi-annual average;

- the maximum or the minimum observed within the period;

- high or low position expected with a given probability;

- the largest fluctuation observed within the period (the difference between maximum and minimum);

- the variance of the statistically analyzed data series;

- the difference between the multi-annual average and the mean value of a shorter time interval (e.g., the mean in a given month).

It is necessary to note that simultaneously observed actual values of the absolute height of the water table or piezometric surface related, e.g., to the sea level, should be used to construct contour maps when the purpose of the investigation is the determination of hydraulic parameters (direction, gradient) for the calculation of groundwater flow. Data not directly observed but calculated statistically (average, variance, expected low or high position) or those showing extreme situations but not necessarily measured at the same time (maximum, minimum, largest fluctuation) characterize only the average conditions, but the actual hydraulic gradient required for calculating the flux of groundwater flow at a given point in time can only be determined from the difference of the absolute heights of the water table (or piezometric surface) measured simultaneously at two neighboring observation wells. On the contrary, the maps showing the depth of groundwater below the surface are generally used to characterize the average conditions of the groundwater regime. It is this reason why simultaneously observed data are hardly applied, but the average or extreme values are preferred when preparing such maps.

Two examples are shown to demonstrate the construction and the application of groundwater maps.

In the first case (Fig. 3-19), two contour maps are compared. Both were constructed from the height of water table (or piezometric surface) above sea level observed at a given day (the date of the first map is October 20, 1938, and that of the second one is October 14, 1904). The two dates

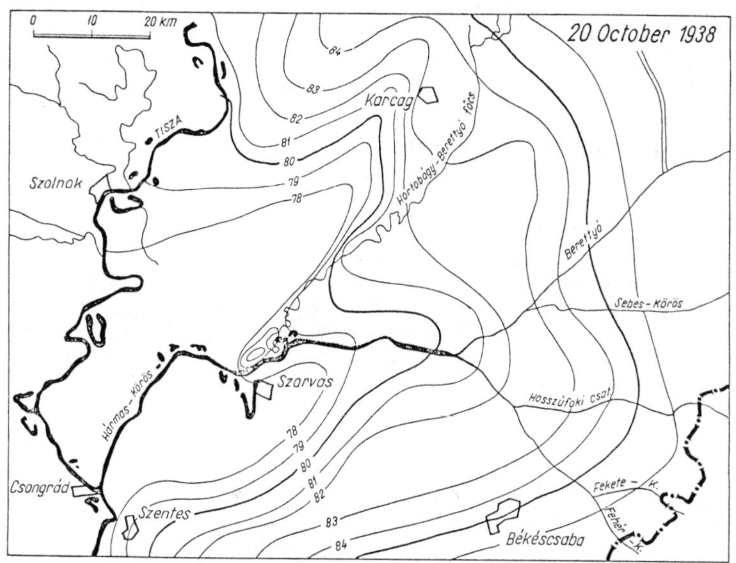

Fig. 3-19a. Regional influence of irrigation characterized by the contour maps of the water table.

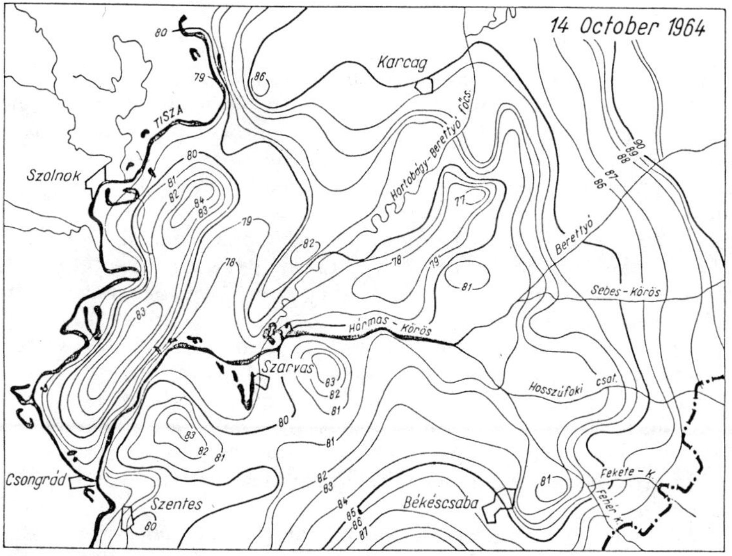

Fig. 3-19b. Regional influence of irrigation characterized by the contour maps of the water table.

were selected by considering that they should approximately characterize the similar climatic conditions, but one should be before the start of the development of irrigation within the area and the second in a year when a considerable period has already elapsed after the complete operation of this activity. The comparison shows the rise of the water table, the change of the direction of flow, and the increase of the recharge of rivers from groundwater due to irrigation. Even the modification of the flux of groundwater flow can be estimated by considering the changes in hydraulic gradients.

The second map shows the multi-annual average of the depth of water table within the same area (Fig. 3-20). The differences in the depth are caused only partly by the different elevations of the terrain. Since the area is completely flat, the absolute heights of the land surface at the wells do not differ considerably. On the other hand, the greater depths of the water table also indicate the differences occurring in the behavior of the covering layer (it was already explained and proved in Part II that the capillary rise is extremely high in loess soils, thus under these terrains a greater depth of the water table is expected). The map is supplemented by isolines showing the position of the monthly mean of data observed in December, 1977 and related to the multi-annual average. The lines demonstrate the considerable rise of the groundwater under irrigated areas and the random scattering around the average where dry farming was further practiced.

There are areas where the local conditions basically determine the direction of groundwater flow, e.g., within the direct zone of influence of the river, the main component of flow is perpendicular to the bank in river plains. In such cases, the representation of the groundwater data in a section take in the direction of flow through the observation wells visualize the water regime better than the construction of maps. In Fig. 3-21, a section is shown running perpendicularly to the Danube River. The first part of the figure shows the hydrographs of the river and three wells, respectively. In the section constructed from the hydrographs the propagation of a flood wave along the river can be followed which has initiated the development of a wave also in the groundwater. The change of both the size and the direction of the gradient can be calculated, and thus the modification can be estimated which occurred during the flood wave in the contribution of the shallow groundwater to the base flow of the river.

There is a special type of the joint analysis of data measured simultaneously at more observation wells when the purpose of the investigation is the determination of the positive and negative accretions of groundwater. The explanation of the method requires, however, more detailed study

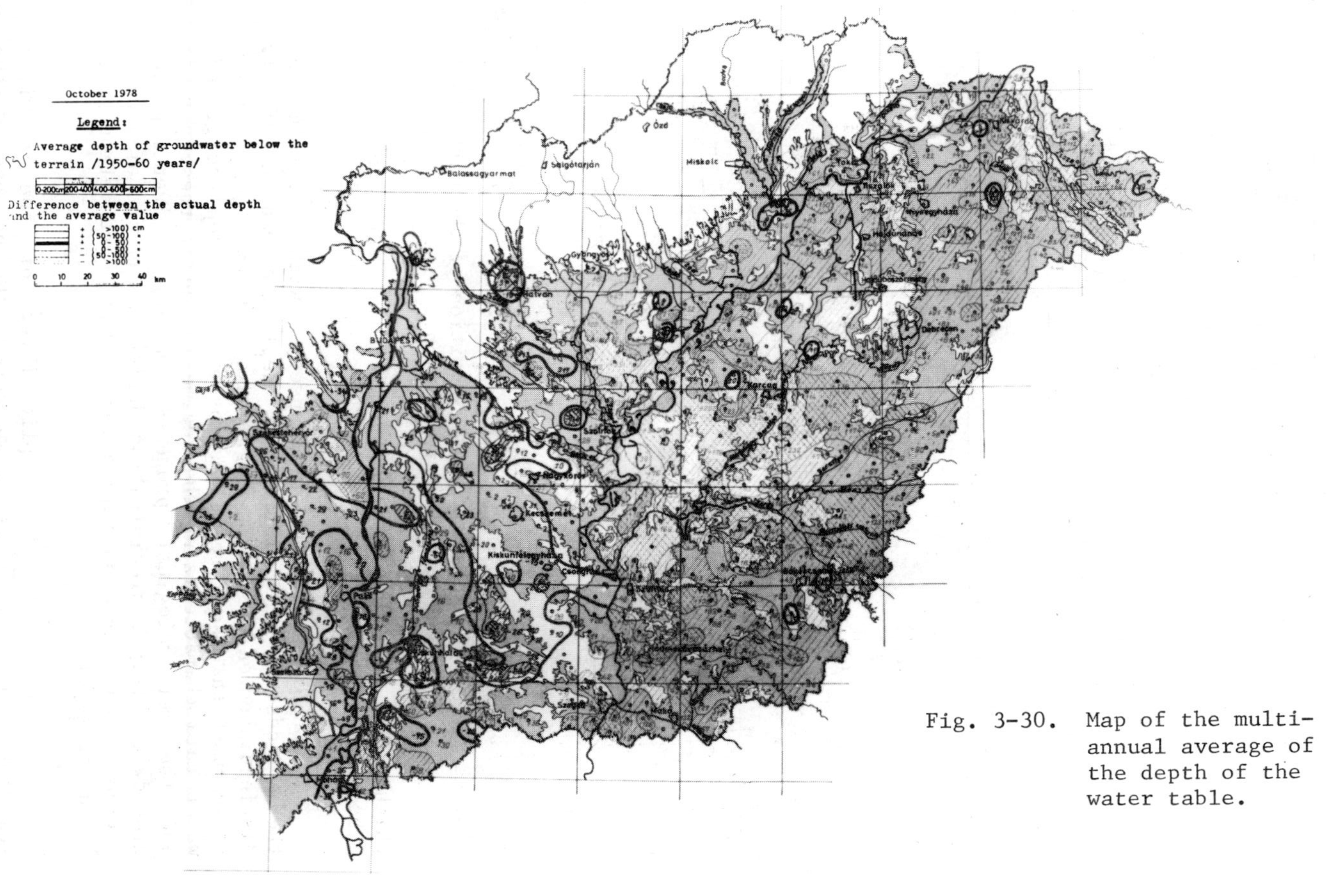

Fig. 3-30. Map of the multi-annual average of the depth of the water table.

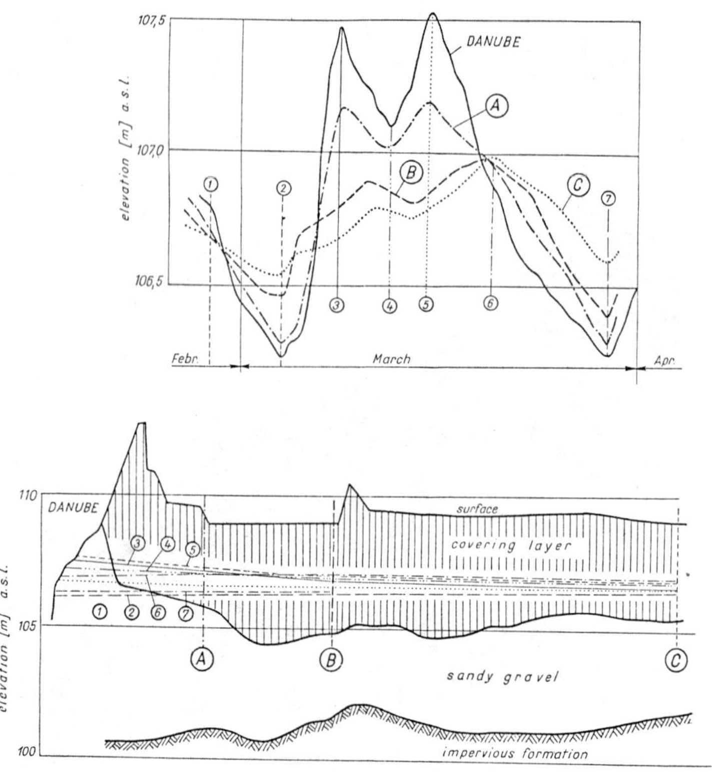

Fig. 3-21. Representation of the position of the water table in a characteristic section.

which will be dealt with, therefore, separately (see Subsection 10-3).

It was already shown in connection with Fig. 3-16 that the periods of accumulation and depletion can be clearly distinguished on the basis of groundwater hydrographs. It may be assumed that some relationships exist between the groundwater data characterizing the regime in one of these periods and the meteorological processes mainly influencing the groundwater (e.g., precipitation during the period of accumulation and evapotranspiration when depletion is characteristic).

To demonstrate the applicability of the previous statement, an example is shown here. The purpose of the investigation was to determine the relationship between the rise of the water table during the period of accumulation and the amount of precipitation received at the same time. The precipitation of the winter half-year was reduced by the probable evaporation assumed to be equal to the potential value and calculated as a function of monthly temperature (this hypothesis seems to be acceptable since potential evaporation is relatively low in winter and there is always water on or near the surface to provide the amount to be consumed by evaporation). The difference calculated in this way was measured on the vertical axis of a coordinate system, while the rise of the water table during the period of accumulation in question was plotted along the horizontal axis (Fig. 3-22). Each year is characterized by one point in this way and it is assumed that there is a linear relationship between the variables. The intersection of the regression line with the vertical axis indicates whether surface runoff reduced the infiltrating amount (if the intersection is positive as in the case of Well No. 3864) or some water reaching the surroundings of the well and coming from higher parts of the catchment increased the infiltration around the well (negative intersection of Well No. 3240).

The amplitude of the fluctuation was further investigated and it was found that it was reduced with the increase of the depth of the water table beneath the surface (Fig. 3-23). It was shown by this correlation that the hydrometeorological processes occurring on and above the land surface, i.e., precipitation, evaporation, transpiration, cannot influence the regime of the groundwater if its water table is overlain by a very thick soil-moisture zone (it is evident that the limiting value of this depth depends on the material of the covering layer; it is the largest in loess and decreases if either finer or coarser texture characterizes the upper stratum).

Where there is a river in the vicinity, the water table is influenced by the fluctuation of the river stage and there may be no correlation between the change of the position of the water table and the seasonal variation of climatic conditions. The extent of the belt influenced by the river can be determined on the basis of hydrographs, if a series of observation wells are located along a line perpendicular to the bank of the river (Fig. 3-24). Comparing the water level of the river to that of a well near the bank, the periods when the river recharges the groundwater and when the latter is drained by the river can be detected. In many cases some numerical parameters characterizing the fluctuation can be related to the distance between the river and the well (e.g., in the figure, the time lag between the development of peaks and the change of the level during a given period) which result may provide further information on the connection between surface and groundwaters.

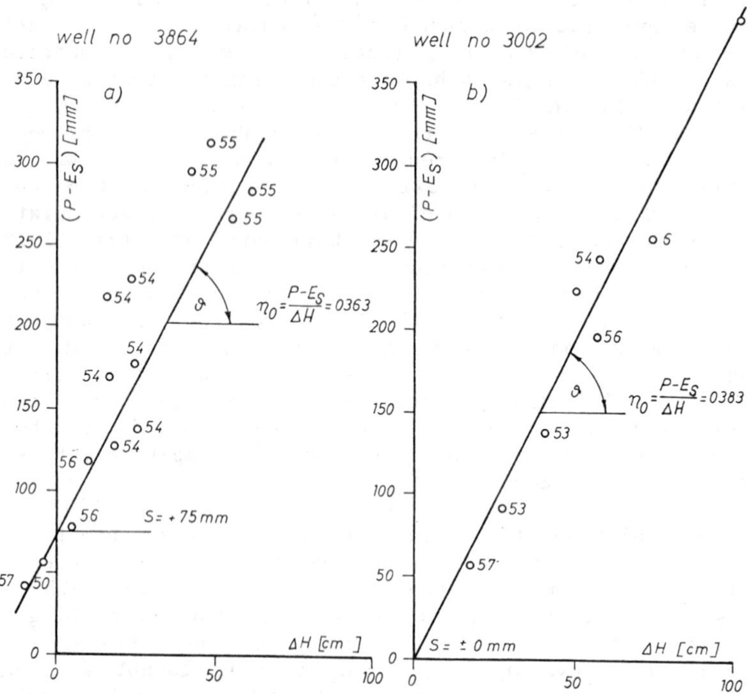

Fig. 3-22a,b. Relationship between the rise of the water table and precipitation in winter.

More examples to show the various application of the correlations existing between the parameters of the shallow groundwater and other hydrological elements will be given in Subsection 10-4 where the use of regression analysis for the exploration of the various connections will be discussed.

10-3 Determination of Water Exchange between Soil Moisture and Groundwater by Using Observation Well Data

An important component of the water balance developing in the groundwater zone is accretion which measures the water exchange through the water table. Its sign is positive if the flow is directed downwards and the drainage of groundwater by capillarity to recharge soil moisture is called negative accretion. The numerical description of this upper boundary condition of the groundwater zone can be solved on the basis of water level data recorded in observation wells. The application of the method requires the investigation of horizontal

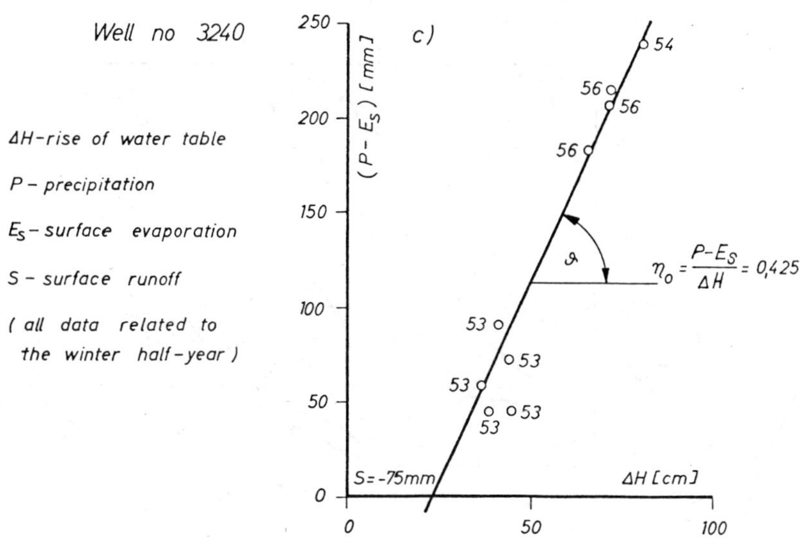

Fig. 3-22c. Relationship between the rise of the water table and precipitation in winter.

groundwater flow. To simplify the hydraulic analysis of seepage, two hypotheses are accepted as starting conditions:

- the seepage velocity is linearly proportional to the hydraulic gradient (Darcy's law);
- the flow is horizontal and, therefore, the velocity is constant along a vertical section, thus the numerical value of velocity can be calculated as the function of the slope of the water table (Dupuit's hypothesis).

Following Kamensky's (1940, 1943) investigations, let us consider the water balance of an elementary part of the gravitational groundwater space, the lower boundary of which is a horizontal impervious layer, and which is bordered in the horizontal plane by a rectangle (Fig. 3-25). There are five observation wells for recording the position of the water table, one inside the oblong, and four outside its border. The wells form a rectangular cross, and the borders are fixed so that they divide the distances between the internal and external wells into halves. Let $b_1, \ldots, b_4$ indicate the length of the different side of the investigated column

$$b_1 = b_3 = \frac{1}{2} (\Delta x_2 + \Delta x_4) \; ; \quad b_2 = b_4 = \frac{1}{2} (\Delta x_1 + \Delta x_3) \quad ,$$

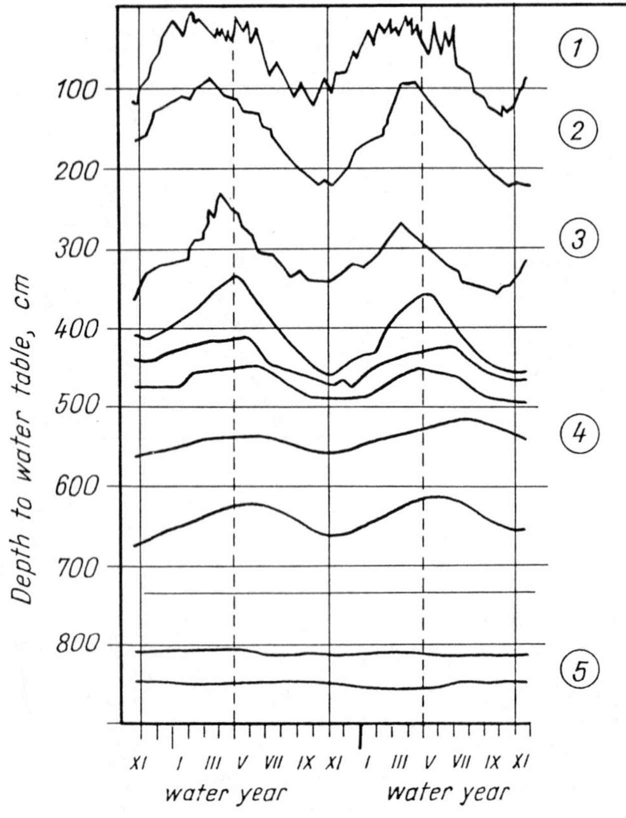

(1) Groundwater recharged from rainfall

(2) G.w. influenced by surplus evapotranspiration

(3) G.w. influenced by infiltration and evapotranspiration

(4) G.w. influenced only by infiltration

(5) Undisturbed groundwater

Fig. 3-23. The amplitude of groundwater fluctuation as a function of the average depth of the water table.

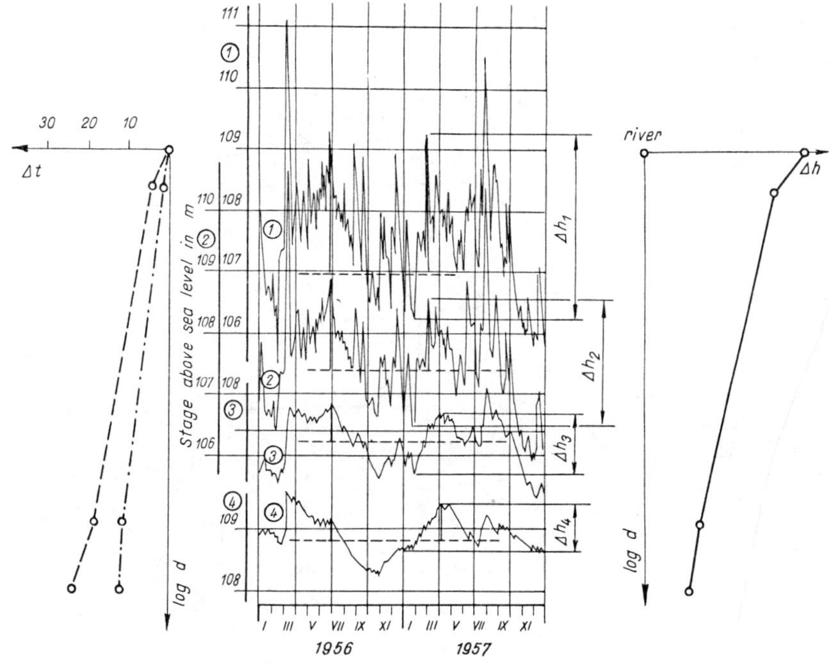

Fig. 3-24. Groundwater hydrographs influenced by a river.
(1) Hydrograph of the Danube River;
(2) Groundwater influenced immediately by the river observed at a distance of 70 m;
(3) Indirect influence of the river observed at a distance of 550 m;
(4) Groundwater table fluctuation independent of the river observed at a distance of 1150 m.

the time dependent area of the vertical cross sections at the borders are

$$A_1 = b_1 \frac{H_o(t) + H_1(t)}{2} \; ; \quad A_2 = b_2 \frac{H_o(t) + H_2(t)}{2} \; ;$$
$$A_3 = b_3 \frac{H_o(t) + H_3(t)}{2} \; ; \quad A_4 = b_4 \frac{H_o(t) + H_4(t)}{2} \; . \tag{3-4}$$

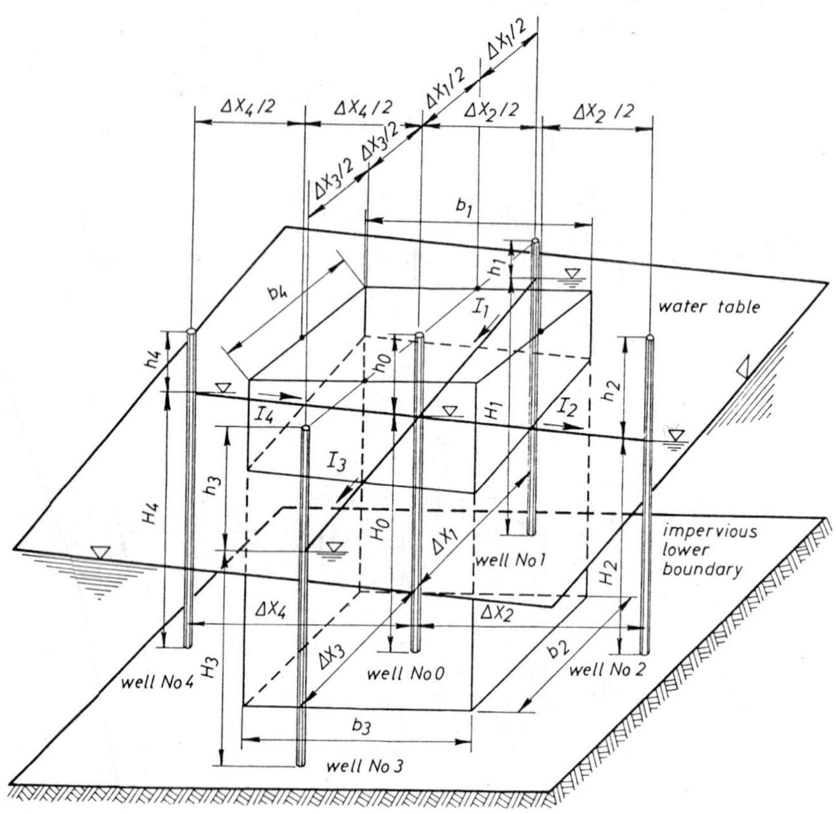

Fig. 3-25. Sketch showing the group of observation wells suitable to determine the positive and negative accretion of groundwater.

where $H_i(t)$ indicates the elevation of the water level in the i-th well above the impervious lower boundary at time t. The average slope of the water table can also be calculated from the data of the observation wells:

$$I_1 = \frac{\Delta H_{1x}(t)}{\Delta x_1} \; ; \quad I_2 = \frac{\Delta H_{2x}(t)}{\Delta x_2} \; ; \quad I_3 = \frac{\Delta H_{3x}(t)}{\Delta x_3} \; ;$$

$$I_4 = \frac{\Delta H_{4x}(t)}{\Delta x_4} \; ; \tag{3-5}$$

where $\Delta H_{ix}(t)$ is the difference of the water levels registered at time t in the i-th well and the center well, respectively $[\Delta H_{1x}(t) = H_1(t) - H_o(t) \ldots \Delta H_{4x}(t) = H_4(t) - H_o(t)]$.

Accepting the two basic hypotheses mentioned previously, the flow rates crossing the vertical boundaries of the investigated prism can be calculated

$$Q_i(t) = A_i K I_i = b_i \frac{H_o(t) + H_i(t)}{2} K \frac{H_i(t) - H_o(t)}{\Delta x_1} . \tag{3-6}$$

There is, however, a fifth bordering surface of the prism as well through which flow can develop, i.e., the water table. Its area is $A = b_1 b_2$, and the flow rate here can be expressed as the product of the area and the specific vertical water exchange $[\varepsilon(t)]$:

$$Q_o = A\varepsilon(t) . \tag{3-7}$$

Investigating a finite time period ($\Delta t = t_3 - t_1$; and $t_3 > t_1$), the water balance of the prism requires that the product of the elementary time unit (dt) and the sum of flow rates summarized for the entire time period Δt should be equal to the change of the amount of stored water. The latter is the change of the position of the water table (characterized with the recorded water level in the central well $\Delta H_{ot} = H_o(t_3) - H_o(t_1)$) multiplied by the horizontal area of the prism and by the specific yield of the layer:

$$\sum_{t=t_1}^{t_3} [Q_o(t) + \sum_{i=1}^{4} Q_i(t)] \, dt = A \, H_{ot} \, n_s . \tag{3-8}$$

Substituting the time dependent variables with their averages, which can be regarded as the value of the function in question belonging to a time point t_2 within the investigated period ($t_1 < t_2 < t_3$), and applying the generally used concept of transmissivity (which parameter is equal to the product of the depth of flow and hydraulic conductivity $T_i(t) = K (H_o(t) + H_i(t)/2)$, the simplified form of Eq. 3-8 can be given as follows:

$$\varepsilon(t_2) + \frac{1}{b_2}\left\{\frac{T_1(t_2)}{\Delta x_1}\Delta H_{1x}(t_2) + \frac{T_3(t_2)}{\Delta x_3}\Delta H_{3x}(t_2)\right\}$$

$$+ \frac{1}{b_1}\left\{\frac{T_2(t_2)}{\Delta x_2}\Delta H_{2x}(t_2) + \frac{T_4(t_2)}{\Delta x_4}\Delta H_{4x}(t_2)\right\} = \frac{\Delta H_{ot}}{\Delta t} n_s \quad . \tag{3-9}$$

The geometrical parameters (b_i and Δx_i), and those calculated from water level records (ΔH_{ix} and ΔH_{ot}) can be determined with relatively high accuracy. The methods applicable for the calculation of specific yield were already discussed. Thus, transmissivity is the remaining variable, the knowledge of which is necessary if Eq. 3-9 is intended to be used for the direct calculation of the time dependent water exchange through the water table. The determination of this parameter is further simplified if the position of the impervious lower boundary is known because only the hydraulic conductivity has to be estimated as a basic data. The equation itself can be used for the calculation of transmissivity by substituting recorded water level data of a time period when the vertical flow is surely negligible [$\varepsilon(t) = 0$] (i.e., in winter, when the frozen top layer hinders both infiltration and evaporation). It is advisable, however, to execute a pumping test (pumping the central well and using the others as observation wells) and determine the transmissivity of the layer from the data collected in this way.

When the cross of the observation wells is fixed so that one line of wells (1,0,3) is parallel to the dip of the water table, and the other (2,0,4) (being perpendicular to the former) is parallel to its strike, there is no flow in the second direction. It is not necessary, therefore, that the data of wells with even numbers (2,4) should be taken into account, and the investigation can be restricted to the analysis of the water balance within a strip of unit width:

$$\varepsilon(t_2) = \frac{\Delta H_{ot}}{\Delta t} n_s - \frac{T_1(t_2)}{\Delta x_1}\Delta H_{1x}(t_2) - \frac{T_3(t_2)}{\Delta x_3}\Delta H_{3x}(t_2) \quad . \tag{3-10}$$

There are several other proposals as well to simplify the calculation (i.e., to choose very short Δt periods and substitute the time dependent variables with their values belonging to the beginning of the period (Kamensky, 1943); to use continuous time dependent functions and divide the solution into three parts because the mathematical functions approximating the separated parts can be determined in closed form by integration in this case (Székely, 1973); to give graphical interpretation of the solution with finite differences (Lebedyev, 1963); etc.). Other authors have made efforts to generalize the method making it applicable for an existing irregular network of observation wells when the geometrical conditions used as the basis of the derivation are not satisfied (Rowe, 1955; Major, 1973). It is the opinion that the solution of this type of the water balance equations with computers does not raise any special problem and, therefore, the most suitable form of the relationship has to always be determined according to the local conditions (distribution of wells within the network, type and reliability of recorded data, information concerning the depth of flow, hydraulic conductivity and transmissivity, etc.).

In an experimental area (Komlósi Imre groundwater Research Station, Hungary), where long groundwater records were available from several wells, detailed investigations were executed by applying the method described in the previous paragraphs. The layer containing the shallow groundwater is relatively homogeneous fine sand, the transmissivity of which is known from pumping tests. The specific yield of the sand can be estimated on the basis of water retention curves measured in the field with neutron probes. Four areas were chosen for which the water balance of six days intervals were determined from the records covering a period of 18 years (1954-71). The extension and the position of their boundaries were determined by the location of existing groundwater observation wells. The time interval between the regular observations is three days, thus the calculation of six days averages was possible ($t_3-t_2 = t_2-t_1 = 3$ days; $t_3-t_1 = 6$ days) (Major, 1975).

Some characteristic data of the areas and the annual sums for the decade 1961-70 of the two components of the vertical water exchange (i.e., I_g positive accretion and ET_g negative accretion) are listed in Table 3-1.

To characterize the seasonal fluctuation of the components of the vertical water exchange, the monthly averages of both I_g and ET_g was calculated from the data determined for area A in a period of 12 years (1960-71) (Fig. 3-26). The basic data calculated for six days intervals in a year selected as an example are represented in Fig. 3-27 also for the area A.

10-4 Application of Regression Analysis to Evaluate Groundwater Data

As it was already explained, the depth of the water table below the terrain recorded regularly in observation wells is

Table 3-1. Annual values of the components of vertical water exchange determined for various areas by using the water balance method.

The sign of the area	A		B		C		D		
The extension of the area (m^2)	10,000		250,000		13,540		9,200		
The amount of forested area (%)	89		78		58		43		
	I_g	ET_g	I_g	ET_g	I_g	ET_g	I_g	ET_g	Annual Precipitation (mm/year)
	(mm/year)								
1961	188	319	196	250	346	139	277	40	421
1962	88	303	112	196	236	27	260	22	478
1963	148	287	183	195	319	19	347	18	616
1964	66	242	103	146	307	13	352	16	669
1965	131	199	191	145	450	15	559	0	652
1966	261	264	313	230	532	43	604	30	844
1967	153	346	231	286	472	37	611	14	618
1968	4	384	30	225	212	32	306	16	384
1969	137	365	188	219	291	60	346	36	593
1970	191	330	292	255	386	89	386	62	690

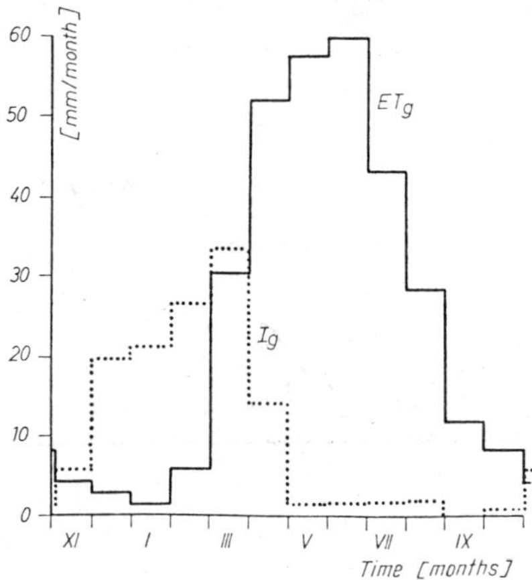

Fig. 3-26. Seasonal fluctuation of positive and negative accretion of groundwater.

the parameter generally used for the hydrological characterization of the groundwater regime. Regarding the set of data observed in a well as a dependent variable, it can be interrelated to similar data sets describing the change in time of other phenomena by using regression analysis. The position of the water level in a river may influence the fluctuation of the groundwater in the floodplains. In other areas, the local precipitation and evapotranspiration may be the dominant factor influencing the development of the groundwater level. There are cases when the governing effects cannot be clearly recognized, but one or more observation wells may be chosen which are suitable to characterize the regime of a larger region and the data sets of the other wells may be related to these standard observations. Thus, either the time series of other hydrological processes or the data sets of other observation wells may be used as independent variables in the regression analysis.

It must also be considered that the effects of different hydrological processes do not occur simultaneously in the data sets, some time lag may be observed even in the fluctuation of the water table at various points. For this reason, the so-called phase displacement (or time lag τ_i) should also be determined between the investigated water level data (P) and each independent variable (S_i) separately. Finally, the relationship between the value of the dependent variable at a time t and the data series of the independent variables can be expressed, therefore, in the form of the following general function:

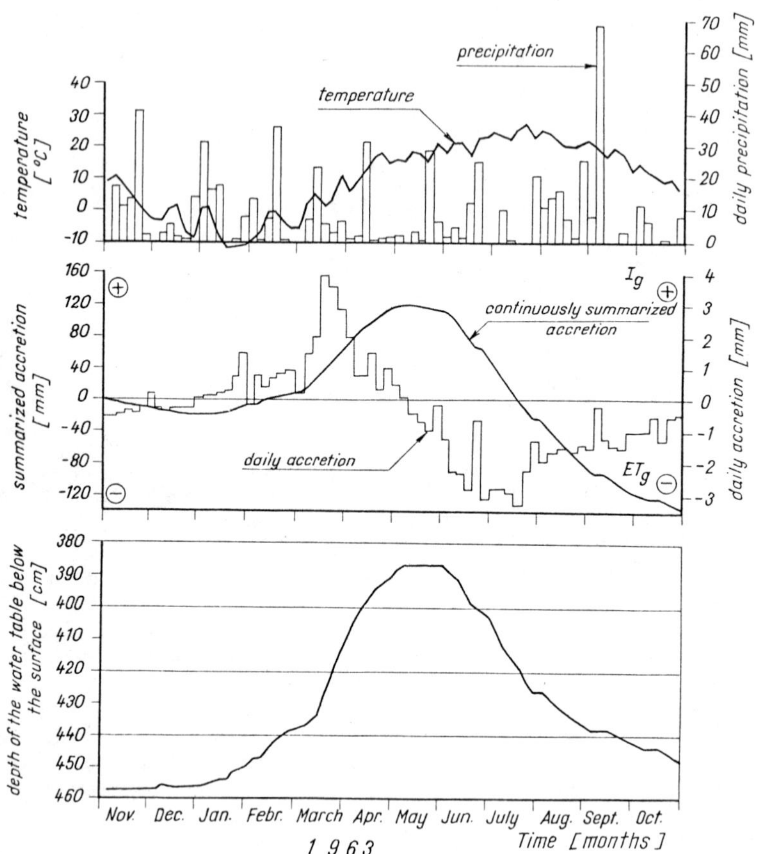

Fig. 3-27. Comparison of the calculated values of accretion with meteorological parameters and the fluctuation of the water table.

$$P(t) = f[S_1(t-\tau_1);\ S_2(t-\tau_2);\ldots\ S_i(t-\tau_i);\ldots\ S_n(t-\tau_n)] \pm \varepsilon\ ; \qquad (3\text{-}11)$$

where ε is a random vector, the expected value of whih is zero, and its variance (σ) is independent of the values of observations and smaller than an arbitrarily chosen limit, but in harmony with both the accuracy required by the practical problem and the reliability of the data used for the investigation.

It is well-known that analytical solutions are available only for the characterization of linear correlation between random variables. Although there are attempts to describe more general relationships (Reimann, 1974), these methods do not provide analytically amenable formulae, and their application is not yet largely extended. This is the reason why the transformation of one or more random variables by using apriori determined functions for the mapping (or choosing these functions on the basis of the trial and error method) may be necessary to perform before the determination of the linear correlation in cases when the supposition of the linear connection between the variables is basically contradictory with the real physical character of the interrelaton. Supposing that each of the independent variables has to be transformed before the determination of the linear correlation by using different mapping functions the simplified form of Eq. 3-11 is as follows:

$$P(t) = a_1 f_1[S_1(t-\tau_1)] + a_2 f_2[S_2(t-\tau_2)] + \ldots + a_i f_i[S_i(t-\tau_i)] + \ldots + a_n f_n[S_n(t-\tau_n)] + b \pm \varepsilon \; ; \tag{3-12}$$

or if the linear approximation is acceptable for all the variables without any transformation:

$$P(t) = a_1 S_1(t-\tau_1) + a_2 S_2(t-\tau_2) + \ldots + a_i S_i(t-\tau_i) + \ldots + a_n S_n(t-\tau_n) + b \pm \varepsilon \; . \tag{3-13}$$

There are many cases when a simple correlation between two random variables gives sufficient approximation and the use of the multiple regression is unnecessary. Naturally, the supposition of a direct linear contact may be sometimes acceptable for the description of simple correlation as well, while the application of a transforming function is necessary in other cases:

$$P(t) = a\, S(t-\tau) + b \pm \varepsilon \; ; \quad \text{or}$$
$$P(t) = a\, f[S(t-\tau)] + b \pm \varepsilon \; . \tag{3-14}$$

This form of the simple correlation is suitable for the interpretation of the role of the parameters b, a, and τ. As it is clearly indicated by Eq. 3-14, b is the difference between the mean values of the dependent and independent random variables, respectively, a is the ratio of their fluctuation, and τ is the time lag between the corresponding stages of the processes characterized by the time series of the two variables. (Naturally, in the case of the second form of Eq. 3-14 the mean and the fluctuation of the transformed parameter is considered instead of those of the original variable.)

Some practical problems will be subsequently listed which are generally solved by the regression analysis of hydrological data.

- There are cases when the observation is not continuous but there are gaps in the record of an observation well. Most of the investigations require continuous time series, and the gaps have to be filled with data. Supposing that a sufficiently close correlation exists between the groundwater levels at two different points (one is the observation well in question and the other one is a well with continuous record) and this relationship can be determined on the basis of data observed in the periods when the depth of the water table was recorded in both wells, the regression can be used for the approximation of missing data.

- The determination of water level data may be the purpose of the investigation, similarly if the influence of some artificial operation has disturbed the natural processes, and the lack of data was not caused by the discontinuity of observation, e.g., in the vicinity of a construction pit or a mine, the discharge necessary for their dewatering causes the drawdown of the water table. The calculation of data characterizing the natural conditions (which are not observable) is required to determine the extension of the zone influenced by the artificial operation.

- Structural functions can be determined between the data observed in neighboring wells (similarly as it was done for meteorological parameters by Czelnay, 1972) which expresses the correlations of the simultaneous observations. These functions are suitable to determine the density of wells within an observation network depending on the accuracy required in the characterization of groundwater conditions and they can also be used to screen the rough observation errors from the time series.

- All the time series observed in the various wells within a groundwater basin may be related to the recorded parameters of one or more other hydrological processes influencing the groundwater regime. Investigating the influence of one external effect, a coefficient of correlation (R) can be determined for each observation point. The mapping of this parameter and the construction of isolines between the mapped values clearly demonstrate the zone of influence where the regime is governed by the investigated process. Repeating the same analysis considering various hydrological processes, the main factors influencing the groundwater balance within the investigated area can be determined. The coefficients of multiple correlation (r_1, r_2 ... r_i ... r_n) can be similarly used to select the most important processes to determine their zones of influence, and to demonstrate the ratio of their effects within the various parts of the investigated basin.

- The mapping of the various parameters of the regression analysis (a,b,τ) can be similarly used to characterize the influence of the various external actions. Correlating the water level in a river and that observed in various observation wells located within the floodplains, the τ parameter indicates the propagation of groundwater waves and the multiplying factor (a) shows their flattening. Thus, the areal distribution of these parameters can also be used to estimate the hydraulic conductivity and the storage capacity of the system.

The examples selected to demonstrate the numerical exploration of the hydrological properties of groundwater systems by using regression analysis and the application of this method for solving some of the practical problems listed in the previous paragraphs was taken from Stol's paper (1969).

In the first investigation the relationship between the water level in River Meuse at Rotterdam (also influenced by tidal effect) and the table of the groundwater within the riparian area was analyzed.

The result of the investigation was directed at the determination of the time lag, and is demonstrated in Table 3-2 where the corresponding values of the supposed time lag (τ) and the calculated coefficient of correlation (R) are listed and were determined for a well at a distance of 150 m from the bank of the river. The closest correlation (R = 0.979) was found by assuming a time lag of half an hour while the largest negative coefficient (R = -0.888) was obtained by substituting τ = 6.5 hours. This result indicates that the time necessary for the propagation of effects from the river along a stretch of 150 m is about half an hour. The negative maximum (6.5 hours) shows the sum of the time of propagation and the length of the period between tide and ebb.

Figure 3-28 summarizes the results of the investigations by representing the areal distribution of the following parameters:

- the simple coefficient of correlation (R) between the river stages and the position of the water table at various points (Fig. 3-28a);

Table 3-2. Relationship between time lag (τ) and correlation coefficient (R) based on 55 half-hour observations in well No. 1 at 150 m from the river bank.

τ (hours)	R	τ (hours)	R	τ (hours)	R
0	0.963	3.0	0.242	6.0	-0.842
0.5	0.979	3.5	-0.009	6.5	-0.888
1.0	0.937	4.0	-0.208	7.0	-0.884
1.5	0.814	4.5	-0.423	7.5	-0.816
2.0	0.665	5.0	-0.529	8.0	-0.704
2.5	0.444	5.5	-0.739		

- the expected amplitude of groundwater fluctuation in cm (Fig. 3-28b);

- the time lag in hours, the substitution of which gives the best correlation between the two variables (river stage and water table) (Fig. 3-28c).

In the second case, the influence of dewatering a construction pit was determined. Using the correlations between the water table data observed in wells being inside and outside the influenced zone, respectively, the probable position of the water table which would have developed under natural conditions without the artificial influence was calculated. Computing the difference between the probable natural levels of the water table (determined by applying the equations obtained from the regression analysis) and the actually observed position of the groundwater surface, the effect of drawdown can be characterized. The results of the investigation are represented by a series of maps showing the development of the cone of depression at various time points (Fig. 3-29). The most striking conclusions drawn from this analysis are as follows:

- the asymmetry of the cone demonstrating the anisotropy or nonhomogeneity of the aquifer, or the nonhomogeneity of the boundary conditions;

- the fact that the position of the contour line showing 35 cm depression remains practically unchanged for a long period (having only some fluctuation) which can be explained by supposing that the excess infiltration within this area and the horizontal flow along the contour of the cone can balance the amount of water discharged;

- the displacement of the center of the cone of depression in a northern direction after stopping the pumping (June, 1968).

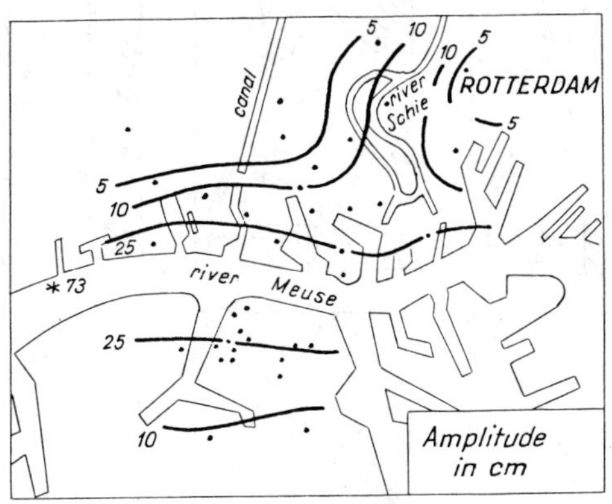

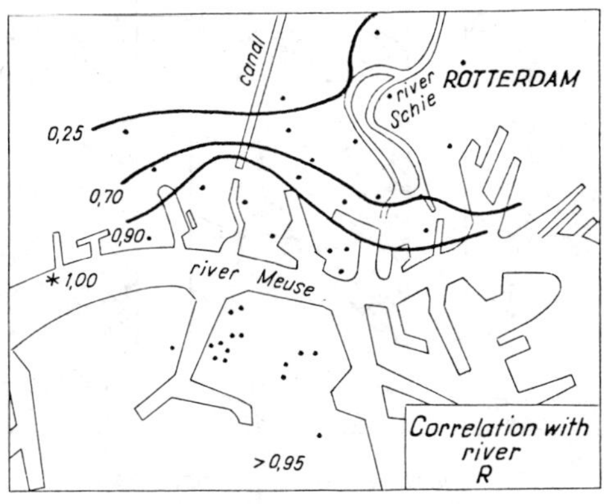

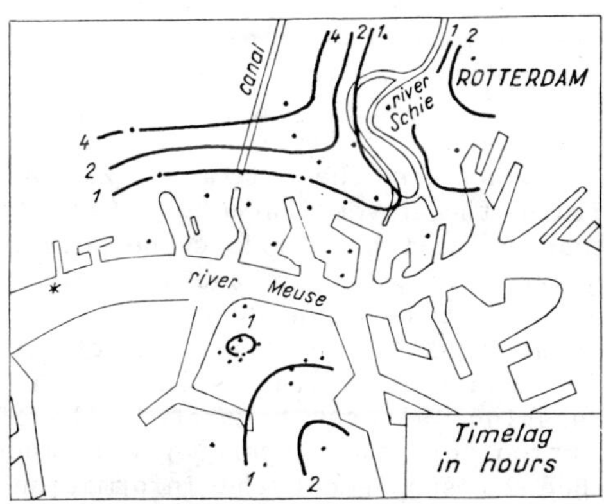

Fig. 3-28. Mapping of the parameters of the regression equation characterizing the regime of shallow groundwater.

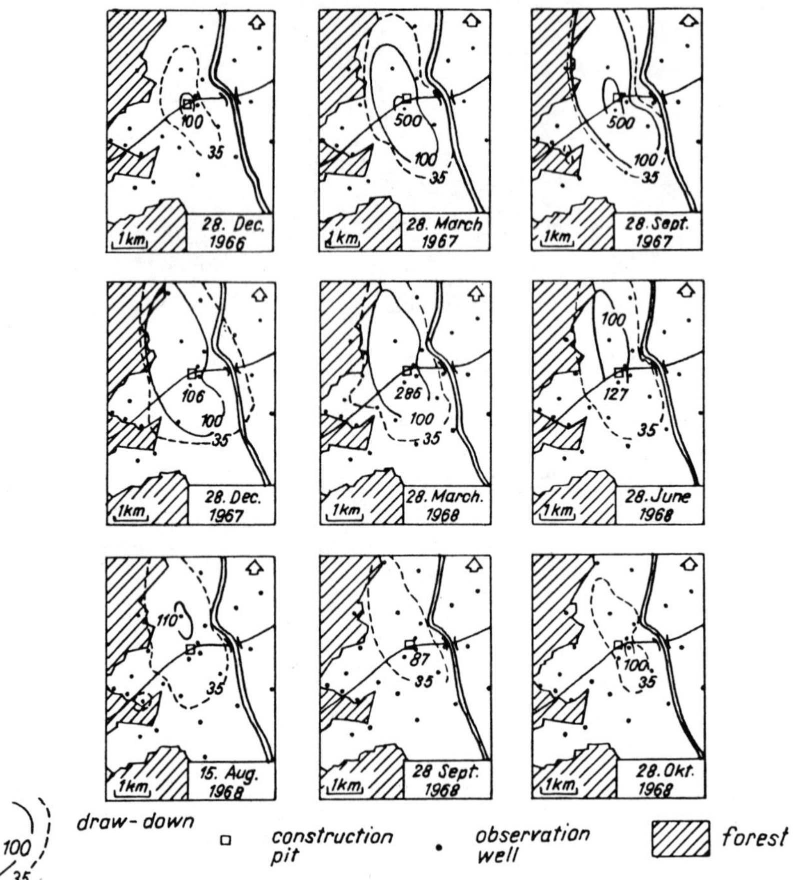

Fig. 3-29. Development of a depression cone in time around a construction pit.

The examples quoted to demonstrate the application of the regression analysis for the characterization of the flow conditions within the investigated aquifer clearly indicate that this method is suitable to describe some general properties of the system (e.g., the propagation velocity and storage capacity in the first case, nonhomogeneity, anisotropy, and the change in accretion as the result of drawdown in the second example) apart from the solution of the practical problems in question. It can be stated, therefore, that the numerical solution of such hydrological models gives an excellent method to supplement the information obtained from hydrogeological explorations.

SECTION 11

HYDRAULIC PARAMETERS CHARACTERIZING GROUNDWATER FLOW

On the basis of surveying the various methods of groundwater observation in the previous section, it can be stated that the position of the water table or that of the piezometric surface if the system is a confined one, is the main information which can be gained by using observation wells. Although there are methods to measure the velocity of groundwater flow, their results are, however, uncertain and their execution is expensive. The flux of flow must be estimated, therefore, in most cases from relationships derived by using only water level data.

The hydraulic gradient initiating and maintaining the groundwater flow can be directly calculated from water level data. The determination of the transported amount of water requires the knowledge of the transmissivity of the system, which parameter is the product of the thickness of the water conveying layer (the depth of groundwater flow in the case of an unconfined aquifer) and its hydraulic conductivity. Thus, the gradient multiplied by transmissivity gives the amount of water transported through a strip of the water-bearing layer having a width of unity measured perpendicularly to the main direction of flow (specific flow rate).

Under natural conditions, the groundwater flow is usually an unsteady process governed by boundary conditions changing in time. The complete characterization of the movement, therefore, requires the consideration of the change of the amount of water stored in the system. As it was already explained in subsection 9-2, the change in storage is calculated as the change of the position of the water table (or that of the pressure head) multiplied by the specific storage capacity of the layers influenced by the hydrological processes. The first parameter is a geometric one which can be calculated directly from the water level data measured in observation wells. The storage capacity has to be determined since it is an important hydraulic parameter of the system.

The natural slopes of the water table or the piezometric surface are very small in most cases. Thus, the seepage velocity maintained by the gradient may remain below the value indicating the lower limit of Darcy's law, thus, the molecular force developing between the water and the solid matrix of the porous medium may have a considerable role among the dominating forces, as it was explained in Part I. There are even cases when the gradient is smaller than the threshold gradient characterizing the layer, and it is not sufficient, therefore, to initiate movement. The pressure difference observed between two points of such systems indicates only the possibility of movement and the accumulation of energy but not the actual development of flow.

Considering the principles summarized in the previous paragraphs, there are three important hydraulic parameters of the layers which should be known for the numerical characterization of the water balance of a groundwater system, i.e.:

- the specific storage capacity of the porous medium;

- the hydraulic conductivity of the saturated layer;

- the threshold gradient characterizing the molecular force acting between the grains and the water and hindering the development of flow. It is also necessary to investigate how the value of hydraulic conductivity changes as a function of velocity and how it is decreased by the molecular force.

It was already explained in Subsection 9-2 that there are two parameters applied for characterizing the specific storage capacity of a layer, i.e., n, the specific yield, gives sufficient result in the case of an unconfined system while the compressibility of water as well as the decrease of pore-volume due to the lowering of pressure should be taken into account when seepage through a confined layer is investigated. The latter two effects can be considered by using S the storage coefficient. The interpretation of specific yield and the way to measure or calculate its numerical value was already given in Part II. The storage coefficient will be dealt with in more detailed form in connection with the analysis of deep groundwaters (Part IV) because shallow groundwater is usually unconfined, and thus, the role of the storage coefficient is much more important when the water balance of and the available resources stored in deep-lying aquifers are investigated.

This is the reason why the discussion of this section starts with the analysis of the hydraulic conductivity of loose clastic sediments. The investigation of the relationship existing between this parameter and seepage velocity is followed by the survey of laboratory and field methods suitable to determine the hydraulic conductivity of natural layers and by the detailed description of the evaluation of pumping tests. The change of hydraulic conductivity due to the considerable influence of molecular force is dealt with separately. On the basis of the investigation of microseepage (flow through porous medium influenced by three dominating forces, i.e., gravity accelerating the movement as well as internal friction and molecular force retarding it) relationships will be given to calculate the threshold gradient from soil-physics data.

11-1 Hydraulic Conductivity of Loose Clastic Sediments and Its Change with Seepage Velocity

The concept of v, the seepage velocity (and that of q specific flow rate or flux being equal to the former) was already explained and used in Part I for the characterization of seepage through a saturated medium; v is the flow rate through a unit area of a section crossing the porous medium perpendicularly to the main direction of flow. Its numerical value can be calculated, therefore, as the ratio of the total discharge transported through the section (Q) and the area of the latter including both the area of the pores and that of the solid matrix (A) (Fig. 3-30):

$$v = q = \frac{Q}{A} \quad . \tag{3-11}$$

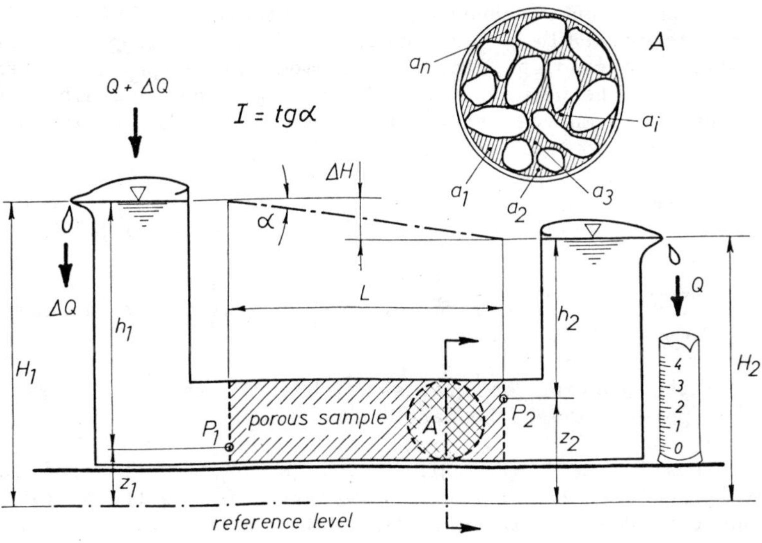

Fig. 3-30. Sketch of an equipment for repeating Darcy's experiments.

It was also mentioned that the actual mean velocity in the pores (v_{eff}) can be approximated by dividing the seepage velocity by the porosity:

$$v_{eff} = \frac{v}{n} \quad . \tag{3-12}$$

The dynamic analysis of seepage has proved that there exists a linear relationship between seepage velocity and hydraulic gradient, if gravity (the force accelerating the movement) is balanced only by internal friction and the other retarding forces (inertia, and the molecular force between grains and water) can be neglected as compared to friction. The coefficient of proportionality is the so-called Darcy hydraulic conductivity (K_D);

$$v = K_D \, I \quad . \tag{3-13}$$

The simple v = KI relationship can be accepted even for the characterization of all types of seepage through saturated media if the concept of hydraulic conductivity is used in a more generalized form instead of Darcy's original parameter. The K_D coefficient is composed of two factors, one characterizing the properties of the solid matrix (k, the intrinsic permeability) and the other describing the behavior of the transported fluid. These two components must be supplemented with a third one, depending on the condition of flow. The third factor should be the function of either seepage velocity or hydraulic gradient, in order to modify the apparent linear relationship between v and I in this way and to achieve the suitable mathematical connection between the two basic variables in each zone of seepage.

Investigating seepage through loose clastic sediments, the intrinsic permeability is influenced by two properties of the solid matrix:

- the size and form of grains which the layer is contructed;

- the packing of grains which determines the size of pores between the particles.

Thus, four components have to be investigated and determined for the complete characterization of the hydraulic conductivity of saturated loose clastic sediments:

- the geometrical parameters of grains including the grain-size and its distribution which two characteristics are expressed in a combined form by using D_h, the effective diameter (see Eqs. 1-12 and 1-13), as well as the shape of particles described by α, the shape-coefficient (see Eq. 1-10);

- the porosity of the layer (n; its interpretation is given in Subsection 3-1 and its connection with soil-physical parameters is dealt with in Subsection 3-4);

- physical parameters of the propagating fluid which is generally expressed as the ratio of specific weight related to dynamic viscosity (γ/μ) or, dividing both parameters by ρ, the fluid density, this property of fluid can be given in the form of acceleration due to gravity over kinematic viscosity (g/ν) (see Eq. 1-85);

- the flow condition of the investigated seepage (considering the dominant forces among those discussed in Subsection 4-1).

The method of dynamic modeling of the flow through porous media was already explained in Section 1. It was seen there that the investigation starts with the determination of a simplified geometric model which is hydraulically equivalent with the actual network of channels composed by the pores. This step is followed by the construction of the dynamic model which describes the flow through the conduits of the geometric model by considering the forces acting. There are cases when the actual size of the pores influences the character of the movement and the simulation of the water transporting network by a geometric model, considering only the average diameter of openings, may cause considerable error. A statistical model should be used in these cases to describe the probability distribution of the sizes of pores.

It can be easily proved by using the continuum approach that the representative elementary volume, the average characteristics of which has only random variation within a homogeneous medium according to the location of the unit investigated, is negligibly small in cases when the solid matrix is composed of evenly distributed grains of loose clastic sediments. There are only a few special problems (e.g., dispersion of pollutants, water transport through unsaturated medium) when the consideration of the pore-size distribution is inevitable. In general, the network of the pores can be substituted by a given number of elements having the same size calculated from the average pore-diameter when the permeability of a loose clastic sediment is investigated. It is not necessary, therefore, in this case to consider the statistical model, but the simple combination of the geometric and the dynamic models provides a conceptual model suitable to describe the water transport through homogeneous solid matrices constructed of individual grains and to determine the resistance against flow in the water conveying network composed of evenly distributed and randomly interconnected pores.

The various types of geometric models being suitable to simulate the flow through saturated loose clastic sediments were investigated in detailed form in Part I (see Subsection 4-3). It was found by comparing the models proposed in the literature that a bundle of straight tubes being parallel to the direction of flow and composed of small stretches with two different diameters is simple enough to be amenable to mathematical treatment, at the same time, it is suitable to correctly reproduce the most important hydraulic processes occurring in such layers. To summarize the results of the investigation, the longitudinal section of the tube (Fig. 3-31) and the equations expressing the relationships between the characteristic sizes of the model tube and the soil-physical parameters are repeated here:

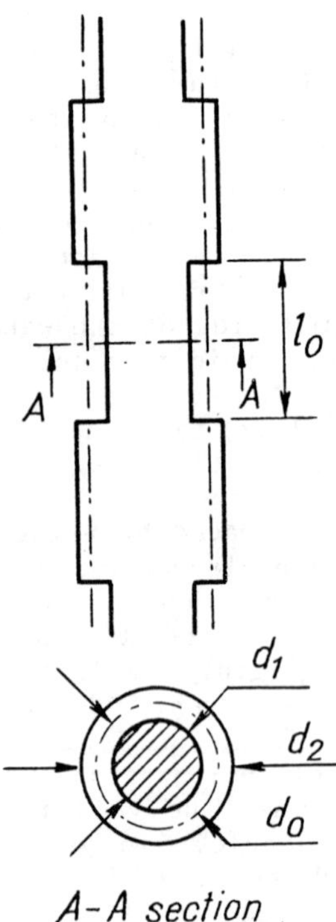

Fig. 3-31. Longitudinal section of the proposed model pipe.

average tube diameter $$d_o = \frac{4n}{1-n}\frac{D_h}{\alpha}\,; \qquad (3\text{-}14)$$

diameter of the smaller stretch $$d_1 = \frac{2.67n}{1-n}\frac{D_h}{\alpha}\,; \qquad (3\text{-}15)$$

diameter of the larger stretch $d_2 = \frac{5n}{1-n} \frac{D_h}{\alpha}$; (3-16)

length of the varying stretches $\ell_o = 2/3 \frac{n}{1-n} \frac{D_h}{\alpha}$; (3-17)

the number of tubes crossing a unit area of the section $N = \frac{(1-n)^2}{4n} \frac{1}{\pi} (\frac{\alpha}{D_h})^2$. (3-18)

It was proved on the basis of the dynamic analysis of the forces initiating and maintaining the seepage developing through saturated porous media (see Subsection 4-1) that gravity is the only one force which should be considered among those accelerating the movement, since the pressure difference can also be expressed as a special type of action of gravity by taking it into account in the form of the weight of an equivalent water column. Among the actions retarding the seepage, internal friction, inertia, and the attraction between grains and water (molecular force) were listed as the main forces to be considered. The weight of their influence changes, however, depending on the velocity of movement, thus, only one or two retarding forces can be simultaneously dominant. The various types of seepage are distinguished according to the dominant retarding forces acting against gravity (see Fig. 1-45):

- turbulent seepage (only the action of inertia is dominant apart from gravity);

- transition zone (gravity is balanced by inertia and friction);

- laminar seepage (friction is the sole retarding force to be considered);

- microseepage (friction and the molecular force should be jointly taken into account as retarding forces);

- static condition (gravity does not initiate movement because its effect does not surpass that of the molecular force).

The latter two conditions will be dealt with separately in Subsections 11-4 and 12-5. Here, the movement equations describing the laminar, transition, and turbulent zones of seepage are analyzed in detailed form. It is evident that the changes of the flow conditions depend, in these cases, on the degree of influence of inertia and friction, respectively. The limits of these zones were expressed, therefore, by given numerical values of the Reynolds number, which dimensionless number was derived in Subsection 4-2 from the quotient of the two forces in question.

In the case of laminar seepage it is assumed, as a dynamic principle, that gravity is balanced by friction. Accepting this hypothesis, the well-known Poiseuille equation can be derived to calculate the water transport through a model pipe of diameter d_o. The basic differential equation expresses the equilibrium of gravity and friction for a cylinder which is concentric with the axis of the pipe and having a radius r and a length ℓ (Fig. 3-32):

$$I\gamma r^2\pi\ell + 2r\pi\ell\mu\frac{dv}{dr} = 0 \quad . \qquad (3\text{-}19)$$

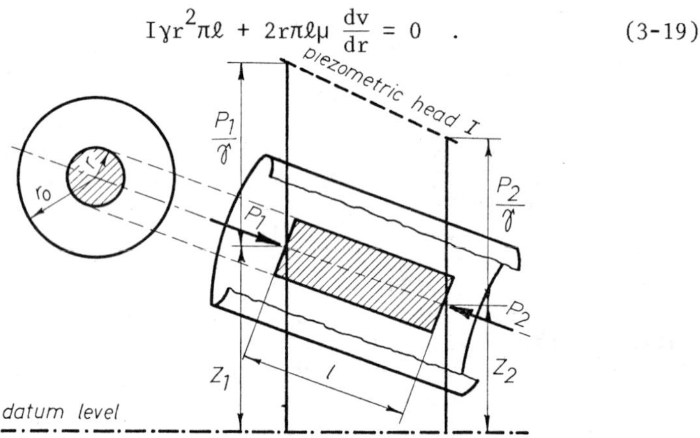

Fig. 3-32. Symbols used for deriving Poiseuille's equation.

After solving the differential equation by considering a boundary condition according to which the velocity at the wall of the pipe is zero (v=0 where $r=r_o$), the velocity distribution within the section of the pipe can be determined. Integrating the product of the local velocity and the elementary area for the whole section, the water amount transported through the pipe is achieved. The ratio of this amount and the area of the cross section is the mean velocity developing in the model pipe of diameter d_o. The amount transported by a pipe should be multiplied by the number of pipes crossing a unit area of the model to get the flux conveyed through the system simulated by using tubes of average diameter (see also Eq. 1-81):

$$q = \frac{1}{32}\frac{g}{\nu}nd_o^2 I \quad . \qquad (3\text{-}20)$$

The flux of the model pipes with constant diameter calculated from Eq. 3-20 is 2.5 times greater than the actual value determined by reliable experiments, as it was mentioned in Subsection 4-3 in connection with the detailed discussion of the establishment of the model system. It was also explained there that the difference can be eliminated if the pipe is constructed from short stretches of diameters d_1 and d_2 (the volume of the pipe remains the same as that of the pipe with diameter d_o and the ratio of d_1^2 and d_2^2 is equal to 1:3 to 1:3.5). The increase of flow resistance (or the decrease of flow rate) can be mathematically proved (Fig. 3-33):

$$Q_1 = Q_2 = \alpha Q_o \quad ;$$

$$\ell/2\,(I_1 + I_2) = I_o \ell \quad ;$$

$$I_1/I_2 = \left(\frac{d_2}{d_1}\right)^4 \quad ; \qquad (3\text{-}21)$$

$$Q_o = \beta I_o\, d_o^4 \quad ;$$

$$Q_1 = Q_2 = \beta I_1\, d_1^4 = \beta I_2\, d_2^4 = 2\beta I_o \left(1 - \frac{I_2}{2I_o}\right) d_1^4 =$$

$$2Q_o \left(\frac{d_1}{d_o}\right)^4 \frac{I_1}{I_1 - I_2} = 2Q_o \left(\frac{d_1}{d_o}\right)^4 \frac{1}{1 - \left(\frac{d_1}{d_2}\right)^4} \quad ;$$

$$\frac{Q_1}{Q_o} = \frac{2\left(\frac{d_1}{d_o}\right)^4}{1 - \left(\frac{d_1}{d_2}\right)^4} \quad ;\text{and}$$

$$\alpha = 0.444 \text{ to } 0.363 \;; \quad \text{if} \quad \left(\frac{d_2}{d_1}\right)^2 = 3 \text{ to } 3.5 \quad \text{and} \quad \frac{d_1}{d_o} = \frac{1}{1.5} \;.$$

On the basis of this exaplanation, the flux transported through the proposed geometrical model under the influence of a hydraulic gradient I can be determined by multiplying the right-hand side of Eq. 3-20 by a constant of 0.4 (see Eq. 1-85). Since it was assumed that the model was hydraulically equivalent with the irregular channels composed of the pores of loose clastic sediments, the flux calculated in this way is equal to the virtual seepage velocity characterizing the flow developing through the saturated solid matrix:

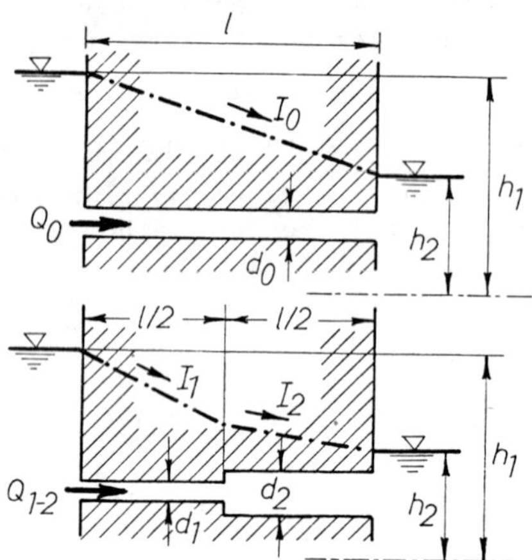

Fig. 3-33. Comparison of flow rate of pipes having constant or changing diameter.

$$v = q = 0.4q = \frac{1}{80}\,\frac{g}{\nu}\,n\,d_o^2\,I = \frac{1}{5}\,\frac{g}{\nu}\,\frac{n^3}{(1-n)^2}\,\left(\frac{D_h}{\alpha}\right)^2 I \quad . \qquad (3\text{-}22)$$

Comparing this result to Darcy's equation, the theoretical value of hydraulic conductivity can be determined. It is in good agreement with the dynamic analysis which has stated that the seepage velocity is linearly proportional to hydraulic gradient in the case of laminar seepage, and includes all the effects of the influencing factors:

$$K_D = \frac{1}{5}\,\frac{g}{\nu}\,\frac{n^3}{(1-n)^2}\,\left(\frac{D_h}{\alpha}\right)^2 \quad . \qquad (3\text{-}23)$$

The variables influencing the parameter can be divided into three groups:

- physical constants expressing the behavior of the fluid [γ/ρ, or using the kinematic viscosity ($\nu = \frac{\mu}{\rho}$) the equivalent ratio is $\frac{g}{\nu}$];

- characteristics of grains composing the layer (D_h/α) which express the size and shape of grains as well as the grain-size distribution;

- the effect of porosity $[\frac{n^3}{(1-n)^2}]$.

To verify the proposed formula (which is the Hungarian version of the modified Kozeny-Carman equation) various ways were investigated in which these groups of variables were taken into consideration earlier in the literature.

The dependency of hydraulic conductivity on the behavior of fluids have been expressed by using the same ratio in the recent publications (Lindquist, 1926, 1943; Kozeny, 1953; Carman, 1956; Chardabellas, 1964). Earlier, the temperature of the water was applied to characterize the same relationship (Havrez, 1974; Seelheim, 1880; Hazen, 1895; Forchheimer, 1924), but the numerical data determined in that way are in good harmony with the change of the viccosity of water depending on temperature. Thus, all the measurements used to determine the hydraulic conductivity as the function of temperature prove the validity of the term expressing the properties of the transported fluid in Eq. 3-29.

Concerning the parameter characterizing the grains, there is one point where all the authors are generally in agreement with each other, i.e., the hydraulic conductivity is proportional to the square of a characteristic diameter, as it is shown in Fig. 3-34, where the results of the measurements published by Rumer (see de Wiest, 1969) and Bear (1972) are summarized. The ratio of the effective diameter and the shape-coefficient is equal to that of the total surface and volume of grains according to the definition of the α parameter (see Eq. 1-10). Thus, the use of the D_h/α quotient in Eq. 3-29 gives the same result as the proposals of Kozeny (1953) and Carman (1956). Although Zamarin (1928) applied another factor to express the different weights of the grains having various diameters, his method leads practically to the same numerical value as Eq. 3-23 does. Even the investigations using characteristic diameters other than D_h, e.g., Paladin, 1964; Beyer (see Juhász, 1976) prove the reliably of the method proposed here, when the numerical results are compared.

The most versatile proposals were published in connection with the consideration of the influence of porosity. There are numerous methods using the same expression as that included in Eq. 3-23. The trend of the change of hydraulic conductivity in the zone of the most probable range of porosity does not differ, however, from that calculated by applying the $n^3/(1-n)^2$ formula (the so-called Kozeny parameter) even if the dependency of hydraulic conductivity on porosity is determined in a different way (Schlichter, 1899; Goldstein, 1938; Ward, 1964; de Wiest, 1969; or Terzaghi,

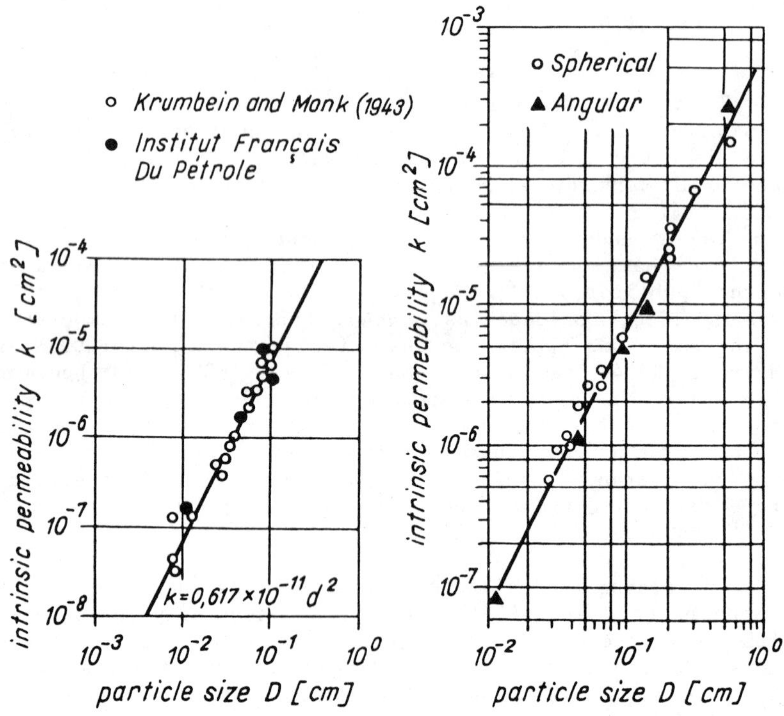

Fig. 3-34. Relationship between intrinsic permeability and grain diameter (after Romer and Bear).

1926, 1943; Chardabellas, 1964; Juhasz, 1967). This fact proves the assumption that the correctness of Kozeny's parameter is acceptable.

Summarizing the foregoing, it can be stated that Eq. 3-23 based on theoretical investigations is in very good agreement with the empirical formulae proposed and with the experimental measurements published previously. Thus, Eq. 3-23 (which is the Hungarian version of Kozeny-Carman equation and provides its simplified and improved form) is one of the most reliable and easily applicable methods to calculate the hydraulic conductivity in the zone of laminar seepage.

It was stated on the basis of the dynamic analysis of seepage that the hydraulic gradient is proportional to the square of the velocity in the turbulent zone where inertia is the sole dominant retarding force balancing the effect of gravity. Thus, Chezy's equation can be applied to calculate the mean velocity in a model pipe of diameter d_o:

$$v_{eff} = \sqrt{\frac{2g}{\lambda_1}} \sqrt{I\ d_o} \quad . \tag{3-24}$$

It is necessary to note here that the λ_1, the resistivity coefficient, cannot be considered in this case with the same numerical value as that generally used to characterize the flow in a straight pipe with constant diameter (which depends on the diameter and the relative roughness of the pipe, and on the velocity of the movement, but which can be approximated with an average value of $\lambda = 0.03$ to 0.04). The model pipe, being hydraulically equivalent with the network of pores, is composed of short stretches of diameters d_1 and d_2. The resistance caused by the changes of velocity and diameter many times surpasses that occurring in a smooth pipe. The λ_1 parameter expresses all the resistances in a combined form, its numerical value should be, therefore, much greater that the average Dupuit parameter mentioned ($\lambda = 0.03$ to 0.04).

The total energy loss through a pipe having a length of ℓ, which is equal to the length of the sample parallel to the main direction of flow, and composed of small stretches with narrow (d_1) and large (d_2) diameters each having a length of ℓ_o can be divided into four components. The numerical values of the components expressed in the form of head losses can be calculated from the following equations:

- losses due to friction in a narrow stretch:

$$h_{v1} = \ell_o \frac{v_1^2}{2g} \lambda \frac{1}{d_1} \quad ; \tag{3-25}$$

- losses due to friction in a large stretch:

$$h_{v2} = \ell_o \frac{v_2^2}{2g} \lambda \frac{1}{d_2} \quad ; \tag{3-26}$$

- losses at the transition from a large stretch to a narrow one (confusor):

$$h_{v3} = c_1 \frac{v_2^2}{2g} \quad ; \tag{3-27}$$

- losses at the transition from a narrow stretch to a large one (diffusor):

$$h_{v4} = c_2 \frac{v_1^2}{2g} (1 - \frac{A_1}{A_2}) \quad ; \qquad (3\text{-}28)$$

where A_1/A_2 is the ratio of the two cross sections, the numerical value of which is 1:3 to 1:3.5 in the case of the model pipe, the constant c_1 varies between 0.06 and 0.4, and the constant c_2 between 0.5 and 1.0 according to the character of the transition between the different stretches (sharp edges or rounded forms; immediate or gradual change of the diameter).

Considering that the number of the stretches is

$$S = \frac{\ell}{\ell_o} \quad , \qquad (3\text{-}29)$$

and half of them are narrow, the other half are large, and taking into account also that the number of both the confusors and diffusors is similarly S/2, the average value of the gradient along the model pipe can be calculated as the sum of the head losses related to the length ℓ:

$$I = \frac{\Sigma h_r}{\ell} = \frac{S}{2\ell} \left\{ \lambda \ell_o \left[\frac{1}{d_1} \frac{v_1^2}{2g} + \frac{1}{d_2} \frac{v_2^2}{2g}\right] + c_1 \frac{v_2^2}{2g} + c_2 \frac{v_1^2}{2g} (1 - \frac{A_1}{A_2}) \right\} . \qquad (3\text{-}30)$$

Knowing the ratios of average tube diameter d_o related to the diameters of the narrow and large stretches, respectively, the gradient can be expressed as a function of both actual mean velocity v_{eff} in the pores and average pore diameter d_o:

$$I = \frac{v_{eff}^2}{2g} \frac{1}{d_o} \frac{1}{2} \left\{ \lambda \left[\frac{d_o}{d_1} (\frac{v_1}{v_{eff}})^2 + \frac{d_o}{d_2} (\frac{v_2}{v_{eff}})^2\right] \right.$$

$$\left. + \frac{d_o}{\ell_o} (\frac{v_2}{v_{eff}})^2 c_1 + \frac{d_o}{\ell_o} (\frac{v_1}{v_{eff}})^2 c_2 (1 - \frac{A_1}{A_2}) \right\} \quad ; \qquad (3\text{-}31)$$

where $(v_1/v_{eff})^2$ = 5.00 to 5.20; $(v_2/v_{eff})^2$ = 0.40 to 0.50; d_o/d_2 = 0.80, and $d_o/\ell_o \sim 1$ are numerical values determined by the geometrical parameters of the model pipe (see Eqs. 3-14, 3-15, and 3-17). Combining Eqs. 3-24 and 3-31 the resistivity coefficient λ_1 expressing the total resistance of the system is as follows:

$$\lambda_1 = \frac{1}{2}\left\{ \lambda[\frac{d_o}{d_1} (\frac{v_1}{v_{eff}})^2 + \frac{d_o}{d_2} (\frac{v_2}{v_{eff}})^2] + \frac{d_o}{\ell_o} (\frac{v_2}{v_{eff}})^2 \right.$$

$$\left. c_1 + \frac{d_o}{\ell_o} (\frac{v_1}{v_{eff}})^2 \; c_2(1 - \frac{A_1}{A_2}) \right\} . \qquad (3\text{-}32)$$

Substituting the realistic scattering of constants c_1 and c_2 (the edges are not sharp but the change of diameter is not gradual) and that of the other numerical parameters, the expected range of the coefficient is $\lambda_1 = 1.2$ to 2.0, $\lambda_1 = 1.6$ being the most probable value. This result is supported by Krischner's (1962) experiments, who has measured the λ_1 value in the transition and turbulent zones by using a sample composed of spheres. The range of the parameter was $0.5 < \lambda_1 < 2.0$, and it can be supposed that the smaller values (between 0.5 and 1.2) were belonging to cases when the turbulence was not yet completely developed. Thus, the following equation can be proposed to calculate the seepage velocity in the case of turbulent flow from the soil-physical data:

$$v = n\, v_{eff} = \sqrt{5g \frac{D_h}{\alpha} \frac{n^3}{1-n} I} \; . \qquad (3\text{-}33)$$

In order to determine the hydraulic conductivity of turbulent flow (K_T) and the multiplying factor (Φ_T) which transforms Darcy's coefficient into this new parameter ($K_T = \Phi_T K_D$), the ratio of seepage velocity and hydraulic gradient has to be analyzed:

$$v = K_T\, I = \frac{1}{v}\, 5g \frac{D_h}{\alpha} \frac{n^3}{1-n} \; ;$$

$$I = K_D \frac{\nu}{v}\, 25 \frac{\alpha}{D_h} (1-n) I \; ;$$

$$K_T = \Phi_T K_D = \frac{\nu}{v}\, 25 \frac{\alpha}{D_h} (1-n)\, K_D \; ; \qquad (3\text{-}34)$$

$$\text{where;} \quad \Phi_T = \frac{\nu}{v}\, 25 \frac{\alpha}{D_h} (1-n) = \frac{100}{Re_P} \; ;$$

$$\text{and;} \quad Re_P = \frac{v}{\nu} \frac{4}{1-n} \frac{D_h}{\alpha} \; .$$

The laminar seepage characterizes the groundwater movement having low velocity (in practice and especially under natural conditions this state is the usually occurring one) and the total development of turbulence is expected only in the case of extremely high velocity. Between these two flow conditions, both friction and inertia have dominant roles in balancing the accelerating action of gravity so that the importance of inertia increases related to that of friction as the velocity increases. There are two different ways generally applied to approximate the relationship between seepage velocity and hydraulic gradient in this transition zone:

- binomial form: $I = a\,v + b\,v^2$;

- potential form: $I = C\,v^m$, or $v = K\,I^{1/m}$.

The dynamic analysis has shown that the structure of the binomial equation proposed by Forchheimer (1924) is in good agreement with the theoretical results (see Eq. 1-63). The scientists dealing with the problem recently have concentrated their effort, therefore, to determine the a and b parameters as the functions of soil-physical parameters. It was found, however, that these parameters are not material constants. Most of the authors proposing various equations to calculate the values in question applied, therefore, the ratio of head losses caused by viscous resistance and kinetic energy dissipation, respectively, as an independent variable in the formulae providing the a and b parameters (Engelund, 1953; Irmay, 1954; Scheidegger, 1957; Sunada, 1965; Ahmed and Sunada, 1969; Valentin, 1970). The gradual change of the constants of the Forchheimer equation is expressed in this way depending on seepage velocity or Reynolds number. Thus, a fixed pair of a and b parameters gives only the approximation of the v vs. I relationship within a range of velocity. The limits of the validity zone of the various equations can be determined in the form of given numerical values of the Reynolds number.

There were attempts also to apply the potential form to achieve a general equation describing the relationship between seepage velocity and hydraulic gradient (Hatch, 1940; Slepicka, 1969; Perez Franco, 1973). It is well-known that the laminar seepage is described with a power m=1. This parameter increases gradually as the effect of inertia becomes stronger, and it achieves a value of m=2 in the case of turbulent flow. It can be stated, therefore, that the parameters of the potential form depend also on the character of seepage. Thus, this approximation can neither be applied with constant numerical parameters, but a validity zone should always be attached to a given value of the m power and to the corresponding multiplying factor.

In order to achieve a general formula which can be applied to characterize the seepage in the whole range of laminar, transition, and turbulent zones without any restriction of the validity, numerous experimental data should be evaluated. Lindquist (1933) proposed a way to represent the data in a coordinate system using the Reynolds number on the horizontal axis and the product of the coefficient of resistivity

$$\lambda = \frac{I}{v^2} 2g\ D \quad , \tag{3-35}$$

and the Reynolds number on the vertical axis. It was found that the Re_s, the seepage Reynolds number, does not express the influences of porosity and the shape of grains. It was proposed, therefore, to make a transformation along the vertical axis to get a Y variable including Darcy's hydraulic conductivity according to Eq. 3-23 which can be achieved by multiplying the λRe_s product with some dimensionless parameters

$$Y = \lambda Re_s \frac{n^3}{(1-n)^2} \frac{1}{\alpha^2} = 10 \frac{I}{v} K_D \quad . \tag{3-36}$$

Following Thirriot's (1969) proposals, similar transformations can also be executed along the horizontal axis to get the best queuing of the points representing the numerous measurements, the data of which were published in the literature. In the example shown in Fig. 3-35, the Re_p parameter is applied along the horizontal axis. This solution gives an excellent result in the case of samples composed of spheres having identical diameters. In Fig. 3-36 all the collected measurements are represented including also heterodisperse samples and grains having a shape other than a sphere. In this case the best queuing was achieved by using the product $Re_p\ n^2$ as the X variable. Approximating the queue of points with a continuous curve, the following equation can be proposed to describe the v vs. I relationship having unlimited validity in the laminar, transition and turbulent zones:

$$I^{3/4} = Av^{3/4} + Bv^{6/4} \quad ;$$

where

$$A = \left(\frac{1}{K_D}\right)^{3/4} \quad ; \text{ and} \tag{3-37}$$

$$B = \left(\frac{2.5n^3}{\sqrt{K_D g \nu n}}\right)^{3/4} \quad .$$

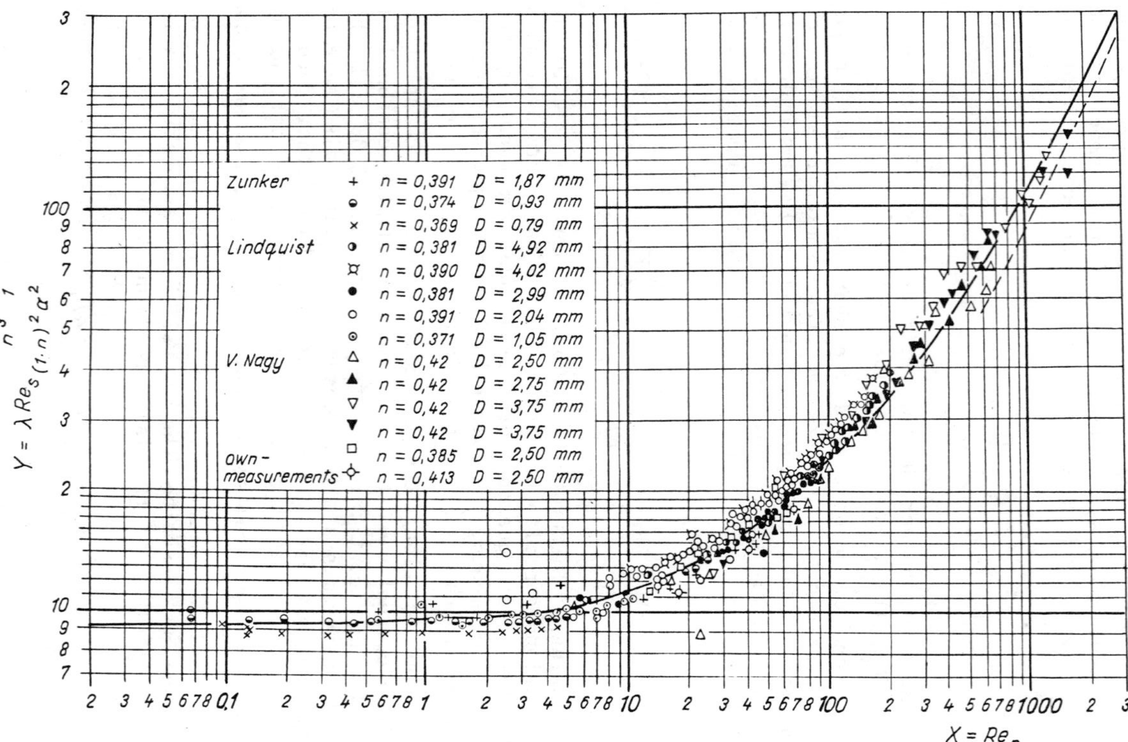

Fig. 3-35. The resistance of homodisperse samples of spheres represented as a function of Reynolds number.

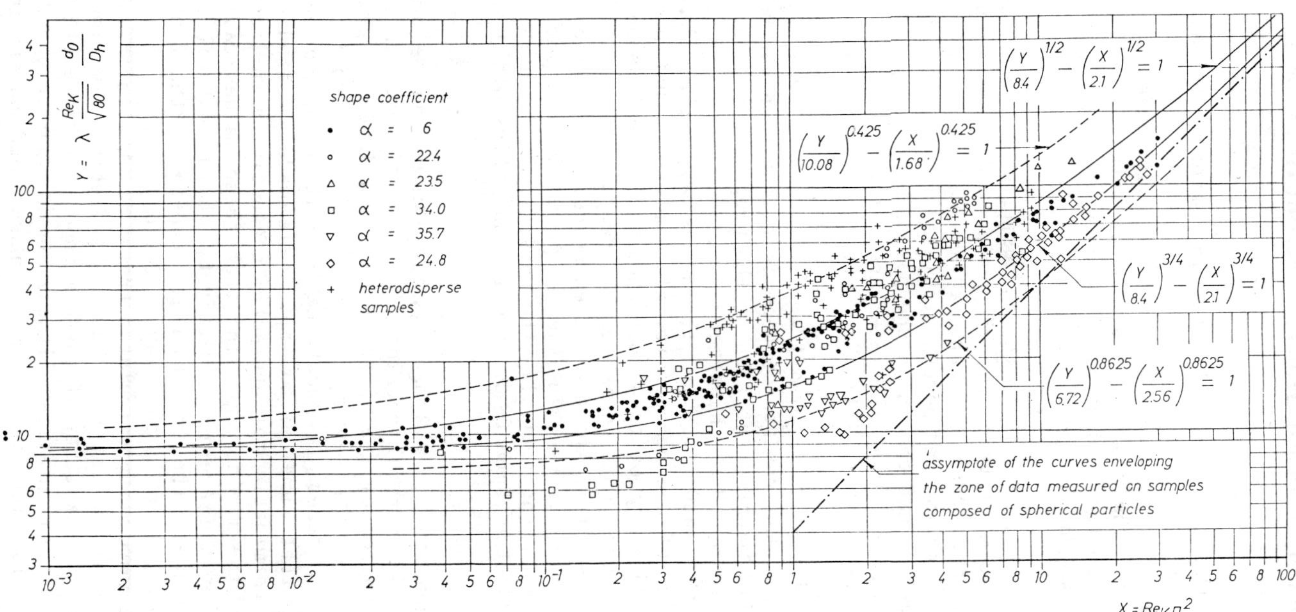

Fig. 3-36. Representation of the relationship between the resistance and Reynolds number on the basis of data measured by using different samples.

11-2 Laboratory and Field Methods to Measure Hydraulic Conductivity

When the resistance against flow through the network of channels composed of the pores of loose clastic sediments is theoretically investigated, it is generally assumed that this resistance is the same at every point within the flow field and it is also independent of the direction of flow, viz., the field is homogeneous and isotropic.

The first condition of the supposition of the homogeneity is the uniform distribution of grains and pores within the layer. In this case the parameters characterizing the size and shape of grains and pores calculated for elementary reference volumes do not depend on the position of the point determined as the center of the reference volume. It follows from this condition that intrinsic porosity is also constant because its numerical value is an unambiguous function of the listed soil-physical parameters. It was already proved, however, that the resistance depends not only on the behavior of the solid matrix, but also on the properties of the propagating fluid and the character of flow and in some special cases on the saturation of the pores as well. Thus, the precondition of the homogeneity of seepage is that both viscosity and density of the fluid should also be independent of the space coordinates (constancy of temperature, salt concentration, etc. within the considered space) and the flow condition must not change in the seepage field. The latter condition means theoretically the constancy of Reynolds number or that of the amount of actual and threshold hydraulic gradients.

Homogeneity of a loose clastic sample can be achieved only with the careful packing of the particles. In nature, the distribution of grains and pores in a layer is generally irregular. In solid rocks, even the use of the model composed of capillary tubes is questionable. It is quite natural, therefore, that numerous attempts were made to find methods for the determination of hydraulic conductivity in laboratories and in the field, supposing that these values can better characterize the average resistivity of the natural layers, and to eliminate the uncertainties caused by the nonhomogeneity of the layer in this way. Previously, when the coefficient of permeability calculated from soil-physical parameters was only a rough estimate due to negleting some important factors, the reliability of these methods was higher than that of the simple calculation, although both laboratory and field tests are also affected by many unknown and uncertain phenomena. At present, when the reliability of the available formulae was considerably improved by various theoretical investigations, there is practically no difference between the correctness of the various methods and sometimes the use of formulae supplies even more accurate data than some laboratory or field tests.

A further advantage of the numerical methods is their simplicity. In most cases nonhomogeneity can be better characterized by numerous data pertaining to different points of the flow space than a single measurement even if the latter describes the in situ condition. The number of the investigated points can be increased, however, when the method used to determine the permeability of the layer is simple, quick, and cheap. From this point of view, the best methods are the numerical formulae.

As it was already explained, the influence of the flow conditions can also be expressed by the various equations. Measuring the parameter in question in laboratory only, one flow condition is investigated in general. In most cases of field tests, even the character of the actual condition is not known or it occurs often that the flow conditions at various parts of the flow space and created by the test (e.g., by pumping) are different. It can be stated, therefore, that the use of laboratory and field methods is reasonable only when parameters must be and can be determined which cannot be calculated from equations (e.g., anisotrophy, development of large channels, direction of recharge, storage capacity, etc.).

The laboratory methods and the field tests directed at the determination of hydraulic conductivity are based on simultaneously measured data of the gradient and flow rate in a system where the flow was created and maintained artificially. Since the uncertainties mentioned in the introduction may cause considerable errors in the direct measurements, only those methods are discussed, the reliability of which is satisfactory and which provide supplementary information on the properties and behavior of the layer apart from permeability. This is the reason why the detailed analysis of the quick field tests is neglected here.

Laboratory tests can be performed with two different equipment. The simple one operates with constant pressure head. This test is practically the repetition of Darcy's experiments (Fig. 3-37). The length of the sample is ℓ and the pressure head is h. In a time interval t, the total amount of water percolating through the sample is V. The discharge (Q=V/t) divided by the cross section of the sample (A) yields seepage velocity, (v), which is proportional to the hydraulic gradient, $I=h/\ell$. The coefficient of proportionality is independent of the gradient if the movement is laminar (the latter condition must be ensured in any case), and this coefficient is equal to the hydraulic conductivity which can be calculated from the measured parameters:

$$K_D = \frac{v}{I} = \frac{V}{tA} \frac{\ell}{h} \quad . \tag{3-38}$$

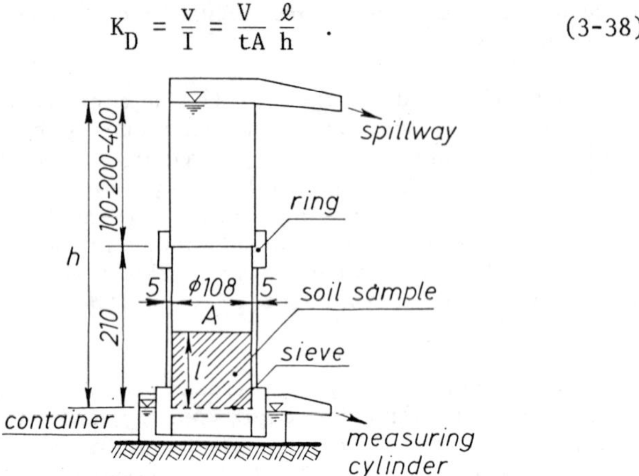

Fig. 3-37. Permeameter with constant pressure head (measurements in mm).

This method cannot be used for measuring the permeability of a cohesive sample because the amount of water percolating through the sample is very small in this case and, therefore, the relative error of measurements may be considerably high. An equipment operating with changing pressure head can be used to measure the permeability of such a sample (Fig. 3-38). The difference between the two methods is that in the second case only the lower water level is constant and the pressure is maintained by a water column, the upper level of which sinks gradually during the experiment in a calibrated glass tube. Both the pressure head and the flow rate can be calculated by measuring the changing water level in the tube. Since the pressure head changes, the movement is unsteady. The flow rate is equal to the change of water stored in the glass tube having a cross section of A_o. This condition can be expressed by the following differential equation:

$$A_o \frac{dh}{dt} = K_D \frac{h}{\ell} A \quad . \tag{3-39}$$

After separating the variables and integrating the equation, a relationship can be established to calculate the hydraulic conductivity:

$$K_D = \ell \frac{A_o}{A} \frac{1}{t_2 - t_1} \ln \frac{h_1}{h_2} \quad ; \tag{3-40}$$

where two boundary conditions are taken into account, i.e., at the time of t_1 the height of the upper water level above the lower constant level is h_1, and the same pressure head at the time of t_2 is h_2.

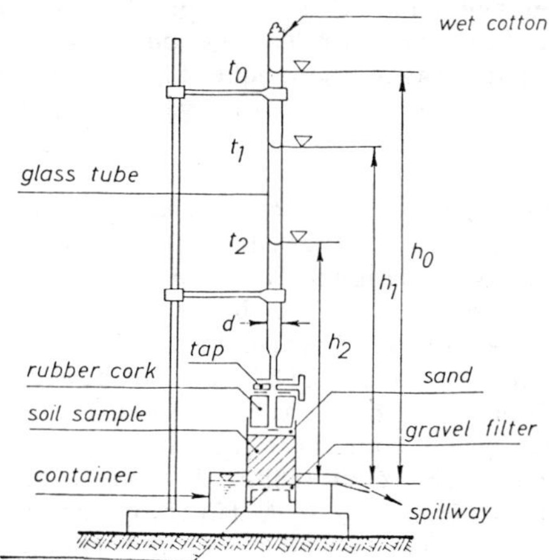

Fig. 3-38. Permeameter with changing pressure head.

Finally, the results of experiments aiming to measure the consolidation of a sample are also sometimes used to determine the permeability of very cohesive clays. The development of compression in time depends also on the permeability of the sample as it is clearly expressed in Terzaghi's differential equation:

$$\frac{\Delta e}{\Delta e_1} = 1 - \sum_{n=0}^{\infty} \frac{8}{(2n+1)^2 \pi^2} \exp (2n + 1) \psi t \quad ;$$

where (3-41)

$$\psi = \frac{\pi^2}{4} \frac{C}{h^2} \quad ; \text{ and } \quad C = K_D \frac{M}{\gamma} \quad ; \text{ and}$$

Δe is the decrease of the void ratio until the time of t, Δe_1 is the total expected decrease, h is the thickness of the layer and M is the coefficient of compressibility. To calculate the consolidation, the index of consolidation (C) is measured by special equipment (oedometer). Hydraulic conductivity can be calculated from this parameter knowing the coefficient of compressibility and the specific weight of the sample.

Laboratory tests have a common feature: a sample has to be taken from the layer and built into the measuring instrument. The disturbed samples are not convenient for this purpose. From the boreholes, the drilling tools bring up the material selected according to the size of grains (from coarse-grained layers) or in kneaded condition with altered structure (from clays). The porosity of the disturbed sample in the measuring instrument depends on the preparaton of the test and not on the natural condition of the layer, it is only an estimated value, similar to the parameter used in the case of numerical formulae. Consequently, the investigation of such a sample which does not preserve any special character of the layer is unjustified.

The determination of a parameter representing some natural conditions of the layer, which cannot be calculated from soil-physical parameters, can only be expected using undisturbed samples which are built together with the core-tube into the measuring instrument. To take an undisturbed sample is, however, very difficult especially from grained saturated sediments. There are special equipment constructed to get an undisturbed core from boreholes or from trenches dug for exploration. The core-tube, however, always causes some deformation along its wall. A further error may be caused by the discontinuity between the core and the wall of the tube. The influence of the water amount percolating along the wall may be significant as compared to the total discharge in the case of clay samples when the total amount percolating through the sample is relatively small. To decrease this possible error, the cross section of the core should be as large as possible.

In spite of all the listed difficulties, the measured permeability may supply some additional information on the natural structure of the layer. Thus, there are cases when this method is one of the most reliable tests to determine some special parameters of the sediment, e.g., anisotropy.

One way to avoid the difficulties arising in connection with taking undisturbed samples is to measure the permeability in natural condition (in situ). For this purpose, various field methods were developed.

The permeability of the unsaturated layer near the surface can be measured by the simplest method which is the so-called infiltration test, the detailed description of which was given in Subsection 6-6.

Field tests to measure the permeability of layers below the water table can be divided into three groups:

- recharge of the borehole;
- short period of pumping observing at first the lowering of the water level and its rise after stopping the pumping;
- pumping test.

The common basis of the methods pertaining to the first group is that an excess pressure is created in the borehole related to the natural water table, and the rate of flow induced by this pressure is measured. Differences between the various methods are caused by the position and size of the surface through which infiltration takes place, and which may be either at the bottom or at the sidewall of the borehole, perhaps at both places. A further difference can be caused by the character of the flow. Sometimes the unsteady state of movement is observed, in other cases the development of the balanced condition is to be awaited, and the parameters of steady seepage measured. Khafagi's and Lugeon's methods belong to this type of tests.

A characteristic test in the second group is Porchet's method. The water from the well or borehole is pumped at a constant rate for a short period. The level in the well is observed during and after pumping until the well is refilled. From the discharge and the graph of the water level changing in time, the expected yield of the well or the hydraulic conductivity of the layer is calculated.

A general error in the listed methods (so-called quick pumping tests) is that the boundary conditions of the flow field are not known and, therefore, neither the condition of flow nor the effects acting can be determined. The lack of this information causes many uncertainties and the reliability of these methods, therefore, is much lower than that of the simple mathematical formulae. In the case of pumping, the uncertainty is further increased by the development of the free exit face, viz., the level in the pumped well is always lower than the pressure head in the layer.

The most reliable method for in situ investigation of the hydraulic conductivity or permeability of layers below the water table is the pumping test using more than two observation wells. The pumped well penetrates the aquifer crossing its entire depth if possible. Using this method, the pumping rate is constant and the hydraulic conductivity is calculated from the observed depression cone. The depression is determined by observation wells generally after awaiting the balanced condition, when the depression curve attained its constant position. There are, however, methods for calculating the K value from the modificaton of the depression using the parameters of unsteady flow.

The basic equation of this calculation is the Dupuit formula according to which the yield of a pumped well can be calculated from the following relationships:

unconfined flow field

$$Q = \pi K_D \frac{H_2^2 - H_1^2}{\ln \frac{R}{r_o}} \quad ; \qquad (3\text{-}42)$$

confined flow field

$$Q = 2\pi K_m \frac{H_2 - H_1}{\ln \frac{R}{r_o}} \quad , \qquad (3\text{-}43)$$

(The meaning of the symbols used in the equations is given in Fig. 3-39.)

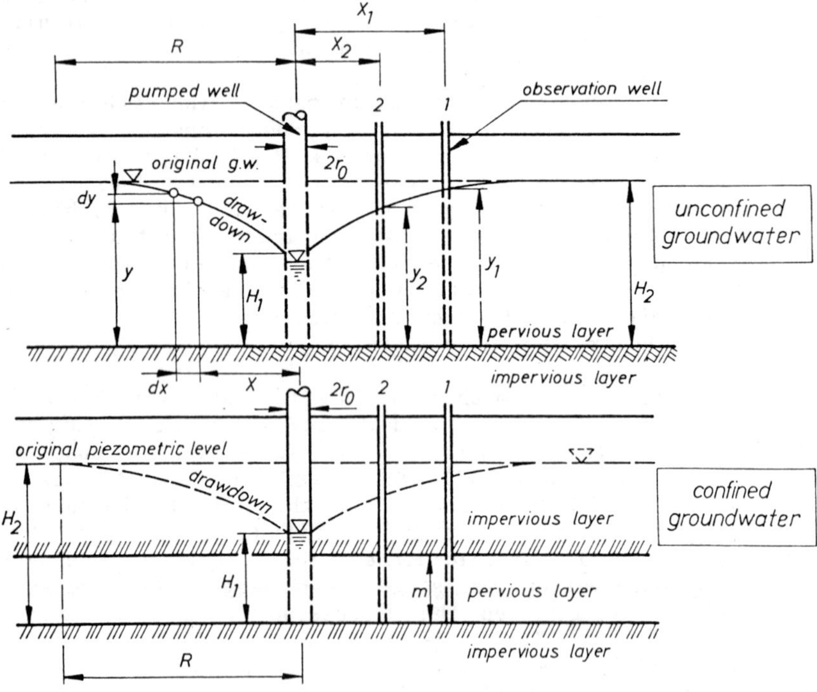

Fig. 3-39. Symbols used for deriving Dupuit's equations to characterize axial symmetrical flow.

There are two problems hindering the application of these equations:

- the development of the free exit face at the screen of the well;

- the fact that the length of the zone affected by the depression (R) is generally unknown.

Both difficulties can be eliminated using two observation wells in addition to the pumped one. Dupuit's equation was derived from the differential equation of movement by double integration. The constants of the integrations were determined by using the water level in the well and that at the end of the depression as boundary conditions. The integration can be similarly executed with other boundary conditions as well. To avoid the errors caused by the uncertain flow conditions around the well and near the end of the depression, the water levels measured at the observation wells have to be used as boundary conditions.

Equations determined on the basis of this assumption, more precisely the hydraulic conductivity expressed from these equations, can be given as follows:

unconfined flow field

$$K_D = \frac{Q}{\pi} \frac{\ell n \frac{x_1}{x_2}}{y_1^2 - y_2^2} \quad ; \qquad (3\text{-}44)$$

confined flow field

$$K_D = \frac{Q}{2\pi m} \frac{\ell n \frac{x_1}{x_2}}{y_1 - y_2} \quad . \qquad (3\text{-}45)$$

Summarizing the discussions about the various laboratory and field tests, it can be stated that there are only three methods whose use is reasonable:

- laboratory tests with undisturbed samples;

- infiltration tests;

- pumping tests with more than two observation wells.

All other methods (including pumping tests executed without or with only one observation well and, therefore, the observed level of the pumped well and the length of the affected zone have to be taken into account as well) evaluate parameters measured in a seepage field, the boundary and flow conditions of which are unknown and, therefore, the hydraulic conductivity determined in this way is very uncertain. This is the reason why tests apart from the previously listed three should be avoided.

11-3 Evaluation of Pumping Tests with More Observation Wells

It can be stated on the basis of the critical evaluation of laboratory and field tests given in the previous paragraph, that the use of formulae are preferable to many types of field tests and the application of experimental methods is reasonable only in those cases when some special feature of the layer can be in this way determined. Among the advisable methods, there is only one field test which is suitable to investigate the permeability of layers below the water table, i.e., the pumping test with several observation wells. The evaluation of the data observed in this way is very difficult because they are influenced not only by the natural conditions of the flow space but also by the special flow conditions created by the pumping itself. Grouping the collected data carefully and evaluating them from hydrodynamic aspects, very valuable information can be gained on the properties of the layer and on the conditions affecting the system.

To achieve the better explanation of the theoretical basis of the evaluation of pumping tests, the geometrical, flow, and boundary conditions have to be at first summarized, which were the fundamental assumptions for the derivation of Dupuit's equations (Eqs. 3-42 and 3-43). These approximations can be listed as follows:

- the flow space is homogeneous and isotropic;

- the aquifer is bounded from below by a horizontal impervious layer, its upper surface is either uncovered or the contact surface of the covering impervious (or semi-impervious) layer is also horizontal;

- the flow space is neither recharged (infiltration) nor drained (evaporation) along its surface;

- the length of the stretch affected by the depression can be characterized by a radius of R which condition is ensured only when the well is situated in the center of a circular island and the pumped water can be recharged without any resistance along the bank of the latter, thus an axial symmetrical flow can develop;

- the seepage remains within the validity zone of Darcy's equation at every point of the flow space;

- the water is in the state of equilibrium before pumping and thus its table is horizontal under natural conditions;

- the seepage is a steady flow, this assumption also requires that the recharging section of the flow space should be within a finite distance from the well and the level of the recharging surface water has to be constant (if any one of these hypotheses is not satisfied the movement is unsteady);

- finally, the last approximation is the so-called Dupuit's hypothesis which assumes that each cylindrical surface, concentrical around the well is a potential surface, the velocity is constant at every point of this surface and proportional to the gradient of the water table or the piezometric surface determined at this distance from the well; in other words Dupuit assumed that the stream lines can be approximated by horizontal lines, and the vertical component of the velocity vector can be neglected.

Analyzing the applicability of these hypotheses in the case of a pumping test, it can be determined how the unsatisfied assumptions react on the observed data. According to Eqs. 3-44 and 3-45, the hydraulic conductivity can be calculated knowing the distances of two observation wells from the pumped well and the drawdowns at these two points. Having more than two observation wells, a parameter can be determined from the data of each pair of wells. If all the assumptions listed above were satisfied, these parameters would be identical within an acceptable scattering caused by the errors of measuring and calculating hydraulic conductivity.

When the differences between the various values of hydraulic conductivity calculated from any pair of wells are higher than that which can be explained by these errors, the nonhomogeneity of the layer may be assumed, or the fact indicates that one or more of the listed hypotheses are not satisfied. If the calculated data are not constant but their change shows some regular pattern, this regular arrangement also gives some ideas on the basis of which of the assumptions was unsatisfied. Some important information can also be gained on the natural conditions of the layer, e.g., the main direction of the flow, the type and amount of recharge, the flow condition, etc.

The nonhomogeneity of the layer is indicated by the irregular but high scattering of the data. A regular arrangement of the variation of permeability may also be caused by nonhomogeneity if the change of the physical behavior of the layer is regular.

The position of the surface of the lower lying impervious layer is determined by using geological and geophysical methods. Thus, its declination from the horizontal plane can be checked directly, or knowing the actual position of the lower boundary, the change caused by the actual thickness of the layer can be considered.

The natural effects influencing the seepage along the flow space (accretion) cannot be avoided. There are methods taking into account these influences as well, thus some corrections can be calculated when the vertical recharge or drainage is significant compared to the horizontal flow. The relationships describing this special condition are, however, more complicated. It is advisable, therefore, to choose a period for the execution of the pumping tests when the accretion is relatively low, e.g., in winter when the frozen surface hinders both infiltration and evaporation. An important aspect which must be considered is that the flow space should not be recharged artificially. The water pumped out during the test should be conveyed through pipes to a recipient.

The exact radial flow and the circular recharge can be ensured only in the laboratory. When the natural recharge is coming from several directions, the uniform distribution of flow around the well can be assumed. In some other cases, when the flow has one main direction (e.g., in the case of a well near the bank of a river or a lake), the calculated permeability coefficients differ depending on the direction where the observation wells are located (Fig. 3-40). Evaluating the measured data, it is assumed that the observed slope of the depression cone has developed under the influence of a uniformly distributed flow. On the contrary, the flow in the main direction is higher than the average and creates, therefore, a steeper slope. Thus, the pair of wells in this direction gives a lower K value than the actual one, while the permeability calculated from the data of other wells is higher than the average. Some special types of nonuniformly distributed flow can be taken into account using such equations for evaluation which were derived by considering relevant boundary conditions.

This analysis draws to attention the importance of the length of the influenced zone (R) as well. The use of empirical formulae to determine some characteristic length (e.g., Sickhart's equation) is unacceptable because such arbitrarily chosen parameters do not describe the actual geometry of the flow space. The fictive values calculated, e.g., on the basis of an assumed permeability, have no physical meaning at all. When there is a river or lake near the well the distance of its bank can be taken into account as R. It is more reliable, however, to choose a method which eliminates the application

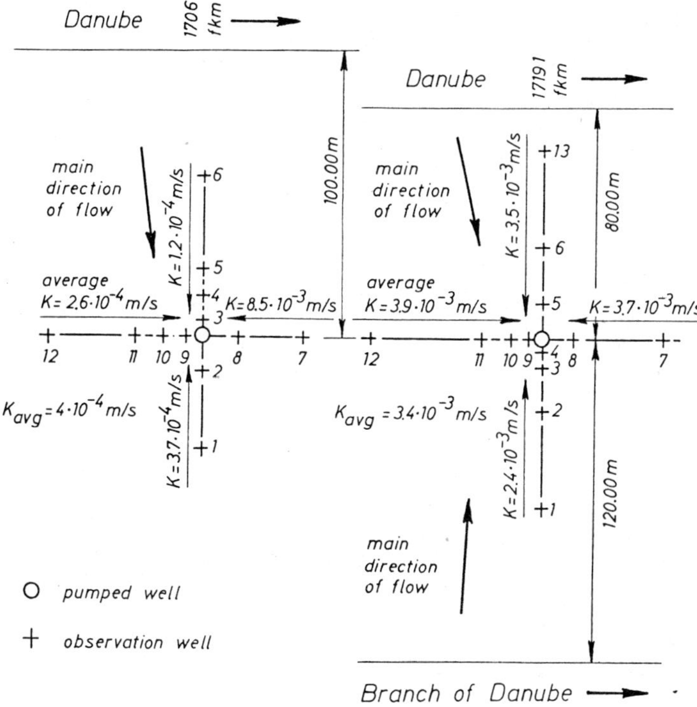

Fig. 3-40. Modification of hydraulic conductivity calculated from the data of pumping tests having uneven recharge.

of this parameter. This is the reason why the use of Eqs. 3-44 and 3-45 is preferable to Eqs. 3-42 and 3-43 which require at the same time the application of a minimum of two observation wells.

When the investigated aquifer is very thick, the drilling of the pumped well down to the lower impervious layer would be costly and the yield of a well perforated along the entire depth of such a layer would be very high compared to a reasonable pumping capacity. In this case, the pumping test may be executed with a partially penetrating well. A general practice is to evaluate the data collected from such a system also by using Dupuit's equations assuming, in most cases, that the influence of the layer below the horizontal plain through the sole of the well can be neglected. The y variables are consequently measured from this plain as a datum of reference (Fig. 3-41). It is evident that a smaller flow rate moves through

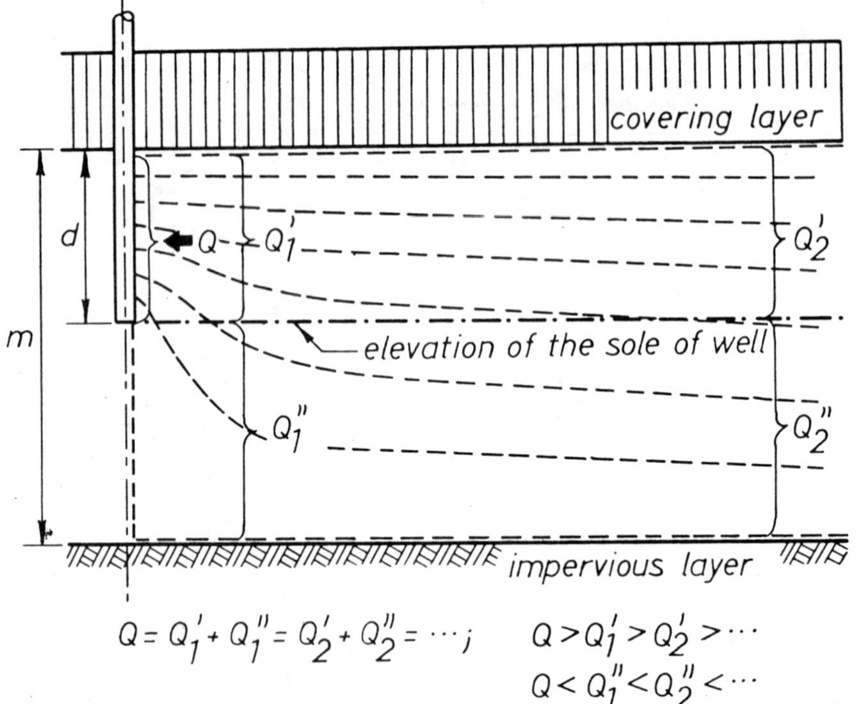

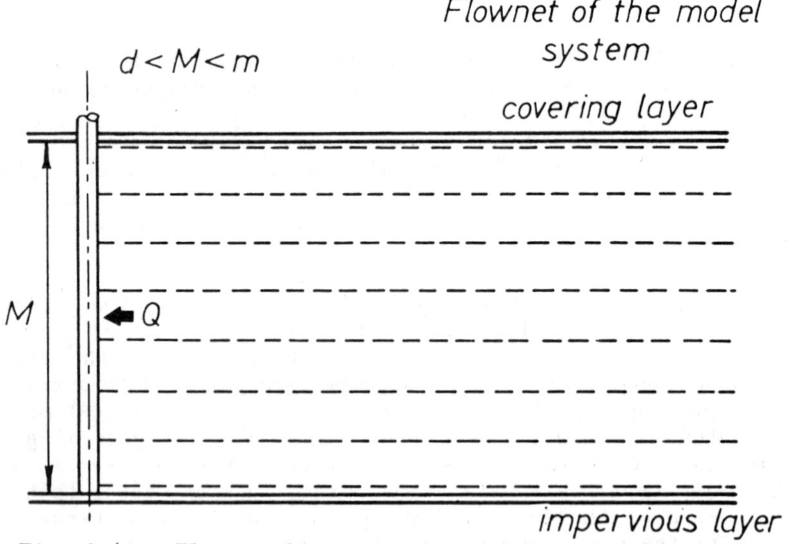

Fig. 3-41. Flow condition in the vicinity of a partially penetrating well in a vertical flow plane.

this considered part of the layer than the total yield of the well and, therefore, the actual slope of the depression cone is smaller than it would be in the case of an aquifer which is actually bordered at this depth by an impervious layer. This is the reason why the calculated permeability is always higher than the actual one independent of the location of the pair of wells used for the calculation. There is, however, a regular arrangement of the determined parameters. The ratio of the flow moving above the horizontal plain of the sole of the pumped well to the total yield increases when coming nearer to the well as it is also shown in Fig. 3-41. The regular error caused by this effect is smaller when the observation wells are near the pumped one and the calculated K values show an increasing trend as a function of the distance from the center of the system.

The case of a pumped well penetrating down to 22 m below the surface of a very thick layer of gravel is shown in Fig. 3-42 as an example. Using a method (Kovács, 1966) suitable to calculate the hydrodynamic parameters of a flow system around a partially penetrating well for evaluating the observed data, the hydraulic conductivity was calculated as a function of the distance from the well, assuming various actual depths of the aquifer. According to the dependency explained previously, the K value increases with the distance and decreases with the assumed actual depth. At the same time, the modification in the horizontal direction becomes less and less when the fictive thickness of the layer is increased. Theoretically, the calculated permeability must be independent of the location of the observation wells used to determine the hydraulic conductivity, if the fictive depth is equal to the actual one. In the example, this condition is attained at a depth of 200 m which is in good agreement with the real thickness of the layer known from some deep boreholes in the vicinity. The comparison of the hydraulic conductivity calculated from the observation wells assuming this depth (the former being practically constant in this case K = 0.6 to 0.7×10^{-3} m/sec) to that determined on the basis of soil-physical parameters (K = 0.9×10^{-3} m/sec) proves also the reliability of both the method of the evaluation and the method proposed for calculating the yield of a partially penetrating well.

Neglecting the development of the free exit surface causes error only when the flow space is unconfined since in the case of seepage under pressure the water level in the well is above the upper surface of the aquifer, thus a free exit surface cannot develop. Excluding the water level in the pumped well from the data used in the calculation, this error is practically eliminated because it can be proved hydrodynamically that Dupuit's equation is an acceptable approximation to determine the yield of a well (Charnyi, 1951). There is,

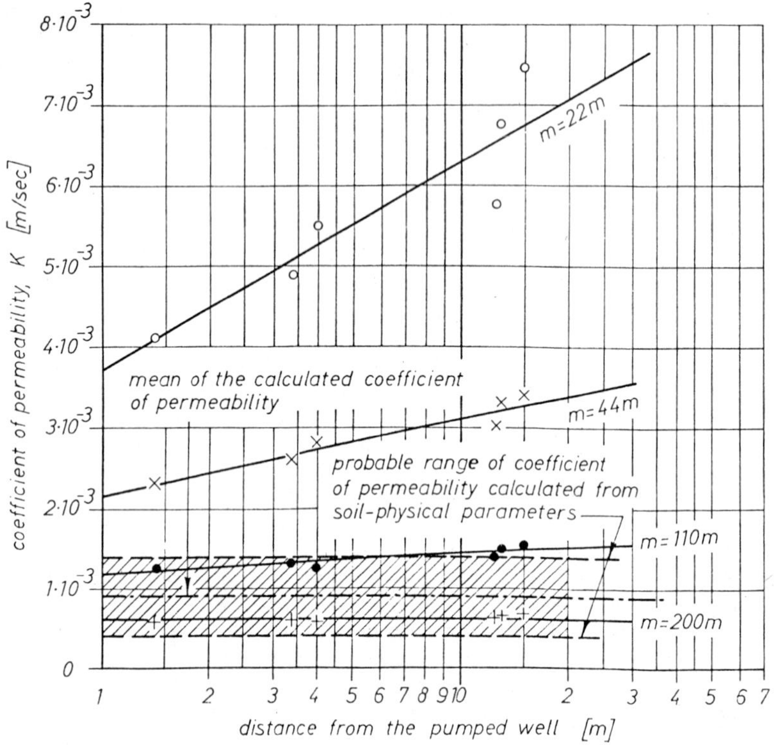

Fig. 3-42. The correction of hydraulic conductivity calculated from the data of a pumping test executed with a partially penetrating well.

however, a small regular error, i.e., the actual yield may be slightly higher than the calculated one depending also on the capillary water conveyance of the layer. The maximum of this difference may reach about 10 percent when the water column in the well is zero ($H_2 = 0$). The error decreases with increasing H_2. Thus, the hydraulic conductivity determined by pumping tests may be 10 percent greater than the actual value in the case of total drawdown.

The next hypothesis is the validity of Darcy's law. To show the possible error caused by not satisfying this condition, the results of a laboratory experiment are given. A well was placed in the center if a circular shaped basin recharged along its outer edge. The sandy aquifer was covered by horizontal glass plates at both sides and the flow remained under pressure during the whole experiment. Thus, all the other hypotheses applied for the derivation of Dupuit's equation were correctly satisfied. Fig. 3-43 shows a vertical radial section of the basin indicating both the total area of the cross ection (A) and the Reynolds number (Re_p) as a function of the distance from the axis. Similarly, the drawdown curve and the hydraulic conductivity calculated from the latter, depending also on the distance, were plotted on the graph. The calculated permeability increases gradually going away from the well and becomes constant where Re_p decreases below 10, consequently where the validity of Darcy's law can be assumed. Outside this section, the average of K determined on the basis of the observed piezometric line is $K = 4.37 \times 10^{-4}$ m/sec which is in good agreement with the parameter calculated from the grain-size distribution curve ($K = 4.5 \times 10^{-4}$ m/sec. The parameter calculated would be only about half of the realistic value (2.4×10^{-4} m/sec) if one of the wells whose data are used to determine permeability were in the zone of nonlaminar movement.

When there is groundwater flow in natural condition as well, the water table has a natural slope. In this case the excess drawdown caused by pumping must be measured. This requirement makes the observations more difficult because the value to be measured is sometimes smaller than the error expected because of the inaccuracy of the measuring methods used. It is a general attempt, therefore, to choose such a location for the pumping tests where the natural slope of the water table can be neglected.

There are many reported pumping tests indicating a constant hydraulic conductivity around the well and a varying value only at a greater distance. If the permeability were really smaller here than the constant parameter, the change could be explained by the development of microseepage, although the effect of adhesion is generally small in coarse-grained sediments in which the experiments quoted were executed. In the reported cases, however, a higher permeability is always indicated at a greater distance from the well (Fig. 3-44) (Szilágyi, 1954; Ubell, 1958). This irregularity can be caused by the fact that the drainage of the stored water does not instantaneously follow the lowering of the water level and, therefore, at the areas far from the well the depletion of the pores is delayed, the water amount drained from these pores becomes a part of the horizontally transported flow only after a considerable time lag has elapsed.

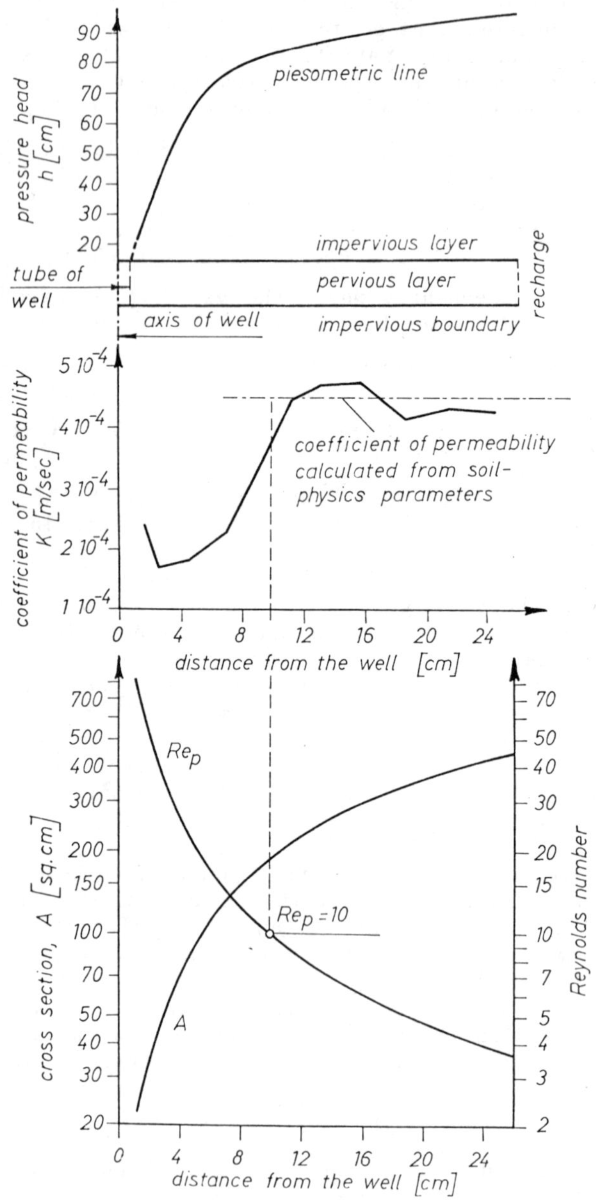

Fig. 3-43. Decrease of the calculated hydraulic conductivity caused by nonlaminar seepage.

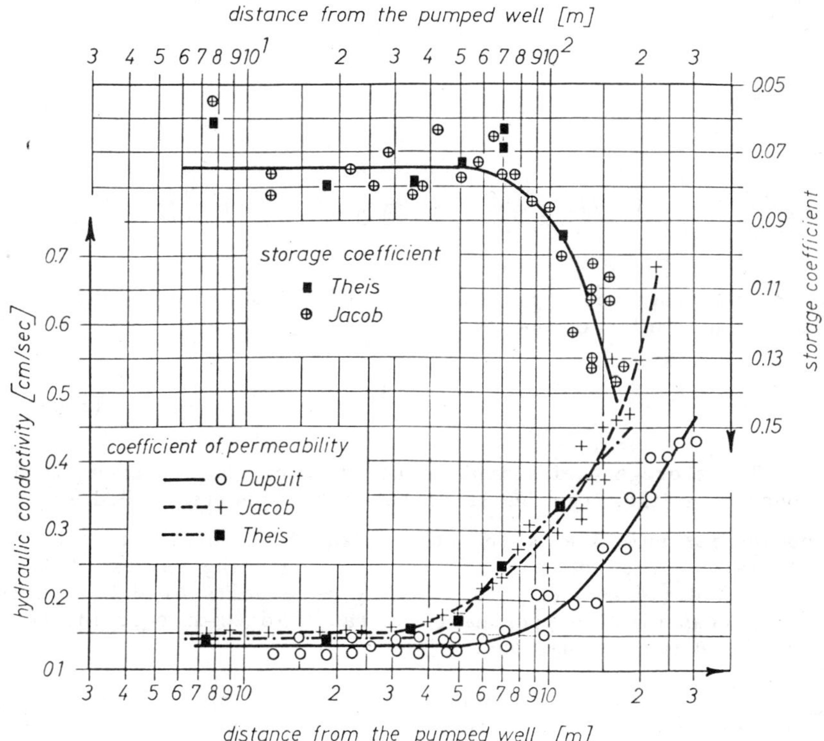

Fig. 3-44. Increase of the calculated hydraulic conductivity far from the pumped well.

To avoid the limit of application caused by the assumption of steady flow, which is a fundamental hypothesis of Dupuit's equation, formulae based on the hydrodynamical analysis of axial symmetrical unsteady flow were developed which can be used for evaluating the unsteady period of movement around the pumped well and to calculate the permeability and storage coefficients from the data characterizing the lowering of the drawdown curve. The basis of the method generally used is the Theis equation (Theis, 1935):

$$K = \frac{Q}{4\pi\ sH} \text{ Ei } (u) \quad ; \qquad (3\text{-}46)$$

where s is the drawdown in an observation well at a distance of r from the well axis after a time t having elapsed from the beginning of pumping, and H is the original depth of the groundwater (or the thickness of the aquifer if the system is confined). The exponential integral function Ei(u) can be expressed by expanding the term to be integrated into a series:

$$Ei(u) = \int_u^\infty \frac{e^{-u}}{u} du$$

$$= -0.5772 - \ell n u + u - \frac{u^2}{2 \cdot 2!} + \frac{u^3}{3 \cdot 3!} - \frac{u^4}{4 \cdot 4!} + \ldots ; \qquad (3\text{-}47)$$

where the u variable is a function of the storage coefficient and the hydraulic conductivity:

$$u = \frac{r^2 S}{4KHt} \qquad (3\text{-}48)$$

The storage coefficient in the case of shallow unconfined groundwater flow can be substituted by specific yield (n_s). Knowing its numerical value, the hydraulic conductivity can be calculated from one pair of the corresponding values of the observed drawdown and time. Having more pairs of data, both storage capacity and permeability can be determined using the listed system of equations.

The calculation cannot be directly executed because of the exponential integral function. Theis applied, therefore, a graphical method. Jacob (1940) explained, however, that the parameter u is inversely proportional to time, thus, the third and further members on the right-hand side of Eq. 3-53 are negligible as compared to the first two members when the time is long enough. Considering this simplification, the corresponding data of time and drawdown at a given distance from the well can be represented in a semi-logarithmic coordinate system (i.e., log t and s) by a series of points fitted to a straight line. Each line belonging to a given distance of r intersects the horizontal axis at a $t_o(r)$ point which is the fictive starting time of the drawdown at the distance in question (Fig. 3-45). On the basis of the slope of these lines, both hydraulic conductivity and storage coefficient can be calculated:

$$K = \frac{Q}{4\pi\, H(s_2 - s_1)} \ell n \frac{t_2}{t_1} ; \qquad (3\text{-}49)$$

$$S = 2.245 \frac{KHt_o(r)}{r^2} . \qquad (3\text{-}50)$$

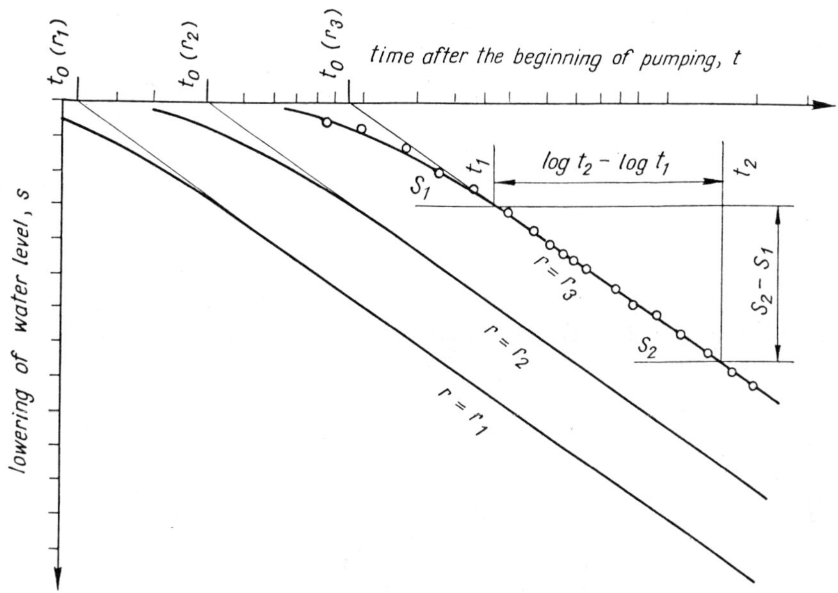

Fig. 3-45. Representation of the relationship characterized by Jacob's equation.

Comparing the various values of hydraulic conductivity calculated from Dupuit's, Theis' and Jacob's equations, it can be stated that the calculation of steady and unsteady movement gives practically the same result in the vicinity of the well where the calculated parameters are constant (in the case of the example shown in Fig. 3-44, this distance is about 50 m). After a short period from the beginning of pumpng, this section of the drawdown curve is lowered parallel to its previous position because the relatively rapid depression and high suction cause a quick emptying of the gravitational pores. The similarity of the results of the two basically different methods can be explained by the fact that their fundamental assumptions are equally satisfied (or can be regarded as acceptable approximations) in the vicinity of the pumped well. In the case of Dupuit's equation, this hypothesis is the steady condition of flow and in Theis-Jacob's method it is the parallel change of the drawdown curve.

Because of the variation of storage capacity in time, the parameter determined by the Theis-Jacob methods for a given time point is also a function of the distance measured from the pumped well. The coefficient near the well can be regarded as a good approximation of the actual value. At a greater distance, water originating from the delayed depletion of the upper lying pores increases the horizontal flow rate compared to the theoretical value. This process results in higher storage capacity than the actual one and makes the calculated parameter very uncertain as indicated in Fig. 3-44 also.

As a summary of this analysis, it can be stated that both steady and unsteady methods are applicable only when data observed near the pumped well (where the drawdown curves are practically parallel to each other in different points of time) are used. Thus, the same pumping test can be evaluated by both methods. Generally, the use of Dupuit's equation is preferable because of its simplicity. The application of the Theis-Jacob method is advisable in cases when the final objective is the study of an unsteady process and parameters (including also storage capacity or specific yield) are to be determined for this investigation by the pumping test.

11-4 Characterization of Microseepage

The ideal (non-viscous) fluid is a theoretical concept. Its molecules are absolutely movable, their displacement related to each other is not hindered by any resistance.

There exists attraction, however, between the molecules of the actual fluids, the result of which is the occurrence of a shearing stress between the layers of molecules if the two layers move with different velocity. The internal friction acting on a unit area separating the two layers is equal to the shearing stress which is proportional to the change of velocity determined perpendicularly to the direction of flow. The coefficient of proportionality is the dynamic viscosity (μ) (Newtonian fluid).

In the case of fluids having dipole molecules (e.g., water) the binding of the negative and positive poles of the dipoles increases the mass attraction of the molecules existing in any types of fluids. The dynamic viscosity of these special fluids characterizes, therefore, the combined action of the two affects, i.e., mass attraction and binding of dipoles. The orientation of the dipoles is not constant. The bonds are disintegrated by Brownian movement and new groups are continuously developing. Thus, the viscosity of fluids having dipole molecules is a probability parameter of randomly bound molecules which always represents the equilibrium between attraction and Brownian movement.

Any external action (e.g., pressure) which increases the probability of the orientation and binding of the dipoles increases the shearing stress of the fluid as well, while those lowering the number of bounds (e.g., increasing temperature) have the inverse effect. Among the influences causing higher shearing stress, the electrostatic charge of the solid wall surrounding the fluid has a special role. The dynamic investigation of this effect considers the action of a second force (i.e., adhesion) apart from internal friction as a retarding effect acting against gravity. The dynamic aspects of the consideration of such molecular force were already explained in Part I. Here some results of the analysis are repeated in a summarized form.

The electrostatic charge of the wall polarizes the water molecules. The dipoles are fully oriented and bound to the wall in the first layer. The polarization can also be observed within the water body inside this double layer. The number of the oriented molecules, however, is inversely proportional to the distance measured from the wall (δ). The adhesive (Van der Waals) force acting perpendicularly to the wall (and expressed by suction which is the force related to a unit surface) is generally described in theoretical physics by a hyperbole of sixth order:

$$p = \frac{C}{\delta^6} \quad . \tag{3-51}$$

Adhesion, being perpendicular to the wall, has no component in the direction of flow. Its retarding effect on movement has to be taken into account in an indirect way considering its influence on viscosity or more precisely on shearing stress. The application of this method means a deviation from the Newtonian fluid and requires the investigation of a material having shearing stress in static condition as well. This kind of media is a special type of Bingham's bodies, the static shearing stress of which is a function of the distance measured from the wall ($\tau_o(\delta)$ non-Newtonian fluid). Applying the total shearing stress (the sum of the static one and that depending on viscosity), the combined retarding effect of friction (S) and adhesion (E) can be described with one equation:

$$\tau(\delta) = \tau_o(\delta) + \mu \frac{dv}{dn} \quad ;$$

consequently

$$S + E = A \left[\tau_o(\delta) + \mu \frac{dv}{dn}\right] \quad ; \tag{3-52}$$

where A is the area between two layers of molecules, the relative displacement of which is investigated.

The existence of the static shearing stress is proved by the fact that in cohesive layers (and also in capillary tubes of very small diameter) there is a limiting value of gradient (I_o, the threshold gradient), below which there is no movement at all, thus, zero velocity belongs to finite gradient. The measured value of the threshold gradient can, therefore, be used to characterize the static shearing stress (Karádi and Török, 1955; Kutilek, 1965).

Bondarenko and Nerpin (1966, 1967) and Bondarenko (1973) have published the results of a series of experiments, the objective of which was the determination of the threshold gradient in capillary tubes. Figure 3-46a shows the corresponding velocity and gradient values measured in tubes with different diameters (the material of the tubes was quartz or glass) by using water as the fluid flowing, while Fig. 3-46b represents the results achieved by applying ethyl alcohol. The data prove that the relationship between the two variables (i.e., v and I) can be well approximated by a linear equation almost in the entire range of velocity. The straight line representing this relationship does not cross the origin of the coordinate system (as it ought to if Darcy's equation were valid), but intersect the horizontal axis at a value of I_1. The I_1 gradient is not equal, however, to the threshold value because the flow starts under the influence of a smaller gradient as well, the linear relationship being not valid in the zone of very low velocities, as it is shown by Fig. 3-46c (which is the enlarged form of the lowest part of a velocity vs. gradient graph). The figure is also supplemented with a theoretical sketch, Fig. 3-46d, showing the interpretation of the threshold gradient (I_o) and the I_1 parameter being the intersection of the linear relationship and the horizontal axis.

Figure 3-46d also represents a third parameter (I_{10}) which is a fictive approximation of the threshold value applied by Bondarenko and Nerpin. They have derived this parameter by solving Buckingham-Reiner's equation in which the static shearing stress is supposed to be constant in the entire water phase, thus, a relationship can also be determined between I_{10} and τ_o:

$$v = K_D I \left[\frac{1}{3}\left(\frac{I_{10}}{I}\right)^4 - \frac{4}{3}\frac{I_{10}}{I} + 1\right] \text{ (Buckingham-Reiner equation);}$$

$$v = K_D I \left[1 - \frac{4}{3}\frac{I_{10}}{I}\right] \text{ (simplified form);} \qquad (3\text{-}53)$$

$$\tau_o = \frac{1}{2} r \gamma I_{10} \quad \text{(constant static shearing stress);}$$

where K_D is Darcy's hydraulic conductivity, r is the radius of the capillary tube, and γ is the specific weight of the fluid.

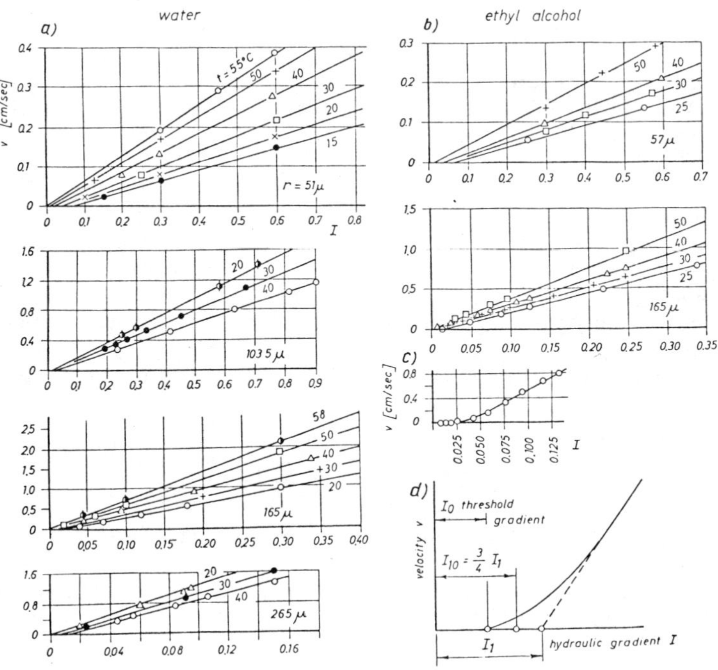

Fig. 3-46. Relationship between velocity and hydraulic gradient in the zone of small velocities (after Bondarenko and Nerpin).

Most of the authors dealing with this problem assume that the static shearing stress is a function of the distance measured from the interface between the solid bordering material and the non-Newtonian fluid (Kovács, 1957, 1958; Juhász, 1958; Karádi and V. Nagy, 1960; Childs and Tzimas, 1971). On the contrary, Bondarenko (1973) argues that this parameter has to be a constant within the entire interior of the fluid because he does not see any physical influence which may modify the property of the fluid. It seems to be proved that the special behavior of the fluid is caused by adhesion. Since the chains of dipole molecules are longer and more stable in the electrostatic field of the wall and this influence decreases gradually going away from the interface, the physical parameter chosen to characterize the non-Newtonian property ($\tau_o(\delta)$, the static shearing stress) should be proportional to adhesion and decrease with the increase of the distance from the wall.

On the other hand, the relationship between adhesion and τ_o does not need to be linear, and a more gentle decrease of the static shearing stress can also be assumed than that of adhesion. The acceptable approximation can be determined by establishing a dynamic model suitable to describe the flow of a non-Newtonian fluid in a straight tube and by calculating the parameters of the model from experimental data. The form of the function does not have to be linked to or influenced by the high power used in Eq. 3-51 to describe the adhesion vs. distance relationship (Thirriot, 1969).

The dynamic model required for the characterization of microseepage (non-Newtonian flow) can be derived from the Poiseuille equation supplemented by considering the influence of adhesion in the form of a static shearing stress expressed as the function of distance of the investigated point from the axis of the tube (r):

$$I\gamma r = 2[\tau_o(r) - \mu \frac{dv}{dr}] \quad . \qquad (3\text{-}54)$$

After differentiating both sides of the equation, the gradient representing the accelerating force (i.e., gravity) can be divided into two parts, one balanced by the static shearing stress and the other by viscous resistance caused by friction:

$$I_\gamma = (I_\tau + I_s)\gamma = 2\left[\frac{d\,\tau_o(r)}{dr} - \mu \frac{d^2v}{dr^2}\right] \quad ;$$

where

$$I_\tau \gamma = 2\,\frac{d\,\tau_o(r)}{dr} \quad ;$$

and

$$I_s \gamma = -2\,\mu \frac{d^2v}{dr^2} = I\gamma - 2\,\frac{d\,\tau_o(r)}{dr} \quad . \qquad (3\text{-}55)$$

Using only a general symbol, for the time being, to indicate the dependency of the static shearing stress on the distance of the investigated point from the wall, or more precisely from the axis of the pipe, the velocity distribution can be determined by double integration (Fig. 3-47a):

$$\frac{d^2v}{dr^2} = \frac{1}{\mu}\frac{d\tau_o}{dr} - \frac{I\gamma}{2\mu} \quad ;$$

$$\frac{dv}{dr} = \frac{\tau_o(r)}{\mu} - \left[\frac{\tau_o}{\mu}\right]_{r=0} - \frac{I\gamma}{2\mu} \quad ; \tag{3-56}$$

because the constant of the integration can be determined by considering the boundary condition according to which dv/dr = 0 where r = 0. The result of the second integration is:

$$v = \frac{I\gamma}{4\mu}(r_1^2 - r^2) - \frac{1}{\mu}\int_r^{r_1} f(r)\,dr \quad ;$$

where

$$f(r) = \tau_o(r) - [\tau_o]_{r=0} \quad ; \tag{3-57}$$

because the boundary condition to be considered now is v = 0 if $r = r_1$.

In the equation, r_1 is the radius of the pipe where velocity is zero because at this distance from the axis, the gradient becomes equal to that value which can be balanced by the static shearing stress ($I_\tau = I$) and, therefore, there is no movement at all.

The part of the gradient consumed to overcome the static shearing stress is proportional to the change of the static shearing stress (Eq. 3-55). At the same time, this parameter can be related to the threshold gradient as well, if the flow through a capillary tube of constant diameter is investigated. In this case, movement can develop only if the actual gradient is greater, at least in a small part of the cross section, than the I_τ value. Accepting the explanation according to which τ_o (and probably its derivative as well) is a decreasing function of the distance measured from the solid wall, it can be assumed that in a circular tube the smallest value of the I_τ parameter belongs to the center of the cross section (r=0). As the final conclusion, it can be stated, therefore, that flow does not start through the pipe if the actual gradient is smaller than that required to overcome the resistance originating from the static shearing stress at the center of the tube, thus, this parameter is equal to the threshold value (Fig. 3-47b). There is no movement in the tube if;

$$I < I_{\tau(r=0)} = \frac{2}{\gamma} \left[\frac{d\tau_o(r)}{dr}\right]_{(r=0)} = I_o \quad . \qquad (3\text{-}58)$$

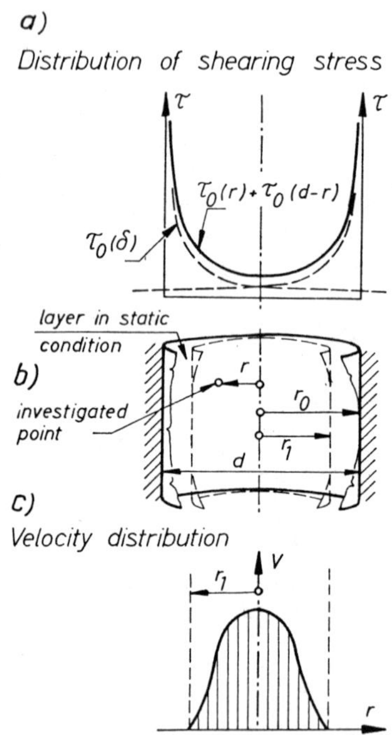

Fig. 3-47. Interpretation of symbols used to derive the relationship describing microseepage.

Another hypothesis is also used at this part of the derivation, i.e., the threshold gradient (I_o) is proportional to the I_1 value characterizing the intersection of the horizontal axis and the straight line approximating the relationship between velocity and hydraulic gradient in the range of higher velocity (see Fig. 3-46).

It was proved, however, by Bondarenko and Nerpin's measurements that the I_1 parameter can be calculated as a B parameter divided by the radius of the tube and the specific weight of the fluid and B (Bondarenko's constant) was found to be a single valued function of temperature. Thus, the threshold gradient in a tube having a radius of r can also be calculated when knowing the physical parameters and the temperature of the fluid:

$$I_o = \frac{B}{2\, r_o \gamma} \quad . \tag{3-59}$$

The combination of Eqs. 3-58 and 3-59 gives a condition which has to be satisfied by the function used to approximate the unknown relationship between the static shearing stress and the distance measured from the solid wall of the tube:

$$\left[\frac{d_o(r)}{dr}\right]_{(r=0)} = \frac{I_o \gamma}{2} = \frac{B}{4\, r_o} \quad . \tag{3-60}$$

The most simple mathematical relationship which fulfills this condition is a hyperbola of first order:

$$\frac{d\tau_o}{dr} = \frac{B}{8\delta} \quad . \tag{3-61}$$

In a pipe, the influence of the two opposite walls has to be taken into consideration because the superposition of the two actions can be assumed (Fig. 3-47c). At the same time, both δ_1 ad δ_2, characterizing the distance of the investigated point from the two walls, can be expressed as the functions of the distance measured from the axis of the pipe (r) ($\delta_1 = r_o - r$ and $\delta_2 = r_o + r$). Thus, the differential quotient $d\tau_o(r)/dr$, and the general form of the $\tau_o(r)$ function can be determined:

$$\frac{d\tau_o(r)}{dr} = \frac{B}{8}\left(\frac{\tau_o}{r_o - r} + \frac{\tau_o}{r_o + r}\right) = \frac{B}{4}\,\frac{r_o}{r_o^2 - r^2} \quad ;$$

$$\tau_o(r) = \frac{B}{4}\tanh^{-1}\left|\frac{r}{r_o}\right| + C = \frac{B}{8}\,\ell n\,\frac{r_o + |r|}{r_o - |r|} + C \quad . \tag{3-62}$$

It can be seen that the minimum of the $d\tau_o(r)/dr$ function (which can be achieved by substituting r=0 value) satisfies Eq. 3-60.

On the basis of the last equation, the velocity distribution from the general formula is given as Eq. 3-57.

$$v = \frac{I\gamma}{4\mu} r_o^2 \left\{ \left(\frac{1}{r_o}\right)^2 - \left(\frac{r}{r_o}\right)^2 - \frac{I_o}{I} \right.$$

$$\left. \left[\frac{r_1}{r_o} \tanh^{-1} \frac{r_1}{r_o} - \frac{r}{r_o} \tanh^{-1} \frac{r}{r_o} + \frac{1}{2} \ln \frac{r_o^2 - r_1^2}{r_o^2 - r^2} \right] \right\} . \tag{3-63}$$

Integrating the product of the local velocity and the elementary area along the entire section, the mean velocity in a capillary tube influenced by adhesion can be calculated:

$$v_m = \frac{I\gamma}{8\mu} r_o^2 \left\{ \left(1 - \frac{I_o}{I}\right)^2 - \frac{2}{3} \frac{I_o}{I} \right.$$

$$\left. \left[\left(1 - \frac{I_o}{I}\right)^{3/2} \tanh^{-1} \left(1 - \frac{I_o}{r}\right)^{1/2} + \frac{1}{2} \ln \frac{I_o}{I} + \frac{1}{2} \left(1 - \frac{I_o}{I}\right) \right] \right\} . \tag{3-64}$$

The characteristic radius r_1 was substituted by the threshold gradient on the basis of the following relationship:

$I = I_\tau$; where $r = r_1$; consequently;

$$\frac{I_o}{I} = \frac{r_o^2 - r_1^2}{r_o^2} \; ; \quad \text{therefore;}$$

$$\frac{r_1}{r_o} = \sqrt{1 - \frac{I_o}{I}} \; . \tag{3-65}$$

It can be assumed that the change of velocity caused by the influence of adhesion i.e., the ratio of the velocity of microseepage and that of laminar flow) is independent of the structure of the solid matrix. Thus, the quotient of the mean velocity determined from Eq. 3-65 related to Darcy's velocity calculated for the same capillary tube can be assumed to be equal to the ratio of the hydraulic conductivity of a soil sample valid in the zone of microseepage to that in Darcy's zone:

$$\frac{v_m}{v_D} = \frac{K_m}{K_D} = (1 - \frac{I_o}{I})^2 - \frac{2}{3}\frac{I_o}{I}$$

$$\left[(1 - \frac{I_o}{I})^{3.2} \tanh^{-1}(1- \frac{I_o}{I})^{1/2} + \frac{1}{2} \ell n \frac{I_o}{I} + \frac{1}{2} (1- \frac{I_o}{I})\right]. \tag{3-66}$$

This ratio tends to unity as the gradient tends to infinity, thus, the difference from Darcy's value becomes negligible in the range of high gradients (Fig. 3-48). Since the basic value related to which the difference is calculated increases with increasing gradient, the absolute difference is practically constant above a given limit and the two relation curves (i.e., the Darcy line and that representing microseepage) become parallel to each other (Fig. 3-48b). This limit can be chosen at a value of $I = 12\ I_o$ which is, therefore, the upper limit of microseepage and the lower one of the laminar zone.

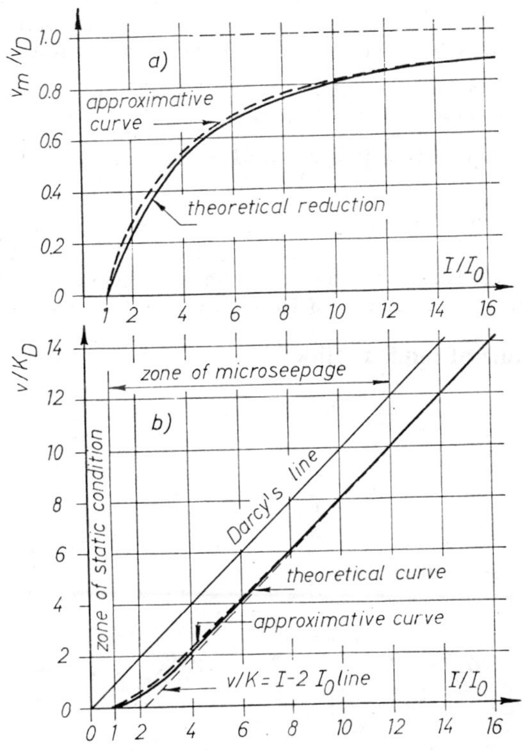

Fig. 3-48. Relationship between velocity and hydraulic gradient in the zone of microseepage.

Above this limit, the second member of the right-hand side of Eq. 3-66 tends to zero, even the quadratic member of the first part can be neglected:

$$\frac{v_m}{v_D} = 1 - 2\frac{I_o}{I} + (\frac{I_o}{I})^2 \sim 1 - 2\frac{I_o}{I} \quad . \qquad (3\text{-}67)$$

The intersection of the horizontal axis and the straight line characterizing the laminar flow was indicated by I_1. The equation of this line is $v_m = K_D(I - I_1)$, the comparison of which with Eq. 3-67 results the $I_1 = 2I_o$ relationship.

It is indicated by Fig. 3-48 that the influence of adhesion ought to be taken into consideration in the laminar zone as well. The correct relation curve is not Darcy's line, but another straight line parallel to the former and intersecting the horizontal axis at the point of $I = 2I_o$.

In practice, the factor given by Eq. 3-65 can also be simplified even in the lower zone of velocity, and approximated by a parabola of vertical axis, the minimum of which is at the point of $v = 0$; $I = I_o$. Another condition which determines the parameters of the parabola is that the tangent of the latter at the point of $v = 10\ K_D\ I_o$; $I = 12\ I_o$, is the line mentioned previously [$v = K(I - 2I_o)$].

Taking the derivation presented in the previous paragraphs into consideration, the following equations can be given to describe the relationship between velocity and hydraulic gradient for laminar movement and microseepage:

in the zone of laminar flow

$$v = K_D\ (I - 2\ I_o) = K_D\ \Phi I \quad ; \quad \text{where}$$

$$\Phi = (1 - 2\frac{I_o}{I}) \quad ; \qquad (3\text{-}68)$$

in the zone of microseepage

$$v = 0.714\ K_D \frac{(I-I_o)^{1.1}}{I_o^{0.1}} = K_D\ \Phi_M\ I \quad ; \quad \text{where}$$

$$\Phi_M = 0.714\ (\frac{I}{I_o})^{0.1}\ (1 - \frac{I_o}{I})^{1.1} \quad . \qquad (3\text{-}69)$$

11-5 Determination of Threshold Gradient and Practical Checking of the Concept of Microseepage

The practical application of the movement equations describing the v vs. I relationship requires the knowledge of the threshold gradient as an important hydrodynamic parameter of seepage. An important result of the experiments of Bondarenko and Nerpin is that the $I_{10}\gamma\ r_o$ product was found to be independent of the material of both the tube and the fluid as well as of the diameter of the tube, the parameter in question being a single valued function of the temperature only. I_1 is proportional to I_{10}, it can be expected that the $I_1\gamma\ r_o$ product behaves similarly. The data determined from the graphs of Fig. 3-46a and 3-46b prove this expectation (Table 3-3).

Bondarenko (1973) has also proved that the non-Newtonian behavior of the fluids is caused by the hydrogen bonds between the molecules. The comparison of the virtual hydraulic conductivity values (the ratio between the corresponding velocity and gradient data) measured with various fluids (Fig. 3-49) testifies that the rapid decrease of this parameter was observed in the zone of small gradients only if the fluid applied for the experiments had dipole molecules. At the same time, the change of the concentration of dissolved materials in the water (KCl in the experiments) did not modify the $I_1\gamma\ r_o$ product, as it is also shown in Table 3-3 (recalculated from the shearing stresses given in the original publication). This result is a negative proof of the previous statement showing that the dissolved minerals do not influence the special behavior of the fluid.

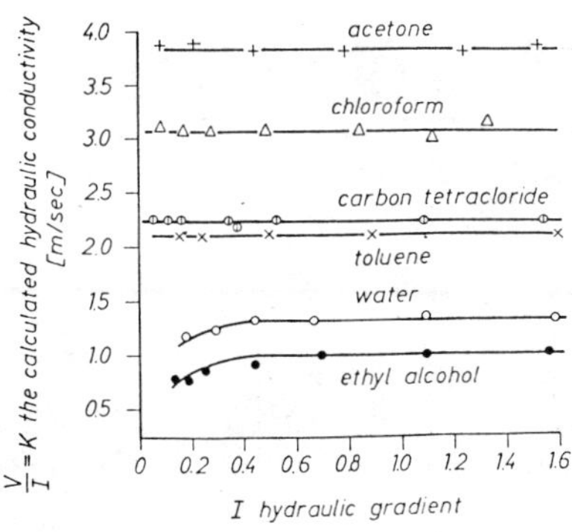

Fig. 3-49. Comparison of the virtual hydraulic conductivities determined by various fluids (after Bondarenko).

Table 3-3. Characteristic I_1 gradients measured in capillary tubes by Bondarenko and Nerpin.

Type of Fluid	Pipe Diameter r	$I_1 \times 10^2$ parameter if temperature in °C								$I_1 \gamma r \times 10^4$ p cm^{-2}								Comments
		15	20	25	30	40	50	55	58	15	20	25	30	40	50	55	58	
distilled water	51	6.4	4.1	-	2.80	1.30	0.40	0.08	-	3.26	2.09	-	1.43	0.66	0.20	0.04	-	
$\gamma = 1.0$ p cm^{-2}	103.5	-	2.5	-	1.26	0.65	-	-	0.00	-	2.59	-	1.30	0.67	-	-	0.00	
	165	-	1.3	-	1.75	0.25	0.15	-	-	-	2.14	-	1.24	0.41	0.28	-	-	recalculated from τ_o
	198	-	1.3	-	-	-	-	-	-	-	2.57	-	-	-	-	-	-	
	265	-	1.0	-	0.59	0.15	-	-	-	-	2.65	-	1.56	0.40	-	-	-	
									average	3.26	2.41	-	1.38	0.54	0.24	0.04	0.00	
ethyl	57	-	-	5.4	3.66	2.22	0.75	-	-	-	-	2.46	1.66	1.01	0.33	-	-	
alcohol	109.25	-	3.3	-	-	-	-	-	-	-	2.88	-	-	-	-	-	-	
$\gamma = 0.8$ p cm^{-2}	165	-	-	1.1	0.68	0.31	0.10	-	-	-	-	1.47	1.12	0.41	0.13	-	-	
									average	-	2.88	1.97	1.39	0.71	0.23	-	-	
KCl solution	198																	
$\gamma = 1.0$ p cm^{-2}																		
10^{-2} N		-	1.2	-	-	-	-	-	-	-	2.38	-	-	-	-	-	-	recalculated from τ_o
10^{-1} N		-	1.3	-	-	-	-	-	-	-	2.57	-	-	-	-	-	-	
10^{0} N		-	1.3	-	-	-	-	-	-	-	2.57	-	-	-	-	-	-	
									average	-	2.51	-	-	-	-	-	-	

It was also found that the heat value of the H-bonds (G) can be related to the shearing stress in the form of an equation having similar structure as those giving this parameter as a function of other properties caused by the dipole character of the molecules:

$$G = A \frac{T_1 T_2}{T_1 - T_2} \ln \frac{(\tau_o - C)_2}{(\tau_o - C)_1} \quad ; \qquad (3\text{-}70)$$

where T is the absolute temperature, A and C are constants, and the two subscripts (1 and 2) indicate the corresponding values of T and τ_o.

Since the G value can be regarded as a constant within a limited range of temperature, Eq. 3-70 can be used also to determine the general pattern of the relationship between the $I_1 \gamma r_o$ product and temperature. Choosing a fixed T_2 value (T_2 = 273°K; i.e., the temperature in centigrades is equal to zero), the dependence of $I_1 \gamma r_o$ on $T_1 T_2$ can be neglected in the zone having practical importance and the ($T_1 - T_2$) difference can be substituted by the temperature expressed in centigrades. The negligence of the $T_1 T_2$ product can be partially compensated by selecting an appropriate C parameter. Thus, a linear relationship can be expected between the investigated variables (t(°C) and $I_1 \gamma r_o$) in a semi-logarithmic coordinate system.

In Fig. 3-50 the data summarized in Table 3-3 were plotted and a straight line was fitted to the points. The constants and the corresponding basic values of the temperature and the $I_1 \gamma r_o$ product were found to be as follows: C = -5×10^{-5}; $G(AT_1 T_2)$ = 1/50; T_2 = 273°C; $(I_1 \gamma r_o)_2 = 1 \times 10^{-4}$ [p cm^{-2}]. It is also shown by the figure that two enveloping curves can be achieved by shifting the line horizontally by ±5°C. These lines border a strip including the whole range of scattering of the measurements. The probable zone of the measured points and the I_1 vs. t relationship within the 10°C < t < 60°C range of temperature can be described by the following equations:

$$\frac{t°C \pm 5°C}{50} = -\left[\log\left(I_1 \gamma r_o + 5 \times 10^{-5}\right) + 1.125\right] \quad ;$$

$$(3\text{-}71)$$

$$I_1 = \frac{1}{r_o \gamma} \{\exp\left[-4.6\, t°C + 56.3) \times 10^{-4}\right] - 5 \times 10^{-5}\} \ .$$

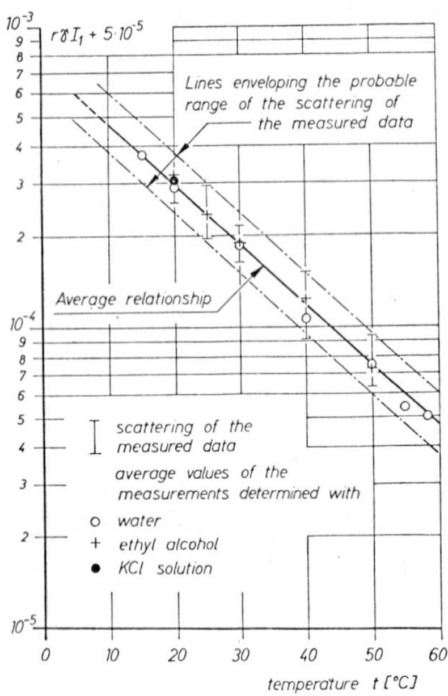

Fig. 3-50. Relationship between the $r\gamma I_1$ product and temperature.

Considering Eq. 3-71 and the relationship between the average diameter (d_o) of the capillary tube (used as a geometrical model) and the soil-physical parameters of the sediment (D_h, Kozeny's effective diameter; n, the porosity; α, the shape coefficient of the grains), the threshold gradient of the layers can also be calculated:

$$I_o = \frac{B}{4\gamma}\,\frac{1-n}{n}\,\frac{\alpha}{D_h} = \frac{1}{4\gamma}\,\frac{1-n}{n}\,\frac{\alpha}{D_h}\,\{\exp[-4.6(t°C+56.3)x10^{-4}]-5x10^{-5}\}\ . \tag{3-72}$$

The applicability, in the practice of the proposed method, was originally proved by comparing calculated graphs with measured data (Kovács, 1969). Kutilek's measurements (1967) were used for the comparison having been, in that time, the most detailed series published in the literature (Fig. 3-51). This investigation has shown that the calculated values are in good agreement with the measured data and the proposed method can follow both the actual character of the relationship and the influence of the water temperature as well.

Only a slight difference can be observed between the measured and calculated data in the zone of very small

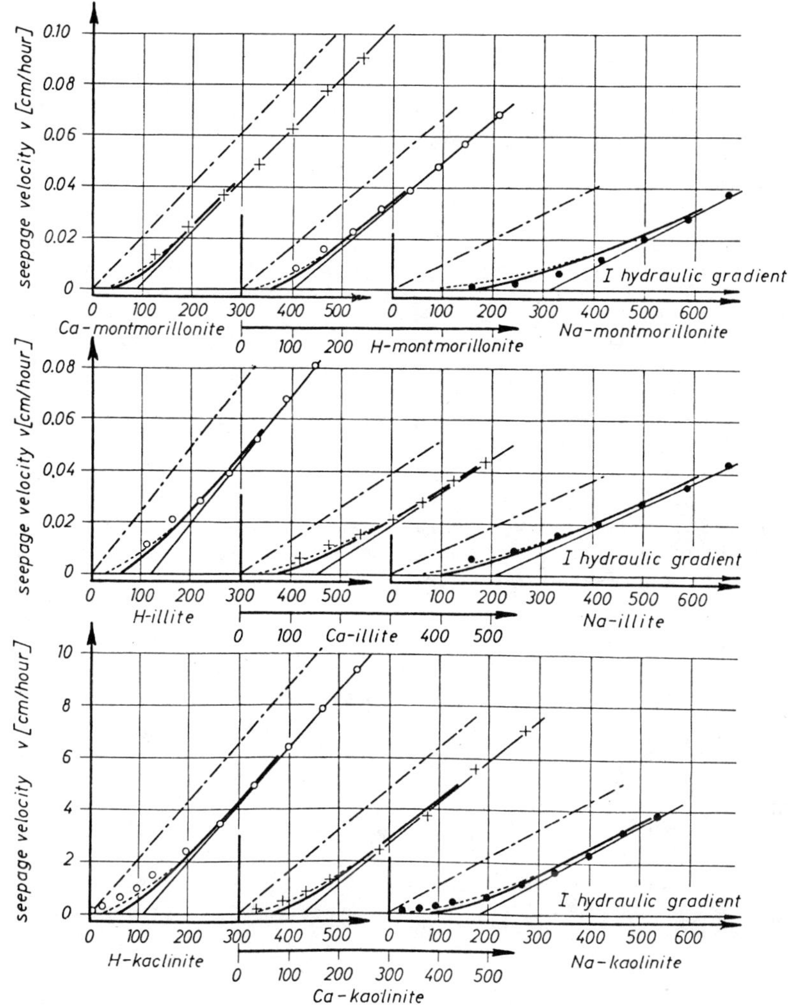

Fig. 3-51. Comparison of the theoretical relationship with Kutliek's measurements.

velocities. Some measured points seem to contradict the existence of the threshold gradient, especially in the case of kaolinite samples as Thirriot (1969) has pointed out. At the same time, the measurements executed in capillary tubes clearly indicate that there exists a limit of the gradient, below which the velocity is zero, there is no flow in the tube (see Fig. 3-46c). This discrepancy between the results of experiments with glass tube and actual soil, respectively, can be explained by the fact that the pores are considered in the derivation with average diameter d_o and thus, the calculated threshold gradient is also an average characteristic of the sample. This parameter can be accepted, therefore, if the phenomenon extends to a volume of the sample large enough to be characterized by average parameters on the basis of the continuum approach. The completely static state can be disturbed, however, if there is only a few larger pores within the solid matrix, through which movement can be initiated even by a smaller gradient than the threshold value. Considering the size of the probable large pores (d_2: the probability of the occurrence of such large pores is still higher than 4 percent according to the statistical analysis of pore-size distribution), some correction can be made (see the dotted line in Fig. 3-52), which eliminates most of the differences. The observed discrepancy between the measured and calculated data draws attention to the fact that the commencement of seepage is governed by a property of the sample which cannot be characterized by applying the continuum approach and, therefore, the derived relationship gives only the general character of the phenomenon and the actual data may differ from the calculated averages.

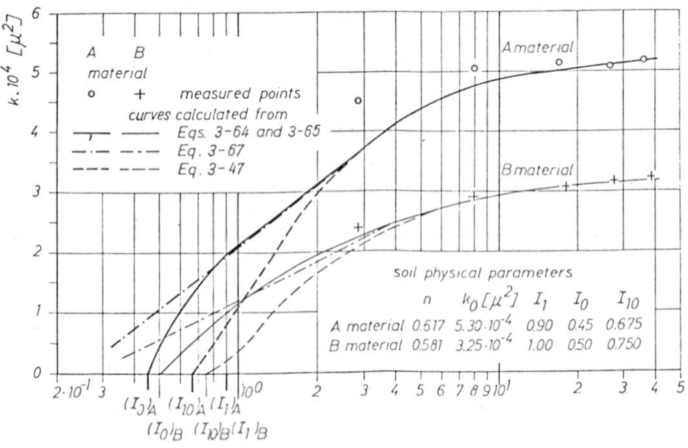

Fig. 3-52. Comparison of the results of various methods on the basis of Habib's measurements.

Bondarenko (1973) has also collected some data concerning the I_1 gradient measured on soil samples. Among the soil-physical parameters, porosity (n) and the specific surface (A/V) are given in the publication (Table 3-4). According to the definition of shape-coefficient, the specific surface is equal to the ratio of the latter and effective diameter (α/D_h). Supposing a given temperature (e.g., 20°C), the threshold gradient can be calculated from Eq. 3-72 as it is listed in the second to last column of the table. Finally, the last column gives the ratio between the measured I_1 and the calculated I_o values. It can be seen that the scattering of this quotient is very large, caused by the uncertainties of the measurements, but the average is in good agreement with the theoretical value (2.02 instead of 2.0).

To give complete information on the methods applicable to characterize the interrelation between seepage velocity and hydraulic gradient in the zone of microseepage, the empirical relationships have to be mentioned. These methods approximate the points determined on the v vs. I field from measured data in most cases by exponential formulae, the general form of which according to Habib (1971) is:

$$v = K_D \{I - I_1 [1 - \exp(-I/I_1)]\} \quad . \qquad (3\text{-}73)$$

Examples for this type of equation are:

$$v = K_D \{I - A [1 - \exp(-CI)]\} \ ; \ \text{(Swartzendruber, 1962)};$$

$$v = K_D \{\frac{1}{B} \ln [A + \exp(BI)] - I_1\} \ ; \ \text{(Kutilek, 1964)};$$

$$v = k_D \frac{g}{\nu} \{I - I_1 [1 - \exp(-I/I_1)]\} \ ; \ \text{(Hansbo, 1960)}; \qquad (3\text{-}74)$$

where the constants (A,B,C) are functions of the threshold gradient.

Habib (1971) compared the data determined by carefully executed experiments using kaolinite samples with the calculated graphs and found that the experimental relationships are more reliable than the theoretical ones because both Eq. 3-53 and Eq. 3-69 gave smaller velocities than the measured values. This contradiction originates, however, from the fact that Habib did not make distinction between the I_o, I_1, and I_{10} values (see Fig. 3-46), but substituted I_1 into the theoretical equations as well. After recalculating the necessary basic data from the published measurements, the same comparison was executed (Fig. 3-52). As it is shown by the figure, there is no considerable difference between the calculated

Table 3-4. Comparison of measured and calculated values of the threshold gradient of soil samples on the basis of Bondarenko's data.

Type of the Sample	Specific Weight γ $[g/cm^2]$	Porosity n	Specific Surface $\frac{A}{v} = \frac{\alpha}{D_h}$ $[cm^{-1}]$	Measured Characteristic Gradient I_1	Calculated Threshold Gradient (Eq. 3-66)	I_1/I_o
medium sandy clay	1.65	0.38	130×10^2	0.8	1.32	0.60
medium sandy clay	1.68	0.37	124×10^2	1.9	1.32	1.44
heavy sandy clay	1.60	0.40	116×10^2	1.9	1.09	1.75
clay	1.40	0.47	118×10^2	4.0	0.82	4.86
light sandy clay	1.53	0.42	92×10^2	0.6	0.79	0.75
medium sandy clay	1.68	0.37	83×10^2	0.7	0.88	0.79
fine rock flour	1.55	0.42	76×10^2	1.9	0.66	2.88
fine rock flour	1.55	0.42	70×10^2	0.8	0.60	1.33
light sandy clay	1.45	0.45	51×10^2	1.0	0.38	2.60
clay	1.57	0.41	131×10^2	3.6	1.18	3.06
rock flour	1.45	0.45	42×10^2	1.1	0.32	3.40
light sandy clay	1.58	0.41	59×10^2	0.4	0.53	0.76
					average	2.02

intrinsic permeabilities, only in the vicinity of the threshold gradient, if the correct interpretation of I_o, I_1, and I_{10} is used. The smallest measured permeability belongs to a gradient being six times greater than the threshold value. The points determined from the observations fit very well to each of the calculated curves.

On the basis of the detailed data published by Habib, it was also possible to check the accuracy of Eq. 3-72. The trend of the I_1 gradient vs. porosity relationship determined from the equation is in good agreement with the position of the measured points (Fig. 3-53). The scattering of the measurements is considerably large. Two curves indicated by dotted lines and also calculated from Eq. 3-72 by using different multiplying factors were constructed to envelop the zone of scattering. The factors of the enveloping curves related to the constant of the average curve indicate an uncertainty (or meauring error) of about ±30 percent. The curves show only the influence of one independent variable (i.e., porosity), while the effects of the others are included into the multiplying factor which is assumed to be constant. Taking into account a temperature of 20°C maintained during the experiments, the probable effective diameter of the kaolinite can be recalculated. The value determined from the constant of the average curve (i.e., ~ 1μ) is smaller than that calculated from the published grain-size distribution curve, but considering the probably high degree of coagulation of the kaolinite, the result of the experiments do not contradict the theory and Eq. 3-62 can be accepted as a realistic approximation of the threshold gradient.

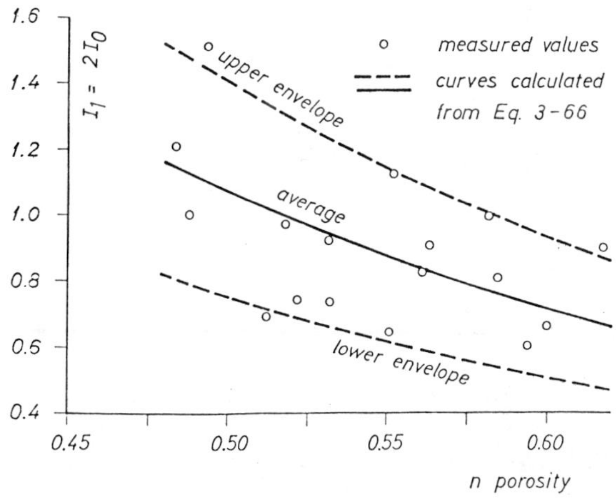

Fig. 3-53. Relationship between I_1 gradient and porosity on the basis of Habib's measurements.

SECTION 12

HYDROLOGICAL ANALYSIS OF DATA OF SHALLOW GROUNDWATERS

The qualitative analysis of the water balance developing in the zone of shallow groundwaters was already given in Subsection 9-3. The various processes composing the versatile system of the external actions were divided into three groups:

- the percolation of groundwater into the beds of surface waters (including the springs as well through which groundwater is raised to the surface and becomes a part of surface runoff), or vice versa, the recharging effect of surface waters forming the input of the groundwater balance;

- water exchange between the soil-moisture and the shallow groundwater which is composed of two parts, i.e., the infiltrating precipitation reaching the water table and the capillary drainage of the groundwater raising a given amount of water above the water table to partly replenish the soil-moisture consumed by evaporation and transpiration;

- seepage between the neighboring aquifers which either drains the shallow groundwater, transporting water to the deeper lying layers, or recharges it, raising water near the surface from the reservoirs of the deep groundwater. Apart from the three types of external processes, the storage, as the internal operation of the system, governs the water regime of shallow groundwaters.

It was already explained in connection with the general description of the four groups of factors influencing the water balance that the surface waters discharge the groundwater in most cases. This process is very important not only from the point of view of groundwaters, but it should also be considered when the task is the characterization of surface water resources because the design value of their available fraction mostly depends on the flow rate developing during long dry periods when the water propagating in the riverbeds originates generally from groundwater reservoirs (base flow). This is the reason why the determination of base flow is a paramount task of the hydrological analysis of groundwaters. Since the rivers are primarily supplied by shallow groundwaters, the investigations aiming to determine the ratio of base flow related to surface runoff are discussed in this chapter.

It was already mentioned that the storage capacity of a groundwater system depends on the character of the aquifer, where it is described by using specific yield if a water table condition prevails in the water-bearing layer, and storage coefficient has to be used in confined systems. The interpretation, the numerical characterization and the method of application of specific yield, were already discussed in Part II. The change in storage due to the compressibility of water and the change of pore volume (the phenomena governing the development of the storage coefficient) may have an important role in connection with the investigation of deep groundwaters. This type of storage process will be analyzed in Chapter 4.

The water exchange between neighboring aquifers due to the differences between pressures prevailing in the layers in question is the most important external action influencing the water balance in deep-lying formations. The accumulation of energy is also a more characteristic process at greater depths than near the surface. This is the reason why the detailed analysis of the group of actions determining the effects of seepage between neighboring layers will also be given in Chapter 4.

Thus, the main topic of the present chapter, remaining to be discussed apart from the determination of base flow, is the characterization of the influence of meteorological processes on the water regime of shallow groundwaters. To meet this goal, a general balance equation should be determined at first considering the infiltrating preciptation reaching the water table and the capillary drainage of groundwater maintained by evapotranspiration. The solution of the equation gives a powerful tool for the general description of hydrological processes in a groundwater basin, i.e., the characteristic curve of groundwater balance. Even the connection between groundwater flow and the migration of salts in the soil profile can be analyzed in this way or the influence of irrigation on the regime of the groundwater can be characterized by using this method.

12-1 Characterization of the Base Flow of Rivers

The flow rate of subsurface runoff is influenced, even within a relatively short stretch of a river, by many factors such as the geological structure of the river plain, the external natural recharge of the groundwater, the vertical effects along the path of the seepage which are affected by the depth of the water table, etc. As it was already explained in Subsection 9-2, the rivers are generally supplied with water from the groundwater reservoirs (positive base flow) but there are circumstances (flood on the river and relatively deep water table) periodically causing the change of the direction of seepage, the result of which is the development of negative base flow, i.e., the storage of surface water in the riparian layers (Fig. 3-54).

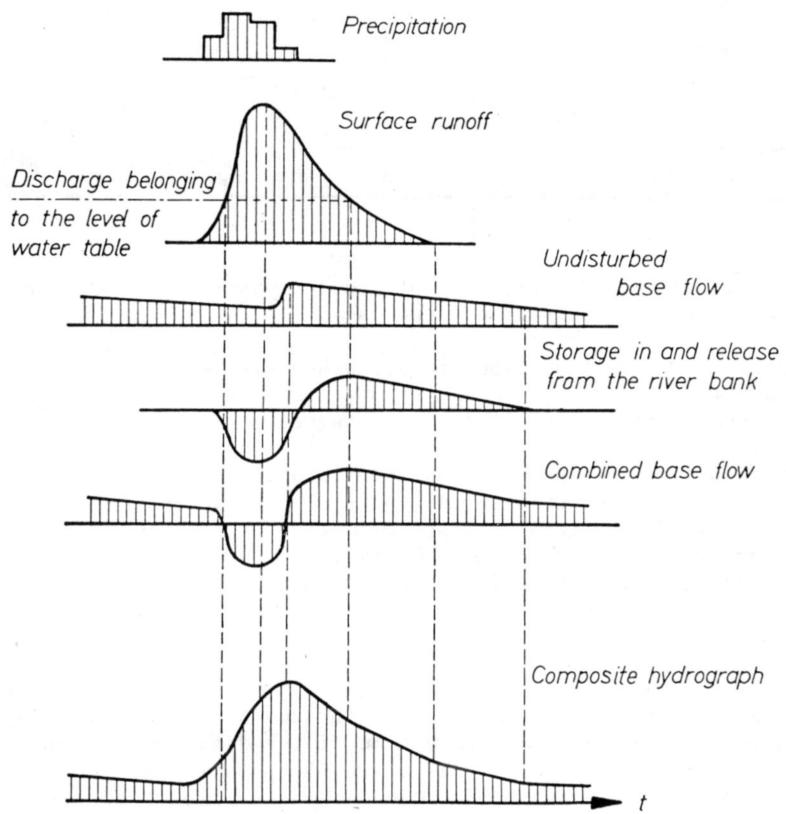

Fig. 3-54. Development of negative base flow during flood periods.

The numerical determination or forecast of the base flow is also hindered by the fact that the subsurface runoff at an investigated section of a river indicates the integrated effects along the whole river system which effects reach the section in question with different time lags. This is the reason why the hydrographs are generally divided into two parts, i.e., surface and subsurface runoff, using the most simple graphical methods (Fig. 3-55) and only that part of the

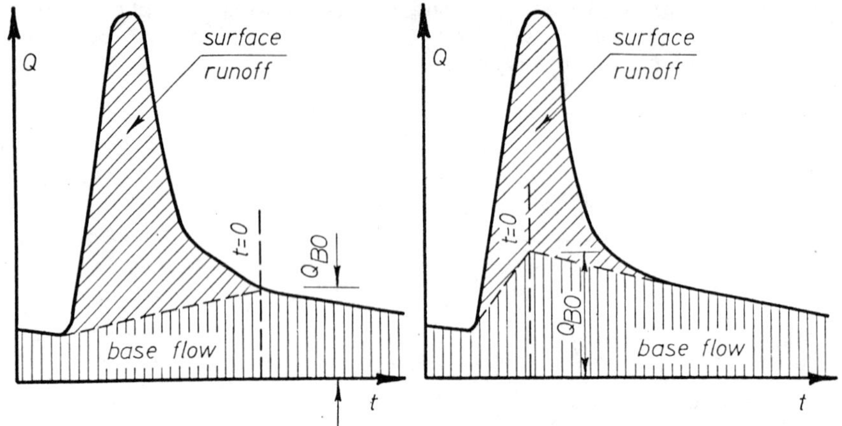

Fig. 3-55. Distribution of hydrographs into surface and subsurface runoffs.

flow rate versus time relationship is approximated with mathematical models which develops after the practical ending of the surface runoff (recession curve).

Because of the numerous uncertainties influencing the run of the recession curves, only very simple models are generally used for the mathematical approximation of the decrease of the flow rate as a function of time. Such models can be constructed by investigating the unsteady movement occurring when a vertical water column is lowered or when a horizontal water bearing layer is drained.

In the first case, the physical model is a cylindrical container from which water is withdrawn at the bottom through a pipe. The detailed derivation of the differential equation was given in Subsection 3-7 where the porosity of solid rocks (especially that of carbonate rocks) was discussed (see Eq. 1-32). Here, only the final result is repeated which is achieved by integrating the differential equation between limits, the lower limit is characterized with the interrelated t_o and q_o value of the starting point of the recession curve while the upper limit is given by the coordinates t and q of the forecasted point. By combining all parameters into one symbol, the solution is as follows:

$$q = q_o \exp\left[-a\,(t - t_o)\right] \quad . \tag{3-75}$$

The form of the other model (which is the solution of the differential equation describing a nearly horizontal unsteady groundwater flow) depends on the method of linearization of the equation and on the boundary conditions considered, e.g., the rate of change of the water level at the exit face. Without going into details related to the correct derivation of this type of relationship, it can be stated that the flow rate is inversely proportional to a function of elapsed time since the beginning of the investigated period:

$$q = \frac{q_o}{f(t-t_o)} \quad ;$$

generally

$$q = \frac{q_o}{a(t-t_o)^n} \quad . \tag{3-76}$$

The natural conditions of a small catchment composed of karstic aquifers are the most similar to those used as hypotheses for the derivation of the first model, as it is evident from the physical interpretation of that model. This is the reason why the recession curves of karstic springs can be better approximated with Eq. 3-75 than those of water courses in other geological regions. Even in the cases of karstic springs, the recession curve is generally composed of two parts (both described with an exponential relationship), the first characterizing the depletion of the large karstic channels while the second that of the thin interstices, as it was shown by Eq. 1-33 and Fig. 1-67.

At a section of a river where the subsurface runoff combines the various effects of a large, nonhomogeneous catchment, the simple exponential approximation is generally not sufficient. In the example given in Fig. 3-56, nine yearly hydrographs are represented for a section of the Tisza River (Vásárosnamény), the catchment of which is 29,000 square kms. It was found that in this case the logarithm of the discharge can be better approximated with an exponential curve than the original data (Böröcz, 1971):

$$\ell n\ Q = 3.167 + 4.833\ \exp[-0.6(t-t_o)] \quad . \tag{3-77}$$

The time in this equation should be substituted in months, measured from that fictive starting point t_o, where the recession curve determined by Eq. 3-77 and fitted to the observed stretch of the hydrograph intersects the horizontal line belonging to the $\ell n\ Q = 8.0$ value. The graphs clearly indicate that the recession curve gives the lower enveloping curve of the actual hydrograph. Equation 3-77 does not mean a new type of mathematical model, the structure of which may be proposed for characterizing the development of base flow. It is only a possible approximation of the recession curve in the

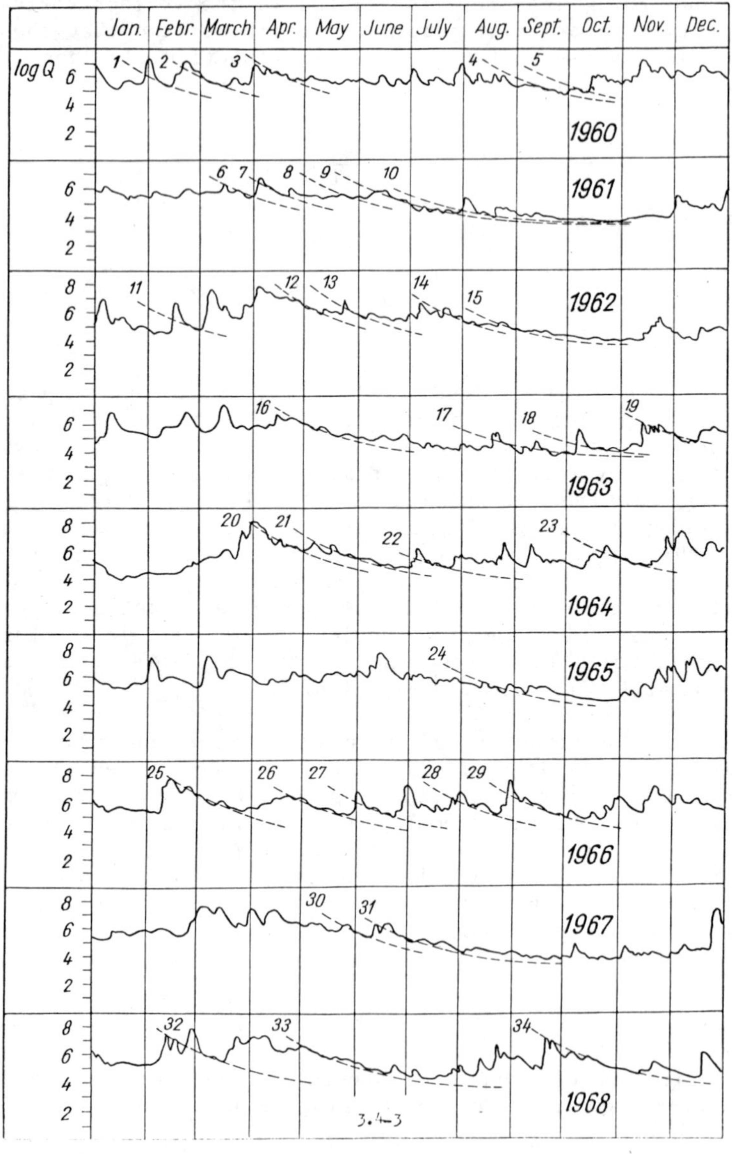

Fig. 3-56. Recession curves in the section at Vásárosnamény of the Tisza River.

special case of the section in question. The example is also suitable, however, to demonstrate the difficulties of the application of any mathematical model for forecasting the expected base flow.

A further example for proving the uncertainties of the determination of the recession curve draws the attention to the versatile physical processes influencing the development of base flow. In Fig. 3-57 the discharge hydrographs of two sections (Zagyva River at Nemti, with a catchment area of 159 square kms and Köszörü Creek at Parádsasvár, with a catchment area of 6.4 square kms) are represented for the summer months of a very dry year (1961). That summer there was no considerable rain after June and there was no rain at all between the 7th of September and 17th of October. In spite of this fact a definite rise of base flow could be observed at the end of September which was in good correlation with the sudden decrease of temperature at that time. The reason of the phemomenon can be explained as follows. In water-bearing slope deposits, a great quantity of groundwater becomes stored during spring and early summer when the precipitation is abundant. This water percolates slowly towards the valleys, continuously recharging the groundwater of the valleys. The recharged amount (due to the low hydraulic conductivity of the water conveying layers) does not or hardly exceed the evapotranspiration occurring at the bottom of the valleys during the warmest months (July and August) and, therefore, the water courses do not receive any appreciable amount of water from this resource in that period. When temperature (and thus evapotranspiration) decreases, the bed becomes the main draining element and the percolating groundwater rapidly raises the base flow. The seasonal fluctuation of base flow (depending on air temperature) can be expected, therefore, in those areas where the geological structure and the topography create suitable conditions for the development of the described process (Kovács, 1963).

As a final result of the analysis of base flow summarized in the previous paragraphs, it can be stated that the general forms of mathematical models proposed for numerical determination of the recession curve can be applied only with great precaution. The first task is always the careful study of the local conditions. The investigation has to be commenced with the collection of data series, representing the water balance in precipitation-free periods along geologically homogeneous stretches of the investigated river. Knowing the changes of discharge along such stretches under different conditions (river stage, season, temperature) as a function of time elapsed after the ending of surface runoff, the dominant processes influencing the development of base flow can be determined and the models can be accepted if the effects of these processes are considered sufficiently. The water

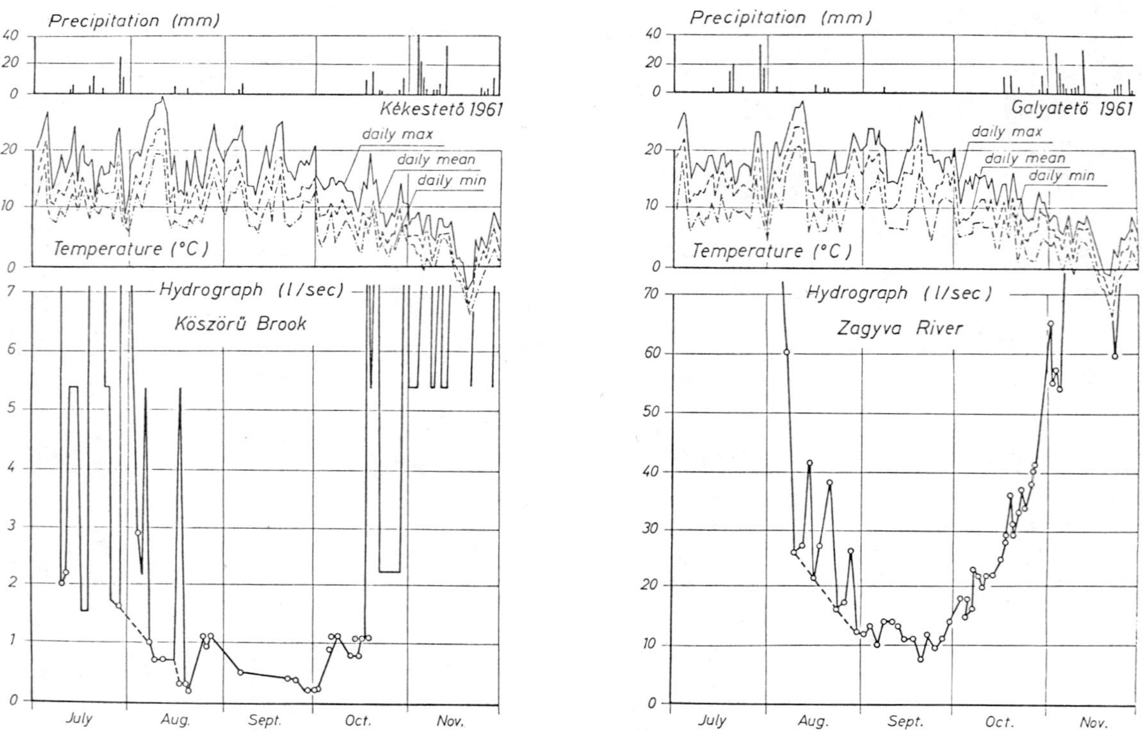

Fig. 3-57. Hydrographs of Köszörü Brook and Zagyva River in the late summer 1961.

balance studies have to be regarded, therefore, as the basis of the establishment of base flow models.

The classical tools for determining the water balance along a river stretch are the discharge measurements. The accuracy of current metering (and thus that of the calculated difference of outflowing and inflowing water amount) is low, its error has sometimes the same magnitude as the base flow within the investigated section. The reliability of such an investigation can be considerably raised by determining the balance of both water and environmental isotopes at the same time. The latter method is based on the difference between the isotope content (deuterium, tritium, oxygen-18) of groundwater and that of surface water. The change of tritium content (expressed in TU's) along the Hungarian stretch of the Tisza River is shown in Fig. 3-58 as an example (Deák, 1974). The example shows the continuous decrease of the TU content at a time when the flow rate was relatively low in the bed. Since the surface waters recharging the river along this stretch have higher TU content than 150 TU, the decrease is caused by the groundwaters percolating to the riverbed and having negligible tritium content. Comparing the water balance and the tritium balance of the stretch investigated, it was estimated that the recharge of groundwater is about 18 m^3/sec long a length of 300 km, thus about 15 percent of the flow rate at the upper border originates from base flow reaching the bed along the stretch.

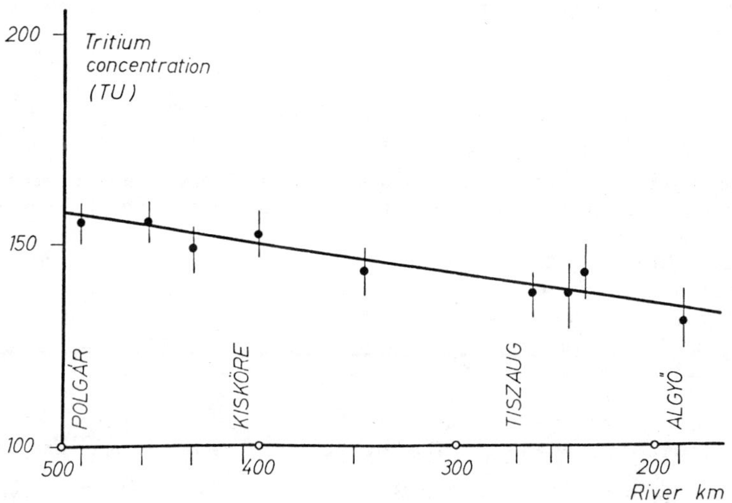

Fig. 3-58. The change of tritium content along the Hungarian stretch of the Tisza River.

12-2 Characterization of the Balance between Horizontal Flow, Vertical Recharge, and Drainage

It was already explained in Subsectio 9-2 that the water table generally exhibits a regular yearly fluctuation in large basins where shallow groundwater is not influenced by rivers. Although it also has longer periodic changes, in most cases, the balanced state can be expected under natural conditions. The assumption of equilibrium means that the change in storage is negligible, if the period under investigation is sufficiently long. It can be stated, therefore, that the average input (positive accretion and incoming groundwater flow) is equal to the average output (negative accretion and groundwater flow discharging the investigated basin) supposing that the mean values of the components are calculated for a long period. For the numerical characterization of the processes acting and for the determination of the average parameters, the water balance of a limited soil column should be investigated in the vicinity of the surface (Kovács, 1971). This column includes a part of the gravitational groundwater space below the water table and the unsaturated zone above it up to the surface (Fig. 3-59).

The equation of continuity, determined for this column, states that the difference between the quantities of water flowing into the column and discharged from it during a given period is equal to the change of stored water therein:

$$\sum_{t_1}^{t_2} [I_s(t) - ET_s(t) + D_{in}(t) - D_{out}(t)] \Delta t = \pm \Delta V \pm \Delta W \quad . \tag{3-78}$$

The members on the left-hand side of the equation are functions of time. $I_s(t)$ is the infiltration through the surface originating from precipitation which can be determined by subtracting surface runoff and water evaporated directly from the surface and retarded by vegetation from the total precipitation. Apart from climatic parameters (amount and distribution of precipitation, factors affecting evaporation) this value depends on the condition of the surface (the degree and direction of slopes covering vegetation) and the type of the soil which influences the process of infiltration.

One part of the infiltrating amount is stored in the unsaturated zone (I_w). The other part reaches the gravitational groundwater space either during the rain or with a longer time lag (I_g positive accretion):

$$I_s(t) = I_w + I_g \quad . \tag{3-79}$$

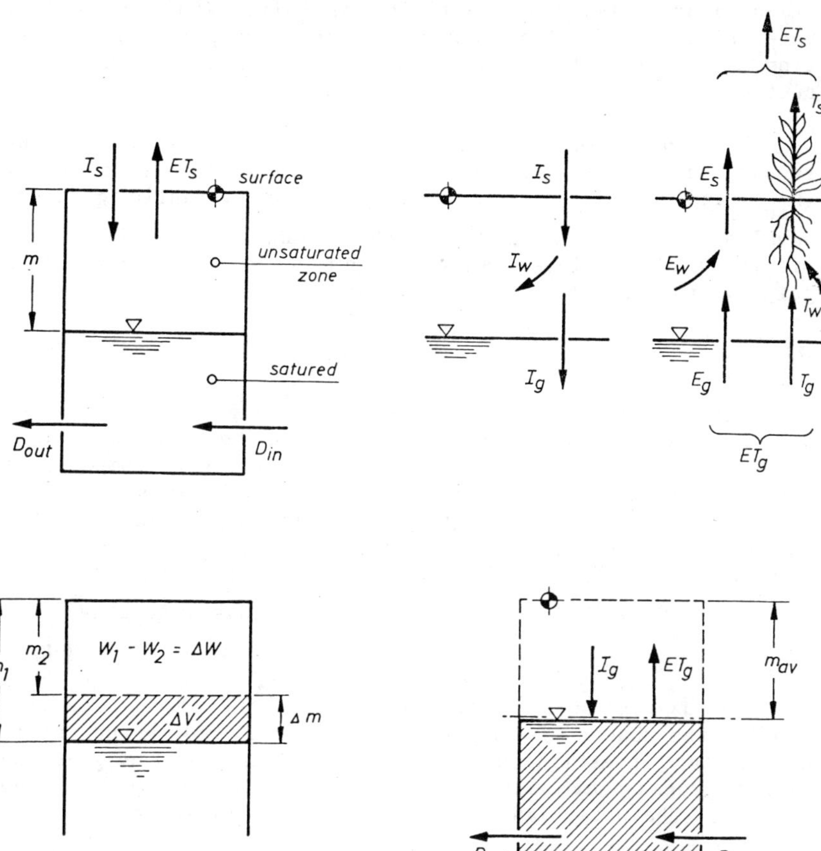

Fig. 3-59. Interpretation of symbols used for investigating the hydrological processes above the water table.

That part of evapotranspiration, $ET_s(t)$, for which the water is supplied by subsurface water includes direct evaporation (E_s) and the transpiration by vegetation (T_s). Both evaporation and transpiration can consume either the moisture content of the unsaturated zone (E_w and T_w, respectively) or the gravitational groundwater stored in the saturated zone and raised above the water table by capillarity (E_g and T_g):

$$ET_s(t) = E_s + T_s = E_w + T_w + E_g + T_g = ET_w + ET_g \quad . \tag{3-80}$$

The amount of water drained from the groundwater by evapotranspiration depends on potential evapotranspiration (temperature, moisture deficit of air, wind velocity), vegetation (depth of root zone, water consumption of vegetation), and the type of soil (water retention capacity, suction, capillarity).

The third influencing effect is the groundwater flow recharging or draining the gravitational space. Both inflow (D_{in}) and outflow (D_{out}) depend on the hydraulic gradient developing in the vicinity of the column and on the transmissivity of the layer. The two effects can be investigated jointly. taking into consideration their difference:

$$\pm \Delta D = D_{in} - D_{out} \qquad (3\text{-}81)$$

The factors listed are functions of time. Investigating a long period, however, their average can be used instead of the varying values. Thus, the sum of the products composed of the variables and elementary time, Δt, can be substituted by the product of the average values and the length of the investigation $(t_2 - t_1)$:

$$[(I_g - ET_g) + (I_w - ET_w) \pm \Delta D]\ (t_2 - t_1) = \pm \Delta V \pm \Delta W \ . \qquad (3\text{-}82)$$

The two members on the right-hand side of the equation express the change of the volume of water stored in the column during the investigated period. ΔV refers to the gravitational groundwater space and is equal to the product of the change of the water table (Δh) and specific yield (n_s):

$$\pm \Delta V = \pm \Delta h\ n_s \ . \qquad (3\text{-}83)$$

The second member (ΔW) indicates the change of soil moisture between time points t_1 and t_2 summarized in the whole unsaturated volume (V_H):

$$\pm \Delta W = \int_{V_H} (W_1 - W_2)\ dV \ . \qquad (3\text{-}84)$$

Since it was the precondition of the analysis that the period under investigation was sufficiently long, thus the balanced state could be accepted as an approximation, the equilibrium of the inputs and outputs of the column can be assumed. It follows from this hypothesis that the change of the water amount stored in the gravitational groundwater space is zero (ΔV=0) because the difference between the water tables at the beginning and at the end of the investigation, respectively, is zero (Δh=0).

The change of the stored soil moisture can also be neglected ($\Delta W=0$), if the starting and closing points of the period are in the same season. The analysis of the water balance can be limited, therefore, to the gravitational groundwater below the average water table (m_{av}). The equation of the groundwater balance is also valid for this case, but the recharge from infiltration refers to the amount of water reaching the water table (I_g) and not to the infiltration through the surface. Similarly, the losses caused by evapotranspiration (ET_g) from the gravitational water are only a part of the total amount of evapotranspiration originating below the surface. The resultant of the groundwater flow (the difference between outflow and inflow) is then equal to the difference of the positive and negative accretions:

$$I_g - ET_g = D_{out} - D_{in} = \pm \Delta D \quad . \qquad (3\text{-}85)$$

Knowing the result of accretion (the difference of the two values on the left-hand side), groundwater balance can be described numerically. Both infiltration and evaporation depend on numerous variables which can be divided into three main groups:

- climatic conditions: amount and annual distribution of precipitation and potential evaporation (the latter includes the effects of temperature, humidity, wind velocity, etc.);

- characteristics of the surface: angle and direction of slopes, method of cultivation, vegetation covering the surface, etc., the resultant of which can be measured in the form of surface runoff;

- behavior of the soil within the unsaturated zone: geological character and structure of the upper-lying layer, water retention capacity, suction, and the instantaneous soil moisture of the top soil.

Investigating only the gravitational groundwater space, a further variable should be included, i.e., the average depth of the water table (m). Thus, the general form of the two functions is as follows:

$$I_g = \Phi_1 \text{ (m; climate; surface; soil)} \quad ;$$

$$ET_g = \Phi_2 \text{ (m; climate; surface; soil)} \quad . \qquad (3\text{-}86)$$

Factors describing the influence of climate, surface, and soil are time and space dependent variables. It was already mentioned that instead of the investigation of their modification in time, the mean values determined for the period are used. Similarly, the equalization of local effects can be assumed in a larger area, thus the individual parameters can be averaged over the area as well. Regarding all variables except the depth of the water table as several constants for the given area, and describing those by their average over the investigated period, Eq. 3-86 can be simplified:

$$I_g = f_1(m) \quad ;$$
$$ET_g = f_2(m) \quad . \qquad (3\text{-}87)$$

These two relationships, their parameters in any case, but perhaps their structure as well, are functions of natural conditions (climate, surface, soil) which were regarded as constant for a given place. Some general rules describing the character of the functions can be, however, determined.

Infiltration through the surface (I_s) is the difference of precipitation (P) and the sum of evaporation from the surface (E) runoff (R), and retention on and above the surface (S_s):

$$I_s = P - (E + R + S_s) \quad . \qquad (3\text{-}88)$$

This amount can be divided into two parts, one stored above the water table in the unsaturated zone and another reaching the zone of saturation (see Eq. 3-15). The quantity of these two values depends on the storage capacity of the unsaturated zone which is influenced by soil physics parameters and the instantaneous water content of the soil at the time of infiltration. Although the amount of the stored water changes from case to case and, therefore, the amount of water stored in a year also varies every year, the average of the yearly retention can be calculated. Thus, graphs can be constructed representing the average yearly storage for various soils as a function of the depth from the terrain as it is shown on the right-hand side of Fig. 3-60 (where the S(m) symbol indicates the yearly average of the free storage capacity of a layer having a thickness of unity at a depth of m).

On the basis of the curve representing the distribution of available storage capacity, the relationship between the depth of the water table and the infiltrating recharge can also be determined. The former function (S(m)) is the derivative of the latter (I(m)) because it shows the probable yearly storage at a given depth which is equal to the average

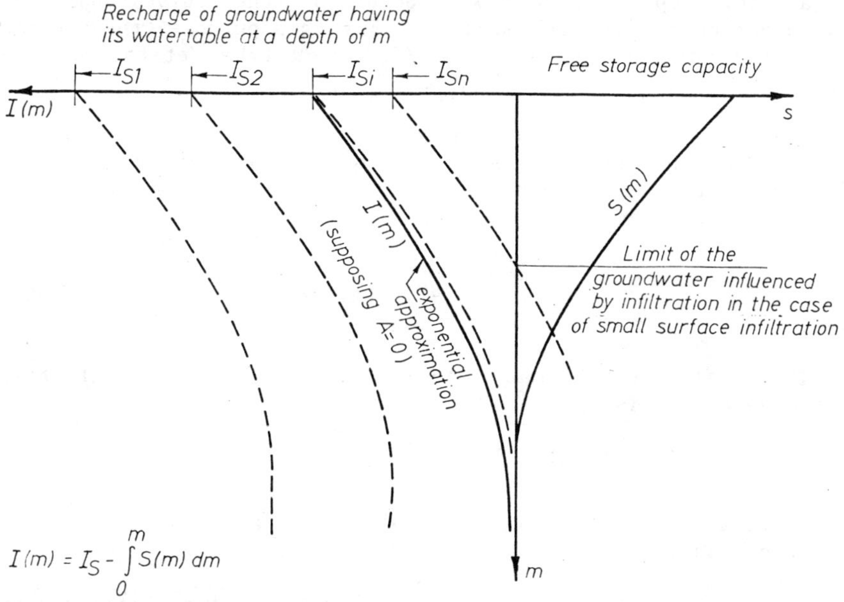

Recharge of groundwater having its watertable at a depth of m

I_{S1} I_{S2} I_{Si} I_{Sn} Free storage capacity

I(m) s

S(m)

I(m) (supposing A=0)

exponential approximation

Limit of the groundwater influenced by infiltration in the case of small surface infiltration

$I(m) = I_S - \int_0^m S(m)\,dm$

m

Fig. 3-60. Relationship between positive accretion and the depth of water table.

decrease of the infiltrating water within an infinitely small layer at the same depth

$$\frac{d\,I_g(m)}{dm} = -\,S(m) \quad ;\ \text{therefore;}$$

$$I_g(m) = I_s - \int_0^m S(m)\,dm \quad . \tag{3-89}$$

For a given soil and in an average year, a series of parallel curves can be obtained by graphical integration to represent the relationship between the infiltrating recharge and the depth of the water table. The parameter of the individual curves is the infiltration through the surface which value gives the intersection of the infiltration curve with the horizontal axis (Fig. 3-60).

To make the mathematical characterization of the relationship easier, the infiltration curve should be approximated by an easily amenable equation. It was found that the exponential function is suitable for this purpose. The water balance of groundwater is very complicated when the water table is near the surface because very small precipitations may also exert influence. It is advisable, therefore, to take another depth as a fixed point of the curve than its intersection with the surface. For this purpose, the average yearly infiltrating recharge (I_o) should be determined at a given level (m_o). The additive member of the equation (A) is the difference of the amount of infiltration through the surface and the total storage capacity of the profile, and characterizes the average positive accretion of a groundwater having very deep water table:

$$A = I_s - \int_0^\infty S(m)\, dm \quad . \qquad (3\text{-}90)$$

Thus, the final mathematical form of the investigated relationship is

$$I_g(m) = I_o \exp\,[-\alpha(m-m_o)] + A \quad ; \qquad (3\text{-}91)$$

where the parameters (I_o, m_o, α, A) are functions of natural conditions.

The curve representing the relationship between the negative accretion and the depth of the water table is very similar to the curve described previously. This also decreases monotonically with depth. It is evident that the amount of water discharged from the groundwater is approximately equal to potential evapotranspiration when the water table is near the surface (the depth is practically zero).

Actual evapotranspiration originating below the surface (ET_s) decreases as the depth of the water table increases and becomes constant at a given depth (which is between 2 and 3 m in Hungary). In this case, the actual value is 50 to 70 percent of the potential evaporation, depending on the type of soil and vegetation. The total amount drained by evapotranspiration must be divided into two parts, one supplied from the soil moisture and the other from the gravitational groundwater (Fig. 3-61). The latter, interesting from the aspect of the recent investigation, is also a function of the depth of the water table, and there exists a level from where no water can be raised to the root zone, and thus, there is practicaly no negative accretion from this depth.

The relationship between the depth of the water table and the negative accretion can also be approximated by an exponential equation having a structure similar to Eq. 3-91

$$ET_g(m) = ET_o \exp\,[-\beta(m-m_o)] \quad . \qquad (3\text{-}92)$$

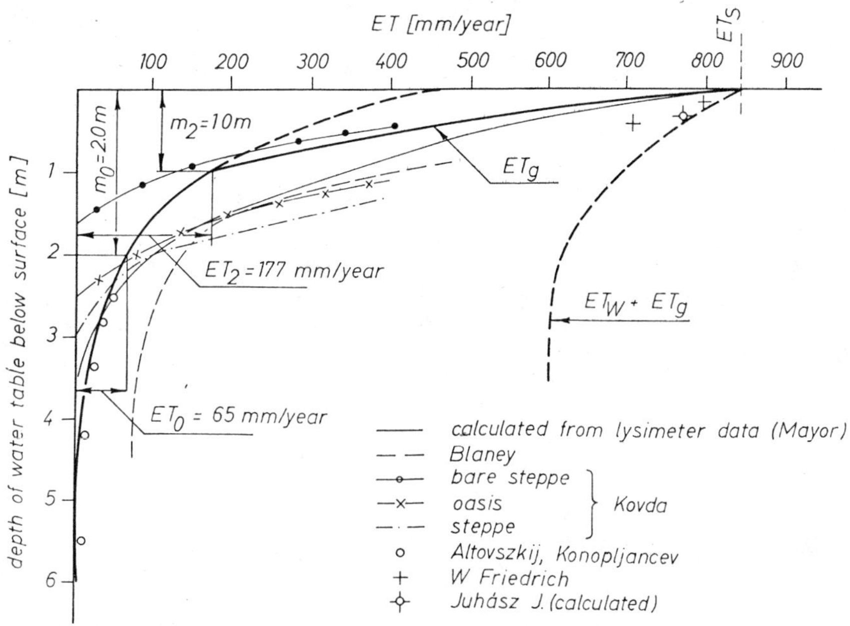

Fig. 3-61. Relationship between negative accretion and the depth of water table.

The process of evaporation depends on the capillarity of the soil, the tension between grains and water, and the influence of vegetation. Thus, the parameters of Eq. 3-92 (ET_0, m_0, β) are also functions of natural conditions.

According to Eq. 3-85, the difference between positive and negative accretion is equal to the resultant of the groundwater flow. Both variables (I_g and ET_g) are given as a function of the average groundwater depth, thus, their difference can also be plotted against the same depth. This curve is the characteristic curve of the groundwater balance (Fig. 3-62).

At the level where the curve intersects the vertical axis, the positive accretion is equal to the negative one (equilibrium level). A water table can develop at this level only if there is no horizontal flow or the inflow is equal to the outflow. Above this level, the zone of horizontally recharged groundwater can be found. In the case of such a water table, the negative accretion is greater than the vertical recharge and the difference is balanced by the excess inflow compared to the outflow. On the contrary: when the investigated column is drained by the groundwater flow

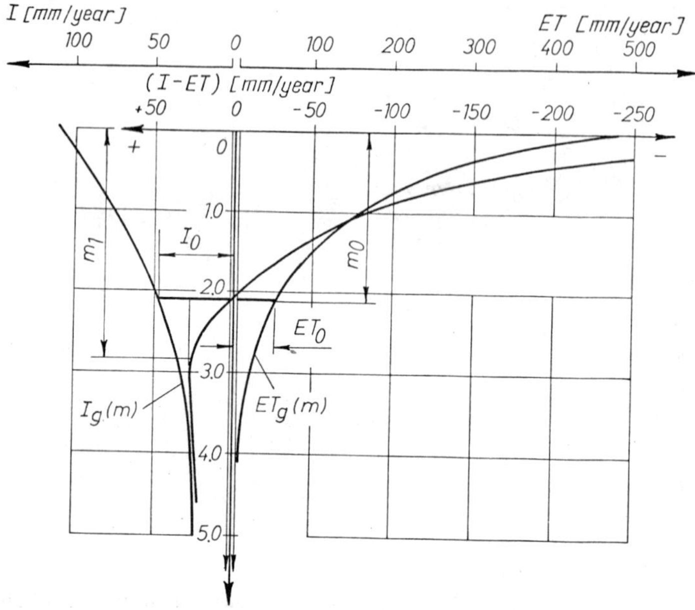

Fig. 3-62. Characteristic curve representing the groundwater balance as a function of the depth to the water table.

(outflow greater than inflow), this amount should be balanced by the excess recharge originating from infiltration, thus the water table can develop only below the equilibrium level.

A further typical point of the characteristic curve of groundwater balance is its maximum which divides the range of the possible water table into two zones. As a result of continuous groundwater exploitation, a new equilibrium can develop only in cases when the original water table was above this level, because at greater depths, the difference between vertical recharge and negative accretion decreases with depth.

Assuming the m_o parameter in Eqs. 3-91 and 3-92 to be equal to the depth of the equilibrium level, the depth of the maximum point can also be calculated:

$$m_1 = m_o + \frac{1}{\beta-\alpha} \ln \frac{\beta}{\alpha} \quad . \tag{3-93}$$

The type of shallow groundwater can be classified based on the characteristic curve. The water table near the surface is influenced directly by precipitation. Below 0.50 to 1.0 meters down to the depth of the maximum point there is the zone of the water table in which groundwater is recharged by flow and infiltration, and drained by evapotranspiration and groundwater flow. The third zone can be characterized by the balance of infiltration and groundwater flow. Finally, at a given depth and below that, groundwater is not influenced by surface effects at all. This is practically the limit of the depth of shallow groundwater (Fig. 3-63).

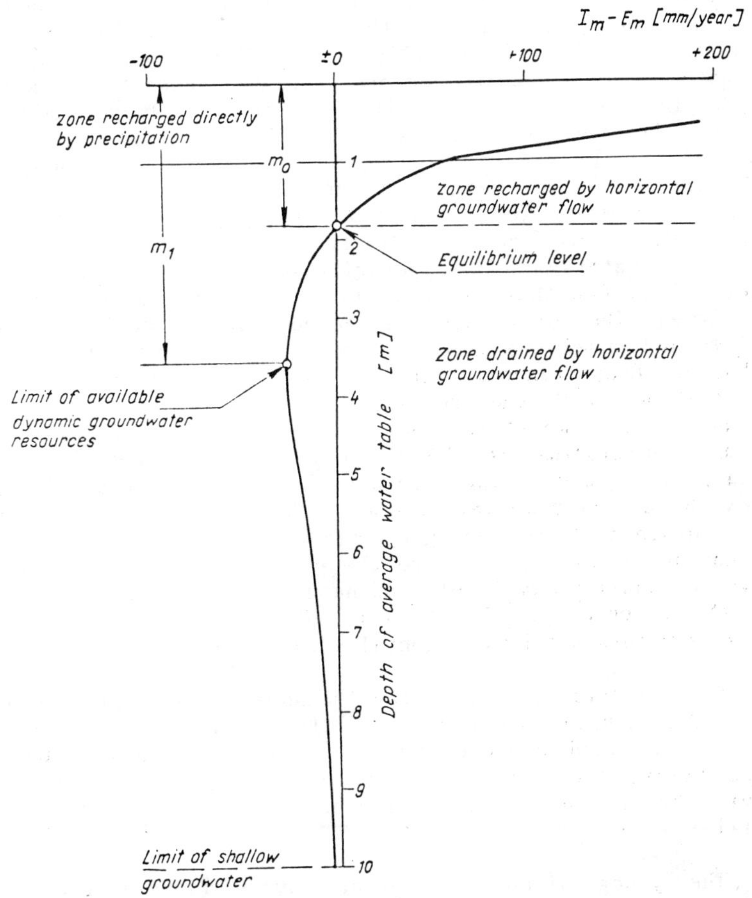

Fig. 3-63. Classification of shallow groundwater according to the influencing effects.

12-3 Hydrological Balance and Migration of Salts in and above Shallow Groundwater

The stretch of the hydrological cycle reaching the shallow groundwater and turning back from it, crossing the soil-moisture zone twice, can be easily investigated by applying the characteristic curve of groundwater balance. Since the movement of water governs, at the same time, the migration of the various salts in the soil profile, the latter process can also be explained in this way.

The water table is not horizontal under sloping terrain, but it generally follows the contour of the surface in such a way that the groundwater depth is greater below the higher areas than in the valleys (Fig. 3-64). Although the slope of the water table fluctuates according to the seasonal variation of the influencing factors, its gradient is always directed towards the deep-lying areas. The flow maintained by this gradient should be recharged and discharged continuously. Where the elevation of the terrain is higher than the average of the catchment, the water table is below the equilibrium level and the groundwater is recharged here by infiltration. The excess water may be temporarily stored by the rise of the water table, but there is a continuous water transport along the slope. The water table is raised near the surface in the valleys and under the areas of low elevation by the permanent recharge. Evapotranspiration is, therefore, higher here than infiltration, and the resultant of the vertical seepage through the soil-moisture zone is directed upwards. The total dynamic equilibrium of the water balance requires that the recharge integrated along such areas, where the water table is below the equilibrium level, should be equal to the amount of water drained from the groundwater by evapotranspiration developing in the areas where the phreatic surface is raised above the equilibrium level (and naturally by riverbeds crossing the catchment). The contact between the two types of areas is maintained by horizontal groundwater flow.

Smaller local fluxes in the groundwater and soil-moisture zones may develop even in absolutely flat plains. Since neither the land surface nor the water table has slope in these areas, the movement is created and maintained not by gravity but by the tension difference prevailing in the soil profile.

The young airborne covering layers (loess) may have different compactness, depending on the fact whether they were deposited on a dry surface or in lakes, pools, or marshes. The same variation can be found in lacustrine sediments following the pattern of the current in the lake. The genetical differences are further increased by the yearly inundations, which is a very characteristic process forming the surface of large

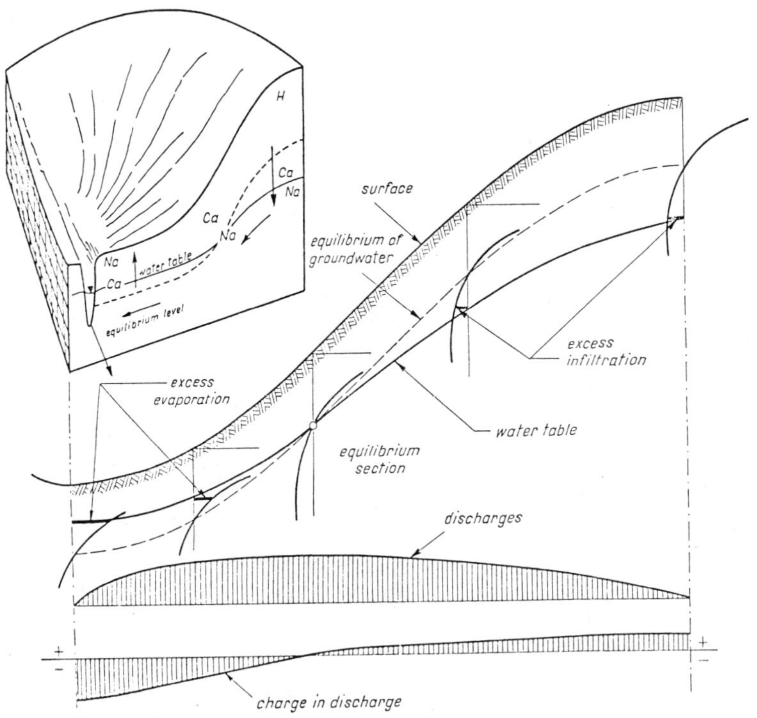

Fig. 3-64. Development of the water table under sloping terrain.

plains. The more compact layers are below the deepest depressions which are covered by water for a longer period than other areas. The very fine particles are also collected here by surface water and upon entering into the layer, clogging increases the impervious character of the sediment.

After protecting the area from floods and starting with agricultural cultivation, the differences are eliminated on the surface, but not in the layer (Fig. 3-65). There is an almost horizontal water table below the surface. Its level develops at a depth which is suitable to maintain the normal water balance (infiltration and capillary rise) under the semi-pervious covering layers, being dominant on plains. At the same time, the soil-moisture zone is almost completely saturated up to the ground surface where the covering sediment is more compact because the capillary rise is higher than the depth of the water table. Here, water cannot infiltrate through the surface during wet months because of the low

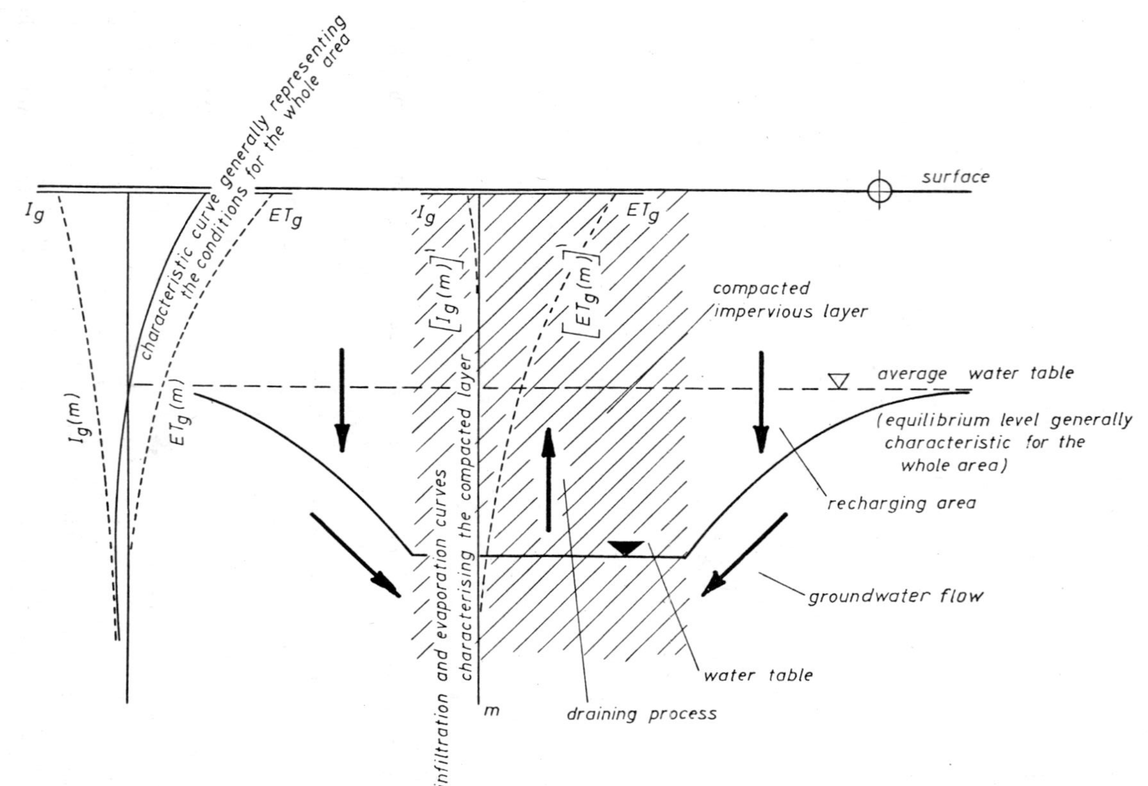

Fig. 3-65. Groundwater transport maintained by tension differences under plains.

conductivity and high tension of the layer, but in dry seasons there is considerable evapotranspiration from the soil which is replenished from the deeper horizons. To maintain the water balance, the amount evaporated is collected from the neighboring areas by horizontal flow either in the shallow groundwater space or in the soil-moisture zone.

The contact between the accumulation and leaching of salts on the one hand and the hydrological process developing in the soil-moisture zone on the other is quite evident on the basis of the characteristic curve of the groundwater balance. Both precipitation and evaporation transport water almost free of salts. Where the multi-annual positive accretion surpasses the negative value, there is a continuous vertical salt movement downwards and the soil-moisture zone is gradually leached above the deep lying water table. On the contrary, the high water table means high negative accretion and a propagation of salts upwards. It is expected that the salt content of the soil is also in a balanced condition (calculating it from the average values of the parameters determined for a long period) if the water table fluctuates around the equilibrium level (Fig. 3-66).

All the factors governing the hydrological processes in the soil-moisture zone have seasonal fluctuations. It is quite natural, therefore, that the same fluctuation can be observed in salt accumulation as well. During wet periods, when the infiltration is dominant, the salts are generaly moving downwards, and in dry months, the salt-content increases in the upper part of the soil profile. These dynamics of salts are proved by the fact that the total salt content is generally at its lowest at the end of the humid season and it reaches its maximum at the end of the depletion period (Fig. 3-67). The layering of the various salts also indicates the probable direction of their movement. Considering that Na is the most and Ca is the less soluble cation among those generally occurring in the soil, sodium can always be found at the front of the movement and calcium terminates the series of salts.

The explained hydrological process and the propagation of salts maintained by it has considerable influence on the genetics of soils as well. A great part of the soil is composed of clay-minerals which were formed by the decomposition of various silicates. Apart from clay minerals, free cations were also produced by this decomposition, a part of which is bound to the free charges of the clay-minerals while the other part can be found in the soil moisture and in the groundwater in a dissolved state. Among the changeable cations of clay-minerals, sodium is leached out first from the soil and calcium remains dominant as it was already explained in Subsection 9-3. In the next stage of the process, the calcium

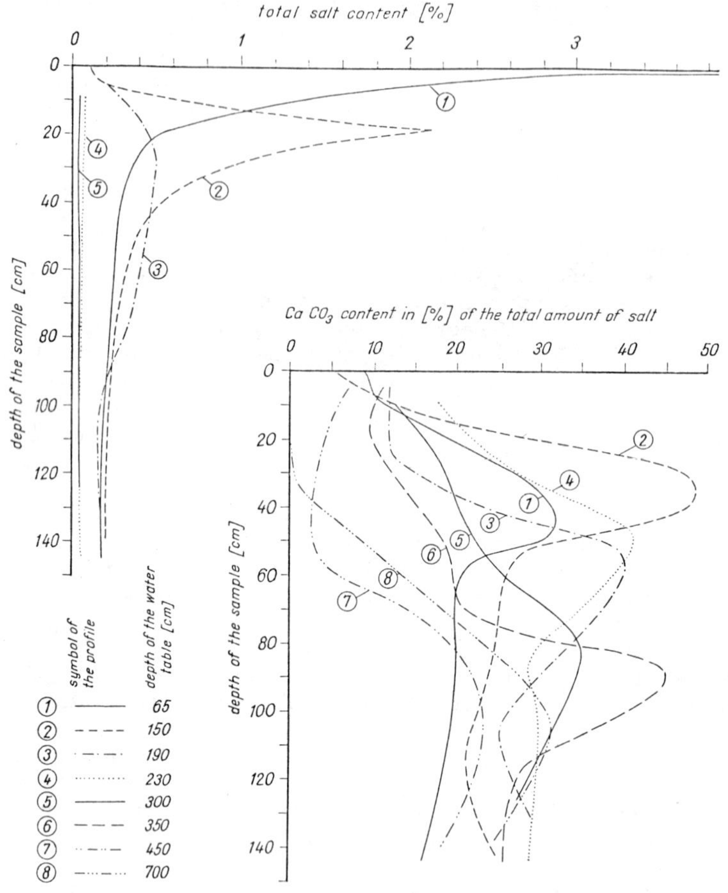

Fig. 3-66. Comparison of salt profiles developed above groundwater with different average depth.

ions will also reach the groundwater, the free electrostatic charges of the clay-minerals will be neutralized by hydrogen ions. Thus, calcereous, later acidic, leached soils develop above the deep groundwater table where the amount of precipitation is sufficient for leaching. Cations washed into the groundwater are carried to the lower lying areas by horizontal flow, and the water having high salt-content (especially

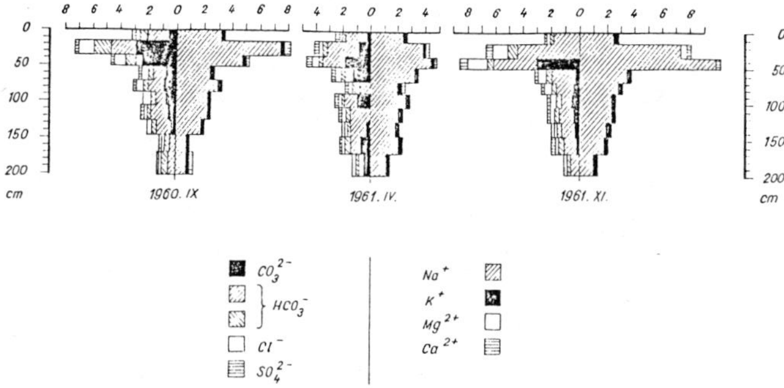

Fig. 3-67. Yearly dynamics of salts in a soil profile.

sodium) is raised near the surface. The vertical movement of water is directed upwards here, and the yearly amount of evapotranspiraton is high. The salts of the evaporated water remain in the soil and their accumulation leads to the development of saline (if Na is dominant alkaline) soils. The best arable soils (black chernozem) can be found between the leached areas and the accumulation zones where the water table is in the vicinity of the equilibrium level.

As it is shown by the explanation, the most important geohydrological parameter, from the point of view of hydrological balance and salt accumulation, is the depth of the equilibrium level. The leaching of the top soil, which is a basic requirement for achieving good production in intensive agriculture, can be expected only if the average water table is kept below this depth. In the other case, the increase of the salt-content in the arable layer is probable, which is extremely dangerous, if the largest part of the salt is Na_2CO_3 (alkalinization). The presence of NaCl causes slower deterioration of the soil.

The depth of the equilibrium level depends on many local factors. However, supposing average conditions and a light well cultivated soil with a layer below it which does not have extremely high capillary rise, some numerical value can be given to show the dependency of this parameter on climatic conditions. On the basis of data showing the salt content of the groundwater (Fig. 3-68), it was estimated that in the arid land of Tafilalet (Morocco), the characteristic depth is around 3.5 m. In Hungary, where the climate is semiarid, hydrological and soil scientific investigations have shown the average position of the equilibrium level to be about 2 m below the surface. Under humid climatic conditions

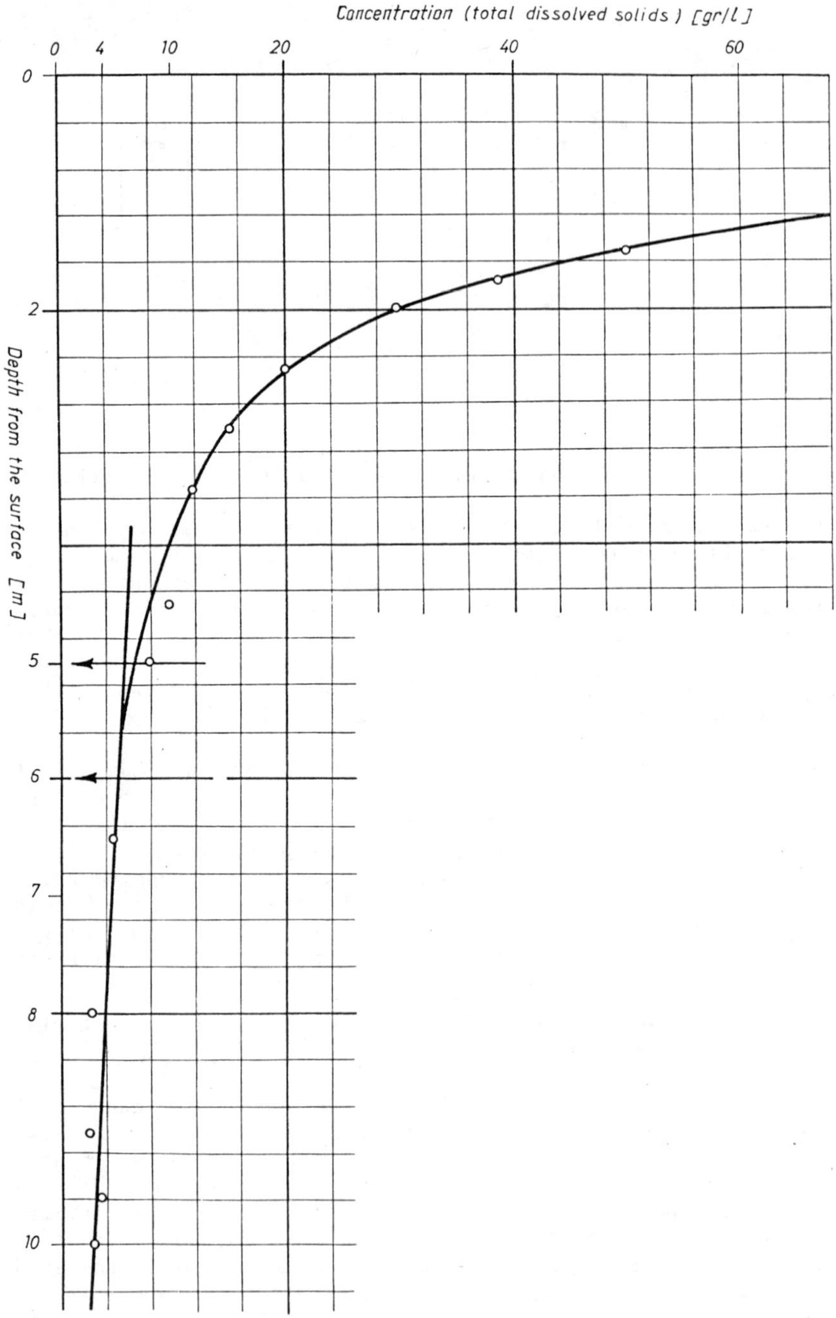

Fig. 3-68. The amount of total salt content in the groundwater as a function of the depth of water table.

(Byelorussia), water balance studies have found a much smaller parameter (0.7 m).

The relationship between the hydrological processes and the migration of salts is the same on plains as in sloping basins. The leached areas and the zones of salt accumulation (dry drainage) are not separated, however, according to the elevation of the terrain but due to the compactness of the covering layer. As it was explained earlier, a water balance develops between the excess infiltration through the semi-pervious covering sediment and the drainage occurring at the places covered by compacted highly impervious layers. The recharging water leaches the soil-moisture zone and carries the accumulated salts to the patches characterized by dry drainage. Here the salts are raised near the surface causing the further deterioration of the soil and the decrease of hydraulic conductivity.

There is practically no water movement directed downwards at the rising wing of the hydrologic cycle occurring in the soil-moisture zones of plains, i.e., where drainage develops through the more compacted sections. This is the reason why the seasonal dynamics of the salts are also missing here. There are no periods when the total amount of salt decreases, its continuous accumulation can be observed and the soil is almost completely saturated by salts at the surface. Another consequence of the lack of infiltration is that the position of the equilibrium level cannot be determined (it is at a practically infinite depth). These soils cannot be reclaimed, therefore, by lowering the water table (or the piezometric surface), but the hydraulic conductivity of the whole medium should be increased to facilitate infiltration through the layer and to initiate the migration of salts downwards in this way.

The complete alkalinization of the very cohesive covering layer near the surface may be improved by surface runoff. Slowly moving water dissolves the salts from the upper few dm of soils and carries it away when the water is drained from the surface. Thus, soils having a high salt content in the profile below 30 cm may have a thin arable layer on the surface. The depth of the improved layer can be increased by deep plowing. However, the water does not carry away the salts from the top soil if it is not drained, but remains on the surface and evaporates from there.

12-4 Change of the Character of Groundwater Balance Due to Human Activity

It is evident that all types of human activities either discharging or recharging the groundwater influences the balance of the latter. Some of these actions must be analyzed separately because they modify the hydrological processes in the soil-moisture zone and thus the curve of the groundwater balance, determined on the basis of the observation of natural processes, cannot be applied to estimate the changes expected as the result of such artificial influences.

In most cases, the water is taken out from or recharged into the gravitational groundwater directly. This artificial effect can be regarded as an additive amount supplementing the groundwater flow (either its inflowing or its outgoing component depending on whether artificial recharge or discharge is investigated). The actions do not change the characteristic curve, the relationship determined originally between the depth of the water table and the $I_g - ET_g$ difference can be used, therefore, to calculate the position of the water table after the development of the new equilibrium in the system.

The method of application of the characteristic curve is shown in Fig. 3-69. In the example represented in the figure, the original water table was below the equilibrium level, thus the area was characterized by excess infiltration (the outflow was higher than the inflow). The size of specific recharge (the amount of water recharged yearly into the groundwater space related to a unit horizontal area) is indicated in the figure in the same dimension as that applied on the horizontal axis. It is evident that the process starts with the gradual rise of the water table when the water amount recharged into the system is used to fill the increased storage capacity belonging to higher water table. New equilibrium develops, the rise of the water table is stopped at this time, when a part of recharge balances the excess outflow and the other is consumed by the excess capillary rise (and evapotranspiration) due to the higher water table. The position of the water table belonging to the new equilibrium can be determined by applying the construction indicated in Fig. 3-69.

This method is not applicable, as it was already mentioned, if the artificial action modifies the hydrological processes occurring in the soil-moisture zone. Such human activity is the change of land use, or that of the plants produced, but the most important type is irrigation which distributes a considerable amount of water over the land surface. When the task is to forecast the change of the water table under such areas (e.g., the expected rise of the water table due to irrigation) the modification of the components of the water balance should be analyzed.

The first step of this investigation is the new interpretation of the two basic curves, i.e., the infiltration curve and the evaporation curve. It is necessary, therefore, that the change of the relationships between the depth of the

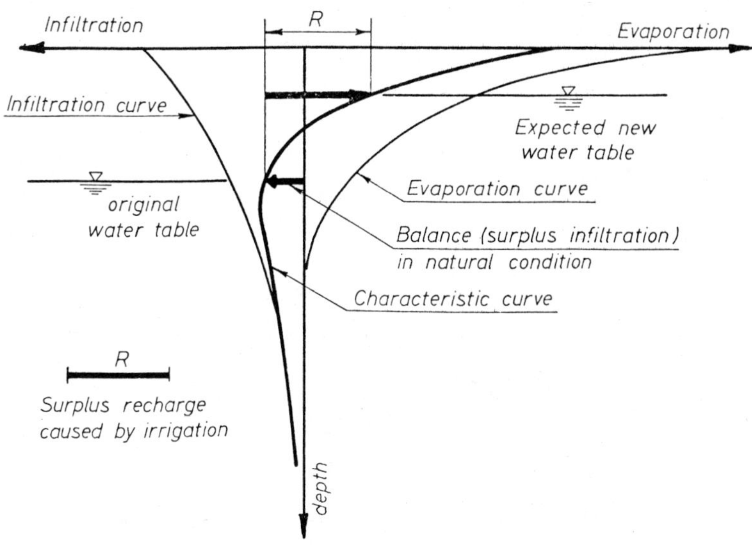

Fig. 3-69. Estimation of the position of the new average position of the water table developing due to the recharge of the groundwater.

water table on one hand and the positive and negative accretion, respectively, on the other should be analyzed. The method of the investigation is explained here by using the example of irrigation, thus the change of the hydrological processes due to the water distributed over the land surface should be determined (Fig. 3-70).

The intersection of the infiltration curve with the horizontal axis (infiltration through the surface) indicates a higher value than the natural recharge. This increase is caused by the infiltrating irrigation water. The curve runs almost vertically downwards from this starting point because the free storage capacity is almost negligible throughout the whole year in the unsaturated zone affected by irrigation. The result of this fact is that the decrease of the recharge reaching the water table is insignificant when this parameter is investigated as a functon of the increasing average depth of the groundwater level.

The starting point of the evaporation curve is practically unchanged at the surface. In this case, the total evaporation hardly depends on the depth of the water table, thus this parameter may be represented by an almost vertical line in the coordinate system. This relationship indicates

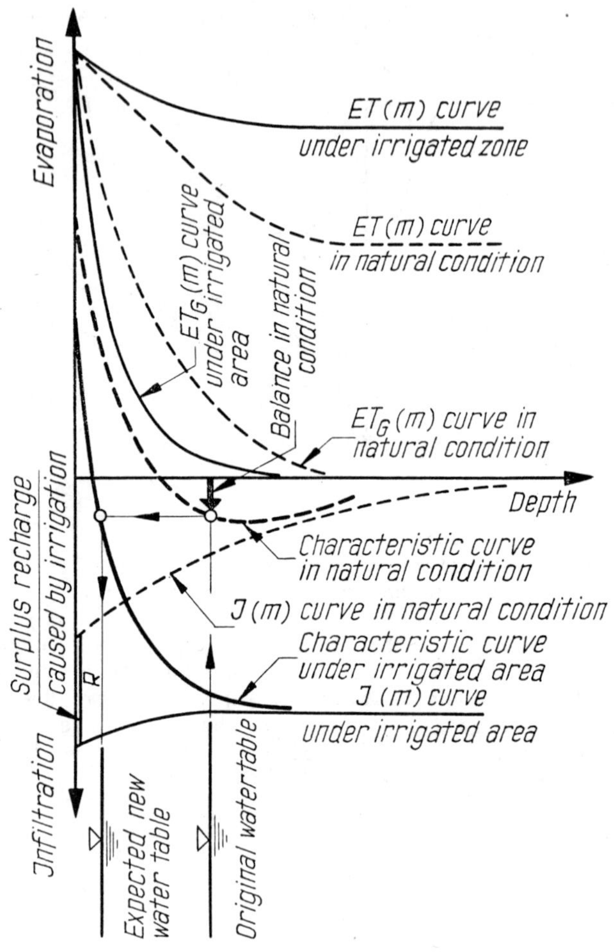

Fig. 3-70. Modification of the characteristic curve of groundwater balance due to irrigation.

that in the case of a well applied and executed irrigation, the plants always have the optimum amount of water necessary for their development. The ratio of the amount of water consumed from the groundwater relative to that supplied from the soil-moisture zone, however, changes considerably. The former becomes smaller than when it was in a natural condition because the capillary suction and the tension difference ensuring the upwards groundwater movement against gravity is relatively small in the top layer, which is almost saturated. Thus, the water consumption from the gravitational groundwater is practically zero at the lower level of the root zone.

From these two curves described above, the characteristic curve of the groundwater balance influenced by irrigation can be easily constructed. This relationship determines the modified resultant of the surface effects (positive and negative accretion) as a function of the depth of the average water table. Thus, the difference of the two curves, mentioned previously, at a given depth indicates the rate of horizontal groundwater flow which is dynamically balanced when the water table develops at that level.

It can be assumed that the horizontal flow is not altered by irrigation because the rise of the water table is limited and, therefore, the change of the gradient in the groundwater space relative to the vicinity of the investigated area can be neglected. Accepting this hypothesis, the numerical parameter of the horizontal flow can be determined as the abscissa of the characteristic curve of groundwater balance representing the undisturbed condition at the original level of the water table. This value, projected vertically to the new characteristic curve, intersects the curve at the depth where the development of the new dynamic equilibrium can be expected.

The modified characteristic curve is an important means for investigating the salt balance of the irrigated area as well. The hydrologic balance of the soil-moisture zone is fundamentally changed by the modification of the processes of infiltration and evaporation, respectively. As it is indicated by the curve, the equilibrium level (where the vertical recharge is equal to negative accretion) moves upwards. Thus, there is a zone between the natural and the modified equilibrium levels where the direction of the resultant of vertical water movement changes. In natural conditions this movement was directed upwards, indicating salt accumulation, while in the case of irrigation, the soil-moisture zone above a groundwater having its table in this zone becomes leached. Thus, the leached area can be extended by irrigation which is in good agreement with practical experience that irrigation (combined with sufficient drainage) can be used for leaching salts from the soil which are already accumulated there.

All explanations and interpretations given previously are valid only in the case when the rise of the water table occurs directly beneath the irrigated area. It has been already mentioned, however, that the investigation of the groundwater balance must be always extended to the entire catchment because the recharged and drained areas within the basin construct a comprehensive system whose elements are connected to each other by horizontal flow and their separate investigation does not supply sufficient information for the evaluation of the whole process. This is the reason why the investigation of areas surrounding the irrigated field is also considered to be necessary in almost every case.

REFERENCES FOR PART III

Ahmend, N. and Sunada, D. K., 1969: Nonlinear Flow in Porous Media (in English), Proceedings of ASCE, HY6, November.

Banks, R. B., Jersates, S., and Owsten, R., 1962: Studies on Dispersion in Porous Media Flow (in English), Technical Report Northwestern University and SEATO School of Engineers, Bangkok.

Bear, J., 1972: Dynamics of Fluids in Porous Media (in English), New York, London, Amsterdam.

Bondarenko, N. F., 1973: Physics of the Movement of Groundwaters (in Russian), Leningrad.

Bondarenko, N. F. and Nerpin, S., 1966: Influence of Viscous-Plastic Properties of Water on its Equilibrium and Transfer (in English), IASH Symposium on Water in Unsaturated Zone, Wageningen.

Böröcz, I., 1971: Low-Water Forecast on the Tisza River (Manuscript in Hungarian), Budapest, VITUKI, No. 1906.

Brown, R. H., Konoplyantsev, A. A., Ineson, J., and Kovalevsky, V. S., 1972: Groundwater Studies (in English), UNESCO, Paris.

Brownell, L., Dombrowski, H., and Dickey, C., 1950: Pressure Drop through Porous Media (in English), Chemical Engineering Progr., Vol. 46.

Carman, P. C., 1956: Flow of Gases through Porous Media (in English), London.

Charnyi, J. A., 1951: The Proof of the Correctness of Dupuit's Formula in the Case of Unconfined Seepage (in Russian), Dokl. Akademii Nauk, USSR, No. 6.

Chardabellas, P., 1964: Specification of Field Tests by Standardizing the Grain-Size Distribution Curve of Water-Bearing Clastic Sediments (in German), Mitteilung des Institutes für Wasserwirtschaft.

Childs, E. and Tzimas, E., 1971: Darcy's Law at Small Potential Gradients (in English), Soil Science, No. 3.

De Wiest, R. J. M., 1969: Flow through Porous Media (in English), New York, London.

Darcy, H., 1856: Public Water Supply of Dijon (in French), Paris.

Deák, J., 1974: Use of Environmental Isotopes for the Investigation of the Connection of Surface and Subsurface Waters in the Nagykunság Area, Hungary (in English), IAEA Symposium on Isotope Techniques in Groundwater Hydrology, Vienna;
1976: Study of the Recharge of Deep Groundwaters and Their Connection with Shallow Groundwaters Using Environmental Isotopes in the Nagykunság Region, Hungary (in English), IAH-IAHS Symposium on Hydrogeology of Great Sedimentary Basins, Budapest.

Engelhardt, W., 1960: Porosity of Sediments (in German), Berlin, Göttingen, Heidelberg.

Engelund, F., 1953: On the Laminar and Turbulent Flows of Groundwater through Homogeneous Sand (in English), Technical University of Denmark, Bulletin No. 4.

Forchheimer, Ph., 1924: Hydraulics (in German), Leipzig, Berlin.

Goldstein, S., 1938: Modern Developments in Fluid Dynamics Volume II (in English), London, New York.

Habib, J., 1971: Flow-Rate of Seepage of Water through Clay and Darcy's Laws (in French) (Doctor thesis), University of Toulouse.

Hagen, G., 1869: Handbook of Hydraulic Structures (in German), Berlin.

Hansbo, I., 1960: Consolidation of Clays with Special Reference to Influence of Vertical Sand Drain (in English), Proceedings of Swedish Geotechnical Institute, No. 18.

Harleman, D. R. F., Mehlhorn, P. F., and Rumer, R. R., 1963: Dispersion-Permeability Correlation in Porous Media (in English), Proceedings of ASCE, HY2, March.

Hatch, L. P., 1940: Flow through Granular Media (in English), Journal of Applied Mechanics, ASME, Vol. 62.

Hazen, A., 1895: The Filtration of Public Water Supplies (in English), New York.

Irmay, S., 1954: On the Hydraulic Conductivity of the Unsaturated Soils (in English), Transactions of American Geophysical Union, No. 1.

Jacob, C. E., 1940: On the Flow of Water in Elastic Artesian Aquifer (in English), Transactions of American Geophysical Union.

Jáki, J., 1944: Soil Mechanics (in Hungarian), Budapest.

Juhász, J., 1958: An Investigation into Percolation (in Hungarian), Hidrológiai Közlöny, No. 1;
1959: An Investigation into Percolation (in English), Acta Technica Academiae, Tom. 24;
1967: Hydrogeology (in Hungarian), Budapest;
1976: Hydrogeology (in Hungarian), Budapest.

Kamensky, G. N., 1940: Differential Equations of the Non-steady Groundwater Flow and Their Application to Calculate Backwater Effects (in Russian), Izvestija Akademii Nauk, USSR, No. 4;
1947: Investigation of Groundwaters (in Russian), Moscow.

Karádi, G., 1963: Hydraulics of Linear Drainage Systems (in Hungarian), Khartoum, Budapest (Thesis; Manuscript).

Karádi, G. and Török, L., 1955: Applicability of Seepage-Law (in Hungarian), Hidrológiai Közlöny, No. 5-6.

Karádi, G. and V. Nagy, I., 1960: Investigation of the Validity of Darcy's Law (in Hungarian), Conference on Hydraulics, Budapest.

Kovács, G., 1957: Theory of Microseepage (in Hungarian), Hidrológiai Közlöny, No. 3;
1958: Theoretical Investigation into Microseepage (in English), Acta Technica Academie Scientiarum Hungaricae, Tom. XXI, No. 1-2.

Kovács, G., 1963: Relationship between Autumn Low-Water Discharge of Small Water-Courses and Groundwater Conditions (in English), IASH Symposium on Surface Waters, Berkeley;
1966: Dynamic Investigation of Seepage by Invariant Numbers (in English), Symposium on Seepage and Well Hydraulics, Budapest;
1966: Yield of Partially Penetrated Wells (in English), Symposium on Seepage and Well Hydraulics, Budapest;
1969: General Characterization of Different Types of Seepage (in English), 13th IAHR Congress, Kyoto;
1969: Relationship between Velocity of Seepage and Hydraulic Gradient in the Zone of High Velocity (in English), 13th IAHR Congress, Kyoto;
1969: Seepage Law for Microseepage (in English), 13th IAHR Congress, Kyoto;

1971: Salt Accumulation in Groundwater and in Soil (in English), IAHS Symposium on Groundwater Pollution, Moscow;
1975: Interaction between Rivers and Groundwaters (in English), IAHR Symposium on Groundwater, Rapperswil.

Kozeny, J., 1954: Hydraulics (in German), Wien.

Krischer, O., 1962: Development of Material Transport through Earthworks and Porous Media by Diffusion, Molecular Movement as well as Laminar and Turbulent Flow (in German), Chemie-Ing. Techn., No. 3.

Kutliek, M., 1964: The Filtration of Water in Soils in the Region of Laminar Flow (in English), 8th Congress of ISSS;
1965: Influence of Interface on Filtration of Water in Soil (in French), Science du Sol (Prague), No. 1;
1967: Temperature and Non-Darcyan Flow of Water (in English), International Soil Water Symposium, Praha.

Kutliek, M. and Salingerova, J., 1966: Flow of Water in Clay Minerals as Influenced by Adsorbed Quinolinium and Prydinium (in English), Soil Science, Vol. 101, No. 5.

Lebedev, A. V., 1963: Methods to Investigate the Balance of Groundwater (in Russian), Moscow.

Léczfalvy, S., 1966: Water Exploitation and Water Supply (in Hungarian).

Lindquist, E., 1933: On the Flow of Water through Porous Soil (in English), 1st Congress of ICOLD, Stockholm.

Major, P., 1973: Investigation of the Balance Parameters in Plains (in Hungarian), VITUKI Scientific Reports, No. 2152 (Manuscript);
1975: Investigation of the Process of the Accretion of the Groundwater in an Experimental Station (in French), IAHS Symposium on the Hydrological Characteristics of River Basins, Tokyo.

Mosonyi, E. and Kovács, G., 1952: Model Law by Joint Consideration of Gravity and Friction (in Hungarian), Hidrológiai Közlöny, No. 7-8.

Mosonyi, E. and Kovács, G., 1956: Model Law of Filtration (in French), Congress of IAHR, Dijon.

Paladin, I. A., 1964: Determination of Permeability Coefficient of Grained Non-cohesive Soils (in Russian), Gidrotechnicheskoie Stroitelstvo, No. 3.

Perez Franco, D. F., 1973: Non-Darcy Flow of Groundwater towards Wells and Trenches, Particularly within the Turbulent Range of Seepage (in English), Technical University of Budapest (Manuscript).

Raimondi, P., Gardner, G. H. F., and Petrick, E., 1959: Effect of Pore Structure and Molecular Diffusion on the Mixing of Miscible Liquids Flowing in Porous Media (in English), Symposium on Fundamental Concepts of Miscible Fluid Displacement, San Francisco.

Reimann, J., 1974: Mathematical-Statistical Analysis of Extreme Values of Flood Waves (in Hungarian), VSzSzI. Budapest (Manuscript).

Reiner, M., 1949: Deformation and Flow (in English), London.

Rowe, 1955: Difference Approximation to Partial Derivative for Spacings in the Network (in English), Transactions of AGU, No. 36.

Scheidegger, A. E., 1957: The Physics of Flow through Porous Media (in English), New York.

Schlichter, C. S., 1899: Theoretical Investigation of the Motion of Groundwater (in English), Annual Report of the U.S. Geological Survey.

Seelheim, 1880: Zeitschrift für analitische Chemie (in German), Wien.

Slepicka, F., 1969: The Linear and Nonlinear Regimes of the Filtration Law and Their Consequences in Geohydraulic Problems (in English), 13th IAHR Congress, Kyoto.

Stol, Ph. Th., 1969: Use of Computers for the Investigation of the Hydrologic Properties of an Area, Exemplified with the Temporary Drawdown of Groundwater Levels (in English), ICID Bulletin.

Sunada, D., 1965: Turbulent Flow through Porous Media (in English), Water Resources Center Contribution No. 103, University of California.

Swartzendruber, D., 1962: Modification of Darcy's Law for the Flow of Water in Soils (in English), Soil Science, No. 1; 1968: The Applicability of Darcy's Law (in English), Proceedings of the American Society of Soil Scientists, No. 1.

Székely, F., 1973: Determination of Groundwater Accretion and Seepage Parameters on the Basis of Groundwater Observations (in Hungarian), Hidrológiai Közlöny, No. 5.

Szilágyi, Gy., 1954: Yield of Wells Calculated with Variable Hydraulic Conductivity (in Hungarian), Vizügyi Közlemények, No. 3.

Terzaghi, K., 1926: Soil Physical Bases of Soil Mechanics (in German), Wien;
1943: Theoretical Soil Mechanics (in English), New York, London.

Theis, C. V., 1935: The Relation between the Lowering of the Piezometric Surface and the Rate and Duration of Discharge of a Well Using Groundwater Storage (in English), Transactions of American Geophysical Union.

Thirriot, C., 1969: Hydrodynamics of Flow in Porous Media (in French), 13th IAHR Congress, Kyoto.

Ubell, K., 1954: Comparison of Methods Serving to Determine the Water Yielding Capacity of Aquifers (in Hungarian), Vizügyi Közlemények, No. 2;
1958: Practical Application of Theoretical Well Hydraulics (in Hungarian), Vizügyi Közlemények, No. 3.

Valentin, F., 1970: Nonlinear Resistance of Porous Media (in German), Mitteilungen, Institut für Hydraulik und Gewässerkunde, München, No. 6.

V. Nagy, I., 1967: Laboratory Measurements to Determine the Coefficient of Permeability of Samples (oral information) (in Hungarian), Budapest.

Ward, J. C., 1964: Turbulent Flow in Porous Media (in English), Proceedings of ASCE, HY5, September.

Zamarin, J. A., 1928: Calculation of Groundwater Flow (in Russian), Trudey I.V.H. Taskent.

Zauberei, I. I., 1932: On the Problem and Determination of the Permeability Coefficient (in Russian), Investia VNIIG, Leningrad, No. 3-5.

Zunker, F., 1930: Behavior of Soils in Connection with Water (in German), Berlin.

PART IV

HYDROLOGICAL PROCESSES OCCURRING IN DEEP-LYING CLASTIC SEDIMENTS

by

J. Gálfi, G. Kovács and N. Pataki

SECTION 13

HYDROLOGICAL PROCESSES OCCURRING IN DEEP-LYING CLASTIC SEDIMENTS

DEVELOPMENT OF FLOW SYSTEMS IN DEEP-LYING AQUIFERS

The decomposition and erosion of solid rocks and sediments covering the surface, the transport of the eroded material from higher terrains to deeper lying valleys and basins as well as the settling of the transported grains there, is a continuously acting geological process. It tends to eliminate, or to at least decrease, the differences in the elevation of the surface and counteracts, in this way, the other geological forces causing the vertical movement of the crust (faulting, volcanic eruptions, etc.). One of the results of the process, being the most important from a hydrogeological point of view, is the development of large aquifer systems which store a great amount of water which is one of the most important resources for water supply. For the utilization of this resource, it is necessary to consider that the aquifers compose a natural storage system. The inventory of the available water can be determined, the expected changes in the system as the result of the exploitation of water can be predicted, therefore, only by investigating the flow and storage processes in the aquifers. It follows from this requirement that groundwater hydrology forms the scientific basis of the development of groundwater resources.

Groundwater hydrology, according to its definition, analyzes the processes governing the flow of water below the surface and the change of its amount stored in the aquifers. In the present case, when our purpose is the hydrological investigation of phenomena characterizing deep-lying clastic sediments, the research does not include the entire subsurface section of the hydrological cycle, but it is limited to one part of the continuous system. It was explained, however, as a basic criterion of hydrological studies, that the movement of water along the hydrological cycle should always be investigated and the continuity should be substituted by considering the interactions developing between the separated parts of the system, if any section of the cycle is investigated independently. When distinction was made between shallow and deep groundwater, the contact of the water-bearing formations with the hydrological and meteorological processes occurring on and above the surface was mentioned as the basic aspect of this grouping. Deep groundwaters are not directly influenced by these processes; the water regime of deep-lying aquifers does not depend, therefore, on the influence of surface phenomena.

The boundary conditions to be taken into account in the hydrological investigations of sedimentary basins and other deep-lying aquifers are composed mostly of water exchanges between various layers. Sometimes deeply penetrating beds of rivers or lakes may also have contact with these layers and the recharge or drainage through the beds should be considered as well.

The aquifers and the boundary conditions characterizing their continuity with the environment compose a system being in balanced conditions. It follows from this fact that the difference between inflow and outflow summarized for an investigated period should be equal to the change in the amount of water stored in the system. The fluctuation in storage commonly observed under natural conditions, as wet and dry periods succeed each other, indicates the change of the ratio of recharge and drainage. It can also be stated, however, that the wet and dry seasons counterbalance each other if the interval investigated is long enough. An important hydrological consequence of the state of equilibrium of the system is that the change of the water amount stored in the aquifers is equal to zero for such long periods. Any groundwater exploitation decreases, therefore, the natural draining processes. When the water resources of deep-lying aquifers are utilized, the amount extracted from these layers is balanced by decreasing the resources of shallow groundwaters and surface waters.

Knowing that the greatest part of groundwater resources is not independent but a unified system is composed of the various groundwater horizons and surface waters, a question can be easily raised: are groundwater resources of any importance? The answer is naturally in the affirmative and the arguments verifying this statement are as follows:

- surface water can be utilized only along the watercourses and rivers whereas the wide extension of water-bearing layers makes water available over large areas;

- the quality of groundwater, especially from a biological viewpoint, is generally suitable for human consumption, while surface waters need special treatment in most cases;

- during flood periods, a great portion of surface water reaches the seas and oceans without utilization; a part of this excess water can be recharged into the aquifers, thus artificially increasing in this way the retention of freshwater on the continents.

It is well-known that the hydrological investigation of subsurface water should be built-up from three subsequent steps:

- the determination of the geometry of the interconnected layers in which the transport and storage of water may take place;

- the estimation of physical parameters characterizing the flow through and the water retention capacity in the system;

- the analysis of the dynamic conditions governing the transport and the storage of water and to calculate, in this way, the numerical characteristics suitable to describe the natural processes and those influenced artificially developing in the system of the interconnected layers investigated.

The characterization of the position and size of deep-lying aquifers does not differ from any other geological exploration aiming to determine the structure of the crust.

Most of the parameters describing the flow conditions of the investigated layers in this chapter are also identical with those characterizing any other loose clastic sediment. It is not necessary, therefore, to repeat here the detailed analysis of hydraulic conductivity (or intrinsic permeability), threshold gradient, and porosity. The interpretation of storage capacity may differ, however, depending on the fact whether shallow or deep groundwater is investigated. In the first case, the system is generally unconfined (or the covering layer is semipermeable) and, therefore, the development of storage is primarily determined by the specific yield of the layer in question, while the storage coefficient of deep-lying confined aquifers depends on the compressibility of both the solid matrix and water.

The dynamic analysis of the process requires the investigation of the forces acting, the continuous movement of water and the accumulation of energy content in the system. The forces maintaining and retarding the water transport in the deep-lying clastic aquifers are those analyzed in Part I. It should be considered, however, that the gravity term includes a part of the weight of the overlying formations taken over by the water in the form of pressure (neutral stress). The total energy head is composed, therefore, of two factors, viz., the elevation above the reference level (potential energy) and pressure head. The investigation of water transport requires the introduction of hydrodynamic principles into hydrogeological studies. The entire system of the interconnected aquifers and the aquitards separating the former ones should be regarded as a unified flow field. The determination of the probable flow pattern in this way may provide us with sufficient information required to estimate the contact between the various components of the basically unified water resources system (surface water, shallow and deep groundwater, soil moisture) and to forecast the changes expected as the results of planned exploitations. Finally, the investigation of energy balance (energy content of water recharged into and drained from the system, energy consumed by water transport, and latent energy determined by the resistance of the system belonging to a static condition), apart from that of mass balance, has special importance when the hydrological processes occurring in deep-lying clastic aquifers are analyzed, because the influence of latent energy is generally prominent in these systems.

Considering the aspects listed as those characterizing the special behavior of the systems composed of deep-lying aquifers, three topics are discussed in the introductory section of this part:

- the classification of the aquifers from the point of view of the development of flow systems;

- the application of hydrodynamic principles for characterizing these systems;

- the investigation of consolidation of water-bearing formations, since this process governs the storage capacity of the system and determines the amount of water available as the result of decreasing the pore volume.

13-1 Classification of Aquifer Systems

The various aspects of the classification of the different types of subsurface waters were already listed in Part I (origin of the water, water temperature, chemical substances being dissolved in the water, etc.). It was found, on the basis of the detailed analysis of the possible ways of grouping, that there are two classifications which are the most important from a hydrological point of view, those which can be formed according to:

- the character of the water-bearing rocks;

- the contact between the type of the subsurface water in question and the meteorological as well as hydrological events occurring on and above the land surface.

Considering the first aspect, the loose clastic sediments and the solid rocks were distinguished as the two most important groups of water-bearing layers because the structure of the solid matrix and that of the water transporting channels basically differs in the case of these two formations. In the first case, the solid matrix is constructed of small grains and the pores, encircled by the grains, form a network of channels almost equally distributed and randomly interconnected. A similar structure hardly occurs in the interior of solid rocks (sandstone may be mentioned as an exception), but water percolates through these layers along fissures and fractures. Since one of the most important aspects to be considered in hydrological studies is the movement of water along the hydrological cycle, the structure of water transporting channels, which determines the resistance of the solid matrix against the propagation of water, has an extremely dominant role in the classification. Only the processes occurring in loose clastic sediments are analyzed in this part (as is indicated by the title), and the problems concerning the hydrological investigation of solid rock aquifers are dealt with in Part V.

It also follows from the basic definition of hydrological investigations, viz., the movement of water as a part of the hydrological cycle should be analyzed in each case, that the other important aspect of classification is the contact of the water horizon investigated with other hydrological events. From this point of view, basic distinction was made between soil moisture and groundwater within the large group of various types of water stored in loose clastic sediments. It was also mentioned that the interface between the two zones (i.e., water table) is only a fictious surface because saturation difference does not exist there. The main difference occurring in the behavior of groundwater and soil moisture (which justifies at the same time the separation of the two groups) is the pressure prevailing in the water, viz., in groundwater the pressure is higher than the atmospheric value, while in soil moisture the same parameter is lower than this characteristic limit.

A further distinction was also made by dividing the groundwater stored in loose clastic sediments into two parts, i.e., shallow and deep groundwaters. According to the definitions given in Part I:

- shallow groundwater is stored near the surface below the water table and above the first continuous impervious layer which is largely extended within the investigated area; it follows from this position that this water horizon is directly influenced by meteorological and hydrological events occurring on or above the land surface;

- deep groundwater saturates the pores of water-bearing aquifers lying below a continuous impervious formation which hinders the direct contact between the surface processes and the groundwater over a very large area; the deep groundwater, is, therefore, recharged or drained only through the shallow groundwater and sometimes by deeply penetrating beds of rivers and lakes.

It was also explained that clear distinction cannot be made between these two types of groundwater. Shallow groundwater can be partly covered by an impervious layer, but the groundwater below it may be recharged by precipitation and drained by evaporation if the covering layer is not continuous or large enough. The same processes may be characteristic in the case of semipermeable (leaking) covering formations. In these cases the effects of meteorological and hydrological events can be observed in the regime of groundwater. It is also possible that the aquifer covered over a large area by impervious formations and consequently containing deep groundwater is raised near the surface far from the investigated area. There, the aquifer becomes unconfined and the groundwater stored in it is influenced by precipitation and evaporation. Thus, the same groundwater being in deep position at the place of investigation may be a shallow one not too far from this point.

It is not possible, therefore, to draw a sharp border between the two types of groundwaters. There are several transition forms and our efforts may be concentrated only to find the dominating factors, i.e., to investigate whether the groundwater space is influenced mostly from the direction of the surface, or the regime may be interrelated with the parameters of another water horizon. The strength of surface effects affecting the groundwater depends on the extension and thickness of the formations covering the aquifer in question as well as on the conductivity of both the water-bearing and the overlying layers. The evaluation of the groundwater data (water table and piezometric level) observed at various depths is the way to determine the type of groundwaters and to estimate the expected recharge and drainage, the latter two being the most important factors from the viewpoint of exploitation.

Although the distinction based on the estimation of the influencing processes, and indicated by the adjective "shallow" and "deep", refers to such behavior of the water-bearing system which has a practical importance in connection with the utilization of groundwater resources, it is not yet commonly applied in hydrogeology. Many other terms are also used to characterize the deep position of aquifers or some special behavior of deep-lying water-bearing layers, e.g.:

- closed groundwater (its aquifer is completely enveloped by impervious layers);

- artesian water (the pressure in the aquifer is higher than the depth below the surface and, therefore, the well draining the layer provides water flowing out freely on the surface);

- confined aquifer (refers to the hydraulic character of movement, i.e., the pressure along the upper contour of the flow field is higher than the atmospheric value, therefore, the water table does not develop in this system);

- connate (or fossil) water (although the term refers to the origin of water, i.e., water in which the sediment was deposited and which preserved its original condition, especially chemical composition, the fact that the original character of water was not considerably modified indicates the poor water exchange between the aquifer in question and the neighboring layers).

The terms listed here are used by many authors and indicate the attempts to divide deep groundwater into several subgroups. The purpose of this further distinction is to simplify the versatile system of deep-lying aquifers and to separate such units in which the number of processes dominantly influencing the water conditions is relatively low. It is expected that models can be established to simulate the water horizons separated and simplified in this way. Considering, once again, the basic requirement of hydrological studies (according to which the dynamic character, the processes of flow and energy accumulation in the system, should be always taken into account), it is quite natural that deep groundwater should be classified according to the flow conditions characterizing the system.

The classification established by considering hydrodynamic principles and the energy content of the system can be clearly summarized on the basis of papers dealing with this topic and presented at the IAH-IAHS Symposium on Hydrogeology of Great Sedimentary Basins (Kovács, 1976). The most important common feature of these papers is that they always assume a discintion between the upper layers where water transport governs the groundwater regimes, and the lower part of the sedimentary formations, influenced by internal forces (i.e., consolidation and thermal effects). Considering the two dominating factors, Kissin (1976) even divides the lower part into two zones: elisian (the dominating action is consolidation) and abyssal zones (thermal dehydration being the most important effect). Bogomolov et al. (1976) as well as Rogovskaya and Bezrodnov (1976) also make distinction in the hydrodynamically influenced zone between unconfined and confined systems divided by the upmost regionally extended impervious formation (which is in good accordance with the separation of shallow and deep groundwaters). The paper prepared by Jarvin et al. (1976) also deals separately with the layers having recharge from the neighboring formations maintained by groundwater flow and with those whose resources originate from consolidation. The complete system of classification and the dominating processes may be summarized in the following form:

Table 4-1. Groundwater classification.

Shallow Groundwater			Hydrological and Meteorological Events on the Surface
	Hydrodynamic Zone	--------------	Water Transport
deep groundwater	zone of energy accumulation	elisian zone	consolidation
		abyssal zone	thermal dehydration

The most important consequence of the classification is that basically different approaches have to be applied when investigating the various regimes. The influence of infiltration and evaporation has to be taken into account and these processes have to be combined with groundwater flow in the zone of shallow groundwaters, as it was explained in the previous chapter. The water transport capacity of the system, the resistance acting against the flow, and the water exchange between the contacting formations are the parameters to be determined for the characterization of the water regime of the hydrodynamic zone of deep groundwaters. The dynamic balance of mass and energy provides the basic principle for the analysis and the determination of flow systems developing within the unified seepage field composed of permeable and semipervious layers, as it will be discussed in the next part of this chapter. In the lower zone of deep groundwaters, the static condition of water is the basic hypothesis of hydrological studies and the accumulation of energy as well as the change in the amount of water stored in the pores as a function of pressure prevailing in the system should be investigated. The interrelation between storage capacity and pressure depends on the process of consolidation, which will be dealt with in Subsection 14-3. In the upper part of the zone of energy accumulation (which is sometimes called the elisian zone), consolidation may be regarded as the sole dominating process, while at very geat depths (in the abyssal zone), the influence of thermal dehydration should also be considered.

There is one problem which still remains open, i.e., the determination of the limit between the hydrodynamic zone and that influenced mostly by consolidation. In the papers quoted previously, only some slight indications can be found concerning the depth of this border: a depth of 300 m is mentioned by Jarvin et al. (1976), below which the dominant role of resources originating from consolidation is expected during exploitation, while the upper value of this depth is 300 to 500 m, and its average is 1000 m according to Kissin (1976), but he has also mentioned that in extreme cases the border may be as deep as 6000 m. A difficulty is caused by the fact that sharp distinction cannot be made between the two different regimes since gradual transitions are generally characteristic for any changes in nature, and only the human brain tries to simplify the processes and to describe them by using generalized and deterministic relationships. One of the most important consequences of this gradual change is the fact that man's activity may basically modify the weight of the factors acting and, therefore, this condition has to always be taken into account in hydrological investigations of deep groundwaters.

Two methods can be proposed to eliminate the obstacles hindering the application of the classification explained here, i.e.:

- to determine the physical behavior of the layers and to give numerical data for the characterization of the developing processes (giving more sensitive quantitative description of the processes in this way instead of the sharp and rigid qualitative distinction);

- to use parameters other than those characterizing the mass balance of water (chemical composition, environmental isotopes, heat balance, energy content) to prove the existence or the absence of movement and to assist the determination of the numerical parameters mentioned in the previous paragraph.

It is necessary to consider that the accumulation of energy is a continuous process in most cases. There is no water transport in a system, where the flow is hindered by impervious layers, until the energy content has achieved the upper limit of the latent energy determined by the resistance of the system belonging to its static condition (threshold gradient). When the stored energy surpasses the limit as a result of continuous accumulation, flow develops in the system. Thus, the imperviousness of a layer is not an absolute term, but it is related to the instantaneous energy content of the whole system.

The comparison of actual hydraulic gradient and the threshold gradient of the layer in question is the way to determine whether the system is in static condition or the dynamic balance should be investigated. The physical interpretation of threshold gradient and its calculation from soil-physical data was given in Chapter 3. As it was explained, the recent investigations (Bondarenko, 1973; Kovács, 1976) have shown that the threshold gradient decreases with increasing water temperature and its value may be smaller in great depth than near the surface. This condition may be compensated by the effect of higher pressure, but the determination of this second influence requires still further investigations.

The other important result of recent research (Kovács, 1976) is that the pore size is a random variable and the probability of a pore having twice a larger diameter than the average pore is still 1-4 percent. Thus, the development of flow is still probable under a gradient only half of the theoretical threshold value. The interpretation of the threshold gradient as a random value is also well demonstrated by the evaluation of data measured in mines (Keserü, 1976).

13-2 Theory of Local and Regional Flow Systems

In connection with the use of hydrodynamic aspects in groundwater hydrology, Maxey's (1969) lecture delivered at the first International Seminar for Hydrology Professors (Urbana, Illinois) can be quoted. He stated there that the most important new feature of recent hydrogeological investigations is the combination of hydrological and dynamic methods. He recalled the results presented by Tóth (1962; 1963) which are represented in Fig. 4-1. In these investigations, Tóth considers the whole mass of loose clastic sediments within a basin as a unified flow system and determines the flow net in this field. The boundary conditions along both the surface and the lower impervious contour of the basin can be determined. The impervious boundary is always a flow line, while the potential conditions at the surface can be characterized by the position of the water table. The entry and exit faces of the flow space can also be separated if the main recharging and draining effects are known. Geological and geophysical explorations can provide us with information on the nonhomogeneity, the layered character, and the anisotropy of the field. Taking into account these flow conditions together with the boundary conditions, the probable flow net characterizing the seepage developing through the entire basin can be constructed by applying, e.g., analog models.

The considerable change of the local potentials along the surface within relatively short distance, as well as the nonhomogeneity and the anisotropy of the field may result in

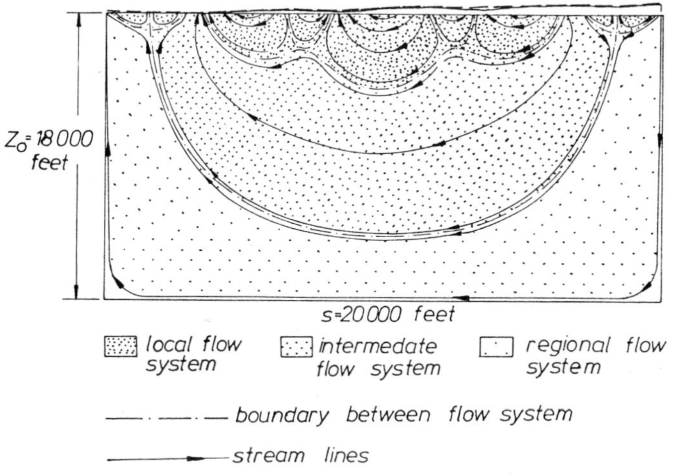

Fig. 4-1. Development of regional and local flow systems in large sedimentary basins (after Tóth).

the development of separated flow systems independent of each other within the unified flow space. Local and generally shallow flow fields develop near the surface, directed from the higher terrain towards the valleys. The paths between the recharging and draining areas are relatively short in these systems and the flow rate is, therefore, high in most cases as compared to the deeply penetrating part of the whole water movement. The flow lines following the contour of the basin may contact regions located far away from each other. These regional flow systems generally start from the edges of the basins and the water transported along these paths (where the hydraulic gradient and, therefore, both velocity and the flow rate are usually low because of the considerable length) reaches the surface at the main draining points. In most cases, there are intermediate systems between the local and regional ones enveloping a few local flow spaces and penetrating some hundred meters. The development of the local systems is generally governed by the morphology of the surface, while the extension of the intermediate and regional systems are mostly determined by the geological conditions.

A similar way of thinking was developed and published by Almássy (Schmidt, 1962) for the characterization of the pressure conditions of deep groundwaters within the Carpathian Basin. Using the data of numerous deep wells, he has constructed the equipotential lines in some characteristic sections of the basin (Fig. 4-2). The potential lines characterize the flow conditions similarly as the set of

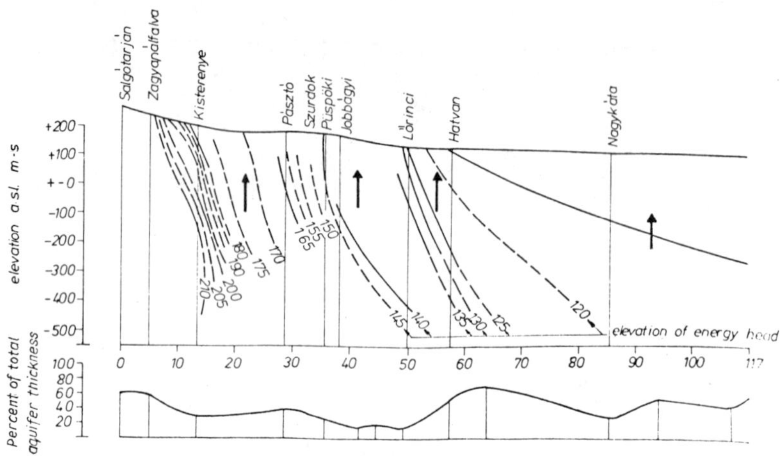

Fig. 4-2. Probable potential distribution in sedimentary formations of the Hungarian Basin (after Almássy).

streamlines, since it is well-known that the two types of curves compose an orthogonal system. Although the streamlines visualize the flow system better (they indicate the path of water) than the potential lines and the set of streamlines contains more information (the density of streamlines is proportional to the strength of the flux), the construction of potential lines has also some advantages:

- piezometric level data measured directly are used for the determination of potential lines; their reliability is better than that of streamlines, since some assumptions concerning the homogeneity and isotropic behavior of the flow field have to be accepted for the construction of streamlines;

- the potential lines are valid not only in the hydrodynamic zone, but also in the zone of energy accumulation where a static condition prevails; the potential differences indicate only the possibility of flow, thus the permeability of the layer should also be investigated to determine whether flow can develop in the system or not; streamlines cannot be applied if there is no water transport through the field, while the potential lines represent the accumulation of energy in this case;

- knowing the normal distance between equipotential lines representing given potential values, the hydraulic gradient prevailing at various points of the field can be calculated; the hydraulic gradient compared to the threshold value provides direct information about the dynamic condition of the system, and if the actual value is higher than the threshold gradient the former multiplied by the hydraulic conductivity of the layer in question gives the estimated value of flux characterizing the water transport of the system.

Almássy's work was continued by Erdélyi (1972, 1976), who has supplemented the evaluated data by adding the information collected from boreholes drilled in the last decade. He was able, in this way, to extend the investigation to the whole internal basin (Great Hungarian Plains) instead of some selected sections. Calculating the pressure gradients between 50 and 100 m as well as between 100 and 400 m at several points and considering also the chemical character of water, he has distinguished the parts of the basin where the potential gradient is directed downwards from those where the possibility of percolation upwards is given in the system, because the gradient is directed towards the surface. The areas discharged and recharged, respectively, were separated in this way and the so-called neutral zones were also determined, where the vertical distribution of the pressure is approximately equal to the hydrostatic pressure distribution, and the gradient is equal to zero. Erdélyi's four figures are repeated here as the summary of his results. There are two maps, the first shows the location of the areas recharged and discharged as well as the neutral zones within the basin (Fig. 4-3), while the second one represents the territorial distribution of one chemical component (the areas are indicated where Cl content of the groundwater stored in the layer between 150 and 350 m is higher than 30 mg/ℓ) (Fig. 4-4). The other two figures are characteristic sections; the first is a SW-NE section through the basin in which the probable flow directions are also estimated (Fig. 4-5), the second is a part of the former one, crossing the recharging area situated at the NE part of the basin (Fig. 4-6). The detailed comparison of the land surface and the piezometric levels of various water horizons can be found on the upper part of the last figure and the lower part shows the geological structure.

The investigation draws attention to the importance of the fact that the entire flow system should always be determined and analyzed. It is a well-known and generally accepted basic requirement in seepage hydraulics that the geometry of the entry and exit faces as well as that of the other contours of the flow field should be determined because the position of the two bordering potential surfaces and the boundary conditions influence not only the numerical values of the parameters describing the water transport, but even the character of flow developing through the field. Although this principle is applied unanimously in hydrodynamics, in hydrogeological studies its importance was hardly recognized in the past. The new approach and the recent investigations explained in the previous paragraphs have testified that only the exploration

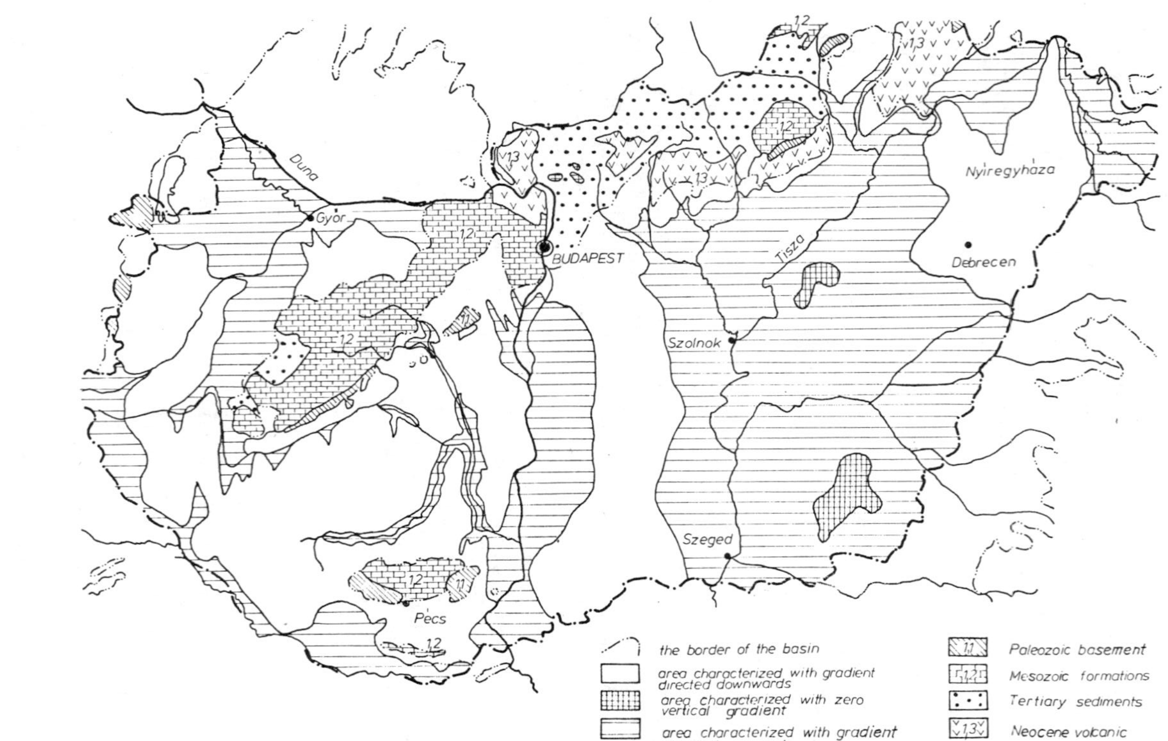

Fig. 4-3. Territorial distribution of the recharging and discharging areas in the Hungarian Basin (after Erdélyi).

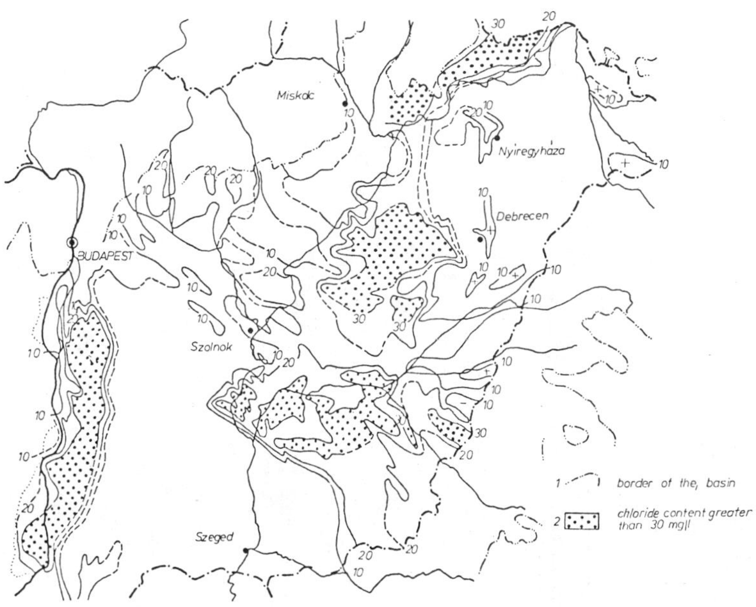

Fig. 4-4. Territorial distribution of the chloride concentration in the deep groundwaters of the Great Hungarian Plains (after Erdélyi).

of the complete flow system can lead to the clear understanding of the hydrological processes governing the regime of deep groundwaters. It is inevitable, therefore, that hydrological studies should start with the determination of the entire interconnected flow field distinguishing:

- its recharging areas;
- the water transporting stretches;
- the places and types of the various draining processes.

The complete description of the flow system requires not only the geometrical exploration of the entire field, but also the estimation of the amount of water transported at the various stretches of the system. Some quantitative conclusion may be drawn up from the comparison of the two maps (Figs. 4-3 and 4-4). It can be assumed that the high salt content of

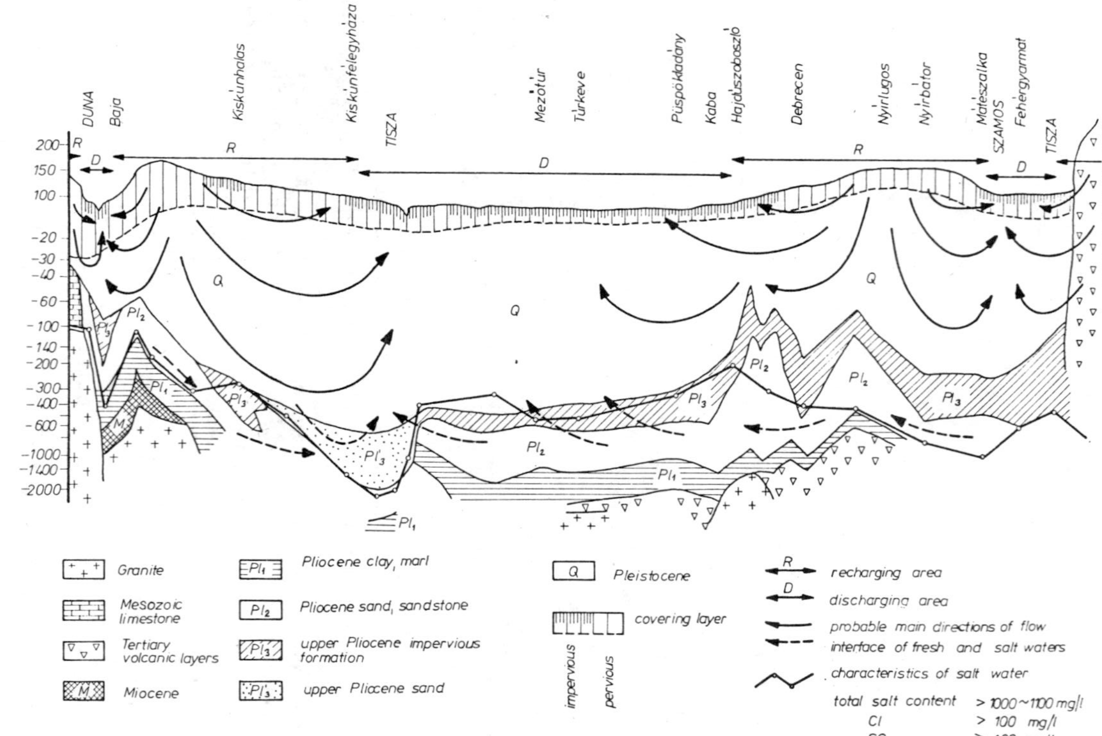

Fig. 4-5. Generalized flow pattern in a section through the Great Hungarian Plains (after Erdélyi).

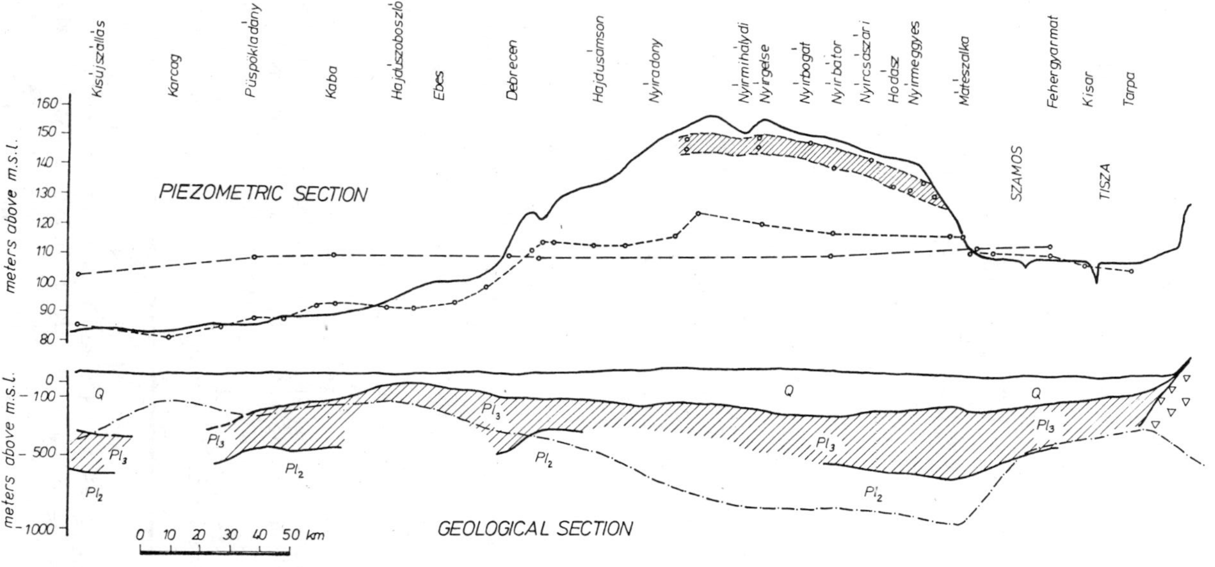

Fig. 4-6. Pressure distribution along a section through the Nyirség area of the Great Hungarian Plains (after Erdélyi).

groundwater indicates the absence of considerable groundwater flow and, therefore, it can be supposed that the amount of water infiltrating at the edges of the basin and reaching its internal part is negligible. This hypothesis is justified by the existence of high chloride content and by the development of a neutral pressure condition here. It can be stated, therefore, that the main drainage areas form belts of 30 to 40 km in width around the recharging areas and the greatest amount of water transported by horizontal groundwater flow is discharged within these belts either by rivers or by evapo-transpiration (Fig. 4-7) (Kovács, 1973).

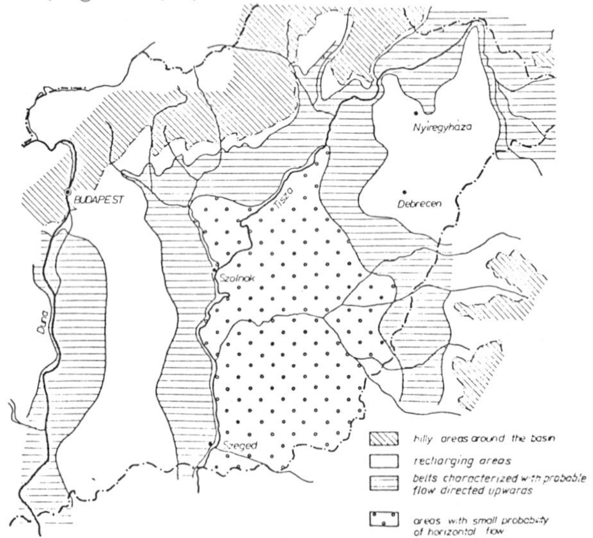

Fig. 4-7. The probable development of the belts influenced by horizontal flow around the recharged areas.

Knowing the gradient of flow and the transmissivity (product of aquifer thickness and hydraulic conductivity) of water transporting layers, the rate of the water conveyance of the separated subsystems can also be determined. In the case of the example represented in Fig. 4-6, the gradient can be calculated as the slope of the water table and the piezometric surface, respectively. Considering also the depth of the various subsystems as well as the conductivity and anisotropy of the layers, the water transport along the various stretches of the system can be estimated. It was found that the greatest amount percolates along the local subsystems, which basically develop in the zone of shallow groundwater. The flow rate of these subsystems can be characterized by an average annual flux of 3 to 10 mm distributed almost evenly within the area of vertical drainage. The intermediate subsystems contact the main recharging areas and the belts of discharge surrounding the former ones. Their flow field occupies the zone of quaternary sediments. The total amount of water transport through this formation is approximately equal to that of local subsystems, but it is distributed over a larger drainage area. The average specific flux is smaller, therefore, than the contribution of the flow in the zone of shallow groundwater to the water balance of a unit area (near the recharge area the component of water balance originating from the intermediate subsystem is 3 to 10 mm/year related to a unit horizontal area similarly as that of shallow groundwater, while this parameter decreases to 0.1 mm/year at the external perimeter of the discharging belt). Finally, the continuous slope of the piezometric level of water contained in Pliocene formations indicates that the regional flow starting from the edges of the basin and directed towards its center is independent of local recharging effects. The contribution of this subsystem to the water balance of a unit area is, however, negligible (less than 0.01 mm/year) since the gradient is very small, the porosity and permeability of the compacted layers is also smaller than the same parameters near the surface, and the transported water is distributed over a large area.

The nonhomogeneity and the anisotropy of the formations may disturb the development of the flow net and may influence the transport of the system. It is necessary, therefore, that some hypothesis should be applied concerning these two behaviors of the seepage field in order to achieve an acceptable simulation of the actual conditions. The example of the Tagus basin (Spain) is analyzed here to represent the construction of streamlines in a layered system (Llamas and Cruces de Abia, 1976).

A section of the basin is represented in Fig. 4-8 showing the contour of the basement of the basin which is composed of tertiary sediments. Considering the relief of the terrain, the development of local, intermediate, and regional flow systems can be assumed by following the principles derived by Tóth, as it is indicated in the figure. The layered and nonhomogeneous character of the field, the alteration of permeable and semipermeable lenses, should be taken, however, into account as well. Figure 4-9 shows, e.g., the influence of sand lenses on the development of the flow pattern in a semipermeable medium. The layered aquifer system would be practically equivalent to a homogeneous anisotropic field having higher horizontal permeability than the vertical one. In the case of the example analyzed now, the sediments might be regarded to be a multilayered system composed of a series of aquifers and aquitards supposing, however, that all the layers are homogeneous and isotropic. Figure 4-10 shows that the upmost layer is an unconfined aquifer which is further followed by two confined water-bearing layers. The three water horizons are separated by two aquitards and the bottom of the basin is impervious. The flow pattern determined by

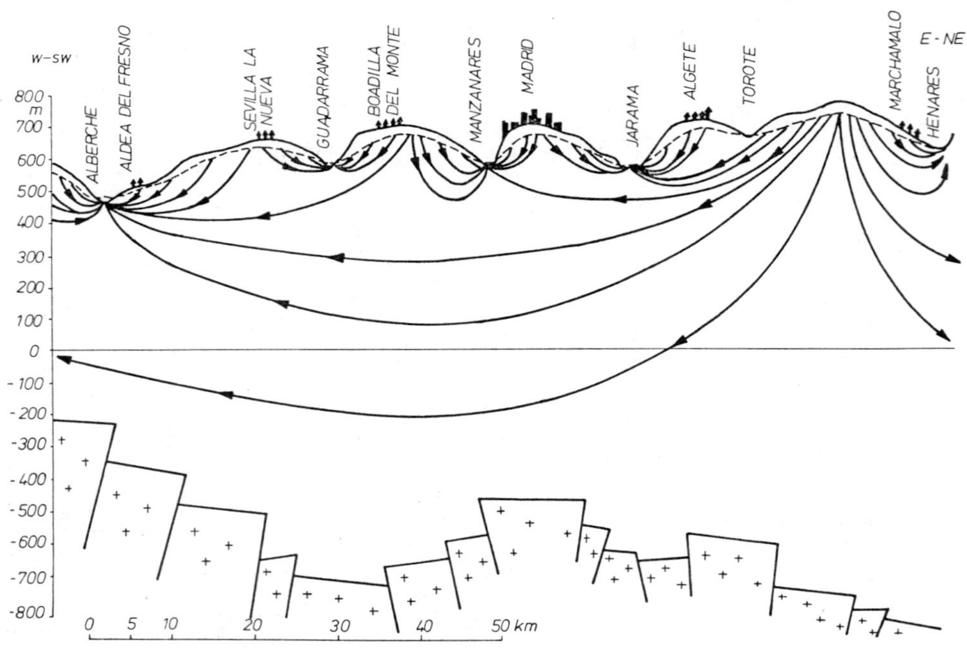

Fig. 4-8. Local, intermediate, and regional groundwater flow systems in the Togus Tertiary Basin (after Llamas and Cruces de Abia).

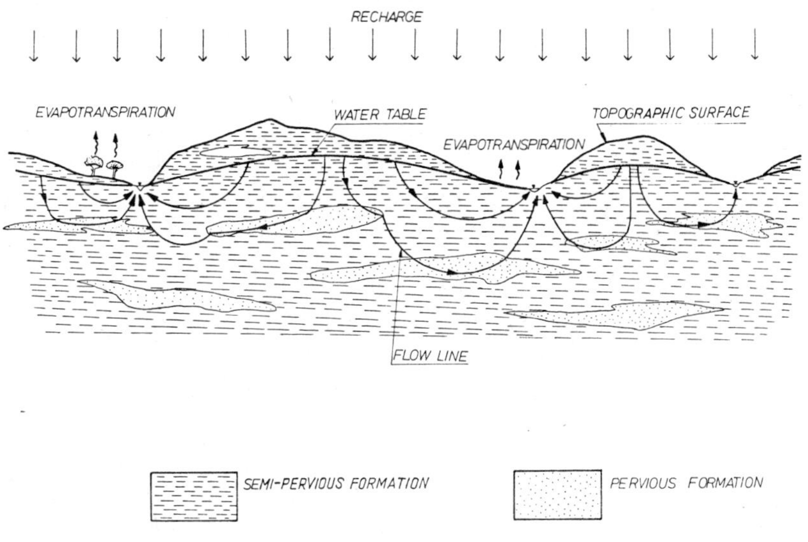

Fig. 4-9. Local groundwater flow system in heterogeneous media (after Llamas and Cruces de Abia).

considering the explained structure of the basin and represented in Fig. 4-10 was in good agreement with the data derived from various observations, and thus the assumed field was applied as the basis of the mathematical model serving the determination of the available groundwater resources.

13-3 The Process of Consolidation and Water Resources Gained by the Decrease of Pore Volume

The development of pressure conditions and the amount of the available groundwater resources depend on the consolidation of the layers in confined systems. This is the reason why the storage capacity of the water-bearing formations has to be investigated at first to survey the close interrelation between the change of pressure, consolidation, and the storage process in deep aquifers.

The storage capacity of a layer is generally characterized by the volume of water released from, or taken into, storage as a result of a unit change in pressure. The volume is related to the horizontal area of the layer and thus it has a dimension of length [L]. This amount depends on:

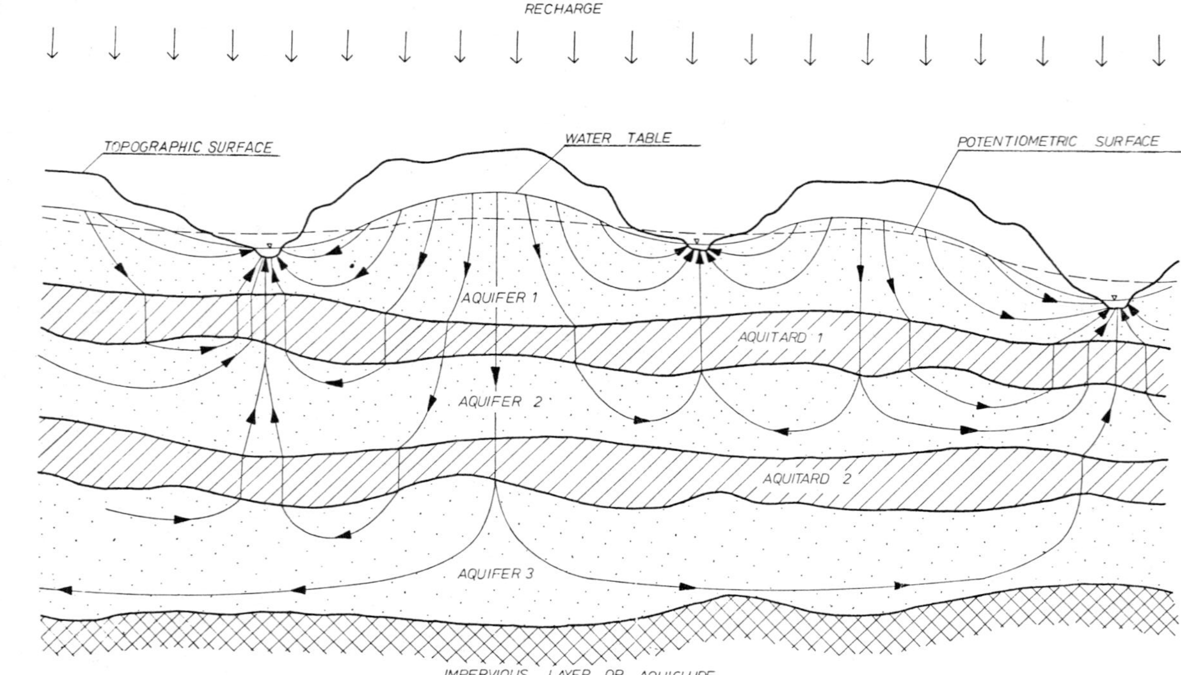

Fig. 4-10. Scheme of a multilayered aquifer system (after Llamas and Cruces de Abia).

- the change of the ratio of the saturated and unsaturated thickness;

- the compressibility of both solid grains and water;

- the change of porosity of the saturated layer.

The influence of the first component occurs only in unconfined systems. The change in storage is characterized in this case by the specific yield of the layer, the detailed explanation of which was given in Chapter 2. Since deep lying aquifers are investigated in this part and the pores remain saturated in these confined aquifers, the specific yield should not be considered in these investigations.

The compressibility of solid grains is very small compared to the other two components influencing the storage capacity of confined aquifers. Its influence is negligible, therefore, in any case. This is the reason why the storage process in confined aquifers is characterized by a parameter called the storage coefficient (S), which depends on the consolidation of the layer and on the compressibility of water. This value represents the total storage capacity of the layer having a thickness of b. Sometimes this parameter is related to the unit volume of the aquifer (water released from, or taken into, storage in a volumetric unit of the layer). This specific storage (S_s) multiplied by the aquifer thickness provides the storage coefficient:

$$S = b\, S_s \tag{4-1}$$

According to its definition, the storage coefficient is composed of two parts: the change of the volume of water caused by the modification of pressure, and the change of the pore volume where the water is stored:

$$S = S_W + S_P \quad . \tag{4-2}$$

The compressibility of water is expressed by the change of density with pressure which is proportional to the density itself:

$$\frac{d\rho}{dp} = \beta\rho \quad ; \tag{4-3}$$

where β (the coefficient of compressibility) is the factor of proportionality, and its numerical value for pure water having a temperature of 20°C at atmospheric pressure is β = $4.6\text{x}10^{-9}\text{m}^2/\text{kp}$; or $4.6\text{x}10^{-8}\text{cm}^2/\text{p}$. The mass being unchanged, the modification of volume of water V_W can be expressed from Eq. 4-3:

$$V_W = \frac{M}{\rho} \quad ;$$

$$\frac{dV_W}{d\rho} = - \frac{M}{\rho^2} \quad ;$$

$$\frac{dV_W}{dp} = \frac{dV_W}{d\rho} \quad ; \qquad (4\text{-}4)$$

$$\frac{d\rho}{dp} = - \frac{M}{\rho} \beta = -\beta V_W \quad .$$

The volume of water in a saturated layer is equal to the volume of pores, thus in a unit volume of the aquifer it is equal to the numerical value of porosity. In a column of unit area of a homogeneous layer having a thickness of b, the total change of the water volume can be calculated considering the total pore volume of the product nb because all the parameters are constant:

$$\left[\frac{dV_W}{dp}\right]_o^b = -\beta nb \quad ;$$

$$(4\text{-}5)$$

$$[dV_W]_o^b = -\beta nb\ dp \quad .$$

The storage coefficient measures this change if the pressure difference expressed as an equivalent water column is unity: $dp/\gamma = dh = 1[L]$. Combining this condition with Eq. 4-5, the first component of storage coefficient can be determined

$$S_W = -\rho\gamma nb \quad . \qquad (4\text{-}6)$$

The process of compression, which is well-known in soil mechanics, may be used for the determination of the second component of the storage coefficient. For simplifying the mathematical description of the phenomenon, some approximative hypotheses have to be accepted. The first among them is the assumption of the linear vertical compression of the layer neglecting the possible horizontal displacement of particles as well as the shearing stress caused by the unequal compression along two neighboring vertical sections.

Loading the sample gradually, the relationships between the load and various soil-physical parameters (specific linear compression, dry specific weight, water content, porosity, void ratio) can be observed and represented (Fig. 4-11) (Kézdi, 1972). It is necessary to note here that the pore volume is generally characterized in this investigation by the void ratio instead of porosity because the latter is given as the ratio of pore volume related to the bulk volume, and thus compression modifies this basic value as well, while void ratio is the ratio of pore volume and the volume of the solid matrix, the latter being not influenced by compression. On the basis of experiments, Terzaghi (1926; 1943) has established an empirical relationship according to which the change of void ratio over the change of load is inversely proportional to the value of load (p) increased with a surplus load (p_o) determined experimentally (Fig. 4-12):

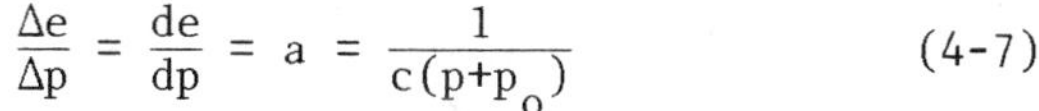

$$\frac{\Delta e}{\Delta p} = \frac{de}{dp} = a = \frac{1}{c(p+p_o)} \qquad (4\text{-}7)$$

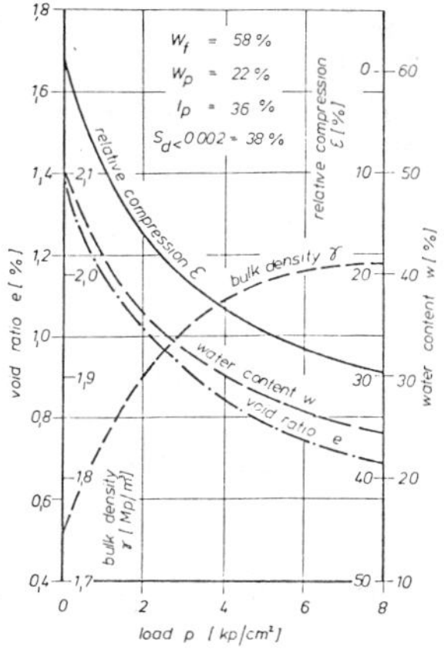

Fig. 4-11. Graphs representing the results of experiments investigating compression (after Kézdi).

The coefficient of consolidation, a, can be expressed as a function of the modulus of compressibility (M):

$$a = \frac{1 + e_o}{M} \quad ; \qquad (4\text{-}8)$$

where $M = dp/d\varepsilon$; and $\varepsilon = \Delta h/h$ is the relative change of the vertical length of the loaded sample (the interpretation of Δe and Δh, as well as the comparison of the two types of curves of compression is shown in Fig. 4-13, while some informative value of M is given in Table 4-2. The integration of Eq. 4-6 provides Terzaghi's equation of compression.

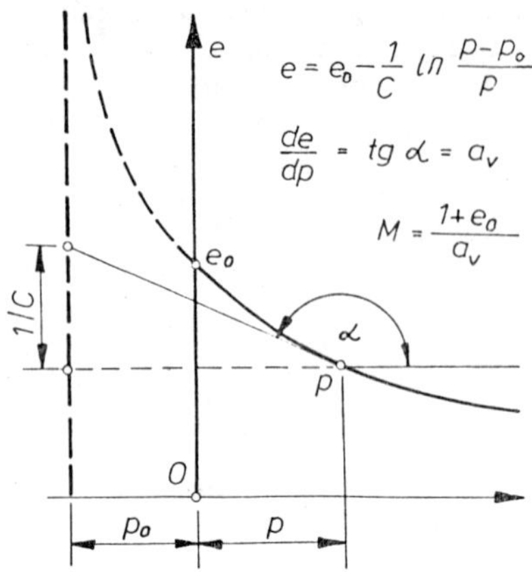

Fig. 4-12. Sketch showing the symbols used in the interpretation of the relationship between load and compression.

The next problems to be discussed is the method of application of the listed equations in order to calculate storability. In a loose clastic aquifer, the weight of the overlying layers (the latter related to a unit horizontal area is characterized by the total vertical stress σ) is transferred only partly from grain to grain (this ratio of the total stress is the effective stress, $\bar{\sigma}$), the other part is balanced by water pressure (neutral stress, u). Creating a depression (Δh) by draining the layer, water pressure decreases and, therefore, the effective stress has to be increased, while the total stress remains unchanged:

$$u_2 = u_1 - \gamma \Delta h \quad ;$$

$$\sigma = u_1 + \bar{\sigma}_1 = u_2 + \bar{\sigma}_2 \quad ; \tag{4-9}$$

$$\bar{\sigma}_2 = \bar{\sigma}_1 + \gamma \Delta h \quad ;$$

(suffixes 1 and 2 indicate the condition before and after draining, respectively). Equation 4-9 shows that the decrease of the pressure head of water increases the effective vertical stress and the change can be regarded as an external load causing compression:

$$\Delta p = \Delta \bar{\rho} = \gamma \Delta h \quad . \tag{4-10}$$

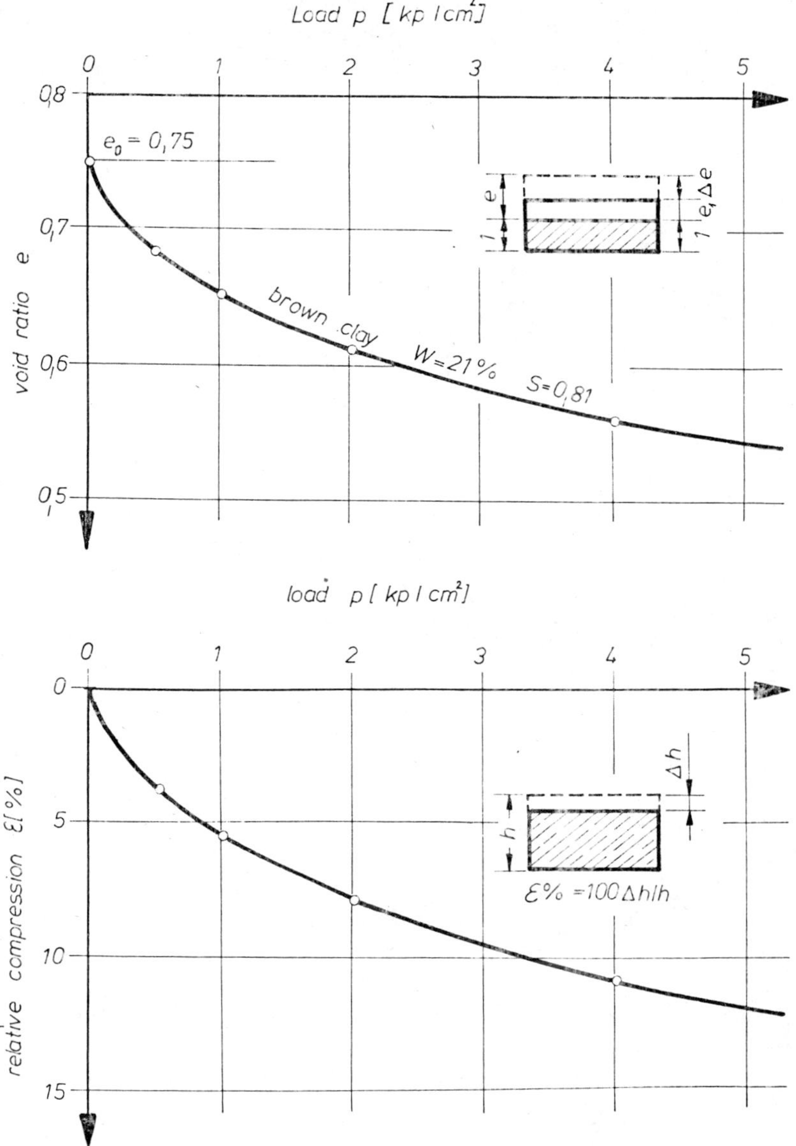

Fig. 4-13. Comparison of the representation of compression depending on either void ratio or the change of linear size.

According to the definition, the second component of the storage coefficient is the change of the total pore volume in a column having a unit horizontal area and a height equal to

Table 4-2. Informative values of the modulus of compressibility (after Kézdi).

M [kp/cm^2]

Type of Sample	Condition of the Sample					
A) Coarse grained sediments	loose		medium		compacted	
Sandy gravel	300 - 800		800 - 1000		1000 - 2000	
Sand	100 - 300		300 - 500		500 - 800	
Very fine sand (mo)	80 - 120		120 - 200		200 - 300	
B) Cohesive sediments	soft	plastic		hard		very hard
Silty very fine sand	50 - 80	100 - 150		150 - 200		200 - 400
Silt	30 - 60	60 - 100		100 - 150		150 - 300
Light clay	20 - 50	50 - 80		80 - 120		120 - 200
Heavy clay	15 - 40	40 - 70		70 - 120		120 - 300
Organic silt			5 - 50			
Organic clay			5 - 40			
Peat			1 - 20			

the thickness, b, of the layer, if the pressure head is modified by unit height. In a unit volume of the layer, the pore volume is equal to porosity. Its change over the change of pressure can be expressed from Eq. 4-7:

$$\frac{dn}{dp} = \frac{dn}{de}\frac{de}{dp} = \frac{(1-n)^2}{c(p+p_o)} \quad ; \quad dn = \frac{(1-n)^2}{c(p+p_o)}\,dp \quad ;$$

or for unit change of pressure head

$$(dn)_1 = \gamma\,\frac{(1-n)^2}{c(p+p_o)} \quad ; \tag{4-11}$$

where the pressure is equal to the effective stress, $(p=\bar{\sigma}_1)$ in the initial state. Thus, the final form of S_p is;

$$S_p = \gamma\int_o^b \frac{(1-n)^2}{c(p+p_o)}\,dz = \gamma\int_o^b \alpha(z)dz \quad ; \tag{4-12}$$

where

$$\alpha(z) = \frac{[1-n(z)]^2}{c[p(z)+p_o]} \quad ;$$

$\alpha(z)$ is a special variable, combining all the depth dependent parameters in one factor. Considering the numerous uncertainties influencing both the process itself and the parameters, it is generally accepted as a first approximation to assume this factor to be independent of the depth. The form of the storage coefficient usually published in literature (Bear et al., 1969; Bear, 1972) is based on this hypothesis:

$$S = \gamma b(\alpha - n\beta) \quad . \tag{4-13}$$

In highly compressible layers, the effect of water compressibility can be neglected but the change of both porosity and effective stress along a vertical section has to be considered. This can be done either by integrating Eq. 4-12, or in tabulated form, assuming the constancy of the parameters within short stretches of the vertical section (Kovács, 1972).

A further problem is caused by the fact that the layer is not completely elastic. Thus, the extension following the increase of neutral stress is not equal to compression caused by the decrease of water pressure of the same size, and the first loading creates greater compression than reloading (Fig. 4-14). Although the definition of the storage coefficient supposes that the amount of water released from, or stored in, the layer as a result of a unit change in pressure head is equal to one another, it is quite evident from the figure that this statement is only a very rough approximation. When solving practical problems in seepage hydraulics, it is impossible to analyze the whole history of loading and reloading of the layer and, therefore, an average parameter is acceptable to characterize the storage capacity of aquifers.

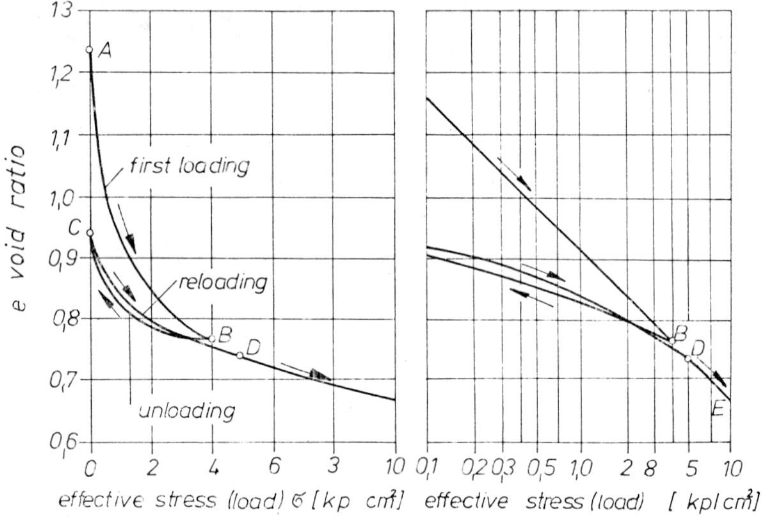

Fig. 4-14. Compression curve represented in arithmetic and logarithmic scale.

The detailed analysis of compression and the direct application of soil mechanical results are also hindered by the time lag occurring between loading and the total development of compression. The curves representing the relationship between changes of both load and void ratio (see Figs. 4-11 and 4-14) were achieved with gradual loading, waiting after all changes of the load until no further compression or extension could be observed. Investigating the time dependent process of consolidation, many further parameters and conditions have to be considered. The mathematical treatment of the problem needs, therefore, the application of further simplifying approximations.

For exhibiting the complexity of the problem which ought to be solved, the representation of the most simple example of the linear vertical compression will be continued here, assuming that the α parameter in Eq. 4-13 is constant, and thus the relationship between the change of porosity and that of effective stress can be expressed from Eq. 4-11 considering also Eq. 4-9:

$$\frac{dn}{d\bar{\sigma}} = - \frac{dn}{du} = \alpha \quad . \tag{4-14}$$

(see Tables 4-3 and 4-4).

Accepting the validity of Darcy's law, the specific flow rate in the vertical direction crossing a horizontal area of unity can be calculated;

$$q = v = KI = -K \frac{\partial(u+z)}{\partial z} \quad ; \ [LT^{-1}] \quad ; \tag{4-15}$$

because the hydraulic gradient can be determined as the change of pressure and potential energy along the z-axis (the negative sign indicates that the flow is directed upwards, while the z-axis is positive downwards). The modification of the flow rate within an elementary stretch, dz, expressed from Eq. 4-15 is;

$$\frac{\partial q}{\partial z} = - \frac{K}{\gamma} \frac{\partial^2 u}{\partial z^2} \quad . \tag{4-16}$$

The flow rate can increase only if there is a source in the elementary volume, and in the case just investigated, only the decrease of the pore volume can be such a source. Thus, the change of porosity in time should be equal to the change of flow rate along an elementary length. This condition combined with Eq. 4-14 provides the following relationship:

$$\frac{\partial q}{\partial z} = - \frac{K}{\gamma} \frac{\partial^2 u}{\partial z^2} = \frac{\partial n}{\partial t} = -\alpha \frac{\partial u}{\partial t} \quad ; \text{ therefore;}$$

$$c \frac{\partial^2 u}{\partial z^2} = \frac{\partial u}{\partial t} \quad ; \text{ where} \tag{4-17}$$

$$c = \frac{K}{\gamma \alpha} \quad .$$

Thus, finally a differential equation is gained for describing the change of neutral stress and consolidation, which can be solved by integration if the boundary conditions are known at both horizontal boundaries of the layers. This condition increases the complexity of the process, and indicates that the investigation of consolidation together with that of storage capacity cannot be limited to one layer

Table 4-3. Calculation of parameters characterizing the water resources gained by consolidation.

A. Clay with high porosity

(n_1 = 0.500; C = -.4-; σ_o = 1.2 kg/cm^2)

Depth m	Before drainage				After drainage					T m	n_1-n_2	K_1 m
	n_1	e_1	σ1 kg/cm^2	γ'tl t/m^3	n_2	e_2	σ2 kg/cm^2	γt2 t/m^3	γ't2 t/m^3			
0	0.500	1.000	0.000	0.800	0.500	1.000	0.000	1.300	-	4.888		
10	0.476	0.910	0.820	0.840	0.464	0.870	1.350	1.400	0.860		0.012	0.400
50	0.422	0.730	4.270	0.914	0.414	0.720	4.950	-	0.930		0.008	0.375
100	0.387	0.630	9.020	0.980	0.380	0.620	9.750	-	0.990		0.007	0.600
200	0.338	0.510	19.500	1.060	0.333	0.500	20.050	-	1.070		0.005	
										ΣT=4.880m	ΣK=1.375m	

B. Fine sand

(n_1 = 0.375; C = 0.06; σ_o = 1.2 kg/cm^2)

Depth m	Before drainage				After drainage					T m	n_1-n_2	K_1 m
	n_1	e_1	σ1 kg/cm^2	γ'tl t/m^3	n_2	e_2	σ2 kg/cm^2	γt2 t/m^3	γ't2 t/m^3			
0	0.375	0.600	0.000	1.000	0.375	0.600	0.000	1.630	-	3.710		
10	0.368	0.584	1.005	1.010	0.364	0.577	1.640	1.650	1.020		0.004	0.120
50	0.358	0.559	5.070	1.025	0.356	0.554	5.750	-	1.032		0.002	0.075
100	0.351	0.541	10.200	1.038	0.350	0.539	10.950	-	1.041		0.001	0.100
200	0.343	0.624	20.630	1.048	0.342	0.523	21.400	-	1.050		0.001	
										ΣT=3.710m	ΣK=0.295m	

because the neighboring formations may also recharge or drain the aquifer in question.

Table 4-4. Comparison of water amounts gained by water release from the pores and by consolidation, respectively.

Total Available Water	[m]	Sand 4.00	Clay 6.25
Water stored above the level of depression	[m]	3.71	4.88
	% of the total available water	92.8	78.1
Water gained by consolidation	[m]	0.29	1.37
	% of the total available water	7.2	21.9
Expected land subsidence	[m]	0.35	1.50

This short summary of consolidation clearly indicates that the storage coefficient, influenced by the compressibility of the layer as well as the development of pressure conditions in deep-lying aquifers, depends mainly on the following factors:

- compressibility of water (β);

- compressibility of the solid matrix (α, being a function not only of material constants, but also of parameters describing the initial state of the layer which may change with depth of the investigated point, and thus, the α = constant hypothesis is only a rough estimation);

- history of previous loading of the layer;

- time elapsed between the start of consolidation and the time for which the storage capacity is determined;

- hydraulic conductivity of the layer (K, the effect of which can be characterized with the time dependency of absolute or relative compression, the former being the change of the height of the investigated sample, and the latter is the same value related to the total compression after the complete development of consolidation (Fig. 4-15);

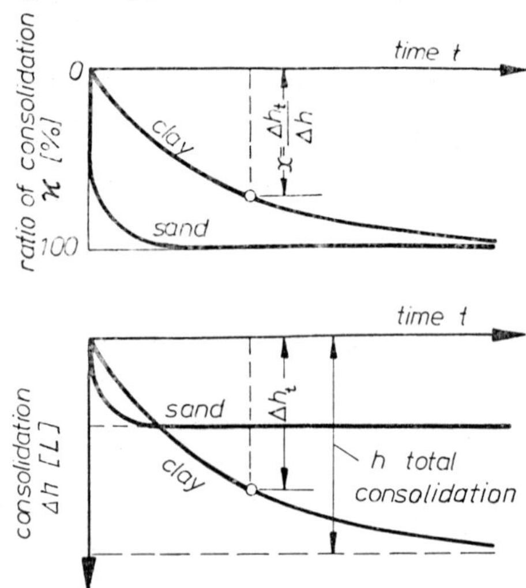

Fig. 4-15. Characteristic curves showing the development of consolidation.

- boundary conditions (hydraulic conductivity and pressure conditions of the neighboring layers) and initial condition (pressure distribution in the layer in question) prevailing along the boundaries of and within the investigated layer. (For demonstrating the last influence, Fig. 4-16 shows the time dependency of relative compression considering three different boundary and initial conditions on the basis of Kézdi's (1972) investigations.)

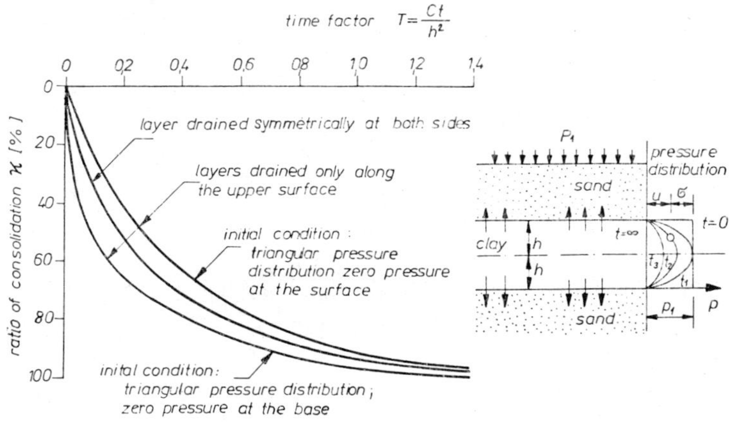

Fig. 4-16. Relative compression depending on time for different boundary and initial conditions (after Kézdi).

SECTION 14

EXPLORATION AND EXPLOITATION OF DEEP GROUNDWATERS

There exists many groundwater problems waiting for solution in connection with deep-lying loose clastic sediments in many different parts of the world (inventory of groundwater resources, water supply for communities, industry or irrigation, etc.). The tasks and methods to be applied in exploration may vary considerably both in their geographical-hydrogeological environment and in their general objectives.

From the point of view of hydrogeological exploration, the most frequently encountered typical regional groundwater schemes are as follows:

- large sedimentary basins;
- coastal plains;
- deltaic areas;
- alluvial plains or great river systems.

Characteristic geological features of these areas were already discussed in Subsection 2-4. The geological structure is constructed of various kinds of clastic sediments deposited sometimes in great thickness on impermeable substratum. Both the horizontal and the vertical extension of the whole formation can be on the order of a hundred to a thousand meters. The water-bearing formations are buried in the strict sense of the word, irrespective of their sedimentary or structural character. In sedimentary basins and in great plains, the topographical forms, if there exist any, are destructional ones depending upon the relative resistance of the materials eroded. The underlying structure is of secondary account although it generally governs the shape and distribution of the topographic elements. The featureless topography renders the geomorphological and surface geological studies rather difficult, sometimes even impossible.

The main objective of a regional groundwater project is to produce the desired amount of groundwater in suitable quality available where and when required. This purpose may be achieved through the construction of structures suitable to withdraw water from the aquifers, i.e., using wells of various types in most cases. The secondary purposes of drilling and prospecting wells, other than their primary economic goal (i.e., to provide groundwater for various users), can be divided into three groups:

- to determine the stratigraphic or rock sequence penetrated by the well;
- to locate the water-bearing layers within the section penetrated and determine their hydrogeological parameters;
- to provide data for correlating the lithologic sequences from hole to hole in order to facilitate the plotting of underground stratigraphy and structure.

Hydrogeologists working with exploration programs should also become as familiar as possible with the practical and theoretical aspects of drilling and well construction.

The groundwater development phase, the final phase of the project, must be based on the inventory of groundwater resources of the region in order to find out whether the desired amount of groundwater is available or not and where is the most suitable location of exploitation. This is the groundwater survey phase of the project.

The survey phase can be carried out either prior to the development phase or simultaneously with the earlier stages of the development phase (or at least considerably overlapping the latter). The approach to be selected depends on such factors as the amount of data available at the beginning of the project, the urgency of the final phase, i.e., of supplying water, etc.

A consecutive approach is preferable where relatively few data are available on the subsurface structure and hydrological conditions. The advantage of this method of prospecting is that the results of a fairly complete investigation (including definite recommendations concerning the most convenient zones for groundwater exploitation and the estimation of exploitable quantities) are available before the initiation of a large drilling program. The disadvantage is, however, that such an investigation does not provide a generally adequate amount of drilling of large-diameter wells, and the results will remain, therefore, largely untested, or based on partial conclusions from small-diameter exploration holes.

The simultaneous approach takes advantage of the possibility to step-up water supply gradually and concurrently with other advancements in the development of the region. It needs continuous and mutual adjustment of the investigation and drilling programs through feedback of forthcoming results which are more conclusive and more adequately tested in this way.

The survey phase of the groundwater project starts with the preliminary assessment of hydrogeological conditions based on pre-existing surface and subsurface data. The work of the geologist is confined mainly to create some working hypothesis of an aquifer model based on data known previously from the interpretation of air photos and of regionale geological studies. The later stages of the investigation will have to prove or disprove the model.

The regional exploration of deep-lying sediments also requires the use of geophysical methods at various stages of the survey. Their purpose is mainly the determination of the subsurface structure and the location of suitable drilling sites. It may also assist in the solution of special problems (e.g., tracing fresh and saline waters in aquifers, location of regions of groundwater outflow in the sea) and in well construction.

Geophysical investigations, as an integral part of regional groundwater surveys, must be carried out in close and continuous coordination with geological studies and drilling since the geophysical methods are indirect by nature and subject to hydrogeological interpretation. It is desirable to include geophysical measurements at an early stage of the survey and these results should be correlated with reliable data of well documented boreholes. Where such boreholes do not exist, test holes (slim holes) should be drilled. The sites for these drillings should be selected so that the largest amount of information could be obtained besides the geological-geophysical correlation. They should permit the observation of water levels in various aquifers as well as the sampling of water quality, and they should be drilled with a diameter suitable for the performance of pumping tests. Finally, it is always desirable that the test holes could be used later as production wells.

In the later phase of the survey, the hydrogeological and geophysical investigations made on the surface are reduced, and emphasis is given to the application of hydrogeological and geophysical methods of subsurface type, using the records of data (called logs) obtained in test holes and wells. The geological log is constructed using samples of rocks removed from their original place by various kinds of drilling and sampling methods. The samples, secured in different methods, are likely to vary considerably in their reliability. The rock core, cut by an improved type of core bit or by sidewall sampler (Subsection 14-3), in ordinary rotary drilling, is by far the best index of the penetrated formations. The samples obtained as cuttings in cable percussion drilling are fairly reliable when a suitable bailing method is used. The least satisfactory record is that secured from the cuttings of rotary drilling because they are mixed with the drilling mud.

Drilling is an expensive means of access to the lithosphere and that of sampling the rocks and fluids penetrated, but these are the only means to study directly the characteristic features of buried strata. The expenses of geological core sampling can be considerably reduced by using geophysical well logging methods which provide further information on the layers in a relatively inexpensive way.

On the one hand, the geophysical logs can provide continuous objective records with values that are consistent from hole to hole and from time to time in contrast to the geological log which is sometimes subjective and greatly dependent upon personal skills. On the other hand, samples of rock or fluid almost never provide continuous data. Even if a hole would be cored with 100 percent recovery, laboratory analysis of the cores involves the selection of point samples. Continuous coring and subsequent analysis of enough samples to be statistically meaningful are much more expensive than most geophysical logging programs.

Geophysical well logging, also called borehole geophysics, includes all techniques of lowering sensing devices in a borehole and recording some physical parameter that may be interpreted in terms of lithological characteristics of the rocks and the flüid contained in it, such as:

- geometry, formation factor, bulk density, porosity, and moisture content of rocks;

- data related to the permeability and specific yield of water-bearing strata;

- source, movement, chemical and physical characteristics of groundwater.

Log data can be used in the construction of wells (e.g., collar locator), in testing and economic development of groundwater supplies, and recharge and disposal systems.

For parameters to be measured by well logging methods and a summary of log applications, see Tables 4-5 and 4-6.

Subsurface hydrogeological methods applied in the study of deep-lying aquifers are characterized by the need to arrive at a conclusion from interpolation between subsurface data, and to construct subsurface geological maps (Subsection 2-3). In order to use the records of boreholes for determining subsurface geological structure, the strata penetrated by drilling must be correlated, i.e., the formations must be recognized and traced from hole to hole. In this process, the geophysical well logs are indispensable methods.

Table 4-5. Parameters measured by well logging methods.

Parameters to be Measured		
lithologic	fluid	technical
Specific resistivity	Specific resistivity	Hole diameter
Formation factor	Chemical composition	Collar detection
Effective inter-granular porosity	Temperature	Cement bond
Total intergranular porosity	Movement	Casing quality
Secondary porosity		Quality of filters
Density		
Moisture content		

On designing and implementing drilling work for water prospecting and exploitation, the selection of the most productive and economical drilling process is determined by such factors as:

- the geological characteristics of the region;

- the depth of drilling;

- the depth of the exploration hole;

- structural dimensions of the producing well.

Among these factors the hydrogeological properties of the area play the most important role since the others, the drilling depth and constructional aspects, are very closely related to them. This is the reason why the degree to which a given region is already explored is a significant factor determining the technology to be employed in the implementation. Using a consecutive approach, a fairly complete hydrogeological report is at the disposal of the designer; in case of a simultaneous approach, part of the hydrogeological data will be obtained during slim hole drilling which calls for continuous adjustment of designing the wells to the actual hydrogeological data.

In any case, the design of deep wells, either for prospecting or for exploiting water, includes the most purposeful solution of tasks such as:

- selecting the drilling equipment;

- working-out the drilling technology;

Table 4-6. Summary of log applications (Keys, 1971).

Required Information	Available Logging Techniques
Lithology and stratigraphic correlation of aquifers and associated rocks	Electric, sonic, caliper logs in open holes; nuclear logs in open or cased holes
Total porosity or bulk density	Calibrated sonic log in open holes, calibrated neutron or gamma-gamma log in open or cased holes
Effective porosity or true resistivity	Calibrated long normal resistivity log
Clay or shale content	Natural gamma log
Permeability	No direct measurement. May be related to porosity, injectivity, sonic amplitude log
Secondary permeability, fractures, solution openings	Caliper, sonic, or borehole television log
Specific yield of unconfined aquifer	Calibrated neutron log
Grain size	Possible relation to formation factor derived from electric log
Location of water level or saturated zones	Electric, temperature, or fluid conductivity log in open hole or inside casing. Neutron or gamma-gamma logs in open hole or outside casing
Moisture content	Calibrated neutron log
Infiltration	Time interval neutron log, or radioactive tracers
Direction, velocity, and path of groundwater flow	Single-well tracer, point dilution, and single-well pulse
Dispersion, dilution, and movement of waste	Fluid conductivity and temperature logs, gamma logs for some radioactive wastes, fluid sampler
Source and movement of water in a well	Injectivity profile, flow meter or tracer logging during pumping, temperature log

Table 4-6 (continued)

Chemical and physical characteristics of water (salinity, temperature, density, viscosity)	Calibrated fluid conductivity and temperature logs in well. Multi-electrode resistivity. Neutron chlorine log outside casing
Guide to screen setting	All logs providing data on the lithology, water-bearing characteristics, correlation and thickness of aquifers
Cementing	Caliper, temperature, gamma-gamma logs. Acoustic log for cement bond
Casing corrosion	Under some conditions caliper or collar locator
Casing leaks and/or plugged screen	Tracer and flow meter log

- elaborating the casing draft;

- selecting the well screen and filtering method;

- planning the aquifer test;

- designing the pumping plant.

A short description of the drilling and well construction methods will be found in Subsection 14-3.

At the end of the drilling operation, or when a phase of the drilling is finished, well construction is necessary to ensure the undisturbed drilling operation, exclude aquifers not intended for use, and to construct testing and final screens (filters).

The exact position of aquifers to be screened will be determined by the use of well logging techniques, thus eliminating the uncertainties due to the use of drilling mud in aquifer detection. When the length and position of the depth intervals to be screened is determined, the length of casing pipe and screens can be calculated, built together on the surface, and lowered into the hole, enabling rapid casing work.

Various porous and rubble layers behind the casing are separated by cementing. With the drilling and development technologies actually employed, cementing is the only means for preventing the motion of water behind the casing sections of several hundred meters length, for separating aquifers of different pressure containing water of various temperatures and chemical composition, and for the directionally regulated connection of such aquifers.

Correct position of cement bond may be checked by the use of temperature and nuclear logging (Subsection 14-2).

The most important elements of water producing wells are the screens (i.e., perforated tubes or rod structures) and the gravel filters replacing a part of the aquifer around the well.

In practical construction work, the requirements concerning the filter structures are as follows:

- optimum water yield with minimal detector;

- the water should enter the well from the largest surface of the aquifer at the lowest inflow velocity;

- although the screen (filter) increases the total resistance of the well, the increase should be reduced to a minimum;

- the water produced during operation should be free from sand;

- the screen material should have mechanical strength and resistance against corrosion.

Although there are screens and filter structures in an enormous variety, the common principles governing their design may be determined considering the process of suffusion around the screened well. The design of filters will be discussed in Subsection 14-4.

The last step in the well construction, the aquifer cleaning, has double objectives, such as:

- the mud cake formed on the walls of the borehole while drilling and the substances penetrated into the aquifer and adversely affecting the permeability must be removed, i.e., clay is to be eliminated;

- seepage conditions more favorable than those in the original state should be provided by flushing other non-desirable grains contained in the aquifer (i.e., movable particles due to suffusion).

Using either a jetting or compressor operation, the water of the well will be free of sand when the cleaning pumping is completed. The construction of the well is finished and ready for test pumping followed by routine water production.

14-1 Exploration of Geological Structures by Applying Geophysical Methods on the Surface

The fundamental principles and the general procedure are essentially the same for all geophysical surface methods, depending upon the following basic facts:

- the subsurface is composed of rocks having certain physical parameters differing from each other;

- the natural or artificial force fields within the lithosphere are affected to different degrees by these various physical properties;

- the extent to which force fields are affected depends, among other factors, on the magnitude of the particular physical properties and on the sizes, masses, and arrangement of the subsurface rocks;

- the effects produced by the subsurface variations in physical parameters upon certain force fields can be measured at the surface of the ground;

- the data obtained by geophysical methods of explorations can be interpreted in terms of the geological or hydrological problems to be solved.

The last statement emphasizes that the relationship of the various physical parameters to problems of practical hydrology is the main factor determining the applicability and the choice of methods for exploring any area. In order to obtain information about the character of this basic relationship, Fig. 4-17 displays, in a simplified representation, a comparison between the specific electric resistivities of waters, aquifers, and impermeable beds. It may be seen that the specific resistivity is a rather sensitive parameter in the hydrogeological range. It is also shown that there is a marked difference between the specific resistivity of fresh-water-bearing gravels and that of the impermeable clay. This unique relationship serves as a basis for solving problems such as gravel detection and location of clay substratum. On the other hand, the non-uniqueness of the basic relationship, the so-called hydrogeological equivalence, may be also seen in Fig. 4-17. Clays do not differ in their specific resistivity from saltwater-bearing sands, and massive limestones may have the same resistivity as sandy layers saturated by fresh water. Another characteristic geophysical parameter, the seismic (acoustic) wave velocities, are also given in Fig. 4-17 in parentheses. Using both parameters, specific resistivity and wave velocity, the geological equivalence may be narrowed down: saturated gravels and massive limestones differ remarkably in their wave velocity and there is a difference in the wave velocity of clays and gravels, at least locally.

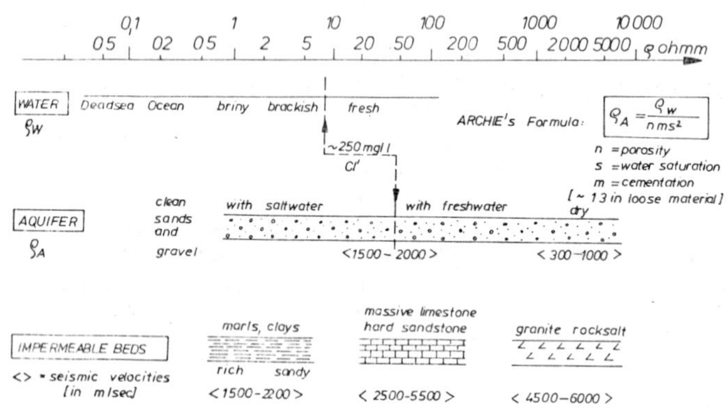

$$\rho_A = \frac{\rho_w}{n^m s^2}$$

Fig. 4-17. Specific electric resistivity and seismic wave velocity ranges of waters, aquifers, and aquicludes (Flathe, 1967).

Regarding the hydrogeological (lithological) interpretation of geophysical data, the exploration methods may be grouped under headings such as stratigraphical and structural prospecting. The aim of stratigraphical prospecting is to determine the depth and thickness of the aquifer, disregarding whether it is saturated by water or not. This objective requires a fair contrast in the physical parameters of aquifers and aquicludes and the determination of the continuity of these properties in the lateral direction.

Structural prospecting attempts to find locations which are favorable for the occurrence of water. Although it is a purely indirect method of prospecting, it is very effective in areas where structural relation exists between the occurrence of water and definite key horizons. It should be noted that supplies of groundwater cannot be located directly by geophysical methods. The reason is that the physical parameters of water-bearing sediments are generally predetermined by the interstititial water and will be significantly altered by saturation merely in exceptional cases.

A wide range of geophysical methods are available today for application in groundwater exploration. Surface electrical resistivity methods are still the most widely used for general and detailed subsurface reconnaissance because of the low expenses involved and the wide range of applicability. Electromagnetic methods have been introduced recently and they are now employed at an increasing rate for locating fractured zones and the freshwater/saltwater boundaries. Relatively inexpensive seismic refraction methods are used at present alone or in combination with resistivity methods to study structures and formations lying in greater depth. Gravimetric methods could be useful for structural analysis of the deep-seated basement and for the determination of the boundaries of aquifers overlying hard rocks. Magnetometric methods might be useful in a number of special cases. These methods will be discussed later (Subsection 18-3). Thermometric methods have been applied with success recently in hydrogeological investigations (e.g., infrared imagery) and in the analysis of hydrological processes.

ELECTRICAL METHODS

Two properties of rocks are of primary concern in the application of electrical methods:

- the specific electrical conductivity of rocks, i.e., their ability to conduct an electrical current;

- the induced polarization which occurs when an electric current is passed through them and then interrupted suddenly.

The electrical conductivity of the Earth's materials can be studied by measuring the electrical potential distribution produced at the surface by an electric current (i.e., supplied into the ground by means of electrodes and called the resistivity method) or by detecting the electromagnetic field produced by an alternating current introduced into the ground (called the electromagnetic method). The study of the decaying potential difference, when the current was turned off, is known as the induced polarization method.

The study of natural electric phenomena, such as:

- spontaneous polarization;

- streaming potentials;

has also found application in hydrogeophysical investigation but only to a small degree.

Resistivity methods of electrical prospecting. In making resistivity surveys, a commutated direct current (DC) or alternating current (AC) of very low frequency (less than 1 cps) is introduced into the ground via two electrodes (A and B in Figs. 4-18 and 4-19) and the potential difference between two other electrodes (M and N in Fig. 4-19) is measured. The current and the potential distribution within a homogeneous and isotropic rock in a vertical plane containing the current electrodes is represented in Fig. 4-18.

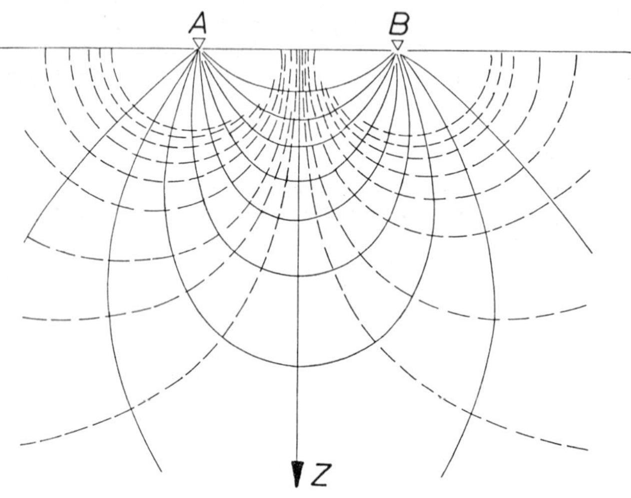

Fig. 4-18. Current distribution (-) and equipotential lines (--) in a plane through the current electrodes.

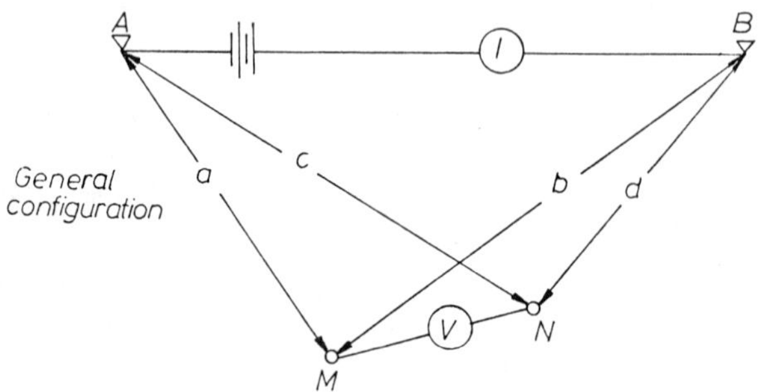

Fig. 4-19. General electrode configuration used in resistivity prospecting.

The electrical field may be described by using Ohm's law. Its elementary and familiar form is valid for thin and long conductors (wires):

$$E/I = R = \frac{\rho L}{A}$$

where E is the potential difference between the ends of the conductor, I is the intensity of the current through it, R is the resistivity of the wire, L and A are the length and the cross-sectional area of the conductor.

The parameter ρ, which characterizes the material of the conductor, is called specific resistivity. The unit of the specific resistivity can be determined using Ohm's law according to which the unit is given as ohm.m or ohmm in the MKSA (meter-kilogram-second-ampere) system of units.

The reciprocal quantity is called specific conductivity and is measured in S/m or in ohmm^{-1}.

Ohm's law may be applied to any infinitesimal part of a physical body, as;

$$i = I/A = \rho^{-1}(dV/dL) \quad ;$$

where i is the current density and dV/dL the potential gradient. The integration of this equation gives a relationship for a semi-infinite medium, which is the simplest model of the lithosphere, having current and potential electrodes placed at the surface as point electrodes:

$$V = \frac{\rho I}{2\pi r}$$

where r is the distance on the surface between the positive current electrode (e.g., A in Fig. 4-19) and a potential electrode, M. When two current electrodes, one positive (A) and one negative (B) are used and the potential difference (V) is measured between two potential electrodes (M and N), the repeated use of the above equation will give

$$V = \frac{\rho I}{2\pi}\left(\frac{1}{a} - \frac{1}{b} - \frac{1}{c} + \frac{1}{d}\right) = \frac{I\rho}{K} \quad ; \qquad (4\text{-}18)$$

hence;

$$\rho = \frac{kV}{I} \qquad (4\text{-}19)$$

Equation 4-19 is a fundamental equation in DC electrical prospecting. The factor K, which depends on the electrode arrangement, is called the geometric factor (Fig. 4-19).

If the measurement of ρ is made over a semi-infinite homogeneous and isotropic material, then the value of ρ computed from Eq. 4-19 will be the true resistivity of the material. However, if the medium is nonhomogeneous and/or anisotropic, then the specific resistivity computed from Eq. 4-19 is called an apparent resistivity, ρ_a. The value of the apparent resistivity is a function of several variables (the true resistivities and other characteristics of the subsurface materials, such as thicknesses, angles of dip and anisotropic properties of the layers) besides the geometry of the electrode array. The value of ρ_a in Eq. 4-18 depends on four distance variables. If ρ_a is made to depend on only one distance variable, then the interpretation of the measured values can be greatly simplified. Electrode arrays invented to fulfill this goal may be grouped as linear arrays and dipole-dipole arrays. In the linear arrays (Fig. 4-20), the four electrodes are placed at the surface of the ground along a straight line in symmetrical position to the center of the current electrode interval. The configuration of the symmetrical type used the most widely in electrical prospecting is that introduced by Schlumberger. The interpretation procedure will be explained for this array in the following pages. In a dipole-dipole array (Fig. 4-21), the distance between the current electrodes (current dipole) and the distance between the potential electrodes (measuring dipole) are significantly smaller than the distance r between the centers of the two dipoles.

Electrical sounding and horizontal profiling. The difference in the electric resistivity of water-bearing layers and that of the impervious rocks (Fig. 4-17) suggests the applicability of resistivity methods. In fact, this method has proved to be successful in stratigraphic as well as in structural prospecting of groundwater.

The method of horizontal profiling is used to determine the variations of the apparent resistivity in the horizontal direction within a pre-selected range. For this purpose, a fixed electrode spacing is chosen, preferably on the basis of studying the results of electrical soundings, and the whole electrode array is moved along a profile after each measurement is made. The value of apparent resistivity is plotted, generally, at the geometric center of the electrode array. The method of horizontal profiling is very suitable to study hard rock terrains, for this reason it will be discussed in Part V.

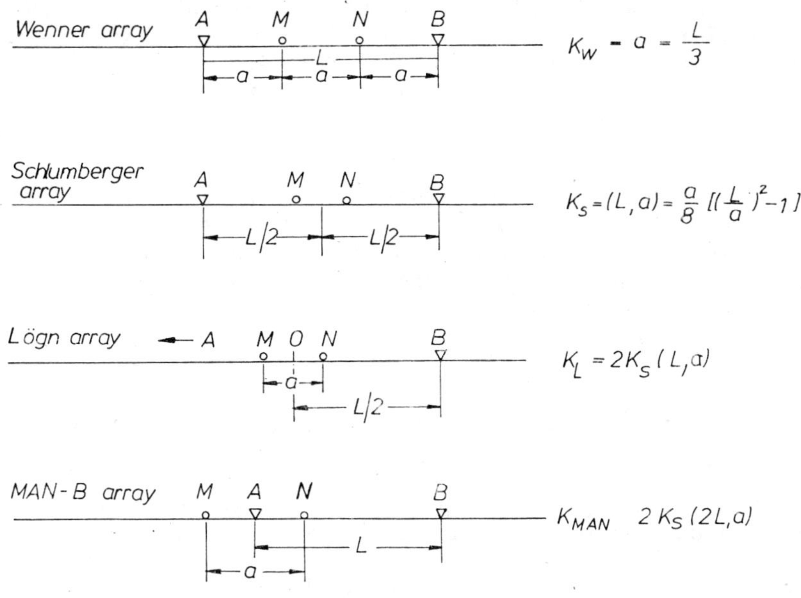

Fig. 4-20. Linear arrays used in resistivity prospecting.

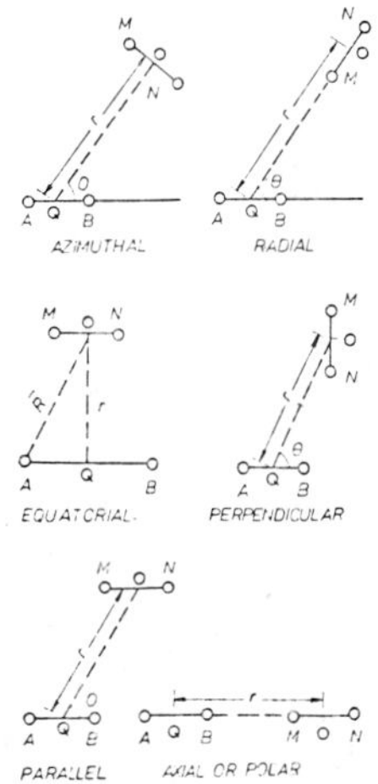

Fig. 4-21. Dipole-dipole arrays used in resistivity prospecting.

Electrical sounding is the process by which depth investigations are made. The basis for making an electrical sounding, irrespective of the electrode array used, is that the farther away from a current source the measurement of the potential or the potential differene is made, the deeper the probing will be. The depth of probing depends on the distance between the two current electrodes (this condition is not valid when the sounding is made with dipole-dipole array). When using the Schlumberger array for sounding, the distance between the current and potential electrodes is increased with the increase of the distance of the current electrodes.

In electrical sounding with the Schlumberger array, the electrode spacing (AB/2) is increased at successive logarithmic intervals and the value of the apparent resistivity is plotted as a function of the spacing on bi-logarithmic coordinate paper. The curve of

$$\rho_a = f(AB/2)$$

is called the electrical sounding curve and it constitutes the basis of the interpretation of resistivity measurements.

The advantages of representing the sounding data on logarithmic system of coordinates are as follows:

- field data can be compared with pre-calculated theoretical curves for given models by simple translation of the curves;
- the form of an electric sounding curve does not depend on the resistivity of the first layer, provided that the ratios ρ_i/ρ_1 and h_i/h_1 remain constant from model to model, where ρ_i and h_i denote the resistivities and thicknesses of the layers, respectively;
- the effect of variations of the thicknesses of layers at large depth will be suppressed and that at shallow depth will be enhanced.

The final resistivity sounding curve is obtained in practical field work by graphical equalization of the series of ρ_a data points. The sounding curve consists of ascending and descending smooth branches as represented in Fig. 4-22.

Geoelectric parameters. A geological section differs from a geoelectric section when the boundaries between geological layers do not coincide with the boundaries between layers characterized by different resistivities. In this respect, the correlation between lithological parameters and geoelectric parameters should be studied.

The specific resistivity of rocks (Fig. 4-23) varies within wide ranges. It is a very sensitive parameter, but

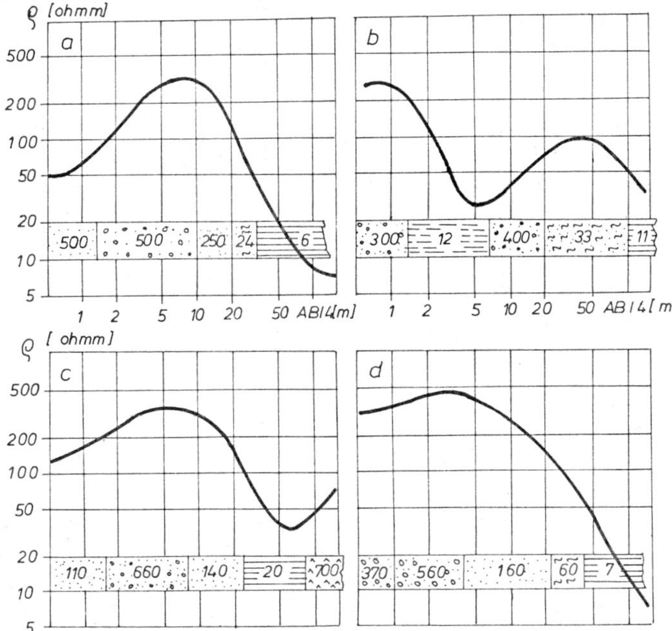

Fig. 4-22. Sounding graphs obtained over a gravel cone. The layers are from left to right: (a) silt, dry sand, water-bearing sand, impervious clay; (b) the same as (a) but covered with dry sand; (c) as in (a) but the bottom clay is underlain by hard rock; (d) as in case (a) but the silty cover is missing.

there is no general correlation between the types and resistivities of the rocks. In most rocks, electricity is conducted electrolytically by the interstitial fluid, and resistivity is controlled more by porosity, water content, and water quality than by the resistivities of the solid matrix (see Archie's formula in Fig. 4-17). Clay minerals, however, are capable of conducting electricity electronically and the flow of current in a clay layer is both electronic and electrolytic. For the present purposes, the rocks of the basin fill can be regarded as an aggregate of grains of very high resistivity compacted to a macroscopically homogeneous body. One important constituent of this skeleton structure is the water film enclosing the individual grains which is bound by adhesive forces and cannot be removed under natural conditions. This attached water film forms an electrolyte of low resistivity even in the case of minimal clay content. This condition gives a relatively good electric conductivity even to the "dry" skeleton which will be altered but slightly by

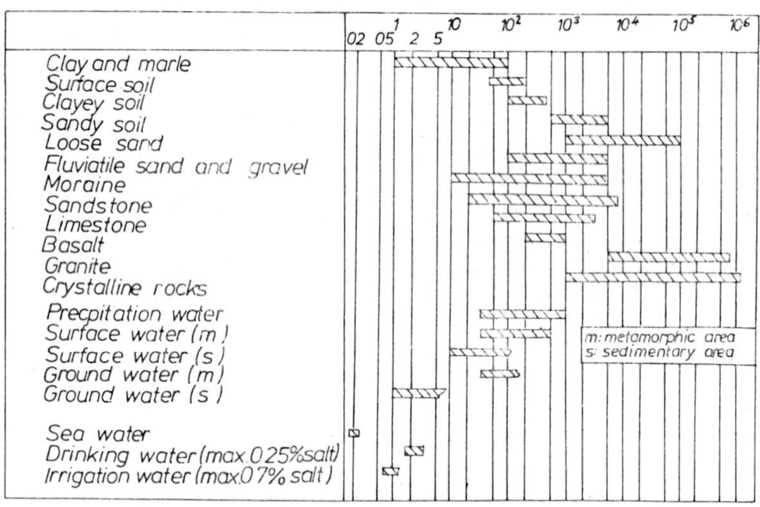

Fig. 4-23. Specific electric resistivity of rocks (ohmm).

the saturation of the pore volume by groundwater. The water film is very thin in the coarse-grained sands and pebbles, thus the effect of saturation is remarkable on their resistivities. In fine-grained clays, however, where the water film fills the pore volume almost completely, no substantial deviation is to be expected in resistivity. Since the specific resistivity is not a definite lithological parameter for rocks, it is not possible to identify a common rock solely by means of the resistivity value, in general, however, there are some rules which constitute a basis for most cases in water prospecting:

- sediments are less resistive than igneous rocks;
- their basic types are better conducting than the acidic ones;
- clayey rocks possess lower resistivity than sandy types;
- stratified and schistous rocks (Subsection 2-3 and 18-1) display electric anisotropy, i.e., their specific resistivity is higher in the direction normal to the plane of schistosity than parallel to it.

The resistivity method is based in principle on the contrast in resistivities of rocks rather than on absolute resistivity values. Resistivity contrasts, sharp enough for practical prospecting purposes, exist locally in most cases in spite of wide-range regional variations.

The geoelectric layer is described by two fundamental parameters: its specific resistivity (discussed above) and its thickness, denoted by ρ_i and h_i, respectively (Fig. 4-24). Other geoelectric parameters are derived from the layers' resistivity and thickness. These are the following:

- longitudinal unit conductance $S_i = h_i/\rho_i$
- transverse unit resistance $T_i = h_i \cdot \rho_i$
- longitudinal resistivity $\rho_L = h_i/S_i$
- transverse resistivity $\rho_t = T_i/h_i$
- anisotropy factor $\lambda = \rho_t/\rho_L$

The secondary geoelectric parameters are particularly important in the description of a geoelectric section consisting of several layers (Fig. 4-24). For n layers, the total longitudinal unit conductance and the total transverse unit resistance are as follows:

$$S = \frac{h_1}{\rho_1} + \frac{h_2}{\rho_2} + \ldots + \frac{h_n}{\rho_n}$$

$$T = h_1 \cdot \rho_1 + h_2 \cdot \rho_2 + \ldots + h_n \cdot \rho_n \quad . \qquad (4\text{-}20)$$

The total thickness of the column is the sum of each thickness:

$$H = h_1 + h_2 + \ldots + h_n \qquad (4\text{-}21)$$

Using the quantities S, T, and H, the average longitudinal resistivity, ρ_L, the average transverse resistivity, ρ_t, and the factor of anisotropy, λ, may be defined for the column as: $\rho_L = H/S$, $\rho_t = T/H$, and $\lambda = \rho_t/\rho_L$, respectively. The physical meaning of the parameters defined above may be seen in Fig. 4-24 where a column of unit cross-sectional area is represented cut-out of a group of layers of infinite lateral extent. If current flows vertically only through the column, then the total resistance of the column is the same as T, and if the current flows parallel to the bedding, the conductance of the column will be given by S.

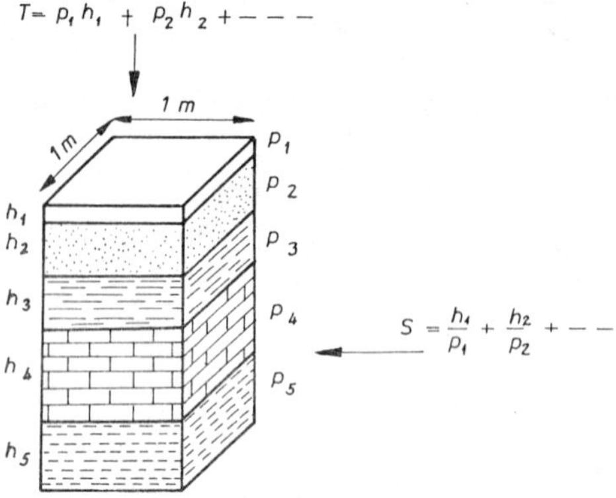

Fig. 4-24. Columnar prism used in defining geoelectric parameters of a section.

In the interpretation of multilayer electrical sounding curves, the evaluation of S or T is sometimes all that can be uniquely determined. The study of these parameters is also the basis of important graphical procedures (e.g., auxiliary point method) for the interpretation of electrical sounding curves. It is interesting to note that the quantity

$$T_i = \rho_i \cdot h_i$$

is analogous to the transmissivity (used in groundwater hydrology)

$$T_i = K_i \cdot h_i$$

where K_i and h_i denote the hydraulic conductivity and the thickness of the ith layer, respectively. This analogy is frequently used in interpolation of transmissivity data for areas where the scarcity of transmissivity data hinders the construction of transmissivity maps.

Electrical sounding over horizontally stratified lithosphere. An earth's model, which is composed of homogeneous and isotropic layers of resistivities ρ_i and thicknesses of h_i separated by horizontal plane surfaces of infinite lateral extension, approximate very well the basin fill in the great sedimentary basins. This is the reason why this kind of electrical prospecting will be discussed more in detail.

The ρ_a sounding curves represented in Fig. 4-22 were obtained over horizontally stratified earth. The qualitative interpretation of sounding curves of this type is based on the simple fact that any increase in the dimensions of the electrode array increases the depth of penetration of the electric field. It is evident that a sounding curve measured over homogeneous isotropic half-space is a straight line parallel to the AB-axis of coordinates. Any change in the direction of the tangent of the sounding curve indicates that the electric field has reached the boundary of a stratum of different resistivity. Thus, any turning point on the sounding curve may be interpreted as a resistivity boundary and its depth may be estimated (very roughly) as the AB/2 value of the turning point. This method of interpretation serves for orientation only and must be followed by quantitative evaluation.

The quantitative interpretation of the sounding curve with the ultimate aim to determine the thicknesses, h_i, and the true resistivities, ρ_i, of the individual layers in the ground, is carried out by means of master curves which are sounding curves of ρ_a (AB/2) type, calculated and presented in bi-logarithmic scale considering various arrangements of horizontally stratified media.

For practical purposes, the master curves are drawn in a bi-logarithmic system of coordinates. The number of master curves can be reduced if the values of variables and parameters are related to selected model parameters. For the sake of convenience, the resistivity ρ_1 and thickness h_1 of the superficial (first) layer are taken as units for the other parameters of the system. The form of the master curve has not been changed by this process since the multiplication or division of variables means only a translation on a logarithmic scale. Due to this principle, any measured sounding curve can be compared by translation (e.g., superposition) to the corresponding master curve, even if the values of ρ_1 and h_1 are different, assuming that both curves are represented in the same system of logarithmic coordinates.

Collections of master curves, constructed for the Schlumberger sounding may be found in the literature (e.g., Orellana, 1966). A set of master curves for the two-layer case is represented in Fig. 4-25. In this case, $\rho_1 = 1$, $h_1 = 1$, according to the convenience, $h_2 = \infty$, and the only changing quantity, $\rho_2/\rho_1 = u$ is given for each curve in a circle. Thus, for the interpretation of the two-layer case, one set of master curves is sufficient. A set of master curves for solving the three-layer case is shown in Fig. 4-26.

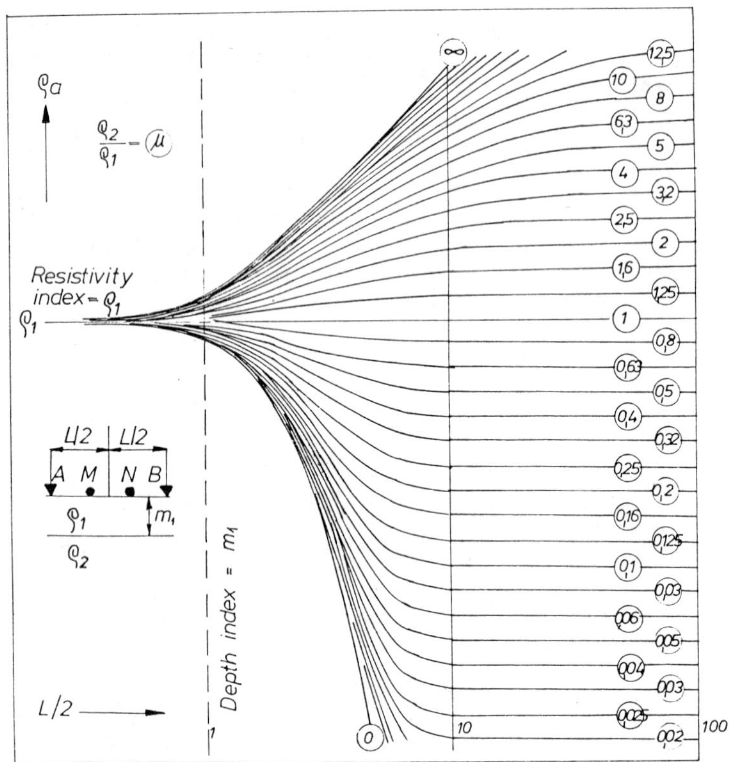

Fig. 4-25. Master curves for the two-layer case.

They are constructed on the same basis as the two-layer curves, but the number of changing parameters is greater. These are the relative resistivities of the second and third layers as well as the relative thickness of the second one. It is obvious that no complete set of master curves can be given when the number of layers is more than two. In the three-layer case, each set is constructed for a given distribution of resistivities (marked as ρ-values) and the relative thickness of the second layer is marked (in squares) on each of the master curves (CGG, 1963; Orellana, 1966). Master curves for the four- and five-layer cases are constructed, but only for a few distributions of resistivities and thicknesses encountered in hydrogeological prospecting (Orellana, 1966; Flathe, 1963).

The procedure for the interpretation of the sounding curves in the two- and three-layer case is similar. The measured curve is plotted on a transparent sheet using the coordinate system of the master curves, and then is placed on

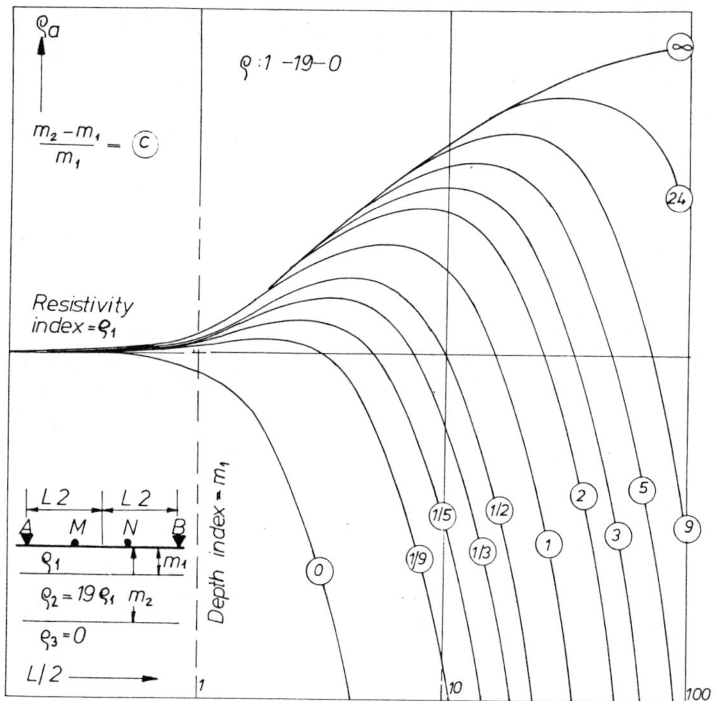

Fig. 4-26. Set of master curves for the three-layer case. The ratio of resistivities (1-19-0) is the same for all curves, the thickness of the second layer varies.

the top of the set of master curves. The working sheet is moved keeping parallel the axes of the two systems of coordinates to a position where the measured curve coincides (is matched) with one of the represented or interpolated master curves. The coordinates of the "cross" (i.e., point (1,1)) read off from the working sheet correspond to the true resistivity ρ_1 and to the thickness h_1 of the first (superficial) layer. The other parameters can be determined following the simple rules given on the sheet of the master curves.

The approximate interpretation of the multilayer type sounding curves is usually carried out with the so-called auxiliary point method. The first step is to determine the parameters of the first layer as it was described above. In the second step the two-layer master curve is matched to the second branch of the measured curve as it was done in the case of the first layer with the restriction that the cross must be moved along an auxiliary curve determined by the resistivity of the second layer, and so on. The auxiliary curves are published in the collections of master curves (e.g., Orellana, 1966). The true specific resistivities of the individual layers are obtained during the matching process, while the true thicknesses by appropriate corrections carried out by means of auxiliary diagrams.

The accuracy of the interpretation may be increased, especially in the multilayer case, by using a special set of master curves constructed for the problem in question. It is necessary to presuppose, at first, a number of probable resistivity and thickness values for each layer obtained, e.g., by the auxiliary point method, and then a special set of multilayer curves is constructed, either by graphical methods (Matveev, 1964) or using digital computers (CGG, 1963). The measured sounding curve is matched with this special set of curves, and the parameters of the best fitting curve will be regarded as the best mathematical solution of the problem.

The problem of equivalence. The interpretation of a multilayer sounding curve is not unique, even theoretically, since a given electrical sounding curve can correspond to a variety of subsurface distributions of layer thicknesses and resistivities. When differing sets of parameters provide the same sounding curve, within 5 percent of deviation, these layerings are said to be equivalent. Examples of different kinds of equivalence are represented in Fig. 4-27. Among them, the equivalence of the three-layer curves of maximum type (the curves marked by a, b, e, and f) is the most important since these types are frequently encountered in water prospecting. This type of equivalence is called equivalence by T, for the transverse resistivity $T = \rho_2 h_2$ is the same (within the error of measurement) in both of the pairs of equivalent layerings. This type of equivalence may be constant for the three-layer curves of a monotonic descending type. Three-layer curves of minimum type (curves marked by c and d in Fig. 4-27) and with monotonic ascending branches are called equivalent by S, where $S = h_2/\rho_2$.

The problem of equivalence, this limitation to the interpretation, should not be discouraging to the hydrogeologist or geophysicist working in the field of water prospecting. There are problems where the equivalence of layers should not be considered (e.g., to get an idea or estimation about the distribution of the transmissivity in an area, the interpretation of the maximum type three-layer curves is needed, regarding the transversal resistivity in the second, water-bearing layer, only and disregarding the ambiguity due to the equivalence). The most complete information on the vertical succession of layers can be obtained from a sounding carried out in the immediate vicinity of boreholes which data is

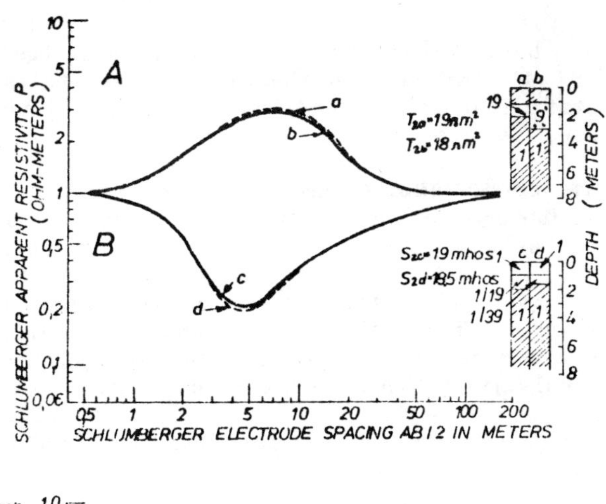

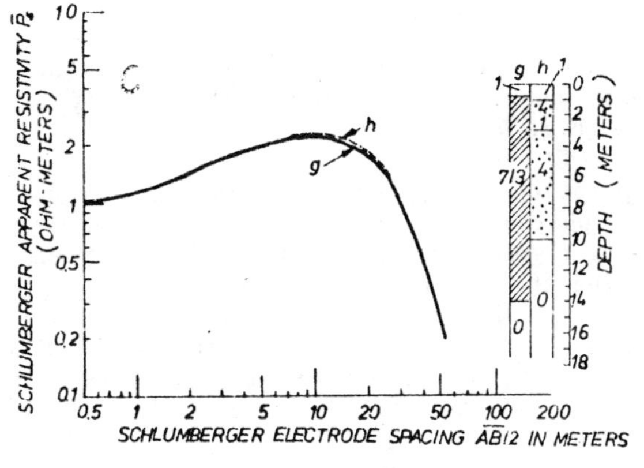

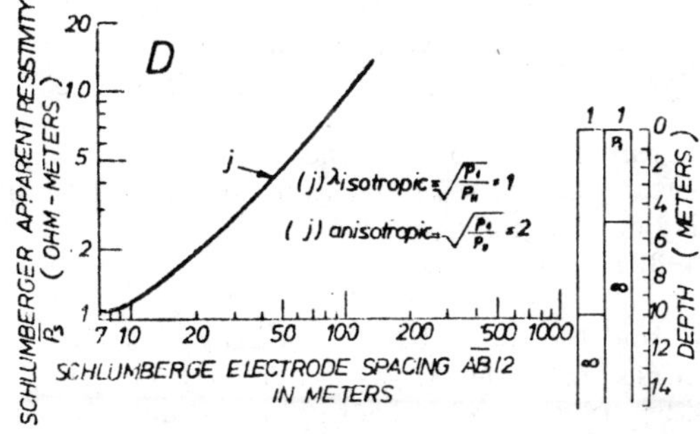

Fig. 4-27. Examples of different types of curve equivalence: (A) according to transverse resistance with the same number of layers; (B) according to longitudinal conductance with the same number of layers; (C) different number of layers; (D) isotropic-anisotropic layer (Zohdy, 1974).

compared with the geological log in order to accurately determine the layer thicknesses. Then, using the resistivity parameters determined with the aid of layer thicknesses, the interpretation of the sounding curves will give the layer thicknesses in areas where drilling information is lacking.

Areal representation of electrical sounding data. The final result of the interpretation of sounding curves may be considered as a log, i.e., the succession of resistivity data related to the vertical line in the center of the measuring array. It is formally the same as the geological log obtained by drilling. The methods of preparing geological cross sections and maps can be applied, therefore, to the construction of resistivity cross sections and maps.

The areal representation of sounding data involves the following:

- study of the types of the sounding curves obtained, and the representation of the areal distribution of these types, possibly depicting them on a map of the survey area (Fig. 4-28);

- preparation of apparent resistivity maps by plotting the apparent resistivity value as registered on the sounding curve at a given electrode spacing, and contouring the results (Fig. 4-29);

- preparation of apparent resistivity sections in the vertical by plotting the apparent resistivities, as observed along vertical lines located beneath the sounding stations, on the chosen profile (Fig. 4-30);

- preparation of maps showing the areal distribution of resistivity parameters (e.g., transverse resistivity) (Fig. 4-31);

- preparation of subsurface contour maps for the representation of well defined resistivity units (e.g., contour map of the resistive or conductive substratum) (Fig. 4-32);

- preparation of block diagrams showing the characteristic lithologic features of the area interpreted by means of electrical sounding (Fig. 4-33).

Application of resistivity prospecting in groundwater studies. Buried stream channels and water-bearing, sandy-gravelly layers covered by comparatively loose overburden are favored targets for resistivity exploration, owing to the great difference between the aquifer and the resistive or conductive substratum. The examples given in Figs. 4-22, 4-28,

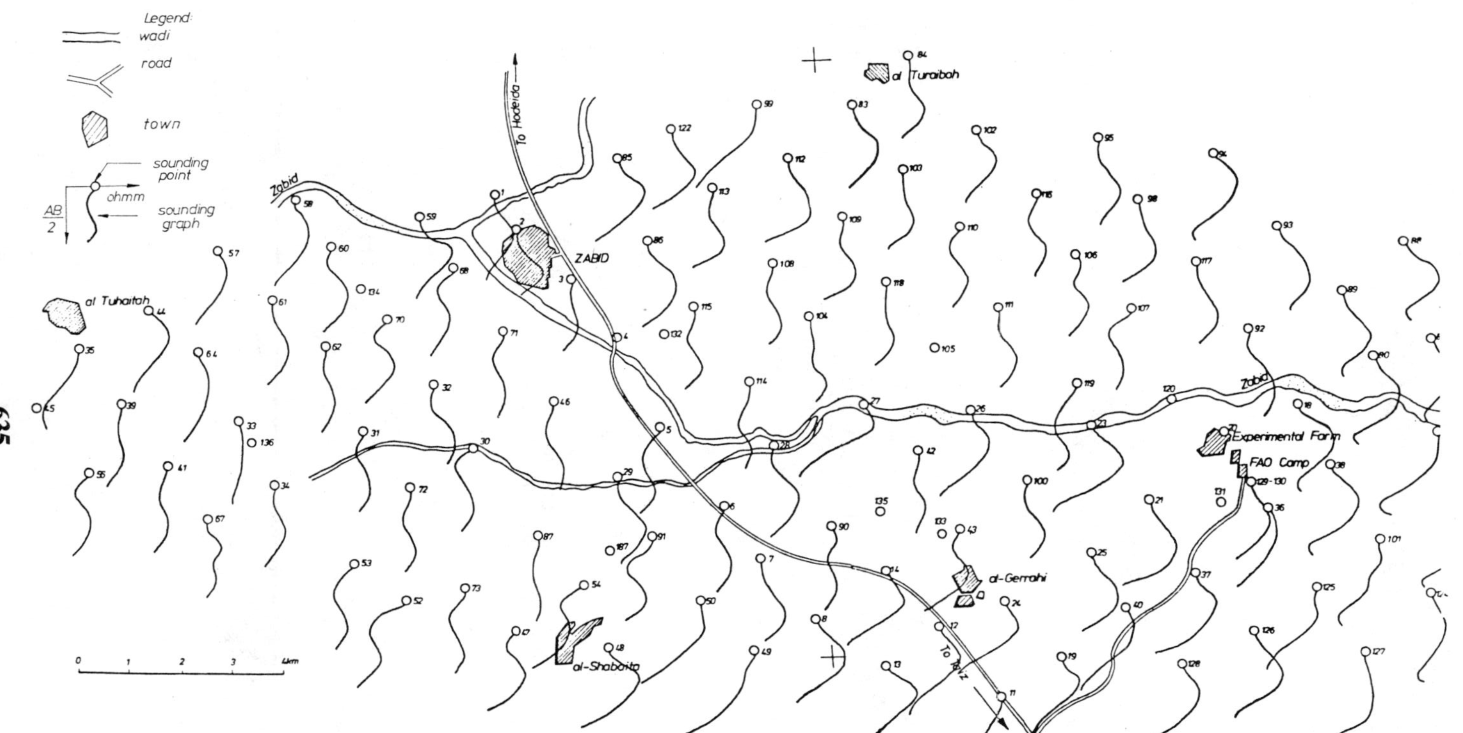

Fig. 4-28. Areal representation of resistivity sounding curves measured on the gravel cone of the Zabid Wadi.

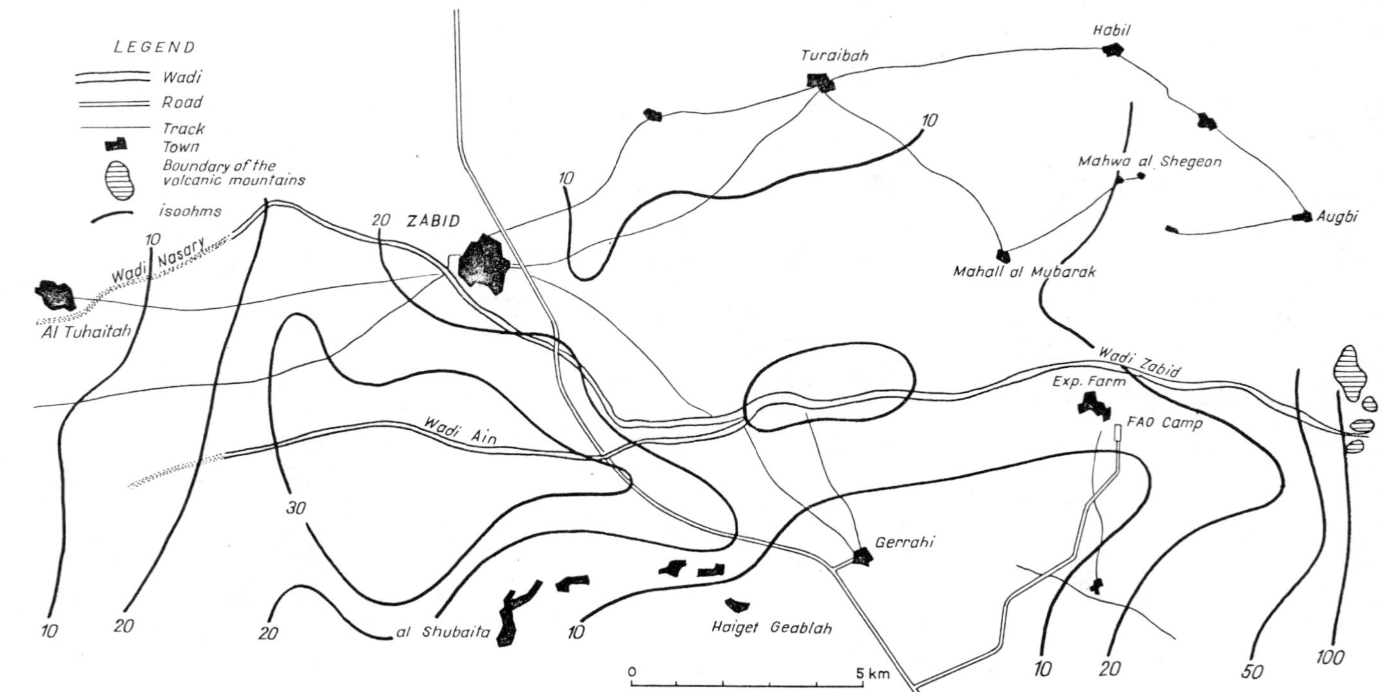

Fig. 4-29. Apparent resistivity map of the Zabid Wadi area. The apparent resistivity values were determined at AB = 500 m electrode separation.

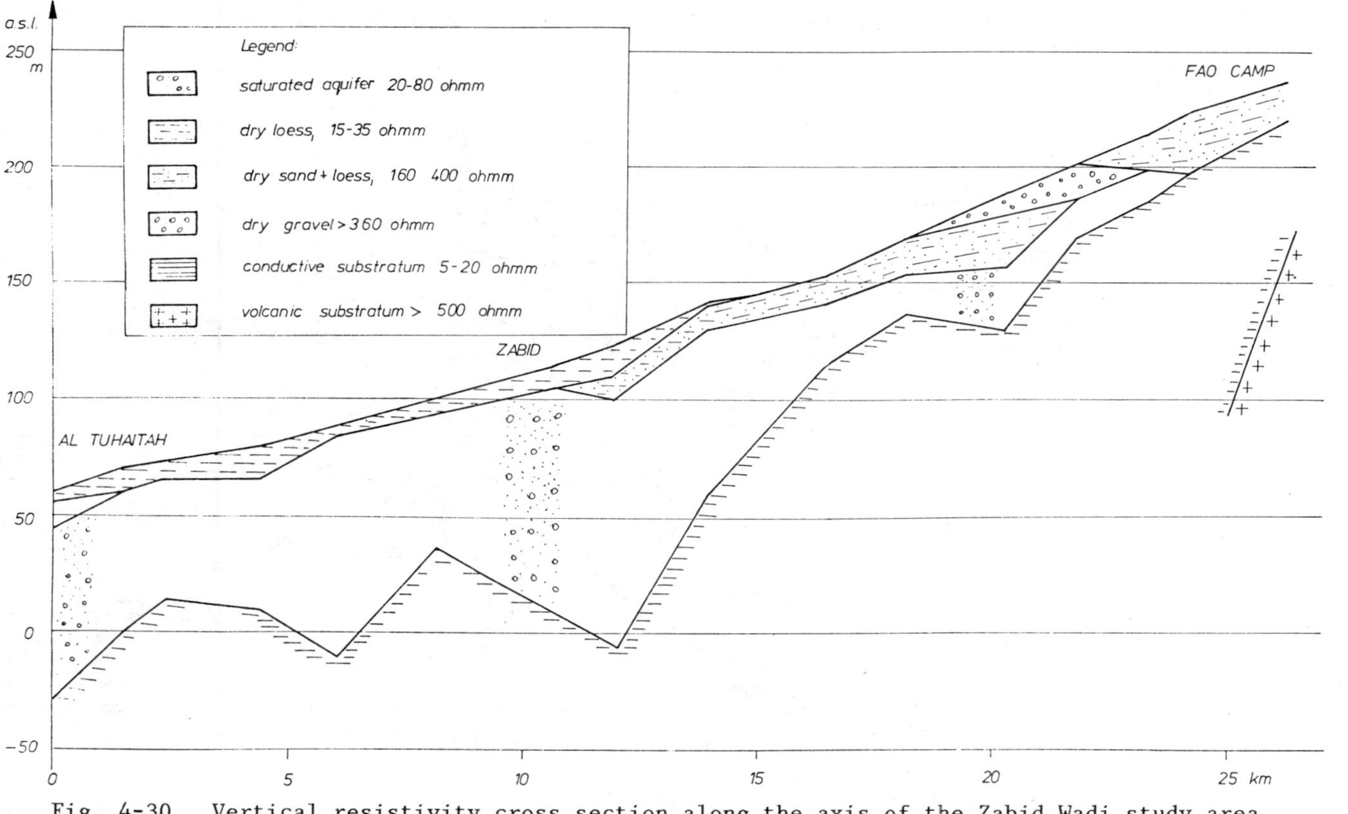

Fig. 4-30. Vertical resistivity cross section along the axis of the Zabid Wadi study area.

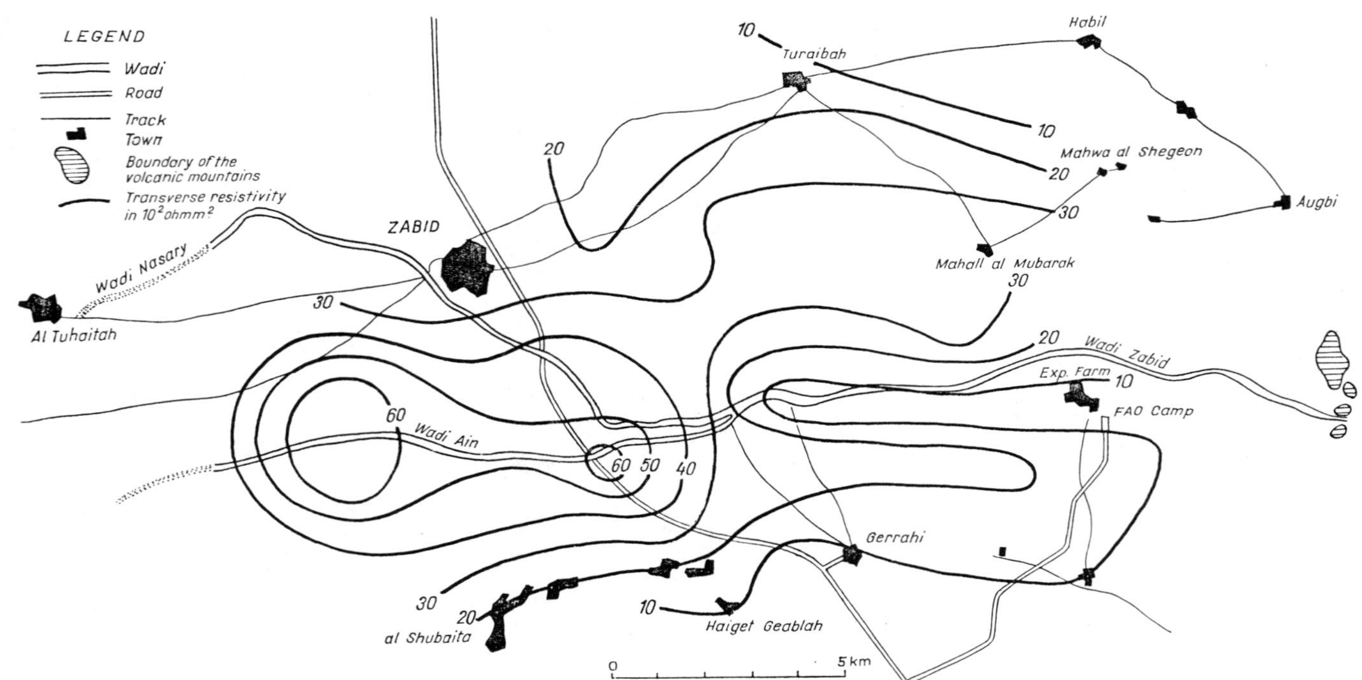

Fig. 4-31. Map of the transverse resistivities of the aquifer in the Zabid Wadi study area.

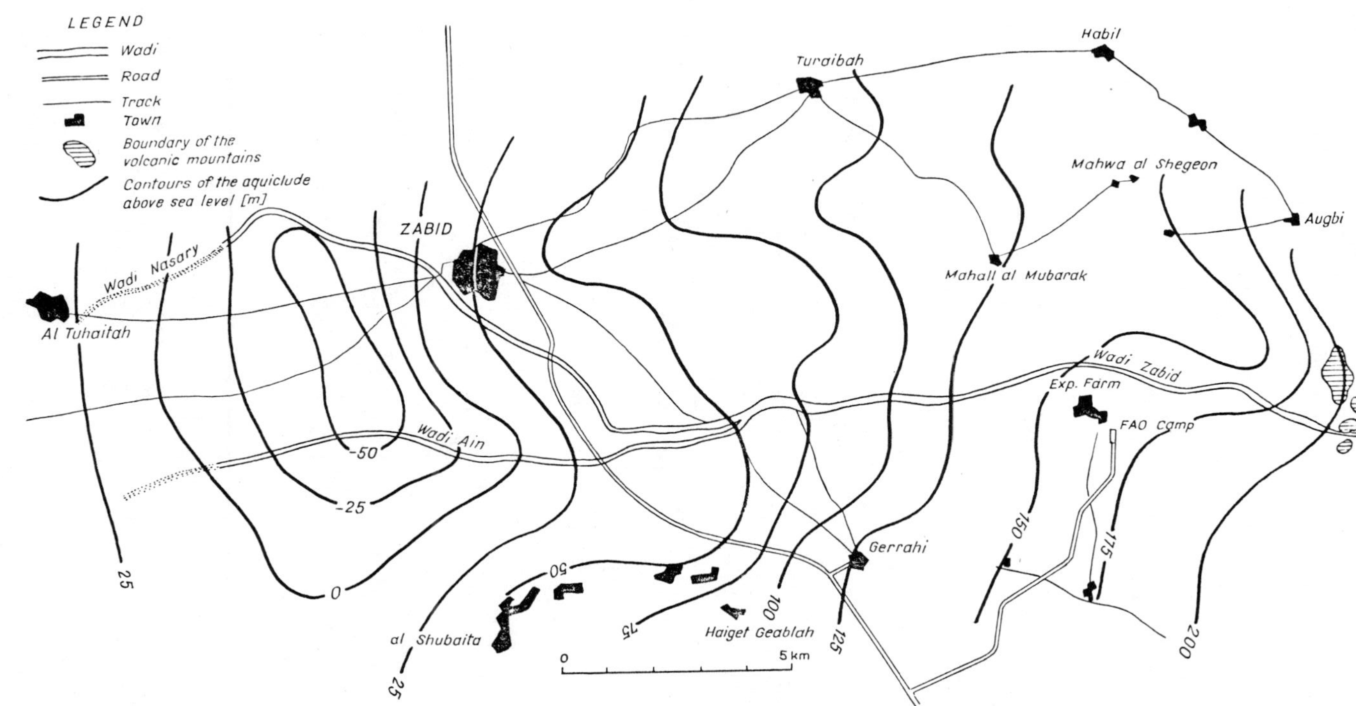

Fig. 4-32. Map of conductive substratum in the Zabid Wadi study area.

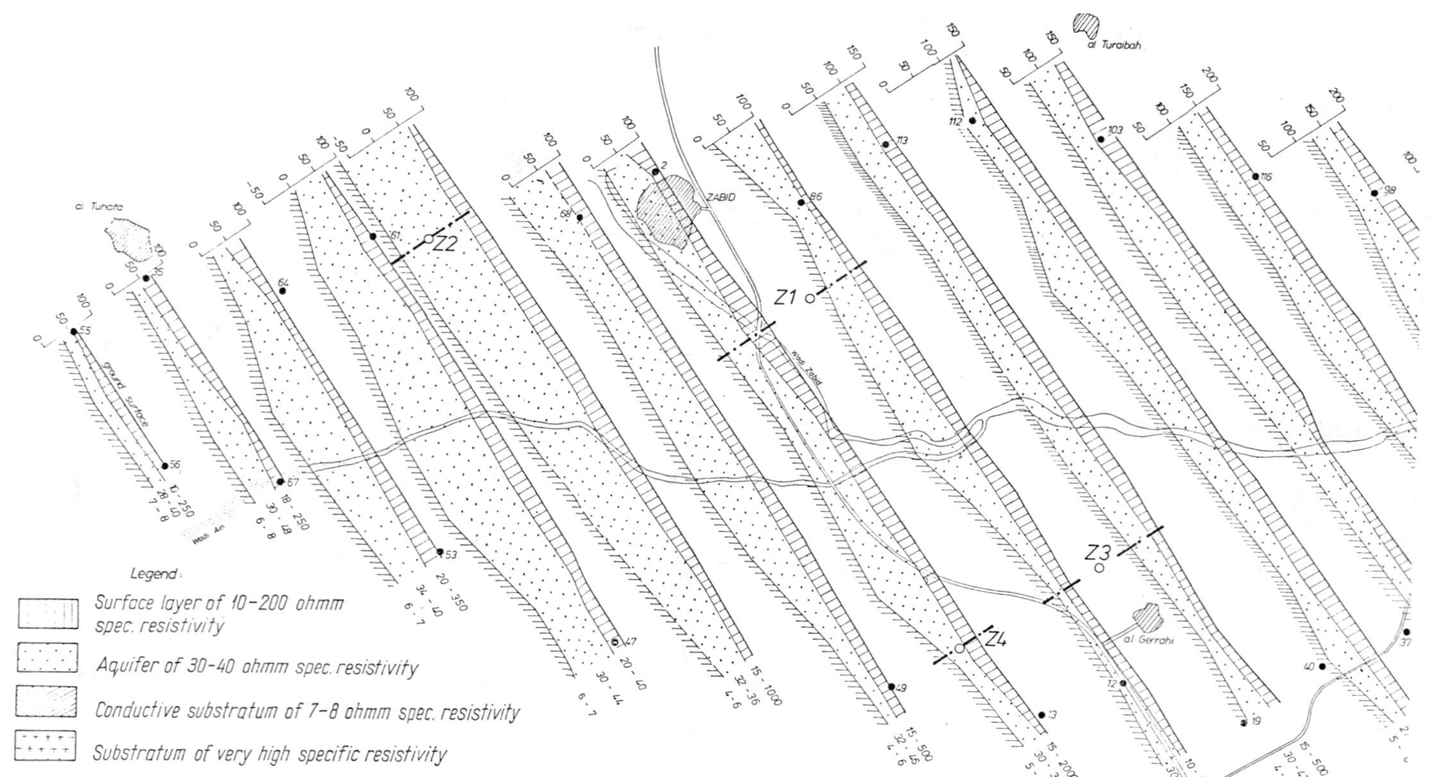

Fig. 4-33. Block diagram of the aquifer and aquiclude layers in the Zabid Wadi study area.

4-29, and 4-32 are the results of this kind of exploration. Thick clay layers separating two aquifers usually can be detected on a sounding curve, but they cannot be detected by other methods. Electrical methods are unique in furnishing information concerning the depth of the interface between freshwater and saltwater in coastal areas especially where no clay interbeddings are present (Fig. 4-34). Clay layers, which have low resistivity values in general, may have the same influence on the sounding curve as sandy layers saturated by brackish water. Some of these layers may be differentiated when investigated by the induced polarization method. Unlike the mapping of the freshwater-saltwater interface, the determination of the depth to water table is generally a more difficult problem. Wherever the water table is overlain and underlain by several layers of different resistivities, its detection on the sounding curve may be virtually impossible (Deppermann, 1965). Under favorable conditions, the water table can be detected on a sounding curve as a conductive layer. This situation is realized in very dry areas where the unconfined water table forms a resistivity boundary in coarse-grained sediments of high effective porosity. In the dry part of these sediments, the adhesive water film is very thin, therefore the resistivity of such a layer would be effectively increased by saturation.

The electromagnetic methods of prospecting. Electrical surveys are also made using a time-varying electromagnetic field as an energy source. The magnetic field is produced by passing an alternating current through a wire loop. When this primary field reaches the lithosphere, a flow of electrical current begins. The amount of current flow depends on the conductivity (resistivity) of the earth's materials. This current flow produces a secondary magnetic field which can be measured at or above the ground by means of another loop of wire, the receiver, in which voltage is induced.

Electromagnetic surveys can be made either on the ground or from a low flying aircraft. The effective depth at which conductive bodies can be detected with electromagnetic methods is dependent upon both the frequency and spacing between the transmitter and receiver loops. Thus, electromagnetic measurements can be used in the same manner as resistivity measurements to obtain horizontal profiles and depth soundings.

Among the various types of electromagnetic prospecting methods, a transient airborne method, the INPUT system, has found a wide-range of application. This system operates on the principle of generating eddy currents in the ground by current pulses in the primary field. The eddy currents will collapse to zero instantaneously in highly resistive media, but in a medium or low resistivity, the collapse will be

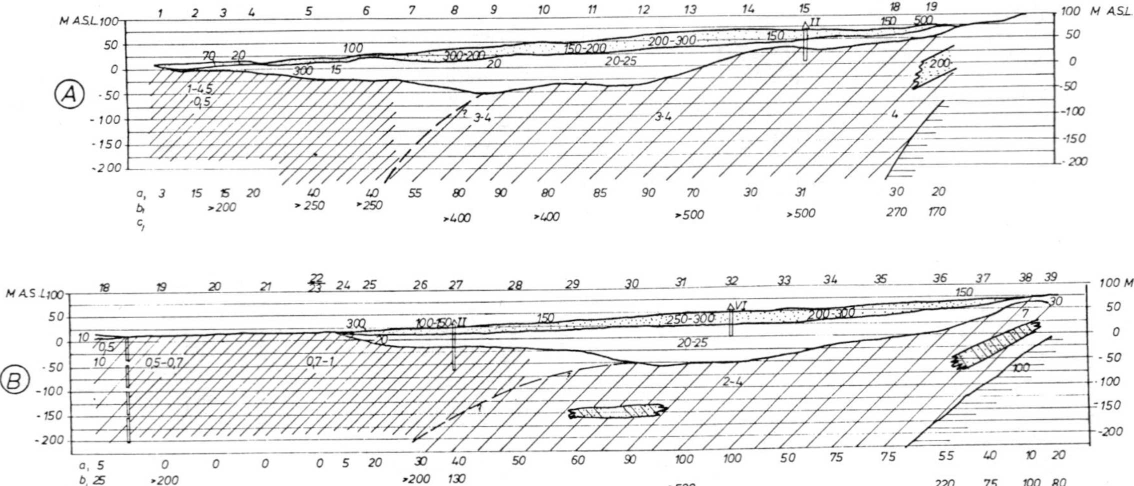

Fig. 4-34. Electric resistivity cross section perpendicular to the seashore showing saltwater intrusion. Layers of 0.5 - 1 ohmm resistivity are saturated by seawater, 2 - 4 ohmm denote the conductive substratum (clay), layers of high resistivity values are saturated by freshwater.

delayed by inductive effect (Fig. 4-35). The secondary field, i.e., the field of the eddy current, is sampled in time to determine the exact rate of decay (e.g., the time-constant) of the eddy current. This time-constant is in inverse proportion to the resistivity of the ground under consideration (Fig. 4-36).

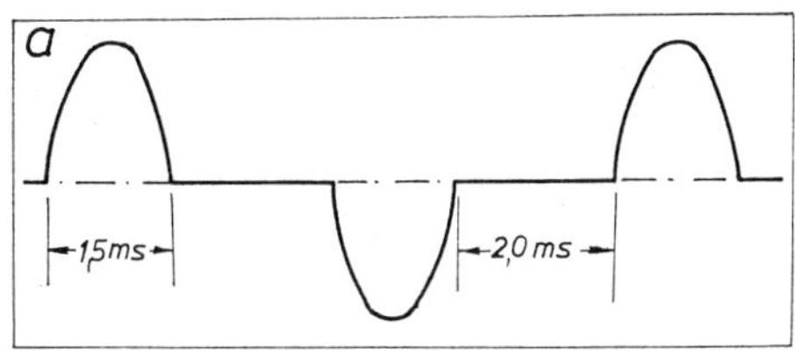

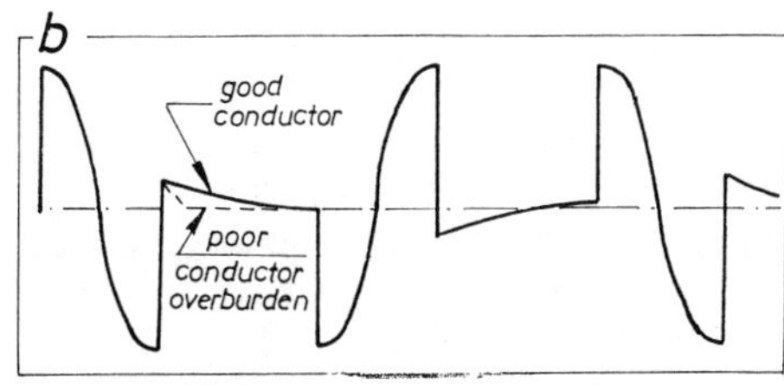

Fig. 4-35. The transmitted signal (a) and the received one (b) in the INPUT system.

Induced electrical polarization method. This phenomenon may be observed at a sudden interruption of direct current circulating in the ground. When the flow of a direct current, which has been circulated in the ground for a long interval of time, is abruptly interrupted, the electric field will drop instantaneously to a value E_p which is a small fraction of the original static field, E_s, then it will decay slowly over a longer period of time ranging from minutes to hours. Certain layers in the ground became electrically polarized forming a battery when energized with an electric current. Upon turning off the polarizing current, the ground gradually discharged and returned to equilibrium.

The procedure for measuring induced polarization response in the field is very similar to that used in the direct current resistivity prospecting. Polarization is excited by a galvanic current of a square-wave form driven into the ground by two current electrodes and the effect of polarization is measured noting the potential decay measured between two potential electrodes following the end of the exciting pulses

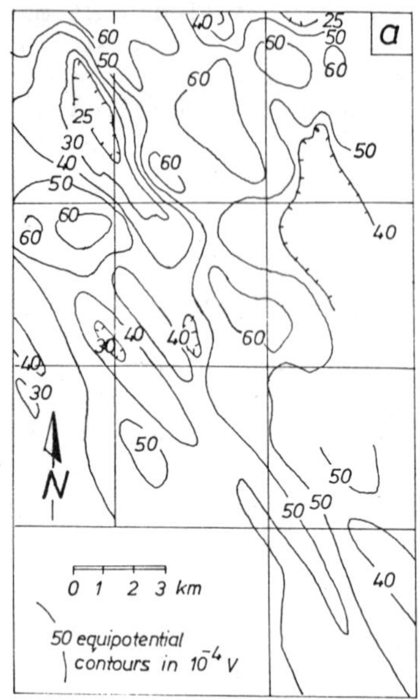

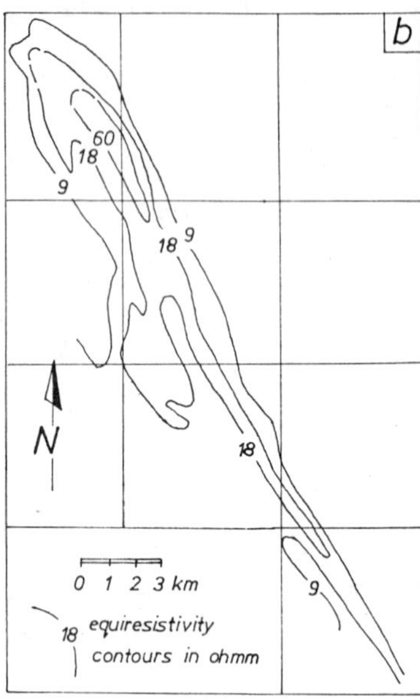

Fig. 4-36. Prospecting over a gravel body by INPUT system, (a), and by horizontal resistivity mapping using the Wenner array, (b). The gravel bed is characterized by comparatively high apparent resistivity (about 60 ohmm). The decay of the secondary magnetic field proceeds very suddenly in the resistive gravel, it is characterized therefore by small voltage values in map (a) (Collett, 1967).

(Fig. 4-37). To avoid the coupling between the current and potential measuring circuit, the dipole-dipole array is generally used. Groundwater studies generally have been made with "time-domain IP". In this method, the potential decay is sampled and then several indices are used to define the response of the medium. Seigel (1959) defined the "chargeability" (in seconds) as the ratio of the area under the decay curve (in millivolt-seconds) to the potential difference (in millivolts) measured before switching the current off. Komarov (1966) defined the "polarizability" as the ratio of the potential difference after a given time from switching the current off. The polarizability is expressed as a percentage.

The apparent chargeability (or polarizability) versus the electrode spacing (AB/2) is plotted on logarithmic coordinates and is interpreted by curve matching procedures (Seigel, 1959).

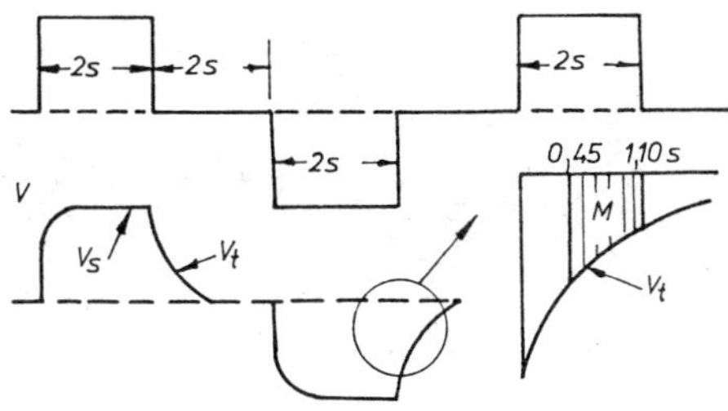

Fig. 4-37. Pulse transient method of induced polarization. The exciting current (I) and the measured voltage (V_t) is shown. The area marked by M is a measure for polarizability.

An induced polarization sounding curve can be of significant value in complementing a resistivity sounding curve. Figure 4-38 illustrates the case where the alluvial deposits in the study area were poorly differentiated by their resistivities, but three horizons were clearly distinguished by their polarizabilities (Kuzmina, 1965).

The method of equipotential lines. This method is treated here as a special method although it is a variant of the horizontal resistivity method discussed previously.

In the mapping by linear current electrodes, long straight wires serve for both current electrodes laid out parallel to each other in the field. Equipotential lines obtained over homogeneous ground will therefore be equidistant lines parallel to the wire electrodes. If the potential measurement is carried out over nonhomogeneous ground, the potential differences may vary from the value of the homogeneous Earth. The electrical structures characterized by a change in apparent resistivities will be located, therefore, by the change of the spacing between equipotential lines. Over conductive bodies, the equipotential lines become denser, and they are rare over resistive soil.

The idea of the mise-a-la-masse method is to use a subsurface conductive mass itself as one current electrode of a pair by connecting it directly to one pole of a battery, the second current electrode being as usual a metal rod, but placed at the surface of the ground at a great distance. If it is considered that the mass represents an electrically charged body, the shape of the equipotential lines will, in some measure, reflect the geometry of the conducting body and should be expected to give clues with regard to its extent, dip, etc. It is obvious, of course, that at least a part of

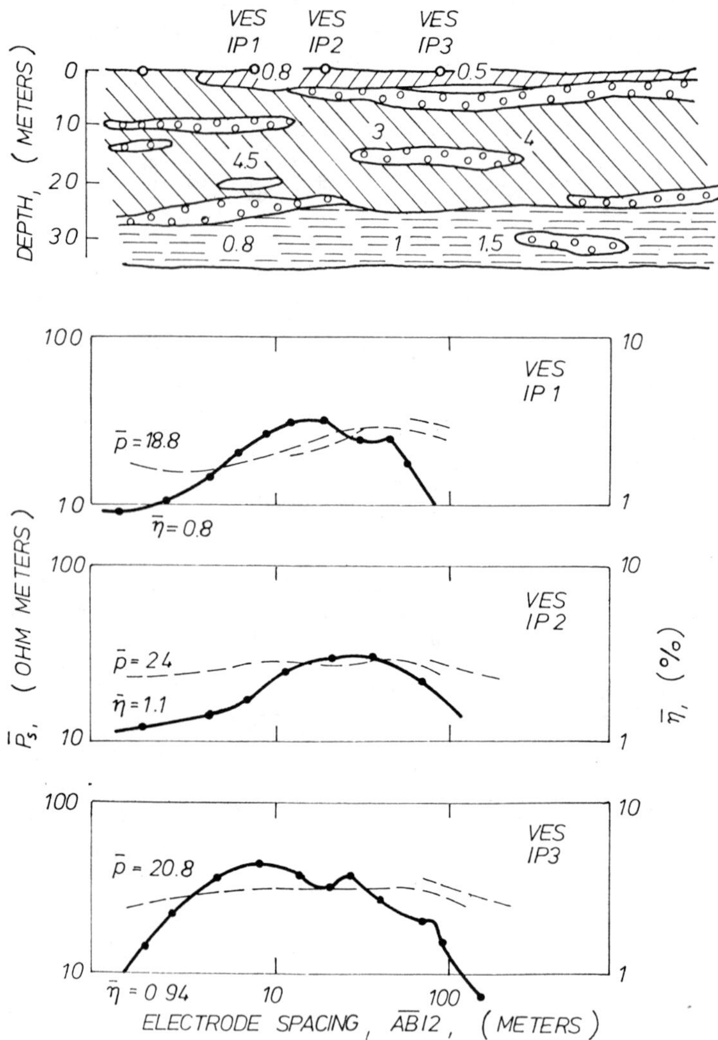

Fig. 4-38. Geoelectric section, vertical electrical sounding curves, and IP sounding graphs of alluvial deposits in Crimea (Kuzmina, 1965).

the mass concerned must be accessible so that an electrical earthing can be made in it.

A unique but important method is the detection of seepages by potential measurement. The free liquid in the center of rock pores is usually enriched in ions of one sign which represent an electric current when the water moves (filtrates) through the pores. The location of the current flow may be detected by the usual method of equipotential lines.

SEISMIC PROSPECTING

The seismic prospecting has, as its basis, the timing of artificially generated pulses of elastic energy propagated through the ground in the form of seismic waves and picked up by electromechanical transducers (seismometers) operating as detectors. These detectors respond to the motion of the ground and their response, transformed into electric signals, is amplified and recorded on magnetic tape or film on which timing marks are also placed.

The geophysicist is interested in two parameters which affect the elapsed time of transmission of the seismic wave: the propagation velocity and the geometry of the propagation path since both of them are influenced by the elastic properties of the subsurface materials encountered. Theoretically, it could be possible to make some deductions about the nature of the subsurface if the arrivals of the seismic waves can be observed and timed at the surface of the ground.

Elastic wave energy can be imparted to the ground in a variety of ways. The most commonly used method is that of firing a charge of explosives in a hole. For very shallow work (about 20 m or less) adequate energy can be generated sometimes by a hammer flow on a steel plate. Analogous sources of energy in larger amounts are produced by weight dropping or by impacting the ground with a hydraulically driven plate. In any case, as the disturbance moves out from the source, it comes within a few meters in the range of linear elasticity. From this point on, the theory of elastic wave propagation may be applied. The theory of elasticity, on which the laws of elastic wave propagation are based, treats materials as homogeneous and isotropic. Although naturally occurring rocks do not fit either specifiction very well in many areas, this theory has proved very useful in understanding seismic phenomena.

According to the theory of elasticity, when a disturbance is produced in an unbounded homogeneous isotropic body being in equilibrium originally, two types of waves will spread out from the place of the disturbance, namely:

- a dilatational-compressional wave with velocity V;

- a torsional or shear wave with velocity v, less than V.

When the disturbance is produced by an explosion, as usual in seismic prospecting, the shear deformations produced in the elastic medium are very small as compared to the change in volume, hence in this case most of the energy will propagate in the form of a compressional wave. This is the reason why compressional waves are used in hydrogeology exclusively. The term seismic wave will be used in this sense.

The surfaces in which the disturbances are of the same value are called wave fronts and the normals to it are referred to as rays. Drawing the normal trajectory to a selected point of the wave front at any time, the path of the wave will be obtained. The time in which the wave front travels from the source to one point of the wave path is called travel time. The velocity of the seismic wave (i.e., the velocity of the wave front) is characteristic for the medium in which the wave propagates and depends on the elastic moduli and on the density of the homogeneous isotropic material. It is obvious that the propagation of a disturbance may be described equally by using either the term wave front or the term wave path.

Taking into account the surface of the earth, an elastic half-space must be considered. A disturbance excited on the surface will travel in the form of half spheres. The wave fronts observed on the surface are spreading out, thus, in the form of concentric circles about the source. This wave, called the direct wave, travels parallel to and just below the earth's surface.

Where the earth is layered, i.e., it is composed of homogeneous and isotropic layers of different (compressional) wave velocities and/or densities, then the energy of the wave, when it reaches the interface between two distinct media, will be divided into two parts (neglecting the generation of shear waves): some of the energy may be reflected towards the surface, and some is transmitted through the boundary. The paths of four rays emanating from a point source of energy at the surface are shown in Fig. 4-39. The model is a horizontally layered earth where the seismic wave velocity V_1 in the upper medium is less than the velocity of the underlying medium V_2. These four rays are as follows:

1. The direct ray defined above.

2. The totally reflected ray which is generated when a ray strikes the boundary at an angle of incidence i_1 greater than the critical angle i_c, and all of the energy is reflected back toward the surface.

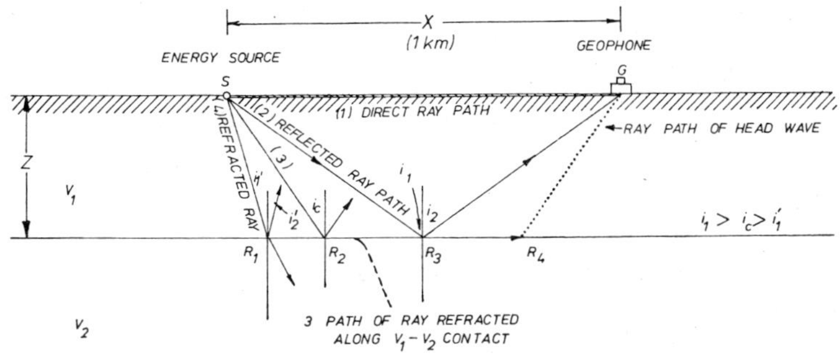

Fig. 4-39. Schematic ray path diagram for seismic waves generated at source S and received at seismometer G.

3. The critically refracted ray which strikes the boundary at precisely the critical angle of incidence i_c, part of the energy being reflected back to the surface and part refracted, the latter traveling parallel to the interface with velocity V_2. As this wave travels along the interface, it continuously generates waves within the upper medium and these travel back toward the surface. The wave fronts of the critically refracted waves are cones which are coaxial with a perpendicular dropped from the source to the interface.

4. A ray striking the interface at an angle of incidence i_1', less than critical angle, part of the energy being reflected toward the surface and part being refracted in the lower medium away from the normal to the surface at angle r. Division of the incident energy between reflected and refracted waves depends on the angle of incidence and the contrast in velocities and densities of the two media.

The angular relationships among the various parts of the ray paths just described are as follows:

The law of reflection, valid for ray path 2;

$$i_1 = i_2 \quad ; \qquad (4\text{-}22)$$

and for the reflected branch of ray path 4,

$$i'_1 = i'_2 \quad ; \tag{4-23}$$

Snell's law of refraction according to which for the refracted branch of ray path 4,

$$\frac{\sin i'_1}{\sin r} = \frac{V_1}{V_2} \quad . \tag{4-24}$$

At the critical angle of incidence, the angle of refraction is 90° and sin r = 1. Thus, the critical angle can be defined in terms of the two velocities as

$$\sin i_c = V_1/V_2 \quad . \tag{4-25}$$

These simple equations, plus that expressing the relationship between velocity, distance, and time constitute the basis for the interpretation of seismic data.

Regarding the technique of seismic prospecting, in all common arrangements, the seismometers are spaced at equal intervals along a straight line containing the shot point too. From all travel time values measured in a seismic profile, the so-called travel time diagram is constructed by using the distances between the shot point and seismometer as abscissa and the corresponding value of travel time as ordinate in a linear scale suitable for the clear representation of the time data.

The interpretation of seismic measurements is based on the travel time diagram since its qualitative and quantitative characteristics are determined by the distribution of wave velocities as well as by the position of seismic boundaries in the underground.

Two methods described as reflection and refraction techniques are employed in seismic prospecting. In both methods, travel times of seismic waves are measured from the source to selected points on the surface. The basic difference lies in the character of the waves whose times of arrival are utilized. In the reflection method, observations are made of the times of arrival of seismic waves which have been reflected from seismic boundaries in the ground. In the refraction method travel time data are obtained from seismic waves which have been critically refracted on boundaries separating two media of different wave velocities.

The reflection method of seismic prospecting. The travel time diagram of a reflected seismic wave will be obtained when the reflected ray path (2) in Fig. 4-39 is considered. The relationship between the velocity V_1, the length of the propagation path, $\overline{SR_3G}$, and the transmission time, t, is:

$$V_1 = (\overline{SR_3} + \overline{R_3G})/t \qquad (4\text{-}26)$$

Using the law of reflection, according to which $i_1 = i_2$ and thus $\overline{SR_3} = \overline{R_3G}$, the equation for V_1 can be rewritten as

$$(V_1 t/2)^2 = (X/2)^2 = Z^2 \quad ; \qquad (4\text{-}27)$$

which represents a hyperbola in the time distance system (Fig. 4-40). The distance X in this equation is predetermined by the location of the seismometer, and the transmission time (t) is read from the travel time diagram. Thus, if the velocity V_1 of the upper medium is known, the depth to the interface (Z) may be calculated. The velocity V_1 is obtained either from independent measurement or by calculating the reciprocal of the tangent of the asymptote drawn to the time distance hyperbola in Fig. 4-40 which is equal to the velocity V_1.

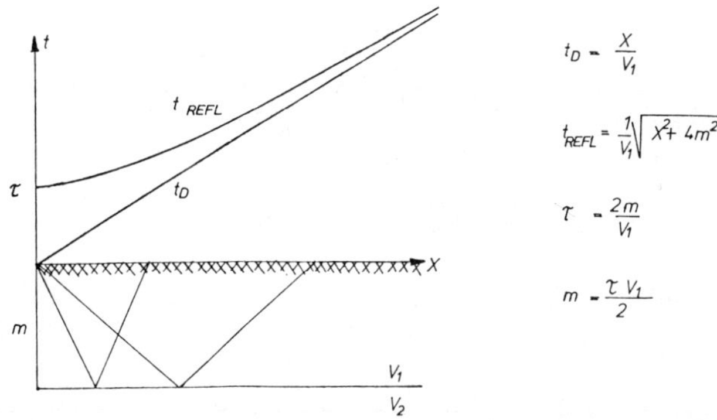

Fig. 4-40. Travel time graph of reflected seismic waves over a horizontal reflecting boundary.

Contrary to the simple technique of depth measurement by the reflection method, which is essentially the same in the case of multilayered earth, this method is used in water prospecting work only exceptionally. The reason is that the reflection technique is not suited to the shallow exploration taking part in the prospecting for water because the waves reflected from shallow horizons reach the surface too soon after the arrival of the direct and critically refracted waves so that their effect is overshadowed by the vibrations excited by these earlier arrivals. The selecton of the reflection arrivals requires sophisticated equipment and technique too expensive compared to the budget of a typical groundwater study. Thus, the discussion of the reflection method may be omitted here.

The refraction method of seismic prospecting. In this method measurements are made of the elapsed time between the shot and the arrival of the critically refracted wave, the path of which is marked by (3) in Fig. 4-39. The significance of the symmetry of ray path (3) is that the right-hand branch, $\overline{R_4G}$, is a mirror image of the left-hand or incident branch $\overline{SR_2}$. It follows that the relationship between the velocities V_1 and V_2, the length of the propagation path $\overline{SR_2R_4G}$, and the transmission time, t can be written as:

$$t = 2\overline{SR_2}/V_1 + \overline{R_2R_4}/V_2 \quad . \qquad (4\text{-}28)$$

The geometry of the figure allows some substitution which leads to;

$$t = X/V_2 + (2Z/V_1) \cos i_c \qquad (4\text{-}29)$$

where $\cos i_c = \sqrt{1 - \sin^2 i_c} = \sqrt{1 - (V_1/V_2)^2}$. An important characteristic of the travel time diagram may be deduced from this equation when differentiated:

$$dt/dX = 1/V_2 = \text{const}_i \qquad (4\text{-}30)$$

which means that the travel time diagram of the critically refracted wave is represented by a straight line having the tangent of its slope equal to the reciprocal value of the wave velocity in the lower layer (Fig. 4-41). The travel time diagram, represented in Fig. 4-41 needs some additional remark. If the observing point is near to the source, only the direct wave (wave marked by (1) in Fig. 4-39) propagating with velocity V_1 reaches the seismometer. It is obvious that part (1) of the travel time diagram, due to the direct wave, is a straight line, the tangent of which is equal to the reciprocal value of the wave velocity in the upper layer (V_1). The critically refracted wave travels along the boundary partly

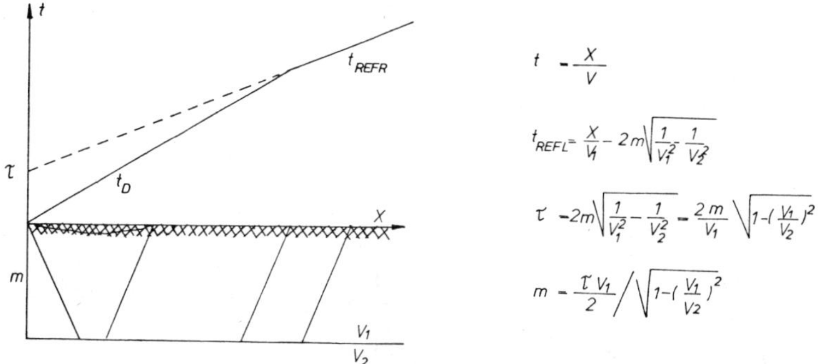

Fig. 4-41. Travel time graph and ray path in the case of one horizontal refracting boundary. Refraction seismic method.

with a speed V_2, higher than V_1. There will be, therefore, one point of observation (X_c) where the direct wave and the critically refracted one arrive at the same time. From this point, the critically refracted wave will go before the direct wave producing part (3) of the travel time diagram.

When a seismic refraction measurement is carried out over a horizontally layered earth composed of two homogeneous and isotropic strata, the obtained travel time diagram consists of two straight line segments. The velocities in the two layers may be determined by calculating the reciprocals of the tangent of the slopes, successively. Then the time intercept (t_o) is constructed by extrapolating the second branch of the time distance diagram to the time axis. An equation for t_o may be obtained by putting X = 0 in the travel time equation (Eq. 4-29);

$$t_o = (2Z/V_1) \cos i_c \qquad (4\text{-}31)$$

After rearranging the terms and substituting an expression containing V_1 and V_2 for $\cos i_c$, a formula for the depth Z results as:

$$Z = (t_o \cdot V_1/2) \cdot (1 - V_1^{\,2}/V_2^2)^{-1/2} \qquad (4\text{-}32)$$

If a seismic refraction measurement is carried out over dipping beds, the travel time diagram is composed of straight line segments, as in the case of horizontal layers. In such a case, one travel time diagram is not enough for the determination of the layering. This problem may be solved by generating seismic energy on both ends of the setup. Using this pair of time distance diagrams, the velocities, the dip of the bed, and the vertical depth can be simply calculated as it is shown in Fig. 4-42.

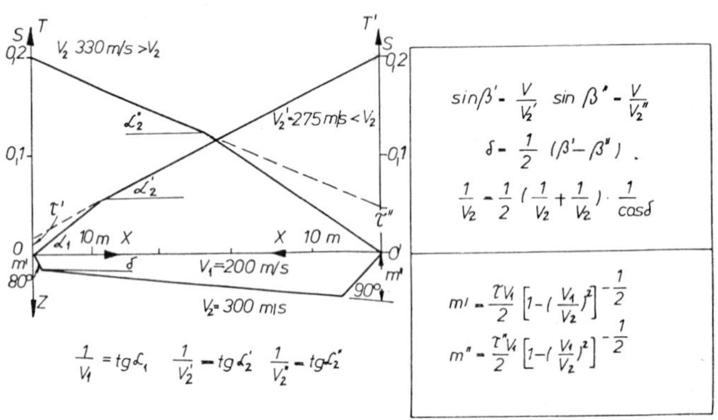

Fig. 4-42. Travel time graph and ray path in the case of one dipping refracting boundary. Refraction seismic method.

Seismic refraction measurements in hydrogeology. The models represented in Figs. 4-41 and 4-42 are highly simplified. The complication of the real systems are illustrated by some hypothetical models and the corresponding time distance diagrams in Fig. 4-43.

The multilayered models may be solved by applying the method of successive decomposition. In the first step, the thickness and wave velocity of the layer underlying the surface is determined using the formulas given in Fig. 4-41 or in Fig. 4-42. Knowing the geometry of this layer, parts of the travel times due to this layer will be calculated and subtracted from each travel time value. By this method, the first layer is scraped off, and the number of layers is reduced by one.

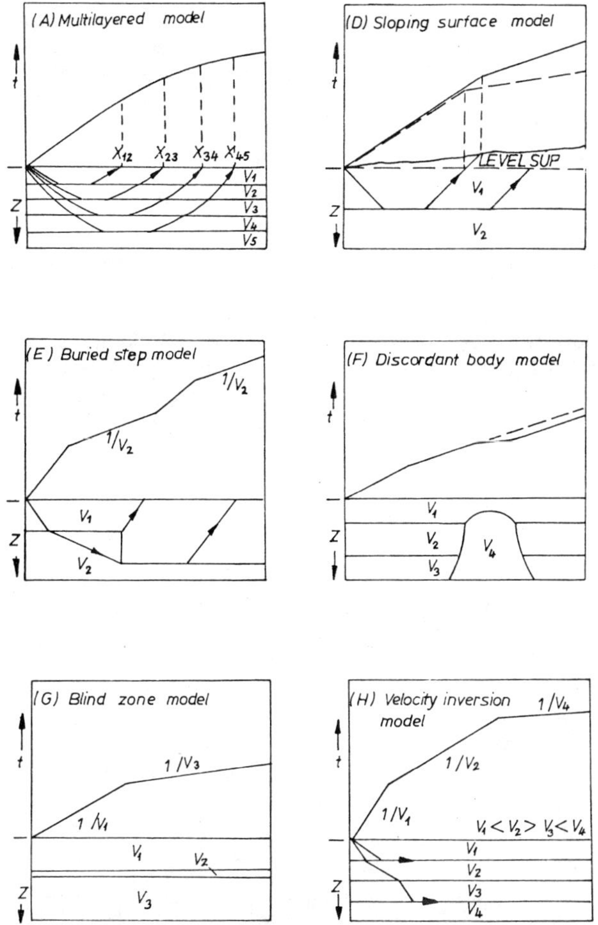

Fig. 4-43. Schematic travel time graphs for typical nonhomogeneous geological models: (a) multi-layered model, (b) sloping surface model, (c) buried step model, (d) discordant body model, (e) blind zone model, (f) velocity inversion model (Zohdy, 1974).

A similar process may be used to reduce the travel time diagram to a plane if the measurement were carried out in rugged territory or if velocity nonhomogeneity is present in layers near to the surface.

The problem of velocity inversion and that of the blind zone should be discussed more in detail.

It follows from the basic character of the critical refraction that a necessary condition for the travel time diagram, which must yield complete information about subsurface, is that the velocity of the seismic waves in each layer must be higher than in the layer immediately above it. The layers that violate this condition are said to exhibit velocity inversion and they are missing from the travel time diagram.

In an earth, where the velocity of seismic waves increase step by step with the depth, critically refracted waves will be generated at each boundary which appear on the surface as first arrivals, generally if the distance between the source and seismometer is increased. There are cases in which a critically refracted wave appears on the surface but always in the form of late arrivals. The so-called blind zones, producing this kind of critically refracted waves are very thin layers, underlain by a thick layer of comparatively high velocity. To overcome the difficulties due to velocity inversion and blind zones in the interpretation of travel time diagrams, a well shooting program (velocity logging) may be advisable.

The velocity of seismic waves in rocks. The distribution of seismic wave velocities in rocks which are important in hydrogeological prospecting is given in Fig. 4-44. The overlapping character of the ranges of wave velocities suggests that a unique correlation cannot be expected between the type of rock and the wave velocity therein. In spite of this phenomenon, a local relationship may be obtained in many cases.

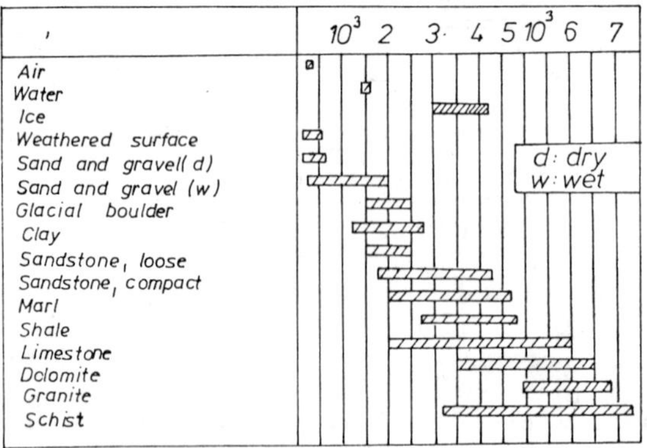

Fig. 4-44. Velocity of seismic waves in rocks (m/s).

A general inference, drawn from Fig. 4-44, is that the wave velocities in dense, non-weathered rocks are much higher than in the fractured variety. Thus, the seismic refraction method may be used for the determination of bedrock topography, the feature of which may indicate locations which are favorable for the occurrence of water. Seismic refraction prospecting may be considered, therefore, as a structural prospecting method applicable for the detection of bedrock channels for the location of buried valleys, faults, and dislocations.

A further conclusion is that, in coarse-grained layers saturated by water, the seismic wave velocity of the particles also is accentuated depending on the size of the grains. Water-bearing layers composed of gravel and debris which are fragments of materials (e.g., granite) of high wave velocity exhibit wave velocities higher than that of the fine-grained environment. Thus, the seismic refraction method may be applied as a gravel detector, or generally, as a stratigraphic prospecting method. It can be used for the location of water-bearing gravelly layers, channels, or lenses covered by sandy-clayey sediments.

Seismic refraction measurements of semi-consolidated or unconsolidated clastic deposits reveal that the (compressional) wave velocities increase abruptly at the water table. The reason being that in the dry bed, above the water table, the wave velocity will be characteristic on a macroscopic scale of the grain skeleton, while in the wet layer, below the water table, a velocity of about 1500 m/s, characteristic of water, will be expected. The above conditions hold best primarily for coarse-graind beds. In the case of fine-grained layers, such as the clays, the inherent water film filling the greatest part of the pore volume also gives enough compression elasticity even to a dry bed which reveals itself in the velocity of waves. Thus, only a minor change in the wave velocity can be observed at the water table. If layers of sand are interbedded with clay under the level of groundwater, the waves refracted critically on the water table and those on the boundary between sand and clay beds can interfere with one another. Clay lenses interbedded and overlying the water table may screen refraction arrivals from the water table. The travel time diagram obtained over an unconfined water table is presented in Fig. 4-45.

One of the most important regional application of the seismic refraction studies is their use for the determination of the gross stratigraphy of large sedimentary basins. If some of the velocity discontinuities in unconsolidated or semiconsolidated deposits represent stratigraphic breaks in the sedimentary section, seismic refraction measurements can be used to unravel the gross stratigraphy of the deposits. If these breaks further represent significant hydrological boundaries, such as those between aquifers and aquicludes, the

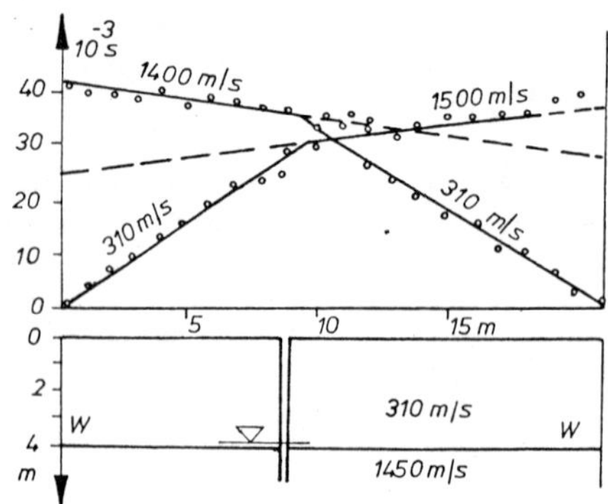

Fig. 4-45. Travel time graphs over an unconfined water table. The calculated refracting interface is marked by W, the observed water table by triangle.

seismic work may have special hydrogeological interest. Regarding the applicability of the seismic refraction method in sedimentary basins, it may be stated that the basic assumption of applicability, i.e., the increase of wave velocity with depth, is generally satisfied due to the compaction of layers. A typical application of seismic refraction measurement in a littoral basin is illustrated in Fig. 4-46.

In this kind of application, the seismic refraction prospecting exhibits considerable advantages when compared to the resistivity exploration: these are the results of the non-integrating character of the method. The basic principle is the same in both cases: the methods are based on differences in the physical parameters of the ground (i.e., in the wave velocity and in the specific resistivity, respectively). Values of the wave velocities may be obtained independently from one another in interpreting the travel time diagrams, but not so in resistivity prospecting. The calculation of depth is, therefore, more evident in a seismic refraction measurement than in the interpretation of resistivity graphs. This is the reason why the interpretation of the travel time diagrams, although not unique, allows comparatively few possible variations.

GRAVITY AND MAGNETIC PROSPECTING METHODS

Both methods may be used for the determination of the gross configuration of an aquifer as rapid and inexpensive studies. Gravimetry is useful in locating areas of maximum aquifer thickness, in tracing the axis of buried channels, and in locating a buried bedrock high (that may impede the flow of groundwater), provided that an adequate density contrast

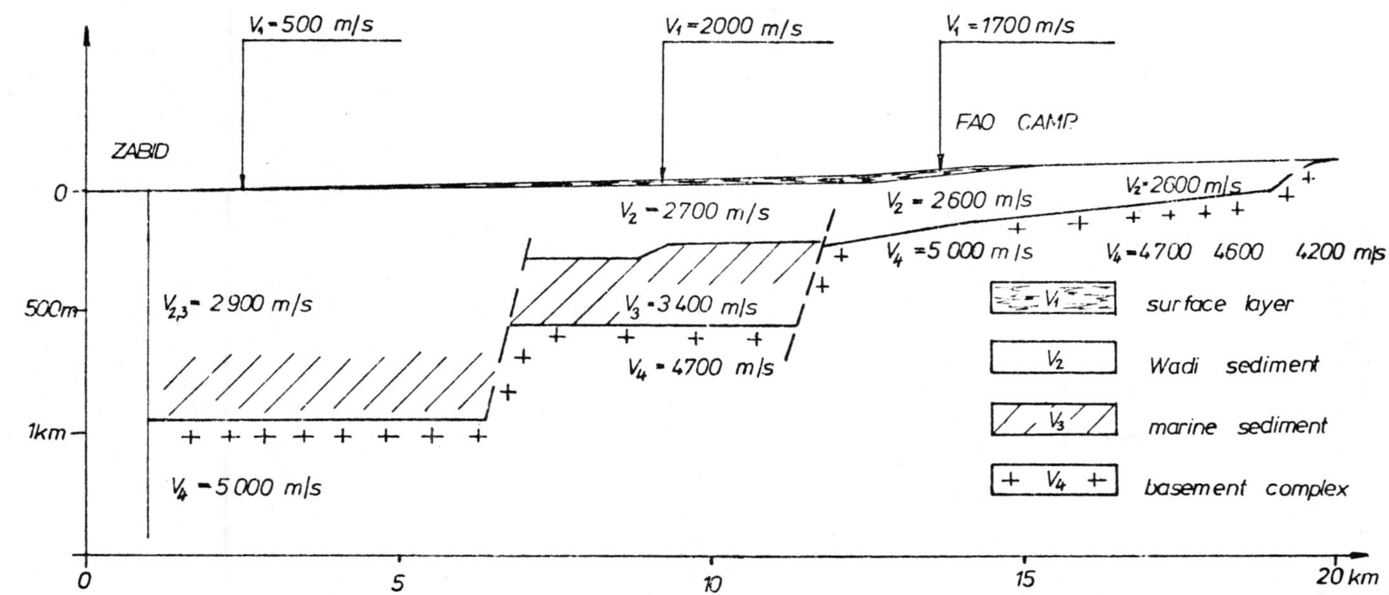

Fig. 4-46. Result of refraction seismic prospecting along the axis of the Zabid Wadi study area.

between the aquifer and the underlying bedrock exists. Where the sedimentary basin is underlain by magnetic basement rock (e.g., basalt) magnetic surveys may be an effective tool in studying the structure of the basin. Both methods will be discussed in Subsection 18-3.

14-2 Borehole Geophysics

In borehole geophysics (called also well logging), surface geophysical methods are generally applied for measuring physical parameters under special condition, i.e., in drilled holes (Fig. 4-47). The physical principles of application are the same in well logging as in the corresponding surface measuring system. Thus, the discussion of the basic principles may be omitted here. Logging techniques that have been utilized in water prospecting and those that have potential application to hydrologic problems are as follows:

- electric (detailed in Table 4-7);
- nuclear (detailed in Table 4-8);
- acoustic (detailed in Table 4-8);
- temperature;
- technical.

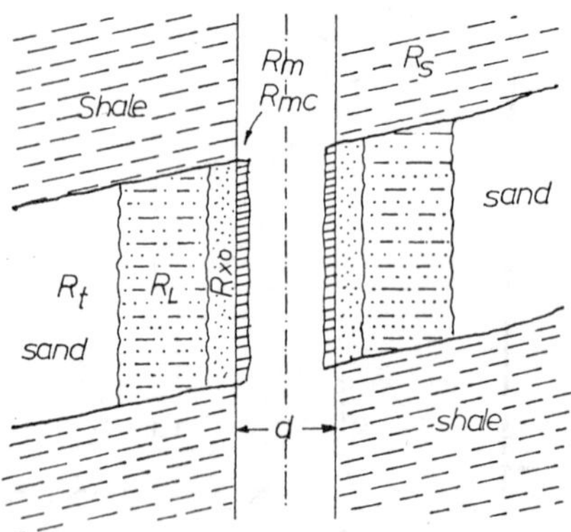

Fig. 4-47. Cross section of a well drilled by rotary system. R_t, R_i, R_{xo}, and R_{mc} denote the unaffected zone, the zone of invasion, the flushed zone, the mud cake, respectively.

Table 4-7. Hydrological applicability of electric logs.

Properties to be Investigated	Type of Log								
	SP	Single Point	Short Normal	Long Normal	Late Normal	Wall Resistivity	Lateral Log	Micro Focused	Induction
Lithologic correlation	x	x	x	-	-	-	x	-	x
Bed thickness	x	x	x	-	-	x	x	x	x
Formation resistivity (low R muds)	-	-	-	x	x	-	x	-	x
Formation resistivity (fresh mud)	-	-	-	x	x	-	-	-	x
Invaded zone resistivity	-	-	x	-	-	-	-	-	-
Flushed zone resistivity	-	-	-	-	-	x	-	x	-
Mud resistivity	-	-	-	-	-	x	-	-	-
Formation water resistivity	-	-	-	x	x	-	x	-	-
Formation water quality	x	-	-	-	-	-	-	-	-
Clay content	x	-	-	-	-	-	-	-	-
Effective porosity	x	x	x	x	x	-	-	-	x

Table 4-8. Hydrologic applicability of nuclear and acoustic logs.

Properties to be Investigated	Type of Log					
	Natural Gamma	Gamma-gamma	Neutron-gamma	Neutron Thermal Neutron	Neutron Epithermal Neutron	Acoustic
Lithologic correlation	x	-	-	-	-	x
Bed thickness	x	-	-	-	-	x
Clay content	x	-	-	-	-	-
Density	-	x	-	-	-	-
Primary total porosity	-	x	x	x	x	x
Secondary porosity	-	-	-	-	-	x
Moisture content	-	-	x	x	x	-
Chlorine content	-	-	x	-	-	-

The most widespread uses of logs in groundwater hydrogeology at the present time are to define the geometry and lithology of aquifer systems and to estimate the quality of the contained water. In order to properly interpret the geophysical logs and phenomena in well hydraulics, technical logs are carried out. The most important parameters measured by well logging methods and those which may be interpreted from the measured data are summarized in Table 4-5. These parameters, defined already, will be used here without further explanation. Information about the application of the various logging methods, to obtain data on the properties of rocks, fluid, wells or the groundwater system, is given in Table 4-6.

Logging equipment. Although the variety of instruments in use is very large, any well logger can be divided into the same basic components, as sensor, signal conditioners and recorders or indicators (Fig. 4-48). In a well logging system, the sensor, which is lowered into the hole, is contained in a watertight probe or sonde. It generally receives power from the surface and transmits a signal to the surface through the logging cable. The cable also serves to position the probe in the hole by means of a winch. The signal is processed by electronic equipment at the surface and is changed to a varying direct current voltage that moves the recorder pens or optical galvanometers horizontally. Chart paper or photographic film is moved through the recorder as a function of the vertical position of the sonde in the hole. The vertical scale of the record may be selected by changing a gear ratio in the recorder. This graphic presentation of geophysical logs allows rapid visual interpretation and comparison at the well site. Decisions on where to set screen and on testing procedures can be made immediately. Geophysical logs can be digitized in the field or in the office and are then amenable to computer analysis and the collation of many logs.

Log interpretation. This refers to all processes of obtaining qualitative or quantitative information from geophysical logs. To make quantitative analysis of logs, the equipment must have been properly calibrated and standardized at first, the extraneous effects of petro-physics and fluid characteristics must be considered, and all corrections for borehole and geometric effects must be made. Calibration of logging equipment refers to the processes ensuring the correctness of the values on the curve. This is generally accomplished in models that simulate in-hole environmental conditions or in test holes which were drilled in a well studied environment. Most of the environmental calibrators are too large for transportation in logging vehicles, yet it is highly desirable to run checks on the reproducibility of logs in the field. These are the standards (e.g., reference signals) which provide a means of comparing the response of

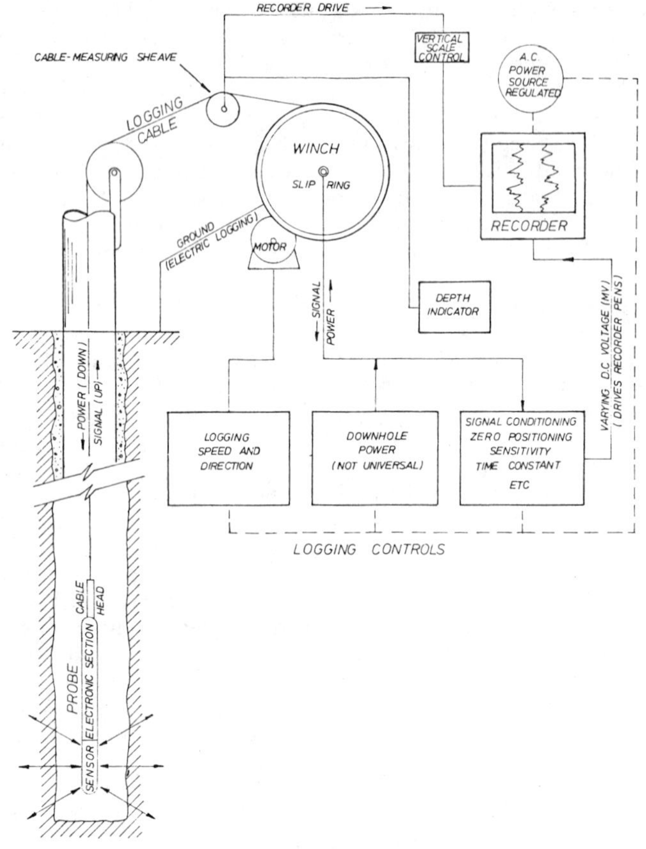

Fig. 4-48. Schematic block diagram of geophysical well logging equipment.

the logging equipment from time to time and from place to place.

As a general rule, the more types of geophysical logs available for a well, the better the hydrological interpretation that can be expected from logging. The importance of the composite interpretation is due to the fact that each type of log actually measures a different parameter, and when several are analyzed together each will tend to support or contradict conclusions drawn from the others. Figure 4-49 illustrates how the relationship of various logs can be utilized to determine hydrologic characteristics. The simplified response of the six logs represented in Fig. 4-49 is listed in Table 4-9.

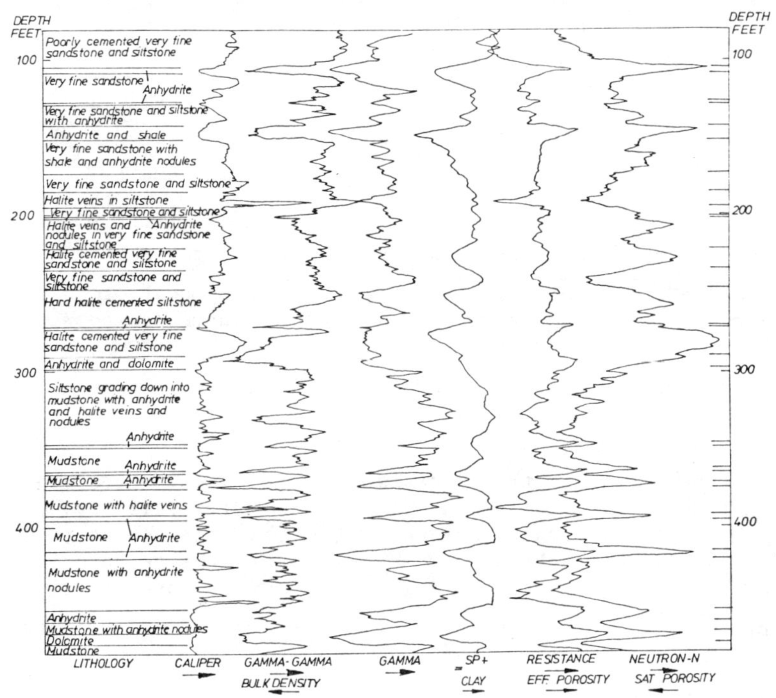

Fig. 4-49. The relationship of six different geophysical logs to lithology (Keys, 1971).

Corrections for geometric and borehole effects. The basic relations are derived for sondes embedded in an infinite and homogeneous medium. This simple situation is never realized in practice for the earth is layered and drilling disturbs the homogeneity near to the well. Anomalies in logs due to layering and observed by the logging device are called geometric effects and must be considered in the interpretation of all geophysical logs (e.g., when a thin bed is logged, the volume being investigated may include a part of the adjacent rocks so that the sonde does not measure the same value that it would in a thick bed of the same lithology).

One of the most common extraneous effects present on logs is caused by changes in the diameter of the hole. It is obvious that any change in the thickness of the material between the sonde and the rock being measured is likely to

Table 4-9. Simplified response of six logs presented in Fig. 4-49.

Type of Log	Parameters Measured and Direction of log Deflection	Parameters Inferred and Direction of Log Deflection
Caliper	Hole diameter (increases to right)	Cementation and effect of drilling
Gamma-gamma	Scattered and attenuated gamma photons (radiation increases to right)	Bulk density (decreases to right) Hole diameter (increases to right)
Natural gamma	Natural gamma radiation (increases to right)	Clay content (increases to right)
Spontaneous potential (SP)	Natural electrical potentials (positive to right)	Clay or shale content
Single-point resistance	Electrical resistance of hole and adjacent rocks (increases to right)	Hole diameter and water salinity (decreases to right), effective porosity (increases to right)
Neutron-neutron	Neutrons slowed down and scattered by hydrogen (radiation increases to right)	Saturated porosity (decreases to right)

affect the log response. Since the thickness of the mud sheet, surrounding the sonde, is determined by the hole diameter, the latter should be considered, and when necessary a caliper log should be made. The effect of drilling is visualized in Fig. 4-47. Inasmuch as both the mud cake and the zone of filtrate invasion vary in thickness, logging devices having different volumes of investigation provide a means of measuring these factors. Geophysical logs are generally carried out in uncased wells, only the nuclear logs are routinely used for logging through casing. Casing of consistent thickness and material introduces a constant error that will shift the log response, but it generally does not change the character of the log. The effects behind the casing (e.g., cementing, hole diameter changes) may be considered using radiation devices having different volumes of investigation. The volume of investigation is here considered to include that part of the formation contributing 90 percent of the measured signal. The radius of investigation (defined alike) ranges from centimeter to meter, depending on the construction of the sensor.

ELECTRIC WELL LOGGING METHODS

Types of electric log and their hydrologic applicability are summarized in Table 4-3. Schemes of logging and the arrangements of electrodes in various sensors are represented in Fig. 4-50. Spontaneous potential (SP) logs are records of the natural potentials developed between the borehole fluid and the surrounding rock materials. The spontaneous potential is used chiefly for geological correlation, determination of bed thickness and separating non-porous from porous rocks in shale-sandstone and shale-carbonate sequences. In its simplest form, an SP logging device consists of a movable (E) and a ground electrode (G) connected with a sensitive millivolt meter (M) as it is represented in Fig. 4-50. The most important source of SP arises in the electrochemical potentials produced at the junction of dissimilar materials in the borehole, symbolized by arrows in Fig. 4-50. When the formation water is much more saline than the mud, the current flows in the direction of the arrows, entering the mud column from the shale and moving into the sandstone. As the SP electrode moves upward through the bottom shale, it senses a decreasing potential with an inflection at the boundary and attains its maximum negative potential at midpoint in the sand. If, on the other hand, the formation water is fresh compared with the mud, the polarity of the SP curve is reversed and the reciprocal of the log in Fig. 4-50 is produced. The following equation may be useful to calculate the SP values provided that NaCl is the most dominant among the dissolved salts:

$$SP = -K.\log(R_m/R_w) \quad ; \qquad (4\text{-}33)$$

where SP denotes the log deflection in millivolts, K = 64.25 + 0.24 T, T denoting the borehole temperature in centigrades. R_m and R_w are the resistivities of the borehole fluid and that of the formation water, respectively. Equation 4-33 may be used to calculate the quality of the formation water. In actual practice, the SP deflection opposite a sand bed is read from the log, R_m is measured with a fluid resistivity tool, then R_w can be calculated, from which, using diagrams such as in Fig. 4-51, the quantity of the dissolved salt can be obtained. In the resistance logging, a potential difference is measured between an in-hole electrode (point electrode, E) and a surface electrode (G) and this is converted to resistance when maintaining a constant alternating current (Fig. 4-50). The main use of point resistance logs is the geological correlation determining bed boundaries and changes in lithology and identifying fractures in resistive rocks (Fig. 4-49). The point electrode log is strongly affected by resistivity of the borehole fluid and by changes in hole diameter since the radius of investigation is small, from about 5 to 10 times the

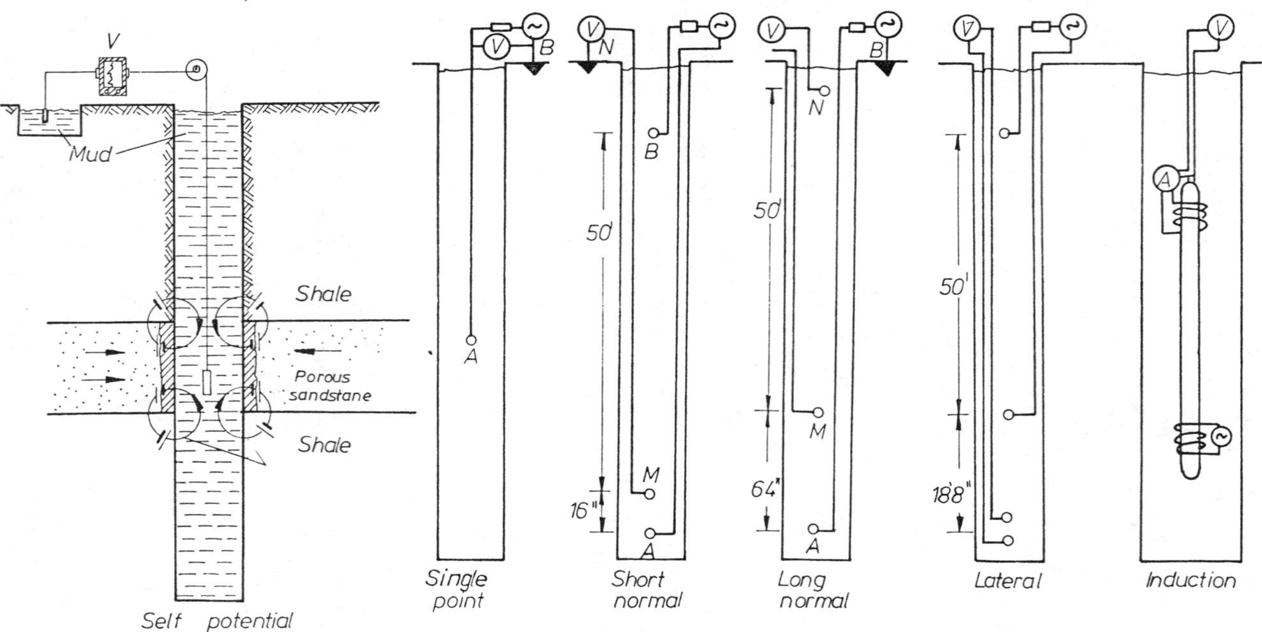

Fig. 4-50. Schematic representation of electric well logging methods.

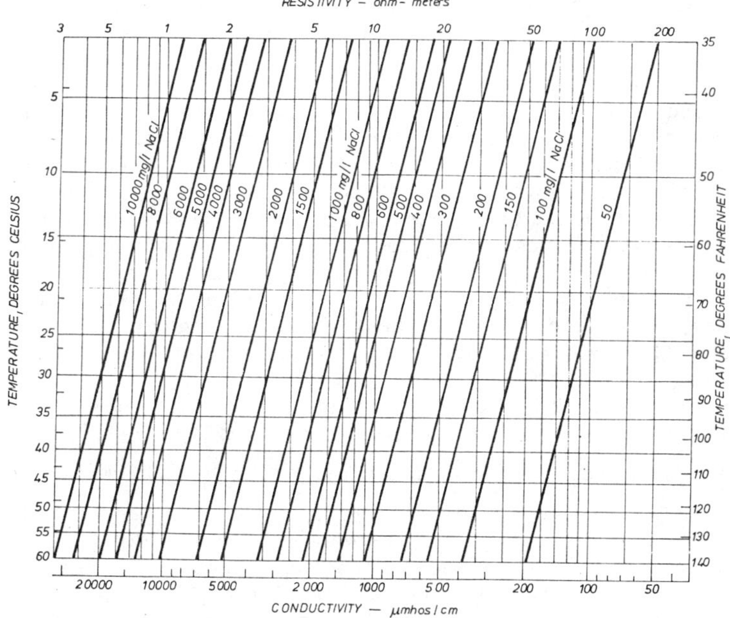

Fig. 4-51. Resistivity of solutions of sodium chloride as a function of concentration and temperature.

electrode diameter. Despite these shortcomings, the point electrode method is a very useful tool inasmuch as any increase in formation resistance produces a corresponding increase in resistance on the log, and the deflections on the single point logs are interpreted to be due to changes in lithology. In fractured rocks, the single point log commonly appears as a mirror image of the caliper log.

Resistivity logging devices measure the specific electric resistivity of earth materials under the direct application of an alternating current or an induced electric current. The measuring techique in resistivity logging is similar to the technique for measuring resistivities in surface resistivity investigation using the four-electrode arrays. Depending on the device employed, resistivity logging can be used for geological correlation although this is not the most important application. Resistivity devices are generally used to determine the resistivity of the formation, mud cake, invaded zone, and fluid, then, using the resistivity data, the formation factor can be determined and hence the porosity. Electrode arrangement of the various resistivity logging devices is represented in Fig. 4-50. Examples of resistivity logs (together with an SP and a natural gamma log) are shown in Fig. 4-52. Multiple electrode resistivity measurements include such curves as the short and long normal, lateral, microlog and focused logs. The electrode arrangements for normal logging (Fig. 4-50) show that the current electrode (A) and the potential electrode (M) are relatively close together compared to the other pair of electrodes. This means that the apparent resistivity is determined primarily by the potential of the measuring electrode, M. The radius of investigation is about the double of the electrode spacing AM. The spacing of the short normal (16 inches) was selected empirically because this spacing gives good vertical details and measures an apparent resistivity that is related primarily to the invaded zone resistivity (Fig. 4-47). The long normal (AM = 64 inches) was selected to investigate the average resistivity beyond the invaded zone for a radius of approximately twice the spacing. Both logs are generally used with charts for interpretation (Schlumberger, 1966) so that invaded zone resistivity (R_i), true formation resistivity (R_t), and depth of invasion (d_i) can be determined. The bed thickness effect is very pronounced in normal logging producing the reversal of the normal curve opposite to resistive beds having thickness equal to or less than the AM spacing. This is the reason why the use of AM spacings longer than 64 inches would not be practical. A quantity, called the formation resistivity factor (F) may be derived from the resistivity data obtained by resistivity logging, by using Archie's (1942) equation:

$$F = R_o/R_w \quad ; \qquad (4\text{-}34)$$

where R_o denotes the resistivity of a formation saturated with water of resistivity R_w. He also established that F is related to porosity (n) and permeability: $F = a(n)^{-m}$, where m denotes the cementation factor. Both F and m tend to be consistent within a lithologic unit in a given sedimentary basin (Table 4-4). One important application of the normal curves is in the estimation of water quality in clastic rocks. A relation between ionic concentration of formation water and long normal resistivity data may be established empirically or Archie's equation can be used for the determination of R_w, when the porosity (n) is derived from a neutron, gamma-gamma or acoustic log. Contrary to this, the porosity may be calculated if R_w is determined from SP log or measured from a control water sample corresponding to the interval logged for R_o. Another potentially useful application of the normal resistivity logs to hydrological problems involves the calculation of the permeability of clastic aquifers. Inasmuch the formation factor is the key to the permeability values, for it

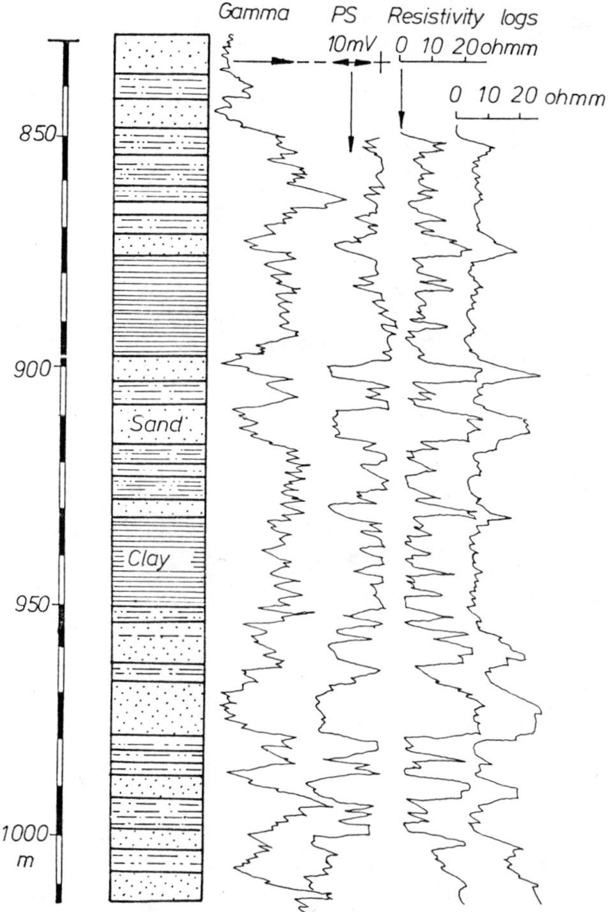

Fig. 4-52. Geological log, natural gamma log, SP-log, and two resistivity logs in a thermal well.

also accounts for tortuosity and cementation, a relation may be obtained between permeability and formation factor (Fig. 4-53). Another method is offered by correlating the transverse resistivities (i.e., R_t multiplied by the thickness of the layer) to the transmissivity values determined in wells, as it was previously explained in Subsection 14-1. The normal devices are the most widely used types in the logging of water wells although they have disadvantageous properties too, e.g., reversal opposite to thin resistive beds.

Various systems have been devised as additional tools to improve the performance of the resistivity logging method. The lateral logging device measures the formation resistivity beyond the invaded zone by use of an electrode spacing large enough to insure that the invaded zone has little or no effect on the response (Fig. 4-50). Its efficiency is poor in highly resistive rock. This disadvantage is eliminated in the focused current devices which are used to measure relatively high formation resistivities through a conductive mud. The good vertical resolution and great penetration of these systems is achieved using guard electrodes or a rather complicated arrangement of point electrodes. Conventional electrode resistivity logging systems require a conductive medium (mud or water) in the borehole to carry the current into the formations; thus, they cannot be used in air filled holes or in wells drilled by resistive (oil base) mud. To overcome this problem, the induction logging system was applied. This method is based on the idea of using electromagnetic induction

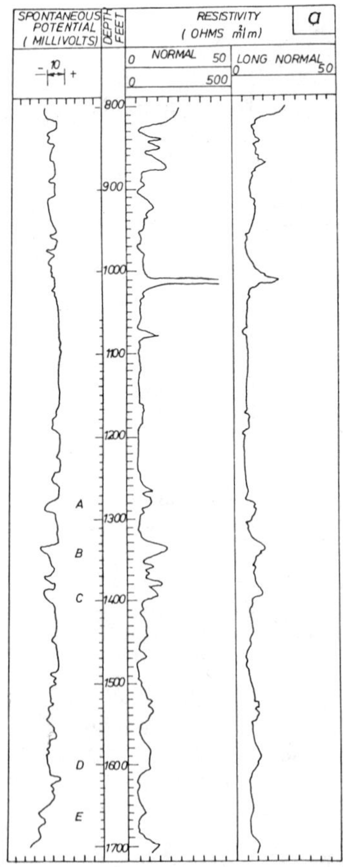

Fig. 4-53a. Electric log of a water well.

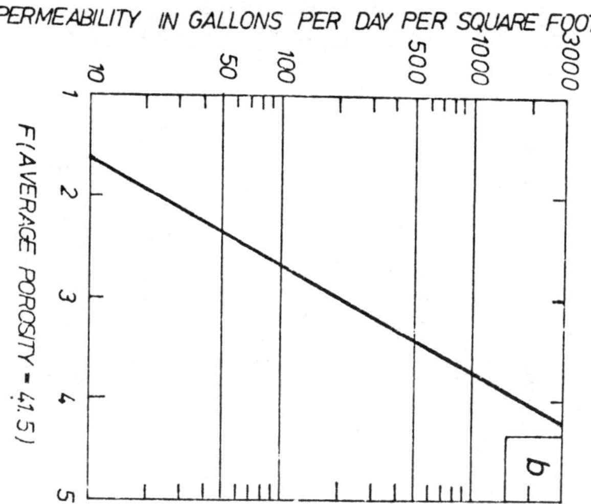

Fig. 4-53b. Graph showing permeability and formation factor (F) (Croft, 1971).

to couple a signal from the logging sonde into the formation and back to the sonde (Fig. 4-50). This method has also proven extremely successful in freshwater muds and is generally the best resistive logging tool in fresh mud except where formation resistivities are much higher than those of the adjacent shale beds. The wall resistivity devices are very short-spaced multiple-electrode arrangements mounted on pads that hold the electrode spacing, they measure the immediate surrounding of the borehole. The micronormal and microlateral sonde give fine lithological detail and are also used to delineate porous versus non-porous formations on the basis of presence or absence of mud cake. The microfocused sondes are designed primarily to measure the resistivity of the flushed zone (R_{xo} in Fig. 4-47) with the objective of obtaining the formation factor (F).

NUCLEAR WELL LOGGING METHODS

Differences both in the intensity of natural radioactivity emitted by rocks and that in the secondary radiation produced by rocks irradiated artificially enable the application of radiation measurements for well logging purposes. Types of nuclear logging and their hydrological applicability is summarized in Table 4-3. The scheme of logging is represented in Fig. 4-54. An important factor in the interpretation of all nuclear logs is the statistical nature of the radioactive emission. Most measurements of

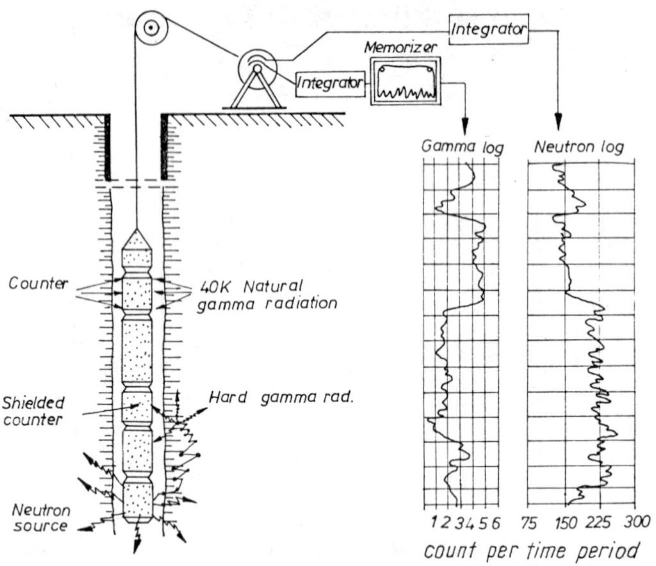

Fig. 4-54. Schematic representation of nuclear logging methods.

nuclear radiation are made in terms of observed disintegrations expressed in counts per second (or counts per minute). The unit of radioactivity is the curie which represents any source that is decaying at a rate of 3.7×10^{10} disintegrations per second. The nuclear logging devices can be standardized to provide similar recorder deflections measured in units such as counts per second, porosity, bulk density in grams per cubic centimeter, etc. An important instrumental parameter is the time constant denoting the time, in seconds, over which radiation pulses are averaged. The time constant and logging speed are usually determined as a result of experience with the logging equipment on the basis of the amount of statistical fluctuation that can be tolerated and on the thickness of the beds that are to be measured.

Quantitative analysis of nuclear logs can be made if the logging equipment is calibrated and corrections are made for the various borehole and instrumental effects present on all logs. Where nuclear logs are used for stratigraphic correlation such corrections are not important.

The sensor of the gamma logging equipment consists of a sodium iodide crystal optically coupled to a photomultiplier tube. The sensor sends electric pulses via an amplifier-integrator circuit to the surface when exposed to gamma or neutron radiation. An important feature of the system is that the height of each pulse is nearly proportional to the gamma photon that produced it. This fact makes it possible to identify radiation both by count rate and according to its energy.

Natural gamma logs are records of the amount of natural gamma radiation that is emitted by all rocks. The gamma emitting radioisotopes normally found in rocks are potassium-40 and daughter products of uranium and thorium decay series. Potassium-40 is probably the most common radioisotope in sedimentary rocks. The total potassium content on the average is about 3 percent for argillaceous sediments and about 1 percent for sandstones. Probably the most important application in groundwater hydrology is the identification of clay- or shale-bearing sediments. Clays tend to reduce the effective porosity and permeability of aquifers and the natural gamma log can be used to empirically determine the shale or clay content in some sediments. Clean sand formations generally show a constant value of radioactivity on the log that is termed "0 percent shale" and the "100 percent shale" line can be determined by maximum radioactivity level (Fig. 4-55). The content of clay in mixed horizons may be calculated proportionally. The radius of investigation of a gamma probe is a function of downhole instrumentation and borehole parameters. Depending on the bulk density of rocks, 90 percent of the gamma photons detected probably originate within 15-30 centimeters of the borehole wall.

Gamma-gamma logs are records of the intensity of gamma radiation from a source in the probe after it is backscattered and attenuated within the borehole and surrounding rock. The main applications of gamma-gamma logs are the identification of lithology, the measurement of bulk density, and porosity of rocks. The logging may be carried out in cased or uncased wells.

The gamma-gamma probe contains a source of gamma photons, generally cobalt-60 or cesium-137, shielded from the sodium iodide detector by lead spacers (Fig. 4-54). Gamma photons from the source penetrate the fluid casing and formation surrounding the probe and are absorbed mainly by Compton scattering. In the Compton range, the gamma radiation absorbed is proportional to the electron density of the material penetrated, which, from the other part, is approximately proportional to the bulk density of the medium. After suitable corrections, the count rate recorded on the gamma-gamma log is inversely proportional to the bulk density. Bulk density may be read directly from a calibrated and corrected gamma-gamma log (called density log) or derived from a chart providing correction factors. Gamma-gamma density is used for the determination of total porosity by application of the following equation:

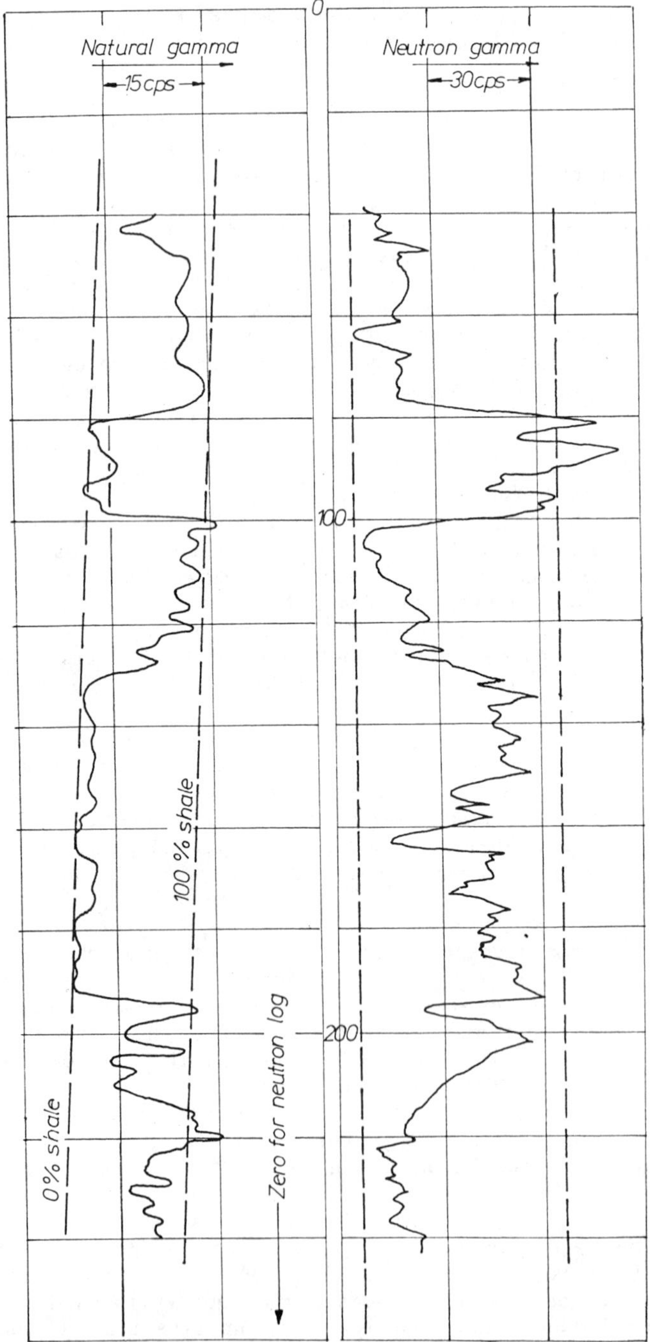

Fig. 4-55. Natural gamma and neutron-gamma logs in a borehole (Keys, 1968).

$$\text{Porosity} = \frac{\text{Grain density(from lab)-Bulk density (from log)}}{\text{Grain density (from lab)-Fluid density (from lab)}} \qquad (4\text{-}35)$$

An approximate value for the porosity may be obtained in quartz sandstones and sands taking a value of 2.65 g/cm^3 for grain density and 1 g/cm^3 for the fluid.

The radius of investigation of gamma-gamma logging devices is reported to be about 15 cm with 90 percent of the signal coming from the material within this volume.

Neutrons, for most logs, are derived from alpha particles impinging on beryllium. The alpha emitter is intimately mixed with beryllium inside a sealed source. The neutrons are then artificially introduced into the rock-fluid system, and the effect of the environment on the neutrons is measured. The neutron has a relative mass of 1 and no electric charge. For this reason, the loss of energy when passing through material is caused by elastic collision. The element most effective in moderating neutrons is hydrogen because the nucleus of the hydrogen atom has about the same mass as a neutron. Thus, the energy of a neutron is largely a function of the hydrogen content of the material through which it passes. In groundwater aquifers it is the groundwater itself that possesses the highest hydrogen content, thus the corrected and calibrated neutron log provides a continuous measurement of moisture content above the zone of saturation and a record of porosity in the zone of saturation. Various types of neutron logs are made by counting the number of neutrons present at different energy levels or counting the gamma photons produced by neutron reactions. The term neutron-gamma log indicates that most of the radiation detected are gamma photons resulting from neutron reactions. The neutron-thermal-neutron probe responds chiefly to thermal neutrons and the neutron-epithermal-neutron sonde responds mostly to neutrons being at a higher energy level (between 0.1 and 100 eV) than the thermal neutrons. Epithermal-neutron measurements provide the highest percentage of response due to hydrogen and are least affected by the chemical composition of rocks and contained fluid. Among the elements generally present in aquifers, boron and chlorine are those which have high neutron-capture cross sections. For this reason, they might interfere with quantitative moisture or porosity measurements. The neutron-gamma probe has been experimentally found to be very sensitive to chlorine content; hence a comparison of neutron-epithermal-neutron logs and of neutron-gamma logs, which are mostly a record of prompt gamma photons, gives a theoretical basis for measuring the chlorine content in place and through casing. Clay horizons will indicate high porosity on neutron logs by virtue of their high water content due to adsorption. This property must be corrected, when logging is made for measuring the effective porosity, by using a natural-gamma log recorded at the same time. A pair of such logs is presented in Fig. 4-55. The radius of investigation of neutron devices is reported to be from 15 centimeters for high porosity saturated rocks to 60 centimeters for low porosity or dry rocks.

ACOUSTIC WELL LOGGING METHOD

All acoustic logging devices contain one or two transmitters that convert electrical energy to acoustic energy which is transmitted through the environment as an acoustic wave. One to four receivers reconvert the acoustic energy sensed to electrical energy for transmission up the cable (Fig. 4-56). The hole must be filled with water or mud in order to permit transmission of acoustic waves from transmitter to formation and from formation to receiver. Two types of measurement may be made in acoustic logging: interval transit time (from which the wave velocity can be calculated) and amplitude (to calculate the attentuation of the acoustic wave in the formation).

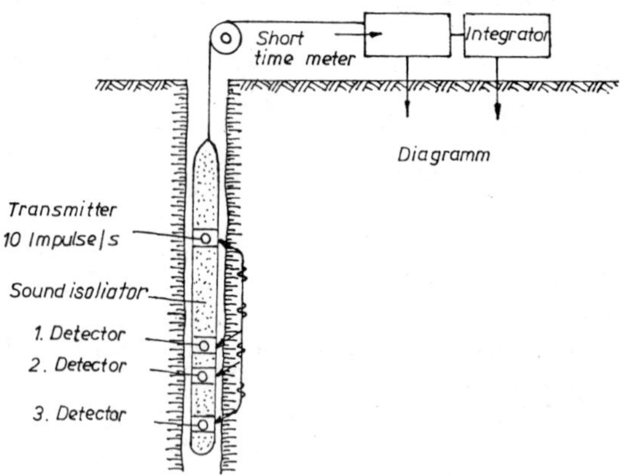

Fig. 4-56. Schematic representation of acoustic well logging method.

At present, the continuous-acoustic-velocity log is primarily used for the measurement of effective rock porosity in open holes. The basis for the porosity calculation from acoustic logs is:

$$1/V_r = n/V_f + (1-n)/V_m \qquad (4\text{-}35)$$

where V_r, V_f, and V_m denote the acoustic wave velocities in the formation (read on the log), in the fluid and in the matrix, respectively, and n denotes the effective porosity.

The effect of secondary porosity on the various elastic wave parameters can be used to locate fractured zones near wells. Fractures may be located as horizons of low velocity, but they may also cause "cycle skipping" on sonic velocity logs due to the attenuation of wave amplitude through the fractured (loose) rocks. Acoustic log data may be used for the identification of lithologic units producing seismic reflections. Reflecting horizons are characterized by the sudden changes in acoustic wave velocities and/or in the bulk density of rocks. The radius of investigation is reported to be about three times the wavelength, which is equal to the wave velocity divided by the frequency. At a frequency of 20,000 cps, the radius of investigation should be about 20 cm in soft rocks and about 1 m in very hard rocks.

TEMPERATURE WELL LOGGING METHOD

Temperature logs are the continuous records of the temperature of the fluid immediately surrounding a sensor in the borehole. The temperature is commonly measured by calibrated electric thermometers. Most temperature logs are made with a single sensor which moves down the hole. The differential temperature logging system consists of two sensors spaced 1 to 2 m apart. The principal advantage of this system (also called the delta log) is that it responds passively to a normal vertical thermal gradient and actively to an abnormal one.

The temperature recorded in temperature logging is that of the fluid surrounding the sensor which may or may not be representative of the temperature in the surrounding rock. A geothermal gradient characteristic for the rocks surrounding the well may be derived from a temperature log when the fluid in the well is in thermal equilibrium with the adjacent rocks and if there is no vertical circulation of fluids in or adjacent to the borehole. Such static conditions are unusual, but where they occur the temperature gradients may be applied to the study of heat balance in sedimentary basins (Subsection 16-3).

Temperature logs may be used to identify aquifers contributing water or gas to the well to provide information on the source of water and to distinguish stagnant and moving water. Seasonal recharge to the groundwater system may be reflected in cyclic temperature fluctuations. Temperature logs can be useful in identifying recharge water or liquid wastes discharged into the ground. Finally, because of the effect of temperature on electrical resistance, temperature logs are necessary for the correct interpretation of resistivity logs.

TECHNICAL WELL LOGGING METHODS

Technical logging includes, in one hand, all techniques which are used to provide data related mainly to well construction such as caliper, fluid conductivity, fluid movement, and casing logs, and on the other hand, the application of hydrogeological well logging methods, listed in the foregoing text to well construction purposes.

Caliper logging. The caliper log is a record of the average diameter of a drill hole. Most caliper sondes consist of one to four bow springs which follow the wall of the hole. Caliper logs are utilized for the identification of lithology and stratigraphic correlation and to correct the interpretation of other logs for hole-diameter effects. The hole diameter changes are functions of lithology, hardness, compaction, and cementing of rocks. In rock formations, which are dense and hard, the diameter of the borehole closely approximates the bit size. Borehole enlargement by the erosion of the drilling fluid is common in sandy or silty clay and shale. The diameter of the hole is seldom appreciably greater than the bit size in most permeable granular formations. Cavities indicated on the caliper log of wells drilled in consolidated rock may be caused by local zones of fractures or solution openings which may have water-yielding possibilities. In well construction, caliper logs are extremely useful for calculating the annulus volume for effective gravel packing (Subsection 15-4).

Fluid conductivity (or resistivity) logs provide a measurement of the conductivity (resistivity) of the in-hole liquid. Any of the resistivity sondes made for microresistivity measurement may be used for fluid conductivity logging if provisions are taken to isolate the electrodes from the wall of the borehole. The fluid resistivity log provides resistivity data required for the quantitative interpretation of resistivity and SP logs. It can often give a general indication of water quality and provide information on the hydraulics of a well. It is useful in determining the depth of saltwater leaks in cased artesian wells and the depth of saltwater aquifers penetrated by uncased wells. Fluid conductivity and temperature logs may also be used to identify aquifers or perforated sections providing water for a pumped well. Water in different aquifers is seldom of the same temperature and/or conductivity. Where two aquifers contribute water of different temperatures and conductivities to a well, it may be possible to estimate the relative contribution of each zone from the temperature and/or from the conductivity logs.

Fluid movement (flow meter) logging. Devices used for this purpose measure the vertical and horizontal flow in water wells. The impeller flow meter transmits pulses which indicate the number of revolutions of the impeller per unit time. The thermal type detects water heated by an element in the sonde (dual thermistor system). Various tracers are injected in the column of water and their movement to a detector is timed. The flow meter log serves for evaluating permeable zones in uncased wells and aquifers and in cased wells having multiple screens. In such cases, a flow meter can be placed in a well beneath the pump, during a test, and the relative contributions of various aquifers can be determined at various discharge rates. An application for in-hole flow measurement is to determine the relative magnitude of permeabilities under an imposed hydraulic test (Fig. 4-57).

Casing logs. The casing collar locator is a useful device consisting of a permanent magnet wrapped with a coil of wire. Changes in the magnetic properties of material moving through the lines of force from the magnet cause a current to flow in the coil which will be registered. The continuous collar log can also be interpreted to accurately locate perforations and screens in a well. Data related to the condition of casing can be obtained by other well logging methods too. Gamma-gamma logs have been used to locate joints and one string of casing outside another. SP logging can be used in casing to locate differences in anodic activity in wells where there is active corrosion, but it cannot be used to determine the extent of corrosion. Both the SP and single point resistance log will generally provide an accurate depth to the bottom of the casing and, under favorable circumstances, they can also be used to locate screens. The caliper log is particularly useful for measuring casing size and for locating joints and screens

Among the various kinds of cement bond loggers, acoustic well logging provides the best means of determining the bonding of cement between the casing and the formation. Sonic velocity in the steel casing is higher than that in the formation, therefore, the arrival of the waves propagating in the casing and in the formation can be recorded separately. When the casing is centralized in the hole and not bonded to the formation, only a symmetrical casing signal will be recorded. Decentralized unbonded casing is characterized by a distorted casing signal. Bonding of cement to the casing but not to the formation will produce a very low amplitude casing signal but produces no formation signal. The amplitude of the casing signal is decreased by good bonding between cement and casing, and the amplitude of the formation signal is increased by good bonding between the cement and formation. Sufficient cement

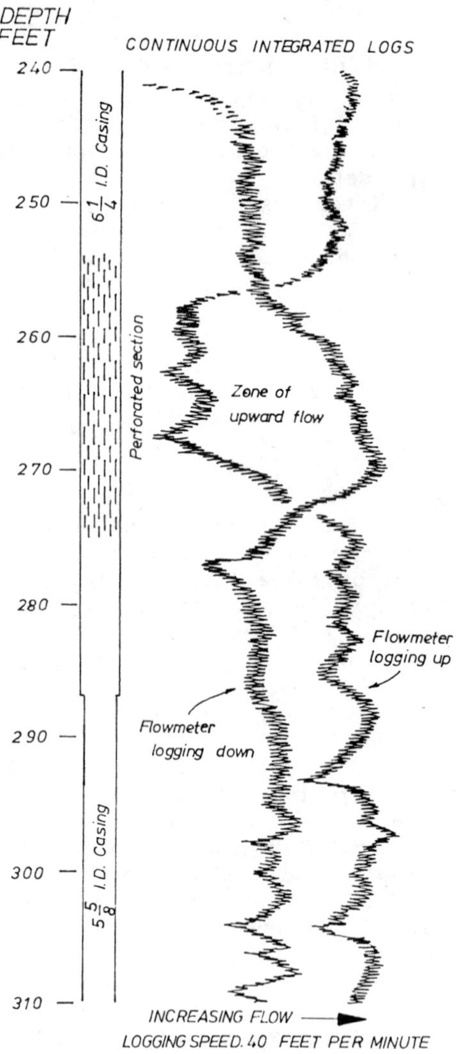

Fig. 4-57. Continuous flow meter logs used to locate the zone of flow (Keys, 1971).

bond allows the recording of a log of acoustic velocities of rock. The gamma-gamma log may be used to locate the position of cement (grout) outside the casing owing to the comparatively high density of the grout. Temperature logging may be proposed to be performed within several days after cementing. Anomalous temperatures due to cement bond may prevail for longer time, but the interpretation of the temperature log becomes more difficult.

Checking of screens and filter structures can be made by logs related to porosity. Procedures used for the development of water wells are intended to increase the porosity near the hole (Subsection 15-5). This is achieved in clastic sediments by removing fine-grained material from between the larger grains. A potential use of neutron, gamma, and gamma-gamma logging is to locate the source of fine-grained material produced during the development of wells by pumping. The difference between a neutron log made prior to the development of a well and a second log made after well development should indicate zones of higher porosity produced by pumping, and so the consequent removal of fines (Fig. 4-58). The same procedure can be used with resistivity logs in and near uncased drill holes or through some plastic well screen. The drilling velocity log, i.e., the record of drilling speed (Subsection 15-3) provides information about the lithology of strata penetrated by drilling.

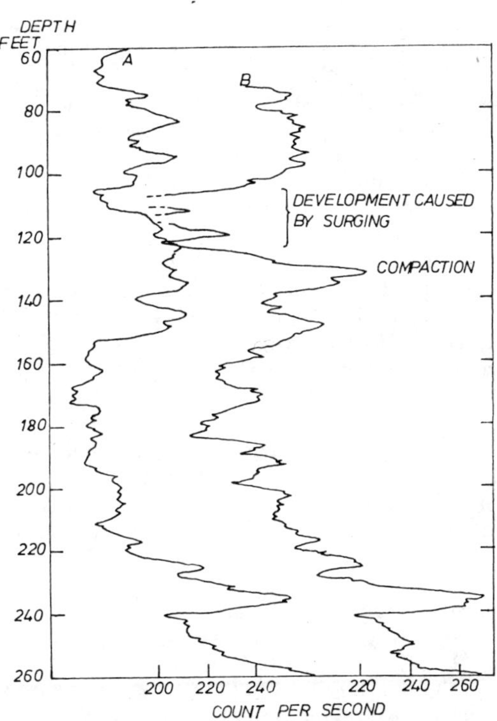

Fig. 4-58. Neutron logs made before perforating the casing (curve a) and made after several days of suring and bailing (curve b) showing the development caused by suring (Keys, 1971).

14-3 Drilling of Boreholes (Pataki, 1972)

There are many possible methods for preparing wells and drilling boreholes. Figure 4-59 gives a general review of the drilling procedures which can be considered in solving water prospecting and well drilling tasks.

In certain regions, particularly in hard rock areas, cable percussion drilling is applied. Exploration holes (slim holes) are drilled mostly by the direct or reverse flushing rotary power method. Portable (i.e., truck-mounted) rotary drilling equipment with modern fittings are also usually supplied with those of percussion drilling. The application of combined drilling methods, in this way, will meet even the highest demands of productivity.

CABLE PERCUSSION DRILLING

The essential elements of the cable percussion drilling rig are shown in Fig. 4-60. The essence of the process is that the bore bit, which is attached to the drilling cable, moves vertically within range of 0.3 to 1 m, and while doing so it applies strokes to the rock forming the bottom of the borehole. The rhythmical motion of the drilling cable and of the bore bit is produced by the spudder. The so-called drag line percussion pulley, accommodated on the spudder which is attached to the eccentric pulley, depresses the cable and causes the bore bit to be raised from the bottom. As soon as the dead center is reached, the cable is released from pressure and the bore bit may drop by gravity to the bottom of the borehole. After dismounting the bore bit, cuttings are removed from the borehole usually with sludgers or gravel bailers.

Cable percussion drilling rigs may be operated either with large lifting height (i.e., 0.6 to 1.2 m) and with lower stroke rate (i.e., 20 to 50 strokes per minute) or with small lifting height (i.e., 0.06 to 0.15 m) and a higher stroke rate (i.e., 60 to 150 strokes per minute).

The design of bore bits and their weight may vary as a function of the borehole diameter and the properties of the rock to be drilled. In hard rock formations, the application of integrally forged bore bits; in plastic formations and in rock slides, that of cross bits and barrel bore bits, gives the best result. The weight of bore bits ranges from 0.1 to 2 tons (Fig. 4-61).

The progress of drilling depends on factors such as the rock drilled, the depth and diameter of the hole, the type of equipment, and the skill of the operator. When drilling

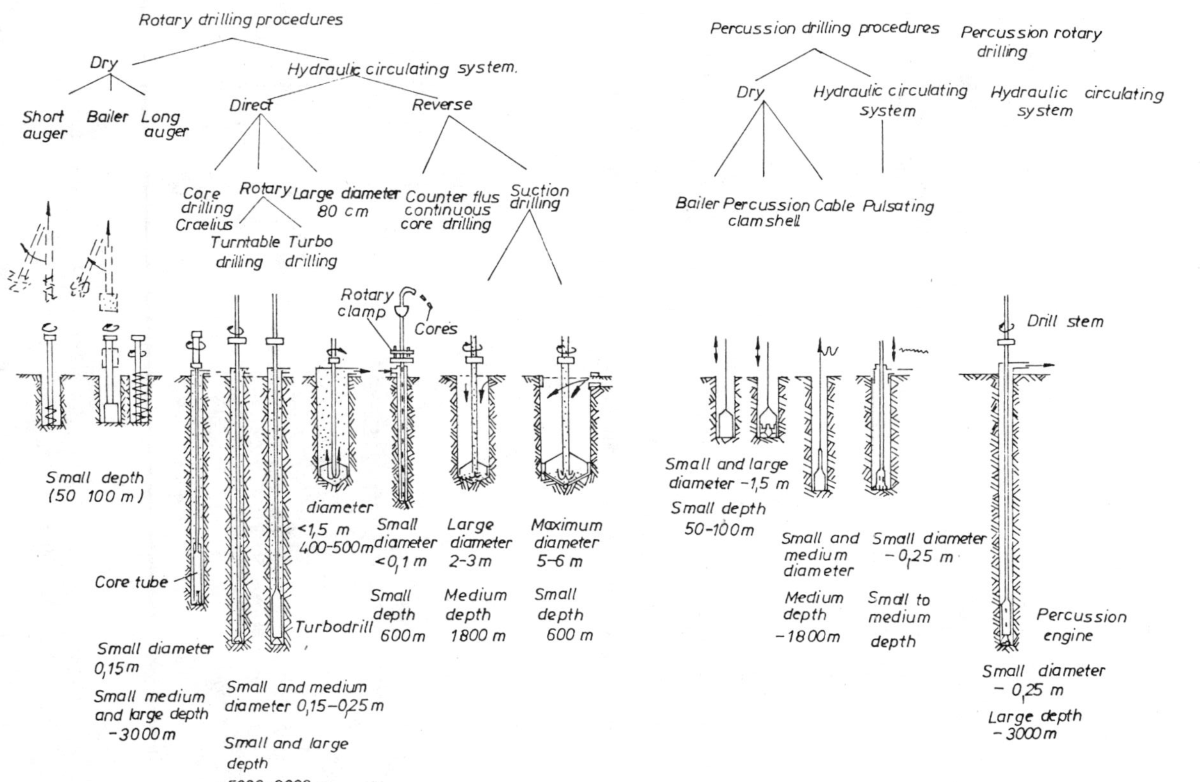

Fig. 4-59. General scheme of drilling procedures.

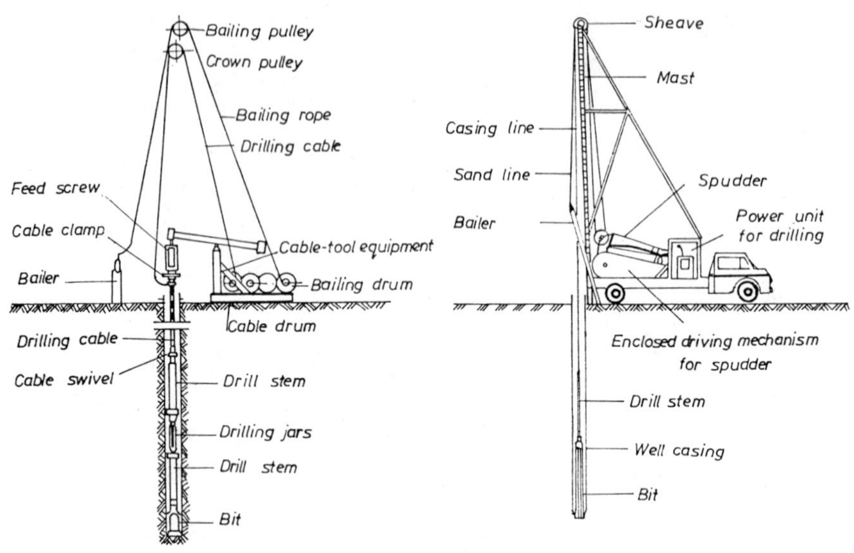

Fig. 4-60. Arrangement scheme of cable percussion drilling.

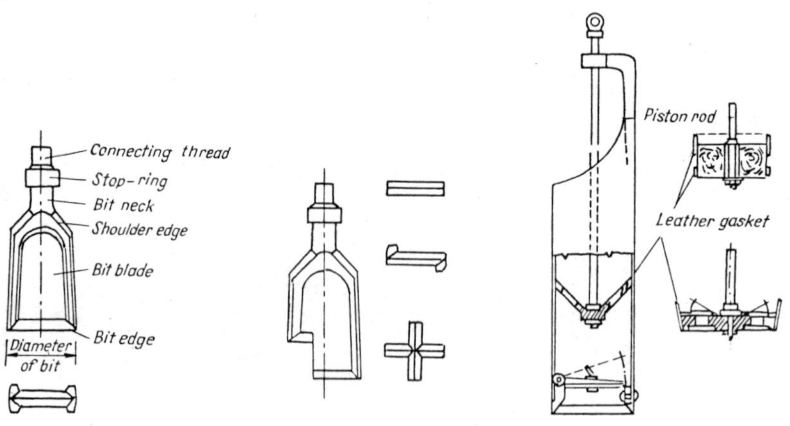

Fig. 4-61. Types of bore bits used in cable percussion drilling.

crystalline rock, a performance of 1.6 to 2.0 m per day may be considered as a very good result. In soft sandstone or in sandy clay, even 15 to 20 m per day can be attained. No difficulties are to be expected in drilling very hard crack-free solid rock. On the other hand, when fissured rocks or rocks of varying slope are drilled, the borehole may incline very rapidly, resulting in the breakage of the drilling cable. Troubles may be caused by loose, fine sand layers since the cleaning of the bottom of the borehole is quite often impossible.

The casing of the wells drilled by percussion method is rather difficult since the tubes must usually be sunk by applying the so-called driving cap which is placed over the casing. Aquifers which were drilled through are difficult to detect and thus electric well logging can be carried out when the hole is filled with mud or water.

The significance of percussion drilling is decreasing at present, owing to the general introduction of the rotary system. In spite of this, percussion drilling rigs are still often operated in certain countries, particularly when shallower holes are drilled, due to the following considerations:

- portable percussion drilling equipment is less expensive than the rotary equipment;

- when drilling in hard rocks to shallow depth, the cable drilling involves comparatively low cost;

- in regions where water is scarce, the supply of water for mud circulation does not cause problems when percussion drilling is applied;

- well construction and connection of aquifers are simpler.

Consequently, the application of cable percussion drilling may be justified when drilling through short sections of disjunctive fissured rocks liable to water losses and when thin layer of crystalline rock lying at the surface are to be drilled through at a rapid rate.

ROTARY DRILLING

Two types of rotary drilling processes are distinguished: the direct and the reverse circulation of mud variants (Fig. 4-62 and Fig. 4-63, respectively).

In the rotary drilling system of direct mud circulation the rock is demolished by bore bits attached to the end of a hollow drill pipe which is rotated by means of a rotary table and by the square kelly of the equipment. The boring tool is suspended by means of the shackle of the rotary swivel from

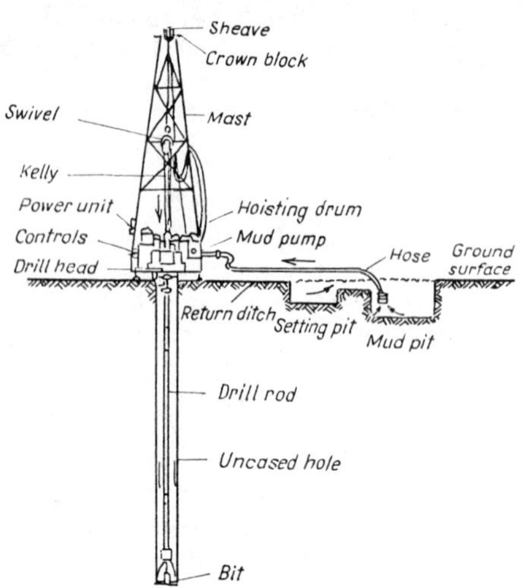

Fig. 4-62. Arrangement scheme of an equipment used for rotary drilling with mud circulation.

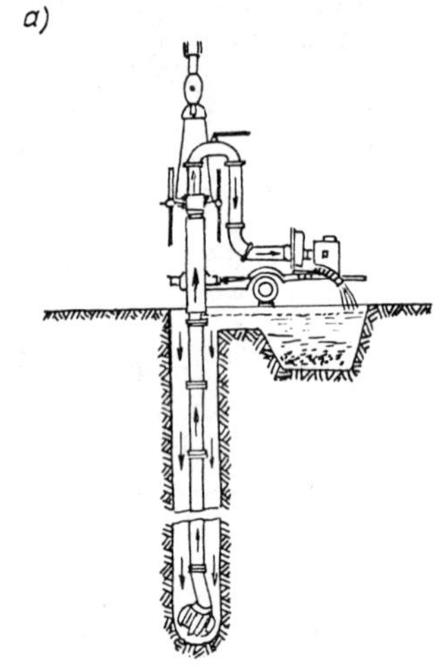

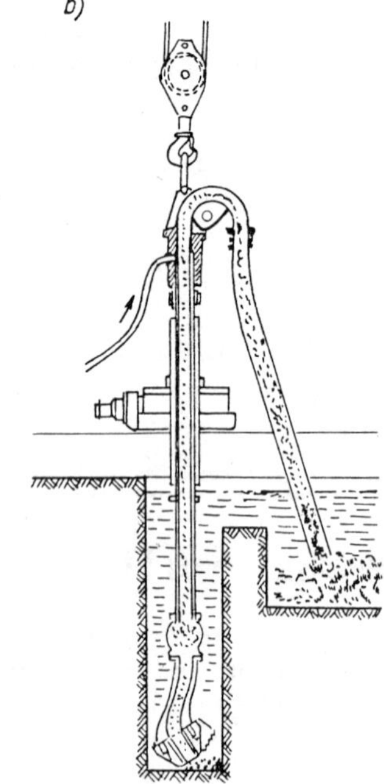

Fig. 4-63. Arrangement of reverse circulation drilling: (a) with centrifugal pump and (b) with air-lift suction.

the drilling cable led over the tower sheave. The other end of the drilling cable is attached and wound over the hoisting drum. The derrick structure of the drilling rig accommodates the crown block and the traveling block, indispensable for both drilling and mounting/dismounting work. The piston mud pump ensures the permanent circulation of drilling mud during drilling. It sucks the mud from a mud pit dug beside the rig and delivers the mud by way of a pressure hose through the rotary swivel and the hollow drill pipes to the bore bit from where it entrains the rock particles (cuttings) dislodged by the bore bits in the annual space between the outside wall of the drill pipe and the inside wall of the borehole into the setting pit which is connected to the mud pit. The hydrostatic pressure of the circulating mud also supports the wall of the borehole and so even longer hole sections can be sunk without casing. When drilling proceeds, the rotary swivel enables the mud circulation, and the square kelly follows the vertical motion of the drill rod. When progress (depending on the length of the kelly) is completed, the drilling tool is lifted from the bottom of the hole and the drill pipes are secured with a pipe clamping wedge, then a new drill pipe is mountained on the kelly, the bore bit again attached and the drilling is continued.

The bore bit types most frequently used with direct circulation system are demonstrated in Fig. 4-64. The winged bits and three-wing bits are chiefly applicable in sedimentary strata when drilling through sand and gravel. Roller bits can be used to drill both sedimentary loose and hard rocks. Long tooth bits are used in sedimentary rocks whereas higher rates of progress are ensured in hard rocks by roller bits with short teeth.

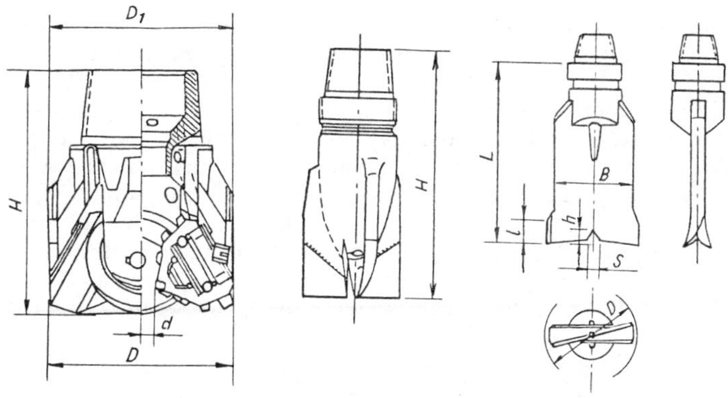

Fig. 4-64. Types of bore bits used in rotary drilling.

The rate of progress depends on many factors, among them the most essential one is the quality of the rock drilled. In soft loose sediments several hundred meters per day can be drilled, but in hard rocks a performance of 10 to 15 meters per day is considered as a good result. The depth of the bore does not influence the rate of progress except when core drilling is performed.

The most considerable difficulties are caused by both porous strata and fissured rocks liable to adsorbing water. When considerable mud loss occurs, the continuous circulation of the mud is interrupted and the cuttings accumulate at the bottom of the hole about the bore bit resulting in the seizure and possibly the breakage of it. Difficulties may also be caused by disjunctive rocks in muddy or loamy strata when the bore bit rotates a piece of rock in its existing position without any progress.

The reverse circulation system of rotary drilling, most widely used in constructing large-diameter (1 to 2 meters) wells, are in principle suction drilling methods classified as:

- suction drilling with centrifugal pumps;

- suction drilling with deep suction head;

- suction drilling with airlift.

This process is distinguished from the "traditional" rotary drilling chiefly by the reverse circulation. The flow laden with cuttings is raised to the surface (represented in Fig. 4-63) through the drill pipe and thus the hydraulic parameters of the ascendant flow can be made independent from the diameter of the borehole. The various types differ from each other only in the means employed for producing circulation.

In suction drilling operating with a centrifugal pump, the circulation is maintained by the effect of a pump of special design since the circulation, full with cuttings, passes directly through the pump (Fig. 4-63a). When operating with deep suction head, the vacuum is produced by water pressed by a centrifugal pump through a nozzle placed in the drill pipe. The vacuum system and the centrifugal pump is substituted by a compressor when operating with airlift (Fig. 4-63b).

Like in the case of the traditional rotary system, the tools of the reverse circulating types include the swivel and the pertinent fittings, the kelly, the drill pipe, the drill collar, and the bore bit. Considering, however, the character of the drilling process and the dimension of the borehole, several differences exist in design, for example the drilling tools are assembled from flanged pipes of larger diameter and when the airlift system is employed, the pipes are also complemented with an air duct. The bore bits used in loose sedimentary rocks are the winged bits and the solid section bits of special shape (called Jumbo bits). For drilling hard rocks, large diameter multi-roller bits are available (Fig. 4-65).

Considerable advantage of the reverse circulating process is that, contrary to the direct circulation system, clayey or specially treated mud is not necessary in the circulation system and that its delivery capacity is very high (e.g., 4 to 9 cu m/minutes), therefore producing a comparatively high rate of flow of cuttings. This accounts for the fact that an average progress of drilling of 1 m per hour can be easily reached in hard dolomite rock interwoven with cracks at a bore diameter of 2 meters.

Factors influencing the drilling work (i.e., the rate of progress) may be divided into groups such as:

- the physical characteristics of the rock to be drilled (compressive strength, abrasion resistance, stress conditions, deformability, water and mud absorption, pore-water pressure, temperature);

- technical factors (the load on the bore bit, speed of rotation, characteristics of drilling mud).

In the following, the technical factors will be dealt with in brief. The rate of progress of the bore bit increases almost in direct proportion to the weight on the borehole bottom. Rates of drilling as a function of the weight on the bit in various rocks are presented in Fig. 4-66. The bottom pressure should be reduced when drilling through the boundary between hard and loose rocks, in creviced and/or collapsible strata, and it should be increased when working in hard, solid rocks. A minimal weight must be applied on the bit otherwise the drilling operation becomes very uneconomical.

With respect to the speed of rotation of the bore bit, two methods of operation can be distinguished:

- drilling with high speed of rotation at low bit weight;

- drilling with reduced speed at increased bit weight.

In water prospecting drilling work, the second method is employed, usually operating in the speed range of 50 to 300 rpm. The ultimate values of bit weights and rotative speeds can be decided on by considering the geological conditions.

Fig. 4-65. Types of bore bits used in suction drilling.

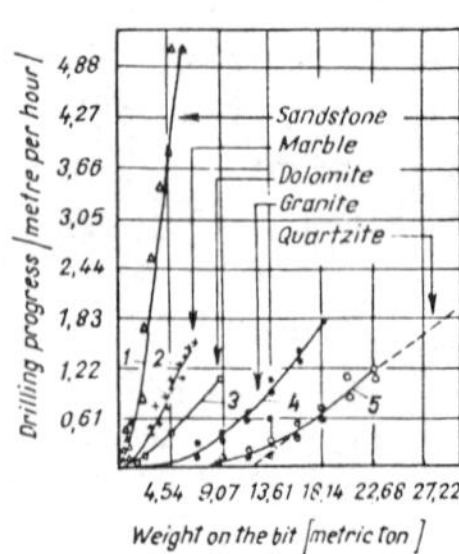

Fig. 4-66. Rate of bore bit progress as a function of rock hardness and bit load.

The circulating mud is usually made of a mixture of water and loam of small sand content. Addition of bentonite increases the swelling ability, therefore, it is recommended only in small quantities to avoid the forming of mud cakes on the wall of the borehole. The ideal density of the drilling mud is 1.1 to 1.2 g/cu cm which may be increased to 1.8 to 2.0 g/cu cm by adding barite or hematite. Along with increasing bore bit weight and rotative speed, the pumping rate of circulation into the bore is also to be increased. This is the only means to ensure the quick disposal of cuttings, otherwise the bottom of the borehole will be filled with cuttings thus limiting the drilling progress.

The modern instruments provide for the continuous checking and recording of the weight on the bottom of the borehole, the speed of rotation, and the quality (specific density, viscosity, sand content, pH value) and pressure of the drilling mud. Using these data, the drilling speed can be reduced to a reference level of weight, of speed of rotation, and of mud conditions. The velocity log constructed from these data provides information about rock properties (Subsection 15-2).

THE PRINCIPLES OF WELL CONSTRUCTION

The technologies applied in well construction are usually distinguished on the basis of whether the well is intended to be built for tapping a known or unknown aquifer, although no sharp dividing line can be drawn between the two technologies since water exploration by means of deep wells is based, to a certain degree, on prospecting work.

As far as the construction technology employed is concerned, well drilling works are classified as follows:

- medium deep wells of up to 500 meters and of initial diameter up to 400 mm;

- deep (thermal) wells the initial diameter of which varies from 14 inches up to 16 inches when drilled to a depth of 2000 m to 3000 m;

- large diameter wells drilled usually up to a depth of 400 meters with an initial diameter ranging from 1 m to 1.5 m.

The technology used for constructing a well intended for exploring a porous aquifer down to a depth of about 300 meters (Fig. 4-67) involves the following steps. A conductor pipe of 10 to 15 meters length is built-in during the first phase of the drilling work. Subsequently a slim hole is sunk to the planned depth using a three-wing or a roller bit of 146 mm diameter. The next stage comprises the well logging and sidewall sampling to study the position and characteristics of the aquifers. The section explored by the slim hole is enlarged to the diameter required by the closing pipe string, then the casing pipe is installed and the work is completed by cementing. According to earlier technological principles, deep wells could only be developed by interconnecting six or seven casing pipes of different diameters. The rotary mud circulating system enables the rapid sinking of long sections, ensures straight holes, and rapid casing work. After installing the closing string of the casing, there follows the enlarging of the hole section to be screened and the installation of the testing screen. After successfully passing the aquifer test, the well can be developed into a producing well. A traditional telescopic tube cutout is made in the final development of the well. The annular space formed where casing joints were released or casings were cut is sealed by stuffing boxes.

As for deep (thermal) wells ranging from 2500 to 3000 meters, drilling is carried out with a similar technology and the casings are perforated at the aquifers (Fig. 4-68).

Both in clastic and hard rocks when drilling large diameter wells, performances meeting economics in every respect can be achieved with both the centrifugal pump suction drilling and the airlift method. The final diameter of drilling is 0.5 m while the initial one ranges from 0.8 to 1.2 m, depending on hydrogeological and drilling problems.

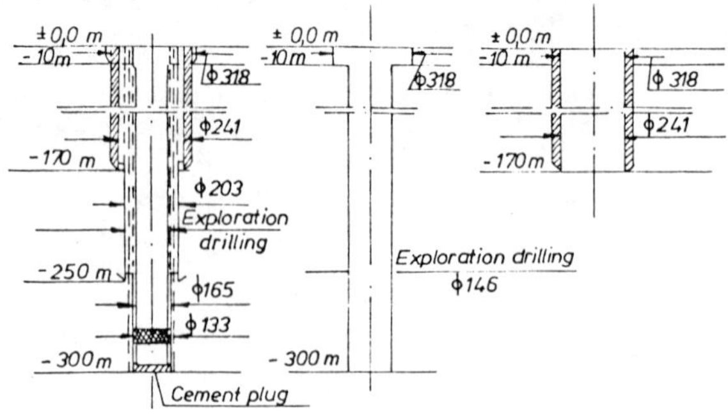

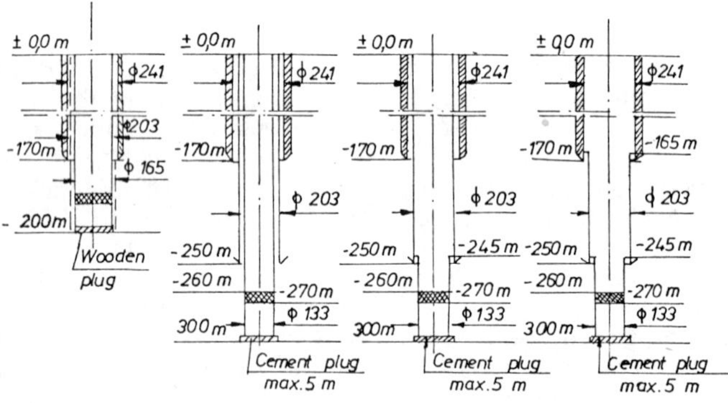

Fig. 4-67. Construction scheme of a water-explorating drilling work planned for a depth of 300 meters.

It has been assumed above that the wells were constructed for tapping a porous aquifer. In hard rock areas, the aquifers may comprise creviced and karstified rocks too. In this respect, first of all limestone and dolomite rocks of various bedding can be taken into consideration. Heavily adhering, packed, and hard sandstone and conglomerate formations call for other technological requirements than do loose sedimentary formations. Slim hole drilling in hard rock is inevitably uneconomical, and thus the borehole is drilled to the final diameter.

Cementing is the only means at present for preventing the motion of water behind the casing thus separating aquifers of very high pressure and for the directionally regulated connection of such aquifers. The cement laitance (also called grout) is pressed into the annular space between the wall of the borehole and the casing using a pressure of 200 to 500 kg/sq cm and a discharge capacity of 1000 to 1500

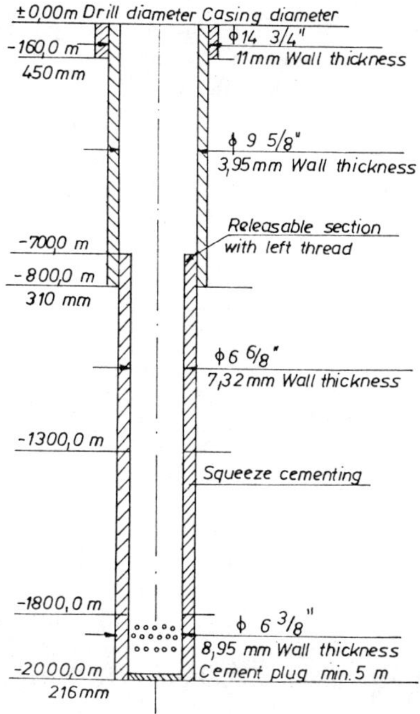

Fig. 4-68. Construction scheme of a deep thermal well.

liters/minute. High-strength Portland cement is generally used, the setting of which may already start two hours after mixing with water. As far as practical, the annular space should not be overdimensioned in order that a suitable high rate of rising can be ensured. It is further advisable to check the diameter of the hole with a caliper log before installing the casing pipe. Technical logs (Subsection 15-2) performed after cementing inform about the position of the cement shell and of the casing in the borehole.

14-4 Design of Well Screens and Filters

Versatile laboratory tests and practical experiments have been carried out in order to develop screen and filter structures ensuring the most efficient drainage of aquifers by wells. The following grouping of the different designs characterize the screens used in practice:

- screens in direct contact with the aquifer such as tubes with perforated or slotted openings, rod-frame structures, and both covered with mesh or spiral-woven wire or without such covers (Fig. 4-69);

- gravel filters covered with screens where the impeding effect occurring in the surrounding of the well and the entrance losses are reduced by replacing the aquifer around the well with an artificial gravel structure of suitable grain size.

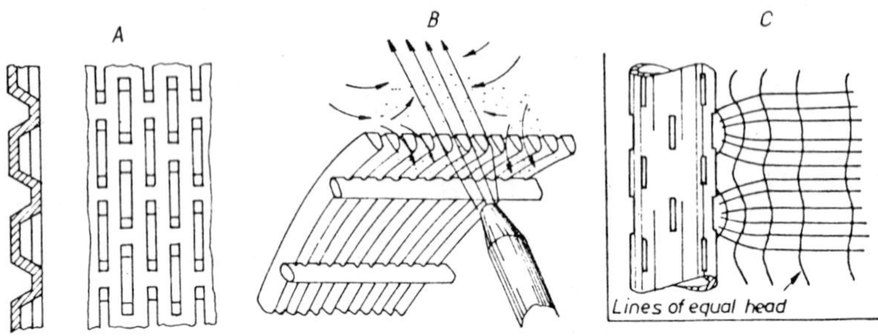

Fig. 4-69. Screen structures: (a) Nold, (b) Johnson, (c) slotted type. The efficiency of decomposing mud by water jetting is shown in the case of Johnson's and the slotted type.

The construction of gravel filters in wells of medium depth may be carried out by methods such as:

- feeding of gravel for building the gravel filter by gravity methods;

- prefabrication of gravel filters before installation.

The problem of filtering may be summarized as follows: the surface where the water leaves the aquifer (i.e., the wall of the well) and becomes a free flow in the well is very often a place where erosion occurs. This is due to the fact that there is a sudden change in the forces and that there are no normal forces acting against the displacement of particles. The water may take the finer particles along and wash them out. This process, the external suffusion is generally allowed and it is even desirable in many cases since washing out fine particles from the surrounding of the well very favorably develops seepage conditions, and thus screens with larger size openings may be used. The suffusion which may be favorable at the beginning should be brought to a stop in its later phase since an increase in the seepage velocity may lead to the destruction of the skeleton formed by coarse particles supporting each other, i.e., to the collapse of the layer around the well. The problem of filtering is reduced, by this reasoning, to the study of suffusion in granular layers.

Concerning the scouring of fine particles, it is necessary at first to decide whether the layer tends to suffusion or the grain-size distribution in the aquifer basically precludes the movement of fine particles, i.e., the layer may be regarded as a stable one (also called self-filtering).

Layers being resistant against suffusion may be defined by comparing the size of pores (d) of the layer with the diameter of the smallest grain (D_m). It is obvious that suffusion is excluded, where $d < D_m$, since even the smallest particle cannot go through the pores of the medium. Taking the minimal pore-diameter (d_m) into consideration, the smallest particle can move in channels wider than d_m, but it will be stopped soon in channels of diameter d_m. It is probable that the medium resists against suffusion. Requirement of higher safety may be proposed for practice using the average pore-diameter (d_o) instead of d_m, formulated as:

$$d_o \leq D_m \quad .$$

Approximating value for d_o may be obtained when simulating the granular media by a bundle of tubes (Kovács, 199) having:

$$d_o = 4\ n\ (n-1)^{-1}\ D_h/\alpha \quad ; \qquad (4\text{-}37)$$

where n and α denote the porosity and the shape coefficient of grains, respectively, and D_h is Kozeny's effective grain diameter (i.e., the harmonic mean diameter of the particles).

A thorough investigation of the problem of suffusion made by Lubotshkov (1965) demonstrated that layers mixed from two media, one of which is coarse-grained and the other is composed of fine particles, may be liable to suffusion although the requirement given above is fulfilled. For demonstration, a mixed layer will be considered which is built of big grains supporting each other, the voids of which are filled by a system of fine particles not liable to suffusion. In such a case, the individual grains in the fine fraction cannot move through the pores, but the assemblage of the fine particles can be washed out through the voids of the big ones. The mixed layer, as a whole, is liable to suffusion. Lubotshkov takes into consideration this possibility in constructing the so-called limiting curves of the grain-size distribution. Layers characterized by grain-size distribution curves lying within the limiting curves of Lubotshkov are not liable to suffusion. Typical Lubotshkov curves are represented in Fig. 4-70.

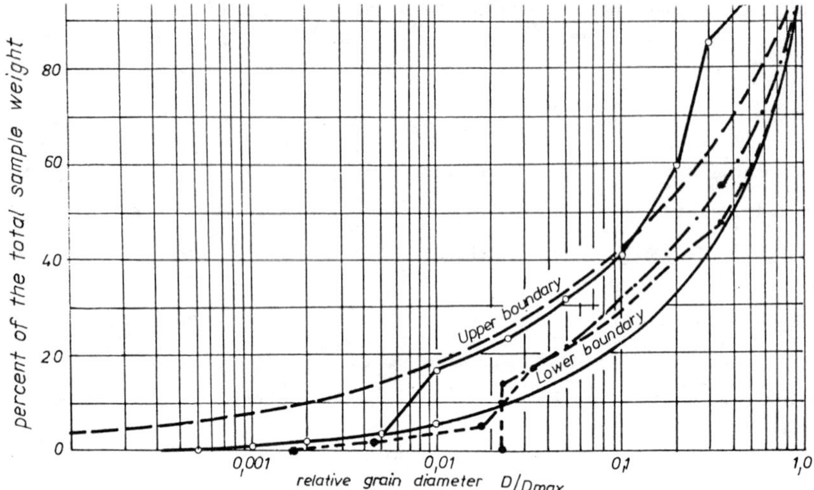

Fig. 4-70. Limiting curves after Lubotshkov showing the upper and lower boundary for soils not liable to suffusion (after Kovács, 1969).

When water is exploited from a layer liable to suffusion, at a certain pumping rate the well starts to produce sand, i.e., at this seepage velocity sand particles are washed from the layer and they move along with water. This phenomenon influences the production of the well by rearranging the structure of the aquifer, therefore, the value of the critical velocity (i.e., the hydraulic condition of suffusion) should be determined.

There are two forces acting on an individual particle (of volume V) being in static condition: its own weight ($V\,\gamma_s$) and the buoyancy ($V\,\gamma_w$) where γ_s and γ_w denote the specific weight of the particle and the water, respectively. When the water starts to flow, there is a third force as well, the seepage force ($V\,I\,\gamma_w$), the direction of which is equal to the direction of the water movement. The particle will start to move (assuming that the direction of seepage is horizontal) when the angle of the resulting force with the vertical is equal to the angle of internal friction of the soil (ϕ), i.e., when;

$$\tan\phi = \frac{V I_{crit} \gamma_w}{V(\gamma_s - \gamma_w)} \quad ; \qquad (4\text{-}38)$$

where I_{crit} denotes the critical value of the gradient producing the seepage. Supposing the validity of Darcy's law, the above equation will give;

$$v_{crit} = K \tan\phi \, (\gamma_s - \gamma_w)/\gamma_w \quad ; \qquad (4\text{-}39)$$

where v_{crit} and K denote the critical seepage velocity and the conductivity, respectively.

Considering now the actual distribution of forces at the well screen, the effect of the screen and that of the rock pressure should be accounted (Fig. 4-71). The continuity of the material of the aquifer at the screened surfaces is substituted by the supporting effect of the screens which means that the forces acting horizontally on the investigated cube from the side of the screen (σ_3) and from the opposite direction (σ_3') may be considered to be equal, therefore, they may be omitted in the study of equilibrium. The suffusion begins when particles move through the openings of the screen, thus the forces should be related to the individual particles. In addition to the forces described above, stresses due to rock pressure (N_1 and N_2) should be considered. The stresses acting vertically (σ_1) and horizontally (σ_2), perpendicularly to the flow direction, compress the particles, thus increasing their resistance against the seepage pressure (P). The force acting in the direction of flow is:

$$P = \pi D^3 I_w/4 \quad ; \qquad (4\text{-}40)$$

and forces perpendicular to this direction are as follows:

$$G = D^3 (\gamma_s - \gamma_w)/6 = \text{the volume weight minus buoyancy} \; ; \qquad (4\text{-}41)$$

$$N_1 = a D^2 h (\gamma_s - \gamma_w) = \text{the vertical rock pressure} \; ; \qquad (4\text{-}42)$$

$$N_2 = a D^2 h (\gamma_s - \gamma_w) \lambda = \text{the horizontal component due to rock pressure} \qquad (4\text{-}43)$$

Suffusion will being (as in the case described above) when;

$$\tan\phi = P/(G+2 N_1+2 N_2) \quad ; \qquad (4\text{-}44)$$

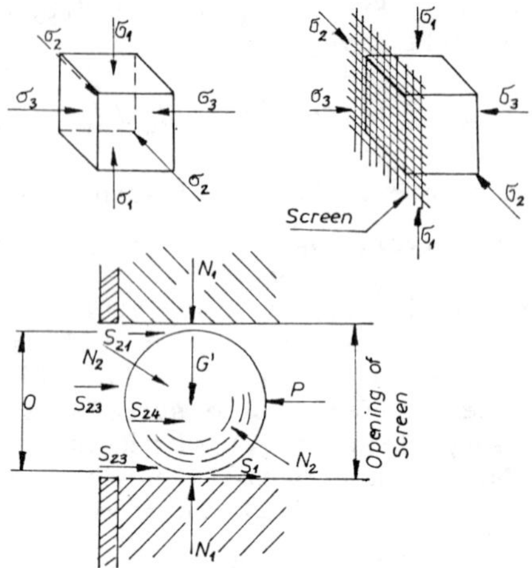

Fig. 4-71. Allowance for rock stresses in stability analyses in the case of well filters (after Kovács, 1969).

from which a relation for the critical seepage velocity may be obtained as follows;

$$v_{crit} = K \tan\phi \frac{\gamma_s - \gamma_w}{\gamma_w} \left[\frac{4}{6} + \frac{8a}{\pi} \frac{h}{D} (1 + \lambda)\right] \quad ; \qquad (4\text{-}45)$$

where h denotes the depth and $\lambda = \tan(45° - \phi/2)$. There is an uncertainty in calculating both the cross section of the column, which transfers the rock pressure to the individual particle, and the effective porosity which decreases the vertical pressure. This uncertainty is expressed mainly by the factor a in Eq. 4-45, where; $a = 1 - n$ when spheres in a cube-packing are considered, and which decreases considerably where the particles form vaulted structures (Kovács, 1969). The formulas derived for the critical seepage velocity will lead to the conclusion that it may be determined as a function of both hydraulic conductivity and friction angle of the aquifer containing the moving particles.

Approximate relations may be found in the literature to calculate the numerical value of the critical seepage velocity, such as:

$$v_{crit} = K/15 \quad ; \qquad \text{(Sichardt, 1952)} \qquad (4\text{-}46)$$

$$v_{crit} = K/31 \quad ; \qquad \text{(Abramov, 1952)} \qquad (4\text{-}47)$$

where both v_{crit} and K are given in m/s units.

In designing the well screens, i.e., tubes with different kinds of openings (Fig. 4-69), the requirement to be considered is that the screen should retain the particles of the aquifer and it should collect the water infiltrating through the aquifer. This requirement will be fulfilled if the openings of the screen are wide enough such that the water can wash out the fine particles, but they are not too wide, thus preserving the skeleton of coarse particles. Recommendations for this purpose are given in the literature comparing the size of the openings of the screen to the diameter of a grain type suitably chosen such as (U.S. Army, 1955):

$$D_{85}/\text{width of slots} > 1.2 \qquad \text{(slotted screen)}$$

$$D_{85}/\text{diameter of performation} > 1.0 \qquad \text{(perforated screen)}$$

It is obvious that this kind of screening meets the requirement when the aquifer is self-filtering or not liable to suffusion after having washed out the fine fraction.

Design of gravel filters. The simplest way to protect aquifers against suffusion, even when the conditions delineated above are not fulfilled, consists of putting a layer of coarse sand between the screen and the aquifer. This construction, the filtering layer together with the screen (called the gravel filter), is used successfully in the practical work. The grain-size distribution of the filtering layer has to fulfill requirements such as:

- the filter must be stable, i.e., resistant against suffusion (condition of stability);

- the voids of the filtering layer cannot be much larger than the finest particles of the aquifer to be protected, otherwise these fines will be washed into the voids of the filter causing the plugging of pores (geometrical condition);

- the filter should have appropriate conductivity, possibly better than that of the aquifer to be protected (hydraulic condition).

According to the first condition, granular media may be used as filtering material which are either self-filtering or where the seepage velocity due to the maximum pumping rate falls short of the critical seepage velocity. The second and third conditions will be fulfilled at the same time if the grain size distribution of the filtering layer obeys the so-called filter law. Among the various filter laws, Terzaghi's equation (1943) should be mentioned here, illustrated in Fig. 4-72. The grain-size distribution curves of the aquifers to be protected lie between the limits of the domain A. Diameter of particles belonging to the percentages S = 15 and S = 85 are determined (such as D^a_{15} and D^a_{85}, respectively) as characteristic quantities of the aquifer. Any kind of granular material may serve as a filter for the material of grain-size distribution curve in the field A in which the diameter of particles belonging to S = 15 percent (D^f_{15}) lies between 4 D^a_{15} and 4 D^a_{85} or $D^f_{15}/D^a_{85} < 4 < D^f_{15}/D^a_{15}$. The field of the grain-size distribution curves fulfilling Terzaghi's condition has been designated by B in Fig. 4-72. The left-hand side of the Terzaghi unequality represents the geometrical condition, its right-hand side determines the hydraulic condition in such a way that the diameter D^f_{15} in the filtering material should surpass that in the aquifer (i.e., the diameter D^a_{15}) at least four times, implementing higher hydraulic conductivity (approximately 16 times) in the filtering material than in the aquifer.

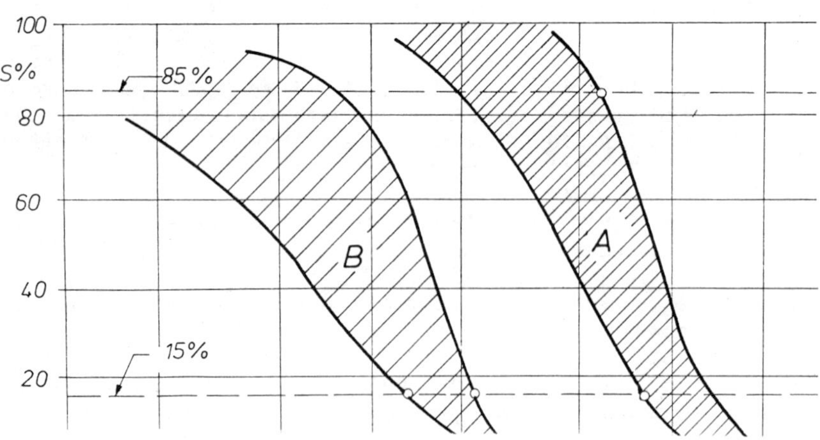

Fig. 4-72. Terzaghi's filter law.

Designing the filters used in water wells, the geometry of the screen (diameter and length) plays a very important role since it must be considered at the drilling work too.

The minimal diameter of the screen may be determined when the planned specific yield (discharge related to the unit length of the screen) and the grain-size distribution in the aquifer is considered. The discharge velocity at the screen (v) can be calculated simply by dividing the specific yield (q) with the surface of the screen of unit length:

$$v = \frac{q}{2\pi r} \quad ; \qquad (4\text{-}48)$$

where r denotes the radius of the screen. The calculated value will give a good average where the hydrostatic stresses at the top and at the bottom of the aquifer do not differ considerably and where the inflow across the screen may be considered as homogeneous. It is obvious that the maximal value of the velocity should be less than or equal to the critical seepage velocity (v_{crit}) to avoid suffusion. Considering also a safety factor, c_s, the above condition may be expressed as:

$$v\, c_s \leq v_{crit} \quad \text{or} \quad r \geq \frac{qc_s}{2\pi v_{crit}} \qquad (4\text{-}49)$$

The critical velocity may relate to the aquifer where it is self-filtering, or it may relate to the gravel filter protecting the aquifer against suffusion. In the latter case, coarse gravel or sand will be selected which fulfill the stability requirements (e.g., Terzaghi's unequality).

An estimation for the minimal diameter (or that for the maximal velocity) may be obtained by a simpler way, accepting two obvious suppositions such as:

- the angles of friction are equal both in the aquifer and in the filtering material involving the proportionality between the conductivities (K) and critical velocities of the aquifer and of the filter, respectively;

- the diameter D^a_{85} of grains may be considered as a minimal diameter (in respect to the basic equation $d_o \leq D_{min}$), since these particles together with the somewhat smaller ones may retain the whole fine fraction when stopped (Cedergran, 1968).

Knowing the proportionality between the hydraulic conductivity (K) and the effective grain-diameter (D_h), the unequality may be transformed on the basis of the first supposition, as

$$c_s \frac{v_{max}}{v^a_{crit}} = \frac{K^f}{K^a} = \frac{(D^f_h)^2}{(D^a_h)^2} \qquad (4\text{-}50)$$

The second supposition results for the filtering material when it is simulated by a bundle of tubes (as it was made at the beginning of this chapter), with the following formula:

$$4 \frac{n}{1 - n} \frac{D^f_h}{\alpha} \leq D^a_{85} \qquad (4\text{-}51)$$

These equations give an estimation for the filter design in a summarized form, applicable for the determination of the minimal diameter of the screen, whether the filter is prepared by feeding-in method (Fig. 4-73) or they are pre-fabricated elements (Fig. 4-74). It is obvious that in case of cemented (bonded) filters, the consideration of stability criteria may be left out.

It may be concluded from the above discussion that the application of up-to-date filter design methods cannot dispense with the examination of the grain-size distribution of the aquifer based on a preliminary sieve analysis. Due to the specific characteristics of the rotary mud circulation drilling method, test specimens (samples) complying with the requirements cannot be furnished from cuttings delivered to the surface by mud circulation since they are graded and mixed with mud contaminants. The gunpowder method and the use of spring-operated sidewall samplers have been introduced into practical well construction as a result of the consideration outlined above. At present, only the sidewall sampler (Fig. 4-75) is employed since it ensures the taking of larger samples with more success from loose formations than does the gunpowder method.

Being acquainted with the method of estimating the diameter of the screen, the length of the screen (m) will be determined as:

$$m = Q/q \quad ; \qquad (4\text{-}52)$$

where q and Q denote the specific and total yield of the well, assuming homogeneity of inflow along the screen.

Discharge measurements performed at various depths over the screened length of operating wells have shown, however,

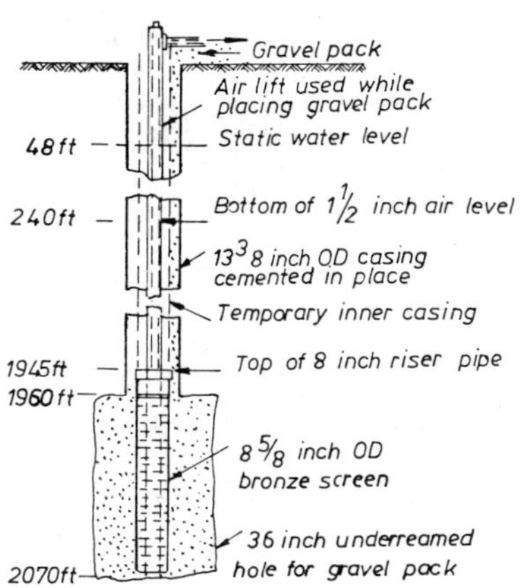

Fig. 4-73. Locating a gravel pack filter by gravity with simultaneous compressor water lifting.

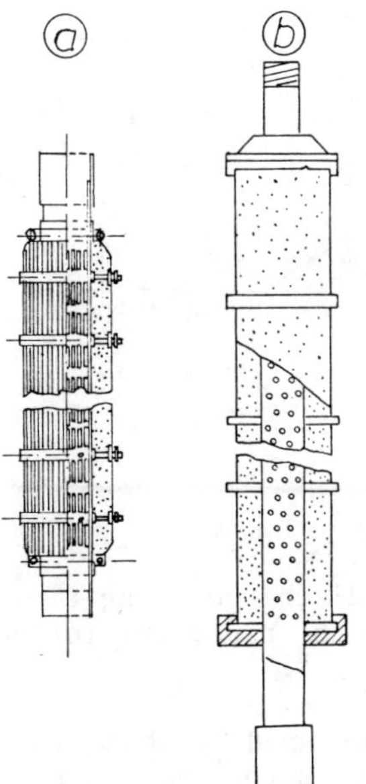

Fig. 4-74. Filter structures. (a) Cased gravel filter. The perforated or slotted filter frame is covered with a square woven wide-mesh filter cloth, and the annular space between the web so formed and the filter tube is filled with gravel of proper grain-size distribution. (b) Block filter is made of blocks of gravel of given sizes and the pre-fabricated elements are placed at the filter frame. When materials such as cement, rubber, plastics are used for bonding the grains of gravel, bonded gravel filter structures will be constructed.

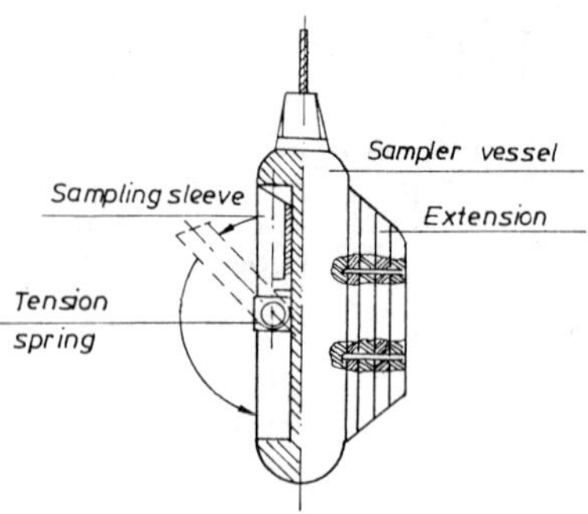

Fig. 4-75. Mechanical sidewall sampling device.

that the inflow is unevenly distributed over the length of the screen (Fig. 4-76); the upper part of the filter takes part in the production of water to a large extent compared to its lower part. The determination of the length of the filter involves, in this case, the estimation of the productive length, i.e., the determination of the rate of discharge produced by definite intervals along the length of filter.

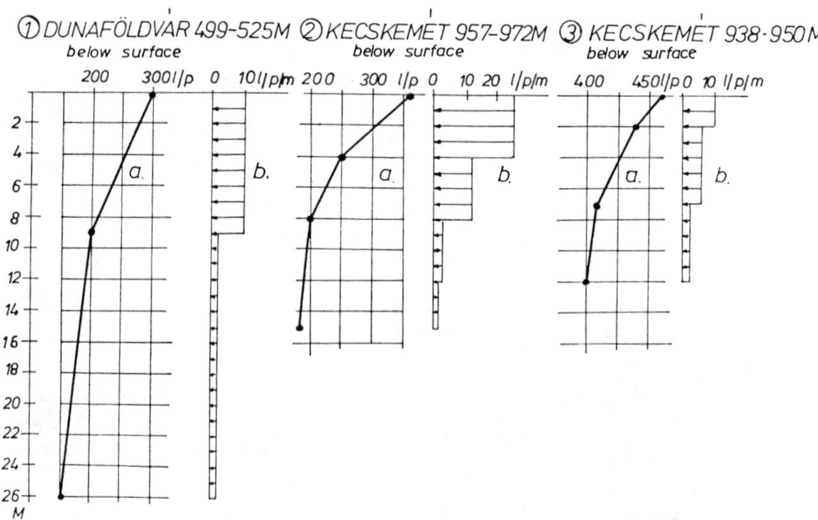

Fig. 4-76. Distribution of discharge along the screen. (a) discharge in the pipe, (b) inflow across the screen.

The reason for the phenomenon described above will be found when perceiving that energy is consumed in conveying the discharge from the aquifer to the well-bore. The head losses characterizing the energy consumption may be calculated according to Kovács (1972) by the following method (Fig. 4-77):

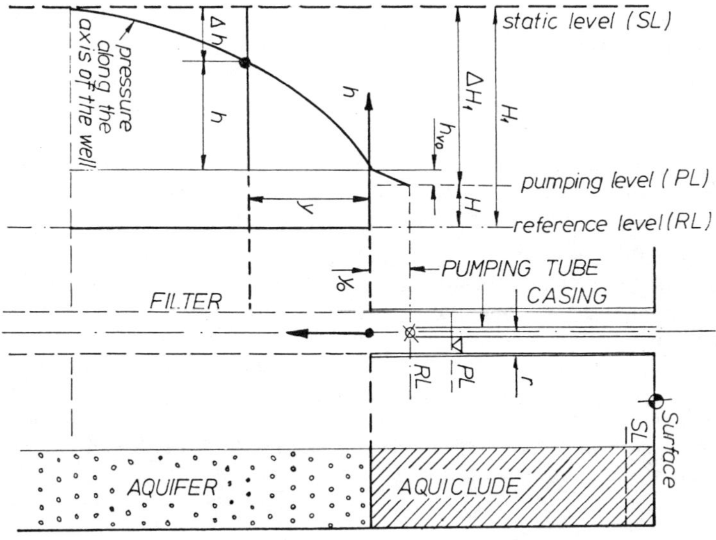

Fig. 4-77. Diagram showing the pressure conditions along the screen.

When the well-known Dupuit equation is used for the description of the hydraulic condition around the well, it is tacitly assumed that the total difference in pressure, i.e.;

$$H_T = H_2 - H_1 = Q/Bm \quad ; \qquad (4\text{-}53)$$

is used to convey the water through the aquifer. Q, B, and m denote here the yield of the well, the resistivity factor, and the thickness of the aquifer, respectively. The validity of this assumption is limited in practice especially if long casing and screen are used in water wells. One part of the total pressure difference (h_{vo}) will be lost due to the resistivity of plain casing and the other part of it (h) along the screen owing to the reason mentioned above.

For the plain casing, Chézy's formula may be applied, as;

$$h_{vo} = A_o \, Q_o^2 \, y_o \quad ; \qquad (4\text{-}54)$$

where Q_o and A_o denote the constant yield and the resistivity factor in the casing, respectively.

Over the length of the screen, the phenomenon is more difficult to analytically describe. An obvious assumption is that Chézy's formula may be applied only for infinitesimal intervals along the screen, as;

$$dh = A\, Q^2(y)\, dy \quad ; \tag{4-55}$$

where Q denotes the inflow rate at the distance y measured from the upper edge of the filter. The resistivity factor, A, used here, drastically differs from that in a smooth pipe (A_o) due to the fact that the flow across the screen is perpendicular to the mean direction of water flow in the well. The development of a boundary layer along the screen is thereby prevented and radial flow tends to make the movement of water appreciably more turbulent than it would be in the case of plain casing. The pressure loss (h) between the upper edge of the screen and a cross section lying at a distance y measured from it may be obtained by integrating the latter differential equation, as;

$$h = \int_o^y A\, Q^2(y)\, dy \quad . \tag{4-56}$$

The seepage across the screen will be supported, then, by the remaining pressure difference; $\Delta h = \Delta H_T - h_{vo} - h$. Applying Dupuit's equation for this case;

$$dQ = B\, \Delta h\, dy \quad ;$$

or using its first derivative, as;

$$d^2Q/dy^2 = -B\, d(\Delta H_T - h_{vo} - h)/dy = B\, dh/dy \tag{4-57}$$

Assuming an infinite thickness of the layer and that of the screen, and going downward along it, a section may be reached where the sum of resistances acting along the overlying screen section becomes large enough to balance the full depression, i.e., no inflow is likely to occur underneath. Equation 4-57, applied for this case together with the modified Chézy formula, will result in the following differential equation:

$$d^2Q/dy^2 = A\, B\, Q^2 \quad . \tag{4-58}$$

Using proper boundary conditions, the solution of this differential equation, i.e., the discharge, Q, as a function of the distance y measured along the screen from its upper edge, will be expressed, as:

$$Q = Q_o \ (1 + \sqrt{Q_o} \ \sqrt{A \ B/6} \ y)^{-2} = Q_o \ (1 + y/L)^{-2} \quad . \tag{4-59}$$

The discharge velocity (v) at the screen will be obtained when calculating, at first, the specific yield, $q = dQ/dy$, then dividing it with the perimeter of the screen:

$$v = \frac{1}{2 \ r \ \pi} \frac{dQ}{dy} = \frac{1}{2 \ r \ \pi} \frac{Q_o}{L} \ (1 + y/L)^{-3} \quad . \tag{4-60}$$

The maximal value of the velocity, v, to be used in the design of filters with respect to the hydraulic condition will be found at $y = 0$, i.e., at the upper edge of the filter, as:

$$v_{max} = \frac{1}{2 \ r \ \pi} \frac{Q_o}{L} \quad . \tag{4-61}$$

The physical meaning of the quantity L, introduced above, may be obtained from this formula: L would be the length of a filter along which the discharge is homogeneous with constant seepage velocity of value v_{max}.

Using Eq. 4-59, the percent change (p) of Q can be calculated along an interval y_p, such as:

$$p = \frac{Q_o - Q}{Q_o} = 1 - (1 + y_p/L)^{-2} \quad . \tag{4-62}$$

Expressing y_p as:

$$y_p = L \ (\sqrt{\frac{1}{1-p}} - 1) \quad ;$$

an estimation will be obtained for the productive length of the screen since y_p denotes (according to the above definition) that upper part of the filter along which the part of the discharge in a given percentage of the whole yield will be produced.

The formulas with respect to yield, discharge velocity, and productive length of filters are given for filters of infinite length. It seems, however, that for estimation, the same reasoning may also be applied in the case of long filter tubes.

14-5 Aquifer Cleaning

Even if the structures of filters to be inserted into wells are design satisfactorily, they will be efficient if the structure of the aquifer, which served as a basis for the filter design, has not been changed during drilling, or if so, the original structure will be restored by cleaning the aquifer. When deep wells are constructed, the removing of the mud and the flushing-out of the well by pure water is followed by the aquifer cleaning process.

Many different methods of this work are known which can be divided into mechanical and chemical aquifer cleaning procedures. Of the mechanical processes, the frequently used types are the following:

- bailing and swabbing;

- screen jetting;

- compressor operation.

By the bailing and swabbing method, a water column of some larger thickness is removed by using a bailer or by mowing a swab in the casing pipe. The wide pressure difference occurring when a certain water column is abruptly removed and the vacuum effect results in breaking down the mud cake.

The principle of screen jetting is that penetrated zones are broken down and a small part of the infiltrated mud is also removed .by the impulses of water streaming out from the openings of the jetting tool designed as shown in Fig. 4-78, and moved in any desired section of the screen by also utilizing the pressure drop produced by the contraction of the flow cross section (Fig. 4-69). It often occurs that some sections around the screen become clogged particularly in fine, micaceous sand and so the entrance surface may get reduced considerably. Clogging should be released by reverse water circulation produced preferably by the jetting method. Jetting is in fact necessary in clogged sections. The determination of such sections requires fluid movment logging (Subsection 14-2), otherwise the jetting must be continued for times longer than necessary and should cover the full length of the screen. Particles loosened and broken up by the jetting tool are removed by compressor water lifting. Screen clogging produced during compressor operation can be loosened by putting the jetting tool into operation again.

Water lifting with compressor (air lift operation) is illustrated in Fig. 4-79. This structure incorporates a production pipe and an air duct built in parallel to the former. The bottom ends of the pipe and air duct are

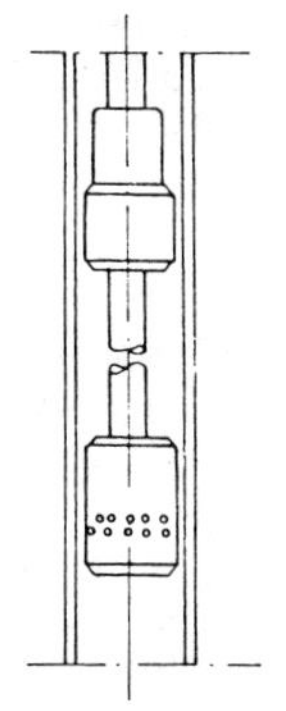

Fig. 4-78. A type of jetting tool used in aquifer cleaning.

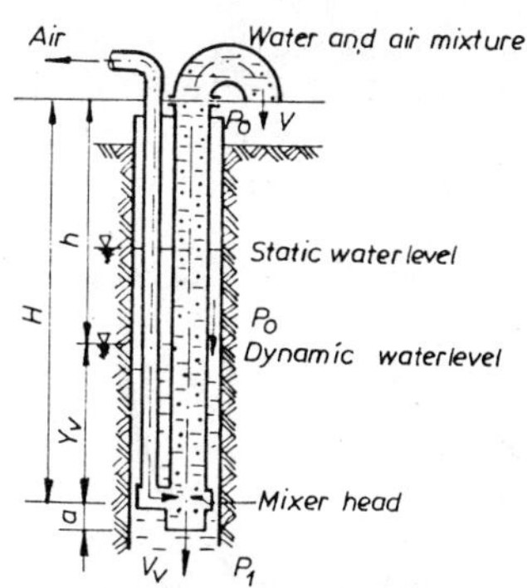

Fig. 4-79. Borsig's arrangement for water lifting with compressor.

connected deep under the water level to a common terminating member, the so-called mixing head. The compressed air fed into the air duct is mixed with water in the mixing head forming bubbles, and furnishes the power necessary for lifting the water by its inherent potential energy. The mixture thus formed, having specific density lower than that of the water, rises in the production pipe. Other systems based on the same principle are illustrated in Fig. 4-80.

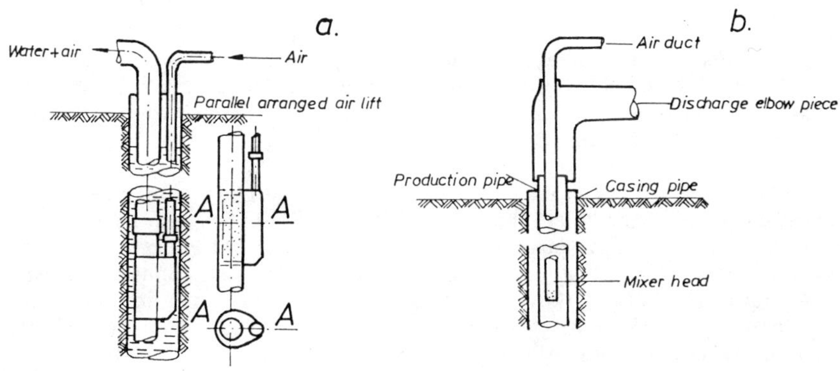

Fig. 4-80. Arrangement of air-water lifting systems (a) in parallel installation and (b) for aquifer cleaning.

Using the fitting shown in Fig. 4-80, the mixer head of the air duct is installed to an extent corresponding to the pressure limit of the compressor so as to provide the highest possible pressure drop in the well to ensure efficient mud cake dislodging. The pressure drop produced in the well is one of the factors playing part in dislodging the mud. Other favorable effect of compressor operation is the non-uniform, intermittent water delivery or the pressure fluctuation.

In the joint application of screen jetting and compressor operation, two openings are used for installation purposes in the compressor fitting as shown in Fig. 4-81. One of these openings is used for installing the compressor air duct to the required depth while the jetting tool is introduced through the other opening. The favorable effect of screen jetting applied jointly with compressor operations provides that mud cake particles broken up by the jetting tool can be removed quickly and efficiently with the simultaneous discharging of the aquifer.

The chemical processes are used where the mechanical methods dealt with in the foregoing are not efficient enough or under special technical circumstances (e.g., wells with artificially deposited gravel packs). The main methods applied for chemical aquifer cleaning are as follows:

- chemical decomposition of the mud cake mixing hydrofluoric acid or fluorate in the fluid circulation;

- replacing the argillaceous drilling mud with a special starch-containing circulation liquid. The watertight starch film formed on the wall of the borehole can be dislodged by some weak flushing;

- chemically treating the clayey, bentonite mud normally used in drilling work using hydrochloric acid to provide mud decomposition to take place after installing the screen.

If the water yield of the well meets, during the cleaning pumping, both the qualitative and quantitative requirements, the so-called test pumping may follow. In this phase, the objective is to get acquainted with the recharge conditions of the aquifer. It must be determined whether the aquifer is capable of permanently supplying the quantity of water that can be produced hydraulically from the well.

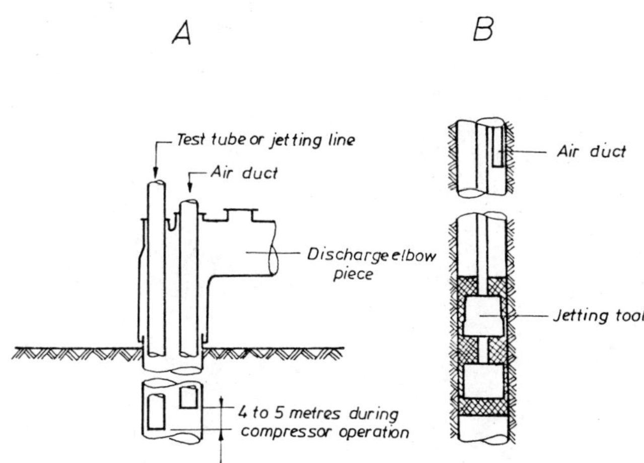

Fig. 4-81. Installation scheme of a compressor air lift system for cleaning pumping (a) with the jetting tool (b) to be attached.

SECTION 15

ANALYSIS OF HYDROLOGICAL PROCESSES INFLUENCING THE REGIME OF DEEP GROUNDWATER

The basic principle used in this analysis is the theory of transport phenomena expressed by Onsager (1967) at the first time in a universal form. According to this theory, the physical quantities may be divided into two groups such as extensive quantities and intensive ones. The extensive quantities are changing in proportion to the dimensions of the physical system to which they relate. For instance, the mass is considered as an extensive quantity since the mass M_1 of a homogeneous material within volume V_1 increases to a mass $M_2 = p\ M_1$ when its volume increases to $V_2 = p\ V_1$ or $M_1/M_2 = V_1/V_2$. The intensive quantities do not change when the dimensions of the physical system are increased or decreased. For example, the hydrostatic pressure in a mass of water is the same whether a smaller or greater volume is taken into consideration.

The process of equalization between intensive quantities will always be carried out by the transport of extensive quantities. For example, when the pressures between two points are equalized, a mass of fluid is transported from the place of greater pressure to the place of smaller one.

In order to obtain a mathematical formulation of the theory of transport, transport equations should be constructed describing the quantitative relation existing among the extensive and intensive quantities (e.g., Darcy's equation of fluid transport, Fourier's equation of conductive heat transport). The parameters involved in the transport equations are only very slightly dependent on the transport process, they may be considered, therefore, as invariable quantities in view of the process, characterizing the material in which the transport takes place (e.g., hydraulic conductivity (K) or heat conductivity (λ)).

The rigorous and complete description of the hydraulic transport system of equations which should be solved simultaneously, are:

- mass (fluid) transport;

- heat transport (conductive and convective);

- chemical transport.

The solution of this system of equations would be very difficult and over-complicated for practical purposes. In order to attain a practical mathematical formulation, the chemical transport equations will be omitted generally from this group of equations, and considered as a qualitative description of phenomena supporting or denying the result obtained by the simultaneous solution of the equations of the mass and heat transport. Even the latter problem is complicated and is solved numerically by the use of computer programs (Lasseter, 1975). The computer program called SHAFT (simultaneous heat and fluid transport) solves the equations of mass and energy transport describing their simultaneous transport in a case of one- or two-phase fluid moving in a porous media for transient or steady-state conditions in one, two, or three dimensions.

There are cases where the flow system taking part in deep-lying clastic sediments may be properly approximated by simple physical models, rendering the analytical solution also possible.

Extensive and intensive quantities as well as transport equations playing part in the description and analysis of hydrological processes occurring in deep-lying clastic sediments are summarized in Table 4-10.

An important consequence of the transport theory is illustrated by Table 4-10, viz., the form of the transport equations is always the same irrespective of their physical content. Based on this characteristic, the equations of the transport of heat will be used in the following discussion without any verification since the foundation and analysis of the fluid transport equation were given already in Section 11-3.

The objective of the present study is limited to the deep part of the subsurface section of the hydrological cycle (Section 13-1). There are reasons, of a methodological nature, that the complete flow system developed in porous aquifers should be divided into parts when investigated, but the continuity of the fluid transport should be kept in mind. This property of transport will be preserved in the following consideration by ways such as:

- the validity of transport equations governing the flow of deep groundwater should be extended over the entire subsurface section of the hydrological cycle (e.g., in the case of the thermal delineation of recharge and discharge areas);

Table 4-10. Basic relations applied for the description of various transport phenomena.

	Fluid Transport	Heat Transport	DC Electric Transport
Basic equation	Darcy's equation: $q = K \text{ grad } H$	Fourier's equation: $f = \lambda \text{ grad } T$	Ohm's law: $i = (1/\rho) \text{ grad } V$
Continuity equation	$\text{div } q = S\, \partial H/\partial t$	$\text{div } f = (\gamma/\lambda)\partial T/\partial t$	-
Steady state condition	$H = 0$	$T = 0$	$V = 0$
External quantities	seepage velocity: q	heat flow: f	current density: i
Internal quantities	pressure, well head: H	temperature: T	potential: V
Parameters	Hydr. conductivity: K Storage coeff.: S Transmissivity: $K\,h$	heat conductivity: λ heat capacity: -	specific resistivity: o - transverse resistivity: o h
	Simultaneous fluid and heat transport		
Basic equation	$f = -q\, \gamma\, T + \lambda \text{ grad } T$		
Continuity equation	$-\gamma \text{ div } qT + \text{div } (\lambda \text{ grad } T) = \gamma\, \partial T/\partial t$		
Parameter	heat capacity $\gamma = n\, \gamma_{fluid} + (1-n)\, \gamma_{solid}$		

- continuity should be substituted by suitable boundary conditions between the separated parts of the system, when only one section of the cycle is investigated separately.

The detailed analysis of the flow system should involve the determination and physical characterization of the recharging and discharging areas as well as that of the water transporting stretches.

The same problems faced in describing the behavior of shallow groundwater were discussed in Chapter 3 of this study. It was demonstrated that a suitable method can be obtained by using:

- the fluid transport equation as the main quantitative relation, supported by

- the chemical (dissolved salts, isotopic content) transport, considered qualitatively, and

- the heat transport, as a mainly qualitative tool.

It is obvious that the same system of transport studies should be applied to investigate the deep part of the subsurface hydrological cycle but with some deviation due to the differences in conditions when deep aquifers are analyzed.

The shallow groundwater systems have easy access by wells, and the pressure data collected by this way are reliable ones in respect to the description of the flow field. On the contrary, in the case of deep aquifers, the construction of observation wells is very costly and from the pressure data collected in wells generally under operation, the reconstruction of the undisturbed pressure field is a successful process only exceptionally (Subsection 16-1).

This is the reason why other quantities than those involved in the mass transport must be used widely to characterize the fluid transport. The practical experience, according to which collection of thermal data is an easy process and the thermal data collected are of good reliability, suggests the use of the heat transport equations in quantitative form as an aid to the determination of fluid movement.

The method to be applied for the investigation of the deep groundwater will be, therefore, the analysis of three kinds of transport phenomena:

- mass (fluid) transport;
- heat transport;
- chemical transport.

The mass and heat transport will be studied quantitatively using the transport equations (Table 4-10) in more or less simplified form, the chemical transport (dissolved salts in and isotopic content of groundwater) will be considered in its qualitative relations only.

Owing to the essential identity of the system of investigation applied in the study of both the shallow branch of the groundwater and its deep branch, it is needless to discuss the methods already mentioned in connection with the hydrology of shallow groundwater (Part III). Here, the results obtained there will be summarized and the differences in the use of methods, if there exists any, will be emphasized.

Considering the fluid transport, the joint use of pressure data and the values of hydraulic conductivity of aquifers has been proved as a successful method for the characterization of the flow field in the case of deep clastic aquifers (Subsection 13-1). The pressure conditions existing in deep sedimentary basins will be discussed, therefore, in Subsection 15-1 to provide a basis for characterizing the recharge and discharge areas by the use of data both of fluid pressure and that of the earth's temperature distribution. The scarcity of pumping tests and, as its consequence, the small number of reliable data of hydraulic conductivity and transmissivity related to deep aquifers was already mentioned. As an aid, the analogies existing between the fluid and electric transport will be used: the hydraulic conductivity and transmissivity data will be supported by the corresponding electric quantities (i.e., by the specific electric conductivity and transversal resistivity, respectively) calculated basically from well logs (Subsection 15-1).

The chemical composition and the environmental isotope content of groundwater are used to delineate the flow path and to estimate the age of the groundwater. The applicability of the chemical and isotopic methods are not limited to the shallow groundwater, although they were described in this relation (Section 9-3). Chebotarev's relation is valid for the entire subsurface water cycle and the ^{14}C method can be used for the age determination in the same way as it was done in the case of shallow groundwater. Some additional data related to the thermal history of deep groundwater may be obtained considering the quartz solubility and the Na, K, Ca ratios (Subsection 15-2).

The problem of heat transport will be taken into account to an ever increasing degree all around the world. One of the reasons was mentioned already, viz., the role of the study of heat transport as an aid in the investigation of the flow field. This kind of study takes an important part especially in the regional hydrological investigation of sedimentary basins where the increasing exploitation of groundwater due to industrial or agricultural demand renders the collection of pressure and yield data rather difficult.

The other reason, of no less importance, is the research for the utilization of new energy sources. It has been proved that the heat energy of thermal waters can be used economically not only in "hot" areas (where the thermal energy is exploited in the form of steam), but in the "cool" regions too where groundwater of comparatively low temperature (60-80°C) can be exploited by wells (areas of low enthalpy). The "hot spots" existing only in a few places of the earth are confined to comparatively small areas, but sedimentary basins of low enthalpy having enormous reserves of heat energy can be found everywhere (e.g., the Hungarian Basin, the Valley of Po). The planning of the reasonable exploitation of the heat energy necessitates the careful study of heat transport within the upper part of the lithosphere, and the flow of groundwater, the latter being the natural carrier of the thermal energy.

Any effect of the heat transport is manifested in the earth's thermal field. The constructon of geothermal maps means, therefore, the first step in the thermal investigation. The analysis of these maps will give information about the nature of the thermal anomalies superimposed on the regional thermal field. It may be that they are caused by variations in the thermal properties of the water-bearing rocks or by the heating (cooling) effect of the water flowing in the rocks. The interpretation of the thermal maps, namely, the delineation of recharge and discharge areas, and the determination of the flow paths can be carried out more easily in the case of deep-lying aquifers than in the case of the shallow ones, due to conditions such as:

- the disturbing thermal effects (thermal background noise) caused by insolation and by the unequal distribution of the heat conductivity are decreased in the regime of the deep groundwater flow compared to the shallow one;

- the deep-lying aquifers can be well approximated by simple and easy-to-interpret thermal models.

It should be emphasized that the determination of the flow system cannot be done even by a very thorough analysis of the thermal field. Hydrological data must always be used for the characterization of the flow system, except for a few, indicating the advantage of the thermal studies.

There are but a few general rules to be applied in the interpretation of geothermal problems. The well-known diversity of the hydrological problems requires the use of "ad hoc" methods that allows the simplification of the interpretation at the same time. A more or less general method and its application to the study of a deep groundwater circulating in clastic sediments will be discussed in Subsection 15-3. Thermal investigations applied to the study of fissured rock aquifers will be presented in Subsection 18-4.

15-1 Investigation of the Pressure Condition in Sedimentary Basins

The geological structure constructed from pervious and impervious layers is a closed system in which the flow cannot be observed or visualized directly. The velocity of seepage is relatively very small, its measurement, therefore, is almost impossible by the methods generally used for such investigations. In spite of the small velocity, the quantity of water transported by this flow is quite considerable, partly because of the very large area of the cross section perpendicular to the main direction of seepage (i.e., the entire horizontal extension of the layer) and partly because the process acts over geological ages. It is necessary, therefore, to choose a method for indicating the water transport between layers by which the effect of seepage summed over a long period can be measured.

There are parameters (e.g., pressure) integrating in time all previous effects caused by water exchange and thus giving information on probable processes taking place prior to the investigation. Among the various methods using such kind of parameters for the hydraulic investigations, the evaluation of the vertical pressure distribution will be discussed here.

After a coarse-grained sediment is deposited in water, its grains contact each other and the weight of the overlying ones is transferred downwards, grain by grain, at the contact points. The water contained in the pores forms, therefore, a separate independent medium whose internal pressure is produced only by its own weight (Fig. 4-82) as:

$$p = h \gamma_w \quad ; \qquad (4\text{-}63)$$

where p denotes the excess water pressure (i.e., the pressure of water above atmospheric pressure) at a depth h below the water table, and γ_w denotes the specific weight of water.

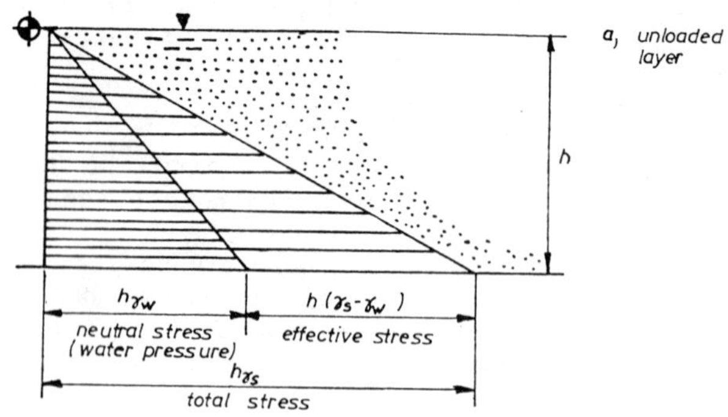

Fig. 4-82. Water pressure distribution in an unloaded aquifer.

When loading the surface with an excess pressure of p_1 (Figs. 4-83 and 4-84), two extreme cases may occur. When the skeleton of the grains was absolutely rigid, all this load would be taken over by the grains and water pressure would remain unchanged. In nature, however, the layers are compressible. When such a system is absolutely closed (i.e., the water cannot percolate out of the aquifer), after the solid struture has been compressed, the whole excess pressure is transferred to the water which is practically incompressible (Fig. 4-83). This second condition is also an extreme one because the layers surrounding the aquifer are never absolutely impervious. Where the hydraulic gradient at the borders of the aquifers surpasses the so-called threshold gradient (Subsection 11-5) of the impervious layers, the movement of water will start. The flow rate of seepage discharging the system depends on the pressure difference, the length of the path, and the permeability of the surrounding layers. Any significant small discharge is enough to drain the aquifer if the time elapsed since the loading is long enough, and thus, the load will be gradually transferred to the grains while the water pressure decreased towards its original static pressure (Fig. 4-84). This process is called consolidation (see Subsection 13-3).

When investigating the natural system of an aquifer, the excess pressure, however, is generally caused by a new layer deposited over the formation already covering the layer in question. Where all overlying strata are pervious, pressure differences will be equalized in a short time and the whole deposit can be characterized by a hydrostatic pressure distribution (Fig. 4-85). Where the investigated aquifer is covered by an impervious layer followed by an upperlying aquifer, there are again two extreme possibilities. In the case of a rigid skeleton, water pressure does not change in the lower

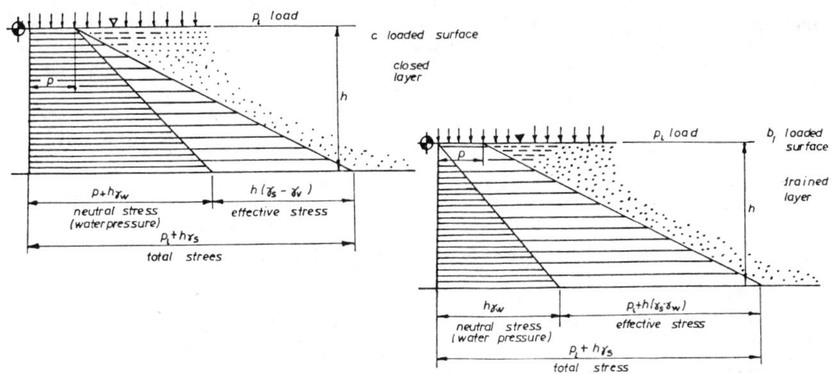

Fig. 4-83. Water pressure distribution in a closed aquifer when loaded at the surface.

Fig. 4-84. Water pressure distribution in a drained aquifer when loaded at the surface.

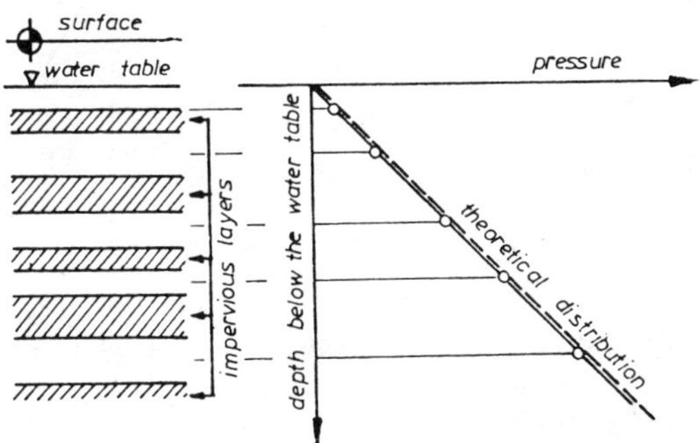

Fig. 4-85. Hydrostatic pressure distribution in a multilayer aquifer system.

aquifer at the first instant. There exists a pressure difference between the two aquifers which is equal to h γ_w (Fig. 4-86) since, using the contact surface of the lower aquifer and the covering layer as a reference level, the pressure of the lower system is zero and that of the upper one is equal to the height of its water table above this level. A flow is induced by the pressure difference through the covering impervious layer, the final result of which will be the equalization of pressures and the development of the hydrostatic pressure distribution as it is represented in Fig. 4-86.

The other extremity is the case when the total weight of the overlying system (i.e., the weight of water and solid grains together) is taken over by the water in the lower aquifer. This total load is

$$p_t = h\,[\gamma_s(1-n) + \gamma_w\, n] \tag{4-64}$$

where γ_s denotes the specific weight of the solid grains. The total load p_t being greater than the pressure of the upper water-bearing layer due to the water contained in it and referred to the contact surface $(h.\gamma_w)$, there exists a hydraulic gradient directed upward which endeavors to equalize the pressures and causes a seepage through a covering impervious layer resulting, finally, in a hydrostatic pressure distribution quite similar to the former case but approaching the static condition from the other direction (Fig. 4-87).

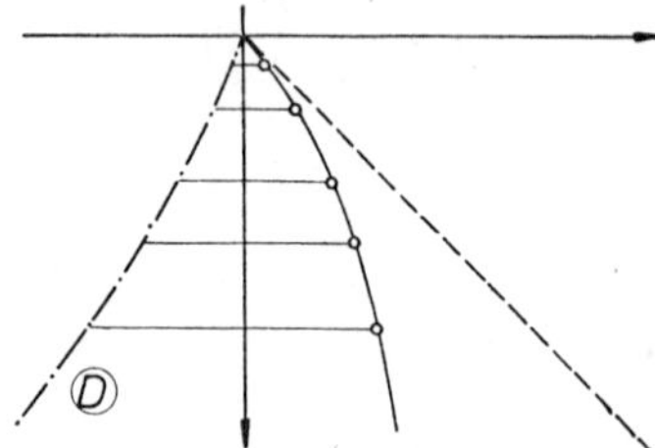

Fig. 4-86. Pressure distribution in a multilayer aquifer system where the load is supported by the rigid skeleton at the beginning. This pressure distribution is characteristic for descending water, i.e., for recharge areas.

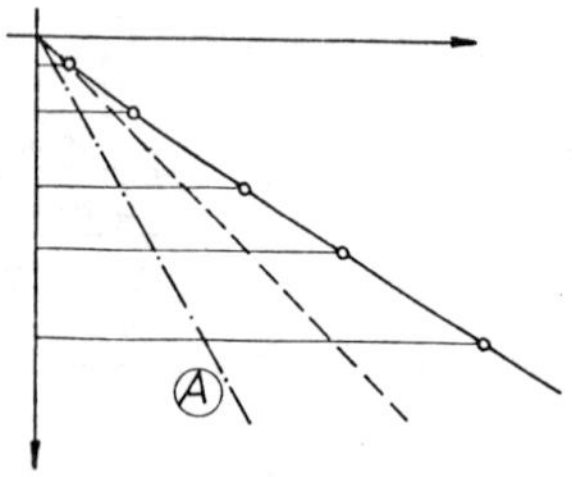

Fig. 4-87. Pressure distribution in a multilayer aquifer system where the load is taken over by the water in the underlying aquifer. This pressure distribution is characteristic for ascending water, i.e., for discharge areas.

The pressure distribution curves are represented in Figs. 4-85, 4-86, and 4-87, denoting the thick (-) line the measured pressure; by dashed (--) line the hydrostatic distribution;

and by dotted (-··-) line the deviation of the measured pressure from the hydrostatic one.

Summarizing the foregoing, two possible zones of actual water pressure can be discerned at the top of an aquifer depending on the fact whether the pressure is approaching the static condition from the zone of lower pressures or from that of higher values. In the first case, the possible pressure (p) may vary between the limits such as:

$$0 < p < h \gamma_w \quad ;$$

the range of variation being $h \gamma_w$. In the second case, the possible pressure is limited in such a way as:

$$h \gamma_w < p < h [\gamma_s (1-n) + \gamma_w n] \quad ;$$

and the range of variation is $h(\gamma_s - \gamma_w)(1-n)$.

By drilling a borehole, the water pressure of each aquifer penetrated by the well can be measured separately. Thus the actual pressure distribution can be determined from the obtained data. If there are several wells in the investigated area, each of which penetrates into different aquifers, the curve of the pressure distribution can be similarly plotted from the corresponding data of the depth of wells and the water level therein, respectively. The actual curve compared to the hydrostatic distribution (dotted curves in Figs. 4-85, 4-86, and 4-87) indicates the possibility and direction of vertical water movement.

Where the two curves are identical (Fig. 4-85) the pressure suits the static condition in all layers, therefore, the possibility of vertical seepage can be excluded. The scatter of the observed pressure data around the line representing the hydrostatic condition indicates the irregularity of sedimentation. It can be expected in this case that the layer separating the various aquifers is highly impervious and the whole series of sediments is relatively young, thus pressure differences could not be yet equalized. It is also possible that there are different draining effects (natural or artificial) influencing the aquifers and causing irregular pressure differences. In any case, vertical seepage can be expected between the neighboring water-bearing layers in the direction of the existing hydraulic gradient.

The regular declination of the pressure distribution curve from the hydrostatic line should be dealt with separately. It may indicate either a higher (Fig. 4-87) or a smaller (Fig. 4-86) pressure at a deeper level than the static value. This phenomenon, however, can be caused not only by unfinished consolidation because a similar pressure distribution also occurs as an effect of continuous vertical flow. Where in an overlying aquifer there is excess pressure as compared to the lower layers, vertical movement can develop from shallow groundwater to deep aquifers. Consequently, this kind of distribution curve (Fig. 4-86) will indicate the area of descending water that is the recharge area of the deep groundwater. On the other hand, in areas characterized by distribution curves representing the lack of pressure at a higher level as compared to a greater depth (Fig. 4-87), the possibility of an upward vertical (i.e., ascending) movement is given and the deep groundwater is probably drained by the shallow one.

It should be emphasized that differences in pressure indicate only the possibility of vertical water movement which, in turn, depend on the permeability of layers. It is not sure at all that continuous water exchange exists there where the hydrodynamic possibility is given. This phenomenon can also be caused by the very slow process of consolidation in which case the same observation refers to the low permeability of the impervious layers. This comparison of pressure distribution curves is, in any case, a useful means for determining the drainage and recharge areas of deep groundwater. It should always be borne in mind, however, that it is only a first step taken for indicating the existence and the probable character of the vertical movement of water and it should be followed and supported by further investigation.

The actual vertical movement of groundwater is controlled by the threshold gradient and by the hydraulic conductivity (K_f) of layers (Subsection 11-1).

In principle, it is possible to calculate the vertical hydraulic conductivity of the strata using the data obtained from pumping tests. For deep aquifers, however, the number and density of pumping tests are usually insufficient due to the varying nature of the layers separating the aquifers. For this reason, attempts have been made to estimate the value of hydraulic conductivity (K_f) and transmissibility (T_f) using borehole data consisting of lithologic, resistivity, and natural gamma logs (Subsection 14-2).

It was stated earlier (Subsection 14-2) that the lithologic type and hydraulic conductivity of clastic rocks is often indicated by their specific resistivity and their natural gamma radiation. A very effective method was worked out by Besenecker (1978) for the determination of the hydraulic conductivity of impermeable and semi-permeable layers based on data mentioned above.

It is obvious that a comparison of well log data measured in neighboring boreholes is possible when the logging instrument is calibrated alike at each measurement. To avoid the deviations due to the differences in calibration, dimensionless parameters will be defined such as:

- the dimensionless parameter of resistivity, K_R, defined as the difference between the specific resistivity (x) of the layer in question and that of "pure" clay (a) related to the difference between the specific resistivity of coarse-grained "pure" sand (b) and of "pure" clay: $K_R = (x-a):(b-a)$;
- the dimensionless gamma activity parameter, K_G, defined similarly to K_R as the difference between the natural gamma activity (number of counts/sec) of the layer in question (y) and that of the "pure" clay (e) related to the activity of the latter: $K_G = (e-y):e$ (Fig. 4-88).

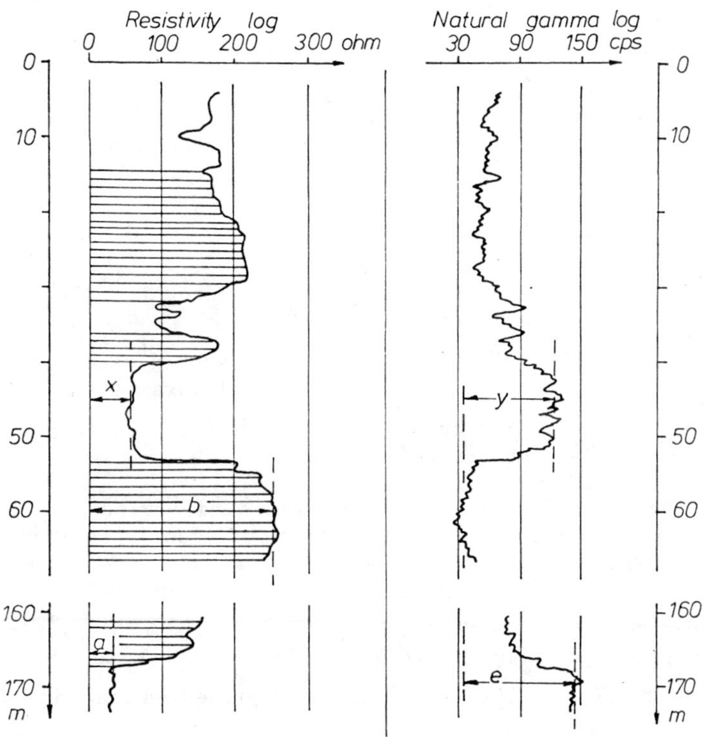

Fig. 4-88. Resistivity and natural gamma activity logs showing the method of calculation of parameters K_R, K_G, and T_R (after Besenecker, 1978).

Figure 4-89 represents the statistical distribution of points constructed from pairs of K_R and K_G values calculated for layers separating aquifers in approximately 300 boreholes penetrating into clastic sediment. The points, plotted with K_R as the abscissa and K_G as the ordinate, form three separated groups (approximated by rectangles) with centroids differing significantly which correspond to three types of aquiclude sediments as silty clay, clayey silt, and sandy silt, respectively.

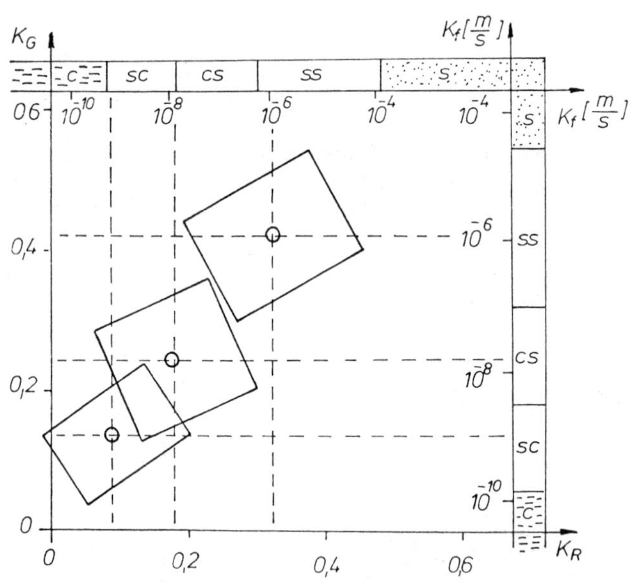

Fig. 4-89. Lithologic types represented with the corresponding values of parameters K_R, K_G, and hydraulic conductivity (K_f) (after Besenecker, 1978).

The hydraulic conductivity values of various grain-size distributions present in the aquifers of clastic sediments can be determined (Subsection 11-1). Since the distribution of the K_R, K_G values makes possible the determination of the grain-sizes of sediments forming slightly permeable layers, it seems to be justified to relate the values of K_R and K_G to the values of K_f. This relation is represented in Fig. 4-89, depicting the lithologic types and the K_f values for the corresponding centroids.

The horizontal movement of the groundwater is controlled by the K_f values and thicknesses of aquifers comprising the vertical sedimentary sequence. The regional flow of groundwater in thick sedimentary sequences may be described properly, therefore, using the transmissivity (T_f) of the sequence as a characteristic parameter (Subsection 13-2). It was demonstrated already that the hydraulic conductivity, K_f, can be related to the specific resistivity ρ of clastic aquifers ranging from fine sand to gravelly sand and gravel. Considering the thickness of the aquifer, the cross-resistivity or transverse resistivity $T_R = \rho\ h$ may be correlated to $K_f h = T_f$, that is to the transmissivity of the aquifer. In multilayered aquifers, where the transmissivity of the sequence can be calculated by summing the T_f values of each aquifer, the same relation is valid for the transverse resistivity of the sequence: it is the sum of the individual transverse resistivities.

The study of the areal distribution of aquifer transmissivity in deep sedimentary basins is supported especially when considering the distribution of transverse resistivities where the water-bearing strata are composed of lenses, coalescent sand beds, and of an alternating sequence of thin sand and clay layers (Fig. 1-16).

The calculation of the transverse resistivity can be easily carried out using resistivity well logs, when after leaving out the impermeable or semi-permeable layers (with the aid of the method described above), the area under the resistivity graph is measured (shaded area in Fig. 4-88). It is obvious that the transverse resistivity distribution can be interpreted in terms of transmissivity where the saturating groundwater shows only minor deviations from an average conductivity value in the study area. The value of the transverse resistivity may be determined with a smaller error using the vertical electrical sounding than that which is produced when determining the resistivity and thickness of a layer separately.

The possibility of representing by a single value (T_R) all of the data obtained from a resistivity log or from a resistivity sounding makes T_R extremely suitable for the preparation of maps showing the areal distribution of this parameter and, by the aid of correlation, that of average transmissivity. Assuming an average value of resistivity (correlated with porosity) for the study area, the porous aquifer system may be modeled as a system of constant porosity, but of varying thickness. This kind of "thickness" map, although in simplified form, may give a clear indication about the flow pattern of the deep groundwater (Fig. 4-90).

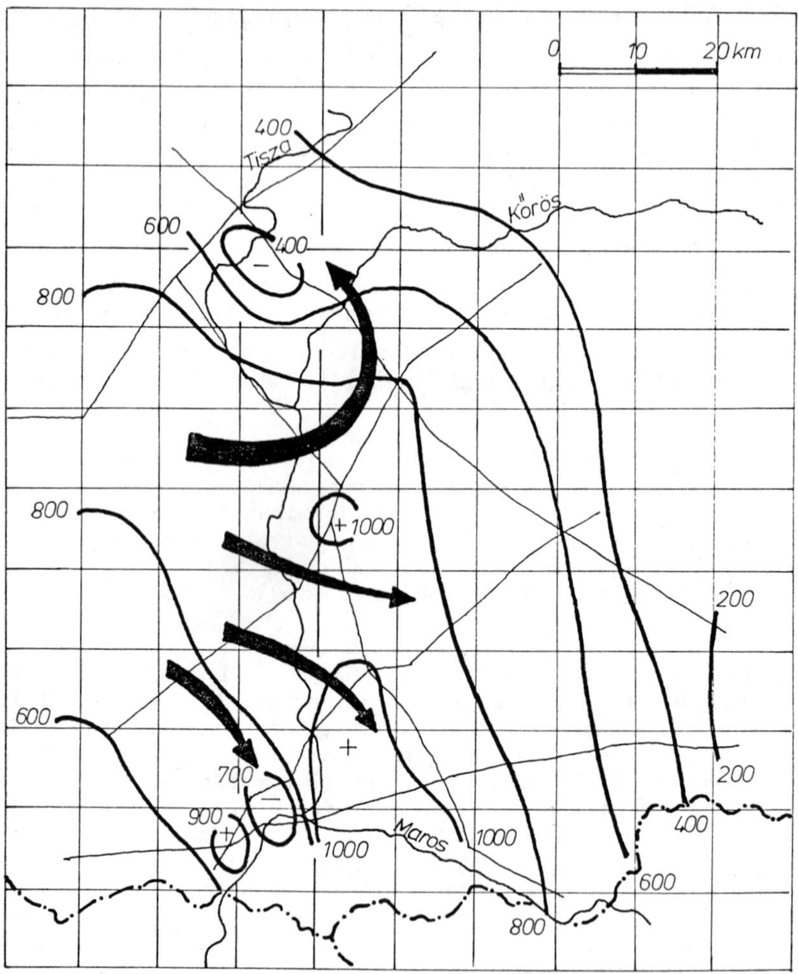

Fig. 4-90. Isopach map of the sand layers down to the lower boundary (top of lower Pannonian) of the groundwater flow system. The thick lines denote equal thicknesses of the sand layer in meters, the arrows show the probable direction of flow.

15-2 Application of the Investigation of Chemical Composition in the Study of Deep Groundwater

The chemical composition of waters is generally considered as an important tool to answer the questions related to the origin and movement of waters occurring in deep sedimentary aquifers. The rapidly developing isotope techniques have indicated new directions for research as far as the origin of formation water is concerned.

The chemical investigation applied to the study of the movement of deep groundwater is based on the worldwide observed phenomenon according to which the mineralization of waters increases with increasing depth. The increase of mineralization is accompanied by the tendency to changes in the chemical composition of water according to Chebotarev's rule (see Subsection 9-3). The subsurface water, regardless of the material through which it percolates, starts as a bicarbonate water, due to precipitation, and changes through a sequence of bicarbonate plus chloride to chloride plus bicarbonate to chloride plus sulphate or sulphate plus chloride finally to a perdominantly chloride water approaching the composition of the seawater.

This sequence forms the basis of the vertical hydrochemical zonality theory according to which the regime of the groundwater may be divided into three zones:

- the upper zone of intensive circulation and exchange (ions: HCO_3^- and SO_4^-);

- intermediate zone of hindered exchange (ions: SO_4^- and Cl^-);

- lower zone where stagnant and extremely slowly moving water occurs (ions: Cl^-).

In terms of a flow system, the chemical succession of Chebotarev may be expected to closely coincide with the chemical changes of groundwater within the system. The sequence of alteration depends on the magnitude of the flow of water (Brown, 1967). Accepting this concept, it follows when measuring the chemical composition of groundwater, its flow path may be traced (Fig. 4-91). The areas where the deep groundwater is discharged can be delineated, therefore, as a local maxima of chloride content in shallow groundwater. With respect to the total concentration of salt in groundwater, areas of recharge are characterized by its low value, and those of discharge by its high value, respectively (Fig. 4-92).

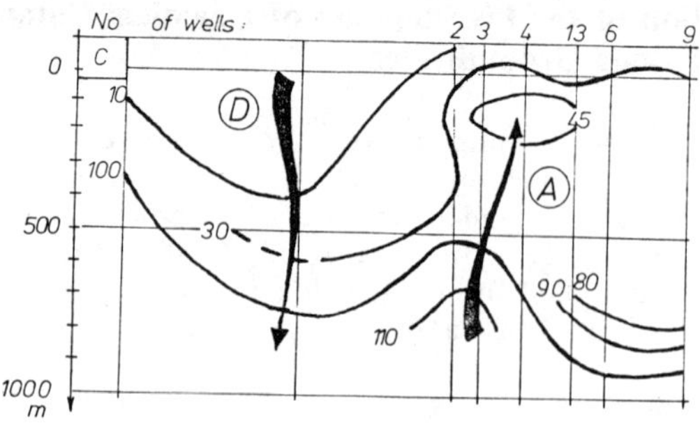

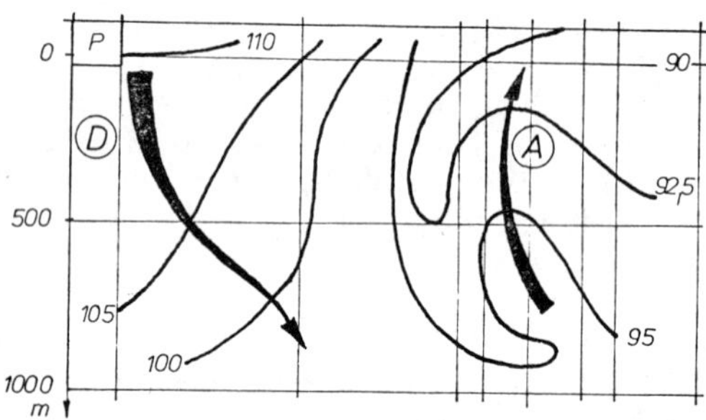

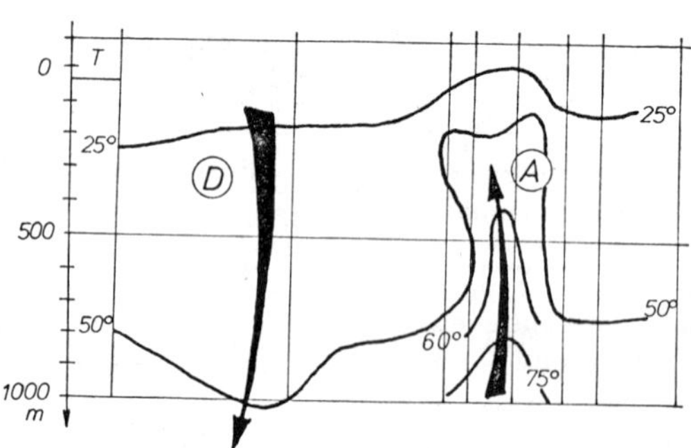

Fig. 4-91. Vertical cross sections through a thermally anomalous area shown in Fig. 4-97. Figure C represents the isoanomaly lines of chloride content in mg/1. For comparison in Fig. P, lines of equal pressure are shown in meters and in Fig. T the isotherms in centigrades (after Erdélyi, 1978).

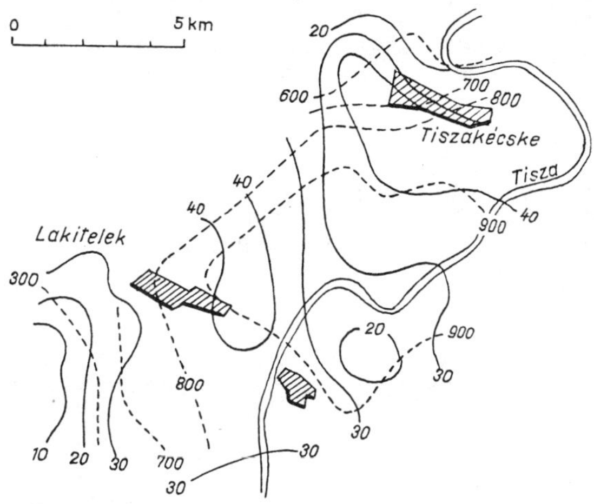

Fig. 4-92. Results of a chemical investigation made on water samples obtained in medium deep wells of the study area shown in Fig. 4-97. The thick and dotted lines represent equal chloride and total dissolved salt content, respectively, numbered in mg/l units.

The hydrochemical sequence discussed above by no means answers the question concerning the origin of waters occurring in deeper zones except the upper one where the atmospheric origin of groundwater is obvious. For example, brines occurring within the lower zones may be both marine connate waters and meteoric ones, respectively, the latters having higher concentration of Cl^- due to leaching of evaporites. The answer may be given by isotopic studies which provide an appropriate way to estimate the age and to trace down the history of groundwaters (Subsection 9-3). In the case of deep (thermal) groundwater, the determination of its original (reservoir) temperature may aid in the disclosure of its origin. Among the chemical methods the most commonly used, "geothermometers" are those based on quartz solubility and on the Na-K-Ca ratios.

The silica geothermometer is based on the experimentally determined variation in solubility of quartz in water as a function of temperature and pressure. Provided the activity of water is not greatly diminished at a given temperature, the solubility of quartz is independent of the local mineral suite, gas partial pressures, and other dissolved constituents commonly found in deep waters (Fournier, 1970).

The geothermometer using Na/K alkali ratios is based on ion-exchange processes since the base exchange or partitioning of alkalies between solution and solid phases is dependent on temperature. Thus, alkali ratios in deep (thermally) groundwater may give indications of subsurface temperature (Ellis, 1970).

Experimental curves allow the calculation of the temperature of the last balance in a depth of the water with quartz on the one hand (Fig. 4-93) and with soda and potash on the other (Fig. 4-94).

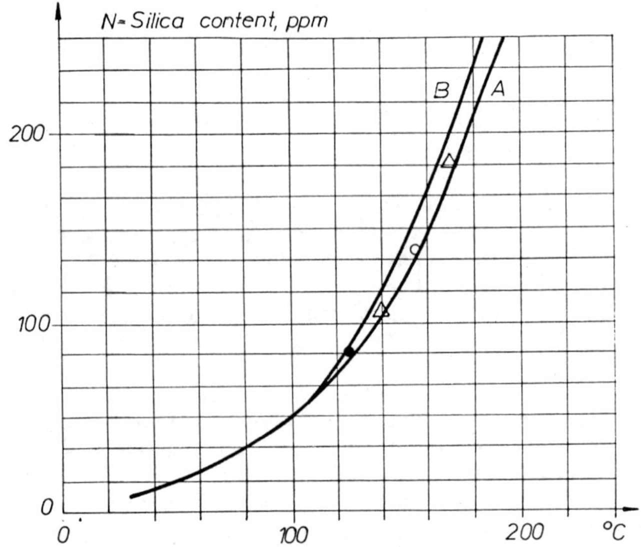

Fig. 4-93. Silica geothermometer. Lines represent silica concentration in waters vs. estimated temperature of last equilibrium. Curve A applies to waters cooled by heat conduction, curve B when cooled entirely by adiabatic expansion (Fournier, 1970).

Subsurface temperatures estimated by the silica methods may differ from those obtained by using Na/K ratios due to the differences in the chemical processes. The deviation of or agreement in the temperature values found by each of the chemical geothermometers is demonstrated by the results of the measurements carried out in the Central French Massif to study the springs emerging from the basement (Risler, 1976). The temperature values determined by the silica and alkali methods are represented in Fig. 4-93 and in Fig. 4-94, respectively,

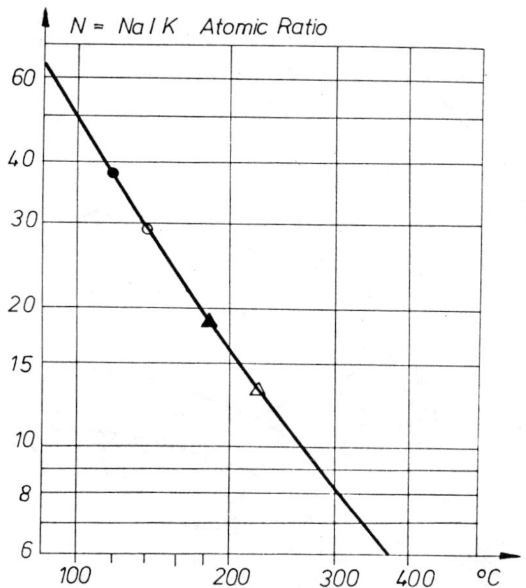

Fig. 4-94. Alkali geothermometer. The line represents the Na/K atomic ratio vs. estimated temperature of the last equilibrium (Ellis, 1970).

denoting the values related to the same spring by the same symbol.

The silica geothermometer was successfully used to study the origin of the thermal springs of Budapest. In Fig. 4-95, where the differences between the temperature values estimated by silica thermometer and the temperatures of the outflowing waters are represented, the great differences (25-30°C) refer to springs with a deep reservoir, the smaller ones indicate springs originating from a "stratum reservoir" (i.e., the reservoir which lies in the depth calculated from the temperature of the outflowing water) (Gölz, 1978).

In using chemical geothermometers, one must be aware of the limitations inherent in the assumptions that chemical equilibrium should exist among the reactants involved in a given temperature-dependent reaction in the conditions of the geothermal reservoir at depth and no re-equilibration should occur after the reactants have left the reservoir. Where mixed with other groundwater, this should be evaluated by chemical or isotopic methods.

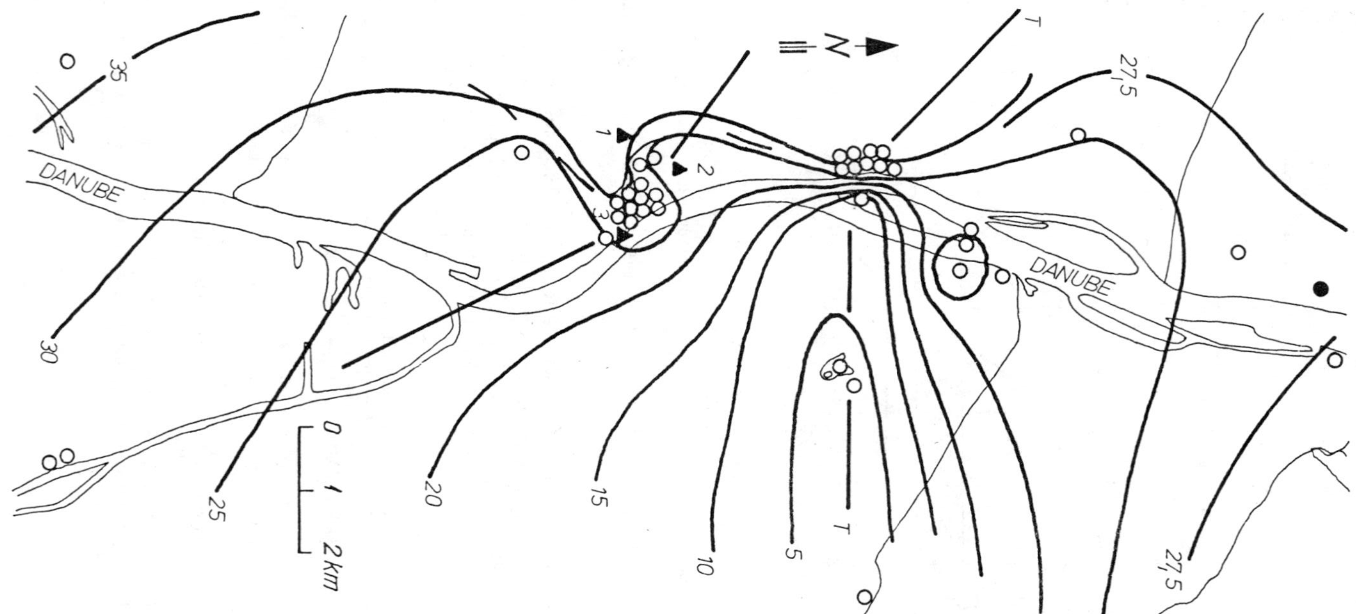

Fig. 4-95. Application of the silica geothermometer to study the origin of the thermal springs of Budapest. The isotherms (--) represent the difference between the temperature values estimated by silica thermometer and the temperature of the outflowing water. High values (south from the "thermal dislocation," marked by T) refer to a deep reservoir, small values (to the north from the dislocation) refer to a "stratum reservoir" (Gölz, 1978).

15-3 Heat Balance under Recharged and Drained Areas

The temperature of the earth is governed by many factors. The practical geothermal measurements aimed at the study of deep groundwaters are chiefly concerned, however, with only one component, viz., the dissipation of the heat coming from the interior of the lithosphere which affect the temperature on the earth's surface.

This steady state thermal field is seriously disturbed at the earth's surface due to the radiation of the sun which causes both diurnal and annual variations. The speed and depth of penetration of the periodic variations depend on the thermal properties (i.e., heat conductivity and specific heat) of the uppermost layer (Gálfi, 1961; Gordienko, 1976) which, in turn, is influenced by the moisture content of the ground. The study of this disturbed zone will be omitted here since it relates exclusively to the regime of shallow groundwater (Subsection 9-3).

Thermal measurements to study deep-lying strata are carried out in holes deep enough to exclude the upper part of the ground subject to radiation from the sun. The thermal field below one meter is not influenced by the diurnal variations, and below several meters depends only upon the internal flow of heat.

The geothermal field in the lithosphere is determined regionally by the heat flow distribution at the bottom of the earth's crust. Irrespective of the origin of the internal heat, whether it is conducted or transported convectively (Stegena, 1975), it manifests itself as a component of deep origin in the thermal field of the lithosphere. It may be assumed, and it is supported by observations, that the regional thermal field is in equilibrium in the lithosphere or at least in its elements of great extension (e.g., in deep sedimentary basins). In respect to the deep-lying clastic sediments, the regional geothermal field may be described, therefore, as a steady state one and homogeneous (i.e., characterized by a constant value of heat flow) in the vertical direction.

There are local anomalies in the regional flow field due to disturbing effects within the lithosphere such as:

- heating effect due to masses (e.g., that of volcanic origin) embedded in it;

- heat transport or removal due to groundwater movement.

The following discussion will be confined to the interpretation of the latter effect although the reasoning may be similar in both cases (Alföldi, 1966).

The recharge areas can be characterized thermally as negative anomalies in the regional heat flow distribution due to the cooling effect of the descending water, on the contrary, the discharge areas represent maxima or positive anomalies in the regional flow field due to the heat transport by the ascending deep groundwater.

The description of the geothermal field may be carried out by various ways using thermic data such as heat flow (f), temperature (T), and vertical thermal gradient ($gg = \partial T/\partial z$) which parameters are interconnected by the basic relationship (Table 4-10) expressing that the vertical heat flow is equal to the product of thermal conductivity and vertical thermal gradient.

The means of representation of the data commonly used in geothermal prospecting are the so-called vertical gradient (gg) maps and sections, respectively, showing the distribution of the $\partial T/\partial z$ values either in horizontal planes or in vertical sections. The regional geothermal field is characterized by the constant value of the vertical component of the thermal gradient. Changes in thermal conductivity produce corresponding changes in the vertical gradient (called conductive anomaly) if it is assumed that the heat flow is constant. The same effect can be produced by changes in heat flow due to convective heat transport of moving water (called convective anomaly). Variations in the vertical gradient are associated, therefore, either

- with variations in the subsurface materials;

- with the movement of groundwater.

The conductive anomalies may be eliminated from the gradient field where the distribution of the heat conductivity (λ) is known. The gradients will be referred to an average value of heat conductivity (λ_a) obtaining by this way the field of the reduced gradients (gg_r) as:

$$gg_r = gg\ (\lambda/\lambda_a) \quad ; \qquad (4\text{-}65)$$

where gg and λ denote the local values of the vertical thermal gradient and of heat conductivity, respectively. The map of reduced gradients can be transformed into a heat flow map when both the gradients and heat conductivities were measured exactly. This represents the relative distribution of the vertical heat flow values where the gradients were measured exactly but only the relative values of the heat conductivity were determined. It is obvious that both the heat flow map and the map of reduced gradients only contain convective anomalies, thus they can be used to study the groundwater flow field. The distribution of the temperature (T) is influenced by the same effects which affect the geothermal gradients. The regional temperature field is characterized by temperature values increasing linearly with depth. The temperature anomalies due to conductive or convective disturbances are superimposed upon the regional field.

Characteristics of vertical temperature distribution in areas of heat transport and removal are visualized in Fig. 4-96 showing temperature logs calculated for a three-layer system where water is moving in the second layer. The heat transport due to water flow produces both positive anomaly in the vertical heat flow and a deviation to the higher temperature values on the temperature log (marked by + in Fig. 4-96). The heat removal, on the other hand, manifests itself by a relative maximum in the heat flow and by decreasing the values of the temperature log (marked by - in Fig. 4-96).

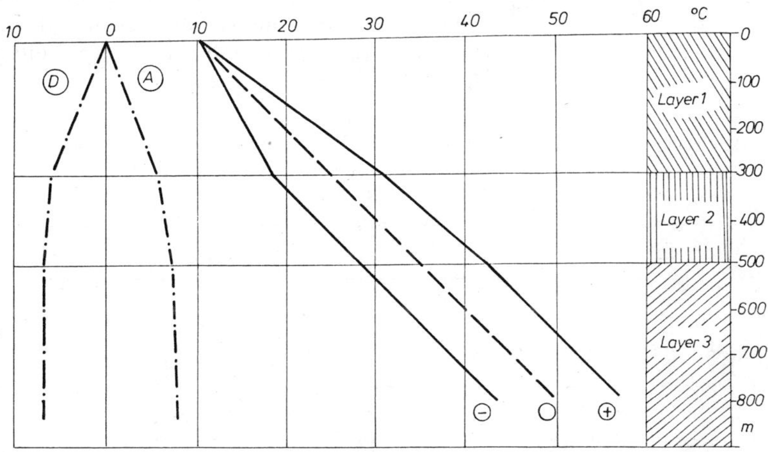

Fig. 4-96. The temperature logs (-- lines marked by + and -) were calculated for a three-layer system of various thermal properties listed in Table 4-9. The average temperature vs. depth line is marked by 0, the deviations of the logs from the average temperature are represented by dotted (-•-) lines.

A comparison made between the graphs representing the deviation in pressure versus depth (dotted lines in Figs. 4-86 and 4-87) and those in the temperature versus depth (dotted lines in Fig. 4-96) suggests that the areas of discharge and recharge are represented by positive values and negative values in the deviation graph with respect to both pressure and temperature, respectively.

It was stated above that the anomalies in temperature (T) and in the gradient (gg) field may be interpreted equally as convective or conductive ones. Both interpretation possibilities are summarized in Table 4-11 where data corresponding to "convective" and "conductive" interpretation are given in columns marked I and II, respectively. It should be emphasized that the interpretation in the flow field is unambiguous: it is characterized by local extremes in the case of convective heat transport.

Table 4-11. A three-layer system characterized by four different sets (-I,-II,+I,+II) of the following thermal parameters: specific heat production 0 (milliW/m^3), heat conductivity (W/m-°C) and heat flow at the surface f (milliW/m^2). Case I and II represent convective and conductive anomalies, respectively. The thermal logs calculated are represented in Fig. 4-96.

Thermal log marked by -						No. of Layer	Thermal log marked by +					
0				f			0				f	
I	II	I	II	I	II		I	II	I	II	I	II
0	0	2	3.3	60	100	1	0	0	2	1.5	140	100
-0.2	0	2	2.1	-	-	2	+0.2	0	2	1.7	-	-
0	0	2	2.0	-	-	3	0	0	2	2.0	-	-

As an example of geothermal methods applied to study the movement of deep groundwater, the investigation of the local thermal maximum at Tiszakécske (Fig. 4-97) will be discussed since it represents, at the same time, the advantage of the simultaneous use of different kinds of transport considerations.

In the first step of the exploration, chemical analysis of the groundwater was carried out using water samples taken from wells of medium depth (about 200 m). The areal

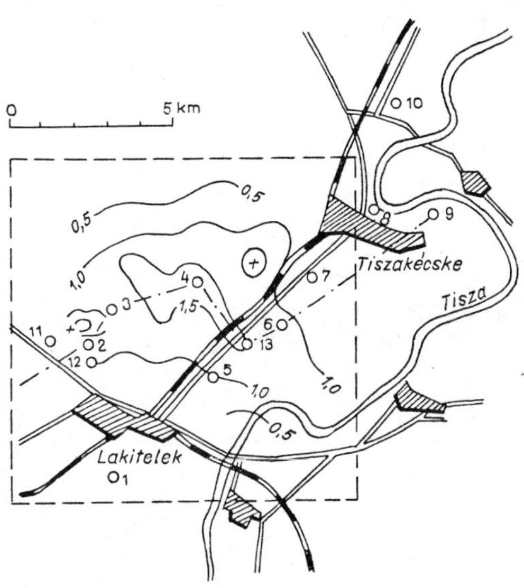

Fig. 4-97. Map of the study area at Tiszakécske. Well numbers (e.g., 0,12) from 1 to 10 denote deep thermal wells; numbers from 11 to 13 denote shallow boreholes. The thick isogradient lines, numbers in °C/10 m delineate the thermal anomaly. The dotted (-•-) lines denote the cross section shown in Fig. 4-91.

distribution of the chlorine and total dissolved solid content (Fig. 4-92) clearly shows the discharge character of the area (Subsection 15-2).

In the second step, an areal temperature measurement was prepared in shallow boreholes at depths ranging from 20 m to 50 m, and vertical thermal gradients (gg) were calculated using the temperature data. The areal distribution of the gg values is represented in Fig. 4-98. To eliminate the conductive anomalies, the relative values of the heat conductivity were estimated within the study area (Fig. 4-99) and considered when preparing the map of reduced gradients (Fig. 4-100) which represents the relative heat flow map at the same time. Comparison of the gradient and relative heat flow map shows that the variation in heat conductivity has not seriously disturbed the general maximum character of the thermal anomaly, but it concealed some details which are important in the final interpretation. Figure 4-100 clarifies the thermal conditions in detail. The steep maximum of the relative heat flow is divided into secondary maxima arranged along two crossing lines suggesting the existence of convective thermal transport due to rising thermal water along crossing planes of good hydraulic conductivity.

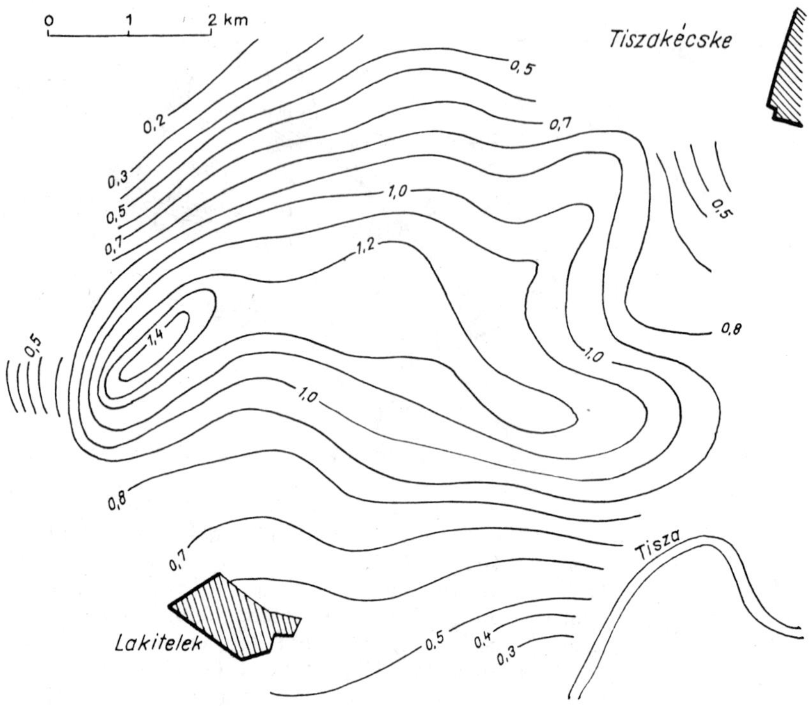

Fig. 4-98. Map of the vertical geothermal gradients (gg) with isogradient lines numbered in °C/10 m units.

The preliminary exploration was followed by an intense drilling campaign. A total of 13 water wells were located in the study area (Fig. 4-97) providing opportunity to investigate the flow field in vertical cross sections.

Data of formation temperature measured in boreholes or calculated from the temperature of outflowing water characterize the vertical structure of the thermal field (Fig. 4-101). It may be concluded, considering the data in Fig. 4-96, that temperature values at the right side of the line showing the average increase of temperature vs. depth mark the ascending branch of a local water circulation system, and those at its left side belong to the descending branch.

The character of the fluid transport can be studied using well head (pressure) data which (when suitably corrected) represent the data governing the flow of the groundwater. The

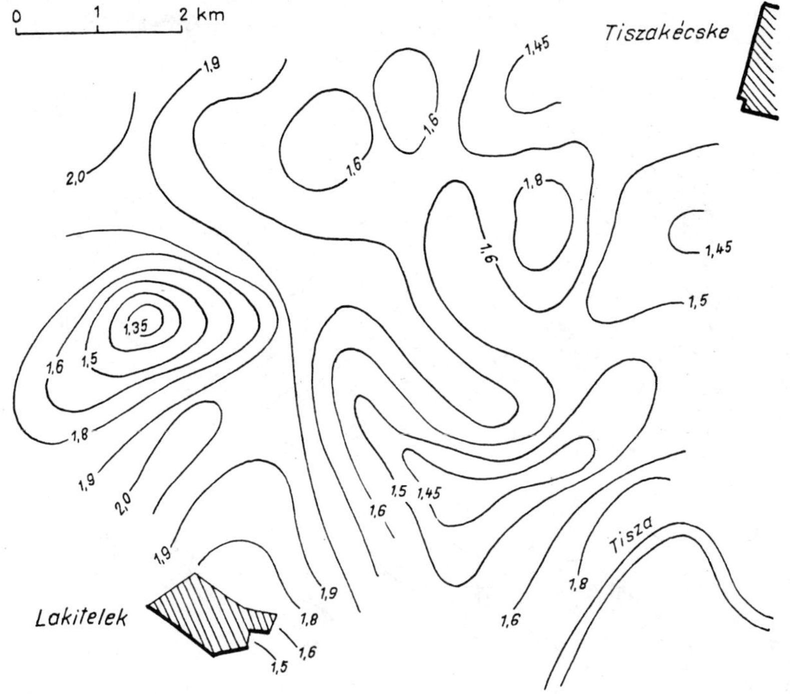

Fig. 4-99. The relative distribution of the heat conductivity (λ) in the study area. The isoanomaly lines are numbered in W/m-°C units.

pressure data represented in Fig. 4-102 delineate the ascending and descending branches of the circulation system when a consideration is carried out, as above, based on Figs. 4-86 and 4-87.

The velocity of flow can be estimated, using thermal data, within areas of homogeneous water flow with constant velocity v. The value of the vertical heat flow is composed in this case of two terms, a convective and a conductive one (Table 4-8) as:

$$f = -\lambda \; gg + \rho \; c \; T \; v \quad . \tag{4-66}$$

Any finite difference Δ in one term must be compensated by the difference in the other since the flow (f) is constant along the vertical axis:

$$\Delta(gg\lambda) = v \; \Delta(\rho \; c \; T) \quad . \tag{4-67}$$

The study of the thermal conductivity, λ, has not shown any marked change in its value within the upper 1000 meters and

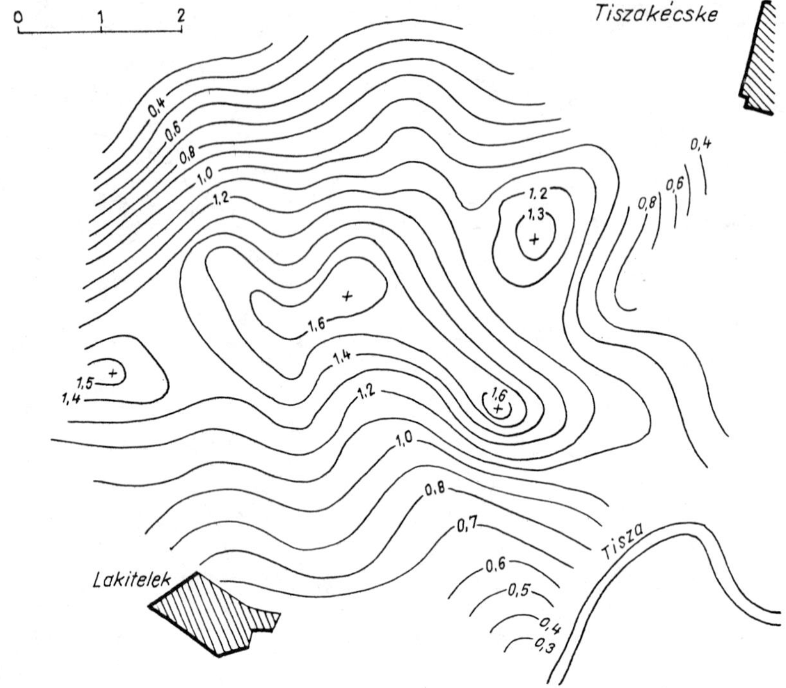

Fig. 4-100. Contour map of the reduced thermal gradients showing the distribution of the relative heat flow values in the study area. The reduction was made to an average value of heat conductivity of 1.45 W/m-°C.

the same may be assumed for the average value of density (ρ) and heat capacity (c). The vertical heat flow equation may be transformed, therefore, as:

$$\lambda \, \Delta gg = v \, \rho \, c \, \Delta \, T \quad . \tag{4-68}$$

The velocity v may be expressed as:

$$v = (\lambda/\rho c) \, (gg_2 - gg_1)/(T_2 - T_1) \quad ; \tag{4-69}$$

where the indices 1 and 2 denote the values measured in depth h_1 and h_2, respectively. After performing this calculation, a flow velocity of about $5x10^{-9}$ m/s was obtained.

The seepage velocity of the flow may also be calculated hydraulically by using Darcy's equation since the difference between the vertical hydrostatic pressure distribution and

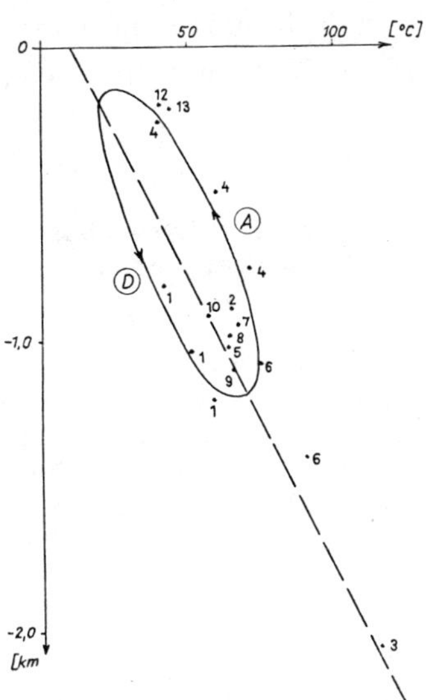

Fig. 4-101. Formation temperature vs. depth. Numbers at measuring points refer to the numbering of wells given in Fig. 4-97. The average increase of temperature vs. depth is shown by broken line. The letters D and A denote the areas of descending and ascending groundwater, respectively.

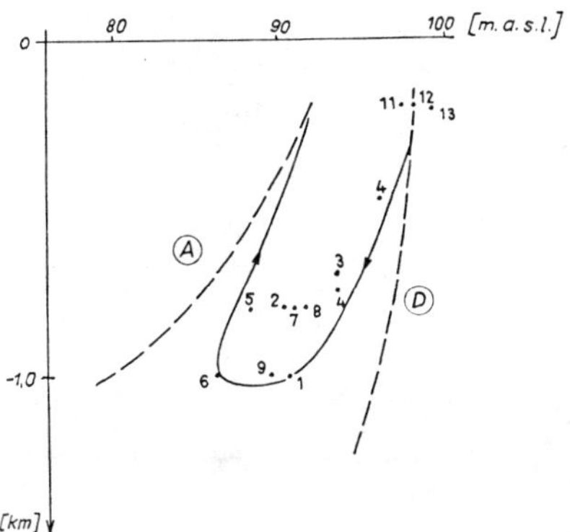

Fig. 4-102. Formation pressure vs. depth. The dotted lines represent the hydrostatic pressure vs. depth, the solid lines equalize the measured values in areas of descending (D) and ascending (A) character, respectively.

the measured one have indicated the existence of the groundwater flow. The hydraulic conductivity in the vertical direction was estimated as 5×10^{-6} m/s by using the result of measurements on core samples. The measured vertical gradient maintaining the movement of water was estimated by considering the data given in Fig. 4-102 and was found to have the order of 10^{-3}. Thus, a value of 10^{-9} m/s obtained for the order of seepage velocity is in good agreement with the value calculated on a thermal basis.

The common interpretation of the two kinds of investigations, based on fluid and heat transport, delineates the local groundwater circulation system in detail, marking a recharge area near to well No. 1 and the discharge area at wells No. 4, 11, 12, and 13, the latter being indicated by chemical methods too.

SECTION 16

HYDROLOGICAL EVALUATION OF OBSERVED DATA

The character of hydrological data observed in deep-lying aquifers composed of loose clastic sediments do not differ essentially from those characterizing the regime of shallow groundwaters. The position of such water-bearing formations, especially the fact that the aquifers are covered by thick layers, causes special problems, and requires the consideration of special aspects in connection with the evaluation of data collected concerning deep groundwaters.

The more limited information content of the data should be mentioned first of all among the problems arising in connection with the hydrological analysis of deep groundwaters. The construction of observation wells is more costly and laborious than the drilling of shallow wells. This is the reason why the observation network is usually not as dense as that providing information about the regime of the first water horizon. It is not possible, therefore, to observe relatively small local changes in the position of the water table or piezometric surface, only the regional character of the development of pressure conditions and that of their fluctuation can be studied in this way.

The small number of wells constructed for the definite task of observation and not used for any other purpose requires that the data collected from other wells, under operation and serving some of the groundwater users, should also be utilized in the investigation. The pumped wells do not indicate, however, the real pressure values in the layer but their data depend on the size and method of exploitation. It is necessary, therefore, to find methods suitable to reconstruct the probable pressure characterizing the static conditions and to repeat the investigation regularly to estimate the effect of artificial drainage on this parameter. The lower reliability of the observation in pumped wells, due to uncertainties of the necessary corrections, should also be considered.

The great depth of the wells may also cause difference between the actual pressure in the layer and that measured at the surface, even in observation wells where the system is in a static condition. Such deviations are due to the difference of the specific weights of water within the pipe of the well and that stored in the layer. The change of the specific weight may be the result of the different pressure, temperature, gas content, and the concentration of solid media dissolved in the water. The investigation of these influences and the method of correction needed to determine the true pressure in the layer is the starting point of the evaluation of water level observations.

It is necessary to consider also that the recharging and discharging processes cannot be observed in the case of deep-lying aquifers and even velocity measurement (e.g., by using tracers) are hardly available because of the thick covering formations. Thus, the water level data provide practically the sole hydrological information on the regime of such layers. This fact emphasizes the great importance of the observation of pressure conditions, and stresses the need for supplementary investigations of the balance of chemical components and isotope contents as well as the heat balance as it was explained in the previous section.

Apart from the many problems listed and originating from the deep position of the water-bearing formation, there are some advantages as well due to the covered character of the aquifer. The random and rapidly changing hydrological events do not influence its regime. All the recharging and draining effects can be characterized as water exchange between the layer in question and the surrounding formations. These processes can be well described by using physical relationships having the pressure gradient as the sole independent variable if the flow parameters (conductivity and storage capacity) of the layers are known. Thus, the application of the movement equations of seepage hydraulics is easier in the case of deep aquifers than that for the investigation of the first water horizon. It follows from this condition that the systems composed of deep-lying loose clastic sediments can be well simulated by the mathematical models derived on the basis of hydrodynamic principles as it will be demonstrated in the third part of this section.

16-1 Determination of Pressure Conditions from Water Level Data Observed in Deep-lying Aquifers

The consideration of the dynamic character of processes was mentioned as one of the main principles of groundwater hydrology. The requirements of this aspect may be satisfied by investigating the flow of water and the accumulation of energy in the system. Both phenomena are closely related to the pressure conditions prevailing on the water stored in the layers since the seepage velocity is proportional to the pressure gradient, and the energy content is generally characterized by the pressure itself. The most important parameters of the simulation models are also the various pressure values: the greatest part of the input (boundary conditions) are expressed in the form of pressure prevailing at the perimeter of the system in most cases, the results to be determined are the change of pressure in time due to human interference and the comparison of the actual pressures measured under natural conditions to the calculated values are used to calibrate the model. At the same time it was also mentioned that the pressure, more precisely the water level which indicates the upper level of the water column assumed to be equivalent with the pressure, is the sole directly measurable hydrological data. Thus, the requirements and the possibilities are in good harmony in this case.

The equivalent water column is always calculated as the pressure divided by the specific weight of water

$$h\ [m] = p\ [Mp/m^2]/\gamma\ [Mp/m^3]\ ; \quad or$$

$$h\ [cm] = p\ [p/cm^2]/\gamma\ [p/cm^3]\ ; \tag{4-70}$$

or vice versa the equivalent water column is measured and the pressure is determined by multiplying the height of the water column by the specific weight. Since the specific weight of water is roughly equal to unity (if it is expressed in a dimension of p/cm^3, kp/dm^3, or Mp/m^3) this multiplication is often neglected stating that the water column measured in m directly gives the pressure in Mp/m^2 dimension or one should take the height in cm to get the pressure in a p/cm^2 value. This approximation does not cause considerable error in the investigation of shallow groundwaters, but its deviation from the true value is not negligible when a water column of several hundred m balances the pressure, and even the specific weight itself may change along the column. The difficulties are further increased if the water is not in a static condition but the well is under operation. The flow developing in the pipe of the casing may produce a quite irregular distribution of the specific weight along the vertical axis of the well. There is, however, another more serious disturbing effect of flow, i.e., a part of the available energy is consumed by the resistance against the flow and the pressure measured at the surface gives, therefore, the difference of the true pressure and the head loss.

Since the information required for characterizing the energy and flow conditions is the pressure acting on the water stored in the layers, the most direct way of observation is the measurement of the actual pressure in the well instead of the water level. The task can be solved technically by closing the pipe used as casing just above the screen with a water tight packer through which the measuring device can be introduced into the closed stretch of the pipe which is in direct contact with the layer and thus has the same pressure.

Although this measurement can be used without any further correction, the observation of the water level is still more generally used than the direct determination of pressure. The probable reason of the limited application of measuring the pressure directly is the more complicated and expensive equipment required for such observations. There are even scientists arguing against pressure measurements because the measuring error may be higher (e.g., if the closing is not absolutely water tight, or if the calibration of the measuring head is influenced by the great depth and high temperature) than that of the data derived from the height of the water column observed in the well and corrected according to the conditions prevailing in the pipe. It should also be considered that pressure can only be measured by closing the casing when the well is not in operation.

The height of the water column measured in the well can be easily transformed into pressure by applying Eq. 4-70 if the water is in a static condition and its specific weight as well as the vertical distribution of this parameter are known. The deviation of the specific weight, γ, from its normal value (which is regarded to be unity under atmospheric pressure and at a temperature of 4°C if its dimension is p/cm^3, kp/dm^3, or Mp/m^3) may be caused by various factors. Either the pressure or the temperature may differ from the basic values determined in the definition. Salts dissolved in water raise the specific weight of the latter while the gas content decreases the parameter. There are many publications and reports proposing practical methods to determine the actual weight of the water column (Lakatos, 1975; Juhász, 1976). Investigating the factors acting separately, the following procedures may be proposed.

The pressure and the specific weight are interrelated variables since the pressure is caused by the weight of the water column overlying the section investigated, at the same time, the specific weight at a given elevation is determined by the compression of water due to the pressure prevailing there:

$$p_{ex}(z) = \int_o^z \gamma(z)\ dz = - \frac{\ln(1-\gamma_o\beta z)}{\beta} = \frac{1}{\beta} \cdot$$

$$[\gamma_o\beta z + \frac{(\gamma_o\beta z)^2}{2} + \frac{(\gamma_o\beta z)^3}{3} + \ldots] \ ; \text{ and}$$

$$\gamma(z) = \gamma_o \ \exp[\beta\ p_{ex}(z)] = \frac{\gamma_o}{1-\gamma_o\beta z} \ ; \text{ because}$$

$$\frac{d\rho}{dp} = \beta\rho \ ; \quad \text{and} \quad \gamma = g\rho \ ; \tag{4-71}$$

where the excess pressure (p_{ex}) is the difference of the total pressure (p_{tot}) and the atmospheric pressure (p_o); β is the coefficient of compressibility which can be approximated by a numerical value of $\beta = 4.6 \times 10^{-8}$ [$cm^2 p^{-1}$]; although β is not an absolute constant, but the proposed value is an acceptable average of the parameters measured between 0° and 100°C as it is shown in Fig. 4-103; g is the acceleration due to gravity; ρ is the density of water, and γ_o is its specific weight under atmospheric pressure, but this parameter depends on temperature, its numerical value of 1[$p\ cm^{-3}$] is at a temperature of 4°C. The change of the specific weight of water as a function of both the total pressure and temperature is summarized in Table 4-10.

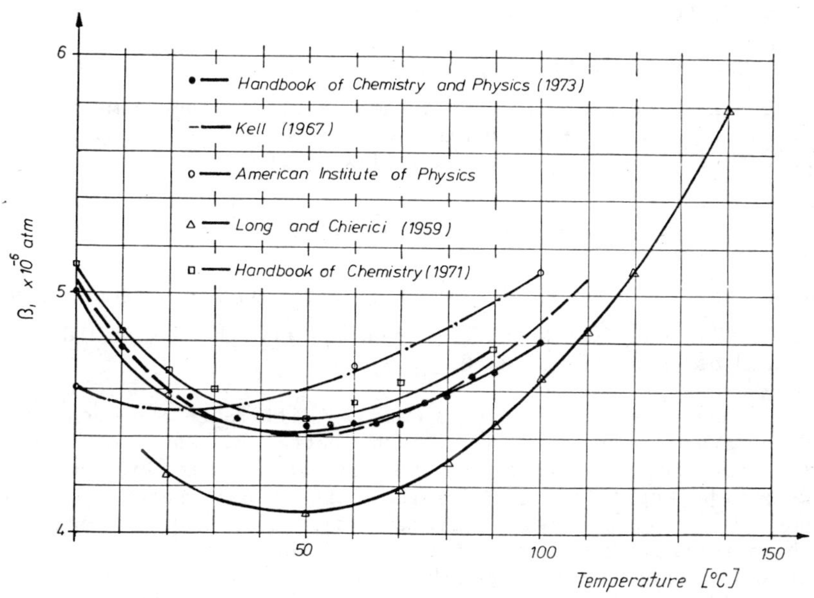

Fig. 4-103. Change of the coefficient of compressibility with temperature.

It is clearly indicated by Eq. 4-71 that the pressure vs. depth relationship can be well approximated with a linear expression until a given limit of depth. The error caused in this way is less than 2.3×10^{-2} percent, if z = 100 m, while it increases to 0.23 percent at z = 1000 m (which means an error in pressure head of about 2.3 cm and 2.3 m, respectively). Considering this approximative value of the probable error, a simple equation may be proposed to calculate the pressure as the function of depth:

$$p = z\gamma_o \; (1 + 2.3x10^{-6} \; z) \; ; \qquad (4\text{-}72)$$

where $p[Mp \; m^{-2}]$, $\gamma_o[Mp \; m^{-3}]$, and $z[m]$ dimensions must be used since the constant is not dimensionless.

The specific weight of water belonging to the atmospheric pressure (γ_o) is the coefficient of proportionality between pressure and depth when the linear relationship is applied. The next problem is caused by the fact that this basic value depends on the temperature of water and thus it may also change along the length of the well. The measurement of temperature is needed, therefore, to provide supplementary data for the calculation of pressure.

One relatively simple equation approximating the γ_{oT} (specific weight of water at a temperature T under atmospheric pressure) vs. T relationship, within a range of temperature between 15°C and 100°C, is as follows (Juhász, 1976):

$$\gamma_{oT} = \gamma_{15} - 4.1x10^{-6} \; (T - 15)^2 \; ; \qquad (4\text{-}73)$$

where γ_{15} is the specific weight of water having a temperature of 15°C under atmospheric pressure, and the temperature T should be substituted in °C.

When there is only a small difference between the temperature values observed at the bottom of the well and at the surface, respectively, the use of a constant specific weight belonging to the average of the two temperatures does not cause considerable error. This condition is usually characteristic for wells under operation because the cooling of water entering from the layer and having high temperature is negligible in most cases because the flow in the pipe and reaching the surface does not require too long a time. It is sufficient, therefore, in such wells to measure the temperature at two points. When the water is in a static condition in the pipe (this is the case of observation wells), the vertical distribution of temperature follows the geothermic character of the surrounding layers since a heat balance develops horizontally at each elevation. A slight difference may occur because of the different thermal conductivity of the layer, the water, and the pipe. In these cases, a more detailed heat profile should be determined since the change in temperature must be considered. The development of the heat balance is also a transient process, which fact draws the attention to the possible error caused by using wells under operation but closed to perform the measurements for the observation of pressure. In such wells, reliable measurements can be expected only after the development of both dynamic and thermal balances.

Up to now, only the specific weight of pure water (and its change with temperature and pressure) was analyzed. In nature, however, there is no pure water at all since water always contains solid and gaseous media dissolved in it. The solid media generally have higher specific weight than water and the opposite relation is usually valid for gases. It is evident, therefore, that the greater the total amount of dissolved salts, the greater the specific weight of water while the latter decreases with increasing concentration of gases in most cases except CO_2 which increases the specific weight of water.

The structure of the molecules and the molecular weight of the constituents ought to be taken into account for the precise consideration of the influence of the dissolved salts on the specific weight of water. Sufficient results may be achieved, however, within a range including most of the values usually occurring in practice, even if some considerable simplifications are accepted. Selecting the most simple relationship from the literature (Juhász, 1976), the following equation may be proposed to calculate the specific weight of water γ_{oT} [p cm^{-3}] having a temperature of T(°C) and a concentration of C (total weight of salts dissolved in a unit volume of water) [p cm^{-3}]:

$$\gamma_{CT} = \gamma_T + A\,C \quad ; \qquad (4\text{-}74)$$

where γ_T is the specific weight of distilled water [p cm^{-3}] at the same temperature and A is a constant, the numerical value of which can be approximated as; A = 0.6 - 0.7.

The calculation of specific weight modified by the gas content of water is more complicated than the characterization of the influence of dissolved salts because the gases may be present in the water not only in dissolved form, but they may form a mixture with the water as well (free gas content). The total amount of gas which may be dissolved in water has an upper limit (saturated condition) above which the system is composed of two phases (i.e., saturated water and free gas). This limit (bubbling point) depends on many factors. The most important ones are:

- the type of gas;

- the salt content of the water;

- the temperature of the water;

- the pressure prevailing in the system.

Since the last two parameters may change considerably along the length of a well (especially in the case of operating wells), the gas-water conditions also generally differ as a function of depth.

In less complicated cases, when the gas is present only in dissolved form, some information can be given on the change of specific weight caused by the gas content of water. Subbota (1972) has published graphs showing the change of specific weight ($\Delta\gamma$) depending on the original specific weight of water (which is supposed to characterize the temperature, the salt content, and the pressure of water in a combined form) for three different gases being most commonly dissolved in water (i.e., CO_2, N_2, and CH_4), assuming the concentration to be 1 m^3/m^3 (that means the volume of the gas dissolved in 1 m^3 of water is also 1 m^3 under atmospheric pressure. These graphs are repeated here in Fig. 4-104.

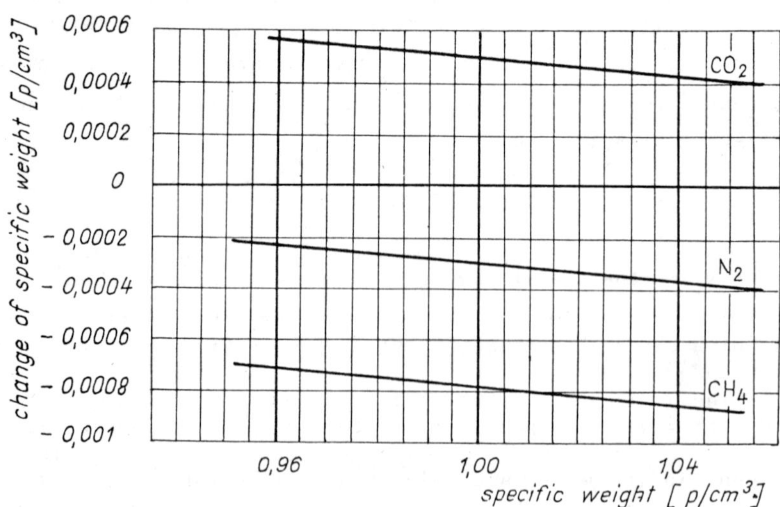

Fig. 4-104. Change of the specific weight of water due to its gas content (assuming 1 m^3/m^3 amount of dissolved gases) as a function of the original specific weight for three different gases.

Another way to characterize the influence of gases dissolved in water is the comparison of two probable vertical distribution curves of specific weights in a water column in a balanced condition, both dynamically and thermally, one determined for pure water and the other calculated by assuming water is completely saturated by the gas in question. An example is shown in Fig. 4-105, determined by Juhász (1976), assuming an increase of temperature with a gradient of 1°C per 20 m and investigating water saturated by methane.

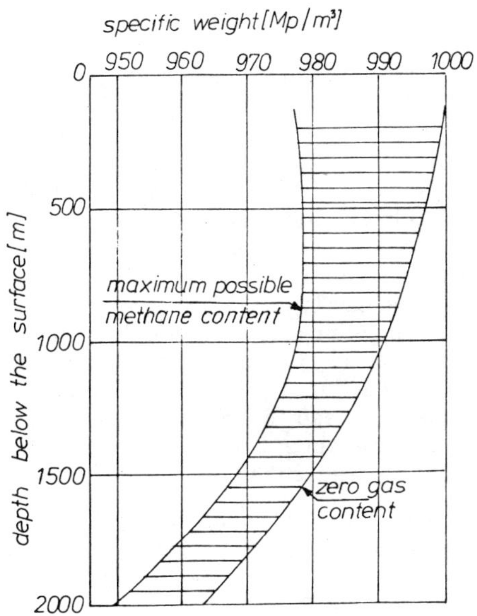

Fig. 4-105. Vertical distributions of the specific weight of pure water and water saturated by methane, respectively.

When the changes of the phases are expected within the system, the detailed investigation of the threefold relationship between pressure, temperature, and volume is required for the sufficient characterization of the change of specific weight of water. This analysis is further complicated when more different gases are dissolved in the water because their bubbling points differ from each other. In the case of high temperature, it should be taken into account that even the water itself may change its phase above a limit value of temperature, the latter depending on the pressure prevailing in the system.

A further correction is necessary when the pressure prevailing in the layer has to be determined from data measured on the surface at the head of wells under operation. It is well-known that a considerable amount of energy is consumed by the resistance against flow (friction) when water is transported in a pipe. The head loss developing along the pipe depends on the length, the diameter, and the roughness of the pipe as well as on the flow rate conveyed through the system. It is evident that the pressure head prevailing at the upper edge of the screen inside the pipe (h_m) is equal to the sum of the pressure heads related to the same reference level and measured at the surface (h_s) and the head loss in the pipe (Δh):

$$h_m = h_s + \Delta h \quad . \tag{4-75}$$

The characteristic curve of a well showing the drawdown vs. yield relationship can be constructed in two different ways. In the first case, the yield is related to the depression observed directly at the surface (s) and this curve characterizes the whole well system (including the vertical pipe transporting the water as well). When the purpose of the investigation is the analysis of the water yielding capacity of the aquifer, the corrected values of drawdown should be used to construct the characteristic curve (s') because this parameter describes the energy conditions at the level of the screen. The pressure head developing in a static condition and related to the reference level chosen at the upper edge of the screen should be indicated with the symbol of h_o. The contact between the two different drawdowns can be expressed by the following equations:

$$s = h_o - h_s \quad ;$$

$$s' = h_o - h_m \quad ;$$

$$s' = s - \Delta h \quad . \tag{4-76}$$

The head loss in the pipe can be calculated from the well-known equation of pipe hydraulics:

$$v = \sqrt{\frac{2g}{\lambda}} \sqrt{d \frac{\Delta h}{\ell}} \quad ; \quad \text{and therefore}$$

$$\Delta h = \lambda \frac{\ell}{d} \frac{v^2}{2g} \quad ; \tag{4-77}$$

where ℓ is the length and d is the diameter of the pipe (if the pipe is composed of more stretches having different diameters, the head loss should be calculated for each stretch separately and the total loss can be obtained by the arithmetic sum of the parts); v is the mean velocity in the pipe (which is the ratio of yield to the area of the pipe cross section $v = 4Q/d^2\pi$), and λ is the coefficient of resistivity which depends on many factors such as the diameter and roughness of the pipe, the condition (laminar, transition, turbulent) and velocity of flow. There are numerous methods published in the literature for the calculation of the λ parameter, among which, the equation derived by Colebrook and White may be proposed for the characterization of turbulent flow in rough pipes:

$$\frac{1}{\sqrt{\lambda}} = 1.74 - 2 \log \left(\frac{2k}{d} + \frac{18.7}{Re\sqrt{\lambda}}\right) \quad ; \qquad (4\text{-}78)$$

where k is the absolute roughness of the pipe (some numerical values are listed in Table 4-12) and $Re = vd/\nu$ is the Reynolds number of the fluid flow. The solution of the equation requires iteration, but it is theoretically well based and gives good practical results when the relative roughness, $\varepsilon = k/d > 10^{-6}$. In cases when $Re\sqrt{\lambda}\ k/d > 200$, the second member in the logarithmic expression is negligible (smaller than 5 percent of the first member), thus the equation can be simplified:

$$\frac{1}{\sqrt{\lambda}} = 1.74 - 2 \log \left(\frac{2k}{d}\right) \quad ;$$

$$\lambda = \frac{1}{\left[1.74 - 2 \log \left(\frac{2k}{d}\right)\right]^2} \quad . \qquad (4\text{-}79)$$

Another simplified form can be used in the case of hydraulically smooth pipes ($\varepsilon < 10^{-6}$) in the range of $3.5 \times 10^3 < Re < 10^5$ (Blasius' formula):

$$\lambda = 0.316\ Re^{-1/4} \quad ; \qquad (4\text{-}80)$$

while in the case of laminar flow ($Re < 2400$), the coefficient of resistivity is:

$$\lambda = \frac{64}{Re} \quad . \qquad (4\text{-}81)$$

A numerical example demonstrating the application of the explained method and comparing the measured and calculated characteristic curves, respectively, is shown in the next chapter (see Fig. 5-61).

Table 4-12. Roughness of pipes (absolute roughness, k, for Colebrook's formula).

Pipe Material	Condition of Pipes	Absolute Roughness (m)
drawn copper, aluminum	smooth	$0 - 1.5x10^{-6}$
drawn steel	new	$1 - 5x10^{-5}$
welded steel	new	$5 - 10x10^{-5}$
	slightly rusted	$1.5 - 2x10^{-4}$
	strongly rusted	$- 3x10^{-3}$
cast steel	new with coating	$0 - 1.2x10^{-3}$
	new without coating	$2.5x10^{-4}$
	rusted	$- 1.5x10^{-3}$
	strongly rusted	$- 3.0x10^{-3}$

When the gas content of water is higher than the limit of the saturated condition, uncertainties are also caused in the computation of the head loss developing in discharged wells. The velocity of the gas bubbles from that of water particles (slipping) and the resistivity is also modified by the bubbles. The determination of the drawdown corrected by considering the energy consumption in the pipe of a well requires, therefore, a more detailed investigation when a medium composed of two phases is transported in the system.

The upper edge of the screen was selected as the reference level in the previous calculation and the head loss developing between this level and the surface was always determined. The flow conditions prevailing below this depth, within the screened stretch of the tube, are more complicated than those within the closed part of the pipe. The collision of water particles entering horizontally into the well causes high losses in energy, surpassing many times the resistance above the screen. The results of this process is the gradual decrease of drawdown acting on the layer going downwards along the screen. The resistance inside the screen acting against the flow developing in the direction of the axis of the well related to the resistance of the layer against horizontal flow influences the active length of the screen. The detailed investigation of this process was discussed, therefore, in connection with the design of screening (Subsection 14-3).

16-2 Investigation of the Fluctuation of the Piezometric Level

The maps representing the horizontal pressure distribution in an aquifer (the contour maps of the piezometric surface) can be constructed for deep groundwater by using the data recorded simultaneously and corrected as it was done similarly in the case of shallow groundwaters. The task is, however, more complicated in the present case. The difficulties are caused not only by the lower number of the available data, but also by the fact that the determination of the hydraulic gradient in a horizontal plain does not provide sufficient information, but the spatial (three-dimensional) investigation of pressure distribution is most important. The uncertainties are further increased since, in most cases, the gradients are relatively low. Thus, the distinction between the hydrodynamic zone and the zone of energy accumulation is very difficult. The consideration of the properties of water other than its pressure condition (chemical composition, isotope content, temperature) was already mentioned as a possible way to collect further information about the development of groundwater flow. It is necessary, however, that the largest possible information content should be utilized from the recorded pressure values as well. This objective may be achieved if not only the pressure heads observed simultaneously at various points of the system are compared, but the fluctuation of the pressure at each observation well is also investigated.

Rónai (1978) has distinguished four different types of fluctuations by analyzing the continuously recorded water level data of deep observation wells. Both the magnitude and the character of these fluctuations are different. They can be distinguished according to the length of their periods which parameter also provides information, in some cases, on the process causing the fluctuation:

- fluctuation with a period of 12 or 24 hours caused by the tidal effect of the earth;

- fluctuation having a period of several days following the changes in the atmospheric pressure;

- seasonal fluctuation with a period of one year and having similar character as the accumulation and depletion of the shallow groundwater;

- long-term fluctuation which seems to be correlated to the wet and dry groups of years.

The amplitude of the fluctuation caused by tidal effects and observed in an aquifer, composed of fine sand which lies at a depth of 870-884 m below the surface, is only 1-2 cm. The change in piezometric head shows practically the mirror image of that of the gravity field (Fig. 4-106). Since the fluctuation is very small, it can hardly be observed where other stronger effects influence the change of the pressure head, as it is shown by Fig. 4-107, in which case a wave caused by the change of atmospheric pressure masks the tidal fluctuation. The figure also demonstrates that the tidal effects are negligible at smaller depth although the material of the layers observed by the other wells is almost the same as that of the deepest aquifer investigated.

The first conclusion drawn from this comparison is the assumption that the fluctuation is not caused by the movement of the mass of water (as in the case of the tides of oceans), but by the deformation of the crust (which is measurable, e.g., in karstic caves where the width of fissures changes with the change of the gravity field (Maucha, 1975) since the movement of water would cause larger fluctuations near the surface. It can also be supposed that any change of pressure is taken up by the water in more compacted layers (at greater depth) while near the surface the deformation of the looser solid matrix responds to the change of pressure acting on the layer without causing observable modification of pressure prevailing on the water stored in the pores.

The fluctuation caused by the change of the atmospheric pressure is more considerable than that due to the tidal effect. Its amplitude was 2-15 cm in the analyzed cases, depending not only on the change in the atmospheric pressure but also on the depth of the aquifer as it is shown in Fig. 4-108. In the case represented by the figure, the atmospheric pressure decreased by 14.5 mm of mercury during three days. At the same time, the water levels rose in the wells observing the deep-lying aquifers. The rise in the deepest well (870-884 m) was 8 cm, at a depth of 351-358 m it was 4.5 cm, and in the aquifer being drained between 193-210 m only 3.5 cm. Similar fluctuation occurs in wells observing shallow groundwater if the water-bearing layer of the first water horizon is confined. During the three days of the period investigated, only a rise of 1-2 cm was observed in a confined shallow groundwater while in unconfined systems there was not any considerable change.

The well and the aquifer may be regarded physically as two intercommunicating vessels. When the pressure condition changes in one of them, the corresponding changes should develop in the other as well. Theoretically, the atmospheric pressure acts on both parts of the system. This action is a direct one in the case of the well (the air mass lies directly on the water column) while the weight of the air is taken up by the upper layers which transfer the pressure only partially

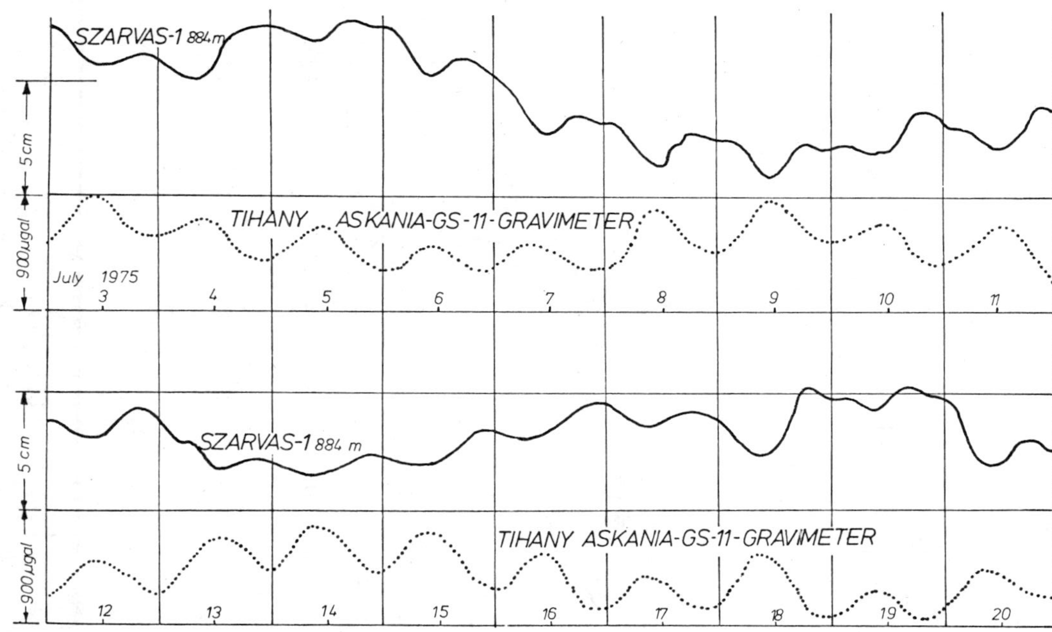

Fig. 4-106. Fluctuation of pressure head due to tidal effect compared to the changes in gravity field.

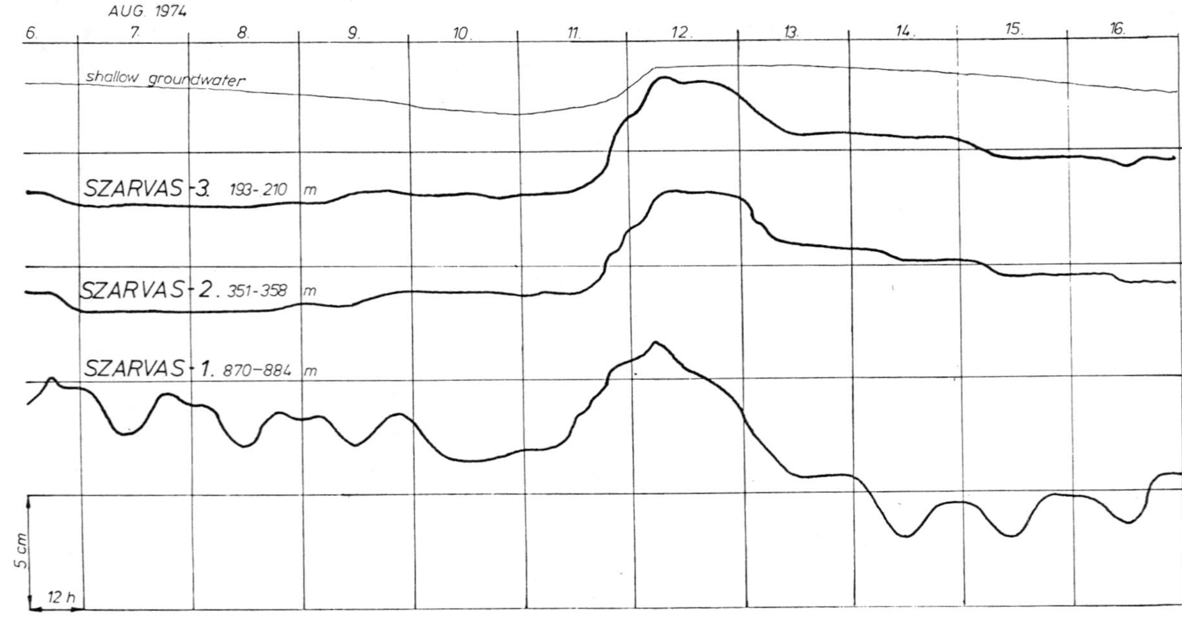

Fig. 4-107. Comparison of fluctuations due to tidal effect and change in atmospheric pressure, respectively, on the basis of records of wells having different depths.

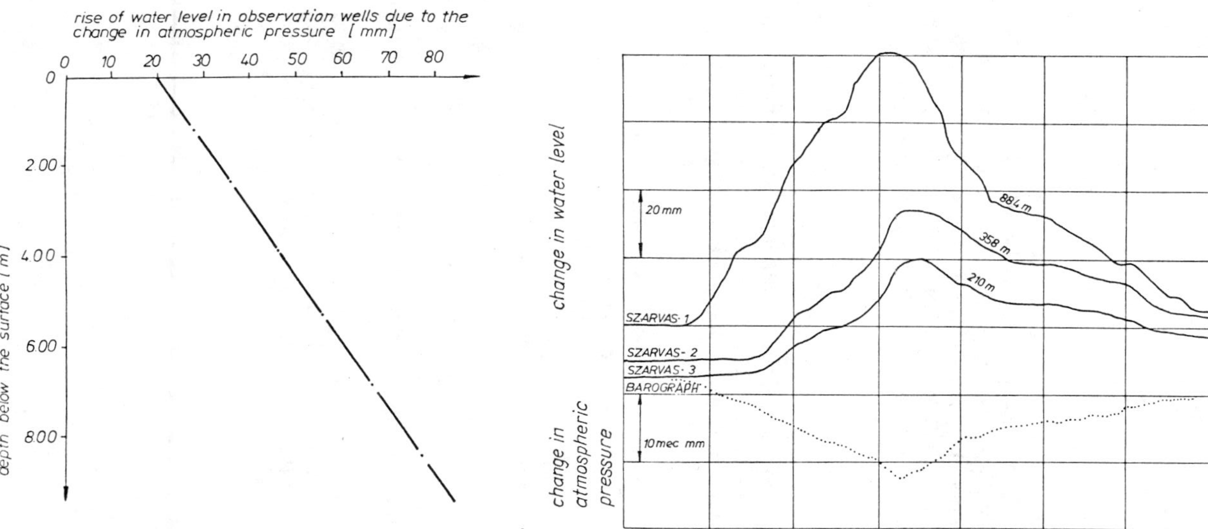

Fig. 4-108. Relationship between the fluctuation of the atmospheric pressure and that of the water level in a well observing deep aquifers.

to the water stored in the pores. The water level would remain unchanged in the well if the decrease of pressure over the water column of the well and in the water content of pores were equal to one another. The observation according to which the rise of the water level in the well increases as the depth of the point investigated in the system increases indicates that the greater the thickness of the overlying layer the more the change in surface pressure is consumed by the deformation of the formation. In the case represented in Fig. 4-108, the greater part of the change of pressure (which is equivalent to a water column of 19 cm) effects the shallow groundwater (maximum 10 percent of it taken up by the solid matrix). At a depth of 300 m, the ratio of the role of the water and the skeleton, respectively, carrying the load is about 4:1 while this ratio decreases to 1.5:1 at a depth of 800-900 m (20 percent of the load is carried by the solid matrix at the depth of 300 m and 40 percent in the deepest aquifer).

It can be stated as the most important conclusion of this analysis that any change in the load over the invetigated aquifer causes the change of the prevailing pressure in the water stored in the pores of the layer almost without any time lag. One part of the load is taken up, however, by the solid matrix of the porous medium, thus the change in the energy content (which is measured by the change of the load) also causes the deformation of the formation and only the remaining part is observable in water pressure. It is evident, and it was also proved by the data represented in Fig. 4-108, that the influence decreases as the thickness of the layers lying between the plain where the change of the load takes place and the aquifer investigated increases.

By comparing the hydrographs recording the pressure conditions at a given section in various aquifers lying below each other and the fluctuation of the water table of the shallow groundwater, it can be clearly demonstrated that there exist some interrelationship between the water regime of the first horizon of groundwaters and the pressure conditions prevailing in the water stored in deep-lying layers even if the depth of the investigated aquifer is more than 1000 m. The existence of this connection is proved by the fact that all the hydrographs recorded in deep horizons are very similar to those of the shallow groundwaters. There are even cases when the fluctuation of the pressure at greater depth is identical with the changes of the position of the water table. The similarity is demonstrated by Fig. 4-109 in which the hydrographs observed in three vertical sections (a. Kecskemét; b. Szarvas; c. Kerekegyháza) at different depths are compared.

When the purpose of the hydrological investigation is the determination of the available groundwater resources, the first question to be answered is whether the aquifer has

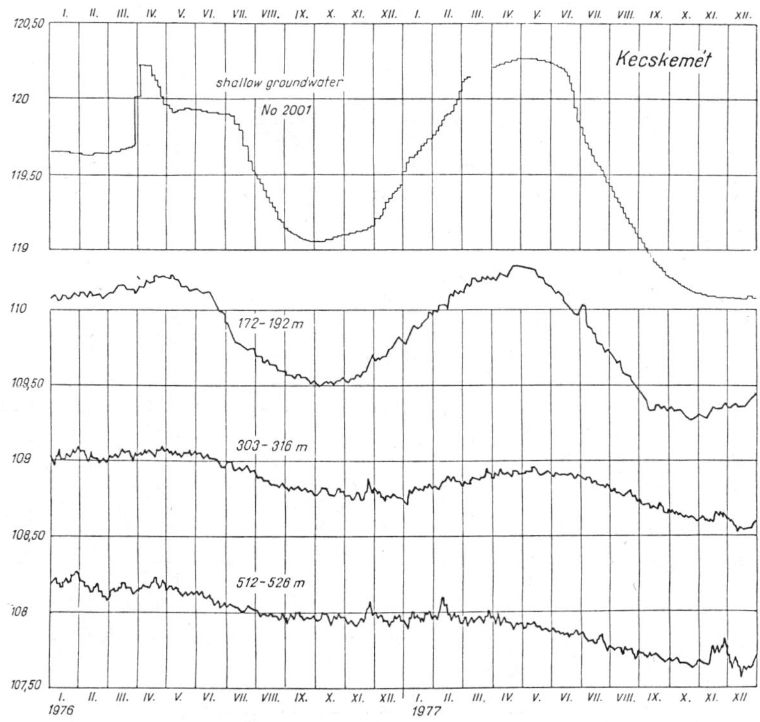

Fig. 4-109a. Comparison of groundwater hydrographs recorded in the same vertical sections at different depths: a) Kecskemét; b) Szarvas; c) Kerekegyháza.

natural recharge and drainage; it is a part of a hydrodynamic system in which the water is in a continuous movement, or it belongs to the zone of consolidation where the accumulation energy is the characteristic process governing the system and exploitaton causes the gradual depletion of the stored resources. It was explained, however, in Subsection 13-1 that the position of the border separating the hydrodynamic zone and the zone of consolidation is very uncertain. A large range (between 100 and 1000 m) is indicated in the literature as the probable depth of this border, and the influence of the local conditions is always emphasized as a basic aspect. The practical importance of the problem and the uncertainty of the distinction between the two zones can be mentioned as the main reason behind the sharp debate concerning this topic. There are scientists completely denying the development of

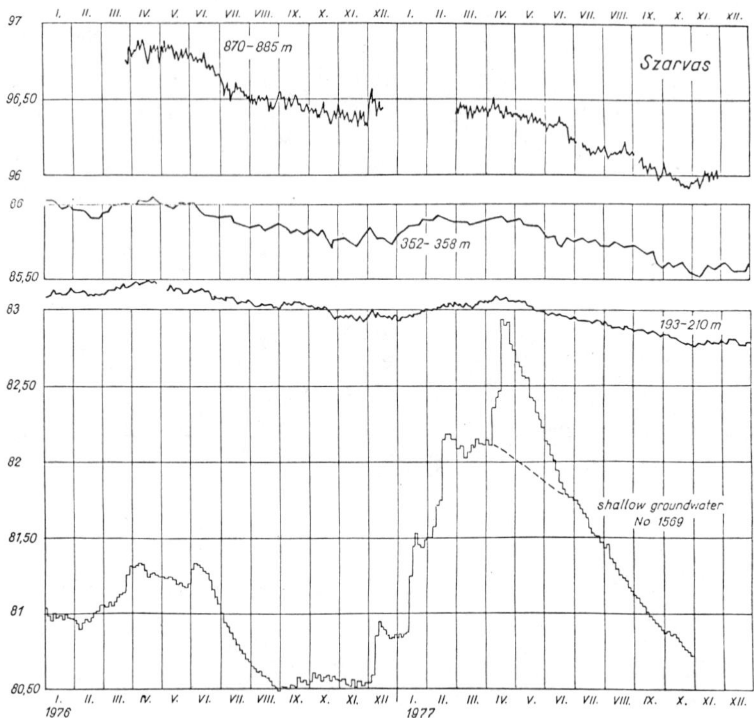

Fig. 4-109b. Comparison of groundwater hydrographs recorded in the same vertical sections at different depths: a) Kecskemét; b) Szarvas; c) Kerekegyháza.

groundwater flow below shallow groundwater (Balló, 1976; Urbancsek, 1976), others apply the hydrodynamic principles even for the investigation of very deep aquifers, assuming that the movement of water is the dominant process governing the water regime everywhere in sedimentary basins.

Rónai has made an attempt, in his paper already quoted, to explain the similarity of the fluctuations observed at different elevations by assuming a vertical groundwater flow is the most important process creating a connection between groundwater horizons using, at the same time, the fluctuaton of pressure to prove the existence of groundwater flow. The basic hypothesis of his theory is to assume the continuous recharge of the deep-lying aquifers from the direction of the edges of the basin. This input is balanced by the vertical

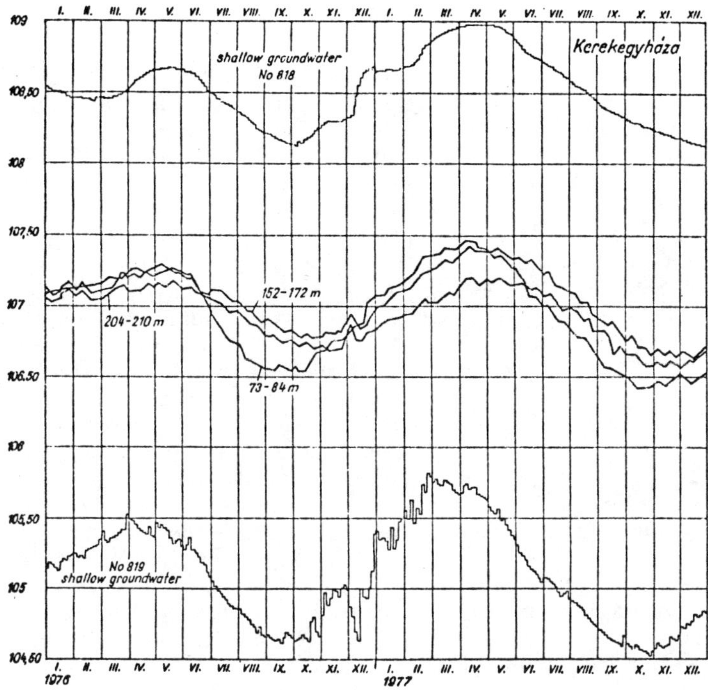

Fig. 4-109c. Comparison of groundwater hydrographs recorded in the same vertical sections at different depths: a) Kecskemét; b) Szarvas; c) Kerekegyháza.

flow developing in the middle of the basin and directed upwards. The water raised near the surface is drained by evapotranspiration. Evaporation consumes a considerably smaller amount of water during wet and cold periods, thus the flux of the vertical flow is also negligible compared to that developing in the warm and arid summer half-year. The smaller flux requires a smaller gradient in the system. Since Rónai has assumed a constant draining level, the decrease of the gradient should cause the increase of pressure, according to this model, in all horizons which take part in water transport.

Theoretically, the process of water transport could perform a role of transmitting actions along the system and controlling the pressure condition in deep aquifers. It is also an acceptable argument that the similarity of the hydrographs would be an a priori expected result if the model were verified since both the depletion of the shallow groundwater and the drainage of the deep aquifers are governed by evapotranspiration. The process of verification cannot be applied, however, in the opposite way: the similarity might be the result of water transport, but it does not prove the existence of flow because there are other processes also resulting in the same control of pressure. Rónai's model is applicable, therefore, only in those cases when the development of water transport was already proved by some other methods (e.g., by investigating the mass balance of water, its chemical constituents, or its isotope content). It is necessary, however, to analyze the other processes which could control the water pressure in deep-lying aquifers.

It is well-known that the change of pressure at an internal point of a water body is proportional to the change of water level (Fig. 4-110a). This linear proportionality is valid even in cases when a part of the space, where the water level fluctuates, is occupied by solid grains (Fig. 4-110b). Thus, the pressure difference does not depend on the weight of water above the investigated point, but on its height only (hydrostatic paradox). This relationship is basically altered, however, when the water body is divided into two parts by an impermeable but elastic membrane. The pressure is recorded in the lower closed space and the fluctuation takes place in the upper part. The pressure is proportional, in this case, to the specific load transmitted by a unit area of the membrane to the water. In a cylindrical container, the specific load (the weight of the water column having a basis of unit area) is equal to the product of the height of the column and the specific weight of the water, thus the change of the pressure head is identical with the fluctuation of the water level (Fig. 4-110c). When the upper part is filled with solid grains and water, the original load is composed of the weight of the two materials. The change of the pressure head is equal to the change of the weight of water, in this case, and the latter can be approximated as the product of the change of the water level and porosity (Fig. 4-110d). The relationship is further complicated when the whole container is filled with a loose clastic sediment. A part of the load is temporarily taken up, in this case, by the solid matrix, the deformation of which is a slower process than the transmission of pressure, thus, some time lag is expected in the development of the corresponding pressure values compared to the action of the load.

It can be stated as a final result of this analysis that the fluctuation of the pressure head is expected to be simultaneous and identical with that of the water table if the water saturating the pores forms a continuous body. There

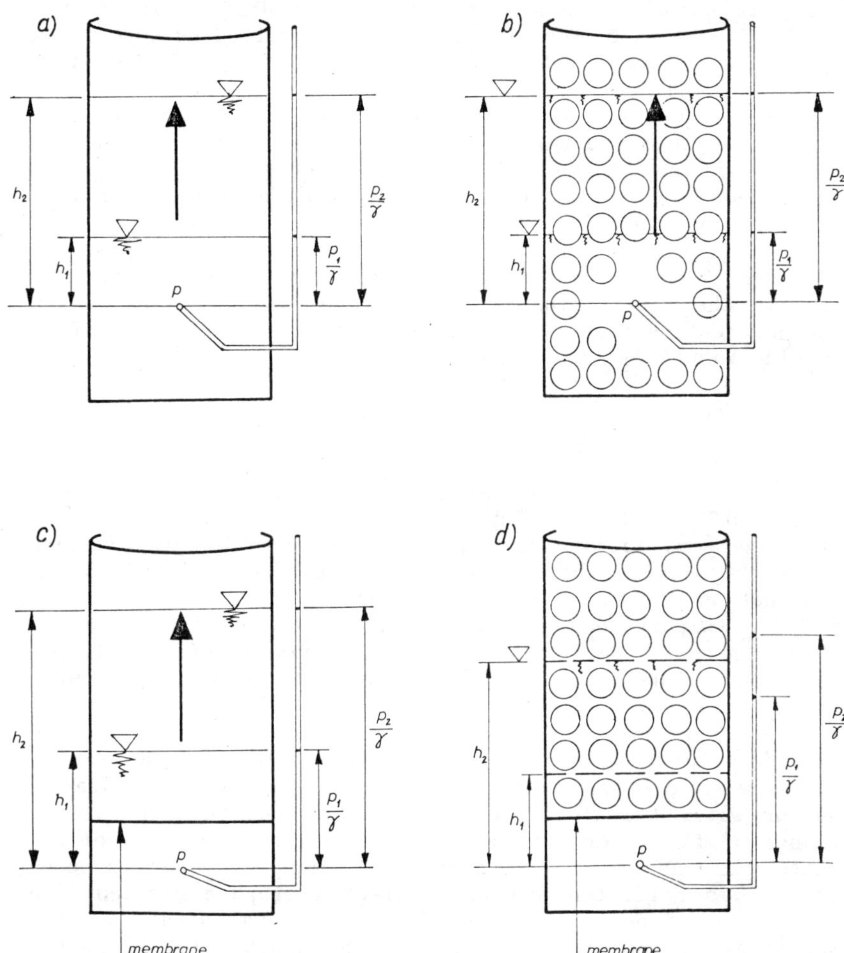

Fig. 4-110. Relationship between the change of pressure inside a water body and the change of the water level.

exist at least some larger channels through which the transmission of pressure is not hindered by retarding forces. The preportion of the amplitude of the seasonal fluctuation of pressure head related to the same parameter of the water is unity (some scattering may be caused by the deformation of the layers, by the local changes in the vicinity of the observation, by the different measuring techniques, etc.) until a depth above which this direct contact characterizes the formation. The identical run of the hydrographs (without any considerable time lag between the corresponding points) is also expected in this zone. The fluctuation of the water table may also influence the pressure conditions at greater depths if the change of the load is transmitted by impervious but elastic layers. The amplitude of the yearly hydrographs constructed from the pressure head data recorded in deep observation wells is smaller than the seasonal fluctuation of the water table and also the graphs are elongated (the development of the minimum of pressure head shows considerable delay compared to the start of the period of accumulation in shallow groundwater).

The examples represented in Fig. 4-109 were selected to compare the fluctuation of pressures at different depths in the vertical sections where the hydrological conditions differ considerably from each other.

At Kecskemét (Fig. 4-109a) the total energy head (the sum of the elevation above a common reference level and the pressure head) continuously decreases when the depth of the investigated aquifer increases. The fluctuation of pressure at a depth of 180 m follows a similar pattern and has almost the same amplitude as the hydrograph of shallow groundwater. Below this level, the amount of the amplitude related to the fluctuation of the water table is about 0.3-0.4 and seems to be almost independent of the depth. There is a time lag of about three months between the development of the lowest position of the water table and that of the minimum pressure.

At Szarvas (Fig. 4-109b), the gradient is directed upwards within the entire section (the shallow groundwater has the lowest energy head). All the hydrographs recorded in deeper water horizons have smaller amplitude than the yearly change of the water table (even if the rapid increase of the latter in April 1977, caused by some local influencing factor, is neglected and the corrected graph indicated with the dotted line is considered), and the magnitude of the amplitudes increases slightly with the depth. (It is necessary to note that the first deep aquifer observed is at a depth of 200 m. The change of pressure in the upper lying sandy layers may have fluctuations identical with that of the position of the water table. The general trend of energy head decreasing in time in deeper formations, which is also observable in the section at Kecskemét, may indicate either a stretch of a long-term fluctuation or the influence of the exploitation of deep groundwaters by water works.)

In the section at Kerekegyháza (Fig. 4-109c), the energy heads of deep aquifers do not differ, practically, from each other. The differences are smaller than 10-20 cm which value does not exceed the possible error of the measurements and corrections (e.g., the influence of temperature was neglected, the readings were corrected only by considering the change of specific weight due to the increase of pressure at greater depth). Thus, the possibility of vertical flow is excluded between these aquifers. The hydrographs of two wells observing the shallow groundwater are represented in the figure which indicate a difference of about 3.5 m in energy head between the two points, proving the statement that the water table roughly follows the relief of the terrain. The energy head of the deep aquifers is approximately equal to the average of the two values observed in the shallow wells. It is probable, therefore, that there is no vertical water exchange between the shallow and the deep groundwaters either. The form and the amplitude of the hydrograph recorded at a depth of about 80 m is almost identical with those of the hydrographs of the water table while a time lag of two months and a decrease of the amplitude characterized with a ratio of about 0.5 are observable at the two deeper horizons.

The results achieved by analyzing the hydrographs represented in Fig. 4-109 seem to contradict Rónai's model in two points:

- there is no vertical gradient in the section at Kerekegyháza, thus the process of water transport cannot be the transmitter of pressure fluctuations;

- the rapid change of the character and the amplitude of the hydrographs (observed at Kecskemét between 180 and 300 m as well as at Kerekegyháza between 80 and 160 m) cannot be explained on the basis of this model.

It is necessary to note also that the fluctuation should decrease in the direction of flow where the transport creates a connection between the pressures developing at various depths. Although this relationship is satisfied in the lower part of the section at Kecskemét and in the whole range observed at Szarvas, but the change of the rate of amplitudes is very small, it does not exceed the uncertainties of this parameter indicated by the large scattering of data and caused, e.g., by the fact that the pressure conditions are also governed by the consolidation of the layer or by the gas content of water. Thus, the trend of the depth vs. fluctuation relationship does not contradict the model, but it does not provide sufficient evidence to prove the validity of the method.

At the same time, a clear distinction can be recognized according to the two possible groups of hydrographs explained in connection with Fig. 4-110. The change of pressure in the first deep horizon follows the same pattern as the fluctuation of the water table at Kecskemét and Kerekegyháza while the other graphs have an elongated form (the amplitude is smaller and the development of the extreme values is delayed). The ratio of amplitudes were plotted as a function of depth for each observation point where more observation wells were available along a vertical section. The purpose of this investigation was to determine the border between the zone characterized by the two different types of pressure transmission (either directly through water bodies or through elastic impervious layers).

Figure 4-111 represents the W-E geological section between the Danube and the Tisza River supplemented by the graphs showing the amount of amplitude vs. depth relationship. The character of the formation (distinguishing the aquifers, the aquitards, and the layered formations composed of sandy and silty layers) and the boundary between the Quaternary and Pliocene sediments are indicated in the figure among the geological data. The lower bed of the Quaternary formations seem to be almost identical with the lower boundary of the zone within which the pressure is transmitted through continuous water bodies.

In Fig. 4-112a, the depth vs. magnitude of amplitude relationships are represented for three sections located at the northern part of the Great Hungarian Plains. The first deep horizon also shows direct contact with the shallow groundwater at each section, but the layer bordering the upper zone is at a higher elevation than the lower boundary of the Quaternary sediments. Thus, the pressure conditions in the lower part of the latter are governed by pressure transmission in the form of loads through elastic layers. The highest layers observed in the middle of the basin are in a relatively deep position and hydrographs identical with the fluctuation of the water table were observed nowhere (Fig. 4-112b). There is only one exception (Öcsöd at 60 m) where a direct connection is observable in the first deep horizon in spite of the impervious upper layers.

The analysis of the data proves that the model assuming the transmission of the changes in energy content partly in the form of hydrostatic pressure (through a continuous water body) and partly through elastic but impervious layers is applicable in large parts of sedimentary basins. The model is also suitable to distinguish the two zones, those characterized by hydrostatic pressure transport and by the transmission of loads, respectively, and even to determine the boundary between them. The position of this border also provides important information in connection with the possibility of the development of water exchange between the aquifers. It is quite evident that the gradient indicates flow through the layers in which the continuous water body transports the change of pressure without any obstacle since this type of transmission requires the existence of a network composed of pores where the adhesion between water and grains does not

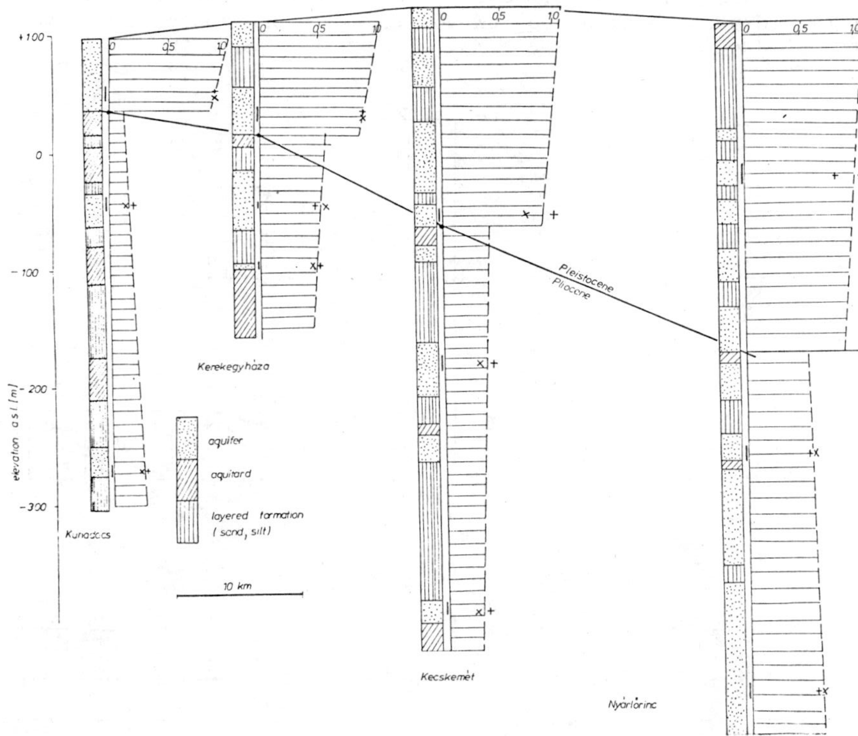

Fig. 4-111. Geological section between the Danube and the Tisza River.

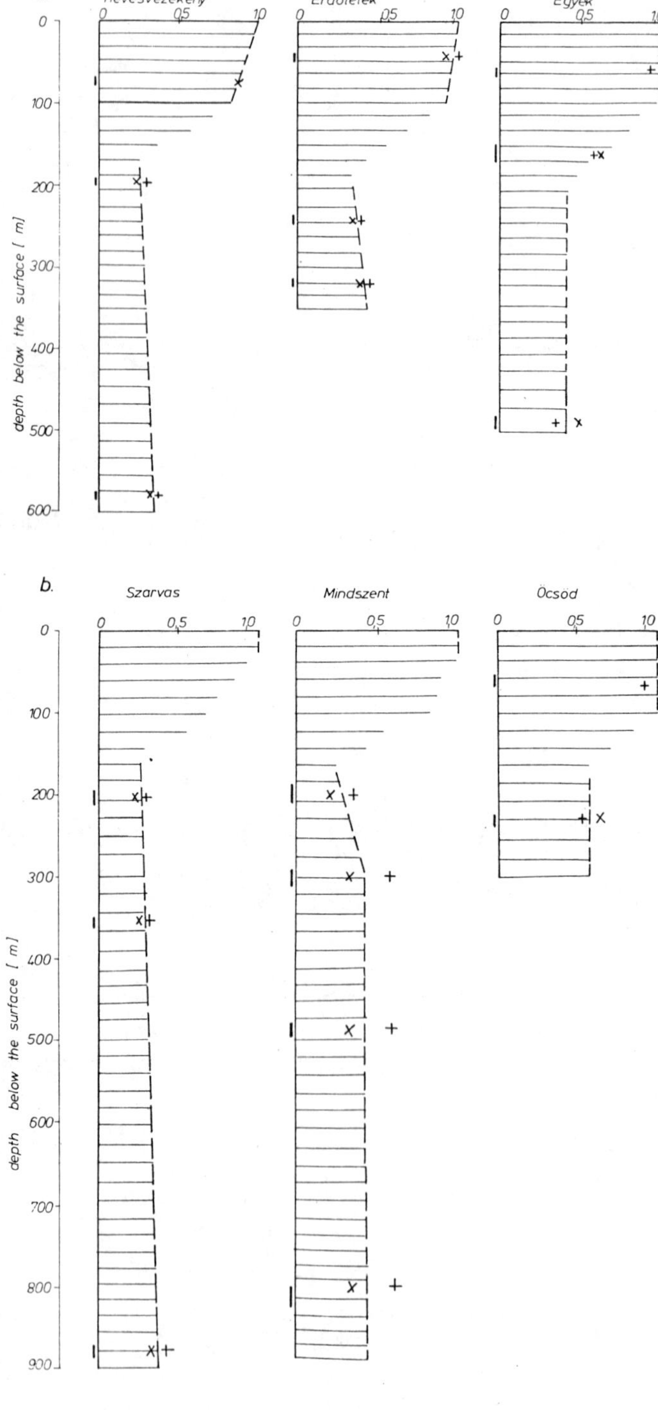

Fig. 4-112. Graphs showing the magnitude of amplitude vs. depth relationship: a) in the northern part of the Great Hungarian Plains; b) in the middle of the basin.

hinder even the propagation of pressure. It follows from this condition that the upper formations of sedimentary basins, where the fluctuation of pressure is identical with the hydrograph of the water table, should belong to the hydrodynamic zone and the zone of energy accumulation may be located only at greater depth. However, this statement is not reversible: it does not mean that the border between the two types of fluctuation should be the upper boundary of the zone of energy accumulation, but the actual gradient may remain below the threshold value with high probability below this elevation. Thus, only the detailed investigation of the mass balance of water, chemical constituents, or isotope content may give an answer to the question whether there is water transport in this zone or not.

Where the principles of pressure transmission through water or elastic layers are applicable and the recorded data prove the reliability of the theory, separation of the formations according to the character of the fluctuation of energy head provides us with valuable information for the construction of the hydrodynamic models of aquifers. It can be seen, e.g., that the hydrogeological unit lying between the Danube and the Tisza Rivers can be simulated with a multilayered system. The Quaternary sediments form a unified formation which is divided from the shallow groundwater only by an aquitard having relatively low resistance. The water exchange between the shallow groundwater and the first deep horizon is significant. The development of flow between the Quaternary aquifer and the Pliocene water-bearing layers is hindered considerably. The aquitard forming the lower bed of the Quaternary formations should be taken into account with a much larger vertical resistivity than the semi-pervious layer separating the shallow and the deep groundwaters.

There are, however, several sections where the model based on the transmission of pressure is not applicable either. It is necessary, therefore, to always investigate all the possible processes which may create the fluctuation of the energy content in deep aquifers and to select the most probable effects governing the water regime and pressure conditions. It is impossible to apply only one model at each part of a sedimentary basin.

The hydrographs recorded at the deepest section observed within the basin are analyzed in the first example to demonstrate a case where the simple model of pressure transmission is not applicable (Fig. 4-113). Apart from the fluctuation of the water table, the graphs show the change of the energy head in four deep aquifers (204-241 m; 428-445 m; 642-655 m; 1029-1056 m). The position of the water table changes according to the normal pattern showing the accumulation and depletion as a result of infiltration and evaporation,

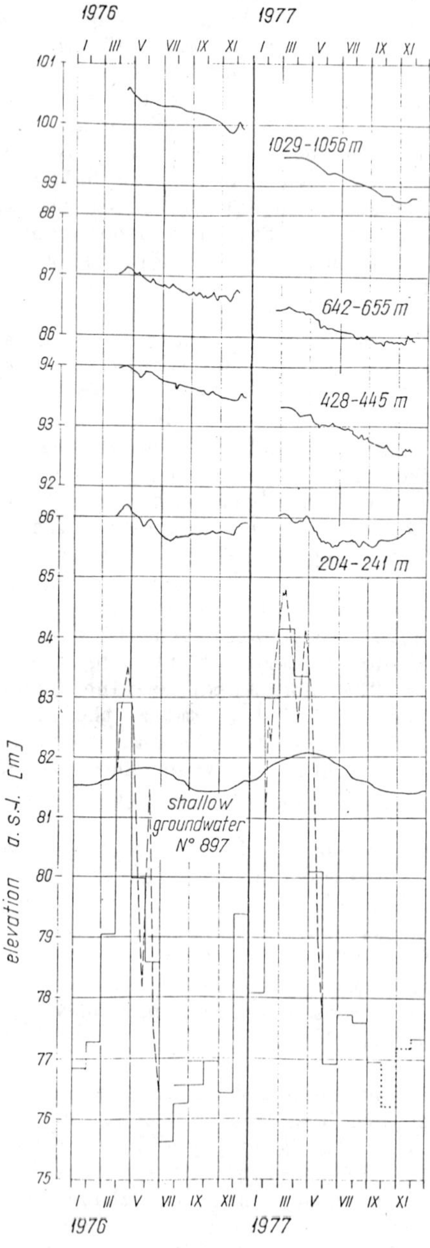

Fig. 4-113. Hydrographs recorded in the various aquifers at Csongrád.

respectively. Seasonal fluctuation was observed in the first deep-lying aquifer (204-241 m) while the other hydrographs indicate the continuous lowering of the energy head. This fluctuation does not show any similarity to the change of the water table either, but it resembles the hydrograph of the water level in the Tisza River which is near to the group of wells. This influence can be observed even in the well being screened between 428 and 445 m although the main character of this graph is its continuous lowering, most probably caused by the over exploitation of the deep aquifers.

The most important conclusion of this analysis is that the versatile system of the different actions does not permit the application of a unified scheme for the characterization of the fluctuation of pressure heads, but the character of the hydrographs assists us to estimate the dominating effects. Another result drawn from the investigation of the hydrographs recorded at Csongrád is that the influence of rivers can be clearly recognized, in some cases, in the regime of deep groundwaters. This observation is in good agreement with the results of earlier investigations based on the analysis of isotope contents of water which have proved that the Tisza River and its tributaries discharge the deep groundwater of the basin and the latter contributes considerably to the base flow of rivers.

Figure 4-114 shows a second example characterizing the influence of surface waters on the pressure conditions of deep groundwaters. The hydrographs, for a period between 1969 and 1976, of four wells located near Szolnok are compared with the average monthly water levels of the Tisza River at Szolnok (Öcsöd 228-233 m and 58-66 m; Tószeg 67-73 m; Szolnok 24-30 m). It is evident that the strongest influence was observed at Szolnok where the well is almost at the bank of the river and the depth of the aquifer is less than 30 m. The energy head fluctuates between 81.50-82.00 m a.s.l. when there is a continuous low water period in the river (e.g., in 1972 and 1973). There is an average difference in energy head of about 2-2.5 m between the river and the aquifer which decreases to 1 m when small flood waves propagate along the river. The large floods (e.g., in 1970, 1974, 1975, and 1976) raised the energy head near or above 85.00 m a.s.l. Floods having long duration hinder the outflow of the groundwater, thus the stored amount is also raised considerably, the depletion of which is a slow process characterized by a slightly lowering recession curve. In this period even a medium flood may raise the water level in the well much higher than the same flood after a low water period (see the comparison of the floods at the beginning of 1969 and 1971, respectively). Similarly, floods following each other with a short time lag may cause a high water level in the well for a long period (see 1974-1976). The average low water level is higher at Tószeg than at Szolnok (between

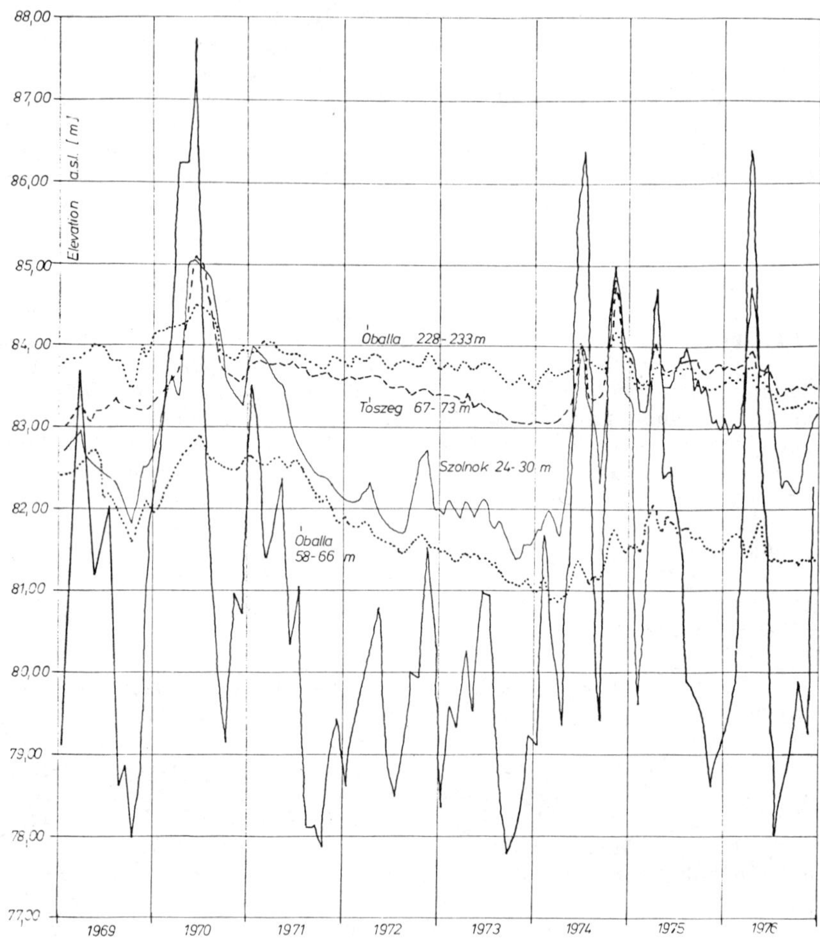

Fig. 4-114. Comparison of the fluctuation of water levels in wells and river, respectively.

83.00-83.50 m a.s.l.). Thus, only the high flood raises the water level considerably in the well. The height of the peaks of the hydrograph are, however, the same as those at Szolnok. The pressure conditions in the two wells at Öcsöd indicate the possible flow from the lower aquifer to the upper one, and from there to the river. The distance from the bank is longer than that in the previous cases, the influence of the river is, therefore, decreased by the larger storage capacity of the system which fact is indicated by the smaller peaks of the hydrographs. The well draining the upper aquifer shows the stronger effect of the river, thus the supposition of a vertical groundwater flow directed upwards and the discharge of the upper horizon by the river seems to be proved.

It can be stated, to summarize the investigation of the fluctuation of the energy head in deep aquifers, that the analysis of this parameter provides valuable supplementary information. It assists to estimate the resistivity of the aquitards separating the aquifers where the model describing the pressure transmission is applicable. At other places the connection between surface waters and deep aquifers can be recognized on the basis of fluctuation and thus the simulation of the flow system is in this way assisted. It is not possible, however, to construct a general method which is suitable to investigate the fluctuation of the energy head. All the processes which may influence the pressure conditions should be analyzed and the models should be constructed by considering the dominating actions. Some aspects of this approach were explained in connection with the examples demonstrated on the basis of hydrographs recorded in deep wells.

16-3 Simulation of Aquifers by Applying Hydrodynamic Models

Both the long-term planning of water resources development and the actual development projects require the numerical characterization of the available water resources. Having long-term records, the design values expected with various probabilities can be determined by statistical analysis of the flow rate data of rivers, and sufficient information can be obtained in this way about surface water resources. To provide similar data on groundwater is a more complicated task.

Difficulties are caused not only by the fact that the aquifers storing and transporting the groundwater are covered (the flow system cannot be observed directly; its geometry is known only at the outcrops and from a few boreholes, thus the information provides us only with rough estimations about the geological structure; the flow parameters are not measurable), it has to also be considered that the groundwater becomes available only by exploitation, the type and the location of the discharging structures as well as their operation influence, therefore, the water amount available for the users. Groundwater is in a balanced condition in nature (the multi-annual averages of recharge and discharge are equal to one another within a region under natural conditions). The artificial drainage (exploitation) causes a drawdown around the draining structures and the whole regional balance is disturbed in this way. Groundwater flow may be initiated from the neighboring areas or from other layers in the direction of the water works.

Evapotranspiration (as one of the most important natural discharging effects) may be decreased by lowering the water table. The other important drainage system of groundwater is the network of rivers. Another consequence of exploitation may be the decrease of the base flow of rivers because the percolation from the aquifers into the rivers is lower if the potential of the groundwater is decreased by drawdown. In many cases, even the direction of seepage is modified, the riverbed may become the recharging place of the groundwater along the stretches where discharging processes were characteristic under natural conditions.

This short summary of the difficulties arising in connection with the determination of groundwater resources clearly indicates that the sole realistic method of the calculation of the available water is to suppose a given average drawdown in the surroundings of the planned water works and to investigate the development of the new regime within the region. All the changes expected as the results of exploitation within the groundwater basin and along its border have to be determined (i.e., lowering of the water table in unconfined or that of the pressure in confined systems; decrease of groundwater provided for the plants by capillarity caused by drawdown; saltwater intrusion as a consequence of the decreased pressure; decrease of base flow in the rivers; the amount of water withdrawn from other basins or recharge from surface water resources; etc.). Evaluating all the effects caused by the different variants of planned exploitations (supposing drawdowns of different sizes at different parts of the basin), some factors can be determined as forming the upper limit of the possible exploitation (saltwater intrusion, deterioration of the environment, etc.). In other cases, the value of the exploited water should be compared to losses caused by withdrawing and consuming it from somewhere else. The economic limits of water withdrawal can be determined in this way. The optimum distribution of the discharging places within a groundwater basin can also be investigated by calculating the drained amounts causing the same effects in the case of the different variants.

The explained concept of the calculation of available groundwater resources requires a large number of computations, the execution of which became possible only after the development of numerical methods and by the use of computers. Horizontal two-dimensional seepage is generally assumed to be acceptable for this analysis in order to simplify the investigations. In this system, however, the boundary conditions prevailing along the various stretches of the flow space cannot be considered in a theoretically correct detailed way. Two hypotheses have to be accepted instead of the determination of entry and exit faces, the free part of the latter, the water table and the impervious boundaries. One of the assumptions simplifies the description of the boundary conditions around the domain while the other substitutes the drawdown in the draining structure, the actual geometry of the latter and the special conditions prevailing along this contour of the seepage space with an average lowering of the potential in the surroundings.

Investigating two-dimensional horizontal groundwater flow, only two types of boundary conditions are distinguished, i.e., the border is either a streamline (impervious boundary or such contour along which the investigated part of the region is separated from the neighboring areas and it is supposed that flow does not develop through this border even after changing the potentials within the investigated domain) or a potential line (entry or exit face through which the investigated field is either recharged or drained). In the second group of boundary conditions it should also be considered whether the flow develops through the stretch of the contour in question without resistance or not, viz., the potential within the seepage field is always equal to that of the contacted surface water body or there exists a difference between the two potential values needed to overcome the local resistance occurring along the boundary. In the latter case it has to be taken into account as well that the potential difference is proportional to the flow rate crossing a unit length of the contour.

The general boundary condition can be described in mathematical form as well by expressing the difference of potential heads prevailing at the two sides of the contour:

$$[\Phi_o(x,y,t)-\Phi_I(x,y,t)]\ \rho_B(t)+q_B(t) - T_B \frac{\partial\Phi_I(x,y,t)}{\partial n_B} = 0\ ; \qquad (4\text{-}82)$$

where $\Phi_o(x,y,t)$ is the potential head at the contour outside the domain (in the contacting surface water body); $\Phi_I(x,y,t)$ is the same value inside the domain; $\rho_B(t)$ is the local resistance at the interface along a unit length of the contour $[LT^{-1}]$; $q_B(t)$ is the specific flow rate crossing a unit length of the contour independently of the pressure conditions $[L^2T^{-1}]$; T_B is the transmissivity of the layer at the contour $[L^2T^{-1}]$; and n_B is the external unit vector perpendicular to the contour.

Any type of the possible boundary conditions may be represented with this general equation by substituting various ρ_B and q_B values into the relationship. Thus, the impervious boundary is characterized with $\rho_B = 0$ and $q_B = 0$ parameters. Entry and exit faces without local resistance can be described by substituting $\rho_B = 0$ and $q_B = 0$ values. Another possible condition is the recharge from a riverbed, the bottom of which is clogged, thus there is a continuous seepage from the river into the layer but its flow rate is independent of the pressure differences because there is an unsaturated zone between the water table and the riverbed. The mathematical form of this boundary condition can be obtained by substituting $q_B \neq 0$ and $\rho_B = 0$ parameters where the numerical value of q_B should be calculated by considering the properties (thickness and permeability) of the clogged layer and the depth of water above the bottom in the river. If the water table reaches the bottom of the clogged riverbed, the recharge depends on the pressure difference, but the influence of the contacting layer having high local resistance should be taken into account. In this case the recharge depends on the pressure conditions ($q_B = 0$) and the local resistance has to be characterized numerically ($\infty > \rho_B > 0$) by considering the properties of the clogged layer.

The continuous domain of the seepage field should be substituted by discrete grid points for the numerical solution of the movement equations. Each point represents the block surrounding it, thus all the effects acting within the block have to be characterized with their average values and these average parameters have to be applied in a concentrated form at the grid points. This is the reason why the geometrical form of a given discharging structure (its actual contour and the boundary conditions prevailing along it) cannot be considered in the computation. The effect of exploitation has to be taken into account in the form of average drawdown estimated for the entire block where the construction of discharging structures is assumed or the flow rates discharged by all such structures (wells or other types of water works) within each block have to be integrated. One of the parameters (the difference between the natural and artificially maintained potential head, i.e., $\Phi_N(i,j) - \Phi_A(i,j) = s(i,j)$ depression, or the sum of the flow rates discharged from the block $\Sigma_{i,j}(q)$) has to be substituted into the numerical model and the other is obtained as the result of computation together with the time variant potential distribution within the domain.

This simplification, which originates from the character of the numerical solution, is in good agreement with the task to be solved when the available groundwater resources have to be determined. In this phase of planning, the precise place and the structure of the wells are not yet fixed. The question to be answered may be raised in two different forms:

- what is the time dependent quantity of exploited water and the change of the potential distribution within the whole domain if a given depth of average drawdown is assumed within the investigated block (the lowering may be constant or changing in time);

- what is the expected time dependent change of the potential head in the block from where a given amount of water (either constant or time dependent, but fixed a priori as a condition of the investigation) is withdrawn, and how does the potential distribution develop in time n the domain surrounding the block discharged by the exploitation.

When the influence of a well (or a group of wells) has to be investigated, the numerical characterization of the domain has to be supplemented by considering the local drawdown around the wells. Any type of numerical methods gives only the average potential value within the blocks and the excess resistance developing around each well should be calculated separately. The operational level in the wells can be determined by adding the excess drawdown to the average depression.

The accretion, either originating from the interaction between soil moisture and groundwater or maintained by the groundwater flow, between the neighboring layers (both actions initiated by the change in potential head within the investigated layer) is a special boundary condition in the case of two-dimensional horizontal flow. It does not act around the contour of the domain, but is distributed over the whole field. The accretion is assumed to be concentrated at the intersections of the grid lines and should be taken into account in the continuity equation determined for each grid point if this action is not negligible.

Investigating the first water horizon (shallow groundwater), the actions reaching the groundwater from the direction of the atmosphere may maintain considerable accretion (infiltration, evapotranspiration). These processes depend, however, on the depth of the water table, the influence of evaporation is negligible, therefore, if the phreatic surface is relatively deep below the terrain while the infiltration becomes constant (independent of depth) in such cases. The other consequence of the relationship between the depth of the water table and the size of accretion is that this parameter has to be considered as a function of the total pressure head at various points of the domain. It is this reason why the computation can be executed only in the form of successive iteration starting with estimated values of accretion, and corrected afterwards by considering the potentials calculated within the preceding step.

The accretion maintained by the groundwater flow developing between neighboring layers has an important role when the influence of exploitation from a multilayered formation is investigated and the three-dimensional flow is approximated by a set of two-dimensional seepage fields assuming that each aquifer may be analyzed as an independent horizontal flow domain and considering accretion through the semipervious layers as the sole action which creates contact between the water horizons. In these cases, the accretion also depends on the change of potentials within the neighboring layers. Iteration has to be used, therefore, in the computation.

In the first example, presented here to demonstrate the investigation of horizontal groundwater flow, the task was the determination of the available groundwater resources within a given stretch of the lowland forming a relatively narrow belt along the eastern shore of the Red Sea. The average width of this belt is about 45 km and the gently sloping terrain starting from sea level reaches an elevation of about 250 m a.s.l. at the eastern side where the lowland is bordered by the steep fault of the Arabian Shield. The depth of the basement is about 100 m at the fault. The thickness of the covering layer increases in the direction of the sea where the impervious Precambrian formation lies at a depth of 400-500 m below the surface (Fig. 4-115). The sediment covering the basement is composed partly from slope deposits and partly from the alluvial fans of the wadis running from east to west in a distance of 30-40 km of each other, thus their deposits form an almost continuous layer. According to its origin, this aquifer is very heterogeneous, mostly composed of sand and gravel but including silts and large boulders as well. The very thick and largely extended layer may be regarded, however, as a continuous, very pervious formation within which the local differences compensate each other, and thus its hydraulic conductivity can be characterized by average values decreasing from the foot of the mountain in the direction of the sea.

The precipitation is very low on the plain, much lower than the potential value of evapotranspiration. Although excess precipitation may occur temporarily, it is stored, however, in the upper soil, there is neither surface runoff nor deep percolation. The water table lies at a depth of some tens of meters almost everywhere under the plain and this depth decreases only near the seashore. Accretion does not influence, therefore, the groundwater regime. The recharge originates from the torrential streams of the wadis. The hydrological investigations have shown that the amount of

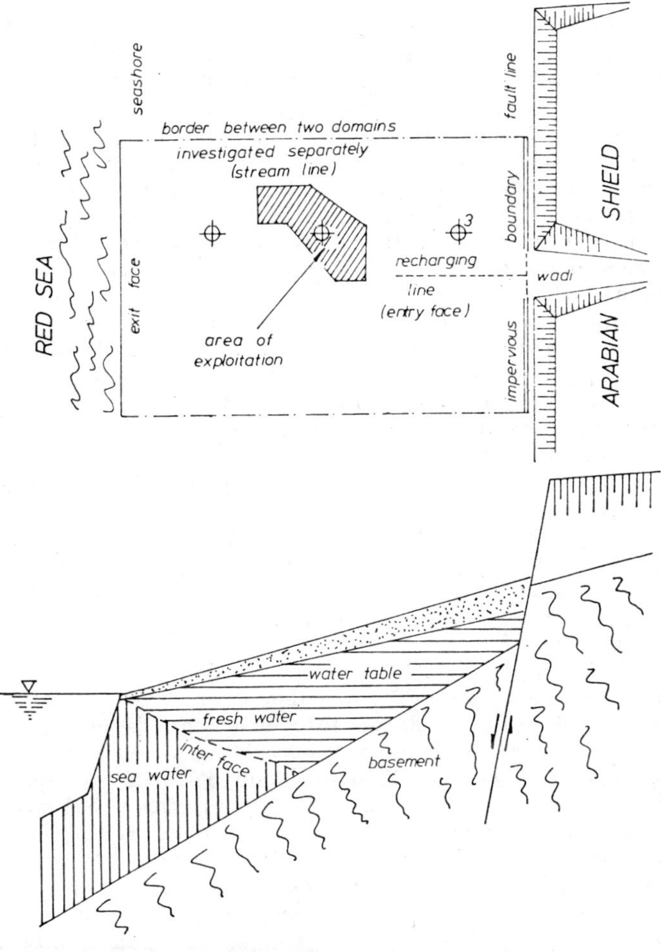

Fig. 4-115. A sketch representing the area simulated with a hydrodynamic model.

water running down from the mountainous areas within the five-month period of the rainy season can be characterized with an average discharge depending on the character of the year. It was found, e.g., that this discharge may be estimated as $Q_d = 3.6\ m^3/sec$ in dry years, $Q_a = 8.4\ m^3/sec$ as the average and $Q_w = 14.8\ m^3/sec$ in wet periods, in the case of the investigated wadi.

The surface runoff does not reach the sea, it infiltrates on the upper part of the plain recharging the groundwater, thus the stretch of the contour recharged is limited to the throat of the wadis and to the upper stretches of the water courses as it is indicated in the figure. The other parts of the eastern border are impervious. Each part of the plain, belonging to one wadi and separated by the lines perpendicular to the seashore and halving the distances between the wadis, can be investigated individually, the dividing lines being streamlines under natural conditions. The separate investigation of this domain is not correct if groundwater is exploited only in one part of the plain since flow commences from the neighboring fields when the potential is lowered by pumping. Supposing the nearly equal utilization of groundwater along the seashore, the approximation is acceptable because almost equal lowering of the water table may be expected in each separate part of the field, thus the dividing lines remain streamlines. The recharge reaching the domain in a fluctuating form, showing not only yearly but longer periods as well, is stored in the aquifer and causes the fluctuation of the water table. It can be assumed that the field is discharged along the seashore, without any local resistance, in an equally distributed form, and the specific flow rate is time independent here.

The continuous water transport from east to west needs the development of a considerable gradient directed towards the sea which is maintained by the sloping water table. Thus, the potential of the freshwater body related to sea level increases going inland. The excess pressure balances the higher specific weight of seawater and depresses the interface between the two types of water into a deep position. The exploitation of groundwater decreases the flow rate reaching the sea. The transportation of the smaller amount of water needs a smaller gradient. The consequence of lowering the water table is the rise of the seawater freshwater interface depending on the yield taken from the groundwater reservoir, and thus the length of time at the end of which the interface will reach the height of the wells can be calculated. The available groundwater resources can be determined, in this way, as a function of the expected length of operation.

The finite difference method was used to solve this problem. A part of the lowland belonging to a given wadi and having an area of about 1200 sq km was covered by a grid. The distance between the grid lines was 1 km in both directions within that part of the area where the exploitation was planned. A less dense network (2x2 km) was applied outside this region. Thus, the total number of grid points was about 700.

The Du Fort-Frankel (1953) approximation was used to express the differential equation in finite difference form. This algorithm has the advantage of yielding explicit and at the same time stable set of equations composed of finite differences, thus it is equivalent to the alternating scheme. The difference between the position of the natural water table $H_o(x,y,t)$ and that of the influenced phreatic surface $H_s(x,y,t)$ may be indicated with a drawdown function depending on time and on the space coordinates (Székely, 1975):

$$s(x,y,t) = H_o(x,y,t) - H_s(x,y,t) \quad . \tag{4-83}$$

Considering the great thickness of the aquifer, the change of the depth of groundwater flow in time is negligible, and the linear form of the differential equation can be used. It was already proved that the accretion can also be neglected in the investigated case. Having a coarse-grained aquifer, the consolidation due to the lowering of water pressure does not have to be accounted for either. Thus, the final form of the simplified differential equation inside the domain is:

$$n_s(x,y) \frac{\partial s}{\partial t} = \text{div}\ [T(x,y)\ \text{grad}\ s] \quad . \tag{4-84}$$

According to the Du Fort-Frankel method, the two differential values included into this equation can be approximated by the following equations of finite differences:

$$\frac{\partial s}{\partial t} = \frac{s(t+\Delta t) - s(t-\Delta t)}{s\Delta t}$$

$$\text{div grad } s = \frac{s(x_{i+1},y_j,t_n)+s(x_{i-1},y_j,t_n)-s(x_i,y_j,t_{n+1})-s(x_i,y_j,t_{n-1})}{(\Delta x)^2}$$

$$+ \frac{s(x_i,y_{j+1},t_n)+s(x_i,y_{j-1},t_n)-s(x_i,y_j,t_{n+1})-s(x_i,y_j,t_{n-1})}{(\Delta y)^2} \quad . \tag{4-85}$$

The boundary condition along the recharged contour is characterized with given specific flow rate q(x,y,t) independent of the pressure conditions while the seashore is a discharging line without local resistance. The other stretches of the contour are streamlines (impervious boundaries). The drawdown is equal to zero everywhere at the beginning of the exploitation which fact determines the initial condition of the system. Considering Eq. 4-82, the mathematical forms of the boundary and initial conditions can be given by the following equations:

$$\frac{\partial s_I}{\partial n_I} = 0 \quad ; \quad \text{(along impervious boundaries)}$$

$$q_R(t) - T_R \frac{\partial s_R}{\partial n_R} = 0 \quad ; \quad \text{(where the field is recharged)}$$

$$s_D(t) = 0 \quad ; \quad \text{(along the seashore)}$$

$$s(x,y,0) = 0 \quad ; \quad \text{(initial condition)}$$

where the suffixes I, R, and D indicate the parameters at the impervious, the recharged, and the discharged contours, respectively.

Relatively little information was available to determine the geometrical and physical parameters of the system. The boreholes reached the basement only at the eastern part of the lowland where the thickness of the aquifer was smaller. Geophysical measurements gave further information about the probable position of the basement. Pumping tests were executed in some boreholes to determine both the hydraulic conductivity and the specific yield of the aquifer. The number of the tests was not sufficient, however, to characterize the average flow conditions within the whole area. The average position and the yearly fluctuation of the water table was observed in some observation wells, and these data were used to check the estimated parameters.

Knowing that the depth and the hydraulic conductivity changes inversely in an E-W section, a constant transmissivity was at first assumed. Starting from a horizontal water table at the elevation of the sea level (initial condition), the influence of the natural recharge during a period of many years was calculated. The output of this computation was the average position of the natural water level which could be compared to the data of the observation wells. Using a trial and error method, the most probable distribution of transmissivity was determined in this way. It was found that assuming a linear change in transmissivity (being 1700 m^2/day at the foot of the mountain and 3300 m^2/day at the sea) the deviation of the calculated heights of the average water table from its observed position was negligible. The seasonal fluctuation of the phreatic surface was also determined by the model at various points and these values were compared to the data obtained from the records of the observation wells. This check proved that a specific yield n_s = 0.22 is acceptable which parameter is also in good accordance with the values estimated from soil physical data and on the basis of pumping tests, respectively.

The model was completely prepared for computation in this way and the influences of the various types of exploitation could be determined and compared. It was found that about 90 percent of the total amount of $3x10^3$ m^3/day of water infiltrating at the upper edge of the lowland was available for utilization in the central part of the project area if the limiting condition was fixed by prescribing that the seawater freshwater interface should not be higher than 100 m below the surface in the region of the wells. The expected development of drawdown in time at three points is indicated on the map in Fig. 4-116. Naturally, the curves demonstrate the influence of a given distribution of pumped wells. The best location of the wells can be investigated by comparing the results belonging to different variants.

The investigations executed to determine the expected effects of the continuously increaing groundwater exploitation to cover the water demand of Madrid on the water regime of the basin of the Tagus River (Spain) was selected as the second example (Llamas and Cruces de Abia, 1976).

The basin is a large graben filled with Tertiary sediments of continental origin as it was already discussed in Subsection 13-2. It is bordered on its northern and southern sides by impervious formations and by karstic rocks at the eastern boundary. Its total area is about 15,000 km. The sediments lie almost horizontally. The relief is generally smooth. It is composed of broad valleys whose bottoms are usually covered by Quaternary alluviums. Low hills and large table lands form the watersheds between the valleys.

The sediments filling up the basin and having a maximum thickness of about 2000 m may be horizontally divided into three lithological facies as it is shown in the geological section (Fig. 4-117). The area where the detailed investigation was executed lies northwards from Madrid. The marginal facies is the dominating formation here within which three main aquifers separated by largely extending semi-impervious

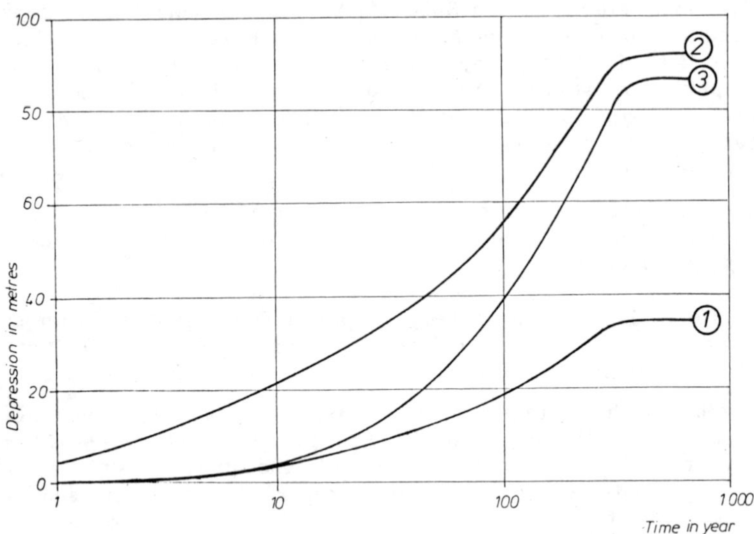

Fig. 4-116. The increase of depression at three points investigated.

layers can be found. The probable paths of groundwater flows governing the water regime of the basin was represented in Subsection 13-2 (see Fig. 4-10).

The large demand for water in the metropolitan area of Madrid and the growing number of wells drilled here have made it advisable to prepare a mathematical model for the simulation of the effects of the different variations of exploitation. Although the first aquifer is unconfined, its thickness is great enough for the change of its transmissivity in time to be negligible. This layer is recharged by the infiltrating rain which is supposed to be constant in space and time. Higher accretion is considered only in the urbanized zones because the leakage from water distribution and sewage networks is estimated to be higher than the decrease of infiltration here. The contact of this aquifer with the rivers through the alluvial deposits is also taken into account. The second and third water-bearing horizons are under pressure. The potential differences between the aquifers maintain groundwater flow through the semi-impervious layers dividing them, and the model has to consider the changes in flow rates caused by the modification of the potentials due to groundwater exploitation.

An area of 30 by 50 km was selected for a detailed investigation and covered a network of uniform squares

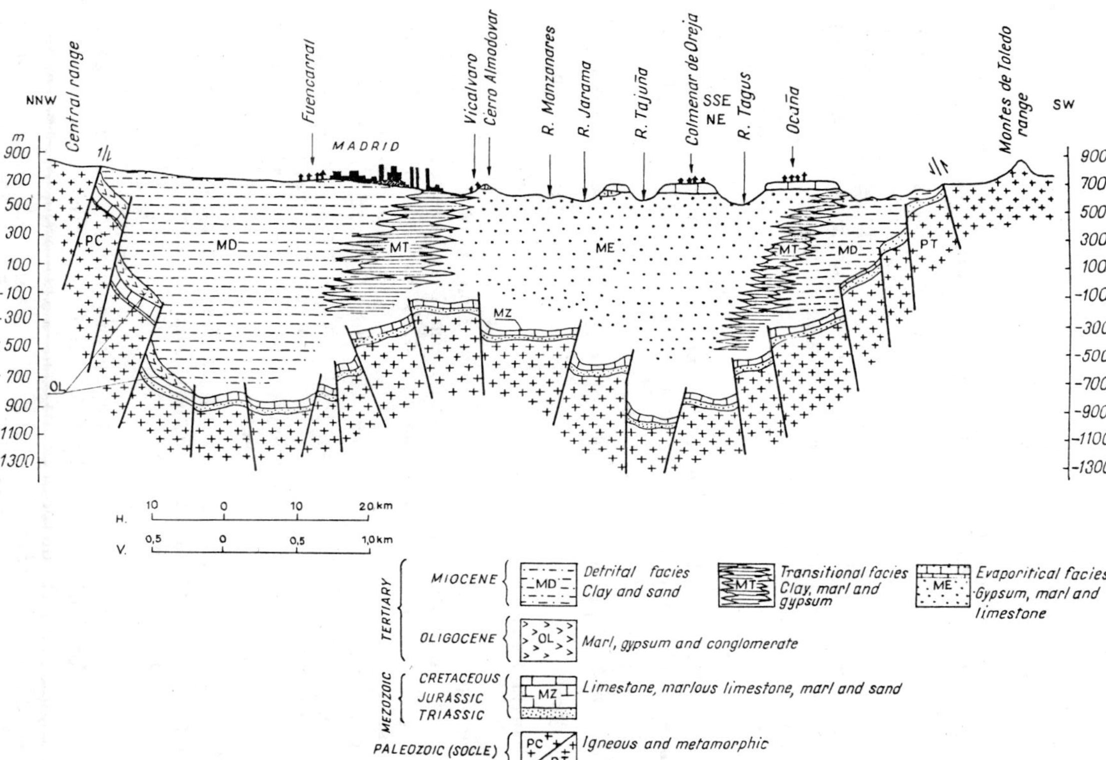

Fig. 4-117. Geological section of the Tagus Basin.

composed of the grid lines. The sides of the squares were 2 km. The three aquifers were regarded as three horizontal flow systems interconnected at the grid points above each other. Thus, the total number of the nodes was 1,125.

The calibration of the model was performed by using previously observed data. It was found that the three aquifers can be characterized with constant transmissivity values: 50, 100 and 50 m^2 day^{-1}, respectively. The storage capacity of the unconfined water-bearing layer (n_s = 0.1) and that of the two confined aquifers (S = 0.0001) were also estimated. The resistance of the aquitards lying between the aquifers was characterized with their leakage coefficient which was the quotient of hydraulic conductivity and thickness. According to the calibration, the parameters of the two separating layers were equal to one another and their independence of the space coordinates might be also assumed. The applied numerical value was K/b = 0.625×10^{-5} day^{-1}. The local resistance between the riverbeds and the unconfined aquifer was taken into account by assuming a clogged layer of one meter thick having low hydraulic conductivity (0.01 m day^{-1}). The recharge originating from rain was considered with an average value of 0.05 m $year^{-1}$ while the same parameter under the urbanized zones was 0.2 m $year^{-1}$.

The differential equations were transformed into finite difference equations by using the implicit iterative method based on the scheme of alternating directions. The calculation results are represented in Fig. 4-118. The first part of the figure (Fig. 4-118a) shows the development of the water table and the piezometric levels under natural conditions as it was determined for the calibration of the model. In the second part (Fig. 4-118b), the influence of exploitation from the second aquifer at six given nodes with a continuous yield of 10^6 m^3 $year^{-1}$ (at each point) is demonstrated showing the development of the potential conditions in each water horizon after 128 years of continuous operation.

The construction of the hydrodynamic models, the application of which was demonstrated here in a summarized form, is the final objective of the collection of hydrogeological data. Efforts should be made, therefore, when evaluating the various data that the geometrical parameters and the flow conditions in the interrelated systems of aquifers should be determined for such models.

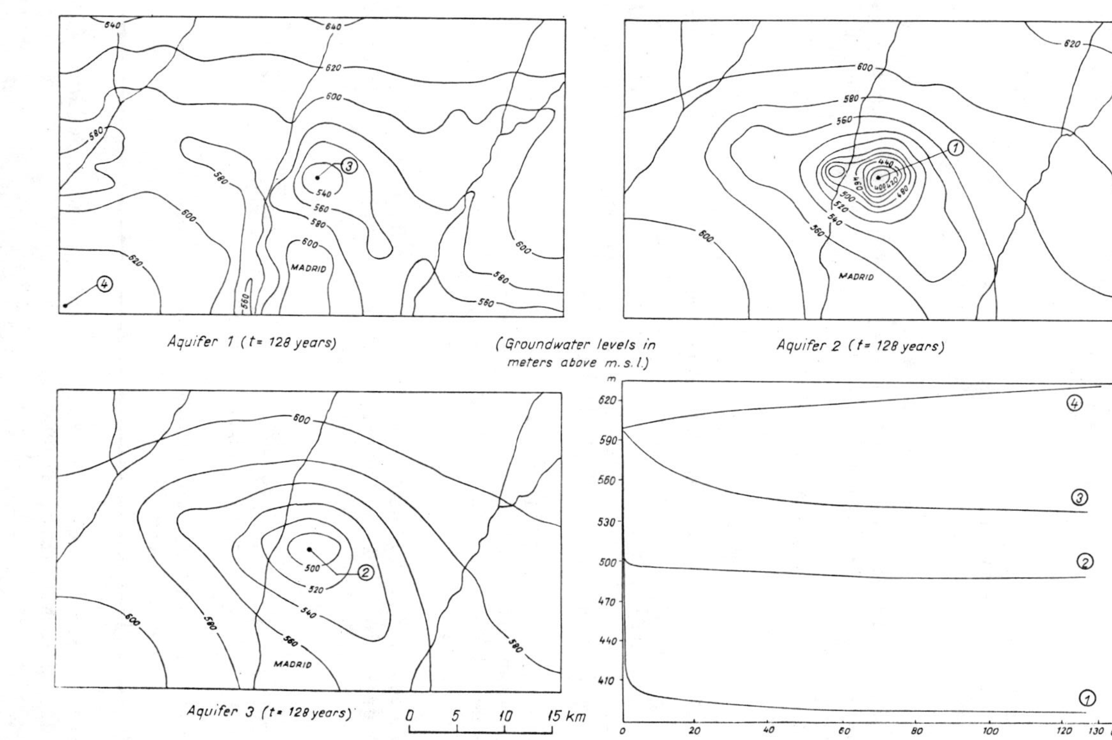

Fig. 4-118. The initial condition and the change of pressures within a layered system composed of three aquifers.

REFERENCES FOR PART IV

Abramov, C. K., 1952: Methods of Selection and Design of Filters to be Applied in Boreholes (in Russian), Moscow.

Alfödi, L. and Gálfi, J., 1966: Hydrogeological and Geophysical Investigations of a Geothermal Anomaly in Hungary (in English), Bulletin of the IASH.

Alföldi, L., Erdélyi, M., Gálfi, J., Korim, K., and Liebe, P., 1978: A Geothermal Flow-System in the Pannonian Basin (in English), Annales Instituti Geologici Publici Hungarici, Vol. 59.

Balló, I. Z., 1976: The Stress-Theory of Water-Bearing Layers Based upon the Experiences in the Great Plain of the Hungarian Share of the Carpatian Basin (in English), IAH - IAHS Symposium on Hydrogeology of Great Sedimentary Basins, Budapest.

Bear, J., 1972: Dynamics of Fluids in Porous Media (in English), New York, London, Amsterdam.

Bear, J., Zaslawsky, D., and Irmay, S., 1969: Physical Principles of Water Percolation and Seepage (in English), Arid Zone Research, UNESCO, Paris.

Besenecker, H. and Lillich, W., 1978: The Determination of Permeability Rates of Aquifer-Separating Layers from Non-hydrological Data (in English), Annales Instituti Geologici Publici Hungarici, Vol. 59.

Bogomolov, J. G., Kudelsky, A. V., and Lapshin, N. N., 1976: Hydrogeology of Large Sedimentary Basins (in English), IAH - IAHS Symposium on Hydrogeology of Great Sedimentary Basins, Budapest.

Bondarenko, N. F., 1973: Physics of the Movement of Groundwaters (in Russian), Leningrad.

Brown, I. C., 1967: Groundwater in Canada (in English), Geological Survey of Canada: Economic Geology Report No. 24, Ottawa.

Cedergran, H. R., 1968: Seepage, Drainage and Flow Nets (in English), New York, Sidney.

Collett, L. S., 1967: Resistivity Mapping by Electromagnetic Methods (in English), Mining and Groundwater Geophysics, Ottawa.

Compagnie Générale de Géophysique, 1955: Master Curves for Vertical Electrical Sounding (in French), DenHaag.

Compagnie Générale de Géophysique, 1963: European Association of Exploration Geophysicists, Leiden.

Croft, M. G., 1971: A Method of Calculating Permeability from Electric Logs (in English), U.S. Geological Survey Professional Papers 750-B.

Deppermann, K. and Homilius, J., 1965: Interpretation of Geoelectric Sounding Graphs in Case of Deep-Lying Water Table (in German), Geologisches Jahrbuch, Vol. 83.

Ellis, A. J., 1970: Quantitative Interpretation of Chemical Characteristics of Hydrothermal Systems (in English), Geothermics, Vol. 1, Part 2.

Erdélyi, M., 1964: Tracing of Subsurface Structures...by Using Indirect Geological Methods (in English), Acta Geologica.

Erdélyi, M., 1972: Hydrology of Deep Groundwaters (in English) (in Kovács, Erdélyi, Major, and Korim: Hydrological Investigation of Subsurface Water; International Post-graduate Course on Hydrological Methods for Developing Water Resources Management), Vol. III(1), VITUKI, Budapest.

Erdélyi, M., 1976: Hydrodynamics of the Hungarian Basin (in English), IAH - IAHS Symposium on Hydrogeology of Great Sedimentary Basins, Budapest.

Erdélyi, M., 1978: Hydrodynamics of the Hungarian Basin (in English), Publications in Foreign Languages, VITUKI, Budapest, in print.

Flathe, H., 1967: Interpretation of Geoelectric Measurements for Solving Hydrological Problems (in English), Mining and Groundwater Geophysics, Ottawa.

Fournier, R. O. and Truesdall, A. M., 1970: Chemical Indicators of Subsurface Temperature Applied to Hot Spring Waters of Yellowstone National Park, Wyoming, U.S.A. (in English), Geothermics, Vol. 2, Part 1.

Gálfi, J. and Stegena, L., 1961: Temporal Variations of the Geothermal Field (in Hungarian), Magyar Geofizika, No. 1/2.

Gálfi, J. and Pálos, M., 1970: Use of Seismic Refraction Measurements for Groundwater Prospecting (in English), Bulletin of the IASH.

Gölz, B., 1978: Study of the Thermal Springs of Budapest by Silica Geothermometer (in Hungarian), personal information.

Gray, D. E., 1963: Handbook of the American Institute of Physics (second edition) (in English), New York, Toronto, London, Vol. 2.

Gribevich, E. A., 1969: Influence of Hydraulic Resistivity on the Flow of Water (in Russian), Riga.

Juodkazis, V. I. and Paltanavicius, J. P., 1967: Filtration Properties of Slightly Permeable Layers of the Baltic Artesian Basin and Methods of their Investigation (in English), IAH - IAHS Symposium on Hydrology of Great Sedimentary Basins, Budapest.

Juhász, J., 1976: Hydrogeology (in Hungarian), Budapest.

Kell, G. S., 1967: Volume Properties of Ordinary Water (in English), Journal of Chemical and Engineering Data, No. 12.

Kesserü, Zs., 1976: The Communication between Subsurface Water Reservoirs in the Light of Mining Experiences (in English), IAH - IAHS Symposium on Hydrogeology of Great Sedimentary Basins, Budapest.

Kézdy, Á., 1972: Soil Mechanics (in Hungarian), Budapest.

Kézdy, Á., 1974: Handbook of Soil Mechanics; Vol. I, Soil Physics (in English), Amsterdam and Budapest.

Keys, W. S., 1969: The Role of Borehole Geophysics in Groundwater Exploitation (in English), IHD Expert Meeting on the Cooperative Role of Hydrogeology and Geophysics in Groundwater Exploration, Budapest.

Keys, W. S. and MacCary, L. M., 1971: Application of Borehole Geophysics to Water Resources Investigation (in English), Technique of Water Resources Investigations of the U.S. Geological Survey, Washington.

Kissin, I. G., 1976: The Principle Distinctive Features of the Hydrodynamic Regime of Intensive Earth Crust Downwarping Areas (in English), IAH - IAHS Symposium on Great Sedimentary Basins, Budapest.

Komarov, V. A., Pishpareva, H. N., Semenov, M. B., and Kholoponina, L.S., 1966: Theoretical Fundamentals for Interpretation of Survey Data Obtained with the Induced Polarization Method (in Russian), Leningrad.

Kovács, Gy., 1969: Relationship between Velocity of Seepage and Hydraulic Gradient in the Zone of High Velocity (in English), XIIIth Congress of IAHR, Kyoto.

Kovács, Gy., 1972: Hydraulics of Seepage (in Hungarian), Budapest.

Kovács, Gy., 1972: The Active Length of Well-Screens (in Hungarian), Vizügyi Közlemények, No. 4.

Kovács, Gy., 1973: Hydrological Investigation of Groundwaters (in Hungarian), Budapest.

Kovács, Gy., 1976: Subsurface Water Movements, Filtration, Hydrodynamics, Models (General Report) (in English), IAH - IAHS Symposium on Hydrogeology of Great Sedimentary Basins, Budapest.

Kovács, Gy., 1976: Determination of the Threshold Gradient in Porous Media (in English), IAHR Symposium on Flow of Water through Porous Media, Kiev.

Kovács, Gy., 1976: Statistical Model of the Pore-Size Distribution of Loose Clastic Sediments (in English), IAHR Symposium on Flow of Water through Porous Media, Kiev.

Kuzmina, N. E. and Ogil'vi, A. A., 1965: On the Possibility of Using the Induced Polarization Method to Study Groundwater (in Russian), Razvedochnaya I Promyslovaya Geofizika, No. 9.

Lakatos, S., 1976: Investigation of Pressure Conditions in Water Yielding Well by Using Analytic Methods (in Hungarian), VIKUV, Budapest (manuscript).

Lasseter, T. J. and Witherspoon, , 1975: The Numerical Simulation of Heat and Mass Transfer in Multidimensional Two-phase Geothermal Reservoir (in French), Bulletin du B.R.G.M. Section III, No. 1.

Llamas, I. and Cruces de Abia, J., 1976: Conceptual and Digital Models of the Groundwater Flow in the Tertiary Basin of the Tagus River (Spain) (in English), IAH - IAHS Symposium on Hydrogeology of Great Sedimentary Basins, Budapest.

Long, G. and Chierici, G. L., 1959: Compressibility and Specific Weight of Deep Groundwater under the Conditions Prevailing in the Formations (in French), 5th World Petroleum Congress, New York.

Lubotshkov, E. A., 1965: Graphical and Analytical Method for the Determination of the Suffusion Properties of Granular Ground (in Russian), Izvestiya VNIIG, No. 78.

Maucha, L., 1975: Study of Tidal Movement of Karst Waters and Karstic Rocks (in English), IUGG General Assembly, Grenoble (published in Annales de Géophysique, 1977).

Matveev, B. K., 1964: Methods of Graphical Construction of Electrical Sounding Curves (in Russian), Moscow.

Maxey, G. B., 1969: Subsurface-Water - Groundwater (in English) (in the Progress of Hydrology), Proceedings of the First International Seminar for Hydrology Professors, Vol. II.

Namiot, A. Ju. and Bondareva, M. M., 1963: Solution of Gases in Water under Pressure (in Russian), Moscow.

Nikolsky, B. P., 1971: Handbook of Chemistry (in Russian), Leningrad.

Orellana, E. and Mooney, M. M., 1966: Master Tables and Curves for Vertical Electrical Sounding over Layered Structures (in English), Madrid.

Pataki, N., 1972: Water-Well Drilling and Construction Technology International Post-graduate Course on Hydrological Methods for Developing Water Resources Management, Vol. III/2, VITUKI.

Risler, J. J., 1976: Application of Geothermometers to Study the Hot Springs of the French Massif Central (in French), Bulletin du B.R.G.M., Section IV, No. 1/2.

Rónai, A., 1976: Essential Hydrogeological Character of the Great Plain of Hungary (in French), IAH - IAHS Symposium on Hydrogeology of Great Sedimentary Basins, Budapest.

Rónai, A., 1978: Analysis of the Results of Observation of Deep Groundwaters on the Great Plain (in Hungarian), Hidrológiai Közlöny, No. 4.

Rogovskaya, N. V. and Bezrodnov, V. D., 1976: Regional Hydrodynamic Regularities of Platform-type Artesian Basins (in English), IAH - IAHS Symposium on Hydrogeology of Great Sedimentary Basins, Budapest.

Schlumberger Well Survey Corp., 1966: Log Interpretation Charts (in English), Houston.

Schmidt, E. R., 1962: Hydrogeological Atlas of Hungary (in Hungarian), Budapest.

Seigel, H. O., 1959: Mathematical Formulation and Type-Curves for Induced Polarization (in English), Geophysics, Vol. 24.

Sichardt, W., 1952: Gravel- and Sand-Filters Used in Foundation and Water Construction (in German), Die Bautechnik.

Stegena, L., 1975: The Development of the Pannonian Basin in the Cenozoic (in Hungarian), Földtani Közlöny, No. 105.

Subbota, M. J., 1972: Methods and Interpretation of Hydrogeological Investigations for the Exploration of Carbon hydrogens (in Russian), Moscow.

Terzaghi, K., 1926: Soil Physical Bases of Soil Mechanics (in German), Wien.

Terzaghi, K., 1943: Theoretical Soil Mechanics (in English), New York, London.

Tóth, J., 1962: A Theory of Groundwater Motion in Small Drainage Basins in Central Alberta, Canada (in English), Journal of Geophysical Research.

Tóth, J., 1963: A Theoretical Analysis of Groundwater Flow in Small Drainage Basins (in English), Journal of Geophysical Research.

Urbancsek, J., 1976: Formation Water Pressure in the Quaternary Sediments of the Great Hungarian Plain (in English), IAH - IAHS Symposium on Hydrogeology of Great Sedimentary Basins, Budapest.

U.S. Army Corps of Engineers, 1965: Drainage and Erosion Control (in English), Engineering Manual of Military Construction, Washington.

Zohdy, A. A. R., Eaton, G. P., and Mabey, D. R., 1974: Application of Surface Geophysics to Groundwater Investigations (in English), Techniques of Water-Resources Investigations of the U.S. Geological Survey 2/D1, Washington.

PART V

HYDROLOGY OF SOLID ROCK TERRAINS

by

Dr. J. Gálfi and Dr. G. Kovács

SECTION 17

GENERAL CHARACTERIZATION OF STORAGE PROPERTIES AND WATER YIELDING CAPACITY OF FISSURED AND FRACTURED ROCKS

There are large areas where water stored in fissured and fractured solid rocks is the only source of water supply, and some layers of this type are regarded to be the best aquifers (e.g., limestone and dolomite) which have an important role as water-bearing formations providing available water resources for regions. In some oil fields, the parent layers of the hydrocarbons are also fissured and fractured rocks, and the production of these fields requires the determination of the hydraulic parameters of seepage through such formations. Thus, great practical importance is attached to the determination of a conceptual model suitable for the characterization of this type of seepage.

A general description of the various fissured and fractured rocks will be given in this section. Here, data are also summarized which were collected from the literature in order to give as much general information as possible on the porosity, permeability, and hydraulic conductivity of the various formations as well as on the water yielding capacity of wells drilled in these rocks.

The fissures and fractures occurring in hard solid rocks are of different form and nature depending on the type of rocks. The main types are the igneous (intrusive and volcanic) and metamorphic rocks, the indurated sediments (sandstone, quartzites, shales, etc.), and carbonate rocks, the latter forming a special type of rock. The form of fractures and fissures remains more or less unchanged after their development in the case of igneous and metamorphic rocks while they are enlarged due to solution and erosion in the case of limestone and dolomites. The porosity, the hydraulic conductivity of the fissures, and the yield from the wells penetrating the fissured structure are different and vary widely depending on the type of rock in which the network of fractures are formed.

After listing the definitions of some basic terms, a general description of the usual network of openings is summarized on the basis of Schoeller's work (Brown et al., 1972) grouped according to the types of rocks.

Difference has to be made between fractures and fissures. A fracture is the result of the break of the rock masses caused by tectonic forces while a fissure is a small slit or a narrow opening of varying length without making any restriction concerning its origin. Faults are planes of fractures where obvious signs of differential movement of the rock masses on either side of the plane can be observed. They may be vertical or oblique. Fault zones are sometimes intersected by a network of other faults, thus forming an interconnected draining system.

In other cases, the movement of rock masses along the faults interposes impervious blocks between the separated parts of an aquifer causing the development of a barrier in the way of continuous groundwater flow. Water circulation in a fault is not uniform and is frequently very localized since faults are not open in all places. Joints are fractures in the rock along which there has been extremely little or no movement.

Fractures and fissures used to be divided into further groups according to the processes creating them. The names of the different types immediately indicate their origin, thus a detailed explanation is not given here, only the simple list of the various fissures and fractures:

- tectonic joints, distinguishing closed shear joints, and generally open tension joints;

- weathering fissures, frequently originating from one of the various types of fissures;

- release fractures or sheet joints.

The degree of fracturing depends on the intensity of the tectonic movement creating the fractures. The development of weathering fissures is related to the climate to which the rocks have been exposed and the length of the time of this exposure. The depth at which fracturing occurs also depends on the fracture origin, i.e., faults may run into very great depth, joints generally transform the solid rocks into water-bearing formations only near the surface (down to a depth of 100 m or less). Both characteristics of fracturing and fissuring (degree and depth) are greatly influenced by the type of the rock in which the network of interstices occurs. It is reasonable, therefore, to summarize the more detailed description of the probable structure of openings according to the main groups of solid rocks. Numerical data characterizing the hydraulic properties of various rocks (porosity, hydraulic conductivity, intrinsic permeability, yield of wells, boreholes, and springs) are also listed in tabulated form according to the main types of rocks.

17-1 Intrusive Rocks

The openings in intrusive igneous rock (granite, syenite, diorite) may have the following nature:

- joints parallel to the surface of the rock mass marking limits between different sheets;
- concentric joints forming conical surfaces and running towards the center of the mass;
- radial joints emanating from the center of the mass.

Table 5-1 gives the parameters that characterize the hydraulic properties of intrusive rocks.

The weathering action at the surface of granite disintegrates the feldspars to form a soft, friable mass of greater porosity and permeability under certain climatic conditions or may disintegrate and form clayey substances of very low permeability. The depth of the weathered zone depends to a large extent on climate and exposure time. In the upper part of the granite mass, where sheet joints are open and intersect the concentric and radial joints, water can flow freely through the channels and an aquifer characteristic is exhibited. These openings become narrower and scarcely distributed at greater depths. In temperate climate the weathering zone may go down 30 m or more and water-bearing joints reach depths of over 100 m.

Although other factors than geological conditions are also influencing the yield of wells (diameter, casing, testing method, etc.), data collected from a given region are suitable to characterize the probability of the expected yield of a given geological formation and its dependency on the drained depth. Turk's (1963) analysis is shown here as an example which is based on the data of 239 wells in the Sierra Nevada (California). Most of these wells are in granodiorite, but a number of metamorphic rocks are also involved. Figure 5-1 shows the frequency distribution of the yields of wells without making any distinction according to the depth of wells. The median (most probable value) is about 34 ℓ/min and the mean 90 ℓ/min. It is worthwhile to note here that the form of the frequency distribution curve suggests: that it can be mathematically well approximated using a logarithmic normal distribution which is generally applied in the literature to characterize the probability of a given yield or hydraulic conductivity in fissured and fractured rocks. About 8.4 percent of the wells were unsuccessful and 16.3 percent yield less water than 14 ℓ/min. Grouping the data according to the depths of the wells (fixing depth intervals arbitrarily), the same statistical analysis can be executed for each interval.

Table 5-1. Parameters characterizing the hydraulic properties of intrusive rocks.

Type of Rock	Location	Porosity (%)			Comments				References
		av.	max.	min.					
granite		0.3	-	-					Brace (1965)
diabase		0.1	-	-					
		Hydraulic Conductivity (cm/sec)							
		av.	max.	min.					
granite		-	2.0×10^{-10}	0.5×10^{-10}					Louis (1967)
weathered granite		-	1.8×10^{-3}	3.7×10^{-4}					Morris and Johnson (1966)
granite	Brazil	-	1.0×10^{-4}	1.0×10^{-6}					Franciss (1970)
		Yield of Wells (ℓ/min)			Av. Depth (m)	Av. Drawdown (m)	Number of Wells	Failure (%)	
		av.	max.	min.					
	South Africa								
old granite	Pretoria, Johannesburg	34	-	-	40	11	62	14	Du Toit (1923)
	North Transvaal	83	-	-	47	12	497	25	
	Rustenburg	65	-	-	61	11	130	50	
	South-West Transvaal	98	-	-	37	12	404	10	
	Mafeking	76	-	-	51	6	202	35	
	Vryburg	61	-	-	48	4	136	37	
	Van Rhysndrop	16	-	-	56	8	16	31	
granite	Connecticut (USA)	48	-	-	27	-	-	-	Ellis (1906)
	Uganda	-	45	0.1	20-60	-	-	-	Gear (1951)
	Brazil	25	-	-	30-40	-	11	-	Suszczynski (1968)
	Mauritania	11	330	-	-	-	-	-	Archambault (1960)
	Sweden	22-43	-	-	-	-	-	-	Meyer and Peterson (1951)
	USSR								
	Dnepr River	3.4×10^{-3}	-	-	10-30	-	38	-	Rats and Chernyshov (1965)
	Selenga River	12.6	-	-	5-60	-	53	-	

Representing either the median or mean values, determined in this way, a function of depth, the decrease of permeability of the intrusive rocks with increasing depth can be well characterized (Fig. 5-2).

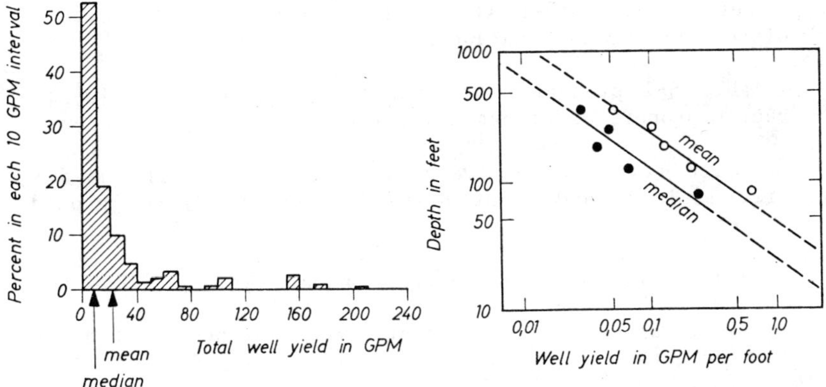

Fig. 5-1. Frequency distribution of yields of 239 wells in Sierra Nevada (California) (after Turk).

Fig. 5-2. Relationship between the depth and the yield of 239 wells in Sierra Nevada (California) (after Turk).

A similar relationship can be achieved between depth and water yielding capacity if the latter is characterized by water injection rate (water take) measured by pressure tests at various depths (Fig. 5-3) (Davis and Turk, 1964). The graph was constructed on the basis of 412 tests executed in granite in California.

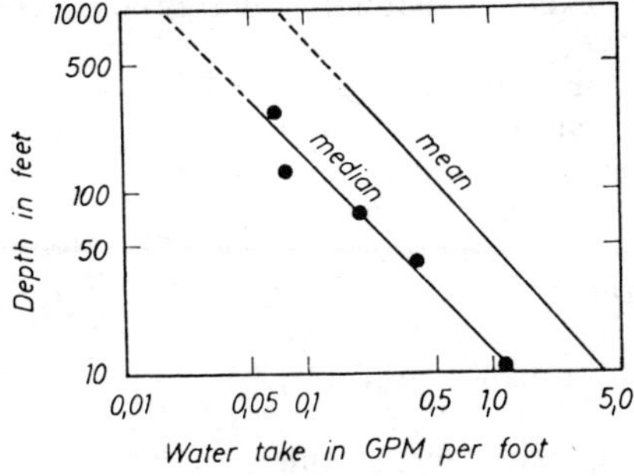

Fig. 5-3. Water injection rates in granite related to depth (after Davis and Turk).

17-2 Extrusive Volcanic Rocks

The most important water-bearing formations are the basalts having low silica and high magnesium and iron content. The mineral composition of the rhyolites has the opposite character while andesites and trachytes form an intermediate type between the other two. The water-bearing capacity of rhyolites, trachytes, and andesites is generally poor.

Table 5-2 presents the parameters that characterize the hydraulic properties of extrusive rocks.

Basalt covers extensive areas in India, North America, Brazil, Ireland, and Great Britain, providing large groundwater resources.

Basalt formations can be grouped as follows:

- as lavas with hardened scoriaceous surface crust due to the escape of gases;
- pahoehoe lavas with smooth, undulating ropy surface;
- pillow lavas formed by outflow below water and composed of blocks (pillows) of a few decimeters in diameter with radially arranged vesicles inside them.

Lava flows may be superimposed on each other varying in thickness and number. There may be a layer of ash, fossil soil, or alluvial deposit between the flows if the dormant phase of the volcano was fairly prolonged.

Basaltic rocks are known as good aquifers due to numerous openings in them. In addition to the interstitial porosity caused by scoriae and breccia flows, the fissure porosity in basalt may be created by the following elements:

- cavities between pahoehoe lava flows;
- shrinkage cracks occurring as fissures normal to the isothermic surfaces brought about by surface cooling and as fissures caused by cooling from beneath resulting in prismatic structure;
- gas vesicles;
- lava tubes within the flow;
- fractures caused by mechanical forces acting on the cooled lava;
- tree moulds.

Table 5-2. Parameters characterizing the hydraulic properties of extrusive volcanic rocks.

Type of Rock	Location	Porosity (%) av.	max.	min.	Comments	References
basalt	USA	17	35	3	94 samples	Morris and Johnson (1966)
basalt (very dense)		0.8	-	-	8 samples	Schoeller (1962)
(porous)		11.4	-	-		
obsidian		0.52	-	-		
phonolite		2.0	-	-		
pumice		87.3	-	-	5 samples	
tuff		31.0	-	-	8 samples	
		Hydraulic Conductivity (cm/sec) av.	max.	min.		
basalt	USA	$1.0x10^{-5}$	$4.7x10^{-5}$	$2.1x10^{-9}$	93 samples	Morris and Johnson (1966)
weathered basalt and fissured compacted ash	Morocco	-	$2.0x10^{-1}$	$1.7x10^{-2}$		Mortier, Quang, Sadek (1965)
basaltic tuffs and slightly fissured compact basalt		-	$7.5x10^{-1}$	$1.1x10^{-1}$		
basalt	Brazil	-	$3.0x10^{-4}$	$5.0x10^{-7}$	156 Lugeon tests	Franciss (1970)
andezite	Mátra (Hungary)	$1.9x10^{-7}$	$2.4x10^{-5}$	$9.2x10^{-9}$	20 wells (35% failures)	Schmieder's data
		Transmissivity (m^2/sec) av.	max.	min.		
andezite	Mátra (Hungary)	$8.5x10^{-7}$	$5.2x10^{-5}$	$1.4x10^{-8}$	20 wells (35% failures)	Schmieder's data
weathered basalt and fissured compacted ash	Morocco	-	$3.0x10^{-2}$	$1.1x10^{-2}$		Mortier, Quang, Sadek (1965)
basaltic tuffs and slightly fissured compact basalt		-	$7.5x10^{-3}$	$1.3x10^{-5}$		

Table 5-2. (continued)

		Specific Yielding Capacity of Wells ($\frac{\ell/min}{m}$)				
		av.	max.	min.		
basalt	Afars and Issas (Africa)	0.7-4.0	33.0	0.1	higher values are doleritic prismoid basalt, the lower ones in weathered basalt	Pouchen (1973)
basalt	Oregon (USA)	-	83.0	7.0		Piper (1932)
		Yield of Springs (m^3/sec)				
Name of the Spring		group of springs		individual spring		
Data group	California-Oregon	40		1.4-3.0		
Snake River Plain	Idaho	110		4.0-16.0		
Thousand Springs	USA	15-20		-		Meinzer (1927)
Kalauaa	Oaluu Island	-		0.4-1.0		
Kaluaoopu	(Hawaii)	-		0.7-1.0		

Horizontal permeability is generally greater than the vertical permeability and pahoehoe flows are more permeable than the aa flows. The latter is reasonably permeable when they consist of breccias. In pillow lavas, water can circulate between the pillows if these channels are not sealed by vitreous material and secondary minerals.

Because of the high porosity and permeability of basalts, the natural infiltration is relatively high. It is estimated to be near 25 percent of yearly precipitation on the island of Oahu in the Hawaiian group, and it can reach 10 percent even in hot and arid climates (e.g., in Afaras and Issas in Africa). This is the reason why the basaltic terrains are known to be as good aquifers as carbonate (karstic) formations with good natural recharge and high available groundwater resources which occur in the form of large springs. Some characteristic data of such springs are also listed in Table 5-2.

In a thick basalt layer covering a large area (trap), the enlargement of the fissures and fractures by weathering can also be characterized by the decrease of permeability with depth. There is, however, another process acting near the surface against this effect of weathering: i.e., the sealing of the openings by clay. The sealing material may be transported onto the surface of the bare rock from the neighboring areas, thus the process was also observed in the upper part of dolomite formations covered by younger marine sediments (Transdanubian Mountain Range, Hungary). The sealing is most common on the surface of extrusive volcanic rocks where clay is locally formed by the decomposition of feldspars as a result of weathering, and in the upper part of indurated sediments, the solid matrix of which is composed of very fine particles (e.g., argillite). The examples shown in Fig. 5-4 to demonstrate the permeability vs. depth relationship summarize observations from the Middle Angara Basin (Far Eastern Region, USSR) (Rats and Chernyashov, 1965). It represents the logarithm of the ratio between yield and drawdown as a function of depth in traps, and in a formation composed of sandstones and argillites. The effect of sealing can be well observed in the latter group.

The extrusive volcanic rocks, other than basalts (rhyolite, trachyte, andesite) are generally very poor aquifers. Their permeability is also the result of the joints dislocating the mass of the rock, thus it decreases with depth. To demonstrate this relationship, permeability data determined by pumping and water injection tests were collected in the andesite layers of the Mátra Mountain (northern Hungary). There were two obstacles hindering the usual statistical evaluation of the data: i.e., the small number of the observations, and the relatively long open stretch of the

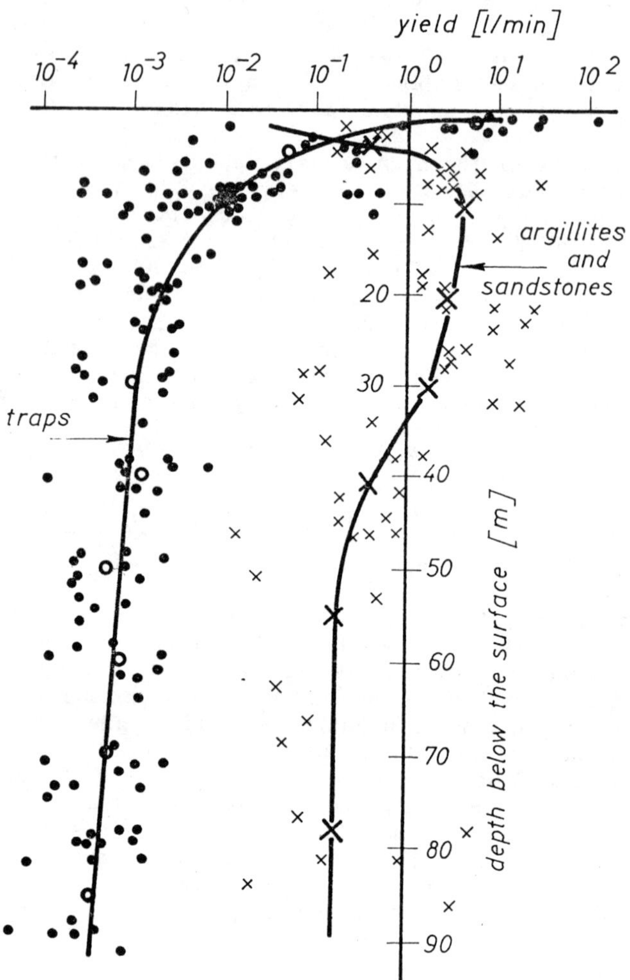

Fig. 5-4. Permeability vs. depth relationship in sandstone-argillite series and in trap of the Middle Angara Basin (USSR) (after Rats and Chernyeshov).

boreholes during the tests (which was necessary to achieve measurable results). In Fig. 5-5 the position of the open boreholes was, therefore, represented as a function of the calculated hydraulic conductivity. The general trend of the hydraulic conductivity vs. depth relationship was determined by averaging the data observed between 0-400 m, 400-800 m, and 800-1200 m. The frequency distribution determined from all observed data can be well approximated by the logarithmic

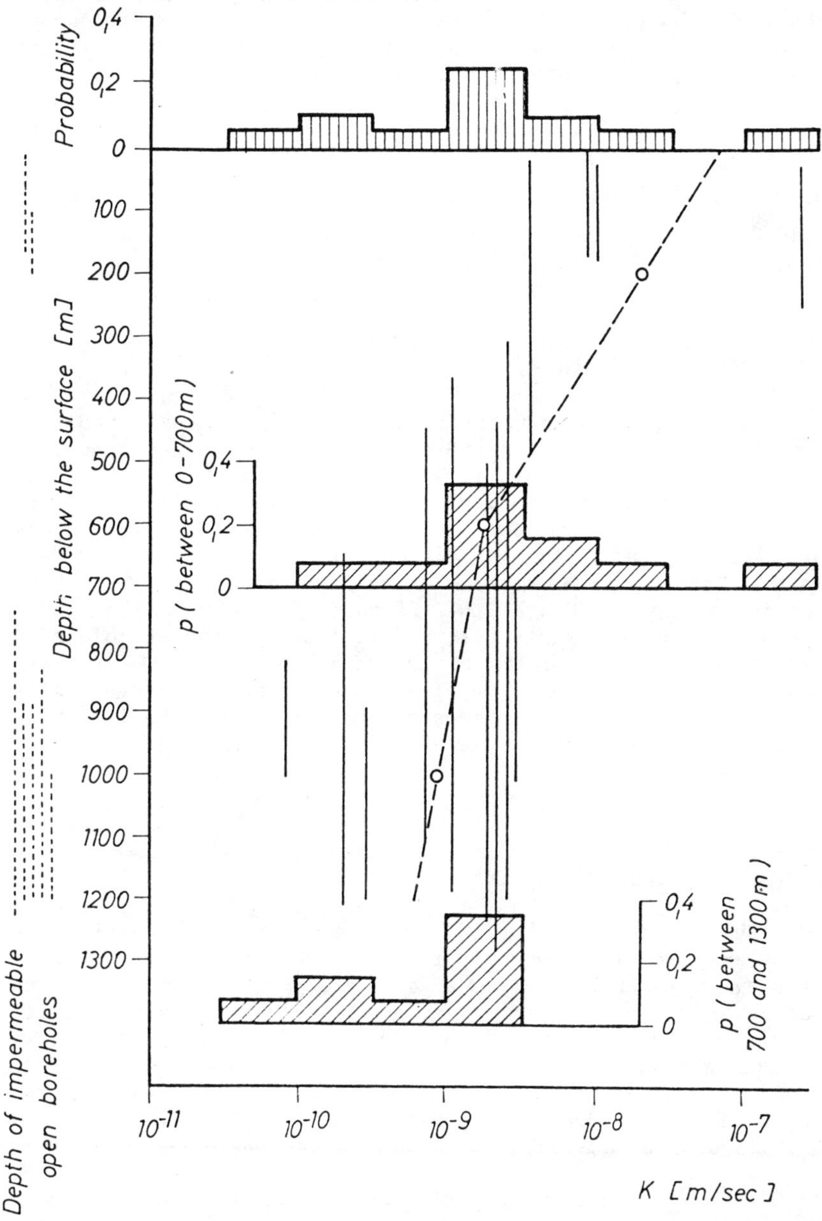

Fig. 5-5. Relationship between hydraulic conductivity and depth in andesite of the Mátra Mountain (northern Hungary) (after Schmieder).

normal distribution function while after separating the data observed above and below 700 m, both empirical frequency distribution curves become asymmetrical even using logarithmic scale on the horizontal axis. It is worthwhile to note that hydraulic conductivity of these rocks is the smallest among the solid rocks investigated (e.g., in a depth of 500 meters, sandstone investigated in connection with Fig. 5-10 has a hydraulic conductivity of $2x10^{-5}$ m/sec and andesite only of $3x10^{-9}$ m/sec). The high rate of impermeable boreholes (30 percent above 700 m and 35% below this level) also indicates the poor water yielding capacity of these rock types.

17-3 Metamorphic Rocks

Gneiss, well crystallized schists, slatey schists, amphibolites, quartzites, and crystalline limestones belong to the group of metamorphic rocks. Fissures occurring in these rocks are due to schistosity (parallel layers of flakes having some petrographic composition), or foliation (parallel layers of flakes of different mineralogical composition), or sheet joints forming a network of fractures normal to the above mentioned ones. Weathering fissures and faults are also common. Comparing schistosity, foliation, and sheet joints of gneiss and slatey schists to the fissures of granite, it can be seen that the openings of the former group are generally smaller. Due to their very steep dip, however, the infiltration is not negligible especially if the openings are enlarged by weathering. Thus, the recharge may be considerable, the number of joints being higher than that in granite.

Table 5-3 gives the parameters that characterize the hydraulic properties of metamorphic rocks.

Sheet joints reslting from decompression, occurring one below another in metamorphic rocks, are known to be capable of storing significant volumes of water. The weathered top mantle is often reasonably permeable allowing greater infiltration of water which can be recovered by digging wells down to the hard rock. At greater depths, the fissures close more rapidly than in granite, they are unidirectional and less interconnected with small possibility of yielding sustained amounts of groundwater.

The water yielding capacity as a function of depth can be similarly characterized as it was done in the case of intrusive rocks. The analysis of water injection data from Oroville dam site (California) is shown as an example in Fig. 5-6 (Davis and Turk, 1964) where the frequency distribution curve of one depth interval (from 24 to 30 m) and the

Table 5-3. Parameters characterizing the hydraulic properties of metamorphic rocks.

Type of Rock	Location	Porosity (%) av.	Porosity (%) max.	Porosity (%) min.	Comments	References
weathered schist	USA	38.0	49.3	4.4	18 samples	Morris and Johnson (1966)
gneiss schist		0.1-3.0	-	-	estimation	Schoeller (see
		0.5-5.0	-	-		Brown et al.)(1972)
Brunswick shale	New Jersey (USA)	$2.5x10^{-2}$	-	-	storage capacity from pumping test	Vecchioli (1965)
slate		3.4	-	-	19 samples	Manger (1963)
marble		0.6	-	-	100 samples	
		Hydraulic Conductivity (cm/sec) av.	**max.**	**min.**		
slate schist	USA	$1.0x10^{-8}$	$5.3x10^{-8}$	$2.1x10^{-10}$	8 samples	Morris and Johnson (1966)
(weathered)		$2.1x10^{-4}$	$1.3x10^{-3}$	$2.1x10^{-9}$	17 samples	
schist	France	-	$1.2x10^{-7}$	$0.9x10^{-9}$		Louis (1967)
slate	Michigan (USA)	$1.3x10^{-9}$	$4.5x10^{-8}$	$5.0x10^{-10}$	9 samples	Stuart, Brown, and Rhodamel (1954)
mica-schist		-	$6.0x10^{-9}$	$2.1x10^{-9}$	2 samples	
gneiss	Brazil	-	$8.0x10^{-4}$	$2.0x10^{-7}$	34 Lugeon tests	Franciss (1970)
granite-gneiss		-	$4.0x10^{-4}$	$5.0x10^{-7}$	72 Lugeon tests	
slate		-	$3.0x10^{-4}$	$5.0x10^{-7}$		
schist	Ukraine	-	$1.6x10^{-1}$	$5.4x10^{-5}$		Rudenko (1958)
ferruginous schist		-	$4.2x10^{-3}$	$7.1x10^{-4}$		
graywacke (fractured)		$4.5x10^{-5}$	-	-	11 samples	Lewis et al. (1966)
metasediment (fractured)		$3.1x10^{-5}$	-	-	6 tests	
quartz-mica schist		$3.3x10^{-5}$	-	-	21 samples	Stewart (1964)
(weathered)		$9.7x10^{-4}$	-	-	field tests	
graywacke		$3.0x10^{-9}$	-	-	5 samples	Stuart et al. (1954)
quartzite		$1.9x10^{-9}$	-	-		
slate		$1.3x10^{-9}$	-	-		
mica schist		$2.1x10^{-9}$	-	-	9 samples	

Table 5-3. (continued)

		Transmissivity (m^2/sec)							
		av.	max.	min.					
Brunswick shale	New Jersey (USA)	1.28×10^{-2}	-	-	pumping test				Vecchioli (1966)

		Yield of Wells (ℓ/min)			Av. Depth (m)	Av. Drawdown (m)	Number of Wells	Failure (%)	
		av.	max.	min.					
gneiss	Connecticut (USA)	46.6	-	-	40	-	73	32	Ellis (1906)
quartzite		27.4	-	-	135	-	3	0	
schist		52.6	-	-	33	-	23	30	
slate		0	-	-	29	-	5	100	
	Sweden								
gneiss	Hallan	46	-	-	-	-	-	-	Meier and Paterson (1951)
	Västergötland	53	-	-	-	-	-	-	
	Bohnslan	18	-	-	-	-	-	-	
	Värmland	40	-	-	-	-	-	-	
	Södermaland	12	-	-	-	-	-	-	
magnetite	Norrland	42	-	-	-	-	-	-	
	USA								
gneiss	Maryland	45	-	-	-	-	-	-	Davis and de Wiest (1966)
	Virginia	64	-	-	-	-	-	-	
	New England	34	-	-	-	-	-	-	
schist	Maryland	91	-	-	-	-	-	-	
	Virginia	45	-	-	-	-	-	-	
	New England	38	-	-	-	-	-	-	
	Brazil								
gneiss	Monteiro	77	-	-	-	-	-	33	Suszczynski (1968)
	Petrolina	26	-	-	-	-	-	46	
slate	Petrolina	23	-	-	-	-	-	20	
slate	Rajasthan (India)	37	47	21	-	-	-	25	Taylor, Ray, and Sett (1954)

relationship between the depth and the injection rate is represented.

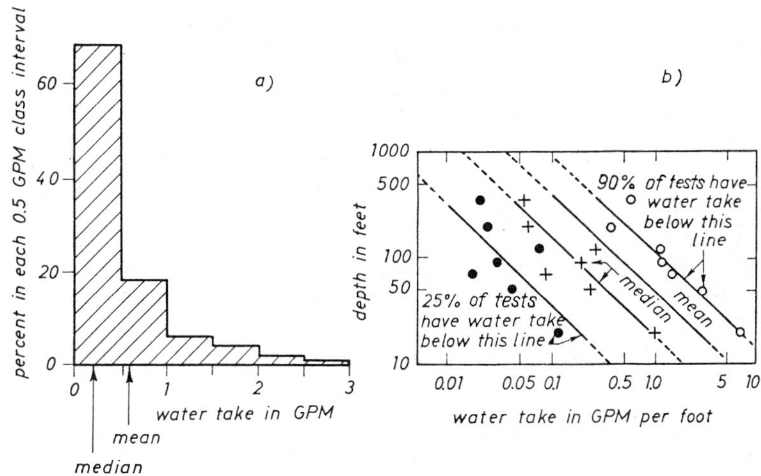

Fig. 5-6. Water injection data from Oroville dam site (California) (after Davis and Turk).
a) frequency distribution of data measured between 24 and 30 m; b) water injection rate as a function of depth.

Although porosity and permeability are supposed to be higher in hard intrusive formations than those in metamorphic rocks, the water yielding capacity of the weathered mantle of the two types of rocks (which two formations are frequently combined as crystalline rocks) does not differ at all. This fact is proved by the comparison of the yield of 1,522 wells in the eastern part of the United States from which 814 drain granite and the others penetrate into schist (Fig. 5-7) (Turk, 1963). The same similarity can be found if the percentage of

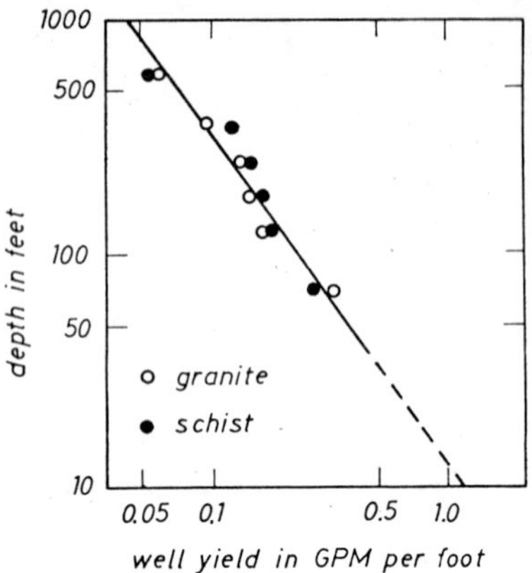

Fig. 5-7. Yield of wells in crystalline rocks in the eastern part of the United States (after Davis Turk).

water injection tests having zero water intake in granite and in metamorphic rocks, respectively, are compared (Fig. 5-8) (Davis and Turk, 1964).

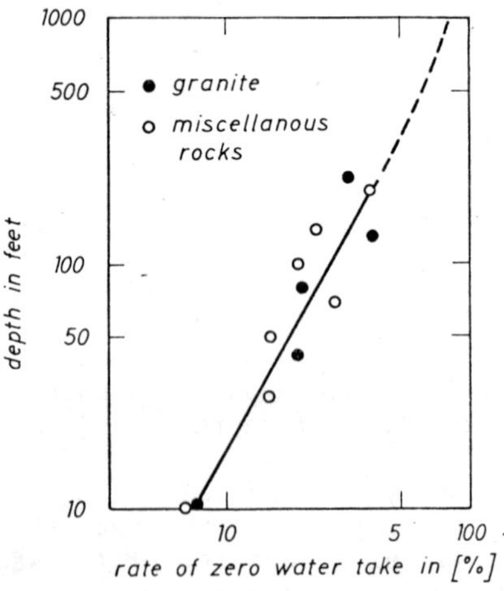

Fig. 5-8. Percentage of tests having zero water intake in granite and metamorphic rocks (after Davis and Turk).

Normal porosity of gneiss is about 0.1 to 3 percent and that of schists is 0.5 to 5 percent. Greater porosity (up to 50 percent) is exhibited by rocks subjected to very high degree of weathering. The relationships of depth vs. porosity and specific yield, respectively, in weathered metamorphic rocks are characterized in Fig. 5-9 based on Stewart's (1962) measurements in Georgia, north of Atlanta.

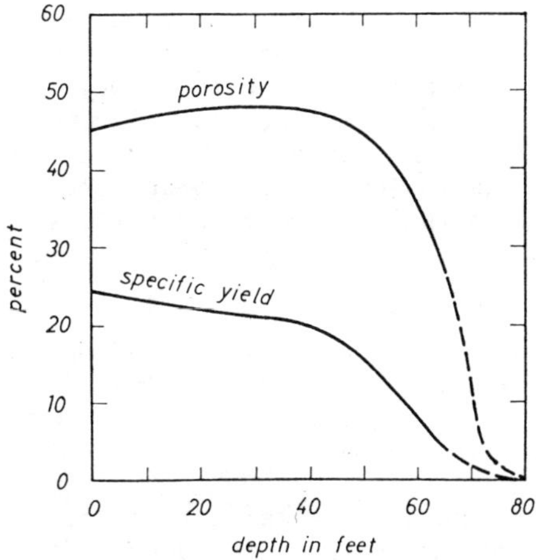

Fig. 5-9. Porosity and specific yield as a function of depth in metamorphic rocks north of Atlanta (Georgia) (after Stewart).

Due to schistosity, metamorphic rocks have highly anisotropic behavior. Permeability along the fracture planes may be many times greater than the average.

17-4 Noncarbonate, Indurated Sediments

Sandstones generally have a double system of interstices. The first part is composed of the original pores between the grains, the second one is created by fissures, fractures, and stratification joints. Primary porosity is highly variable, ranging from less than 1 to 35 percent and depends on the grain-size distribution and the packing of the particles as well as on the material and degree of their cementation. The number of joints is generally high, and they are more open if the rock is more indurated. In most cases, the secondary porosity is lower than the primary one, but the joints are larger than the original pores. Thus, hydraulic conductivity is greatly influenced by the degree of fissuring and fracturing while the storage capacity of sandstones depends mostly on primary porosity.

Table 5-4 gives the parameters that characterize the hydraulic properties of noncarbonate, indurated sediments.

It follows from the structure of the interstices explained in the previous paragraph that wells sunk in sandstones have a high initial yield by draining the network of joints,but the yield characteristic for a long period of operation is determined by primary porosity and hydraulic conductivity (by the parameters representing the continuous blocks without joints). The same role of secondary porosity can be observed in natural condition as well, viz., infiltration and water transport through the joints is rapid which causes a high yield of springs after rains while the second stretch of the recession curve (the graph showing the decrease of the yield of a spring in time after a rainy period) is flat indicating the relatively low permeability and high storage capacity of the solid matrix.

The influence of the double system of interstices decreases with depth as the degree of fissuring is low below a given depth (which is, e.g., 100-200m on the average in Bunter sandstone in the Saar region).

Although the general decreasing trend of yield with depth can be similarly observed as it was done in the case of crystalline rocks, this relationship is not so close in sandstones. The influence of primary porosity and the change of the latter in the layered system can be mentioned perhaps as the reasons of this difference.

In Fig. 5-10, an example is shown based on the yield of wells draining an Oligocene sandstone formation of several hundred meters within a relatively small region in northern Hungary. Three different parameters were investigated: the yield related to the depression measured inside the well (dimension of m^3/sec/m); the former parameter divided by the length of the screen ($m^3/sec/m^2$); hydraulic conductivity calculated from the yield, depression, diameter, and screened length of the well (m/sec) (Lorberer, 1975). Similar relationship between depth and yield was also observed in the Brumundal Sandstone in Norway (Englund and Jorgensen, 1975). There are also attempts to characterize the density of joints and their hydraulic conductivity statistically. It was found by Rats (Rats and Chernyashov, 1965) in sandstones of the Ordovician flysch in central Kazakhstan that the average distance between neighboring fractures (ℓ) in meters and hydraulic conductivity along the fractures perpendicular to the stratification (K_f) in m/sec can be expressed as functions of the thickness of the layer (m) given also in meters:

Table 5-4. Parameters characterizing the hydraulic properties of noncarbonate, indurated sediments.

Type of Rock	Location	Porosity (%) av.	max.	min.	Comments	References
sandstone (fine)		33	49.7	13.7	55 samples	Morris and
(medium)		37	43.6	29.7	10 samples	Johnson (1966)
Miocene sandstone	Tunisia	-	37.0	20.0	2 samples	Schoeller (1962)
sandstone	USA					
Devonian	Bradford	14.8	-	-		Davis and
Missisipian	Berea	19.0	-	-		de Wiest (1966)
Ordovician	Oil Creek	6.7	-	-		
Cretaceous	Woodbine	25.6	-	-		
Pliocene	Repetto	19.1	-	-		
Eocene	Wilcox	15.3	-	-		
conglomerate (coarse)		17.3	-	-		Rima et al.
arkose (fine)		14.4	-	-		(1962)
(medium)		15.6	-	-		
(coarse)		10.9	-	-		
siltstone		9.7	-	-		
sandstone	USA					
	Cromwell	16.6	-	-		Muskat (1937)
	Glicrest	27.4	-	-		
	Prue	11.4	-	-		
	Wilcox	15.6-12.4	-	-		
sandstone	USA					
Cambrian		11.2	-	-	24 samples	Manger (1963)
Pennsylvanien shale		17.4	-	-	587 samples	
Oligocene and Miocene		21.1	-	-	9 samples	
Silurian		5.2	-	-	5 samples	
Permian sandstone	Brumundal (Norway)	15.2	18.8	10.6	9 samples	Englund and Jorgensen (1975)
		Hydraulic Conductivity (cm/sec) av.	max.	min.		
sandstone (fine)		2.6×10^{-4}	1.9×10^{-3}	4.2×10^{-7}	20 samples	Morris and
(medium)		4.1×10^{-3}	1.2×10^{-2}	2.6×10^{-6}		Johnson (1966)
Miocene sandstone	Tunisia	-	1.0×10^{-2}	1.0×10^{-3}		Schoeller (1962)
Oligocene sandstone		-	5.0×10^{-3}	3.0×10^{-3}		

Table 5-4. (continued)

sandstone	USA					
Devonian	Brandford	2.6×10^{-6}	-	-		Davis and de Wiest (1966)
Missisipian	Berea	3.5×10^{-1}	-	-		
Ordovician	Oil Creek	3.9×10^{-6}	-	-		
Cretaceous	Woodbine	4.3×10^{-3}	-	-		
Pliocene	Repetto	3.5×10^{-5}	-	-		
Eocene	Wilcox	2.9×10^{-7}	-	-		
sandstone						
Carboniferous		-	6.0×10^{-11}	2.9×10^{-12}		Louis (1967)
Devonian		-	2.0×10^{-11}	2.1×10^{-12}		
conglomerate (coarse)		3.8×10^{-7}	-	-	vertical	Rima et al. (1962)
		4.9×10^{-7}	-	-	horizontal	
arkose (fine)		1.6×10^{-7}	-	-	vertical	
		1.6×10^{-6}	-	-	horizontal	
(medium)		5.5×10^{-7}	-	-	vertical	
		1.1×10^{-6}	-	-	horizontal	
(coarse)		3.8×10^{-7}	-	-	vertical	
		5.5×10^{-7}	-	-	horizontal	
siltstone		1.6×10^{-7}	-	-	vertical	
		1.2×10^{-7}	-	-	horizontal	
sandstone	USA					
	Cromwell	1.7×10^{-4}	-	-	vertical	Muskat (1937)
		4.1×10^{-4}	-	-	horizontal	
	Oilcrest	5.6×10^{-4}	-	-	vertical	
		8.0×10^{-4}	-	-	horizontal	
	Prue	4.7×10^{-7}	-	-	vertical	
		3.4×10^{-6}	-	-	horizontal	
	Wilcox	3.6×10^{-4}	-	-		
		$\sim 8.5 \times 10^{-5}$	-	-	vertical	
		7.6×10^{-5}	-	-		
		$\sim 8.8 \times 10^{-5}$	-	-	horizontal	

Table 5-4. (continued)

shale						
Cretaceous		$4.0x10^{-11}$	-	-		Gondouin and Scala (1952)
Pennsylvanian		$9.0x10^{-11}$	-	-		
Oligocene sandstone	Nógrád (Hungary)	$3.0x10^{-2}$	$7.8x10^{-1}$	$2.9x10^{-4}$	200 m depth interval, 32 wells	Lorberer (1975)
		$5.2x10^{-3}$	$5.1x10^{-2}$	$1.8x10^{-4}$	200-300 m depth interval, 37 wells	
		$8.7x10^{-3}$	$7.9x10^{-2}$	$1.5x10^{-4}$	300-450 m depth interval, 23 wells	
Permian sandstone	Brumundal (Norway)	$1.4x10^{-4}$	$3.0x10^{-4}$	$1.5x10^{-5}$	9 samples	Englund and Jorgensen (1975)
		Yield of Wells (ℓ/min)				
		av.	max.	min.		
Ordovician argillite	Angara Basin (USSR)	45.7	-	-	5-25 m depth interval	Rats and Chernyshov (1965)
		3.1	-	-	25-55 m depth interval	
Permian sandstone	Brumundal (Norway)	71.3	1250.0	9.0	40-100 m depth interval, 13 wells	Englund and Jorgensen (1975)
		Specific Yielding Capacity of Wells $(\frac{\ell/min}{m})$				
		av.	max.	min.		
sandstone	Saar Region	1.5	12.0	0.5		Seiler (1969)
Oligocene sandstone	Nógrád (Hungary)	$1.7x10^{-1}$	$6.8x10^{-1}$	$3.9x10^{-3}$	10-200 m depth interval, 32 wells	Lorberer (1973)
		$1.0x10^{-1}$	$3.8x10^{-1}$	$6.4x10^{-3}$	200-300 m depth interval, 37 wells	
		$8.1x10^{-2}$	$3.0x10^{-1}$	$3.2x10^{-3}$	300-450 m depth interval, 23 wells	

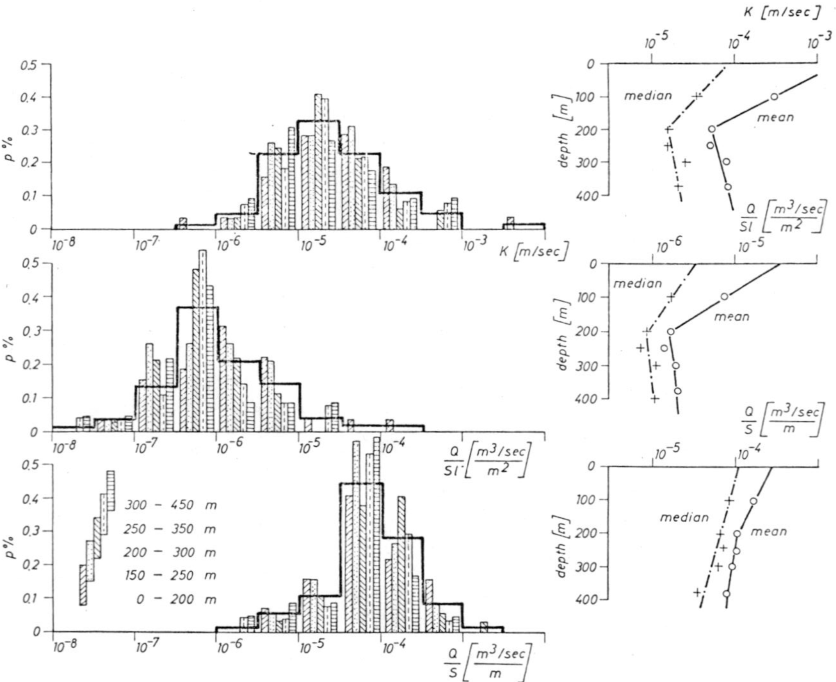

Fig. 5-10. Parameters of wells (amount of yield and drawdown; yield divided by drawdown and the length of filter; calculated hydraulic conductivity) penetrating into Oligocene sandstone in Nógrád (northern Hungary) (after Lorberer).

$$\log \ell = 0.41 \log m + 0.45 \quad ;$$

$$K_f = 0.82 \log m - 0.19 \quad . \qquad (5\text{-}1)$$

Shales usually have very low primary porosity, and thus water circulates mainly through fissures, joints, and planes of stratification. In general, the yield of wells in shales are, therefore, poor. Well developed network of joints can be expected only near the surface due to high degree of weathering. The reported average porosity ranges from 1 percent to 25 percent and hydraulic conductivity was observed to be as low as $5x10^{-11}$ - $4x10^{-9}$ cm/s.

17-5 Hydraulic Properties of Carbonate Rocks

The sediments composed mostly of different carbonates (generally calcium and magnesium carbonates) of chemical or biological origin are called carbonate rocks. The main representatives of this group are the various limestones and dolomites. The term "karstic rock" is used many times as a synonym of carbonate rock although this application is not absolutely correct because the adjective "karstic" indicates the development of chemically enlarged openings in the rock mass which process may occur in noncarbonate sediments as well (in evaporites, e.g., gypsum) while there are carbonate rocks in which the development of dissolved openings is not common (e.g., carbonate marls). The carbonate rocks, and especially those having karstic behavior, are known to be the best aquifers apart from the coarse grained loose clastic sediments. On terrains covered by carbonate rocks, the recharging rate of groundwater is generally high and its evaporation is low because of the wide vertical openings through which precipitation infiltrates rapidly and in which capillarity is negligible. The storage capacity of carbonate rocks depends on their porosity because the compressibility of the solid matrix is practically negligible, thus both important hydraulic parameters (i.e., storage capacity and hydraulic conductivity) are basically influenced by the structure, the volume, and the size of the interconnected pores. This is the reason why the chemical and mechanical processes enlarging the openings considerably increase these parameters.

Similarly as in the case of other indurated sediments, primary and secondary porosity can be distinguished. The former is the volume of the original pores within the rock mass related to the bulk volume. These interstices are the residual pores of the sediment from which the rock was formed, frequently reduced by calcite deposited within the pores during or subsequently from the process of lithogenesis. Primary porosity is generally very low in limestones and dolomites. In young coarse grained limestone, however, primary porosity may be as high as 20 percent. If the dolomite minerals have developed from calcite after the general lithification, the primary porosity of the rock is about 12-13 percent because the transition from calcite to dolomite results in a reduction of volume (Davis, see de Wiest, 1969). Some special carbonate rocks may have even higher primary porosity (chalks up to 50 percent, tuffs up to 60 percent).

Secondary porosity is composed of discontinuities of different origins (fractures due to tectonic movements, bedding planes, expansion fissures). The joints generally compose a dense network throughout the whole mass of the rock. The size of the openings generally decreases with increasing depths due to the burden of the overlying layers. The effects of weathering and exfoliation result in high porosity near the surface. Secondary tectonic movements may cause either the closing or the widening of the joints, the result of which is the nonhomogeneous and anisotropic permeability of the carbonate rocks (the openings are larger and also their number is higher in the fractured zones along cross faults than in the mass of the rock, while the joints are closed, although high in number in compressed zones). Complete or partial sealing of the fissures and fractures may occur near the surface from silt and clay transported by the infiltrating water, and inside the rock mass a similar result is caused by the development of calcite crystals on the walls of the openings from the carbonates dissolved in the percolating water.

Another effect of water percolating through the masses of carbonate rocks (more important than sealing) is the enlargement of the openings and the development of cavities. Water having free CO_2 content dissolves the material of carbonate rocks. This process is more rapid in limestone than in dolomite. Impurities in the carbonate rocks (e.g., clay in carbonaceous marl or quartz in cherty limestone) hinder the development of dissolved openings while suspended materials carried by the water accelerates the development of large cavities because the chemical process is supplemented by mechanical erosion.

The strongest enlargement of secondary porosity can always be observed in the zone of fluctuation of the water table of the groundwater stored in carbonate rocks (karstic water) where the water having continuous contact with the atmosphere always contains free dissolved CO_2, and the water movement is also considerable in the direction of the draining points. Usually gently sloping passages and interconnected galleries are formed here, the intersections of which are enlarged to form chambers and domed caves. Above the water table in the zone of aeration, the dissolved openings are generally vertical or steeply sloping. Shafts, chimneys, and sinkholes are created here by the infiltrating water.

The elevation of the zone of fluctuation related to the mass of the rock might be changed many times during geological ages due to the erosion of the valley crossing the blocks of the mountains or to the vertical tectonic movement of the rock masses. This is the explanation of the development of multi-storied caves having more than one network of galleries above each other interconnected by vertical shafts. The same effect of the changing of drainage level can be observed when the porosity vs. depth relationship is investigated in carbonate rocks especially in limestones. Apart from the generally decreasing trend of porosity, characteristic for any other fissured and fractured rock, local irregularities can be found at special levels where the rapid increase of porosity is observed. An example is shown in Fig. 5-11. The graph is constructed from data observed in the northern part of the Ural Mountain (USSR) (Shevyakov and Mankovsky, 1963), but similar conditions were explored by mining activities in the Transdanubian Central Mountain Range (Hungary) as well (Schmieder et al., 1970).

There are many data concerning the total secondary porosity of carbonate rocks published in the literature although it is well-known that the determination of this parameter is difficult, and thus, the numerical values can be accepted only as rough estimations because of the uncertainties involved in their measurements. The determination is generally based on the investigation of core samples taken from a borehole on the evaluation of pumping tests, on the comparison of the change of water level and the recharge or discharge of the aquifer during a given time period, or on the investigation of the openings along a free wall of the rock either on the surface or in mines. Some collected data are summarized in Table 5-5 for information.

It can be seen from the data listed in the table that the hydraulically active secondary porosity of carbonate rocks is generally between 1 and 3 percent. Schmieder (1970) has proposed that the average porosity of Triassic limestones in the Transdanubian Central Mountain Range (Hungary) might be characterized by n = 1.5 ± 1.0 percent, and the dolomite of the same age by n = 3.0 ± 1.0 percent. There are several investigations, however, showing a much larger parameter. In the same area, Venkovits (1951) has found as high a porosity as 6-23 percent (one local value was 65 percent) using core samples. In the southwestern part of Spain, the porosity of Silurian-Devonian limestone ranges up to 10-12 percent and that of Middle Triassic dolomite to 6-7 percent (Samper and Navarro, 1965). The exaplanation of this very high parameter

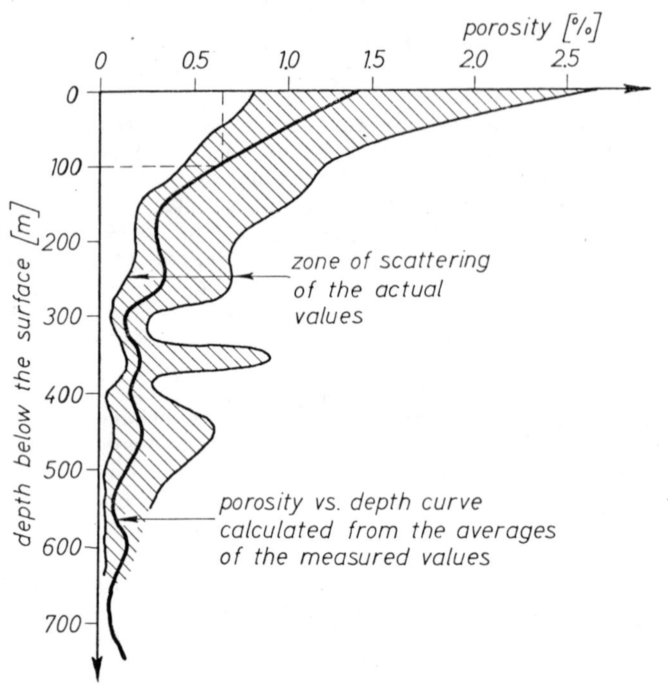

Fig. 5-11. Porosity of limestone as a function of depth (Ural Mountain, USSR) (after Shevyakov and Mankovsky).

can be given on the basis of the investigation of Schoeller and Aigrot (1967). They have found that in the surroundings of the Fontaine de Vaucluse (southwestern France) the porosity of the limestone in the zone of fluctuation of the water table can be estimated as 13-14 percent. Immediately below this zone the parameter is only 0.2-0.8 percet. Considering this observation and also those represented in Fig. 5-4, it can be stated that the average porosity of carbonate rocks is only 1-5 percent near the surface, and it decreases with increasing depth. There are, however, levels where the conditions are or were favorable for the development of karstic formation. Here, extremely high porosity can be observed locally. It is not a rare case that the local maxima may reach 10-15 percent and even as high a value as 30 percent can be expected at some places.

Table 5-5. Porosity of carbonate rocks.

Type of Rock	Location	Measuring Method	Porosity (%) av.	max.	min.	Comments	References
	Transdanubian Central Mountain Range (Hungary)						
dolomite Middle Triassic	Iszkaszentgyörgy	samples	1.5	-	-		Schmieder et al. (1974)
Upper Triassic	Nyirád	samples	4.5	-	-		
		pumping tests	5.2	15.0	1.0		Höriszt (1965)
	Nagylengyel	samples	2.0	-	-		Schmieder et al. (1974)
	Tatabánya	samples	9.0	-	-	at the surface	Gerber (1965)
			2.2-2.5			between 60-80 m	
limestone Upper Triassic	Dorog	pumping tests	1.32	2.50	0.50		Willems (1959)
		samples	0.55	3.87	0.07	areal porosity	
Cretaceous Eocene	Pápa		1.55	-	-		Schmieder et al. (1974)
	Ajka		1.4	-	-		
	Halimba		1.3	-	-		
	Rábasömjén		2.5	-	-		
	Southeastern Spain						
marble Silurian-Devonian			10-12	-	-		Navarro and Samper (1965)
dolomite Middle Triassic			6-7	-	-		
limestone Miocene			0.6-0.7	-	-		
limestone	Fontaine de Vauchuse						Plandrin and Paloc (1969)
Triassic	Southeastern France	pumping tests	1.2				
		fluctuation of the yield		13-14		zone of saturation	Schoeller and Aigrot (1967)
					0.2-0.8	at deeper level	

The extremely wide fluctuation of porosity within a relatively small area needs the investigation of the correct definition and determination of this parameter, especially in carbonate rocks, using the conceptual approach as it was explained in Section 1.

Apart from porosity, the other parameter important for the hydraulic characterization of the seepage field is hydraulic conductivity or intrinsic permeability. There are several data available in various publications giving directly the permeability of carbonate rocks. Some of them are listed in Table 5-6.

The uncertainty of these data is, however, very high. The execution of a correct pumping test with more than one observation well is very expensive and time consuming because of the high expected yield, the great thickness of the carbonate layers, and the high cost of drilling in hard rocks. The simplified field tests are, therefore, applied in most cases, e.g., observation of the lowering of the water level in the pumped well during a short period of pumping, and recording the rise of the water until the original level is recovered; water injection test (water uptake, Lugeon test) in the open borehole; recording the corresponding yield and drawdown data of existing wells operated for any other purposes; etc. There are, howver, attempts to calculate the permeability of the rock even from these indirect data. Borelli and Pavlin (1965) have found, e.g., that Lugeon's number can be expressed in the form of an equivalent hydraulic conductivity

$$1 \text{ Lugeon} = 0.07\text{-}0.15 \ [\text{m day}^{-1}] = 1\text{-}2\text{x}10^{-6} \ [\text{m sec}^{-1}] \quad , \qquad (5\text{-}2)$$

Although the transformation of the directly observed data into the form of hydraulic conductivity or matrix porosity is possible, there are many authors preferring the publication of the original measurements knowing the very uncertain character of the parameters derived from simplified field tests. It is necessary, however, to give information on the conditions influencing the development of the measured process. Thus, the yields of various wells are comparable only if the drawdown, the length of the screen (or that of the open stretch of the borehole), and the diameter of the well is known. In many cases, the yield (Q) related to drawdown (s) is used (q specific yielding capacity of the well) to homogenize the data, or the latter is also divided by the length of open stretch of the well (ℓ) to get the specific yielding capacity of the rock, q' (or double specific value) (Table 5-7):

$$q = Q/s \quad ; \quad q' = q/\ell = Q/s\ell \quad . \qquad (5\text{-}3)$$

Table 5-6. Hydraulic conductivity of carbonate rocks.

Type of Rock	Location	Measuring Method	Hydraulic Conductivity (cm/sec)			Comments	References
			av.	max.	min.		
	Transdanubian Central Mountain Range (Hungary)						
dolomite							
Middle Triassic	Nagyegyháza	water injection	$8.4x10^{-5}$	$1.7x10^{-5}$	$1.3x10^{-6}$		Schmieder et al. (1974)
Upper Triassic	Nyirád	pumping tests	$6.0x10^{-2}$	$2.5x10^{-1}$	$2.0x10^{-2}$	in fault zones	
limestone							
Cretaceous	Nagyegyháza	water injection	$1.5x10^{-4}$	$4.5x10^{-5}$	$7.6x10^{-8}$		
Eocene	Nagyegyháza	water injection	$2.5x10^{-4}$	$1.1x10^{-2}$	$3.0x10^{-6}$		
karstic limestone	Busko Blato (Yugoslavia)	Lugeon tests	$1.5x10^{-3}$	$3.0x10^{-1}$	$1.5x10^{-3}$		Borelli and Pavlin (1965)
gypsum and anhydrite	Jenireh (Syria)	pumping tests	$1.1x10^{-1}$	-	-		Mortier and Safadi (1965)
tertiary	S.Carolina (USA) Beaufort County	pumping tests	$6.10^{-3}-6.10^{-4}$	-	-		Siple (1965)
	Jasper County		$1.10^{-2}-5.10^{-3}$	-	-		
	Savannah		$6.10^{-3}-2.10^{-3}$	-	-		
limestone	Tunisia						Gosselini and Schoeller (1936)
Campanian		pumping tests	-	$8.8x10^{-1}$	$6.5x10^{-2}$		
Eocene		pumping tests	-	$1.2x10^{-2}$	$6.0x10^{-4}$		

Table 5-7. Yield of wells draining carbonate rocks in Hungary.

	Type of Rock d = dolomite l = limestone	Open Length of the Borehole (m)	Depression (m)	Yield (ℓ/min) max.	Yield (ℓ/min) continuous	Specific Yielding Capacity ($\frac{\ell/min}{m}$)	References
	Triassic						
Budapest Városliget	d	54	12.3	470	470	37	Korim, K., verbal information
Budapest Zugló	l	6	9.5	960	960	100	
Bükk-1	d	6	75.0	9000	3500	40	
Harkány-3	l	6	1.5	2500	1000	666	
Harkány-4	l	20	2.0	800	800	400	
Magyarszék	l	138	6.7	1000	950	142	
Miskolc-1	l	28	12.2	1000	923	75	
Miskolc-2	l	40	12.0	4850	4850	404	
Szentendre	l	-	14.0	60	60	4	
Budafok	l	-	1.0	50	50	50	
Leányfalu	l	-	12.2	1040	560	45	
Visegrád	l	-	1.1	585	535	532	
Bia	l	-	34.0	460	460	14	
Komárom-1	l	-	29.0	2500	2500	86	
Tapolca (HGN-39)	l	-	1.3	600	410	31	
Sárospatak-2	l	-	4.9	850	850	173	
Héviz-3	l	-	0.3	1000	1000	3333	
Komló-17	l	-	4.5	610	610	135	
Vác	l	-	6.0	2500	1176	196	
Törökbálint	l	-	3.0	860	860	286	
	Cretaceous						
Pápa-2	l	-	27.1	3250	3250	120	

(It is necessary to note here that all the problems explained in connection with the determination of permeability occur not only in the case of carbonate rocks, but in the investigation of other fissured and fractured rocks as well. It is this reason why data other than permeability were also collected in the previous sections as well concerning the various aquifers composed of hard rocks. The detailed explanation of the problems is given here because carbonate rocks are the most important aquifers.)

The determination of hydraulic conductivity is hindered even in the case when data of correctly executed pumping tests are available or the drawdown of mining activity is observed by more observation wells and the water amount drained through the galleries of the mine is measured (mining can be regarded in this case as a large scale pumping test, and the area of the mine as a fictive well). The difficulties are caused by two factors in these cases:

- the unknown depth of the layer taking part in the transport of water;

- the role of large openings in water conveyance near the draining structures (wells, shafts, galleries).

The thickness of the carbonate rocks is generally very high, a layer of some hundred meters is not rare. The total thickness of carbonate formations composed of more layers and forming a unified reservoir of karstic water may surpass one thousand meters in many cases. At the same time, the decrease of porosity as well as permeability is observed in the carbonate rocks when the depth increases, similarly as in other hard rock aquifers. The approximation of the relationship of hydraulic conductivity vs. depth with an exponential function is proposed by many authors:

$$K = K_o \exp(-\beta z) \quad ; \qquad (5\text{-}4)$$

where K_o is the hydraulic conductivity at the surface and z is the depth. The β factor was found by Borelli and Pavlin (1965) to be 0.0052 in karstic limestone (Busko Blato, Yugoslavia) while from data represented in Fig. 5-12 and characterizing Middle Triassic dolomite (Transdanubian Central Mountain Range, Hungary), a value of β = 0.0030 was calculated. It is also shown by the figure that the probable distribution of hydraulic conductivity can be approximated by using a logarithmic normal distribution function in the case of carbonate rocks as well as it is generally proposed to characterize fissured and fractured rocks although the frequency has some asymmetry. Some other measurements show, however, that there are areas where the distribution can be

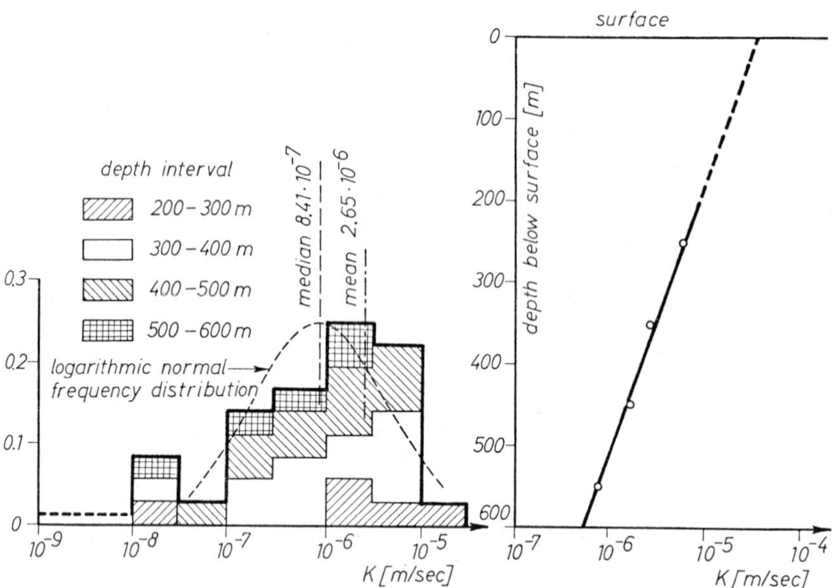

Fig. 5-12. Frequency distribution of hydraulic conductivity in Middle Triassic dolomite (Óbarok, Hungary) investigated as a function of depth (after Schmieder).

better described by a normal than a logarithmic normal distribution function (Fig. 5-13) (Király, 1973).

From Eq. 5-4 an average depth can be calculated, the transmissivity of which (supposing constant hydraulic conductivity K_o) will be equal to that of an infinite layer with the hydraulic conductivity decreasing with depth:

$$T = K_o m_{av} = \int_o^\infty K dz = \int_o^\infty K_o \exp(-\beta z)\, dz = K_o \frac{1}{\beta} \quad ; \tag{5-5}$$

$$m_{av} = \frac{1}{\beta} \quad .$$

It is shown, however, by Fig. 5-11, that porosity may have local high values at elevations where the conditions are at present or were in the past suitable for the development of large dissolved openings. Hydraulic conductivity, being closely dependent on porosity, may also have local changes superimposed on the trend decreasing with depth. Thus, the depth of the flow field estimated or calculated on the basis of any theoretical method is very unreliable and, therefore, many investigators do not try to divide the influence of depth and hydraulic conductivity, but publish the value of

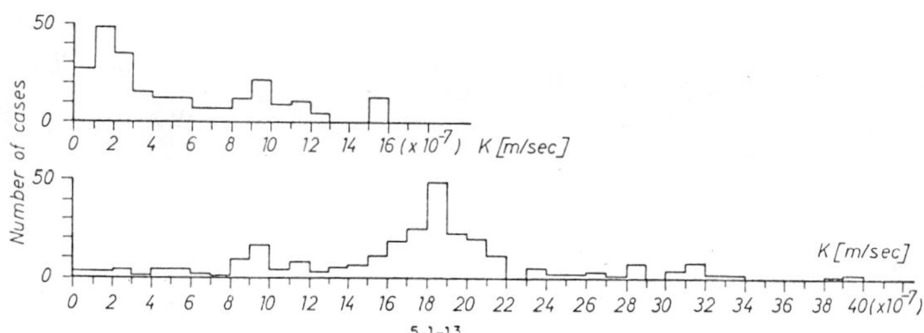

Fig. 5-13. Frequency distribution of hydraulic conductivity of carbonate rocks showing normal and logarithmic normal character, respectively (after Király).

transmissivity (T = Km) as it is calculated from pumping tests or from the depressions developed around mines draining the karstic reservoirs (Table 5-8).

The influence of the large openings around the draining structure was mentioned as the second cause hindering the evaluation of the data gained from pumping tests. Discussing the determination of the representative elementary unit in carbonate rocks (see Part I), it was mentioned that in the close vicinity of wells or shafts water movement cannot be investigated as a three-dimensional seepage extending through the whole space of the fractured rock. The draining structure crosses only a few water conveying elements which may be large channels or fractured zones. The relationship between the discharge and the head loss is determined here, therefore, by the resistance of these openings, and turbulent movement frequently occurs in the channels because the movement of a large amount of water is concentrated into a few openings. Farther from the draining structure, the flow is better

Table 5-8. Transmissivity of carbonate rocks.

Type of Rock	Location	Measuring Method	Transmissivity (m^2/sec) av.	max.	min.	Comments	References
	Transdanubian Central Mountain Range (Hungary)						
dolomite Middle Triassic	Iszkaszentgyörgy	depression of mine	3.0×10^{-3}	5.4×10^{-3}	1.8×10^{-3}		Schmieder et al. (1974)
	Nagyegyháza	water injection	3.4×10^{-5}	8.0×10^{-4}	2.5×10^{-7}		
		pumping test	7.0×10^{-3}	-	-		
Upper Triassic	Iszkaszentgyörgy	depression of mine	7.8×10^{-3}	-	-		
		depression of mine	2.2×10^{-2}	-	-		
		depression of wells	7.7×10^{-2}	-	-	in fault zones	
	Tatabánya	depression of mine	8.0×10^{-3}	2.1×10^{-2}	-		
	Nagylengyel	pumping tests	2.3×10^{-4}		-		
limestone Upper Triassic	Dorog	depression of mine	3.0×10^{-2}	-	-		
	Tatabánya	depression of mine	4.2×10^{-2}	-	-		
Cretaceous	Nagyegyháza	water injection	4.4×10^{-5}	6.3×10^{-4}	2.2×10^{-8}		
Eocene	Nagyegyháza	water injection	4.4×10^{-5}	3.5×10^{-3}	6.0×10^{-7}		
	Ajka	depression of mine	2.0×10^{-4}	-	-		
	Illinois (USA)						
Silurian dolomite	Cook County	pumping tests	9.0×10^{-3}	-	-		Csallány (1965)
	Du Page County		1.7×10^{-2}	-	-		
	Kane County		2.8×10^{-3}	-	-		
	Lake County		1.5×10^{-3}	-	-		
	McHenry County		2.4×10^{-3}	-	-		
	Will County		4.1×10^{-3}	-	-		
limestone	Amphissa-Itea (Greece)						
general		pumping tests	1.8×10^{-5}	3.0×10^{-5}	1.0×10^{-5}		Burdon (1965)
intermediate			1.4×10^{-4}	1.6×10^{-4}	9.1×10^{-5}		
karstified			1.7×10^{-3}	2.6×10^{-3}	1.0×10^{-3}		
gypsum and anhydrite	Jezireh (Syria)	pumping tests	8.6×10^{-2}				Mortier and Safadi (1965)
tertiary limestone	S.Carolina (USA)	pumping tests					Siple (1965)
	Beaufort County		4.3×10^{-3}				
			$\sim1.6\times10^{-2}$	-	-		
	Jasper County		6.0×10^{-2}	-	-		
	Savannah		4.3×10^{-2}	-	-		

distributed, especially if a fractured zone normal to that penetrated by the well creates contact between the water transporting elements. It can be supposed, therefore, that outside a given distance from the well the whole network of the openings take part in water transport in an almost even rate (Fig. 5-14).

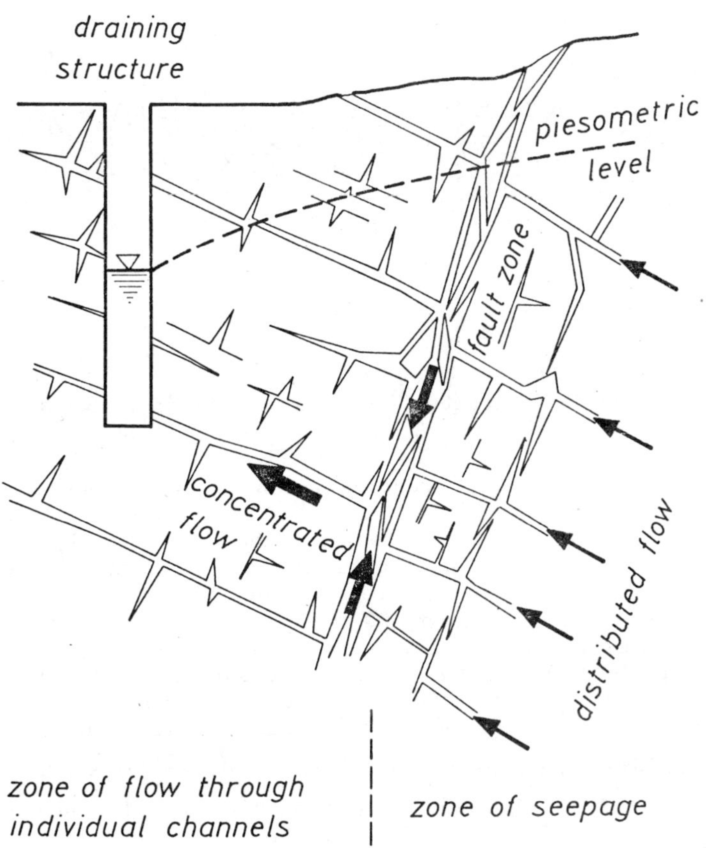

Fig. 5-14. Water transport through large openings in the vicinity of a well.

An example is shown in Fig. 5-15 to demonstrate the influence of the special water transporting elements on the development of drawdown in the vicinity of a pumped well. The drilling crossed a steeply dipping fractured zone. The water uptake was 1600 ℓ/min. After some fluctuation, a constant drawdown of 73.0 m was maintained by pumping 900 ℓ/min (the

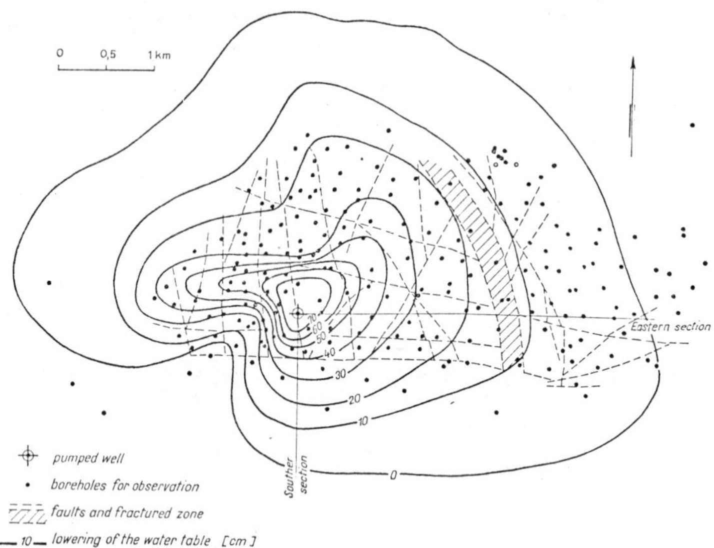

Fig. 5-15a. Results of a pumping test in carbonate rock (after Schmieder).

total duration of the testing period was two months). From the water level data observed in the observation wells located in a relatively dense network around the pumped well, the drawdown cone was constructed as it is shown on the map. About 99 percent of the total head loss occurred within a distance between 150 and 400 m from the well depending on the direction. Only the remaining 0.70 m drawdown caused the general lowering of the piezometric surface of the reservoir in the farther surroundings where the flow can be investigated as seepage. The transmissivity calculated from the outer part of the drawdown cone ($T_w = 7.5 \times 10^{-3}$ [$m^2 sec^{-1}$]) is practically equal to that calculated from the influence of a nearby mine (Tatabánya) ($T_m = 8.2 \times 10^{-3}$ [$m^2 sec^{-1}$]) while the transmissivity of the drained fractured zone ($T_f = 5 \times 10^{-4}$ [$m^2 sec^{-1}$]) is between the maximum and average value of this parameter determined by water injection tests for the same layer (see Table 5-8).

Very similar results were achieved by Király (1973). Investigating the hydraulic conductivity of carbonate rocks in the La Brévine syncline (Switzerland) he has found that the parameter determined by field tests ranges from 5×10^{-6} to

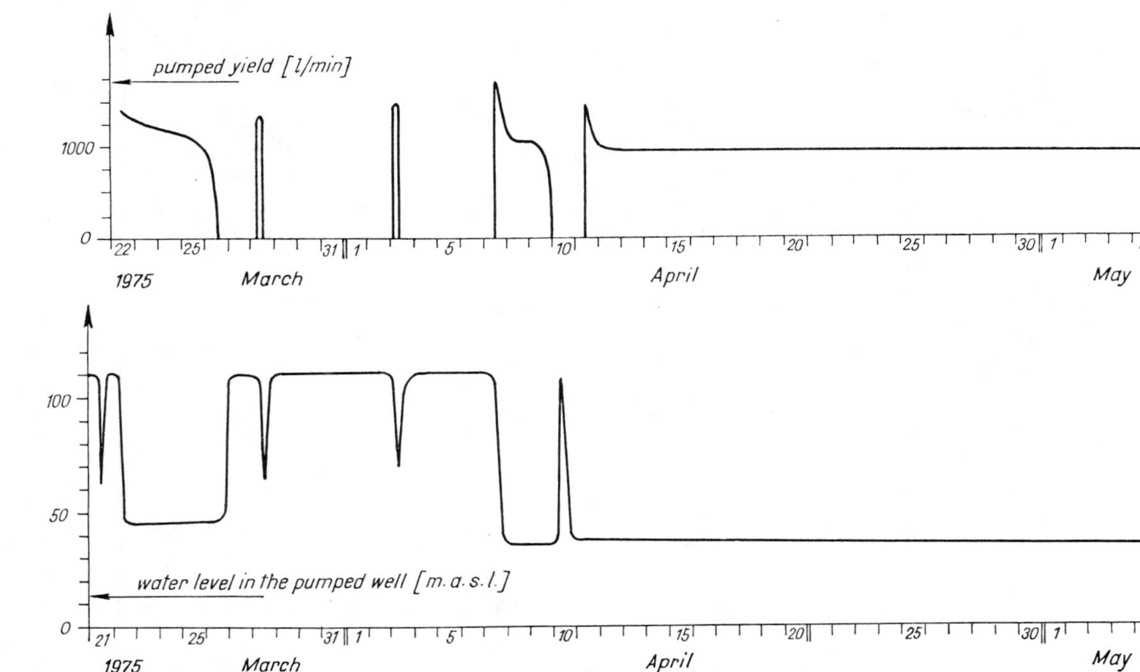

Fig. 5-15b. Results of a pumping test in carbonate rock (after Schmieder).

1×10^{-7} [m/sec] while a value higher than 10^{-3} is characteristic to large regions as it was calculated from the yields of springs and the slope of the piezometric surface. In his opinion, this deviation of about three orders of magnitude or more is caused by the water conveying capacity of large openings spaced in considerable distance from each other, the exploration of which by drilling has very low probability.

The example and the explanation given in connection with the role of individual water transporting elements around the draining structures draws attention to the fact that the water movement cannot be characterized as seepage within a given distance from a well or shaft. It is necessary, therefore, that this limit should be determined for each layer as a further hydraulic parameter supplementing porosity and hydraulic conductivity. A further consequence of this condition is that the total head loss maintaining the flow has to be divided into two parts: i.e., the local loss near the draining structure calculated as the resistance of the water conveying channels and fractured zones crossed by the structure, and the seepage loss along the outer part of the influenced area where the development of a three-dimensional movement through an almost homogeneous flow space can be supposed.

SECTION 18

HYDROGEOLOGICAL EXPLORATION OF SOLID ROCK TERRAINS

The general guidelines in exploration of hard rocks are essentially the same as those for groundwater in general, i.e., the determination of the properties and the geometry of various types of rock. The methods used in their exploration are of geological character supported by geophysical measurements resulting in hydrogeological maps and sections which may be transformed into the hydrological ones representing the rocks according to their porosity and permeability.

The prospecting of hard rock reservoirs means a difficult task for the hydrogeologists: not only the methods of exploration to be used are rather complicated, but also the basic characteristic quantities, even the permeability, can be determined by rather sophisticated processes.

From the point of view of hydrogeological exploration, hard rock areas can be characterized as terrains where the water-bearing character of rocks is controlled by the secondary porosity of the formations due to the fractures, fissures, vesicles, and other kinds of openings within the rock mass. The effective porosity of such rocks is composed of their primary (intergranular) porosity and of the secondary one defined above. The ratio of the latter related to the effective porosity varies considerably according to the types of rocks: it is high, e.g., in crystalline limestones and in sandstones.

Even where the original porosity is relatively high, the original permeability is generally low because the large pores are filled with cement in most cases.

In addition to this common feature, the various types of hard rocks are determined by factors such as: structure, stratigraphy and lithology. The comprehensive analysis of such factors will result in detailed hydrogeological classification of hard rocks. The final objective of the classification should be a quantitative treatment of hard rock aquifers analogous to the aquifers of intergranular porosity. The hydrogeological classification of hard rocks was already given in Subection 2-4. Detailed characterization and description of types of hard rocks will be given in Subsection 17-1, taking into account the point of view of exploration.

In advance, two types of solid rock terrains should be mentioned here, i.e., the deep types and the shallow ones. Both terms were originally applied to karstic forms, the former one for karsts lying under the deeper part of a river valley while the latter one meant karstic rocks, the floor of which were exposed in valley bottoms. Considering the methods of exploration, it is a decisive factor whether a solid rock belongs to the deep type or to the shallow one.

Regions where the hard rocks are overlain by a thick layer of other rock type can be explored by methods applied to study the substratum of thick clastic sediments, previously discussed in Section 14. Contrary to this, in areas where the solid rock is exposed or blanketed by a thin layer, special methods of exploration may be applied since the structural features of the rocks are more or less manifested on the surface, thus providing a good opportunity for the application of proper geological and geomorphological studies.

In the system of exploration of hard rock terrains, the general schedule of hydrogeological prospecting will be considered. The survey will begin with geological field work involving:

- the study and interpretation of rocks, topographic forms, etc.;

- the determination of outcrops where observations are made;

- the mapping of outcrops and other geological data related mainly to structural forms.

An important conclusion to be considered during the exploration concerning the porosity and permeability of hard rocks is that the search must be made for zones of secondary porosity due to fracturing and/or solution processes. Geological prospecting methods especially suitable to study hard rocks will be discussed in Subsection 17-1.

Among the geological (morphological) methods applied to the study of hard rock areas, remote sensing (Subsection 2-5) should be mentioned at first. Although the remote sensing technique has been proved successful in mineral discovery, for water prospecting as well as in basic geological studies its present day usage is mainly confined to the study of hard rock areas. This is the reason why the general characterization and review of remote sensing will be given here in connection with the geological survey of hard rock terrains.

The remote sensing systems which are applicable to hydrogeological exploration and which are either in use today or are well advanced in development are grouped in types such as:

- optical: imaging and spectral correlation;

- electromagnetic: imaging and non-imaging;

- magnetic: both total and vertical field measuring systems.

These will be discussed in Subsection 18-2.

The geological exploration carried out by field methods and by air reconnaissance will be followed by surface geophysical studies based on a proper geophysical model of fractured-fissured rocks. This model, showing similarity with the hydrological model discussed in Section 19, may serve as a basis for the correlation between hydrogeological and hydrological characteristics. According to this model, hard rock terrains are characterized by geophysical features such as:

- different types of igneous rocks may display different magnetic properties;

- primary and chiefly the secondary porosity of hard rocks control their electric resistivities;

- cavities (vesicles) lying near the surface may be detected by gravimetric, magnetic, and electric methods;

- contact zones of hard rocks, especially those of structural character, are also marked geophysical boundaries, thus they may be delineated by geophysical prospecting techniques.

This characterization gives hint at the geophysical methods suitable for the exploration of hard rock terrains, namely:

- electric resistivity mapping;

- airborne and ground magnetic survey;

- gravimetric prospecting.

Both the magnetic and gravimetric prospecting will be discussed generally with respect to their principles, technique, and interpretation, the electric resistivity mapping will be dealt with reference to the general description of the method in Subsection 14-1.

The characterization of the water regime represents the most important step in the general exploration program. It was already demonstrated that the most effective method characterizing the water regime is the joint and complex consideration of transport phenomena, among them the fluid and heat transport represent the two main types in solid rock terrains. The fluid transport is characterized by the comparatively rapid movement of water in the fissures, fractures, voids, and vesicles within the rock mass. The heat transport occurs in very close connection with water movement almost exclusively by convection owing to the lack of primary porosity as well as due to the rapid flow of water.

Data collection, in this respect, relates quantities playing part in the transport processes such as:

- infiltration into hard rocks;

- discharge of groundwater through springs;

- determination of water level in wells;

- measurement of temperature in wells and springs.

There are data relating quantities not involved in the fluid and heat transport which may aid in the description of these processes. This group of data includes the result of studies such as:

- chemical analysis of groundwater;

- determination of environmental isotopes in it;

- use of artificial tracers for flow studies.

The methods of collection of data described above will be reviewed in Subsection 18-4.

18-1 Geological Survey

The solid rock areas are built up of rocks classified according to their genetic types such as:

- igneous;

- metamorphic;

- indurated sedimentary rocks;

whose fissured and fractured parts, vesicles, and solutional openings represent the aquifers of the area. The genetic types as well as the structural forms such as faults, joints, and cleavages were defined and illustrated in Subsection 2-4. Taking into consideration the extreme importance of secondary porosity controlled mainly by structural elements with regard to the water-bearing properties of hard rocks, further details should be given here in respect to the problems of geological exploration.

Delineating both structural relations and the morphological description of hard rock areas, the most important types of rocks and methods of geological mapping will be considered.

The structural elements, with regard to morphology, may be divided into two groups, namely:

- structural elements characterizing the rock mass itself;

- structural elements at the contact of different rocks.

The hard rocks are generally divided into blocks by well-defined cracks. If along such a crack there has been no slipping or if one block has slipped only a very little against the other, the crack is termed a joint. The joints may be classified according to whether they have formed by compression or by tension, both of them having special relation to the genesis of hard rocks.

In igneous rocks, several kinds of primary facture may be recognized which are in connection with the flow structure. For example, cross joints of tension origin are roughly at right angles to the direction of flow. Sheet-like joints, results of vertical expansion, may be frequently seen in granite or other massive and homogeneous rock. These joints are broadly undulating, being roughly parallel to the surface of the ground. The fracturing dividing the rock into flat sheets is believed to be due to the relief of pressure and consequent vertical expansion through removal of the overlying rock by erosion (Section 2).

In stratified rocks, joints may be seen striking in all directions, but among them a large proportion usually fall into two or more distinct sets. These are the master joints characteristic for the stratified rock resulting from compression or from tension, seldom of simple tension.

No individual joints, not even the master ones, are continuous for long distances. When a fracture ends, it is usually replaced on one or the other side often with a few tens of centimeters of overlap by other joints having the same trend.

Metamorphic rocks, due to dynamic metamorphism, display rock cleavage (fissility) which may be defined as a structure having the capacity to part along certain parallel surfaces more easily than along others. The rock cleavage, characterizing the secondary structure of metamorphic rocks, is a common result of plane stresses under which the rock is deformed by fracture or folding (Subsection 2-3).

Cleavage developed in the upper (fractured) part of rocks is called fracture cleavage. In the lower zones, the folding, which is itself a metamorphic process, leads to the origin of flow cleavage, a structure characterized by the parallel orientation of platy and columnar minerals. Flow cleavage is called platy cleavage when it is so perfect that the rock splits into thin plates of uniform thickness, and it is called schistosity when the rock splits easily but not so regularly. Rocks exhibiting this kind of anisotropy are called slates and schists, respectively.

A compact and soluble carbonate rock is called karstified when distinctive surficial and subterranean features appear and are caused by solutional erosion. In such rocks, a part of secondary porosity is produced by solution openings, cavities, caves, galleries, etc. Although the solution openings are of nonstructural origin, minor joints, faults, differences in composition, and small layers of shale may all control their localization in dense limestone and dolomite. The vertical joints that are widened by solution near the surface tend to be filled by the overlying soil. Horizontal openings along bedding planes, on the other hand, tend to remain open.

The contact between two solid rock masses generally display a well-defined boundary mostly characterized by structural elements.

The contact zone in an eruptive rock may exhibit characteristics which are more or less depenent upon the influence of the country rock penetrated by the igneous rock, or in the case of lavas, dependent on the atmosphere. All of these phenomena, such as textural variation, structure with respect to flow, fracture, and vesicles, control the development of secondary porosity along the zone of contact.

First of all, the vesicular structure of lava flow needs some explanation. In sheets of lava, the vesicles (bubbles) are commonly near to the upper surface toward which they become more and more numerous. Vesicles near the lower contact of flow, in many cases, originated at the upper surface or at the front of the moving lava and were then rolled under. The vesicular structure is characteristic of extrusive sheets, but it may be found in rocks which were injected under little pressure. The contact zones of lava flow in contact with the country rock were illustrated in Fig. 1-23.

Contacts characterized by faults represent well-defined boundaries with respect to geological (morphological) as well as geophysical exploration, e.g., the great carbonate reservoirs in Italy where the water-bearing rocks are confined by faulted zones of impermeable formations (flysch).

The fault plane themselves may serve as channels for groundwater while the faults are acting as barriers elsewhere. Among the miscellaneous fracture elements, dikes play an important role in forming reservoirs or a reservoir system, e.g., in France where faults and fractures due to piercing basalt dikes form the reservoir system.

Boundaries of many aquifers are frequently determined by parts of differential development of the rock mass itself. Common example of this lithologic boundary may be found where fractured-fissured carbonate rocks are surrounded by their impervious formation. In this respect, the karstic terrains should be mentioned owing to the extreme importance in the field of groundwater resources.

It is defined here that the term karst is used in this relation as an internal structure form characterized by the abundance of solution openings.

Summarizing the structural features of solid rocks and solid rock terrains, they come under the headline of lineation, from a morphological point of view, a term though obvious, needs some explanation.

In geology, the term lineation refers to a single feature or, more commonly, to a group of parallel linear features. It usually connotes relatively straight or only broadly curving parallel lines, but accidentally highly contorted parallel lines would fall under the geological meaning of the term.

The lines or linear features are called lineaments. There is another connotation also commonly included in the case of parallel lineaments: that the spacing between the parallel lines is relatively much smaller than their length. The lineaments may be microscopic features observed in a microscopic slide or in a hand specimen as well as macroscopic ones such as fracture lines mapped in field work or detected in air photographs.

The attitude of lineation, i.e., its position in the field relative to north and to the horizontal may be defined as it was made in the case of a bedding plane. If the lineation occurs on, and is limited to a horizontal rock surface, its direction is merely referred to north. If it occurs on an inclined surface, its direction is the horizontal direction of the vertical plane in which it lies. Within this vertical plane its inclination, which is the intersection of the vertical plane with the lineated rock surface, is its plunge which like dip is always a vertical angle measured downward from the horizontal. In any case, the attitude of lineaments is determined by angles which, when measured, may be mapped according to methods to be discussed in the following pages.

Actually, lineation embraces a large variety of geological parallel linear features, some of which may be purely superficial ones whereas others relate to rock structure in one way or another. This was already demonstrated for lineaments such as joints, cleavages, fractures, and solution cavities. Where the lineation is carefully mapped, very significant trends may become evident and considered as an index of the geological history of rocks.

Knowing the structural relations of solid rock terrains, the most important types of rocks will be considered in respect to their lithological features.

A classification of solid rock areas suitable for hydrogeological purposes may be compiled under headings such as:

- karstic carbonate rocks characterized by distinctive features having cavities due to solution erosion;

- pseudokarstic rocks showing karstic features but of other origin;

- indurated sediments of the noncarbonate type;

- igneous and metamorphic rocks.

Within the group of carbonate rocks, limestone and dolomite, the two common types originate from a large number of different sedimentary deposits such as inorganically precipitated limey mud, shell fragments, talus deposit, calcite sand, reef masses, and accumulation of remains of small planktonic organisms. Some of the more important changes after sedimentation are caused by compaction, solution of aragonite and calcite, re-precipitation of calcite cement, and formation of mineral dolomite. Some of the limestones are true clastics whereas others are chemical or biochemical precipitates. The dolomite, a modified form of limestone, displays dense, hard, and brittle character as a result of large-scale recrystallization. Chalk, contrary to the dolomite, represents a soft, friable variety of the limestone. Marls are semi-friable mixtures of clay minerals and calcium carbonate in varying proportion.

Various types of cemented clastic rocks belong to the group of pseudokarstic rocks whose binding material can easily be dissolved and washed out such as conglomerates and sandstones cemented with carbonate material, and hypersoluble sedimentary rocks whose most common types are the evaporites formed by gypsum, anhydrite, and rock salt. The pseudokarstic rocks, except sandstones, constitute less than 2 percent of all exposed sedimentary rocks (Avias, 1975).

The water prospecting both in carbonate as well as in pseudokarstic areas involves:

- determination of the type of rock;

- localization of bigger cavities;

- study of the system of joints, fractures, and cavities.

The group of noncarbonate indurated sediments is represented essentially by various kinds of sandstones. The original porosity of sandstones is a function of sorting, grain shape, packing, and degree of cementation, the latter being the most important among these variables. Common cementing materials are clay minerals, calcite, dolomite, and quartz. Sandstones, cemented with soluble materials, belong to the pseudokarstic types since the secondary porosity is due mainly to solution openings. Porous sandstones with minor amount of calcite cement display good aquifer properties. Advanced stages of silica cement can produce orthoquartzites having extreme hardness and very low permeability. Rocks cemented with clay are usually not as firm as other sandstones, showing properties similar to that of clay layers. The secondary porosity of the latter types is produced by fracturing.

In the group of igneous and metamorphic rocks, no more detail is needed than was given in Subsection 2-3. It should be emphasized only that the secondary porosity is of structural origin.

In water prospecting, the morphological studies, although they are rather complicated, may give information about the position of fractured zones both in the case of indurated sediments and igneous and metamorphic rocks.

The geological field work and mapping, the geological exploration of solid rock terrains, is essentially the same as that for groundwater prospecting in general, or rather, it is more suitable for classical geological field studies since there are outcrops in the area.

General morphological studies and detailed structural examination should be emphasized owing to the close relation between structural forms and water-bearing properties of rocks as it was already demonstrated. Tracing outcrops, i.e., the mapping of dikes and stratified rocks, is best accomplished if these rocks are continuously exposed so that they may be easily followed. However, continuous exposure for long distances is rare. It may be observed sometimes where a great dissimilarity shows itself in the resistance of rocks to erosion: the stronger ones constituting scarps or ridges as it is demonstrated in Fig. 5-16 representing a sandstone area. Data collection for morphological studies may be carried out using either field observations or interpreting air photographs.

Fig. 5-16. Air photograph of a sandstone area (Mainguet, 1975). The big, medium, and small eolian crest-corridors are represented by encircled longish areas, thick and thin lines, respectively. The ← and ··· denote the axes of barakhans and erosion patterns, respectively.

The best places in which to look for outcrops are precipices, hilltops, steep hillsides, streambeds and coasts, and first of all, in artificial excavations. The study of outcrops begins with the petrographical description of rock followed by structural determinations. Careful mapping of structural forms with respect to position, orientation, spacing, density of faults, joints, and cleavages as well as their length, shape, and relation to other structures constitute the basis of the whole geological exploration.

The role of hydrographical studies in morphological exploration (Subsection 2-5) of solid rock areas is of higher importance than elsewhere. Valley pattern or drainage pattern depend on the distribution of hard rocks and on the arrangement of surfaces of weakness such as joints and faults. In addition, springs are often very good indications of faults and fault zones. The springs, which are natural outlets of groundwater, mostly occur in the surface, but in many places they are hidden and issue into a body of surface water such as rivers, lakes, and sea or into alluvial cones as well as into caves within carbonate rocks. Marsh, swamp, and moor vegetation or an oasis, i.e., the green vegetation in general, sometimes wet soil only, are good indications of springs.

Data collection with respect to hydrography may be carried out properly using air photographs where the characteristic patterns mentioned above can be detected more easily than in fieldwork. Surface waters, such as rivers, rivulets, lakes, and outflowing springs, can be picked out easily on black and white air photographs as dark lines and spots, respectively. To study the accompanying phenomena such as green vegetation, wet soil, and difference in soil temperature, color air photographs and thermal images are the most suitable, first of all the latter where water content and temperature of soil as well as vivid pattern of vegetation are enhanced at the same time. This is illustrated by a pair of air photographs, one among which was made in visible light (Fig. 5-17) and the other one in the near infrared band (Fig. 5-18). The pattern of rivers and lakes which are rather blurred in Fig. 5-17 are in sharp contrast to the background in Fig. 5-18. The location of springs, observed on the air photographs or mapped in the field and then depicted on the photos, can be correlated with hydrographic features of the investigated area. Figure 5-19 illustrates a direct relationship between the location of high yield springs and wells and major lineament traces in the area represented in Fig. 5-17.

Knowing the concept of lineation, the morphological exploration of solid rock areas may be characterized as the study of lineaments whether examining an outcrop, mapping them in the field, or investigating air photographs. If the geologist detects any evidences of lineation, he should pay

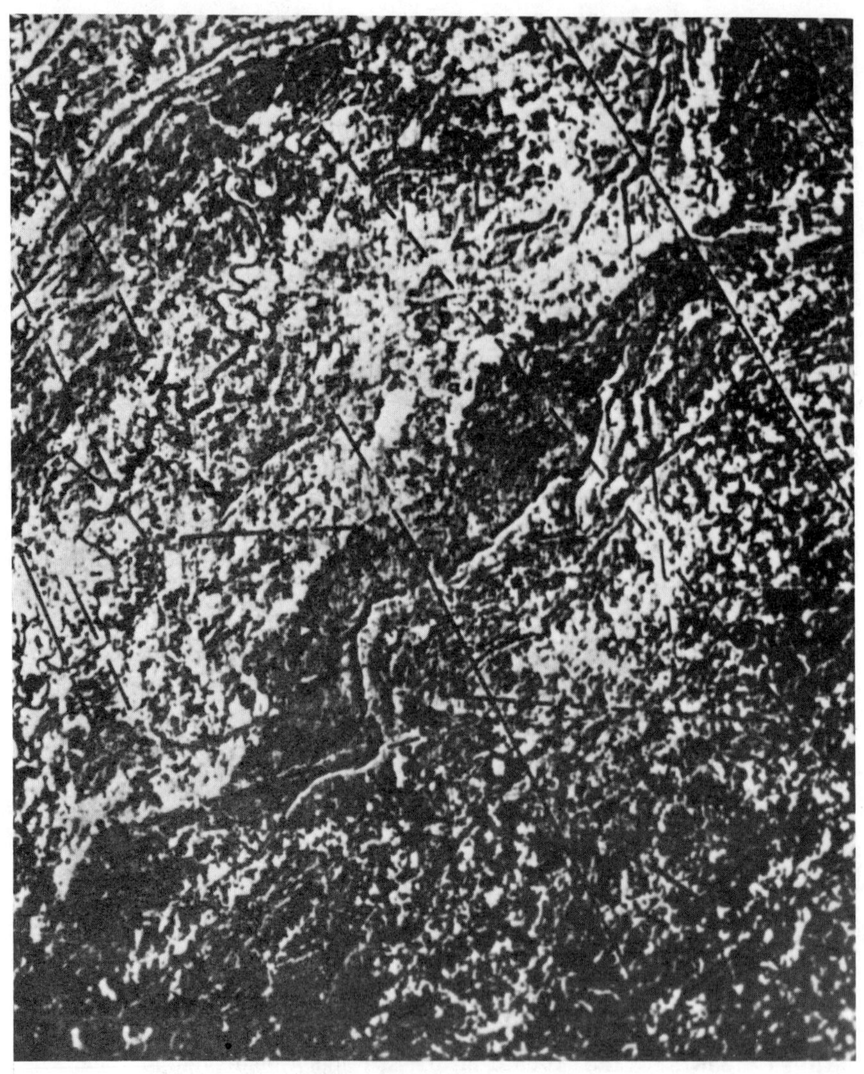

Fig. 5-17. Air photography of a solid rock area made in the visible range of light.

attention to them, ascertain their cause and significance, and correlate them with other geological features of the region.

The instruments used in geological fieldwork to measure the attitude of bedding planes and lineations are comparatively simple measuring devices such as the compass, the clinometer and tape measure.

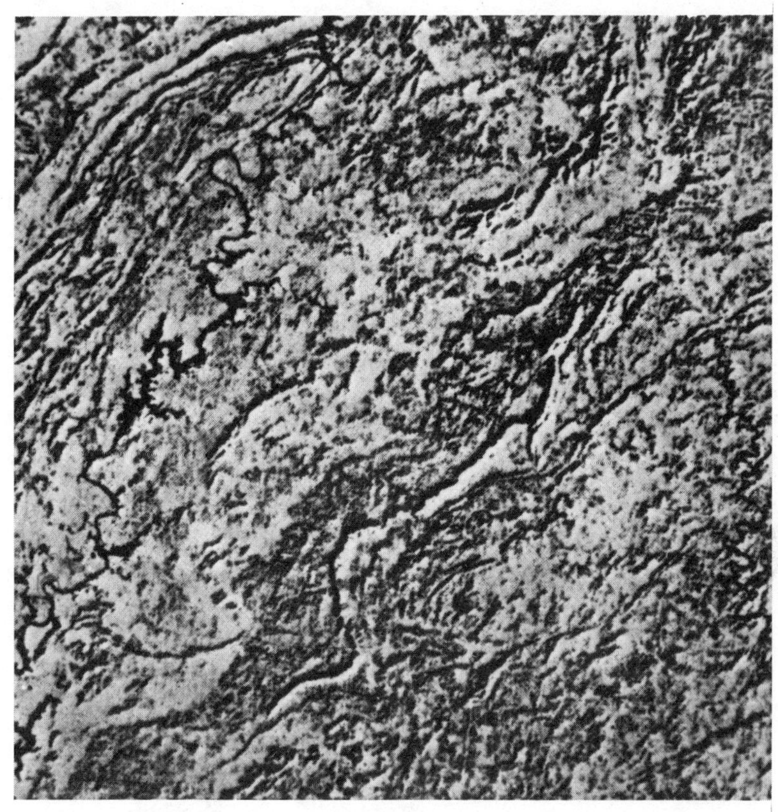

Fig. 5-18. Infrared imagery of the area presented in Fig. 5-17 made in the thermal infrared band (0.7-0.8 μ).

Directions (e.g., position of strike) are measured by means of compass. Concerning the use of this instrument, there are certain rules, namely:

- directions are to be measured and recorded in degrees exactly as one reads the compass;

- all readings are made from north generally to east, sometimes to west (which should be marked on the record), but never from east or west, e.g., N 30° E or N 150° W;

- the needle of the compass gives the magnetic north which may lie a variable number of degrees to the east or west of the true (geographical) north;

Fig. 5-19. Lineaments interpreted in the area shown in Fig. 5-17 using the air photos both in Fig. 5-17 and Fig. 5-18. Locations of springs are marked by circles.

- there is a correction which has to be made for a compass reading to relate it to the true (geographical) north;

- since the amount of this correction is not the same for different localities and since it changes gradually from year to year in any given locality, one must ascertain the proper correction for any place in which mapping is to be performed.

The clinometer is used for measuring vertical angles of slope and dip. The rules to be considered are the following:

- in taking the dip, the instrument is held vertically, parallel to the outcropping edge of the stratum and its face in a plane perpendicular to strike;

- the position of the pendulum gives the angle of dip measured downward;

- recording attitude is to give the amount and exact direction of dip, 45°N 65°E. Strike, always being at right angles to the direction of dip, would then have to be: N 65° + 90°E = N 155°E or N 25°W.

Fracture spacing, ranging from a few centimeters to tens of meters (e.g., in granites) can be measured by steel tapes of mm-division. Very subtle measurements, when needed, may be carried out using a mechanical type micrometer (e.g., measuring the width of fine fissures and joints) or an optical one (e.g., study the schistosity of metamorphic rocks).

Geological maps. In preparing maps of solid rock terrains, the general outlines of hydrogeological mapping (Subsection 2-5) will be followed emphasizing the structural interpretation since fractures, joints, and other structural elements are of particular importance in this respect.

To this end, detailed, often sophisticated, structural maps must be constructed, containing, among other things, all tectonic features of the area with special respect to fractures and joints, irrespective of the origin of data, i.e., whether they were measured in the field, determined in laboratories, or picked up as lineaments in air photographs.

Fracture maps. For the larger fracture system, special maps will be prepared showing some other morphological features of the area together with the fractures, e.g., the distribution of springs. This kind of map is illustrated by Fig. 5-20 displaying hydrogeological features near to Héviz, Hungary where great amounts of thermal karstic water (6500 liters/sec) is issued from a dolomite reservoir through fractured sandstones into a natural depression forming a lake.

Tectonic maps. Data determined on outcrops with respect to the attitude of bedding may be presented as strike-and-dip values. In Fig. 5-21, which is an enlarged part of Fig. 5-20, the attitude of bedding planes on a dolomitic outcrop (marked by D in Fig. 5-20) are represented. The strike and dip directions can be seen on the map, the values of dip are noted at the symbols.

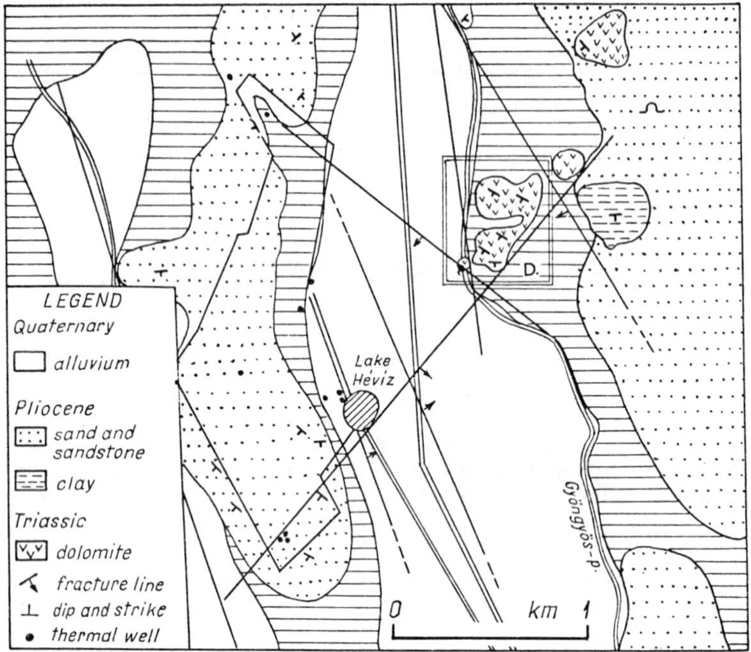

Fig. 5-20. Hydrogeological map of the Héviz (Hungary) area showing structural lines and a characteristic dolomitic outcrop called Dobogóhegy (marked by D) (Korim, 1970).

Presentation of joints. The observed sets of joints may be represented by fan diagrams. For each station the strikes may be plotted on the map as a radiating group of lines which intersect at the station. Along each line, the strike angles as well as the amount and direction of dip may be noted if desired. The joints, observed in the dolomitic outcrop at Héviz (Fig. 5-20), are represented in Fig. 5-22 using this method. All the trends of joints in a region may be presented graphically by plotting all the strikes about a single center lying in the middle of the region.

Rose diagrams. A very general procedure displaying the distribution of lineaments (e.g., strike of joints, fractures, etc.) in a study area may be carried out as follows:

- do the determination of attitude (strike) of lineaments in an area, e.g., which is presented in Figs. 5-17 and 5-18 using either methods or air photographs as it is shown in Fig. 5-19;

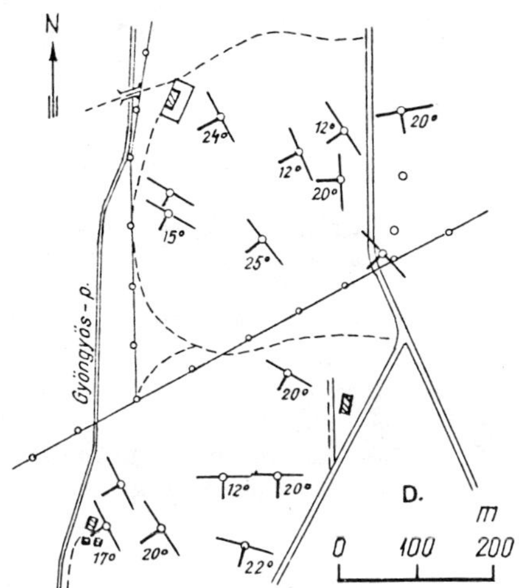

Fig. 5-21. Enlarged map of the area marked by D in Fig. 5-20 showing the attitudes of bedding planes in the outcrop Dobogóhegy.

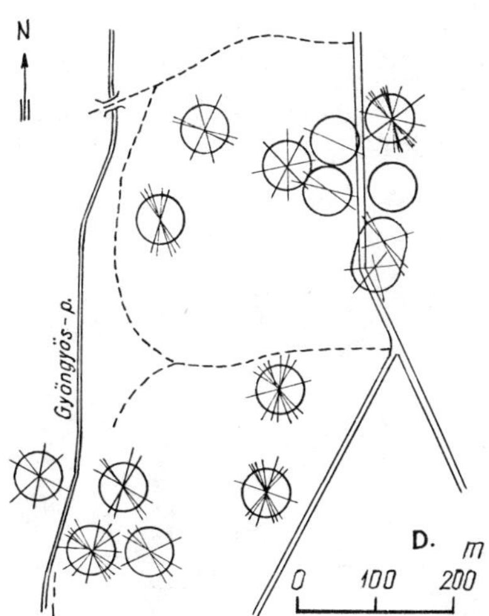

Fig. 5-22. Presentation of joints as fan diagrams determined in the outcrop Dobogóhegy (marked by D in Fig. 5-20). Each of the diagrams relates to the measuring station in the area where the determinations were made.

- tabulate all the strikes in groups of 1°-10° intervals, the size depending on the degree of detail of the presentation required (in the tabulation of data represented in Fig. 5-19, an interval of 5° was chosen);

- having the strikes tabulated, construct a diagram like that in Fig. 5-23, in which the radii are drawn at intervals corresponding to the borders of the intervals (at each 5° in the case of the example);

- the radial distance between the center and the periphery of the circles is measured in any convenient unit and is made proportional to the greatest number of lineaments falling in any one of the groups, for better visualization, the number of lineaments is marked by blackening the corresponding segment outward from the center to a distance proportional to this number.

A.

Strike of lineaments in degrees					
0°	93°	135°	148°	154°	160°
0	96	135	148	154	160
2	98	137	148	154	160
5	112	138	148	154	161
33	122	140	149	156	162
53	123	140	151	156	165
58	126	144	153	157	165
73	131	144	153	158	168
80	133	146	153	158	170
88	133	146	153	159	170

Number of lineaments in groups of 5°			
Group	N	Group	N
0° - 5°	4	120° - 125°	2
30 - 35	1	125 - 130	1
45 - 50	1	130 - 135	5
50 - 55	1	135 - 140	4
55 - 60	1	140 - 145	2
75 - 80	1	145 - 160	6
85 - 90	1	150 - 155	9
90 - 95	1	155 - 160	9
95 - 100	2	160 - 165	4
110 - 115	1	165 - 170	3

B.

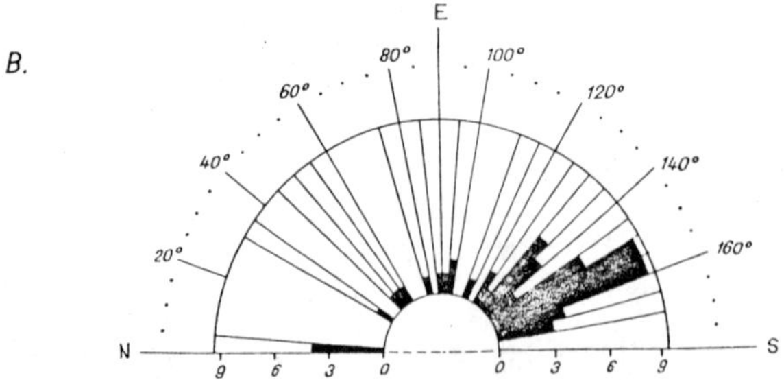

Fig. 5-23. The strike directions of lineaments in Fig. 5-19 are tabulated (A), then presented in the form of a rose diagram (B). The group division is 5°, a radial distance of 1 cm represents three strike values falling in the same interval of 5°.

It is obvious that dips may be plotted in a similar manner. The attitude of fault lines, veins and dikes, even systems of cavities or caves may be illustrated by diagrams similar to that described above when these phenomena are comparatively uniform in their trend (Fig. 5-24).

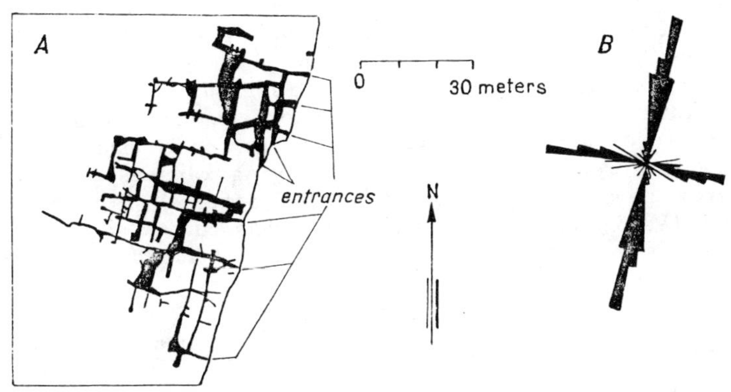

Fig. 5-24. Distribution of axes of a system of caves (A) represented in the form of a rose diagram (B) (after Powell, 1977).

Joint and fracture maps. Toward mapping joint and fracture zones in a large region, methods of interpretation of air photography were developed to take advantage of computer processing. One of these very effective methods will be demonstrated based on the results of a structural investigation carried out in the Edwards Plateau using air photography (Wermund, 1977).

The lineations, or fracture zones, were observed as features of variable intensity on the photos. Some of the known lineations include such phenomena as straight stream segments, darker linear traces of the investigated soil or rock with greater moisture content, as well as relatively straight vegetation alignments. The lineations were identified on quadrangle mosaics of 1:24000 black and white aerial photography (177 pcs).

Two kinds of fractures were identified: short lineations up to 2.5 km and long ones up to 160 km in length. Following the identification, the lineations were traced as lines on overlying transparent paper alike to Fig. 5-19. Each photomosaic was given a six-digit identification number based on an arbitrary system of coordinates. The coordinates of each end point of the short fractures were read in this system. All coordinates together with the identification number of quadrangles were punched onto cards or read onto magnetic tape. In a similar manner, long fractures were punched onto a separate set of cards.

It is obvious that digitalization and computer techniques are essential in interpretation. There are 177 quadrangles averaging 400 short fractures marked by four coordinates. This multiplies to nearly 300,000 data points having three or more integers. There is another set of integers to identify the end point of long fractures. Computer programs were designed to display the detailed distribution of fractures on a variety of maps. The regional orientation of short fractures are displayed as rose diagrams for each quadrangle as it is represented in Fig. 5-25 displaying classes of 10°. One program permits plotting of long fractures (Fig. 5-26). A field test made in 36 quadrangles has shown good agreement between the orientations measured in the field and by office methods.

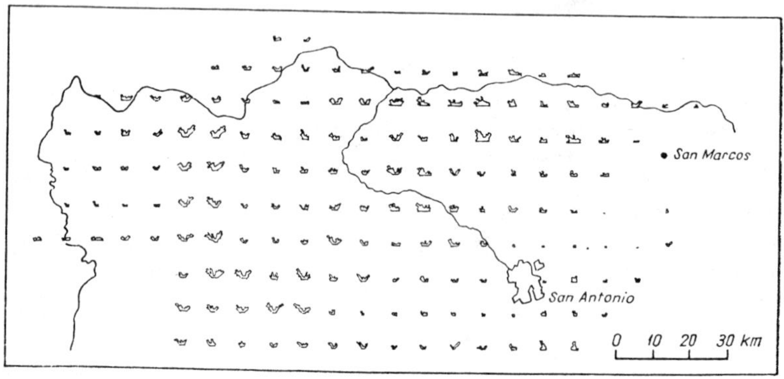

Fig. 5-25. Rose diagrams showing distribution of short fractures (less than 5 km) overlying the Edwards Limestone aquifer.

Considering the classification of hard rock terrains, it is easy to decide that the geological exploration, outlined above, can be applied in the shallow types only which are uncovered or covered by a very thin layer. The geological studies to be applied in areas of deep-lying rocks are the same as used for the exploration of deep-lying sediments (Subsections 14-1 and 14-2).

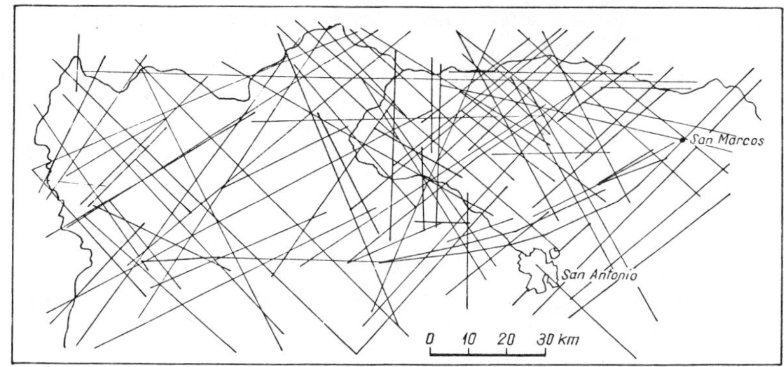

Fig. 5-26. Distribution of the long fracture zones (longer than 25 km) overlying the Edwards Limestone aquifer.

18-2 The Role of Remote Sensing in Hydrogeological Research

"Remote sensing is the measurement of some property of an object without having the sensor or measuring device in direct contact with the object," is stated by a definition which determines the term exactly, but it is too wide in respect of hydrogeological exploration. According to this definition, all types of the surface geophysical methods belong to remote sensing since the objects of exploration, i.e., buried structures, are in contact with the measuring instrument through fields only. In the following discussion of the problem, the concept of remote sensing will be used in a somewhat restricted sense applied for processes where the information transfer from the object to the sensor is carried out by use of emitted, transmitted, or reflected electromagnetic field. The result of remote sensing is represented generally, although not exclusively, in the form of photographs or images resembling photos. For this reason, the term imagery, used frequently for this type of remote sensing, seems justified in most cases.

A few years ago all photographic interpretations used in geology, i.e., the photogeological work, were based on low altitude black and white photographs. The range of altitude increased enormously. It extends from 500 m (e.g., obliques from hills, helicopters and from airplanes) to 20,000 meters (by aircraft) or even more than 1,000 kilometers in the case of photos made by satellites.

The setting of cameras has not been essentially changed. Photographs may be taken with the camera pointing vertically downward or the camera may be set obliquely as the support (the plane) remains horizontally. The former are called vertical images, the latter oblique ones. The high obliques are taken at high angle from the vertical and include the distant horizon. The high oblique photographs may be compared with the view one would have if looking toward the horizon from a high point of a mountain. It gives a good idea of perspective and it shows land forms accustomed to being seen. If the camera is pointed more directly downward, at a low angle from the vertical, low obliques will be made without showing the horizon.

The character of air photos enhancing superficial forms, i.e., the topography, was already discussed in the introduction of Section 2. The most important features observed on the air photos are the lineaments. Surficial features, such as geological structure, particular class of rocks, changes in topography may have a more or less distinct linear appearance on the photos. Whatever the cause may be, this effect is called lineation in accordance with the geological meaning of the term (Subsection 18-1). This is a phenomenon which should be investigated and interpreted or explained. Often they unravel the geological structure, sometimes they are a manifestation connected indirectly to geological features.

The interpretation of air photos was improved considerably in the last few years by using imagery in different spectral bands in the range of electromagnetic waves. The waves of different length suffer absorption in different measure penetrating the atmosphere. Bands used in air imagery are those where the absorption is comparatively low, i.e., the transmissivity is high (Fig. 5-27).

Probably the simplest technique of air imagery, using the visible sector, and one that is generally available throughout the world, is the reconnaissance utilizing black and white photography. This technique can be used with significant result in the hands of experienced operators. On the black and white photos taken during the hydrogeological study of the Edwards Plateau, lineations were identified with fracture zones both directly and indirectly as demonstrated in Subsection 18-1. In the interpretation of these kind of air photos taken on a glaciated area, the type of the covering sediment, i.e., that of the glacial deposit, was indicated on the photos as changes in the tones (Lahermo, 1973).

The use of color film yields some improvement, but another significant use of the visible spectrum concerns the multi-spectral technique. A multi-spectral camera takes several simultaneous photographs of the same area. Each

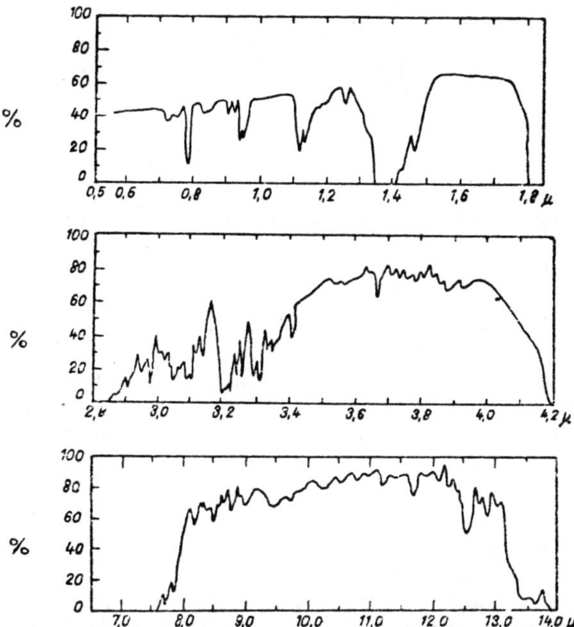

Fig. 5-27. Transmission of electromagnetic waves (in percent) through the atmosphere. The "windows" are marked by high values of the coefficient of transmission represented on the vertical axis. The length of the waves is given in $\mu = 10^{-8}$ cm on the horizontal axis.

photograph records electromagnetic radiation in a different band of the spectrum by a filtering technique. This allows the interpreter to distinguish very subtle differences between objects on a photo that a broad band camera and film might miss. For the interpretation, the black and white transparencies taken in selected bands may be projected on a screen superposed to provide one image, and if the proper filters were used the color quality of the original scene may be reconstituted. Using color filters for this composition, "false color" images can be created. The false color technique enables the observer to see objects of interest much more clearly because the unusual color combination attracts attention. Further, with the multi-spectral technique it becomes possible to attempt lithological and petrographical identification of rocks using their spectral response.

For information about the spectral bands used in multi-spectral technique, the remote sensing system of the satellite ERTS-1 will be delineated. It is equipped with two kinds of remote sensing systems. One is a return-beam vidicon working in the following bands:

- blue-green (0.475 μ - 0.575 μ wavelength);

- red (0.580 μ - 0.680 μ wavelength);

- infrared (0.690 μ - 0.830 μ wavelength).

The other consists of a multi-spectral scanner collecting energy in four bands, namely:

- green (0.5 μ - 0.6 μ wavelength);

- red (0.6 μ - 0.7 μ wavelength);

- infrared I (0.7 μ - 0.8 μ wavelength);

- infrared II (0.8 μ - 1.1 μ wavelength).

The thermography or thermal infrared imagery is particularly useful in hydrogeological studies. Infrared radiation is emitted by all objects as a function of their temperature, thermal conductivity, and inertia or diffusivity. Among the thermal parameters (defined in Subsection 15-3), diffusivisity is the more important being determined mainly by the water content of the rocks. The daytime imagery can be used exceptionally, for the radiation of the sun is superimposed on the reflected radiation of objects. The preferred time of data collection for infrared imagery is in the predawn hours when residual heat patterns are discernible since the heat budget from the previous day's sun radiation has been re-radiated back to the atmosphere.

The windows of the electromagnetic spectrum, i.e., the bands which can be used in infrared imagery, are shown in Fig. 5-27. The windows near to the visible spectrum (0.7 - 1.1 μ) are called solar infrared, the farther thermal infrared.

Based mainly on the data collected in interpretation infrared imageries taken by satellites, the infrared technique is found to be useful in the following applications (Emplaincourt, 1976):

- determination of areas containing groundwater with estimation of availability of groundwater;

- determining areas for water engineering constructions;

- aid in geological, geomorphological, and structural studies.

In order to demonstrate the enhancing character of infrared imagery, two satellite photographs are presented which were taken on the same area, one in the visible spectrum (Fig. 5-17), the other in the thermal infrared band (Fig. 5-18).

A non-imaging remote sensing optical technique, showing considerable promise, has been developed by Lyon (1965). This method operates on the measurement of the infrared emission spectra of rocks and soils in the 8 μ - 13 μ wavelength region where the atmosphere is essentially transparent. It depends upon the shift in wavelength of the silicium-oxygen fundamental vibration transition with changing composition of silicate minerals in rocks. This transition, occurring near 9 μ in the middle infrared region, shifts approximately 2 μ between rock composition of acidic character (e.g., granite) and an ultrabasic one (e.g., dunite). Typical spectra of the wide range of rock types are shown in Fig. 5-28.

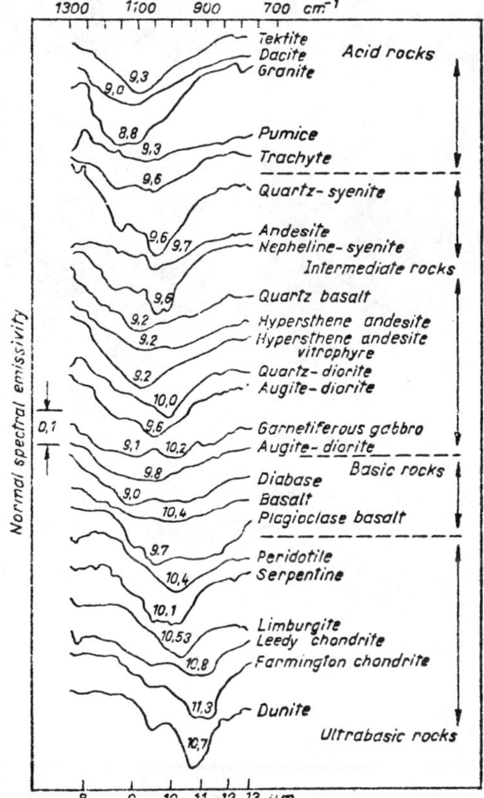

Fig. 5-28. Infrared spectrum of some rocks in the thermal infrared band. The coefficient of surface reflection is represented as a function of the wavelength.

Among the electromagnetic remote sensing techniques using the radio band, the only imaging system, radar imagery, is rather unique since it needs no daylight, clear weather, or thermal emission from the objects when penetrated. Most radars provide their own illumination by means of a transmitter working in the microwave band, from 0.1 cm to 100 cm wavelength. The SLAR (side-looking airborne radar) radiates in a sector of about 35 degrees with the middle plane dipping about 40 degrees to the horizontal obtaining, by the reflected beam, excellent images resembling the oblique air photos (Fig. 5-29).

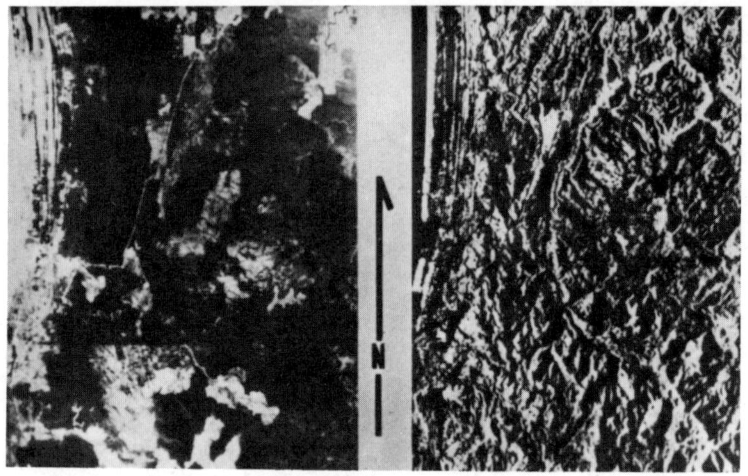

Fig. 5-29. SLAR imagery. In this example the radar image (marked by R) reveals a large arcuate structure that is not shown on the air photo made in visible light (marked by V) over the same area.

The principal advantages of the SLAR may be listed as follows:

- good quality imagery can be obtained through cloud cover;
- accentuates linear features in the topography;
- long, broad strips of imagery can be flown providing continuous unbroken coverage over much larger areas than in common aerial photographs;
- subtle changes in lithology are sometimes depicted which can be further accentuated separating the return beam into two images of different polarization;

- radar signals can penetrate a short distance into vegetation and also into dry soil of low dielectric constant.

Investigations made by SLAR indicate the capability of this method, in general, when used for the study of topographic forms especially in conjunction with other types of air imagery.

The applicability of air photo technique in morphological studies over a karstic area was demonstrated in Figs. 5-25 and 5-26. Two other air photos show lineaments which can be detected only by infrared imagery (Figs. 5-18 and 5-30). The geomorphology so far has concentrated on limestone and crystalline terrains neglecting the sandstones. The classical methods of geomorphological investigation were not very fruitful in sandstone terrains since the observer in the field cannot contract the orientation of surficial forms. Only air photos enable the observer to understand these phenomena as a whole illustrated above by Fig. 5-16.

18-3 Geophysical Model of Fissured Rocks

With respect to the geophysical study of hard rock terrains, the same classification could be used as was made before regarding their lithologic-petrographic characteristics; first of all the distinction of hard rock terrains into deep and shallow types. The main features of the geophysical model and the tasks of the geophysical exploration may be summarized as follows:

- the hard rock terrains are built up of great masses of rocks of differing physical parameters being in contact with each other (e.g., granite intrusion and country rock); the task is to detect the shape and position of the composing rock masses;

- the same rock may have secondary (fissured, fractured zones) and primary porosity differing in some geophysical parameters; the fissured-fractured parts of potential aquifers should be detected;

- special forms due to tectonism or erosion (e.g., cavities) exist on the rock surface or within the rock mass which should be delineated;

- the magnetic properties of igneous and metamorphic rocks differ considerably from those of sedimentary rocks which are practically non-magnetic.

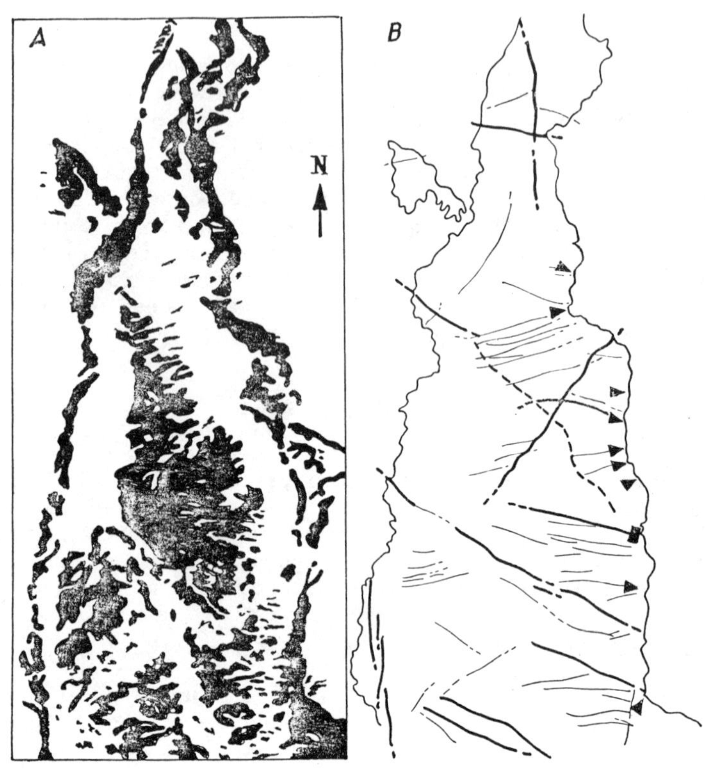

Fig. 5-30. Infrared imagery of the Gramovoussia peninsula near Crete made in the infrared window of 3.5-4.2 μ (A) and the interpretation of the air photo showing the main thermal lines and karstic springs (marked by the triangle) in part B.

Exploration methods suitable for application in deep hard rock areas are essentially the same as those used for the study of deep-lying aquifers in general (Subsection 14-1). Types of geophysical exploration discussed here are those special methods, or the suitable variety of general prospecting methods, applicable to the study of hard rock terrains of the shallow type.

Magnetic exploration. This kind of prospecting involves measurements of the intensity of the earth's magnetic field and the interpretation of variations in its intensity over the study area. Variations, called magnetic anomalies, are the distortions of the regional magnetic field of the earth produced by materials of magnetic properties in the earth's crust. Anomalies of hydrogeological interest are generally of the induced type due to the magnetization induced in a rock by the earth's magnetic field. The anomaly produced is dependent upon the geometry, orientation, and intensity of the earth's field and upon the magnetic property of the rock called susceptibility. Magnetic susceptibility values of some rocks are given in Fig. 5-31.

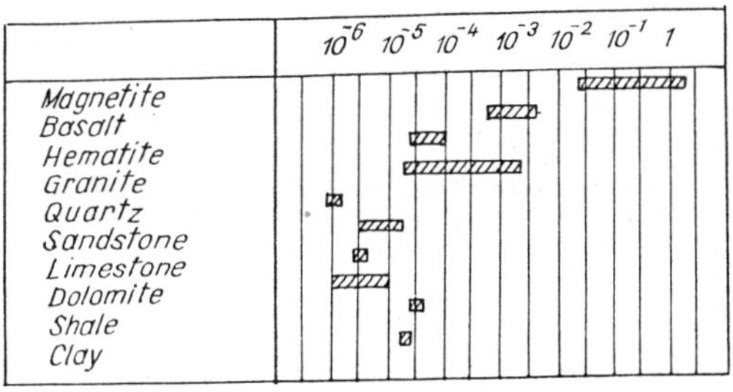

Fig. 5-31. Magnetic susceptibility of some rocks given in cgs units.

The unit of magnetic field intensity (in the cgs system) is defined as the intensity of the field which exerts a unit force (1 dyne) on a unit magnetic pole. In geophysical exploration the gamma is used, one hundred thousandth of the cgs unit. The intensity of the earth's magnetic field is of the order of 10,000 gamma. The intensity range of strong and weak anomalies is 1,000 gamma and 100 gamma, respectively. For most exploration purposes, only the determination of relative magnetic intensities is needed. The measured values are related to an arbitrary 0-level when represented on a magnetic map.

Quantitative interpretation of individual magnetic anomalies yields information about the depth, extent, structure, and susceptibility of rocks. Anomalies due to simple geometrical forms can be interpreted using albums (Vacquier, 1951) or computer programs especially where the forms are more complicated.

The use of magnetic exploration in hard rock terrains is limited practically to the regions of igneous and metamorphic rocks. The sedimentary rocks are non-magnetic and of very small susceptibility. The most common hard rock aquifers, such as the sandstones and carbonate rocks, are thus excluded from the magnetic survey. Igneous and metamorphic rocks display higher susceptibility values. Volcanic rocks may produce high amplitude magnetic variations of very local extent. Igneous intrusions produce anomalies with a wide range of amplitudes but of greater extent than the anomalies associated with volcanic rocks. Metamorphic rocks may produce complex patterns and pronounced lineaments. Although the sedimentary rocks are non-magnetic, a few clastic ones, such as magnetite-bearing sands and gravels, can be studied directly.

In the hydrogeological application of magnetometric prospecting, the study of basalt flows represents an important group of problems. It may have purposes such as:

- location of "windows" in the basalt cover where the sedimentary strata buried by basalt flow can be reached by drilling;

- location of spots covered with a thick basalt layer unsuitable for water wells;

- delineate the water-bearing valleys and drainage system buried by basalt flow.

A thick sequence of lava flow presents rather difficult problems. Generally, basic lavas are strongly magnetic and the magnetic map is a complex one on account of rapid variations in magnetic properties of the rock. A ground magnetic survey may be of little help because the rapid variations in the magnetic properties of lava yield a patternless mass of anomalies. The airborne survey, by flying at a height of 100-300 meters, produces a filtering effect thus rejecting the minor features, the noise from the map. Figure 5-32 represents the airborne magnetic map of such an area. Although the magnetic image is rather complicated, strong lineaments can be observed due to fault zones in the ground covered by the basalt flow (Boyd, 1967). A good example for the application of remote sensing in hydrogeological exploration is illustrated by the magnetic map shown in Fig. 5-32. The usefulness of the ground magnetic survey in a covered igneous rock area is represented in Fig. 5-33 showing the cross section of a barrier structure produced by a diabase dike which was located by means of magnetic prospecting.

Gravity exploration. Gravimetry or gravity exploration involves all kinds of measurements of gravitational

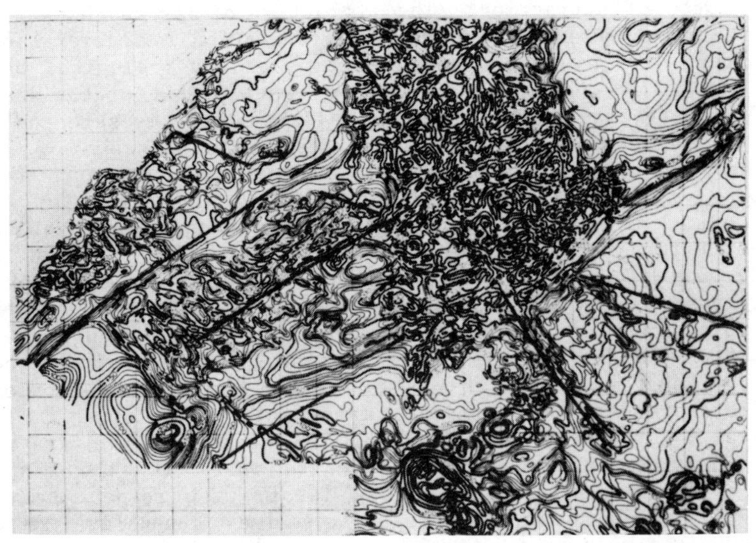

Fig. 5-32. Aeromagnetic survey flown at 300 m above tertiary basalt lava flows illustrating major structural lines which show through the magnetic cover (Boyd, 1967).

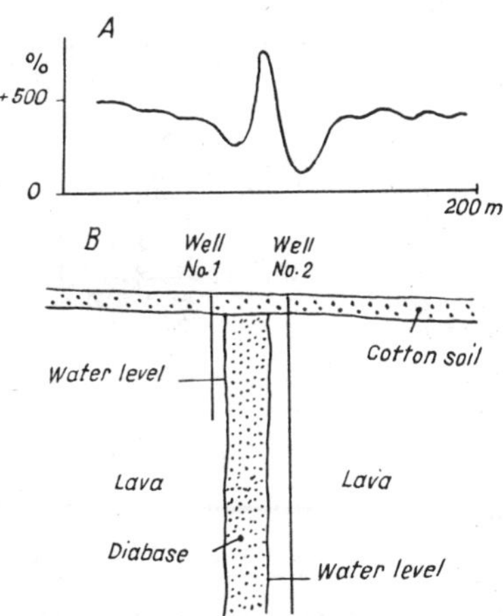

Fig. 5-33. Magnetic survey in Poona (India) showed a strong maximum in the magnetic profile (A) due to a diabase dike which cuts the lava and acts as a barrier illustrated in the geological profile (B).

acceleration. It is based on the law of Newton, according to which there exists a well-defined force between masses, called the force of attraction. The attraction of the earth exerted on a unit mass is defined as gravitational acceleration. Although the gravitational attraction of any geological body is a function of its mass, or more exactly that of the shape and density, the total gravitational attraction measured on or above the surface of the earth represents the sum of the attractions of both the body and the rest of the earth as a whole. In geophysical prospecting, only the effect of the geological body to be surveyed is interesting, therefore, the other effects must be eliminated. This process is called the reduction of gravity data and can be carried out by methods worked out exactly for such purposes. That part of the earth's gravity field which remains after data reduction is known as the gravity anomaly reflecting the effect of the body to be surveyed. According to the process of reduction, different kinds of anomaly maps may be obtained. The so-called Bouguer anomaly map will represent the gravity acceleration on a non-rotating earth where all the masses would be scraped off to the sea level. With respect to the Bouguer anomaly map, the relative values generally are of interest, or, in other words, only the excess or deficiency of mass that the body represents should be considered rather than the absolute values. Under these circumstances, the body can be described quantitatively in terms of density contrast with its surrounding. Density values of various kinds of rocks are given in Fig. 5-34. The gravity acceleration of a sphere and that of two-dimensional bodies of simple geometrical form are represented in Fig. 5-35.

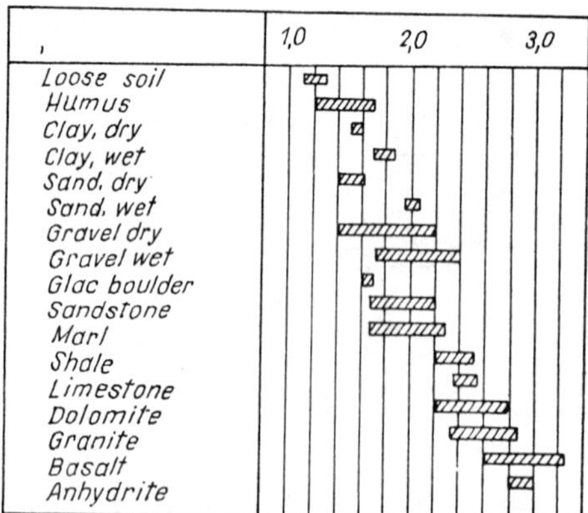

Fig. 5-34. Bulk density of some rocks given in cgs units.

The unit used in gravimetric studies is the gal, equivalent to a unit acceleration, i.e., 1 cm/sec^2 in the cgs system. Since the amplitude (the maximal variation) of an

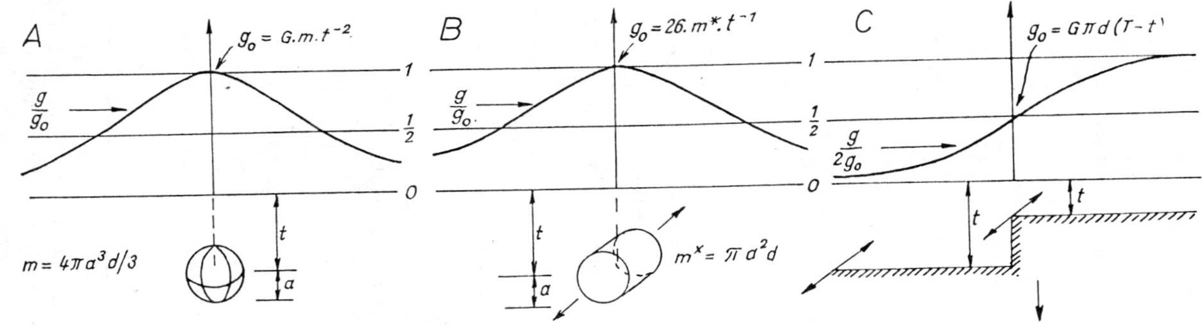

Fig. 5-35. Gravitational acceleration (g) due to a sphere (A), to a cylinder of infinite length (B) and to a step of infinite extension in directions marked by arrows (C). d denotes the relative density of bodies, $G = 6.67 \times 10^{-8} cm^3 gram^{-1} s^{-2}$.

anomaly comes to a fraction of this unit in general, one thousandth of it, the milligal, is used.

Gravimetry can be used for the study of all types of hard rock terrains since gravity acceleration is a common feature of all bodies. The most fruitful field of application of the method are the areas of great density contrast such as big caves in the underground, dikes, and similar forms differing in density from the surrounding. For interpretation, the geological bodies may be approximated with simple geometrical forms, some of which are represented in Fig. 5-35. Owing to the small value of the observed anomalies as well as to the difficulties in making the reduction in hilly areas, only a few applications of gravimetry are known in hard rock terrains.

In the last few years a very sensitive gravimetric method, called microgravimetry, was worked out which is very promising in the study of karstic areas. The essential feature of this method is the use of a highly sensitive gravity meter by which determinations of error less than 0.05 milligal may be carried out (Omes, 1976). The main advantages of the microgravimetry as opposed to electrical and electromagnetic methods are as follows:

- it is not affected by buried metal pipes, stray currents, etc.;

- vibrations due to wind and traffic affect the measurement, but it can be operated in quiet periods;

- conductive soil or overburden which behave as masks in the case of high frequency electromagnetic exploration have no influence on the measurement.

Gravity maps obtained in karstic areas using microgravimetry are shown in Fig. 5-36.

Electric exploration. Considering the basic principles of the electric methods in Subsection 14-1, the horizontal profiling proved to be a suitable method in solid rock terrains since the lateral variations in resistivity of rocks may be detected by this process.

In horizontal profiling, a fixed electrode spacing is chosen and the whole electrode array is moved along a profile after each measurement is made. The values of apparent resistivity, obtained by this method, are plotted at the center point of the array when resistivity profiles or maps are prepared (Fig. 5-37). The capability of the horizontal profiling in solid rock areas may be estimated according to different aspects such as:

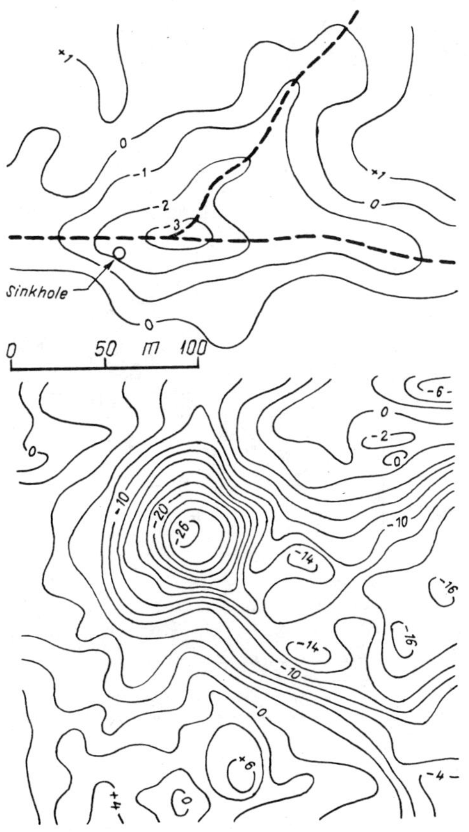

Fig. 5-36. Detection of karstic cavities by microgravity. The values of contour lines (isogals) are given in 10^{-2} mgal units.

- structural prospecting, i.e., delineating great units;
- aquifer prospecting, i.e., detection of fractured zones;
- detection of specific forms, i.e., caves, sinkholes, etc.

Structural forms can be accurately mapped by geoelectric method especially where the resistivity contrast between the rock to be studied and the surrounding rock masses is relatively high. Considering the list of resistivities in Subsection 14-1, this requirement is always realized between the igneous rock and the country rock, and it is mostly fulfilled

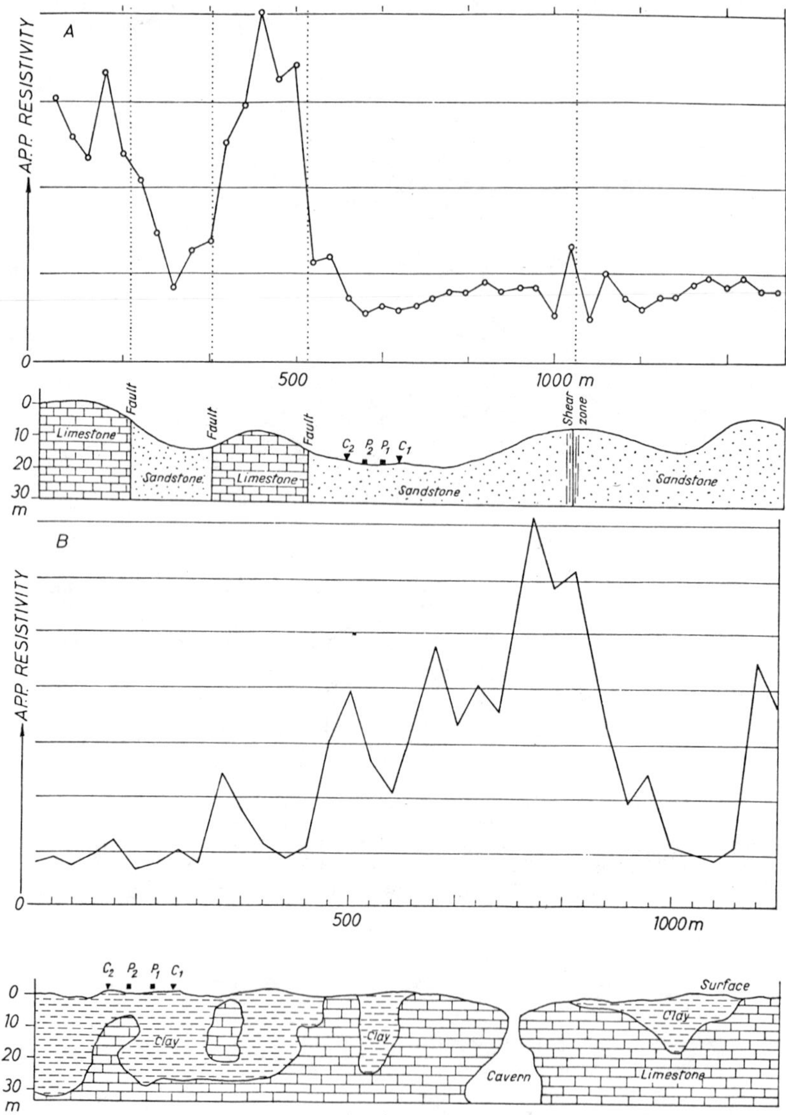

Fig. 5-37. Resistivity mapping over hard rock areas showing apparent resistivity profiles across shear zones of sandstone and limestone fault block (A) and over karst topography (B) with geological profiles. In the electrode pattern, which is shifted along the profile, the distance between current electrodes (C_1C_2) equals 100 m.

between different types of hard rocks being in contact with each other.

The delineation of fractured-fissured zones depends on the resistivity contrast between the rock and its fractured part if there exists any. Data collected by Arandjelovic' (1966) suggest the possibility of distinction between limestones karstified in different measures on the basis of their electric resistivities (Fig. 5-38). The same distinction can be made in volcanic areas based on resistivity values.

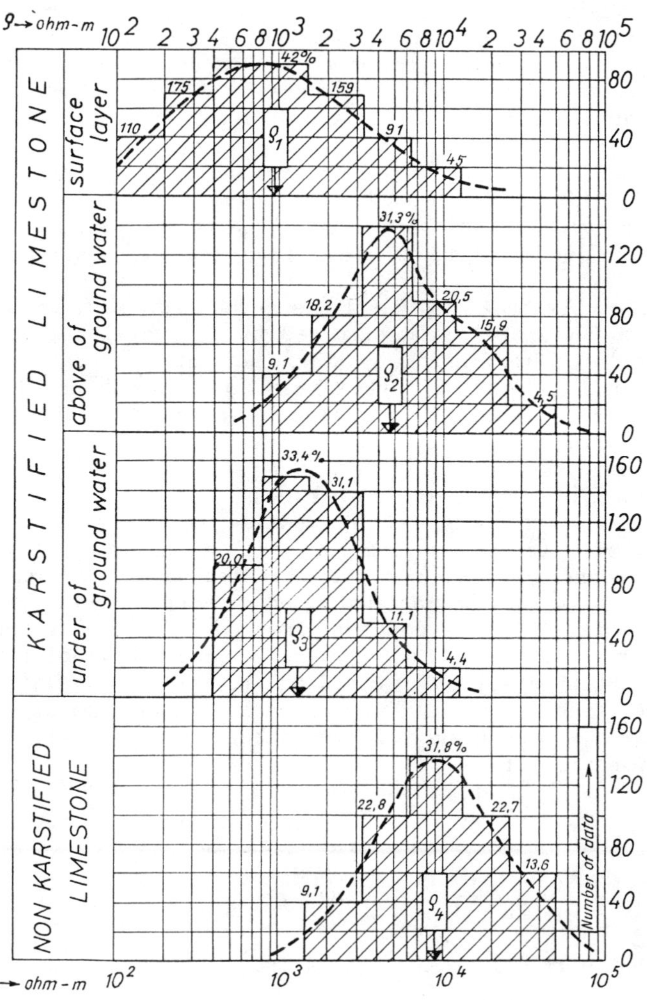

Fig. 5-38. Histogram of electric resistivity values obtained in the Dynaric karst area.

The interpretation of the apparent resistivity profiles can be carried out using master curves calculated for simple forms usually encountered in the study of hard rocks such as: a vertical slab, simulating a dike and a dipping plane discontinuity; half-cylinder and hemispheres, simulating sinkholes; and spheres simulating caves (Fig. 5-39). An observed horizontal resistivity profile over karst topography was illustrated in Fig. 5-37.

The detectability of buried cavities can be estimated by the calculation of the disturbing effect of a perfectly conducting sphere imbedded in a homogeneous medium becase it will have the maximum effect on the measurement of the apparent resistivity for a given geometrical relationship (Fig. 5-40). The curve, for which the depth to the center of the sphere is twice the radius of the sphere, has a minimum where the apparent resistivity varies only 10 percent from the original value of the true resistivity. It would not be safe to depend on finding such a feature when its depth is greater than this value. This result may be extended to bodies of arbitrary shape provided that the variation between maximum and minimum diameter or dimension is less than two (VanNostrand, 1966).

The limiting depth of detectability may be increased using a focalized field produced by a special arrangement of electrodes (Grandinetti, 1967), but at the same time the disturbing effect of nonhomogeneities around the body will grow higher, thus decreasing the average sensitivity of the measurement.

Electric resistivity of fissured rocks. Fissured and saturated hard rocks may be modeled in the first step by a system of alternating parallel sheets consisting of water and rock, respectively. For the calculation of resistivity (or more exactly for the determination of specific electric resistivity of such a system) the same reasoning may be applied as that used in connection with the determination of anisotropy in Subsection 14-1. According to this calculation, the average specific resistivity of such a block, measured perpendicular to the plane of the sheets, may be given by the following formula:

$$\rho_T = (\sum_{iw} b_{iw}\rho_w + \sum_{ir} b_{ir}\rho_r)/(\sum_{iw} b_{iw} + \sum_{ir} b_{ir})$$
$$= n_L\rho_w + (1-n_L)\,\rho_r \quad ; \tag{5-6}$$

and that parallel to the sheets is:

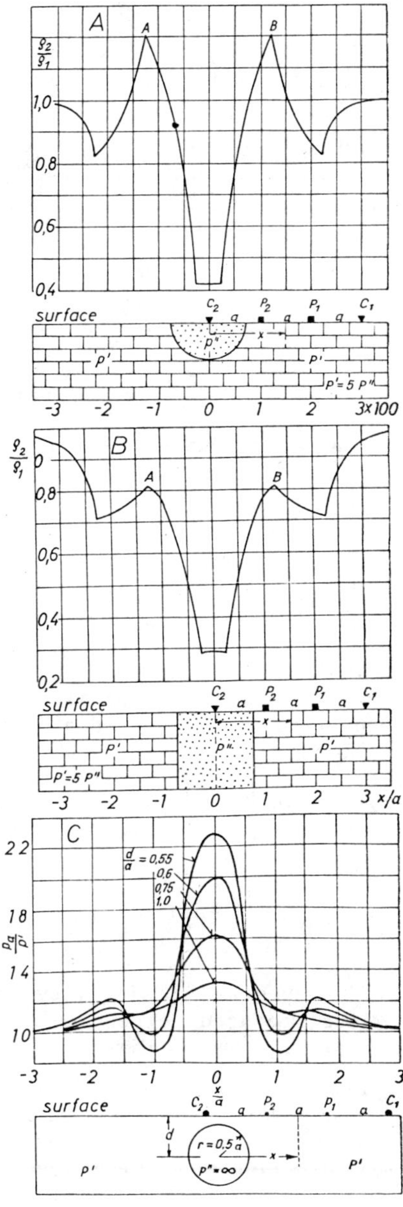

Fig. 5-39. Horizontal resistivity profiles over a hemisphere (A), over a slab (B) and over a buried, perfectly insulating sphere (C) simulating filled sink (A), reef (B), and cave (C), respectively (VanNostrand, 1966).

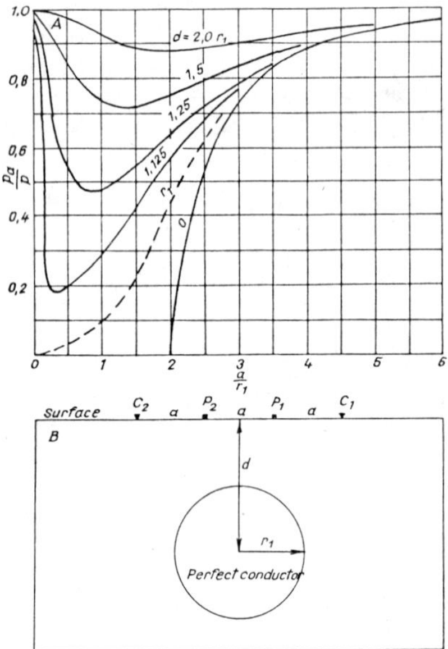

Fig. 5-40. Apparent resistivity profile over a perfectly conducting sphere buried at different depths.

$$\rho_L = \left(\sum_{iw} b_{iw} + \sum_{ir} b_{ir}\right) / \left(\sum_{iw} \frac{b_{iw}}{\rho_w} + \sum_{ir} \frac{b_{ir}}{\rho_r}\right),$$

$$= \left[n_L \cdot \frac{1}{\rho_w} + (1-n_L) \cdot \frac{1}{\rho_r}\right]^{-1} . \tag{5-7}$$

Here, ρ denotes the specific resistivity of the sheets, b denotes its thickness measured perpendicularly to the sheets, n_L is the linear porosity in this direction, the letters w and r relate to the water and rock, respectively. If the sheets of rock consist of a layer with intergranular porosity, then assuming that the particles of rock are non-conductive and the voids are randomly distributed, using the same reasoning as above, the specific resistivity of rock will be given as

$$\rho_r = \rho_w / n \quad ; \tag{5-8}$$

where n denotes the primary bulk porosity of the rock. Using this expression for ρ_r, the average transverse resistivity will be obtained as:

$$\rho_T = \rho_w[n_L+(1-n_L)/n] \approx \rho_w(1-n_L)/n \sim \rho_w/n \quad ; \quad (5\text{-}9)$$

and the longitudinal average resistivity as:

$$\rho_L = \rho_w/[n_L+(1-n_L)n] \approx \rho_w/(n_L+n) \quad . \quad (5\text{-}10)$$

Thus, the longitudinal resistivity of fissured rocks is controlled by both the primary and secondary porosities, but the transverse resistivity is influenced by the secondary porosity only slightly. Considering the direction of the electric field n with respect to the plane of the sheets, the introduction of a factor is necessary, called tortuosity, which equals to $\cos\alpha$ and $1/\cos\alpha$ for ρ_T and ρ_L respectively, α denoting the angle of inclination of the field to the sheets.

There are some limitations of the electric resistivity method in solid rock areas. Most of the inherent limitations of the conventional methods of direct current electric prospecting were already discussed in Subsection 14-1 and earlier in this section too. In addition, the problem of uncovered solid rock areas should be mentioned.

It is known that in the electric prospecting of the direct current type, direct current or alternating current of very low frequency is supplied into the ground by means of metal electrodes. In terrains where the hard rock is exposed, the high specific resistivity of the hard rocks resulting in extremely high contact resistivity between the electrodes and the ground hinders the flow of current into the rock. Where the decrease of contact resistivity cannot be achieved (e.g., by means of salt solution), electromagnetic or seismic methods may be proposed. The latter should be proposed in general since the interpretation of electromagnetic data is rather complicated.

Tracing water-filled galleries by means of electric current is essentially equivalent to the electric prospecting method called mis-a-la-masse, applied sometimes in electric prospecting (Subsection 14-1). In this kind of application, two electrodes are implanted in the water of the gallery and the flow of current, i.e., the position of the gallery, is followed on the surface, measuring the field of current confined in it. The use of alternating current is necessary owing to the commonly high resistivity of rocks.

Seismic prospecting. The conditions for the application of seismic refraction methods in solid rock terrains are generally favorable. The differences in the velocity of seismic waves may be detected between the igneous and country rocks as well as between the different types of hard rocks but in slighter measure. It is an important feature of hard rocks, however, that the seismic wave velocity in the fissured, fractured, and broken part of the rock always falls short of velocities in the unbroken part of the rock. Although the various types of solid rocks cannot be distinguished by seismic methods, the fractured zones and loose clastic fills can be successfully located. Methods, interpretation, and presentation of data were dealt with in Subsection 14-1.

In the exploration of deep-lying solid rocks there are no limitations for the method. The application of the seismic refraction method to shallow studies is somewhat limited. It must be considered that objects, the dimensions of which are equal to or even less than the length of the shortest seismic wave generated, cannot be studied. Taking into consideration that the length of the seismic wave extends from 20 meters to 100 meters, small objects, caves, or cavities may be omitted in exploration.

The geobomb (Arandjelovic, 1969), a special application of seismic measurement, was developed to study wide karstic conduits and galleries. The geobomb, i.e., a lightweight and easy to transport time bomb of spherical shape, is put in the water of a karstic conduit and transported by the water, it will explode after a predetermined time interval. The position of the sphere at the moment of blasting will be determined by simple seismic method and knowing the time lag of the bomb, the average velocity of the water flow can be calculated.

18-4 Data Collection Concerning the Water Regime in Solid Rock Formations

The problem of estimating the hydrological characteristics of solid rock terrains requires the collection of data describing the water regime in the rocks. The methods to be applied are well-known processes used in many fields of hydrological research. Some of the data collecting methods will be discussed here, as it was outlined in the introduction of this section, taking into account the special structural aspects of the hard rock areas.

The principal factor determining the features of data collection is the well-known mobility of groundwater in fissured rocks especially in karstic types. For example, the amplitudes of karstic level fluctuations depending on seepage properties of rocks often amount to 10-15 meters and up to 30-50 meters in some cases. The discharge of karstic springs at periods of maxima are sometimes 30-600 times as large as their discharges at periods of minima (Kovalevsky, 1976).

Considering the mass transport (i.e., the groundwater balance) data related to infiltration, discharge through springs, and position of water table should be collected. Only natural discharge of groundwater will be considered. Where the water is exploited artificially, e.g., through wells or in mines, the data should be acquired on the spot.

The quantity of precipitation of any kind is measured in the form of depth (i.e., the depth to which a flat horizontal surface would be covered if no water were lost by runoff or evaporation). As for snow, it is customary to speak about its (calculated) water depth equivalent.

There are many types of rain gauges for measuring precipitation. Their data has to be reduced by considering surface runoff and evaporation in order to get the quantity of infiltrating water due to precipitation. This study always requires detailed hydrological analysis because the amount of infiltration depends on many local factors such as climate, season, structure, and texture of rocks, covering layer, vegetation, etc.

Springs, the natural outlets or outputs of water transport, may have discharge varying within very wide limits both in time, mentioned above, as well as topographically. The majority of springs of high discharge (springs of first magnitude) occur in karstic rocks such as:

- Warm Spring, Montana (50,000 liters/sec);
- Vaucluse, France (min. 4,500 liters/sec);
- Cantaro, Italy (4,500 liters/sec);
- Héviz, Hungary (6,500 liters/sec).

The category of first magnitude includes only few springs issuing from other types of rocks such as:

- Big Horn, Thermopolis (1,300 liters/sec, sandstone);
- Brandwlei, S. Africa (1,300 liters/sec, crystalline).

There are innumerable second magnitude springs having their discharges vary in wide ranges from 0.3 liters/sec to 100 liters/sec.

For measuring the discharge of springs varying very widely in solid rock areas, special kinds of weirs are used. The weirs have been the earliest known structures of continuous discharge measurement. They can be used in very diversified forms and sizes in open channels, i.e., in the discharge channel of the spring. The weir itself is an obstacle constructed on the bottom of the channel. The water level rises over this obstacle and then the water flows over it. The measuring head is, in the case of free overfall, the water depth above the crest of the weir on the upstream side (i.e., the overfall depth). The measuring weirs to be used for discharge measurement are special types for which the relationship between the discharge and overfall depth has been determined. The quantity to be measured is the overfall depth which can be determined using water level measuring devices.

It is obvious that the relation between discharge and overfall depth is controlled by the cross section of the discharge channel. The wide range measurement of discharge is realized using structures where the narrow lower part serves for measuring the lower discharges and the wide upper part serves for measuring the higher one. There are structures in use constructed collaterally, one of them measuring the low, and the other the high discharges.

The movement of water in the fractured-fissured parts of the rocks is generally studied by determining the water level or head in observation wells. For such purpose, recording gages should be used, i.e., instruments that produce graphic or punched-tape records of water level elevation in time. When selecting the water stage recorders, the following main characteristics should be considered:

- the range of the water level measurement;

- the time scale, i.e., the velocity of the registering device;

- the ratio of the recorded and real value of water stages;

- the inertia of the instrument, i.e., the maximal velocity of water level changes which can be followed in the registration.

Owing to the sudden changes in water level, instruments of low inertia and of high measuring range should be applied.

Water table or piezometric maps constructed from water level data relating on the same date serve as a basis for the study of water regime owing to the following reasons:

- the water table generally defines the zones of greatest circulations;

- it helps in the determination of the general direction of flow, the determination of hydraulic gradient, and it makes possible the delineation of recharge and discharge areas;

- it indicates locally the extent to which caverns or voids are air filled or water filled.

The heat transport and heat balance in solid rock areas is mainly governed by the flow of water or at least where it flows through bigger fissures or cavities. The quantity of heat transported by flowing water may be estimated when the temperature is measured in recharge areas as well as at natural outlets (i.e., in springs). Temperature measurement in observation wells will increase the knowledge about the convective branch of heat transport. It was explained in Subsection 15-3 that in aquifer systems where water is moving through pores of rock at a slow but measurable rate, the temperature of rock and water at depths beyond the effect of insolation, is the result of heat exchange between the rock and water. This conductive part of heat transport, although of minor importance, can be estimated using measurements of the temperature gradient in wells the water of which does not take part in convective circulation.

Collection of temperature data may be accomplished using thermometers of medium, i.e., 0.1°C, sensitivity such as:

- common (mercury) thermometer for making determinations at the surface or in springs;

- electric (e.g., resistivity) thermometer which can be lowered into wells;

- infrared air imagery (Subsection 18-2).

The latter one is especially useful to locate submarine springs (Fig. 5-30). Continuous recording of the temperature is needed only exceptionally. Periodically repeated observations (e.g., to repeat the measurements weekly) are, in general, satisfactory.

The analysis of the chemical composition of water, in addition to the hydrodynamic analysis of graphs of groundwater level fluctuations and to the measurement of temperature and water, reveal the genetic relations to the water regime. According to Shuster (1971), groundwater flow feeding springs in a karstic area may be characterized as diffuse type (i.e., water collected in narrow openings on great areas) or as conduit type (i.e., flowing through wide channels) on the basis of variations of chemistry in time (Fig. 5-41).

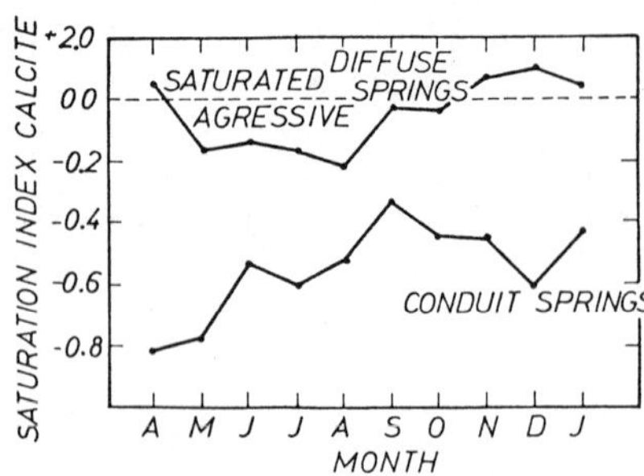

Fig. 5-41. Average seasonal variation in $Si_{calcite}$
The variation of monthly averages differs markedly in the case of diffuse springs and conduit springs.

Environmental isotopes, whose concentration is determined in samples taken from wells or springs, can be used for age determination or for the measurement of the velocity of water moving in the interstices of rock by methods similar to that applied in the case of granular media (Subsection 15-2). It should be noted that recharge of karstic springs through leakage from lakes or rivers could be identified by a high concentration of stable isotopes in the spring water.

The rapid movement of water in joints, fissures, and voids as well as the lack of filter effects in fractured-fissured rocks suggest the application of artificial tracers for purposes such as:

- delineation of the flow field;
- identification of recharge and discharge areas;
- determination of the location of water conduits;
- measuring the velocity of groundwater flow;
- detection of leakage of surface water into the groundwater.

According to the technique generally adopted, the water mass will be marked in an accessible place through wells or natural openings by bringing the tracer material into the water, and then the appearance of tracing material, which is transported by the moving water, will be observed in wells, springs, and occasionally on the surface of the rock at a later time. Knowing the time elapsed between the marking and observing of the tracer, the average velocity of the flow of water can be estimated.

Among the tracers, the salts, mainly sodium chloride and potassium chloride, are the oldest successfully applied materials for karst water tracing. The disadvantage of the method is that salt in the hundreds of kilograms range of magnitude must be applied for successful marking. An advantage of the method is that the output of the injected material can be quantitatively determined by exact chemical analyses. The diffusion of salt can be monitored on the surface of the study area using geoelectric resistivity prospecting methods.

Dyes are very suitable materials for tracing experiments, especially the fluorescein dyes (e.g., uranine, rhodamine, etc.), since they can be detected by spectrofluorometers in water samples even in the case of very low concentrations such as 0.02 mg/m^3. In addition to this, uranine is not dangerous for man and animals even in high concentrations. Artificial radioisotopes can be applied as in the case of clastic sediments (Subsection 15-2). The use of spores and germs is of less importance.

SECTION 19

CONCEPTUAL MODEL TO CHARACTERIZE THE HYDRAULIC PROPERTIES OF SOLID ROCKS

It was already explained in Section 1, where the dynamic principles of seepage hydraulics were discussed, that the generally accepted way to characterize the resistivity and other hydraulic properties of a seepage field is the combination of a simplified geometrical model of the randomly interconnected interstices and a dynamic model. The solution of the movement equation describing the flow through the geometric model can be achieved in this way. There are cases, however, when the absolute size of openings considerably influences the character of the movement and, therefore, the water transporting capacity of the actual channels composed of the interstices of the solid matrix cannot be determined by considering only the average size of the interstices. In these cases, the probability distribution of the actual size of channels should be analyzed and the statistical model derived in this way gives supplementary information to develop a conceptual model necessary for characterizing the flow through the interstices of the solid matrix.

The structure of the interconnected water transporting channels between the solid phase of fissured and fractured solid rocks are basically different from that characterizing the network of randomly distributed pores of loose clastic sediments. This is the reason why a geometrical model different from that generally used in seepage hydraulics and determined by considering the character of porous medium must be applied for the simulation of water transport through hard rocks. At the same time, the dynamic principle of the construction of the model is the same as that applied to derive the relationship between seepage velocity and hydraulic gradient in loose clastic sediments. Thus, a mathematical model suitable to calculate the resistivity of fractures can be easily determined by accepting a slit between two parallel walls as a geometrical model and by combining its geometrical parameters with the movement equation describing laminar flow. The result of this investigation is summarized in Eq. 1-95.

The idea was further developed in Section 1 by applying the model of one slit to calculate the seepage velocity and the hydraulic conductivity of fractured solid rocks although there were some basic hypotheses listed which hindered the practical use of the result of this analysis (walls are parallel and smooth, the openings have equal width, the flow is laminar, and the fractures are parallel with the flow direction). The way to consider the unequality and roughness of walls and to generalize the model for characterizing the water transport through fractures having random position was explained in Subsection 4-4. The first two influencing factors can be taken into account by substituting the average width of the openings and by increasing the numerical constant in the denominator of the expression. Since the fractures have random position, the hydraulic conductivity cannot be characterized by a definite value, but its most probable value (median) and the expected scattering should be determined. On the basis of analyzing the probable position of the fractures and considering the probable distribution of the parameter investigated (which is generally approximated by a lognormal distribution), the median and the limits of the zone of confidence were given in Eq. 1-99, and this equation is repeated here as the basis of the further investigation:

$$K_{min} = \frac{1}{128} K_{max} < K_{med} = \frac{1}{16} K_{max} < K_{max} \ ; \quad \text{and}$$
$$K_{max} = \frac{1}{15} \frac{g}{\nu} (n_{Lx} b_x^2 + n_{Ly} b_y^2) \ ; \qquad (5\text{-}11)$$

where g is the acceleration due to gravity, ν is the kinematic viscosity, there are two measuring lines crossing at a right angle (x and y) fixed on the surface of the rock perpendicular to the flow direction, n_{Lx}, n_{Ly} are linear porosity values and b_x, b_y the average widths of the openings measured along these lines.

It was already explained in Section 1 that the use of the average sizes in this equation hinders the practical application of this model because the character of flow is influenced by the actual width of the fractures. The change of flow conditions according to the size of openings can be considered only if the geometrical and dynamic models are combined with a statistical model characterizing the probability distribution of the fissures and fractures having different width. The analysis aiming to determine this distribution was discussed in Subsection 3-7 and its result (which is given by Eq. 1-39) is also repeated here, this relationship being the other basis of the further investigation

$$\sigma(b;\Delta b) = \sum_{j=1}^{r} \sigma_j = \sum_{j=1}^{r} a_j \exp(-c_j b) \ ; \qquad (5\text{-}12)$$

which relationship states that (after counting the total number of openings along a line S having a length of L, measuring their width b, dividing the whole range of width into Δb intervals, determining the number of widths belonging to each interval S_i, dividing S_i with the interval length to get its specific value s_i, finally relating the latter to the specific value of the total number of openings and achieving the parameter characterizing the probable distribution of the specific number of openings belong to each interval ($\sigma_i = S_i/S$) in this way) the distribution of the number of openings, according to their width [the $\sigma(b;\Delta b)$ parameter], can be well approximated by a series of exponential functions. Each number of the series depends exponentially on the width b and includes two constants, a_j and c_j. It may be assumed that each exponential function characterizes a set of openings having the same origin (e.g., tectonic joints, fissures created by exfoliation or weathering, fractures in faulting zones, karstic openings enlarged by the solution and erosion of rocks, etc.). Among the constants, a_j is proportional to the Δb parameter, chosen arbitrarily as a basic data for the investigation, and c_j can be related to the ratio of total specific number of openings and linear porosity n_L. Both the number of the exponential functions being the numbers of the series and the numerical values of the constants should be determined from the data of openings (number and width) measured in the field by using a trial and error method.

The purpose of our investigation is to find a conceptual model suitable to simulate the flow through the openings of solid rocks. It was explained that this model can be derived by combining the geometrical, dynamic, and statistical models. Since Eq. 5-11 gives the movement equation of laminar flow through the geometrical model, its solution considering the statistical model (Eq. 5-12) simultaneously provides us with the conceptual model of laminar seepage through fissured and fractured solid rocks. Not only the hydraulic conductivity (or intrinsic permeability) of the rock mass can be determined from this model, but the minimum size of flow field necessary for the application of the continuum approach can also be calculated. Finally, the model can be generalized to consider non-laminar seepage as well. These problems are discussed in the following sections.

19-1 Laminar Seepage through Fissured and Fractured Solid Rocks

As it was explained in the previous paragraphs, the hydraulic application of the statistical model (i.e., the combination of Eqs. 5-11 and 5-17) is the necessary step to characterize the laminar water transport through the interstices of a large mass of solid rocks. The measured linear porosities in three orthogonal directions (n_{Lx}, n_{Ly}, n_{Lz}) and the probable distribution of the specific number of openings

$$s_x(b,\Delta b) = s_{xo}\sigma_x \quad ;$$

$$s_y(b,\Delta b) = s_{yo}\sigma_y \quad ; \qquad (5\text{-}13)$$

$$s_z(b,\Delta b) = s_{zo}\sigma_z \quad ;$$

are the required basic data where s_x, s_y, and s_z are the probable specific number of openings within an interval Δb with a mean value b in the x, y, and z directions, respectively; s_{xo}, s_{yo}, and s_{zo} are the total specific numbers of openings in the same directions; the symbols σ_x, σ_y, and σ_z stand for the sum of the determined series of exponential functions [e.g., $\sigma_x = \sum_{j=1}^{r} a_j \exp(-c_j b)$ in the x direction].

The flow rate through the fractures observed along one line of unit length can be calculated as the specific number of openings multiplied by the flow rate of one slit having a width corresponding to the relevant Δb interval:

$$Q = \sum_{i=1}^{m} s_i(b,\Delta b)\; q(b_i) \quad . \qquad (5\text{-}14)$$

Assuming the investigation of laminar flow only and substituting the movement equation accordingly, the following relationships can be given to calculate the water transporting capacity of the fractures in question

$$Q\Delta b = \frac{1}{15}\frac{g}{\nu} I \sum_{i=1}^{m} b_i^3\; s_i(b_i;\Delta b)\; \Delta b = s\,\frac{1}{15}\frac{g}{\nu} I \int_0^{\infty} b^3\, \sigma(b,\Delta b)\, db \;;$$

$$(5\text{-}15)$$

$$Q = \frac{s}{\Delta b}\frac{1}{15}\frac{g}{\nu} I \int_0^{\infty} \sum_{j=1}^{r} b^3 a_j \exp(-c_j b)\, db = \frac{s}{\Delta b}\frac{1}{15}\frac{g}{\nu} I \sum_{j=1}^{r} b\,\frac{a_j}{c_j^4} \;.$$

The method of summarizing the influences of the fractures observed in two orthogonal directions on a surface was already explained when the construction of the geometrical model was discussed and the result of this analysis was repeated as Eq. 5-11. Following the same steps, the most probable (median) flow rate in the z direction through a unit area in the xy plane (i.e., flux) and the virtual hydraulic conductivity or intrinsic permeability in the same direction can be calculated from the following equations:

$$Q_z = \frac{1}{16}\,\frac{1}{15}\,\frac{g}{\nu}\,I\left(\frac{s_{xo}}{\Delta b}\sum_{j=1}^{r} b\,\frac{a_{xj}^{4}}{c_{xj}} + \frac{s_{yo}}{\Delta b}\sum_{j=1}^{p} b\,\frac{a_{yj}^{4}}{c_{yj}}\right) \quad ;$$

$$K_z = \frac{1}{40}\,\frac{g}{\nu}\,\frac{1}{\Delta b}\left(s_{xo}\sum_{j=1}^{r}\frac{a_{xj}^{4}}{e_{xj}} + s_{yo}\sum_{j=1}^{p}\frac{a_{yj}^{4}}{c_{yj}}\right) \quad ; \quad (5\text{-}16)$$

$$k_z = \frac{1}{40}\,\frac{1}{\Delta b}\left(s_{xo}\sum_{j=1}^{r}\frac{a_{xj}^{4}}{c_{xj}} + s_{yo}\sum_{j=1}^{p}\frac{a_{yj}^{4}}{c_{yj}}0\right) \quad .$$

The same parameters in the other two main directions can be calculated by substituting the parameters (s,a,c) determined in the plane normal to the direction in question. The permeability parameters calculated from measured data are listed in Table 5-9. To demonstrate the practical application of this approach, the same field measurements were used which were applied to give the correct interpretation of porosity of fissured and fractured rocks on the basis of the continuum approach and to derive the statistical model characterizing the probability distribution of the size of openings in such formations. The measurements were performed by Balásházy and J. Kovács (1975). They have determined the linear porosity (n_L) and the $\sigma(b;\Delta b)$ distribution function along three measuring lines being perpendicular to each other and fixed on the fresh surfaces of four different carbonate rocks of Triassic age:

- Middle Triassic (Karnanian) dolomite (Óbarok);

- the same rock type in a fractured zone along a cross fault with observable karstic forms (chemically enlarged openings) (Szár);

- Upper Triassic (Norian) bedded limestone (Csolnok);

- Upper Triassic (Dachstein) karstic limestone (Dorog).

The magnitude of the data seems to be very realistic, their reliability can also be proved by comparing them to data measured in the same layer. For this reason Fig. 5-12 is partly repeated here because the field tests represented in the figure were carried out in the same layer as one of those investigated here (Middle Triassic dolomite, Óbarok) (Fig. 5-42). The hydraulic conductivity vs. depth relationship is shown in the figure. The data calculated by this method

Table 5-9. Hydraulic conductivity of carbonate rocks calculated by using the conceptual model.

Type of Rock		Direction	Minimum Length of Seepage Field (m)	Hydraulic Conductivity (m/sec)				
				Supposing Laminar Flow	considering the non-laminar character of flow			
					$I = 1x10^{-5}$	$I = 1x10^{-5}$	$I = 1x10^{-5}$	$I = 1x10^{-5}$
Middle Triassic	x	5°-185°	107.0	$3.43x10^{-4}$	$3.428x10^{-4}$	$3.419x10^{-4}$	$3.371x10^{-4}$	$3.098x10^{-4}$
dolomite (Óbarok)	y	95°-275°	132.2	$2.41x10^{-4}$	$2.409x10^{-4}$	$2.403x10^{-4}$	$2.373x10^{-4}$	$2.238x10^{-4}$
	z	vertical	164.2	$2.85x10^{-4}$	$2.848x10^{-4}$	$2.842x10^{-4}$	$2.805x10^{-4}$	$2.800x10^{-4}$
Middle Triassic	x	120°-300°	2250.0	21.7	17.8	0.32		
dolomite in strongly faulted	y	10°-190°	3650.0	292.5	241.5	4.20	outside the validity zone	
zone (Szár)	z	vertical	326.1	314.2	259.3	4.52		
Upper Triassic	x	69°-249°	253.0	$2.90x10^{-4}$	$2.89x10^{-4}$	$2.87x10^{-4}$	$2.77x10^{-4}$	$2.17x10^{-4}$
bedded limestone (Csolnok)	y	159°-339°	400.3	$4.37x10^{-4}$	$4.36x10^{-4}$	$4.33x10^{-4}$	$4.17x10^{-4}$	$3.26x10^{-4}$
	z	vertical	205.0	$5.07x10^{-4}$	$5.06x10^{-4}$	$5.03x10^{-4}$	$4.84x10^{-4}$	$3.78x10^{-4}$
Upper Triassic	x	24°-204°	2337.5					
karstic limestone	y	non-measurable		0.628	0.611	0.533′	0.087	outside the validity zone
(Dorog)	z	vertical	223.0					

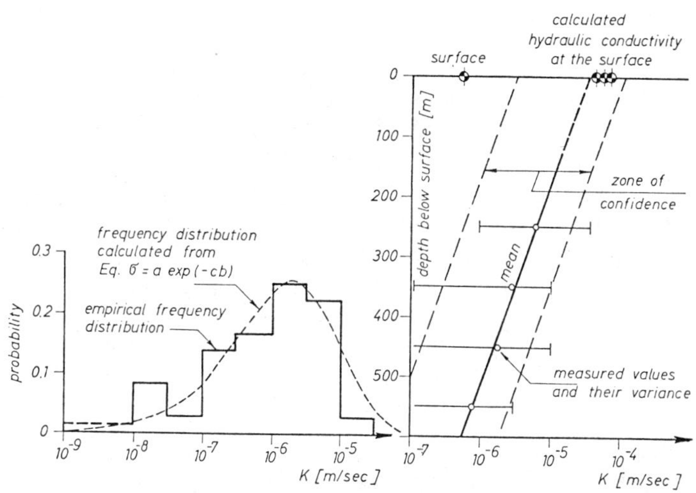

Fig. 5-42. Comparison of measured and calculated values of hydraulic conductivity.

indicated a higher value [K_{cal} = (2.4-3.4)x10^{-4} m/sec] than that extrapolated from measurements carried out at various levels (K_{means} = 3.5x10^{-5} m/sec). If the calculation is repeated by using only the first member of the series of exponential functions (assuming that the second one characterizes a process acting only on the surface, i.e., weathering, exfoliation) a quite good agreement is achieved [K_{red} = (5.0-7.5x10^{-5} m/sec]. The zone of confidence is also indicated in the figure by calculating its limits at the surface and extrapolating these limits downwards by using Eq. 5-4. Comparing this zone to the scattering of data actually measured at various depths, the good agreement between theoretical and practical data can be further proved.

The reliability of the supposed frequency distribution can also be checked in the same figure. Originally the empirical frequency distribution of measured data of hydraulic conductivity was determined. Because the conductivity of a rock mass is proportional to the third power of the width of openings ($K = n_L K_1 \propto b^3$) the b^3 vs. $b^3\sigma$ relationship plotted in suitable scale can be compared to the empirical distribution curve. The figure also proves the good agreement between the two results achieved in different ways.

Király's ideas (1973) concerning the role of very large openings in the development of the relatively high regional hydraulic conductivity of carbonate terrains are also supported by the presently described method. He has found, as it was already mentioned, that it is necessary to assume a network of relatively narrow zones with conductivity higher than 10^{-2} m/sec spaced in a distance of 2-4 km to explain the large flow rate and relatively flat slope in a region where the measured conductivity ranges from $5x10^{-6}$ to $1x10^{-7}$ m/sec. Hydraulic conductivity of the Middle Triassic dolomite was determined by the statistical method at two places: where the rock is relatively intact (Óbarok) and in a strongly faulted zone (Szár). The ratio of the two values is $1:10^5$ which is in good agreement with Király's observations.

19-2 The Minimum Size of Seepage Fields

Explaining the concept of continuum approach and representative elementary unit and also discussing the application of the latter for the characterization of fissured and fractured rocks, it was already mentioned in Section 1 that the flow through the channels can be described by using the seepage equations if the flow field is many times larger than the representative elementary volume. The conceptual model derived by combining the geometrical, dynamic, and statistical models of fractures is also suitable to determine the limit of the size below which the water conveying capacity of individual water transporting elements should be investigated. The basic concept of this analysis is to set up a limit of the flow rate related to the total discharge of the system (e.g., 95 percent) assuming that the water transport through the very large but rare openings is negligible if this amount of water is smaller than the fraction of the total water conveyance being above this limit (e.g., 5 percent).

The error caused by neglecting the very large fractures having a width larger than the limit value (b_h) can be calculated and the requirement is that it should be smaller than or equal to a given value (ε):

$$\frac{Q-Q_h}{Q} \leq \varepsilon \quad ; \qquad (5\text{-}17)$$

where Q is the total flow rate (Eq. 5-15) while the limit value is

$$Q_h = \frac{s}{\Delta b}\frac{1}{15}\frac{g}{\nu} I \int_o^{b_h} \sum_{j=1}^{r} b^3 a_j \exp(-c_j b)\, db$$

(5-18)

$$= \frac{s}{\Delta b}\frac{1}{15}\frac{g}{\nu} I \sum_{j=1}^{r} a_j\left[\frac{b}{c_j^4} - \left(\frac{b_h^3}{c_j} + 3\frac{b_h^2}{c_j^2} + b\frac{b_h}{c_j^3} + b\frac{1}{c_j^4}\right)\exp(-c_j b_h)\right] .$$

Calculating the possible error, the members of the series of exponential functions having high c parameters can be neglected taking into account only the last member with parameters a_r and c_r. Thus, the simplified form of Eq. 5-17 is as follows:

$$\left(\frac{1}{6} b_h^3 c_r^3 + \frac{1}{2} b_h^2 c_r^2 + b_h c_r + 1\right) \exp(-c_r b_h) \leq \varepsilon \quad . \quad (5\text{-}19)$$

After a numerical value has been given to ε(0.05 can be proposed which means to neglect only 5 percent of the total flow), Eq. 5-19 can be solved numerically by expressing the product of $b_h c_r$. Using the limit of 5 percent, the result is:

$$c_r b_h \geq 7.75 \; ; \quad \text{and at the limit } b_h = \frac{7.75}{c_r} \quad . \quad (5\text{-}20)$$

The probable number of the openings having a width of b_h within a unit length is:

$$s_h = s \sum_{j=1}^{r} a_j \exp(-c_j b_h) = s \sum_{j=1}^{r} a_j \exp\left(-7.75 \frac{c_j}{c_r}\right) \; ;$$

(5-21

or neglecting, once again, the members of the series having higher c parameter and considering only the last one:

$$s_h \sim s a_r \exp(-7.75) = 4.3\text{x}10^{-4}\, s a_r \quad . \quad (5\text{-}22)$$

It is necessary to note that the parameter a, and thus the probable specific number of the openings having a width of b_h as well, is a function of the interval Δb, chosen arbitrarily. It is reasonable, therefore, to determine a design value for the interval $[(\Delta b_h)]$ proportionally to the limit width and transform all the results calculated by using any Δb interval to this system. The determination of such a design value is an arbitrary step once again and it can be proposed that the b_h width should be the mean value of the fifth Δb interval. Considering the listed conditions, the probable specific number of openings having the limit width and also that within a length of L can be calculated:

$$(\Delta b)_h = \frac{1}{4.5} b_h = 0.22\ b_h \quad ;$$

$$(a)_h = \frac{a_r}{\Delta b} \frac{b_h}{4.5} \quad ; \tag{5-23}$$

$$s_h \sim s \frac{a_r}{\Delta b} b_h x10^{-4} \quad ;$$

$$S_h \sim Ls \frac{a_r}{\Delta b} b_h x10^{-4} \quad .$$

If a characteristic length is defined in a way that it should be the shortest size of a field regarded as a continuous seepage field, the requirement is that the probable number of an opening having a width of $(1 \pm 0.11)\ b_h$ has to be 1 within this length:

$$S_h = 1 \quad ; \quad \text{consequently}$$

$$L = \frac{\Delta b}{sb_h} \frac{10^4}{a_r} = \frac{\Delta b}{s} \frac{c_r}{a_r} \frac{10^4}{7.75} \quad . \tag{5-24}$$

The data calculated in this way are also listed in Table 5-9 and they are in good agreement with the observations around pumped wells (see Fig. 5-15), and also with Király's hypothesis according to which the spacing of large water transporting formations has to be about 2-4 km.

19-3 Application of the Conceptual Model to Characterize Non-laminar Seepage

The last process, which may cause discrepancy between the actual resistivity of the solid matrix and that calculated from the theoretical model, is the development of non-laminar flow in large openings. The dynamic reason of the development of the various types of seepage in the zone of high velocities is the change of the ratio of the influence between friction and inertia and, therefore, the limits of laminar, transition, and turbulent seepage can be determined with given values of the Reynolds number. In the case of fractured rocks, the mean velocity in the openings is generally used as a characteristic velocity while the characteristic length may be either the average width of the fracture (b) or its hydraulic radius (R) which is equal to half of the width (R = b/2) in the case of a long slit between two parallel walls.

There are several publications proposing slightly different limits on the basis of experiments executed either in laboratories or in the field (e.g., Lomidze, 1951). It is well-known from the investigation of loose clastic sediments that the establishment of a general relationship between velocity and hydraulic gradient is more suitable for the practical characterization of non-laminar flow than the determination of validity zones and different equations to describe this relationship within the zones. This is the reason why experiments were carried out to determine a general movement equation for a slit with parallel walls instead of the investigation of the limits dividing the various zones of seepage (Kovács, 1973).

The equipment used for the experiment was the most simple form of the Hele-Shaw models: i.e., confined horizontal flow of a viscous fluid between two glass plates (Fig. 5-43). The length of the flow field was 20 cm. Two variations of the height were applied, 5 and 10 cm. Eight different widths were investigated between 1.1 and 0.1 mm. After having developed the permanent flow at fixed levels of the head and the tailwater, the flow rate, the temperature of the fluid, and the piezometric head at three points were measured. The next step was the calculation of the corresponding values of the mean velocity and the hydraulic gradient. Supposing a similar relationship between these two variables as that determined previously for loose clastic sediments:

$$\left(\frac{I}{v_{eff}}\right)^{3/4} = A + B\, v_{eff}^{3/4} \quad ; \qquad (5\text{-}25)$$

the data were graphically represented in a coordinate system having $v_{eff}^{3/4}$ on the horizontal and $\left(\frac{I}{v_{eff}}\right)^{3/4}$ on the vertical axis (Fig. 5-44). The graphs proved the reliability of Eq. 5-25. The factors A and B were determined as the intersections of the lines with the vertical axis and the slopes of the lines, respectively. These data were analyzed afterwards as functions of the width of the slit (Fig. 5-45). The results can be summarized in the form of a general relationship between mean velocity and hydraulic gradient:

$$A = \left(12\, \frac{\nu}{g}\, \frac{1}{b^2}\right)^{3/4} = \left(\frac{1}{K_1}\right)^{3/4} \quad ;$$

$$B = \left(\frac{0.03}{gb}\right)^{3/4} = \left(\frac{0.0087}{\sqrt{g\nu K_1}}\right)^{3/4} \quad ; \qquad (5\text{-}26)$$

$$I^{3/4} = \left(\frac{v_{eff}}{K_1}\right)^{3/4} + \left(\frac{0.0087}{\sqrt{g\nu K_1}} v_{eff}^2\right)^{3/4} .$$

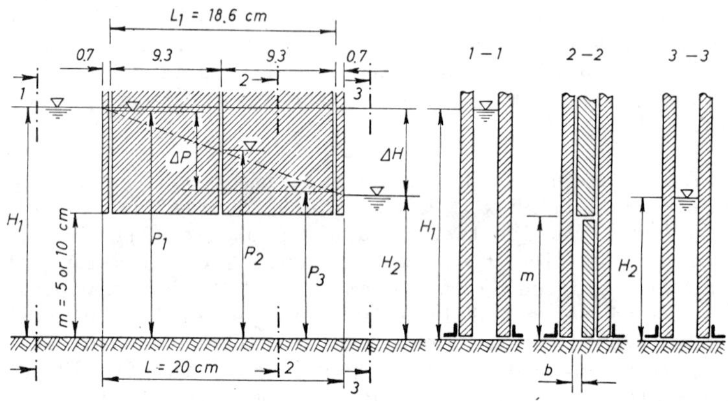

Fig. 5-43. Sketch of the Hele-Shaw model used for measuring the hydraulic conductivity of slits.

The structure of the equation is the same as that determined for loose clastic sediments. The laminar term (i.e., the first member on the right-hand side) is numerically equal as well while the second term (the turbulent one) is smaller in the slit (the ratio is about 1:10) than in pores. This difference may be explained by the fact that the development of turbulence is slower between smooth glass plates, and the role of inertia is smaller than in the pores where the actual velocity changes from section to section. It was felt necessary, therefore, to investigate the character of flow not only in a single slit, but also through a network of fractures. The comparison of the results of such experiments with data determined in a single fracture can characterize the surplus resistance caused by the vertices developing at the edges of openings intersecting each other.

Böcker's experiments (1972) were carried out with cubes of 5 cm edge length and arranged regularly (Fig. 5-46). Three variations were investigated in which the widths of the straight slits between the cubes and forming three orthogonal

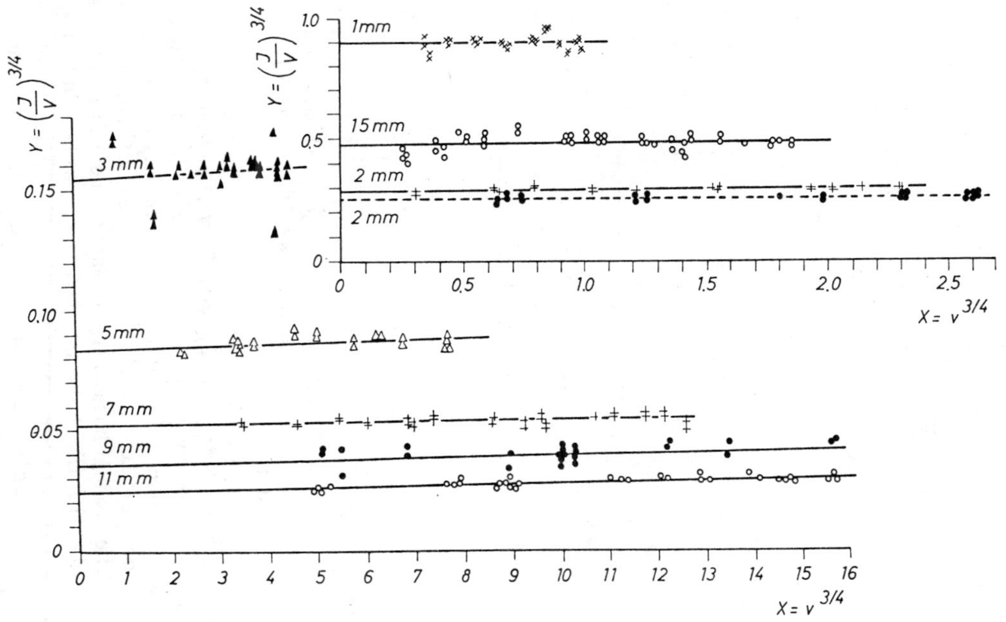

Fig. 5-44. Relationship between velocity and hydraulic gradient in slits with parallel walls.

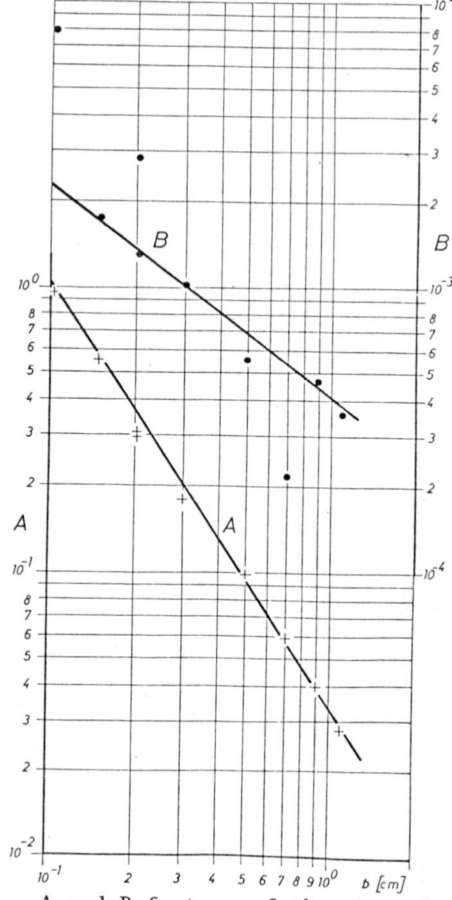

Fig. 5-45. A and B factors of the general movement equation represented as functions of the width of slits.

systems of openings were 1, 2, and 3 mm, respectively. The flow was maintained by constant pressure head between the upper and lower faces of the model. Its main direction was parallel to the vertical slits and normal to the horizontal ones. Thus, the model basically creates the same water movement as that developing between parallel walls, but simultaneously the yield of two systems of slits intersecting one another at right angles are investigated (in the model the water conveyance of two times seven slits). There was no flow in the horizontal slits. They served only to equalize the pressures between neighboring vertical openings and to decrease the smoothness of the walls by intersecting them and forming sharp edges at the intersections.

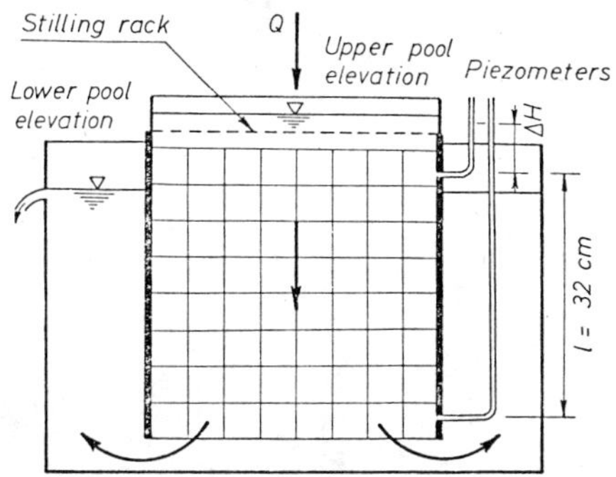

Fig. 5-46. Sketch of the model used for measuring the hydraulic conductivity of three orthogonal sets of slits.

The discharge of the developing steady flow was measured (Q). This value divided by the horizontal area of the vertical openings gives the actual mean velocity and Darcy's seepage velocity can be calculated by multiplying the latter with areal porosity (in this special case $n_A \sim 0.7\ n$). Further measured or calculated parameters were: temperature (to calculate the viscosity of water); pressure head at six points inside the model (to check pressure distribution); and hydraulic gradient (calculated as the amount of head loss to the length of the model).

The interrelated values of mean velocity and hydraulic gradient were evaluated in the form of graphs similarly as it was done in the case of single slits (Kovács, 1973). The results are represented in Fig. 5-47 giving the factors A and B of Eq. 5-25 as functions of the width of slits. Although the scattering of the points is larger than that in Fig. 5-45, the trend of the relationships is the same as that determined for a single slit, and thus the structure of the general movement equation is unchanged, only the numerical values have to be slightly modified (considering the scattering observed in the graphs, the limits of the possible ranges of the factors are given below):

$$\left(12\ \frac{\nu}{g}\ \frac{1}{b^2}\right)^{3/4} < A < \left(15\ \frac{\nu}{g}\ \frac{1}{b^2}\right)^{3/4} \quad ;$$

$$\left(\frac{0.021}{gb}\right)^{3/4} < B < \left(\frac{0.032}{gb}\right)^{3/4} \quad . \tag{5-27}$$

Considering that in nature the walls of the openings are even less smooth than in the applied model, not the average

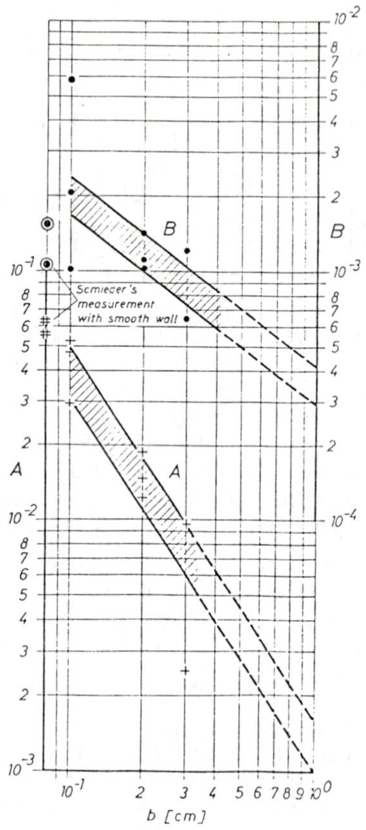

Fig. 5-47. A and B factors as functions of the width of slits in the case of an intersecting network of openings.

values, but the upper limits of the indicated zones can be used for establishing the general movement equation:

$$I^{3/4} = \left(1.25 \frac{v_{eff}}{K_1}\right)^{3/4} + \left(\frac{0.01}{\sqrt{g\nu K_1}} v_{eff}^2\right)^{3/4} . \qquad (5\text{-}28)$$

The final step still missing is the combination of the general movement equation (Eq. 5-28) and the statistical model. Unfortunately, the equation gives the hydraulic gradient as a function of mean velocity while in the model the velocity has to be substituted in explicit form. For expressing the velocity from the general equation in an easily amenable form, some terms have to be expanded into series which requires the application of some new restriction of the validity zone. Thus, the equation loses its general character, but it remains sufficient for the combined characterization of the laminar and transition zones. For this derivation, Eq. 5-28 has to be modified at first

$$v^{3/4} = \frac{A}{2B}\left(\sqrt{1+\frac{4B}{A^2}I^{3/4}} - 1\right) = \frac{I^{3/4}}{A} - \frac{B}{A^3}\left(I^{3/4}\right)^2 + \ldots$$

$$\text{where;}\quad A = \left(\frac{1.25}{K_1}\right)^{3/4} \quad ; \quad B = \left(\frac{0.01}{\sqrt{g\nu K_1}}\right)^{3/4} \quad ; \quad \text{and}$$

$$\frac{4B}{A^2} I^{3/4} \leq 1 \quad ; \quad I < \left(\frac{A^2}{4B}\right)^{4/3} = 20\sqrt{\frac{g\nu}{K_1{}^3}} \quad ; \tag{5-29}$$

is the condition of the use of expanding the term in question into a series. This limit is sufficiently high to ensure the application of the result in solving almost all practical problems.

Following the derivation and expanding the right-hand side once again into series, neglecting the terms with higher powers:

$$v = \frac{I}{A^{4/3}}\left(1 - \frac{B}{A^2} I^{3/4}\right)^{4/3} \sim \frac{IK_1}{1.25} - 0.03\left[K_1\left(\frac{K_1^3}{g\nu}\right)^{3/8} I^{7/4}\right] . \tag{5-30}$$

Consequently, the virtual hydraulic conductivity of a slit depending on hydraulic gradient, if the flow is not laminar, is as follows:

$$K_{trans} = \frac{v_{eff}}{I} = K_1\left[\frac{1}{1.25} - 0.03\left(I\sqrt{\frac{K_1{}^3}{g\nu}}\right)^{3/4}\right] . \tag{5-31}$$

The water transporting capacity of one slit having a width of b is (in the second term its power is considered to be 4 instead of 4.125 for simplification):

$$q = bv_{eff} \sim \frac{1}{12}\frac{g}{\nu} I\left[0.8\ b^3 - 1.83\text{x}10^{-3}\ \frac{1}{\nu^{3/4}}\left(\frac{gI}{\nu}\right)^{3/4} b^4\right] \quad ; \tag{5-32}$$

and that of a set of fractures within a unit length can also be calculated:

$$Q = \frac{1}{12}\frac{g}{\nu} I \frac{s}{\Delta b}\left[\int_o^\infty 0.8 \sum_{j=1}^{r} b^3 a_j \exp(-c_j b)\, db - \frac{1.83x10^{-3}}{\nu^{3/4}}\left(\frac{gI}{\nu}\right)^{3/4}\int_o^\infty \sum_{j=1}^{r} b^4 a_j \exp(-c_j b)\, db\right]$$

$$= \frac{1}{12}\frac{g}{\nu} I\, 6 \frac{s}{\Delta b}\left[0.8 \sum_{j=1}^{r}\frac{a_j}{c_j^4} - \frac{7.32x10^{-3}}{\nu^{3/4}}\left(\frac{gI}{\nu}\right)^{3/4}\sum_{j=1}^{r}\frac{a_j}{c_j^5}\right]. \quad (5\text{-}33)$$

Equation 5-33 provides the flow rate calculated on the basis of fractures measured along one fixed line. Applying once again the geometrical model similarly as it was done to derive the flux in the case of laminar flow, the result is the seepage velocity (or flux), the hydraulic conductivity, and the intrinsic permeability in one of the main directions (after rounding up some figures):

$$Q = \frac{1}{40}\frac{g}{\nu}\frac{I}{\Delta b}\left\{s_{xo}\left[\sum_{j=1}^{r}\frac{a_{xj}}{c_{xj}^4} - \frac{1.0x10^{-2}}{\nu^{3/4}}\left(\frac{gI}{\nu}\right)^{3/4}\sum_{j=1}^{r}\frac{a_{xj}}{c_{xj}^5}\right] + s_{yo}\left[\sum_{j=1}^{p}\frac{a_{yj}}{c_{yj}^4} - \frac{1.0x10^{-2}}{\nu^{3/4}}\left(\frac{gI}{\nu}\right)^{3/4}\sum_{j=1}^{p}\frac{a_{yj}}{c_{yj}^5}\right]\right\};$$

$$K_z = \frac{1}{40}\frac{g}{\nu}\frac{1}{\Delta b}\left\{s_{xo}\left[\sum_{j=1}^{r}\frac{a_{xj}}{c_{xj}^4} - \frac{1.0x10^{-2}}{\nu^{3/4}}\left(\frac{gI}{\nu}\right)^{3/4}\sum_{j=1}^{r}\frac{a_{xj}}{c_{xj}^5}\right] + s_{yo}\left[\sum_{j=1}^{p}\frac{a_{yj}}{c_{yj}^4} - \frac{1.0x10^{-2}}{\nu^{3/4}}\left(\frac{gI}{\nu}\right)^{3/4}\sum_{j=1}^{p}\frac{a_{yj}}{c_{yj}^5}\right]\right\};$$

$$k_z = \frac{1}{40\Delta b}\left\{s_x\left[\sum_{j=1}^{r}\frac{a_{xj}}{c_{xj}^4} - \frac{1.0x10^{-2}}{\nu^{3/4}}\left(\frac{gI}{\nu}\right)^{3/4}\sum_{j=1}^{p}\frac{a_{xj}}{c_{xj}^5}\right] + s_{yo}\left[\sum_{j=1}^{p}\frac{a_{yj}}{c_{yj}^4} - \frac{1.0x10^{-2}}{\nu^{3/4}}\left(\frac{gI}{\nu}\right)^{3/4}\sum_{j=1}^{p}\frac{a_{yj}}{c_{yj}^5}\right]\right\}. \quad (5\text{-}34)$$

The parameters in other directions can be similarly calculated from the data measured in a plane normal to the direction in question.

It is shown by this derivation that the non-laminar character of flow can be considered with a member depending on hydraulic gradient to be subtracted from the parameters determined for laminar movement. For characterizing the effect of inertia numerically, the hydraulic conductivities calculated by assuming various gradient are compared in Table 5-9.

SECTION 20

HYDROLOGICAL ANALYSIS OF DATA CHARACTERIZING THE REGIME OF WATER STORED IN HARD ROCK FORMATIONS

Some of the basic principles accepted to determine the available groundwater resources are repeated here to serve as the starting point of the hydrological analysis of data observed for the characterization of the regime of water stored in fissured and fractured aquifers:

- the surface and groundwater resources are not independent of one another, but they form an integrated unit;

- the groundwater reservoirs are natural storage systems being in a balanced condition, the recharging and draining effects equalize each other if a sufficiently long period of time passes when investigating under natural conditions;

- thus, the concept of available groundwater cannot be applied under natural conditions, these resources become available only by exploitation and their amount depends on the method of artificial draining.

It follows from the principles listed that the theoretically correct determination of available groundwater resources requires the construction of a mathematical model which simulates the flow through as well as the water and energy accumulation in the system composed of the interconnected aquifers. The model describes the water exchange between surface waters and groundwaters under natural conditions. The comparison of the natural water table and piezometrical level, respectively, with those simulated by the model can be used to check the reliability of the model. Introducing the artificial effects planned to be executed in the system as new boundary conditions in the model, the influences of water exploitation (changes in the interaction of surface waters and groundwaters; lowering of both the water tables and the piezometric levels; decreasing the yields of both natural springs and wells operated earlier; changes in the water exchange between neighboring layers; etc.) can be determined.

The importance of the modeling of the aquifers is perhaps greater in the case of hard rock aquifers, and especially when karstic formations are investigated, than for the characterization of the groundwater resources stored in clastic sediments because the flow is more rapid through the large channels or faulting zones and the storage capacity is smaller than in the randomly interconnected pores. More direct contacts can be observed, therefore, between surface waters and groundwaters on hard rock terrains than in areas covered by sediments (e.g., karstic springs are regarded as the natural draining points of the groundwater, but their yield is a part of surface water resources immediately at the time when the water flows out of the layer, their yield has rapid and considerable fluctuation, its increase generally follows the infiltration of precipitation with a relatively short time lag, indicating in this way the limited storage capacity of the system). Since, in most cases, the storage capacity is low and the propagation of the effects is rapid in fractured aquifers, the construction of a new pump well may considerably influence (and in a very short time) the yield of natural springs or existing draining structures. In other cases (e.g., when mines exploiting mineral resources lying under the water table have to be protected against water intrusion by dewatering), the high and unexpected yield of large channels may cause serious problems. The dewatering of mines may also result in the rapid and drastic decrease of the yields at neighboring discharging wells or springs. The sole reliable way to predict the versatile influences of new artificial effects, especially when a large system already discharged at several points is investigated, is the development of the model of the formations.

Accepting the statement that the final objective of any hydrological evaluation of the data collected by the various observations is the construction of the model simulating the flow and storage in the system as well as the determination of the parameters of this model, the suitable ways of data processing can be easily investigated. The first step of this analysis is the survey of information required for the model. The following data are needed to construct a model which gives the reliable simulation of the system.

The geometric parameters of the aquifer (position, depth, thickness, their contact with other water-bearing layers) are determined with the various methods of geological exploration, e.g., surface mapping, drilling, geophysical exploration. Hydrological data are not used directly to characterize the geometry of the system, only their indirect application may be mentioned in connection with this problem: the final checking of the reliability of the model (calibration) is performed on the basis of observed values of water level or pressure and the calibration includes the checking of the supposed geometry of the system as well.

The internal flow conditions of seepage fields are generally characterized with two parameters, i.e., the storage capacity and the hydraulic conductivity of the system. The most important factors composing the storage capacity are the specific yield of the layer and the compressibility of both water and the solid rock matrix. The change of the volume of water is negligible in most cases. The compressibility of solid rocks is also very small and the storage capacity of fractured aquifers originating from the deformation of the solid matrix is hardly considerable. Thus, the decisive factor is the specific yield which is closely related to the effective porosity of hard rock aquifers. There are numerous methods to measure the porosity of these formations. Some of them applied most commonly are as follows: analysis of core samples taken out of boreholes, measuring the number and size of openings on the surface or on the walls of mines, geophysical logging in boreholes, etc. Hydrological observations may also be used to estimate porosity or at least to check its values determined in any other way (e.g., comparison of the fluctuation of water level data and the periodic changes of infiltration or discharge, or the interrelation between the yield of artificial drainage and the lowering of the water table). One of the advantages of the application of hydrological data is that it provides us with the average value of effective porosity (more precisely that of specific yield) over a large area while the other measurements generally give the point values of total porosity which have to be reduced by considering the probable amount of inactive pores and the mean, the median, and the variance should be calculated by regarding the measured values as random variables. At the same time, the probable scattering of storage capacity is also an important information. The combined application of direct measurements of porosity and hydrological observations may be proposed, therefore, as the most effective way to estimate the specific yield of hard rock aquifers. The same conclusion can be stated in connection with hydraulic conductivity. It was explained in the previous section how the expected value and the scattering of this parameter can be calculated by counting the number of openings and measuring their width. The analysis of hydrological data (evaluation of pumping tests, comparison of the gradient of the water table or the piezometric level and the yield of springs, etc.) may provide us with supplementary information giving the most probable average of hydraulic conductivity characterizing larger areas.

The boundary conditions governing the development of seepage through the systems are perhaps the most versatile parameters of the model. Two basically different groups of these conditions should be distinguished:

- the boundary conditions acting along the perimeter of the field, i.e., entry and exit faces (where either the external potential is known or the flow condition is determined in the form of a flux crossing the border and where local resistance developing at the contour or the dependency of flux on the internal potential may cause further varieties of the boundary conditions) and impervious boundaries;

- water exchange within the field through the upper and lower boundaries of the aquifer investigated, i.e., infiltration and the interactions with a neighboring layer (it is necessary to consider that the flow rate of these actions may be independent of the potential prevailing at the given part of the field, e.g., infiltration; or it may be a function of pressure differenes between the aquifer in question and the contacting layer and it changes, therefore, in time as the potential is modified within the field). In some cases, the drainage of the system by evapotranspiration should also be considered as a boundary condition belonging to the second group, but this action is generally negligible in fissured and fractured rocks because the capillary rise in these formations is usually smaller than the depth of the water table.

The determination of the boundary conditions and the estimation of the correct numerical parameters simulating these actions is perhaps the most difficult task in connection with the construction of the mathematical model and only the careful evaluation of hydrological observations can provide us with sufficient guidance in this work.

After surveying the purposes of the hydrological analysis of the data observed in hard rock aquifers, the further discussions have to be grouped according to the character of the observations. The evaluation of the water level data will be dealt with at first. This is followed by the analysis of yields and 'drawdowns. Finally, some water balance studies will be demonstrated.

20-1 Characterization of Water Regime on the Basis of Water Level Data

The pressure prevailing on the water stored in the aquifer in question is indicated by the height of the water table or by that of the piezometric level depending on the fact whether an unconfined or confined layer is investigated. The water level data observed either at natural draining points (springs, bottom of valleys penetrating into the saturated aquifer, moors developing on the surface of unconfined fissured formations) or in observation wells can be used, therefore, to characterize the pressure conditions in the system. Having more simultaneous observations at several points, even the flow direction can be determined and the potential gradient of the system in given directions can be calculated.

The requirements concerning the observation wells are the same as those listed previously in connection with the observation of shallow groundwaters (the well should drain only one aquifer and it must not be used for any other purpose because the water withdrawal disturbs the position of the water level). Including the heights of natural draining points in the investigation, it is necessary to consider that local resistance may hinder the outflow and the water table may be higher in the vicinity of the observation than the level of the spring or that of the bottom of the valley.

The method most frequently used to evaluate water level data is the construction of maps showing the contour lines (curves interconnecting the points having equal heights) of the water table or the piezometric surface at given time points. In other cases, especially if the area is not evenly covered with observation points but they are queuing along the main structural lines of the formation, sections are constructed from the water level data observed simultaneously. Naturally, the combined application of maps and sections provides more information.

Since in hard rock formations the water transporting channels are not always interconnected, it is necessary to first of all check whether the data observed characterize a unified water body or they indicate the pressure conditions of separated systems. There are many cases when some parts of the solid rock is impervious, dividing several independent water conveying networks from each other. This condition frequently prevails above the continuous karstic reservoirs in the so-called descending belt where the springs indicate the outflow of independent water transporting systems. The records of observation wells draining such separated networks cannot be used to construct either maps or sections of the water table, but the discontinuity of the aquifer has to be taken into consideration. Figure 5-48a shows the heights of the outflowing levels of springs draining the descending belt of a karstic formation while the second sketch of the same figure (Fig. 5-48b) represents the development of a unified water table.

The contour map of the water table, combined with the topographic and geological maps of the same area, assists the separation of the recharged and drained areas of the aquifer as well as the determination of the character of the boundary conditions prevailing along the perimeter of the investigated field. The topographic, the geological, and the water table maps of the Transdanubian Mountain Range are compared in Fig. 5-49. The last sketch shows the position of both the areas recharged by infiltration and the draining lines as well as the character of the contours as it was determined on the basis of the maps. Investigating some details of the map,

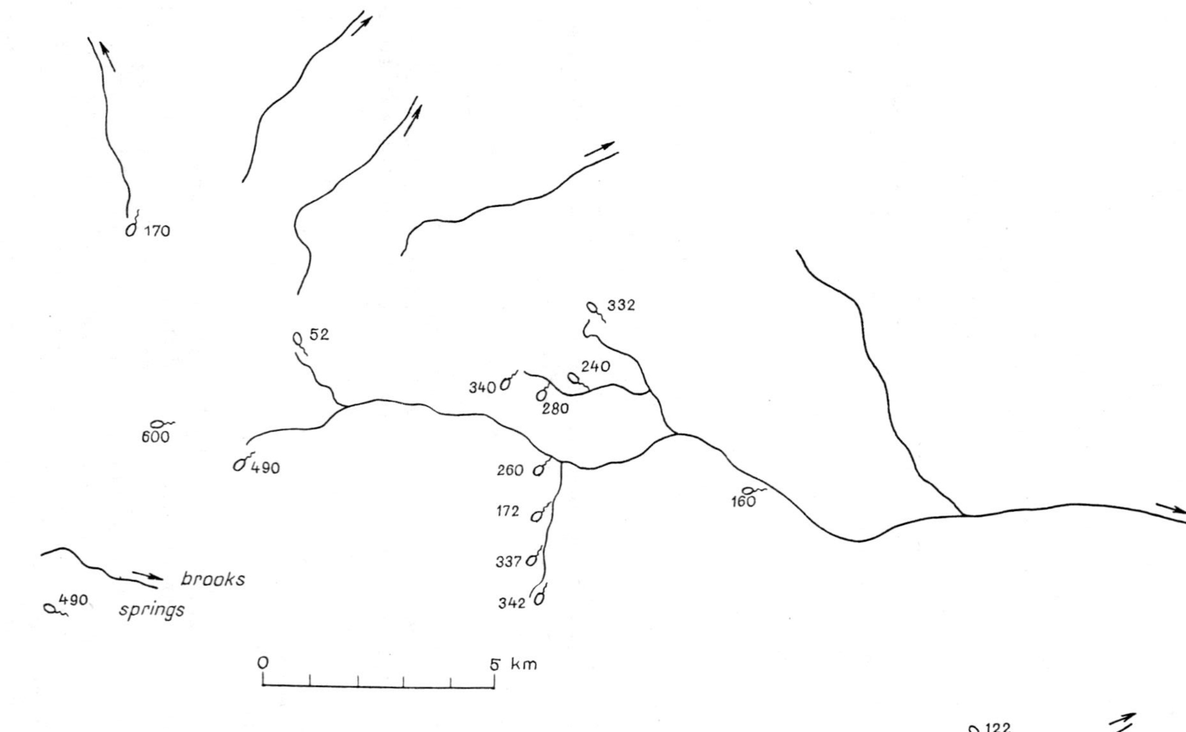

Fig. 5-48. Height of springs; a) in the descending belt of a karstic formation.

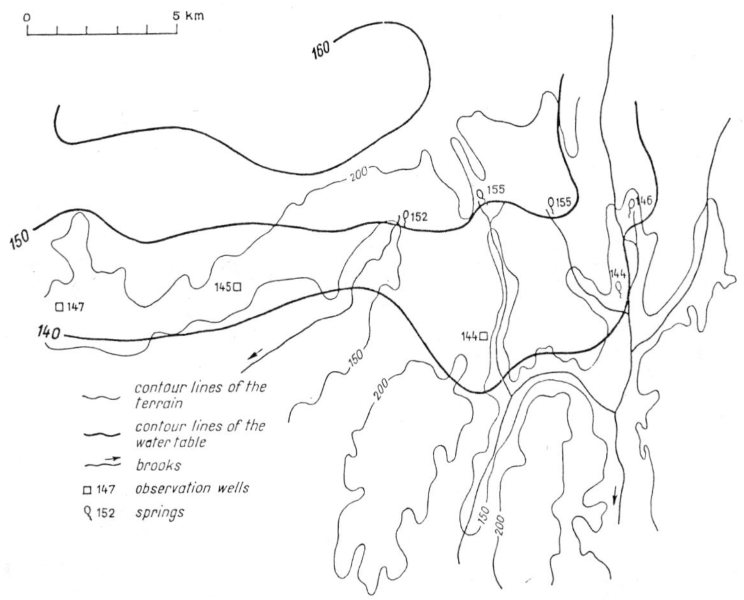

Fig. 5-48. Height of springs; b) within an area having unified karstic reservoir.

further conclusions can be drawn concerning the internal structure and the boundary conditions of the seepage field, e.g.:

Where the contour lines of the water table (or piezometric surface) indicates a very steep slope of the surface, or its discontinuity, the presence of an impervious block interbedded between the water transporting layers may be assumed. Groundwater flow develops around the block, the low permeability of which should be simulated in the model (Fig. 5-50).

In the Danube valley, which borders the area at the northern and eastern sides, the river has contact with the carbonate formations either directly or through the gravel deposits of the river. Not only large springs developed along the faults, performing as the groundwater drain for that valley, but the amount discharged directly into the river is also considerable. This border of the field may be regarded as a contour with constant potential (Fig. 5-51).

Fig. 5-49a. Topographic, geological, and hydrological maps of the Transdanubian Mountain Range.

The contour lines are perpendicular to the impervious boundary of the field. Such borders are expected along faults or other geological structural lines where the aquifer is contacted by impervious formations (in the case represented in

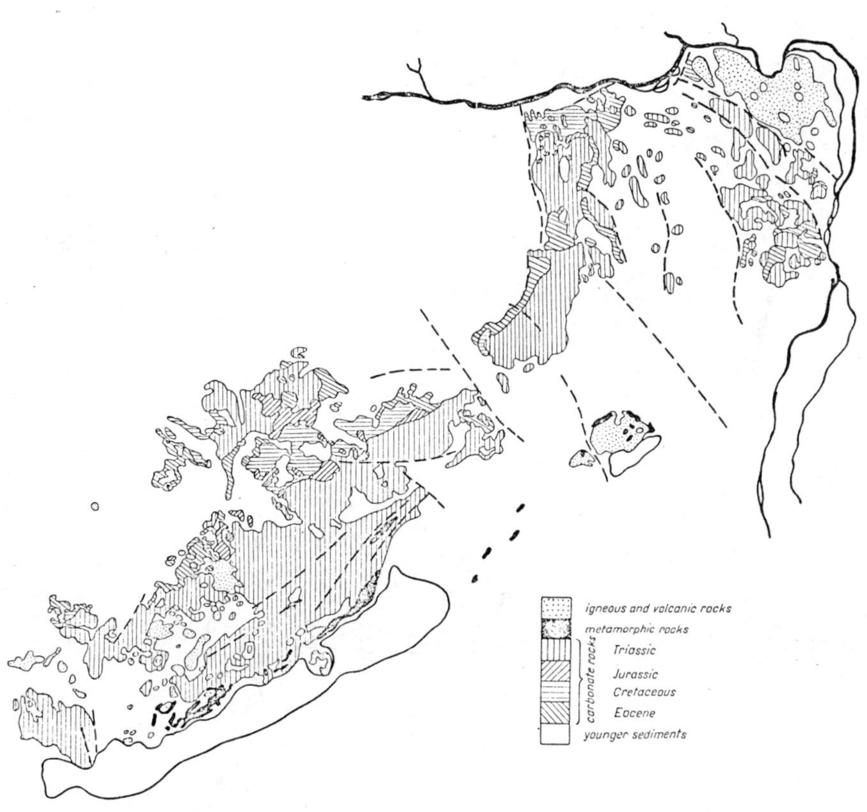

Fig. 5-49b. Topographic, geological, and hydrological maps of the Transdanubian Mountain Range.

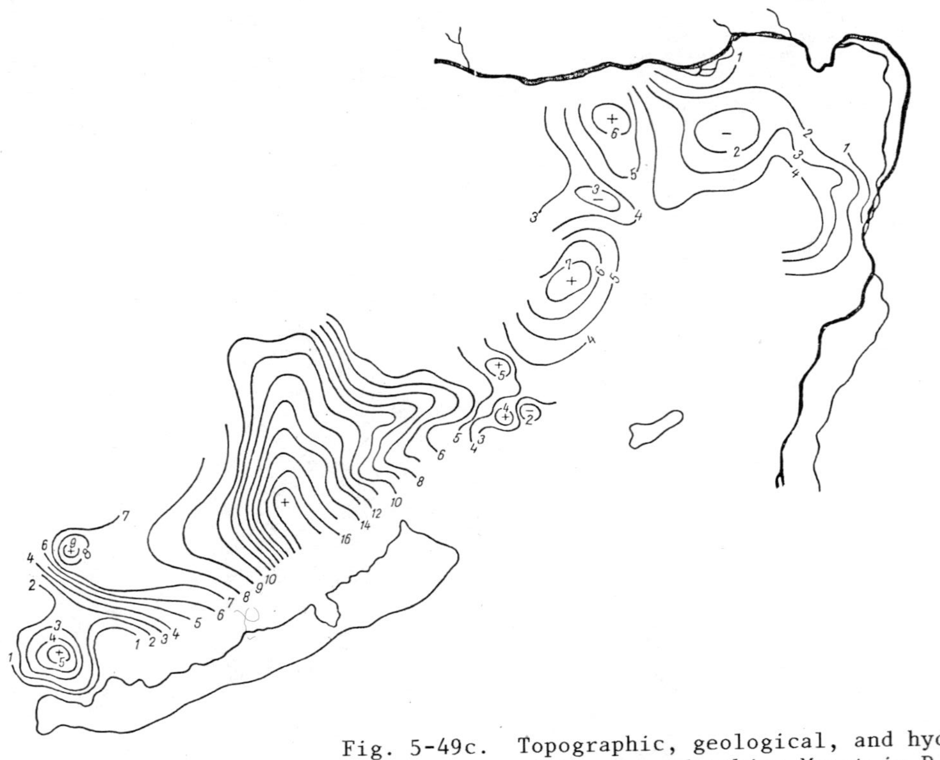

Fig. 5-49c. Topographic, geological, and hydrological maps of the Transdanubian Mountain Range.

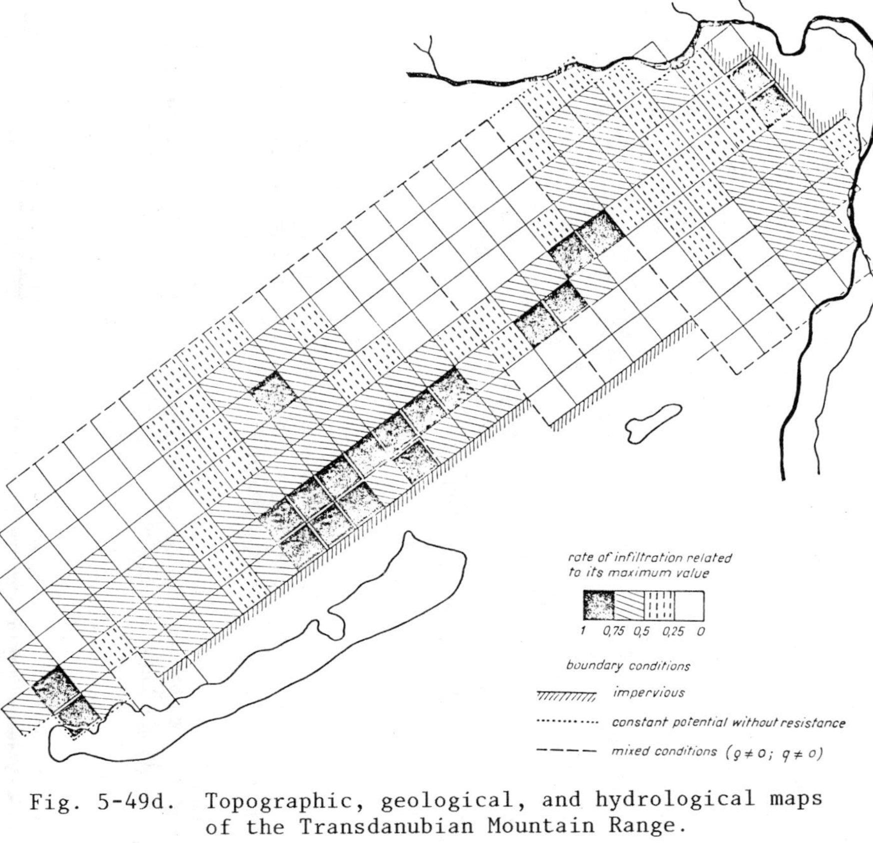

Fig. 5-49d. Topographic, geological, and hydrological maps of the Transdanubian Mountain Range.

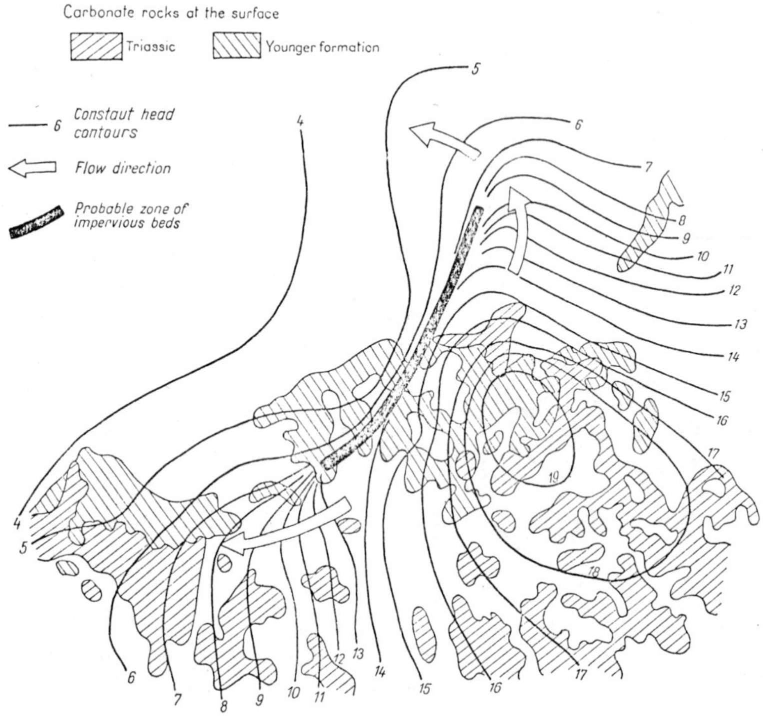

Fig. 5-50. Development of the steep slope of a water table inside a seepage field indicating the presence of an impervious block.

Fig. 5-52 by Miocene andesite, this occurs along the southern border of the investigated formations by Permian slates and very compacted sandstones).

It can be seen also in Fig. 5-52 that the main recharging areas of the aquifers are those where the carbonate formation is not covered by younger sediments. As a result of relatively high infiltration here, local mounds of the water table develop under such terrains. It is necessary, however, to draw attention to the fact that the infiltration through the covering formations within the internal basins is not negligible either because the extension of such areas is large and thus the component of water balance originating from infiltra tion through younger sediments may be considerable although its specific value (crossing a unit horizontal area) is much

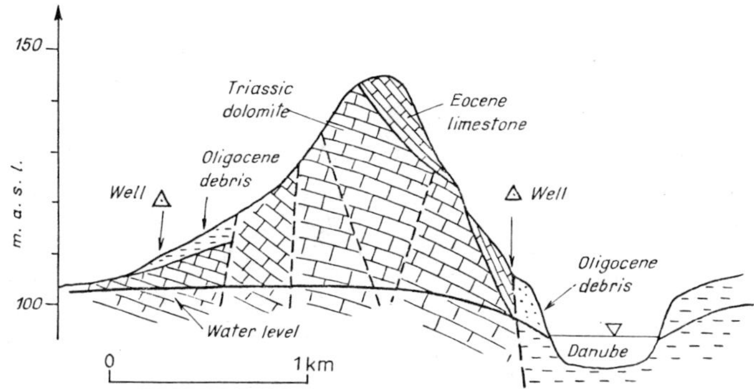

Fig. 5-51. Section at the fault bordering the karstic formation (see Fig. 4-96) (after Böcker, 1967).

less than the same parameter over the bare surface of carbonate rocks. The pressure conditions within the covering layers should always be compared to that in the main aquifers in order to determine whether the internal basins are recharging or draining areas of the karst.

At the borders of the field where the karstic aquifer is contacted by permeable sediments, the boundary conditions may be characterized by the flow rate crossing a unit length of the contour. The direction of flow depends on the pressure conditions prevailing in the contacting layers and the flux of flow is also determined by the pressure gradient developing at the interface of the two formations (and naturally on their hydraulic conductivity). In the example shown in Fig. 5-53, the water pressure is higher in the karstic aquifer than in the contacting loose clastic sediments, this border is, therefore, an exit face of the investigated seepage field.

The deep valleys crossing the mountain and dividing the range into separated blocks drain the karstic water under natural conditions as it is clearly indicated by the form of the contour lines (Fig. 5-54). This drainage should be considered in the model along the lines representing the valleys.

The boundary condition mentioned in connection with Fig. 5-54 changes in time if the water table is lowered due to human activity. In the first period, the valley still remains

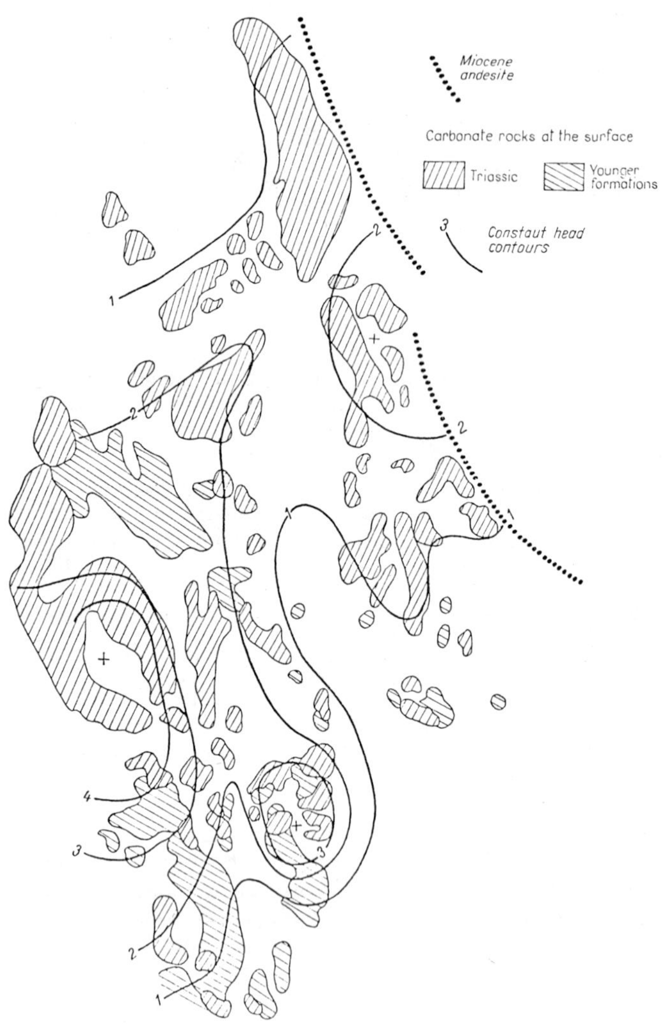

Fig. 5-52. Development of the water table at the NE border of the investigated aquifer system (after Lorberer, 1978).

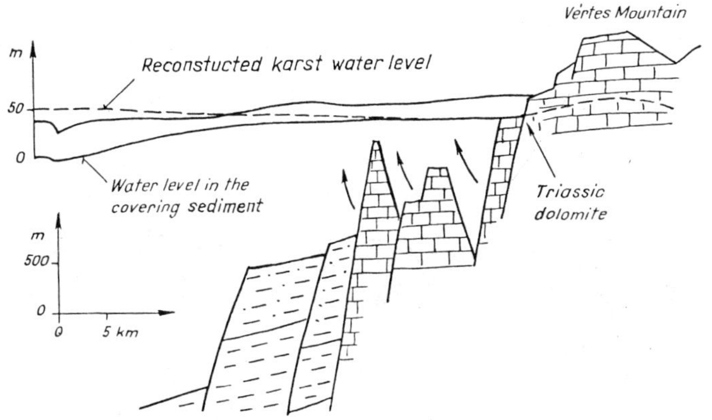

Fig. 5-53. Development of the water table at the NW border of the investigated aquifer system.

a draining line, only the amount of water discharged there from the aquifer decreases, in this way compensating a part of the water drained artificially. When the piezometric level or the water table is lowered below the bottom of the valley, the direction of flow is changed and the valley becomes a recharging line. In this period, this contour can be regarded as a boundary with constant potential. Further change in the character of the boundary condition is expected if the pressure in the aquifer is further decreased, the specific recharge related to a unit length of the lines becomes approximately constant (boundary condition with recharging flow rate independent of the potential conditions). The gradual modification of the conditions along such contours can be well characterized with a section running through the center of an artificial drawdown and being perpendicular to the valley. The development of the water table at various points in time are represented in the sction shown in Fig. 5-55. The numerical value of the flow recharging or discharging the field along such lines can also be calculated in the first two periods when the flux is proportional to the potential gradient (the gradient determined as the slope of the water table perpendicular to the valley should be multiplied by the transmissivity of the aquifer estimated on the basis of other hydrological analyses). The recharging flow rate in the last period depends on the valley characteristics (texture and structure of the alluvium filling the valley, local resistance

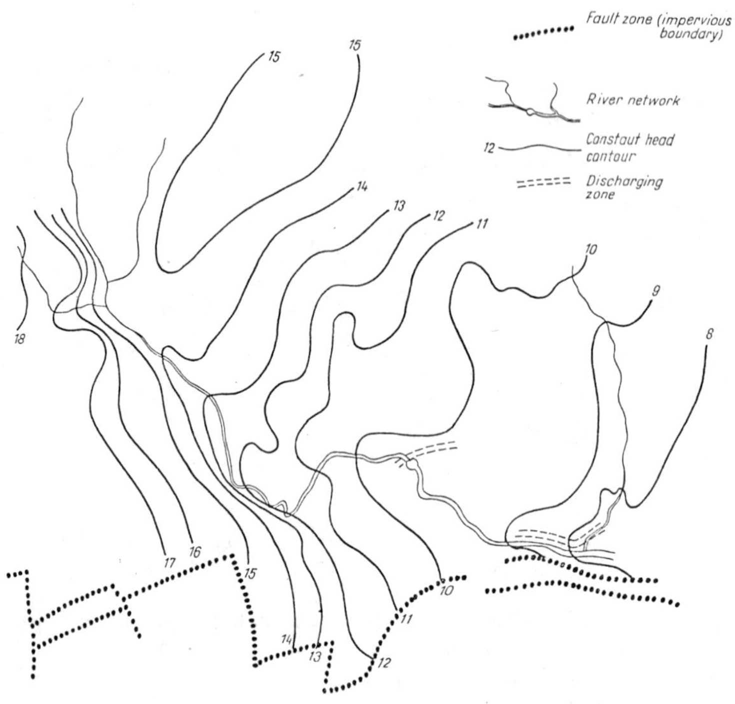

Fig. 5-54. Form of the water table in the vicinity of deeply penetrating valleys.

between the riverbed and the aquifer) as well as on the hydrological parameters of the water course running there (depth flow).

The influence of artificial groundwater exploitation can also be characterized by a series of groundwater maps showing the position of the water table at subsequent points in time (Fig. 5-56). Calculating the volume of drawdown during a given period (as the difference of the water table at the beginning and at the end of the investigation) and multiplying this parameter by effective porosity (or specific yield), the change of the stored amount of water can be determined. Knowing the yield of artificial drainage, estimating the amount of natural recharge or discharge and relating the total amount withdrawn from the aquifer to the total volume of the rock mass dewatered during the investigated period, the average specific yield can also be estimated.

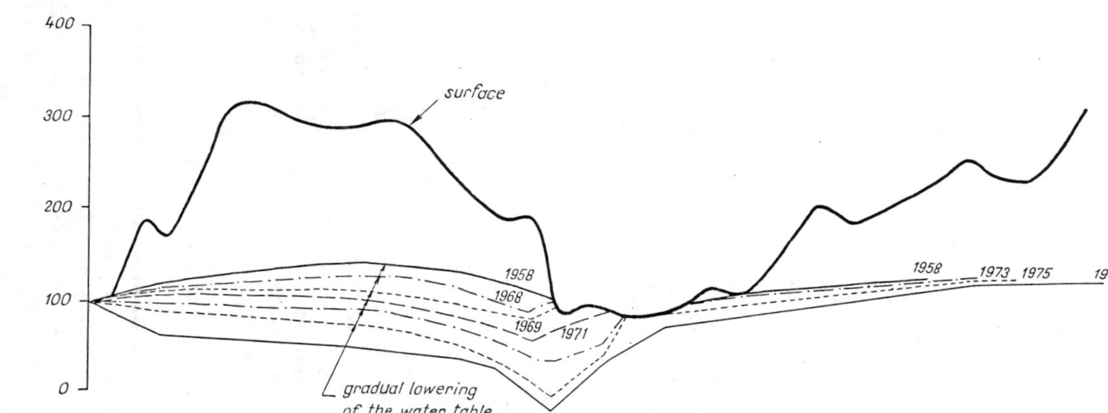

Fig. 5-55. Section through a valley discharging the aquifer under natural condition and recharging it when the water table is lowered due to artificial drainage.

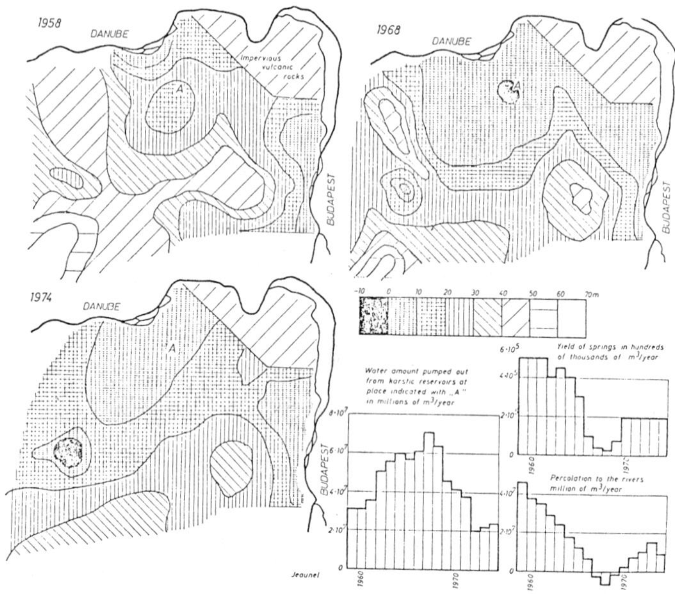

Fig. 5-56. Change of the position of the water table due to artificial groundwater exploitation.

In general, the construction of a groundwater map is repeated yearly using the data observed at the same calendar date in order to exclude the influence of the seasonal fluctuation of the water table. The changes of the stored amount of water within a year may also be characterized by maps, but the sensitivity of the method is not high enough to achieve the reliable analysis of relatively small influences. It is preferable, therefore, to use the water level hydrographs observed at various wells for the characterization of seasonal fluctuations. Where the aquifer is not influenced by human activity, the general trend of the hydrograph is horizontal rising and lowering stretches following each other according to the recharging and discharging character of the seasons (Fig. 5-57). Within areas artificially discharged, the decreasing trend of the hydrograph indicates the effect of water mining and the seasonal fluctuation is superimposed on the descending line (Fig. 5-57). There are even cases where the artificial lowering of the water table is so strong that the natural infiltration cannot raise the level even in recharging seasons (Fig. 5-57).

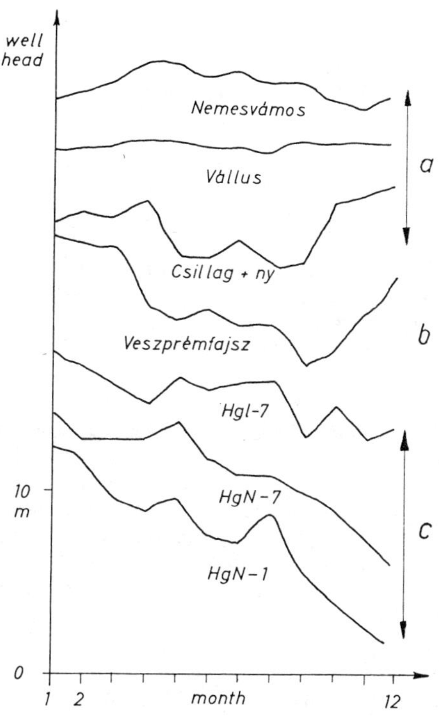

Fig. 5-57. Water level hydrographs at various observation wells (after Böcker, 1969).

The rise of the water table during wet periods is naturally proportional to the change of the stored amount of water and the coefficient of proportionality is the specific yield of the aquifer. The change in storage depends, at the same time, on infiltration which is assumed to be proportional to precipitation reaching the surface of the hard rock terrain during the same period. Some relationships may be established in this way, therefore, between the specific yield of the aquifer and the rate of infiltration (the amount of infiltrating water related to precipitation). It is necessary, however, to consider that many other influencing factors (natural and artificial discharge, water exchange with neighboring formations) may disturb these relationships, the effects of which may achieve or even surpass the magnitude of the two parameters taken into account (i.e., specific yield and infiltration). The rate of infiltration calculated in this way is, therefore, only a rough estimation and its reliability should always be checked by applying other methods as well. This versatile interrelation of the various factors influencing the water regime of the aquifers draws attention to the necessity of a complete water balance analysis. An example of this type of study will be given in Subsection 20-3.

20-2 Evaluation of Artificial Drawdown

Apart from the water level data, the yields of natural springs and artificial water withdrawals are forming the other group of directly observable parameters.

The yield of springs discharging fractured aquifers (and especially those draining karstic formations) generally has rapid changes because the infiltrating precipitation reaches the outflow through large channels in a short time, thus producing high peak of yield. Since the storage capacity of such aquifers is relatively low in most cases, the flow rate decreases rapidly after rains and the springs have small yield during low water periods, or they even dry out, if the interval between two rain events is long. This is the reason why discharge measurements repeated occasionally do not provide us with sufficient information, but the continuous recording of flow rate is required. Since there is usually a great difference between the discharge of flood peaks and the yield characteristic in low water periods, the measuring devices should be suitable to measure low flow rate with sufficient accuracy and to also operate during floods (e.g., Thomson weir, measuring channel with a composite section, etc.), as it was already mentioned.

The most common evaluation of flow rate hydrographs is the use of the various stretches of the regression curve to determine the amount of large and narrow pores as it was explained in Section 3 (see Fig. 1-67). There are also attempts (Gerber, 1975; Maucha, 1975) to correlate the discharge with atmospherc pressure and even with the diurnal cycle of the moon (tidal effects of the crust), demonstrating the change of porosity as the result of the change in pressure and the influence of tidal effect on rock masses (Fig. 5-58).

The flow rate data recorded at natural springs are often used to calculate the rate of infiltration through the bare surface of fractured rocks. The total amount of water discharged during a given period from a block through the springs located around and within the investigated area is generally related to the product of the depth of precipitation belonging to the same time interval and the area of the bare rock surface. This quotient is assumed to be the rate of infiltration. Performing the calculation month by month, the

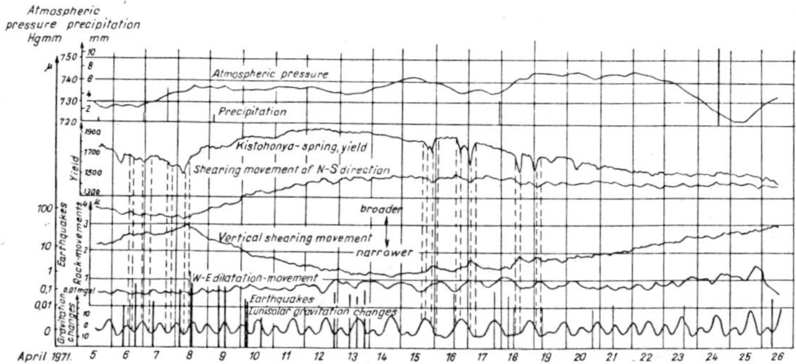

Fig. 5-58. Interrelation between the yield of natural springs and tidal effects.

monthly average infiltration rate can be determined. Two series of the parameter in question are summarized in Table 5-10 calculated for the same mountain. The difference between the two results and the fact that even a rate higher than unity can be found in the table indicate the low reliability of the analyses. The storage of precipitation in the form of snow cover was not considered which fact caused an unreasonable high infiltration rate during the melting periods. Other uncertainties were caused by neglecting both the infiltration through the surface of the solid rock covered by younger layers and the water exchange with the sediments surrounding the investigated aquifer. Although the latter effect is characterized at the area investigated as a discharging effect and thus the two influencing factors neglected in the investigations are acting against one another, their numerical values are not directly measurable and the error caused by the simplification of the water balance equation cannot be estimated. The numbers listed in the table characterize, therefore, only the seasonal fluctuation of the infiltration rate, their absolute values can be accepted only as rough estimations. Since all the processes influencing the infiltration of precipitation depend on the climate of the area and on the structure of the aquifer, such parameters as the rate of infiltration cannot be generalized, they describe only the conditions of the region from where the data analyzed in the research were collected.

Some distinction may also be made in connection with the evaluation of yield discharged artificially according to the fact whether the direct purpose of the withdrawal of water is

Table 5-10. Seasonal fluctuation of the infiltration rate through the bare surface of karstic formations (presented as percent of that months precipitation).

	Jan.	Feb.	March	Apr.	May	June	July	Aug.	Sept.	Oct.	Nov.	Dec.
Kessler (1954)	50	73	122	65	47	28	22	18	16	13	25	50
Maucha (1972)	75	77	87	50	15	17	7	10	5	23	28	50

the determination of some hydrological or hydrodynamic parameters of the aquifer or the water is exploited to serve some other purpose (e.g., water supply, protection of mines, etc.). In most cases, the estimation of hydraulic conductivity of the aquifer is made by the analysis of discharge data independently of the method of data collection. When the thickness of the permeable stratum is not known, the influences of conductivity and depth cannot be separated and transmissivity (the product of the thickness and the hydraulic conductivity of the layer) must be accepted as the result of the investigation. When the time dependent process is initiated by the tests and the transient flow is analyzed, some information can be gained on the storage capacity of the system as well.

The most common method of evaluation of artificially discharged yields is the analysis of data collected from pumping tests executed by using several observation wells. The procedure for this calculation, concerning both cases when the steady state of flow may be assumed and when the transient character of flow is considered, was explained in Section 3 in detailed form. Theoretically, the data produced by pumping tests performed in fractured aquifers can be evaluated similarly as it was discussed in connection with shallow groundwater. In practice, however, many special conditions of solid rocks hinder the analysis, e.g., large water transporting channels in the vicinity of the pumped well or the great thickness of the aquiferous formation. It is necessary to also consider that the cost of the construction of observation wells is considerably higher because deeper wells are required and the specific cost of drilling in hard rocks is also high. This is the reason why, in most cases, only a few observation points are available for the evaluation of pumping tests.

In the vicinity of the draining structure (in the case of pumping tests around the pumped wells) the flow is concentrated in a few large openings and not evenly distributed in the whole mass of the solid matrix. The velocity is generally high in the individual channels and the movement here is, therefore, turbulent causing high potential loss in a relatively short distance. Within this zone (the extension of which may be determined by considering the probability distribution of the size of openings as it was discussed in Subsection 19-2) the seepage law cannot be applied to describe the relationship between velocity and hydraulic gradient, but the resistivity of individual water transporting elements (faults and channels) has to be taken into account. The formulae derived for the evaluation of pumping tests can be used only by substituting the data characterizing the observation points outside this zone. A great part of the energy difference created by the drawdown is consumed, however, within the internal belt and only a small potential difference develops between the neighboring observation wells located within the zone of seepage as it is shown by Fig. 5-15 where the example of a pumping test was used to demonstrate the special flow conditions in fractured aquifers. The transmissivity of the investigated formation is approximately proportional to the difference of drawdowns at two observation points, and if this value is very small, the uncertainties of the measurements may cause considerable error in the result.

The other problem of the evaluation of pumping tests is caused by the great thickness of aquifers. It is not possible, therefore, to cross the whole depth of the permeable formation by the pumped well and by the observation wells although this condition ought to be satisfied for the correct application of the equations derived to calculate transmissivity. The introduction of some corrections, therefore, becomes necessary in order to consider the partially penetrating character of the pumped well. The methods determined for this purpose require, however, the knowledge of the actual thickness of the aquifer which is not available in most cases. The application of any correction increases, at the same time, the uncertainty of the result in any case.

The great thickness of the water-bearing formation is generally associated with its layered character especially in the case of carbonate rocks and sandstones. The nonhomogeneity and anisotropy created in this way hinders the evaluation of pumping tests. The layered field may cause large errors if the partially penetrating observation wells drain different layers, and thus the collected data do not form a homogeneous set. The calculation in these cases may result with contradictory and unacceptable conductivity and transmissivity values.

Since the drawdown related to the depth of groundwater flow is usually negligible in very thick formations, some simplification of the equation is also acceptable in the evaluation of pumping tests performed in such aquifers. The use of simplified methods is also supported by the fact that many other uncertainties hinder the achievement of absolutely correct results, thus the error caused by further approximation does not lower the reliability of the parameter calculated in this way.

The basic idea of the simplification is the use of the equation derived for the characterization of confined aquifers even in the case of a water table condition:

$$Q = 2\pi \text{ Km} \frac{h_1 - h_2}{\ell n \frac{r_1}{r_2}} = 2\pi \text{ Km} \frac{s_2 - s_1}{\ell n \frac{r_1}{r_2}} = 2\pi \text{ Km} \frac{\Delta s}{\ell n \frac{r_1}{r_2}} \quad . \quad (5\text{-}35)$$

The relation is acceptable in deep, unconfined aquifers because both depths (h_1 and h_2) may be approximated with an average depth of groundwater flow:

$$h_1 \sim m \quad \text{and} \quad h_2 \sim m \quad ; \tag{5-36}$$

and thus there is only a multiplying factor of 2 causing a difference between the equations derived in two different ways:

$$Q = \pi K \frac{h_1^2 - h_2^2}{\cdot \ell n \frac{r_1}{r_2}} = \pi Km \frac{h_1 - h_2}{\ell n \frac{r_1}{r_2}} = \pi Km \frac{s_2 - s_1}{\ell n \frac{r_1}{r_2}} = \pi Km \frac{s}{\ell n \frac{r_1}{r_2}} \quad ; \tag{5-37}$$

since $h_1^2 = m\ h_1$ and $h_2^2 = m\ h_2$. (The discrepancy is caused by the fact that the average depth is substituted only after the integration in which the depth is already considered as a variable).

Considering the uncertainties caused by other approximations (the axial symmetry of flow, the substitution of the influenced zone where the drawdown is equal to zero, the assumption of homogeneity and isotropy, the use of Darcy's law) both equations (Eqs. 5-35 and 5-36) can be accepted to evaluate the data of pumping tests. The structure of the two formulae is identical and their constants may be regarded as the limits of the range of this value:

$$\pi < C < 2\pi \quad . \tag{5-38}$$

Applying this general form and assuming that one observation point is selected at the edge of the influenced zone ($r_1 = R$; $s_1 = 0$; $\Delta s = s_2$; and thus the data of the observation well can be indicated without any suffix), the relationship can be expressed between four variables (yield, drawdown, the logarithm of the distance between the pumped well and the observation well related to the radius of the influenced zone, and transmissivity which is the product of hydraulic conductivity and the average depth of groundwater flow $T = Km$). Expressing the drawdown from the equation, the final form used for the evaluation of data is achieved:

$$s = \frac{Q}{C} \frac{1}{T} \ell n \frac{R}{r} = \frac{Q}{2.3C} \frac{1}{T} \log \frac{R}{r} \quad ; \tag{5-39}$$

(the symbols used in the equations are indicated in Fig. 5-59).

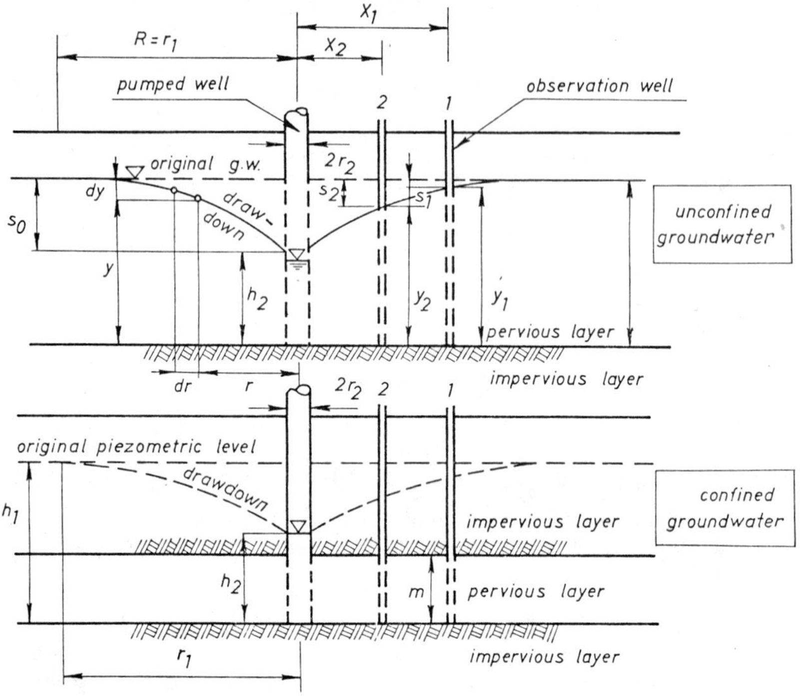

Fig. 5-59. Sketch representing the symbols used in the equations derived for the calculation of transmissivity.

It follows from Eq. 5-39 that the points represented in a coordinate system of s vs. log R/r are queuing along a straight line if the yield is constant. The slope of the line is inversely proportional to the transmissivity of the aquifer. The data of the pumping test, the depression cone of which is shown in Fig. 5-15, were plotted in the coordinate system mentioned (Fig. 5-60). The conditions around the pumped well are not homogeneous (either the hydraulic conductivity or the boundary conditions are nonhomogeneous) which fact causes the distortion of the depression cone. It was, therefore, necessary to execute the investigation in two directions (one southwards and the other eastwards from the pumped well). Considering the constant yield (Q = 900 1/min), the probable values of transmissivity are

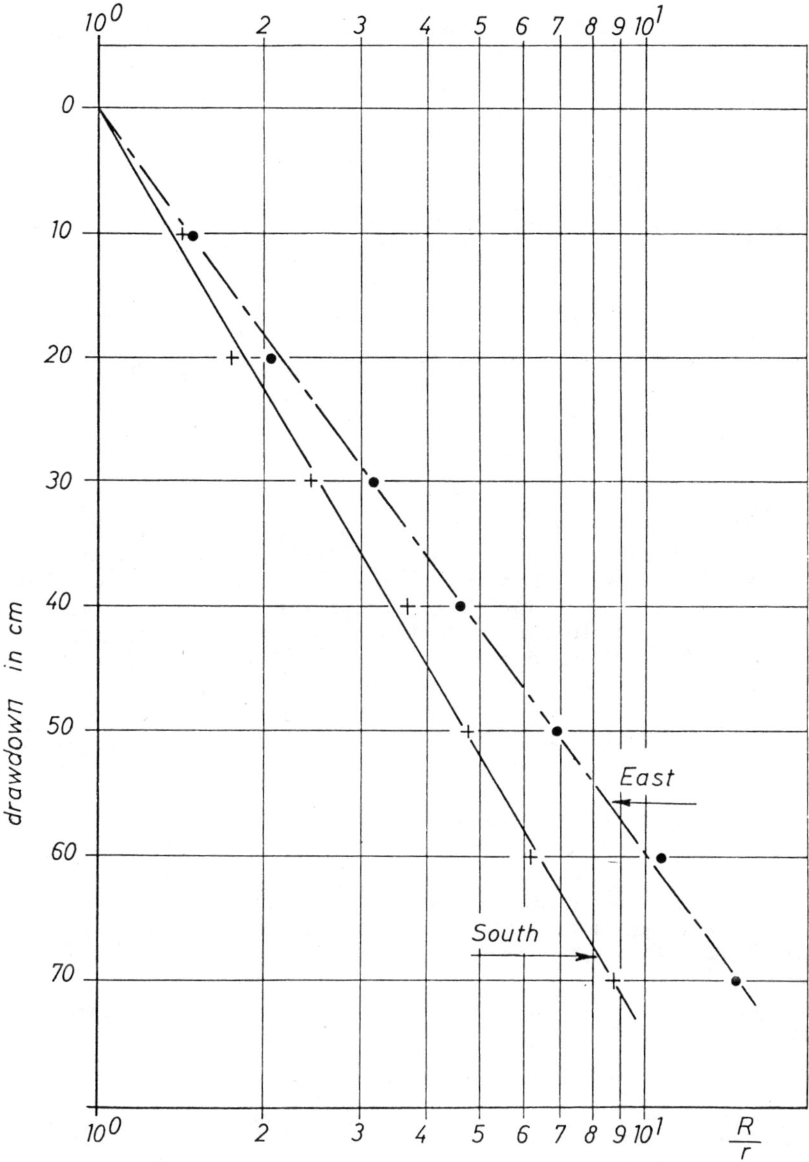

Fig. 5-60. Drawdown data plotted in a coordinate system of s vs. ℓn R/r.

$$T_{South} = 1.41 - 2.82 \times 10^{-3}\ m^2/sec \quad ;$$

$$T_{East} = 1.75 - 3.50 \times 10^{-3}\ m^2/sec \quad ;$$

thus the limits of total scattering are $1.5\text{-}3.5 \times 10^{-3}$ m^2/sec. Combining this value with the hydraulic conductivity of the same formation (Middle Triassic Karnian dolomite, the hydraulic conductivity of which is represented in Fig. 5-42 as a function of depth) the probable thickness of the permeable formation can be estimated. Considering the depth of the upper surface of the formation below the terrain (i.e., about 200 m) the range of scattering of conductivity is $K = 1 \times 10^{-5}\text{-}2 \times 10^{-5}$ m/sec with a mean of about $K = 1 \times 10^{-5}$ m/sec. Thus, the water transporting thickness may vary between 150 and 750 m (probably between 200 and 350 m). At the same time, the average depth of the active layer calculated from Eq. 5-5 is 330 m which is in good agreement with the previous result.

It was already mentioned that the pumping tests aimed at the determination of hydraulic conductivity in fractured aquifers are very expensive. It is a general practice, therefore, that the depression developing around draining structures discharging the groundwater for any economic purpose (e.g., water supply, dewatering of mines) is evaluated similar to the data of pumping tests. The yield of such structures is generally higher than that of a single well, and the period of their pumping is longer. The drawdown is usually deeper, it extends, therefore, to a larger area and the development of steady state flow is more probable. Observation wells are needed anyway for the operation of the structures, thus data being more reliable than those of pumping tests can be gained without any excess cost.

The evaluation of these data is also based on Eq. 5-39. In those cases when the drawdown was gradually increased and the quasi-steady state of flow could be developed after each step, the corresponding data could be represented in an s vs. Q coordinate system. Figure 5-61 shows this relationship for some mines within the Transdanubian Mountain Range indicating the character of the aquifer as well. The evaluation of the graphs is hindered, however, by the fact that the term log R/r_0 is included in Eq. 5-39 as a further variable (the small r_0 indicates, in this case, the radius of the well equivalent with the drained area of the mine). The use of this term causes two problems:

- it would be necessary to know the width of the influenced zone which depends on the time of the investigation;

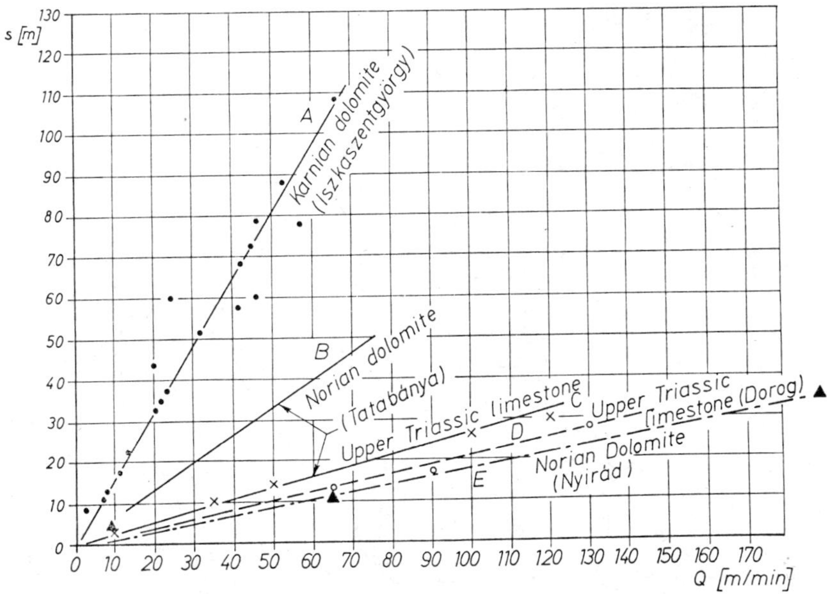

Fig. 5-61. Relationships between drawdown and yield constructed from data observed in mines.

- since the total drawdown is substituted, not only the radius of the equivalent well should be estimated, but the zone near the discharging structures is also included into the field where the application of seepage law is questionable.

The construction of the contour lines of the depression cone around the mine may assist with the estimation of the R/r_o ratio. The cone ought to be circular if the medium of the aquifer and the flow conditions were uniform around the discharged area. The distortion of its shape indicates the nonhomogeneity of the layer or that of the boundary conditions, thus further information is gained by such maps. Figure 5-62 shows the depression cone of the mine, the drawdown of which is characterized by line A in Fig. 5-61. Accepting an $R/r_o = 15$ ratio as an approximation, the transmissivity of the Karnian dolomite can be calculated from Eq. 5-39 by substituting the slope of the line mentioned. The result is $T = 0.85\text{-}1.75 \times 10^{-3}\ m^2/sec$ which is in an acceptable proximity of the previous value although the latter originates from another area of the same formation.

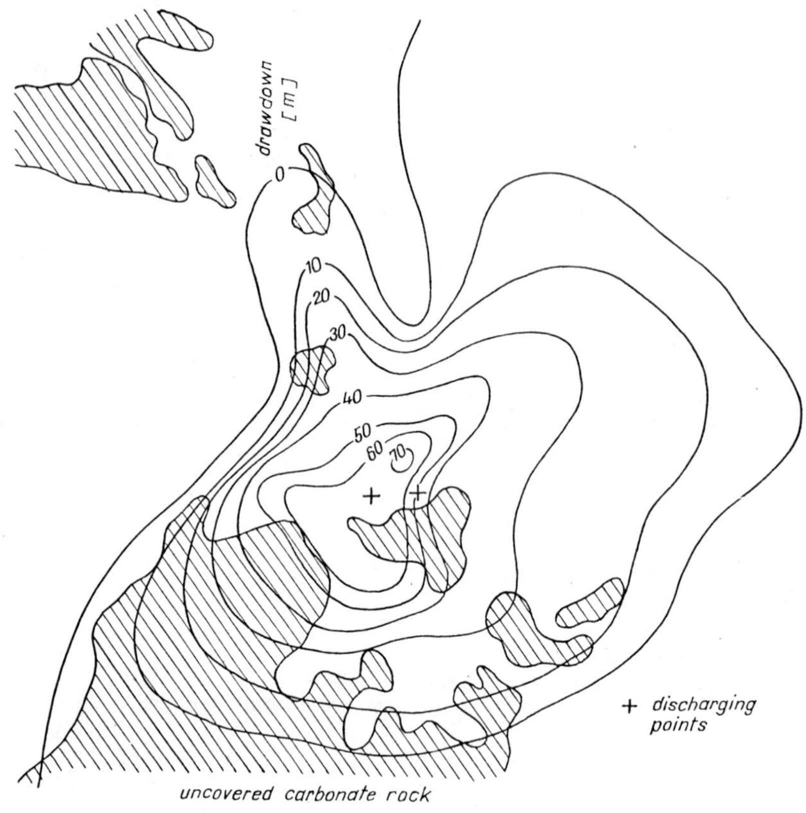

Fig. 5-62. Development of the depression cone around a mining area.

The unknown character of the radius of the influenced zone hinders the determination of transmissivity in these cases also when the data observed around mines are evaluated using the method demonstrated in Fig. 5-60. At the same time, this analysis may also give some guidance for the estimation of the most probable value of the R parameter. An example is shown in Fig. 5-63 where the data collected around one field of a mining area are evaluated in an s vs. log R/r coordinate system and the points determined by the corresponding s and r values are plotted by assuming different R radii. The conclusions drawn from the graphs are as follows:

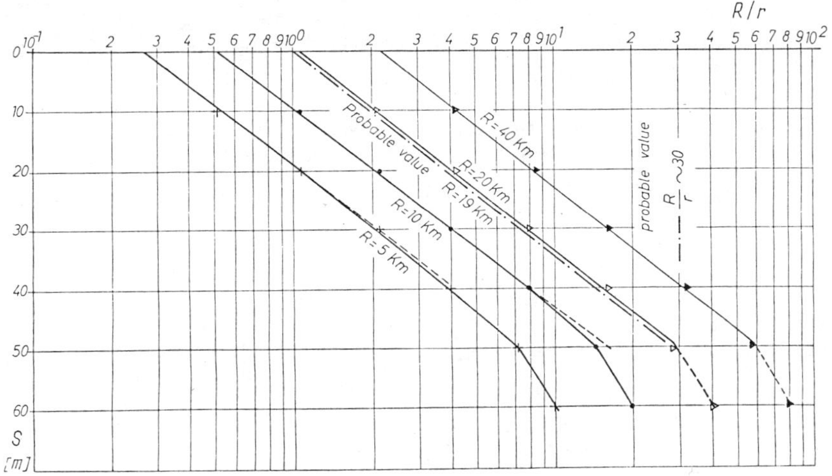

Fig. 5-63. Evaluation of the data characterizing the depression around the field of the X shift at Tatabánya.

- the point representing the largest drawdown (the observation near the discharging place) is not in harmony with the others, indicating the large head loss in the individual channels (the seepage law is not valid in the internal zone, the limit of which is between 500 and 700 m from the center of the depression);

- the points are queuing along straight lines independently of the considered R value (the scattering due to the uncertainties of data is larger than the curvature of the lines); the nonlinear character of the relationship can be observed only in the case of unrealistically small radius (R = 5 km); thus, the probable value of the radius cannot be estimated on the basis of the linearity of curves;

- the slope of the lines does not differ either (the slopes are between 34.5 and 33.5 m depression per one cycle of the logarithmic scale) which fact proves that the result of this method hardly depends on the width of the influenced zone if the latter were estimated within a reasonable range;

- the probable radius may be determined by considering the position of the line having the same slope and intersecting the s = 0 and R/r = 1 point (in the case of the example in question, R = 19 km);

- the lower end of the straight line belonging to this radius indicates the probable value of R/r_o which can be substituted in Eq. 5-39 if the analysis is based on the s vs. Q relationship (this value is $R/r_o = 30$).

Considering all the aspects listed, the transmissivity of the investigated aquifer can be calculated in two different ways from Eq. 5-39: a) by substituting the slope of the straight stretch of the line if the yield discharged from the area is known s:log R/r = 34.5; Q = 1.08 m^3/sec; or b) by substituting the slope of line B in Fig. 5-61 (which characterizes the same aquifer) after accepting the R/r_o radius as it was determined previously (Q/s = 0.025 m^2/sec, log R/r_o = 1.4771). The transmissivity values of the Middle Triassic Norian dolomite at Tatabánya estimated by applying the two methods are in good agreement:

$$T_a = [2.0\text{-}4.0] \times 10^{-3} \ m^2/sec \quad ;$$

$$T_b = [2.5\text{-}5.0] \times 10^{-3} \ m^2/sec \quad .$$

Since the execution of pumping tests is costly and laborious in hard rocks, the simplified field tests, although their results are very uncertain, are applied in greater number for the investigation of such aquifers than for the characterization of loose sediments. The most simple parameter calculated only from the data of the pumped well is the specific yielding capacity of the well which is the ratio of yield and drawdown (q = Q/s $[L^2 \ T^{-1}]$). Similar characteristics are achieved if excess pressure is created in the borehole by pumping water into it and the water recharged in the aquifer is related to a water column (h) equivalent with the pressure (water intake capacity q' = Q/h $[L^2 \ T^{-1}]$). Sometimes these parameters are divided by the open length of the borehole or well also getting quantities having a dimension similar to hydraulic conductivity $[L \ T^{-1}]$. Water intake is characterized in other cases with values measured under special condition (e.g., Lugeon number).

It is necessary to note that these parameters are sufficient only to compare the various formations and to determine their relative conductivity, but numerical values cannot be calculated from them for the description of the flow conditions of the investigated layer. Although there are attempts to numerically interrelate conductivity and water intake (e.g., Eq. 5-2), such relationships only give uncertain information which cannot be accepted even as a rough estimation in the models.

More valuable data can be derived from the analysis of single wells if some pairs of the corresponding yield and drawdown are determined. The usual way of evaluation of such observations is the construction of the characteristic curve of the pumped well which is a graph showing the s vs. Q relationship fitted to the points representing the measured yields and the corresponding drawdown values (Fig. 5-64). It is a general opinion that the curve can be well approximated by a parabola of second order:

$$s_o = A\,Q + B\,Q^2 \quad . \tag{5-40}$$

The same relationship represented in a s_o/Q vs. Q coordinate system is a straight line (as it is shown by the lower graphs of Fig. 5-64) which intersects the s_o/Q axis at at height of A and the slope of which is B:

$$s_o/Q = A + B\,Q \quad . \tag{5-41}$$

Schmieder (1975) has proved that the movement equation describing axially symmetric flow in a confined aquifer results in the same structure as Eq. 5-40 if the relationship between the hydraulic gradient and the seepage velocity is substituted according to Forchheimers' equation

$$I = av + bv^2 \quad ; \tag{5-42}$$

and the change in kinetic energy is neglected along the paths. The final form of his derivation is

$$s_o = \frac{Q}{2\pi\ Km}\,\ell n\,\frac{R}{r_o} + \frac{Q^2}{(2\pi\ K_T m)^2}\left(\frac{1}{r_o} - \frac{1}{R}\right) \quad ; \quad \text{thus}$$

$$A = \frac{1}{2\pi\ Km}\,\ell n\,\frac{R}{r_o} = \frac{1}{2\pi\ 2.3\ Km}\,\log\frac{R}{r_o} \quad ; \tag{5-43}$$

where $K_T m$ indicates the turbulent transmissivity. He has pointed out that even 1/R is negligible compared to $1/r_o$. It is also testified by this analysis that the tangent of the

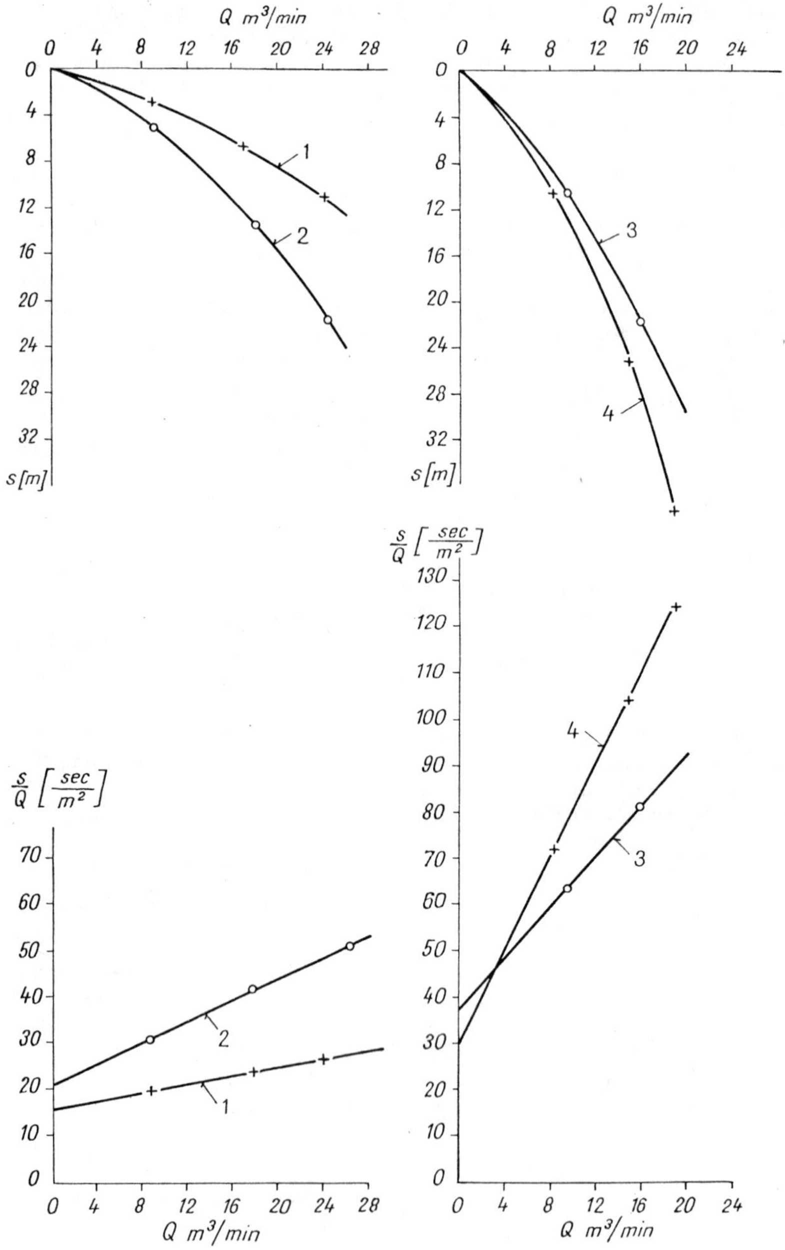

Fig. 5-64. Characteristic curve (graphs showing the s vs. Q relationship) of wells.

characteristic curve at the point s = 0 and Q = 0 (or the height of the intersection of the straight line with the vertical axis if Eq. 5-42 is applied) is proportional to the transmissivity of the aquifer, thus the numerical approximation of the parameter can be achieved in this way even from the data collected by pumping a single well without having observation wells around.

The graphs represented in Fig. 5-64 characterize wells having a diameter of 80 cm and discharging the same formation which was analyzed in connection with Fig. 5-63, but at another place within a strongly faulted zone (Nyirád). According to the characteristic curves, the most probable range of A is 20-40 sec/m^2. The realistic value of log R/r_o is estimated to be 3-4. It should also be considered that a formula having the same structure as Eq. 5-43 can be derived for unconfined deep aquifers, but the constant of 2 is missing from the denominator of A in this case. Substituting the values listed, the scattering of transmissivity expected within the area investigated can be calculated:

$$T = [5.0\text{-}25.0]\text{x}10^{-3}\ m^2/sec \quad .$$

The same parameter can be determined from the slope of the corresponding line in Fig. 5-61 (i.e., line D, the slope of which is about 10 sec/m^2). Assuming a realistic R/r_o ratio (log R/r_o = 1.3), the calculation results in a parameter of

$$T = [9.0\text{-}18.0]\text{x}10^{-3}\ m^2/sec \quad .$$

The transmissivity seems, therefore, to be about three times higher here than that in the same formation at Tatabánya which was indeed expected because of the strongly faulted character of the aquifer at Nyirád.

It was already mentioned that the local resistances around the pumped well (concentration of flow, effect of the structure of casing, the difference in height between the exit point and the water level in the well if the system is unconfined, etc.) may considerably influence the data measured in the well itself, and the results derived from simplified tests are more uncertain than those determined by applying observation wells. The measurements executed in the pumped well are further disturbed in fractured aquifers where the validity of the seepage law is not acceptable in the vicinity of the well. The influence of the development of turbulent flow in the individual channels near the pumped well is also indicated by the nonlinear relationship between depression and yield. It is very important, however, to draw attention to the fact that the characteristic curve must not be constructed from the drawdown data measured at the surface (at the head of the well if the discharging system is closed) because these pressure data do not refer to the conditions at the screen, but they are lowered by the loss of energy consumed by the resistance against the flow between the screen and the surface. The head loss, which can be calculated from the well-known formulae of hydraulics of closed conduits and which increases as the yield increases, should be added to the pressure measured at the surface in order to get the characteristic curve describing the actual conditions at the exit section of the layer. This influence is especially important in the case of deep wells and increases with decreasing diameter.

To demonstrate the influence of the resistance in the pipe on the characteristic curve of wells, the s vs. Q relationship determined by surface measurement in a well having a depth of 1246 m (Városliget I) is shown in Fig. 5-65a (line 1) on the basis of Lakatos' investigations (verbal communication). In part b) of the figure, the casing of the well is demonstrated. Using a simplified form of Chezy's equation and considering the length of the subsequent stretches of casing having different diameter, the relationship between yield and head loss in the pipe was calculated and plotted in part c) of the figure (line 2). The superposition of the head loss over the characteristic curve measured at the surface provides us with the drawdown vs. yield relationship characterizing the actual conditions in the aquifer. The example proves that not only the numerical values but even the character of the curve may be modified by the correct interpretation of the data since the contact between the two variables is linear according to the corrected characteristic curve (line 3) indicating that the flow remains laminar in the aquifer.

The knowledge of the radius of the influenced zone has had an important role in all cases explained here to demonstrate the determination of transmissivity from hydrological data. Some examples were also shown how this parameter can be estimated. The various assumptions give, however, only rough approximations since the radius is a time dependent parameter. More correct information is gained, therefore, on the extension of depression if the transient character of seepage is investigated. The study of nonsteady flow is required in these cases also when the determination of the storage capacity from the hydrological analysis of observed data is desired (which quantity is the other important parameter of the model apart from transmissivity).

The complete form of the differential equation describing axially symmetric flow in a confined leaking aquifer is as follows:

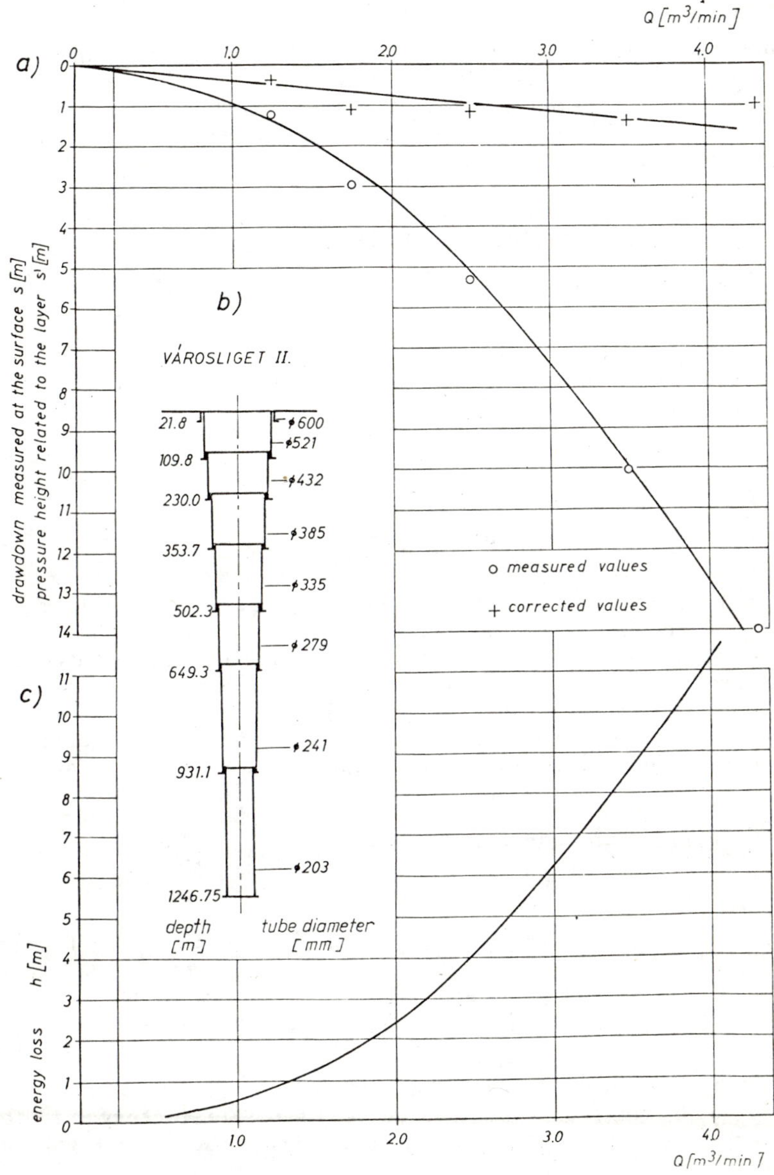

Fig. 5-65. Influence of energy loss consumed by the resistivity against flow in the casing of wells on the characteristic curves.

$$\frac{\partial^2 h}{\partial r^2} + \frac{1}{r}\frac{\partial h}{\partial r} = \frac{S}{2Km}\frac{\partial h}{\partial t} - \frac{\varepsilon(t,r)}{Km} \quad ; \qquad (5\text{-}44)$$

where S is the storage capacity of the aquifer. The layer having vertical recharge along the path is called the leaking system and the amount of the vertical recharge depending on both time and distance from the center of the system is indicated by the symbol $\varepsilon(t,r)$.

The solution of the differential equation leads to the Theis formula (Theis, 1935) if the vertical recharge is negligible:

$$s = \frac{Q}{4\pi\ Km} Ei(u) \ ; \quad \text{where}$$

$$Ei(u) = \int_{\mu}^{\infty} \frac{e^{-x}}{x} dx \ \text{(exponential integral)} \ ; \qquad (5\text{-}45)$$

$$\mu = \frac{r^2 S}{4\ Km\ t} \ ; \quad \text{and}$$

s is the drawdown at a time t and at a distance of r from the axis of the well. Jacob (1940) has proposed to use a simplified form of Eq. 5-45 derived by expanding the exponential integral into a series and neglecting the terms having the μ parameter on the second or higher power. A graphical solution can be applied in this case which was demonstraed in Section 3 (see Fig. 3-26 where the symbols applied here aré also shown):

$$Ei(u) = -0.5772 - \ell n\ \mu - \frac{\mu^2}{2\cdot 2!} + \frac{\mu^3}{3\cdot 3!} - \frac{\mu^4}{4\cdot 4!} + \dots \ ;$$

$$s_2 - s_1 = \frac{Q}{4\pi\ Km} \log \frac{t_2}{t_1} \ ; \quad \text{and} \qquad (5\text{-}46)$$

$$S = 2.245 \frac{Km}{r^2} t_o(r) \ .$$

This approximation was proved to be satisfactory if the interval of $\mu < 0.3$ was excluded from the investigation.

On the basis of the combination of the Dupuit-Thiem and the Theis equations, Muskat (1937) has also published an analytic solution which is applicable for the higher values of time:

$$s = \frac{Q}{4\pi\, Km} \ln 2.245 \frac{Km}{S} \frac{t}{r^2} \quad ; \qquad (5\text{-}47)$$

from which the radius of the influenced zone can be estimated

$$R = \sqrt{2.245 \frac{Km}{S} t} \quad . \qquad (5\text{-}48)$$

The correctness of this approximation can be controlled by analyzing data observed around large draining structures. In the example shown in Fig. 5-66, the depression developing around a mine draining Eocene limestone is investigated (Schmieder, 1975). In the first part of the figure, the drawdown s is represented depending on log r at different time t. Two directions were investigated separately because the propagation of the depression was not axially symmetric due to nonhomogeneity. The corresponding s and r values were determined at three time points (t equal to 3.5, 5.5 and 8.5 years, respectively). Straight lines were fitted to the points which intersect the horizontal axis (s = 0) at the value of log R. After the determination of the radius of the influenced zone in this way, this parameter was represented as a function of time in the second part of the figure. It seems from the data that Eq. 5-48 is an acceptable approximation in the first period, but the last observations (belonging to t = 8.5 years) indicate a smaller radius than that calculated from the equation. This deviation can be explained by the fact that the depression reached, at that time, some area where considerable recharge could develop either in the horizontal or in vertical direction. The transmissivity estimated from other investigations and the storage capacity calculated from this analysis are as follows:

direction 1 $Km = 1.8\times10^{-4}\ m^2/sec$; $S = 0.027$;

direction 2 $Km = 2.1\times10^{-4}\ m^2/sec$; $S = 0.001$.

The determined storage coefficient seems to be realistic although it is lower in the second direction than the expected value. The other problem is caused by the fact that the vertical recharge is unknown and its effect may influence the propagation of the depression.

There are attempts to use the same relationship even for the characterization of leaking aquifers (Schmieder, 1975), but the quantity sometimes called piezometric conducticity (a = Km/S) should be substituted by a time dependent parameter in this case. This value combines not only the transmissivity and storage coefficient, but the effect of the vertical recharge too. Considering the fact that the two physical processes, the description of which is combined into one term

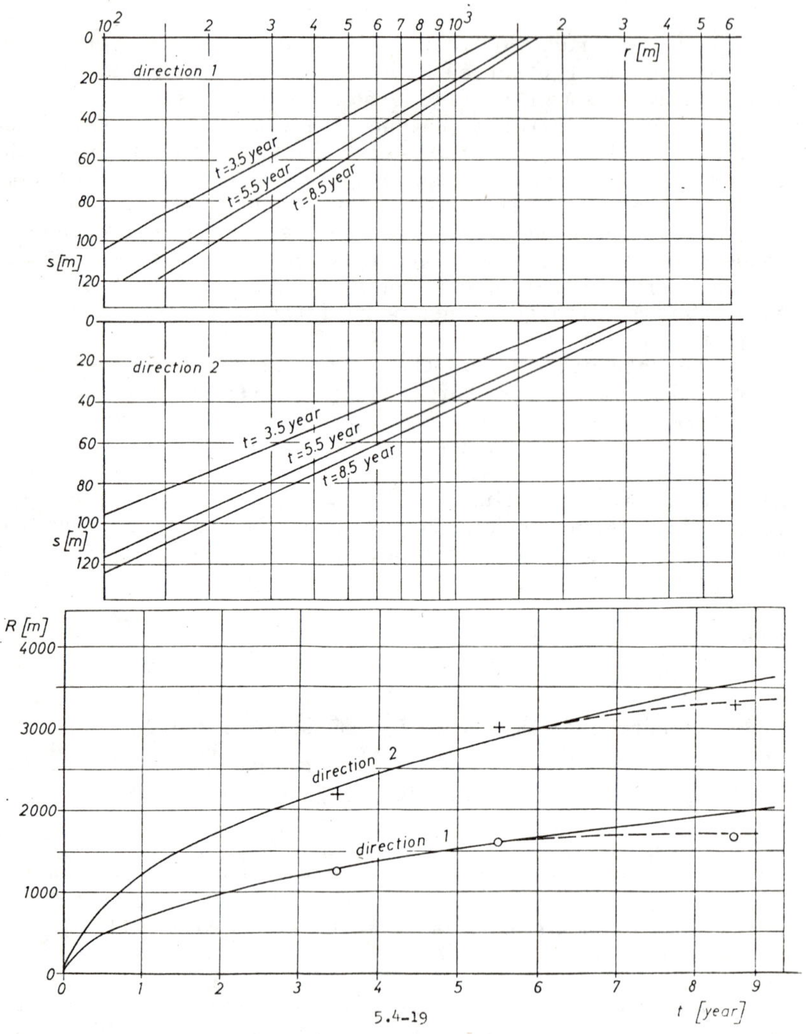

Fig. 5-66. The change of depression in time.

(i.e., storage and vertical recharge) are independent of one another, it is advisable to investigate the original form of Eq. 5-44 and to determine storage capacity, transmissivity, and recharge separately because they cannot be applied in the model in combined form.

20-3 Estimation of Factors Influencing the Water Balance

It was explained in the introduction of this section that one of the main objectives of the analysis of hydrological data is the determination of the mathematical model simulating the water régime of the investigated aquifer system. Three important groups of parameters were distinguished:

- the geometry of the system;
- the flow characteristics of the formation (conductivity, storage capacity);
- the boundary conditions acting around the perimeter and over the area of the field.

The determination of the geometrical parameters (position and size of layers) is a typical geological task. There are various methods to estimate the flow conditions (e.g., investigating core samples, borehole logging, statistical and hydraulic evaluation of the distribution of openings, etc.). It was already demonstrated how numerical information can be achieved concerning transmissivity and storage capacity by analyzing hydrological observations. It is a general rule that as many independent ways should be used as possible because all types of methods provide us only with rough approximations, and having more estimated values, the most probable range of the parameters can be better determined. The most versatile problems arise in connection with the characterization of the boundary conditions. It was already demonstrated that the regular observation of the position of the water table (or that of the piezometric surface) gives valuable information on the character of the boundary conditions and in some cases even numerical parameters can be determined in this way. It was emphasized, however, that these numerical values should be checked by the calculation of the water balance of the interconnected aquifers (or by that of a separable part of the complete system) if possible. An example of this analysis will be given in this section.

The general water balance equation for a hard rock aquifer expresses that the sum of recharging and discharging effects during a given time interval is equal to the difference of water amount stored in the system at the beginning and at the end of the period investigated.

The main source of recharge (Q_R) is infiltration which is usually calculated as the product of the average precipitation (P_a) over the area (A) within the period investigated ($\Delta t = t_2 - t_1$) and the rate of infiltration (β). The latter may be determined as an average for the whole system (β_o) or the infiltration through the bare rock surface (β_B) may be distinguished from that developing in the basins covered by younger sediments (β_C):

$$Q_R \Delta t = \beta_o \; A \; P_a \; ; \quad \text{or}$$

$$Q_R \Delta t = \beta_B \; A_B \; P_{aB} + \beta_C \; A_C \; P_{aC} \; ; \qquad (5\text{-}49)$$

where subscripts B and C indicate the bare and covered surface, respectively (their extension and the average precipitation over the distinguished parts of the catchment).

One of the main components of the naturl discharging effects is the yield of springs. The total amount of water drained from the aquifer by springs (or the product of the average yield of springs and the investigated period $Q_S \Delta t$) can be easily calculated if the flow rate is measured at every spring or at least at the major ones:

$$Q_S \Delta t = \sum_{i=1}^{n} \int_{t_1}^{t_2} Q_i(t) \, dt \; ; \qquad (5\text{-}50)$$

where $Q_i(t)$ is the discharge hydrograph of the i-th spring and n is the number of the springs which were measured.

There is another directly measurable component of the water balance, i.e., the artificial water withdrawal. All water discharging structures operating either by pumping or gravitationally are generally equipped with measuring devices independently of the purpose of exploitation. Thus, the hydrographs are available to characterize the average artificial discharge (Q_A) quite similarly as it was done in the case of springs:

$$Q_A \Delta t = \sum_{j=1}^{m} \int_{t_1}^{t_2} Q_j(t) \, dt \; ; \qquad (5\text{-}51)$$

where $Q_j(t)$ is the hydrograph of the artificial water withdrawal at the j-th point and m is the number of such structures.

The water exchange around the perimeter of the field may be either discharge or recharge depending on the direction of flow. It should be taken into account with a negative sign if it drains the investigated system and its value is positive if the stored amount is increased by the process. Although the magnitude of this component may have a considerable role within the balance, its volume cannot be measured. Water exchange may develop between two aquifers as well where the contour of the field is composed of the interface of two independent formations. The character of the boundary conditions is very versatile because of the many different possible ways of contacts. A border with constant external potential may be assumed where the system is directly in contact with a large body of surface water (river, lake, sea, etc.). Flow may develop without any special local resistance (e.g., where the field is closed by large open faults or the coarse-grained alluvium of the bed directly contacts the permeable layer of the formation). In other cases, interbedded semi-permeable strata or a clogged layer around the bed may decrease the flow rate which fact requires the consideration of local resistivity as a part of the boundary condition at such contours. There are even borders where the flow rate crossing the contour is independent of the potential conditions prevailing within the field, but it is constant or changes in time according to some external effect (e.g., infiltration from the clogged beds of water courses if the water table remains in the coarse-grained alluvium and does not reach the semi-permeable layer because in this case the position of the water table does not influence the amount of infiltrating water).

It follows from the explained conditions that not only the amount of water exchange may change in time as the potential decreases due to exploitation, but even its character, the flow direction, may be altered as well in many cases except at the borders where the flow is independent of the internal pressure conditions. It is necessary, therefore, to find a method for the numerical estimation of the components of the water balance acting around the contour which is suitable to take into account the time dependent water exchange. This demand can be met if the recharge or discharge through the contours is determined from continuously observed water level data by calculating the hydraulic gradient within the field of the fissured aquifer perpendicularly to the contour and multiplying it by the transmissivity of the formation. Thus, the water exchange through a given section (between points L_1 and L_2 of the boundary) at time t is:

$$Q_k(t) = \int_{L_1}^{L_2} T(\ell)\, I(t,\ell)\, d\ell \quad ; \qquad (5\text{-}52)$$

where the symbols indicate that transmissivity $T(\ell)$ depends on the position of the infinitesimally small stretch $d\ell$ only while the hydraulic gradient $I(t,\ell)$ is a function of both space and time.

Supposing that there are r stretches of the closed contour having different character, the total amount of water crossing the boundary of the field ($Q_L \Delta t$) can be indicated by the following symbol

$$Q_L \Delta t = \sum_{k=1}^{r} \int_{t_1}^{t_2} Q_k(t)\, dt = \Sigma\Sigma TI \quad ; \qquad (5\text{-}53)$$

where the symbol $\Sigma\Sigma TI$ indicates the product of transmissivity and hydraulic gradient (both changing in time and along the boundary) summarized for the investigated period and along the perimeter of the field. The form of this equation is quite the same as that of Eqs. 5-50 and 5-51. The basic difference is caused by the fact that in the previous cases the discharge hydrographs were measurable, and now it should be estimated from gradient values substituted into Eq. 5-53 and calculated from hydrological observations. It is quite natural that the gradient directed into the field is considered with positive sign and it is negative when pointing from the field to the perimeter.

There is a further discharging effect which may have a considerable role in some cases depending on the morphology of the terrain, i.e., evapotranspiration. In general, the water table is at or near the land surface only in deep valleys within mountainous areas. The development of small swamps and growing of aquatic vegetation on the flat bottom of the valleys indicate the areas characterized with high evaporation. In other mountains, large moors have developed on the surface of hard rock terrains recharged by groundwater, and in this way drain a considerable amount of water from the aquifer. The numerical value of this discharge may be estimated as the product of the area of moors or that of the flat valley bottom (A_M) and actual evapotranspiration (ET_A), the latter being a function of time. The average discharge drained during the period investigated (Q_E) can be expressed similarly as the terms discussed previously:

$$Q_E \Delta t = \int_{t_1}^{t_2} A_M(t)\ ET_A(t)\ dt = \Sigma A_M\ ET_A \quad ; \qquad (5\text{-}54)$$

where even the change of the extension of the effected area should also be considered as it is indicated by the $A_M(t)$ symbol giving the area depending on time, and $\Sigma A_M\ ET_A$ gives the total amount of water evaporated or transpirated as the product of the two time dependent variables (area and actual evapotranspiration summarized for the investigated period).

The last component of the water balance to be dealt with is the change of the amount of water stored in the system. Knowing the position of the water table at the beginning and at the end of the period investigated, indicating the change in height at a given point with the symbol ΔH, the total volume of rock mass dewatered or saturated can be calculated (ΔV_R). Multiplying this parameter by the storage coefficient of the solid matrix (S which is the average, or n_s specific yield and the storage coefficient S of the confined section of the aquifer) the change of storage can be calculated:

$$\Delta V = S \Delta V_R \ ; \quad \text{and}$$

$$\Delta V_R = \int_{(A)} \Delta H\ dA = \Delta H_o A \ ; \qquad (5\text{-}55)$$

where ΔH_o is the average change of the position of the water table (or piezometric level) over the area A which change is negative if the water level decreases and it has a positive sign in the opposite case. It is also possible to make a distinction between the areas having water table condition (i.e., area with bare surface A_B) and those where the aquifer is confined (A_C)

$$\Delta V = n_s \int_{(A_B)} \Delta H\ dA + S \int_{(A_C)} \Delta H\ dA$$

$$= n_s\ \Delta V_{RB} + S \Delta V_{RC} = n_s\ \Delta H_{oB}\ A_B + S \Delta H_{oC}\ A_C \ ;$$

where the average change of the position of the water table (ΔH_{oB}) is determined separately from that of the piezometric level (ΔH_{oC}).

Combining the terms listed, although the list is not a complete one, but it includes only the most important components of the water balance of hard rock terrains, the final form of the balance equation is as follows:

$$(Q_R + Q_L - Q_S - Q_A - Q_E)\ \Delta t = \Delta V \quad ;$$

$$(\beta_o\ AP_a + \Sigma\Sigma TI - Q_S - Q_A - \Sigma A_M\ ET_A)\ \Delta t = S\Delta H_o\ A \ ; \quad \text{or}$$

$$(\beta_B A_B P_{aB} + \beta_C A_C P_{aC} + \Sigma\Sigma TI - Q_S - Q_A - \Sigma A_M\ ET_A)\ \Delta t = n_S \Delta H_{oB} A_B + S\Delta H_{oC} A_C \quad . \tag{5-56}$$

The first line is the general form of the equation while the other two also indicate the parameters from which the non-measurable terms can be derived. The difference between the last two expressions is caused by whether the confined and unconfined areas were considered separately or not. In the second line, average parameters are determined for the whole catchment neglecting the physical differences between areas having water table conditions and those where piezometric level develops. This is the reason why the third form gives a more correct approximation and why the application of this relationship is advisable in every case when the available data give sufficient information for this more detailed analysis.

In the balance equations, there are variables which are directly measurable. The geometrical parameters form the first part of this group (A total area, A_B area of bare rock or that covered only by a thin blanket, A_C area covered by younger sediments having a considerable thickness where the aquifer is probably confined, A_M area of moors and large flat bottoms of valleys having high water table). The hydrographs of springs and artificial exploitations are also measurable, and the average yields for the period can be determined from the records. The depth of precipitation is also a parameter observed regularly. The basic set of data is the point value of daily precipitation from which the areal average should be calculated either for the whole catchment or for the bare and covered surfaces separately. The sum of daily areal precipitation within the interval ($\Delta t = t_2 - t_1$) provides us with the desired parameters (P_a or P_{aB} and P_{aC}).

The second group of parameters can be derived from the regularly observed data of the water table and the piezometric surface. From maps where the lines of these surfaces are constructed, the average hydraulic gradient (I) can be determined along the perimeter of the field. The difference of the positions of the levels can be calculated on the basis of maps constructed for subsequent times, and the areal average (either for the whole catchment ΔH_o or for the confined and unconfined parts separately ΔH_{oC} and ΔH_{oB}) of the differences provides the basic data required for the determination of the change in storage. The length of the interval for which the water balance is determined can be chosen arbitrarily, but there are several aspects to be considered:

- at the beginning and at the end of the period, hydrological maps should be available;
- the interval should be long enough because otherwise the change of the position of the water table is smaller than the error of the calculation, which fact may result in unreliable data;
- the time dependent boundary conditions should be constant within a period (they do not change their character) which requirement excludes the application of too long intervals;
- the influence of storage processes other than those occurring in the aquifer (e.g., snow accumulation, change of the amount of soil moisture) should be avoided, which target can be achieved by selecting the beginning and the end of the period at the same calendar date when snow cover is not probable over the area.

Considering the aspects listed, the analysis of yearly periods is advisable. The choice of other and longer intervals is necessary only if the hydrological maps are not available for each year.

The parameters still unknown in Eq. 5-56 are those whose determination or checking is the result of the water balance analysis. This group includes:

- the rate of infiltration (β_o the average for the whole catchment, or β_B and β_C if the parameter is determined separately for the surfaces being bare and covered by younger sediments, respectively);
- the transmissivity of the aquifer (T);
- the average actual evapotranspiration from the swampy areas (ET_A);

- the storage capacity of the system which may be characterized either by an average coefficient (S) or by specific yield (n_S) and the storage coefficient of the confined aquifer (S').

The parameters listed in the last group can be determined from hydrological data by analyzing the water balance of the system. Since there are several unknown variables, the balance equation should be set up for subsequent periods, and the set of the parameters to be determined should be looked for which satisfies all the members of the series of equations with the smallest possible error. Naturally, all the parameters should remain within the probable range of the value in question which zone of confidence can be calculated by using other methods already discussed or can be estimated on the basis of the physical meaning of the parameter if other data are not available.

The application of this method is demonstrated by using an example investigated by J. Kovács (1975). The area is located at the northeast part of the Transdanubian Mountain Range represented in Fig. 5-49. Its extension is about 700 km^2. The main aquifer within the area is a carbonate formation, mostly Upper Triassic limestone. The northeast boundary of the water-bearing system is impervious (Miocene andesite). It is bordered at the north and east sides by the coarse alluvium of the Danube (constant potential boundary). The carbonate formation is continued in the west and south directions. The contour of the field was chosen at the groundwater watershed which was indicated by the highest points of the water table. Although the lowering of pressure due to exploitation modified the position of the watershed, and thus water exchange has developed through the contour fixed originally, this amount was relatively low and it was therefore neglected. The carbonate rock forms the surface on about one third of the area and the other two thirds are covered by young sediments. Unfortunately, the available information was inadequate to consider the actual character of the confined and unconfined parts separately and, therefore, the more simple form of the balance equation was applied (second line in Eq. 5-56). The ratio of the two different parts should be taken into account when the reliability of the rate of infiltration and that of the storage coefficient calculated in this way is judged.

A groundwater map was available at the end of 1954, 1958, and 1968. Since that time the records were continuous, thus yearly maps were constructed. The yield of springs and the artificial discharge were known. The drainage at the northern and easten sides was calculated from the average hydraulic gradient determined from the maps by considering a transmissivity of 11×10^{-3} and 7.5×10^{-3} m^2/sec, respectively. The computation is summarized in Table 5-11. Three groundwater maps have been developed together with the change of the average annual values of the yield of springs, the discharge at the boundaries, and the artificially exploited water. The investigation resulted with the most probable rate of infiltration (β_o = 0.115) and storage coefficient (S = 0.027). These parameters are within the expected range of the variables in question, considering that about two thirds of the area is covered by young sediments.

Table 5-11. Water balance in a karstic region (after J. Kovács).

Year	Yield of Springs Q_S (10^7 cm/yr)	Artificial Discharge Q_A (10^7 cm/yr)	Slopes at the Discharging Faults I_1(‰)	I_2(‰)	Discharge (-) or Recharge (+) at the Contour Q_{L1} (10^7 cm/yr)	Q_{L2} (10^7 cm/yr)	Precipitation Pa (10^7 cm/yr)	Infiltration Q_R (10^7 cm/yr)
1955-1958	0.05*	3.00*	-462*	-1.00*	-4.99*	-0.23*	41.7*	4.80*
1959	0.05	3.07	-410	-0.98	-4.43	-0.23	37.9	4.36
1960	0.05	3.06	-333	-0.96	-3.59	-0.22	43.7	5.03
1961	0.05	3.56	-305	-0.95	-3.29	-0.22	31.1	3.58
1962	0.04	5.02	-253	-0.93	-273	-0.21	37.9	4.36
1963	0.05	5.49	-200	-0.92	-2.16	-0.21	45.5	5.23
1964	0.04	5.88	-148	-0.90	-1.59	-0.21	45.7	5.26
1965	0.03	5.59	-095	-0.88	-1.03	-0.20	58.2	6.69
1966	0.01	6.02	-043	-0.87	-0.46	-0.20	53.7	6.18
1967	0.004	7.05	+040	-0.85	+0.43	-0.20	33.1	3.81
1968	0.003	6.32	+076	-0.81	+0.82	-0.19	35.2	4.05
1969	0.008	4.56	+029	-0.83	+0.32	-0.19	41.9	4.82
1970	0.02	4.17	-011	-0.76	-0.13	-0.17	45.3	5.21
1971	0.02	3.76	-053	-0.77	-0.18	-0.18	27.4	3.15
1972	0.02	1.98	-083	-0.68	-0.89	-0.16	42.1	4.84
1973	0.02	2.16	-128	-0.98	-1.39	-0.17	31.2	3.59
1974	0.02	2.37	-075	-0.92	-0.81	-0.16	47.5	5.46

Table 5-11. (continued)

Year		Sum of Recharge and Discharge (10^7 cm/yr)		Change in Storage (10^7 cm/yr) (S = 0.027) in rock	in water	Closing Error (10^7 cm/yr)
1955-1958		-3.47*				
	1955-58		-13.88	-360.4	-9.73	-4.15
1959		-3.42				
1960		-1.89				
1961		-3.54				
1962		-3.64				
1963		-2.68				
1964		-2.46				
1965		-0.16				
1966		-0.51				
1967		-3.01				
1968		-1.64				
	1959-68		-22.95	-974.5	-26.31	+3.36
1969		+0.38		+37.5	+1.01	-0.63
1970		+0.72		±0.0	±0.0	+0.72
1971		-1.38		-129.8	-3.50	+2.12
1972		+1.79		+169.0	+4.56	-2.77
1973		-0.15		-96.0	-2.59	+2.44
1974		+2.10		+129.5	+3.50	-1.40
						-0.26

*annual average of the parameters in the period of 1955-58.

REFERENCES FOR PART V

Arandjelovic, D., 1966: Geophysical Methods Used in Some Geological Problems (in English), Geophysical Prospecting, Vol. XIV.

Arandjelovic, D., 1969: A Possible Way Tracing Groundwater Flows in Karst (in English), Geophysical Prospecting, Vol. XVII.

Avias, J. and Dubretret, L., 1975: Karst Phenomena in Non-carbonate Rocks (in French), Hydrology of Karstic Terrains, Paris.

Bachmat, Y., 1965: Basic Transport Coefficients as Aquifer Characteristics (in English), IASH Symposium on Hydrology of Fractured Rocks, Dubrovnik.

Balásházy, L. and Kovács, J., 1975: Determination of Hydraulic Conductivity of Triassic Carbonate Rocks by Statistical Analysis of Slits (Manuscript in Hungarian), Budapest.

Beer, F., 1972: Dynamics of Fluids in Porous Media (in English), New York, London, Amsterdam.

Böcker, T., 1967: The Interference between the Thermal Wells in Budapest (in Hungarian), Vizügyi Közlemények, No. 3.

Böcker, T., 1972: Theoretical Model for Karstic Rocks (in English), Karszt- és Barlangkutatás, Vol. VII.

Borelli, M. and Pavlin, B., 1965: Approach to the Problem of the Underground Water, Leakage from the Storages in Karst Regions, Karst Storages Busko Blato, Poruca and Kruscica (in English), IASH Symposium on Hydrology of Fissured Rocks, Dubrovnik.

Boyd, D., 1967: The Contribution of Airborne Magnetic Survey to Geological Mapping (in English), Geological Survey of Canada, Economic Geological Report No. 26.

Brace, W. F., 1966: Some New Measurements of Linear Compressibility of Rocks (in English), Journal of Geophysical Research.

Bereton, N. R. and Skinner, A. C., 1974: Groundwater Flow Characteristic in the Triassic Sandstone in the Tylde Area of Lencashire (in English), Water Service, August.

Brown, R. H., Konoplyanisev, A. A., Ineson, B., and Kovalevsky, V. S., 1972: Groundwater Studies, Chapter 9. Analytical and Investigational Techniques for Fissured and Fractured Rocks by M. Schoaler (in English), UNESCO Studies and Reports in Hydrology, Paris.

Bardos, B. H., 1965: Hydrogeology of Some Karstic Areas of Greece (in English), IASH Symposium on Hydrology of Fractured Rocks, Dubrovnik.

Csallány, S., 1965: The Hydraulic Properties and Yields of Dolomite and Limestone Aquifers (in English), IAHS Symposium on Hydrology of Fractured Rocks, Dubrovnik.

Davis, S. H. and de Wiest, R. S. H., 1966: Hydrogeology (in English), New York.

Davis, S. N. and Turk, L. S., 1964: Optimum Depth of Wells in Crystalline Rocks (in English), Groundwater, No. 2.

de Wiest, R. S. H., 1969: Flow through Porous Media (in English), New York, London.

Emplaincourt, L. G., 1974: The Role of Remote Sensing in Hydrogeological Research (in English), Memoires of the IAH Congress in Montpellier, Vol. X.

Englund, J. O. and Inreensen, P., 1975: Weathering and Hydrogeology of the Brumunddal Sandstone, Southern Norway (in English), Nordic Hydrology, No. 1.

Farkasiewicz, J. and Paloc, H., 1965: The Process of Recession of "La Toux de la Vis" - Preliminary Study (in French), IASH Symposium on Hydrology of Fractured Rocks, Dubrovnik.

Franciss, F. ., 1978: Study of Groundwater Movement through Fissured Media (in French) (Doctor Thesis), University of Grenoble.

Gerber, P. and Bodri, B., 1974: Preliminary Results of the Evaluation of Data Concerning Karstic Water Levels and their Relation with the Protection of Mines against Water Intrusion (in Hungarian), Tatabányai Szénbányák Müszaki Közgazdasági Közleményei, No. 4.

Gondouin, M. and Scala, C., 1958: Streaming Potential and the SP log (in English), Transaction of Am. Inst. of Mining and Metallurgist Engineers, Paper 8023.

Gosseline, N. and Schoeller, H., 1939: Observations of the Yield of Artesian Wells (in French), IUGG General Assembly, Washington.

Grandinetti, M., 1967: New Arrangement of Electrodes (in Italian), Bolletino di Geofisica, Vol. IX.

Jacob, C. E., 1940: On the Flow of Water in an Elastic Artesian Aquifer (in English), Transactions of American Geophysical Union.

Jung, K., 1961: Gravity Methods Used in Applied Geophysics (in German), Leipzig.

Kessler, H., 1954: The Available Amount of Water Exploitable Continuously and the Determination of the Rate of Infiltration (in Hungarian), Hidrológiai Közlöny, No. 5-6.

Király, L., 1973: Explanatory Notes for the Hydrogeological Map of Neuchatel County (in French), Bulletin de la Société Neuchateise des Sciences Naturelles, Tom. 96.

Korim, K. and Liebe, P., 1970: Hydrogeology of Hévizfürdö (in Hungarian), Vizügyi Közlemények, No. 3.

Kovács, G., 1973: Determination of the Hydraulic Parameters of Seepage Developing through Homogeneous Earth Dams (in Hungarian), VITUKI, Budapest, No. 6 (manuscript).

Kovalevsky, V. S., Babushkin, V. D., Böcker, T., and Borevsky, B. V., 1975: Regime of Subterranean Water Flows in Karst Regions (in English), Hydrogeology of Karstic Terrains, Paris.

Lahermo, P., 1973: The Groundwater of Central and West-lapland Interpreted on the Basis of Black and White Aerial Photos (in English), Geological Survey of Finland, No. 262.

Leveque, P. Ch., 1976: Study of Karst...Using Remote Sensing (in French), Karst Hydrogeology, Alabama.

Lyon, R. J. P., 1965: Analysis of Rocks by Spectral Infrared Emission (8-25 microns) (in English), Economical Geology, Vol. 60.

Lewis, D. C., Kriz, G. H., and Buray, R. H., 1966: Tracer Dilution Sampling Technique to Determine Hydraulic Conductivity of Fractured Rock (in English), Water Resources Research, No. 2.

Lomidze, T. M., 1951: Percolation in Fissured Rocks (in Russian), Moscow.

Lorberer, Á., 1978: Regional Hydrogeological Cross Sections through Northwest Part of the Transdanubian Central Mountain in Hungary (in Hungarian), VITUKI Report (unpublished).

Louis, C., 1967: Study of the Flow of Water in Fissured Rocks and its Influence on the Stability of Mining Structures are Slopes in Rocks (in German) (Doctor Thesis), University of Karlsruhe, No. 1;
1969: A Study of Groundwater Flow in Jointed Rock and its Influence on the Stability of Rock Masses (in English), Rock Mechanics Center, Imperial College, London;
1970: Three-Dimensional Flow in Fissured Rocks (in French), Rock Mechanics Center, Imperial College, London.

Mainguet, M., 1975: Sandstone Morphology, Paris.

Manger, G. S., 1963: Porosity and Bulk Density of Sedimentary Rocks (in English), U.S. Geological Survey Paper No. 1144-8.

Maucha, L., 1972: Investigation of Natural Processes Influencing the Change of Yield of Springs (in Hungarian), VITUKI Scientific Reports, No. 2162.

Maucha, L., 1975: Study of Tidal Movements of Karst Waters and Karstic Rocks (in English), IUGG General Assembly, Grenoble (published in Annuales de Géophysique, 1977).

Milatovic, B. F., 1970a: Hydraulic Mechanism of Maret Aquifers in Deep-Lying Coastal Collectors (in English), Bulletin of the Institute for Geological and Geophysical Research, No. 7, Beograd;
1970b: A Method of Studying the Hydrodynamic Regime of Karst Aquifers by Analysis of the Discharge Curve and Level Fluctuation during Recession, Bulletin of the Institute for Geological and Geophysical Research, No. 8, Geograd.

Mortier, B., Quens, H. T., and Sadek, M., 1965: Hydrogeology of Volcanic Rock Formations in Northeast Morocco (in French), IASH Symposium on Hydrology of Fractured Rocks, Dubrovnik.

Mortier, B. and Safadi, Ch., 1966: Karstic Phenomena in Gypsum of Dezireh (in French), IASH Symposium on Hydrology of Fractured Rocks, Dubrovnik.

Muskat, M., 1937: The Flow of Homogeneous Fluids through Porous Media (in English), New York.

Van Nostrand, R. G. and Cook, K. L., 1966: Interpretation of Resistivity Data (in English), Geological Survey Professional Paper No. 499, Washington.

Omes, G., 1976: High Accuracy Gravity applied to Detection of Karstic Cavities (in English), Karst Hydrogeology, Alabama.

Öllös, G., 1963: Hydraulic Investigation of Flow Rate in Fractured Rocks by Small-Scale Mdel (in French), Bulletin of IASH.

Powell, R. L., 1976: Joint Patterns and Solution Channel Evolution in Indiana (in English), Karst Hydrogeology, Alabama.

Rate, M. V. and Cherveshov, S. N., 1965: Statistical Aspect of the Problem on the Permeability of the Jointy Rocks (in English), IASH Symposium on Hydrology of Fractured Rocks, Dubrovnik.

Rima, B. S., Meisler, H., and Longwill, S., 1962: Geology and Hydrology of the Stockton Formation in Southeast Pennsylvania (in English), Pennsylvania Topographic and Geological Survey,- Groundwater Report, W-14.

Romm, B. S., 1966: Flow Phenomena in Fissured Rocks (in Russian), Moscow.

Samper, A. A. and Navarro, A., 1965: Problems of the Storage of Water in Southeast Spain (in French), IASH Symposium on Hydrology of Fractured Rocks, Dubrovnik.

Schmieder, A., 1975: Determination of Hydraulic Parameters of the Karstic Aquifer of Transdanubian Mountain Range (in Hungarian), Scientific Report of Bányászati Kutató Intézet (Manuscript).

Schmieder, A., Willems, T., Szilágyi, G., and Keserü, Zs., 1978: Investigation of Laws Governing the Movement of Water and Rocks in Aquifers (in Hungarian), Bányászati Kutató Intézet, Budapest (closing report), Manuscript No. 18-2/69 K.

Schoeller, H., 1948: Hydrogeological Regime of Eocene Limestone in the Synclinal of Dyr-El-Kef (Tunisie) (in French), Bulletin de Society Geological de France, Tom. 18;
1962: Groundwater (in French), Paris;
1965: Hydrodynamics in the Karst (in French), IASH Symposium on Hydrology of Fractured Rocks, Dubrovnik.

Shevyskov, L. B. and Mankovsky, G. I., 1968: Experiences Gained by Dewatering Mining Sites of Productive Minerals having Complicated Hydrogeological Conditions (in Russian), Moscow.

Shuster, E. T., 1971: Seasonal Fluctuation in the Chemistry of Limestone Springs (in English), Journal of Hydrology, Vol. 14.

Siple, G., 1965: Saltwater Encroachment of Tertiary Limestone along Coastal South Carolina (in English), IASH Symposium on Hydrology of Fractured Rocks, Dubrovnik.

Snow, B. T., 1970: The Frequency and Apertures of Fractures in Rock (in English), International Journal of Rock Mechanics and Mining Science, No. 1.

Stewart, J. W., 1962: Water-Yielding Potential of Weathered Crystalline Rocks (in English), Stanford University, Student M.S. Report;
1964: Infiltration and Permeability of Weathered Crystalline Rocks (in English), U.S. Geological Survey, Paper No. 1138-B.

Stuart, W. T., Brown, E. A., and Rhodamel, B. C., 1954: Groundwater Investigations of the Marquette Iron-Mining District (in English), Michigan Geological Survey, Technical Report, No. 3.

Theis, C. V., 1953: The Relation between the Lowering of the Piezometric Surface and the Rate and Duration of Discharge of a Well Using Groundwater Storage (in English), Transactions of American Geophysical Union.

Turk, L. J., 1963: The Occurrence of Groundwater in Crystalline Rocks (in English), Stanford University, Student M.S. Report.

Vacquier, V., 1951: Interpretation of Aeromagnetic Maps (in English), Geological Society of American Memoires, Vol. 47.

Vocchioli, J., 1965: Directional Hydraulic Behavior of a Fractured Shale Aquifer in New Jersey (in English), IASH Symposium on Hydrology of Fractured Rocks, Dubrovnik.

Venkovits, I., 1951: Investigation of Water at Dorog (in Hungarian), Hidrológiai Közlöny.

Wermund, E. G., 1976: Regional Relation of Fracture Zones in the Edwards Limestone Aquifer, Texas (in English), Karst Hydrogeology, Alabama.

Wilson, G. R. and Witherspoon, P. A., 1974: Steady State Flow in Rigid Networks of Fractures (in English), Water Resources Research, No. 2.

INDEX BY SUBJECT

INDEX BY AUTHOR